# Microeconomics

# Microeconomics

Robert B. Ekelund, Jr.
*Auburn University*

Robert D. Tollison
*George Mason University*

**LITTLE, BROWN and COMPANY**
Boston   Toronto

**Library of Congress Cataloging in Publication Data**

Ekelund, Robert B. (Robert Burton), 1940–
  Microeconomics.

  Includes index.
  1. Microeconomics.   I. Tollison, Robert D.
II. Title.
HB172.E39      1985      338.5      85–21401
ISBN 0–316–23125–8

Copyright © 1986 by Robert B. Ekelund and Robert D. Tollison

All rights reserved. No part of this book may be reproduced in any form or by any electronic or mechanical means including information storage and retrieval systems without permission in writing from the publisher, except by a reviewer who may quote brief passages in a review.

Library of Congress Catalog Card No. 85–21401

ISBN 0-316-23125-8

9  8  7  6  5  4  3  2

MU

Published simultaneously in Canada
by Little, Brown & Company (Canada) Limited

Printed in the United States of America


# CREDITS

*Photos:* Page 11: Sharon A. Bazarian/The Picture Cube; Page 53: Courtesy of the Trustees of the Boston Public Library; Page 65: Jerry Gordon/Archive; Page 95: *left* and *right*, The Granger Collection, New York; Page 152: AP/Wide World Photos; Page 204: Historical Pictures Service, Chicago; Page 240: Russell French; Page 257: Wide World Photos; Page 259: *left*, Historical Pictures Service, Chicago; *right*, Peter Lofts Photography, Cambridge, England; Page 320: Courtesy of the Public Services Department, Federal Reserve Bank of Boston; Page 340: *left*, Courtesy of Apple Computer, Inc.; *right*, Courtesy of Atari Industries; Page 370: *left*, The Granger Collection, New York; *right*, Courtesy of Gary Becker; Page 399: Courtesy of GTE SPRINT Communications Corp; Page 419: Photo courtesy of Chicago Transit Authority; Page 453: The Granger Collection, New York; Page 459: *left*, Photo by V. H. Mottram; National Portrait Gallery, London, England; *right*, Courtesy of Ronald Coase; Page 517: *left*, Courtesy of Bettina Bien Greaves, Foundation For Economic Education, Irvington-on-Hudson, NY; *right*, The Bettmann Archive; Page 524: Nicholas Daniloff, *U.S. News and World Report*; Page 535: Kirschenbaum/Stock Boston; Page 539: Wide World Photos. Page 841: *left*, The Granger Collection, New York; *right*, The Bettmann Archive.
*Text:* Page 10: Graph from Gordon Tullock, "The Coal Tit as a Careful Shopper," *The American Naturalist* (November 1969): 77–80. © 1969 by The University of Chicago. Reprinted by permission of The University of Chicago Press; Page 44: Summary of Everett G. Martin and Fernando Paulson, "Victims of a Theory: Chilean Region, Competitive in Few Products, Hits Hard Times Under Rule of 'Chicago Boys,'" *The Wall Street Journal*, May 5, 1983, p. 60. Reprinted by permission of *The Wall Street Journal.* © Dow Jones & Company, Inc. 1983. All Rights Reserved; Page 59: Data for Figure 3–3 from *Facts and Figures on Government Finance* (Washington, D.C.: Tax Foundation, Inc., 1983); Page 60: Data for Figure 3–5 from *Facts and Figures on Government Finance* (Washington, D.C.: Tax Foundation, Inc., 1983); Page 61: Data for Figure 3–6 from *Facts and Figures on Government Finance* (Washington, D.C.: Tax Foundation, Inc., 1983); Page 66: Table 3–1 from Svenska Handelshanken, *Sweden's Economy in Figures*, 1982; Page 67: Table 3–2 from "The Swedish Economy" (Stockholm: The Swedish Institute, 1982); Page 67: Table 3–3 from "The Swedish Economy" (Stockholm: The Swedish Institute, 1982); Page 93: Summary of Frederick W. Bell, "The Pope and the Price of Fish," *American Economic Review* 58 (December 1968):1346–1350. Copyright © 1968 by the American Economic Association. Reprinted by permission

(continued on page 554)

*For My Mother*
*and in Memory of My Father, Bob, and My Friend, Terry*
RBE

*For Anna, April, and My Parents*
RDT

# Preface

In this introduction to microeconomics we stress the interrelationship of theory and economic issues. In each chapter we rely on applications to test the insight of theory, and we rely on theory to provide the context for understanding real-world events. We think this is the best way to prepare students to appreciate the power of economic thinking, to see the vital connection between the logic of economics and the issues confronting them on television, in newspapers, in the voting booth, and in their careers.

To build a solid foundation in theory, we present microeconomics step-by-step, illustrating and reinforcing key concepts at every opportunity. We focus patiently on critical concepts such as supply and demand to make sure students are secure in the basics. We also move strongly into current theories such as public choice, rent-seeking behavior, cartel theory, and modern theories of the firm. Although the range and diversity of topics is wide, we have tried not to be encyclopaedic. Students gain little from exposure to a smattering of topics strung together with little organizing force. We have tried instead to integrate modern perspectives into the book by building upon the fundamentals set forth carefully and consistently throughout.

To connect theory with the world around us, we have woven in hundreds of applications throughout the book, ranging from the current problems facing mass transit planners to the effects of competition in the market for Halloween pumpkins. The diversity of these applications demonstrates the versatility of economic reasoning. (We have a bit more to say about applications under "Special Features" below.)

Since our approach stresses the relationship between theory and real-world issues, the technical presentations in this book are concise and straightforward. We rely heavily on examples to illustrate abstract theory. We describe graphs and graphical relationships in patient detail. And we use an abundance of visual aids to motivate and clarify.

In sum, we feel that the approach of *Economics* reflects our sense that students too often merely receive the facts and theories of economics in an introductory textbook and too often grow uncomfortable with its abstractness. We want our readers to discover economics' intuitive appeal, to share our enthusiasm for a discipline rich in insight.

## Special Features

- Many of the applications in *Microeconomics* are set off in "Economics in Action" boxes that appear at the ends of chapters. Here is a sample of some of the topics explored: "The U.S. Farm Problem," "The Economics of Law Firms," "All That Glitters: The De Beers Diamond Cartel," "Is Urban Mass Transit Worth the Cost?" "The Breakup of AT&T," "If You Are Taxed More, Will You Work Less?" "International Capital Movements and the Dollar." For a full listing of "Economics in Action" boxes, see the table of contents.
- Within each chapter, we have set off brief, interesting perspectives on theory and institutions in "Focus" boxes. Some of the "Focus" topics include "Corporate Takeovers and 'Greenmail,'" "The Pros and Cons of

Advertising," "The Marginal Revenue Product of Professional Baseball Players," "The Soviet Underground Economy." Again, see the table of contents for a complete listing.

- A special historical feature titled "Point-Counterpoint" concludes each Part of the book. Here we offer side-by-side biographies of some of the most important thinkers in economics and compare and contrast their theories. Part III, "Microeconomic Principles of Input Demand and Supply," for example, concludes with a "Point-Counterpoint" on Thomas Malthus and Gary Becker, presenting their different views on the economic roots to population growth. The "Point-Counterpoint" sections are listed in the table of contents.

### Design and Pedagogy

- A broad variety of two-color graphs are integral to our presentation. The graphs are large and easy to read. Captions carefully summarize the major points in each graph.
- Large, flow-chart diagrams and other illustrations and photographs enhance the visual appeal of the book.
- Key terms are printed in the margins and listed at the end of each chapter so that students can review material systematically.
- All of the key terms are also gathered together in a complete glossary at the end of the book.
- Each chapter concludes with a concise chapter "Summary" and a useful selection of "Questions for Review and Discussion."

### Organization

*Microeconomics* is divided into five parts: "The Power of Economic Thinking," "Microeconomic Behavior of Consumers, Firms, and Markets," "Microeconomic Principles of Input Demand and Supply," "Microeconomics and Public Policy," and "International Trade and Economic Development."

### The Power of Economic Thinking

Part I is a general introduction to basic tools of economic analysis, including opportunity cost, marginal analysis, comparative advantage, supply and demand, and market equilibrium. We have packed these chapters with lively, interesting issues and applications. Chapter 2, for instance, looks briefly at the effects of trade barriers between the United States and Japan, and in Chapter 4 we show how usury laws affect the supply and demand for loanable funds.

### Microeconomic Behavior of Consumers, Firms, and Markets

Part II begins the formal presentation of microeconomic theory. It covers elasticity, consumer choice, the firm, the firm's costs of production, and output market structures. Although the sequence and coverage of these chapters is relatively standard (and effective), we do include a number of interesting innovations. Chapter 7, "The Firm," for example, introduces students not only to the various firm structures but also to the broader economic questions of why firms exist, what roles they serve, and why they take the forms they do. Chapter 9, "The Competitive Firm and Industry," introduces students to the concept of rivalrous behavior and to the dynamic process of competition as well as to the more static model of "pure" competition. Chapter 11, "Monopolistic Competition, Oligopoly, and Cartels," explores recent theory about cartel behavior, relating this theory to problems now experienced by OPEC.

### Microeconomic Principles of Input Demand and Supply

Part III covers demand and supply in factor markets for labor, land, and capital. Perhaps the most unusual chapter in this section is Chapter 15, "Rents, Profits, and Entrepreneurship," which, among other things, investigates the modern theory of rent-seeking behavior and the special role entrepreneurial ability plays in our economy. At the end of this chapter we profile the remarkable careers of two well-known computer entrepreneurs, Stephen Jobs of Apple Computer and Jack Tramiel of Atari. Part III closes with a detailed look at the issue of income distribution, including a careful examination of economic discrimination against women and minorities.

### Microeconomics and Public Policy

In Part IV we integrate microeconomic theory and public policy. We cover four broad areas: industry concentration, market failure, taxation, and public choice. Chapter 17, "Market Structure and Public Policy," traces the evolution of antitrust policy from the days of the Standard Oil trust to the present era of decontrol. Chapter 18, "Market Failure and Public Policy," applies the modern theory of externalities to such diverse problems as endangered species, international defense alliances, shrinking oil reserves, as well as the familiar smoke-belching factory. Chapter 19, "Taxation," lays the groundwork for understanding the current plans for tax reform in the United States. Chapter 20, "The Theory of Public Choice," shows how economic analysis yields startling new insights into the political decision-making process. For example, students will discover one of the reasons why lobbyists are such a major part of our political machinery.

### International Trade and Economic Development

In Part V we conclude the book with three chapters on international trade, international finance, and comparative economic systems and development. Chapter 21 applies the basic theory of comparative advantage to an analysis of tariffs and quotas. The "Economics in Action" for this chapter examines how protectionist sentiments extend even to a small industry, domestic flower growers. Chapter 22 covers many of the most pressing international financial problems, including the recent debt problems of several Third World countries. Finally, Chapter 23 analyzes the relationship between a country's economic system of incentives and its economic growth. We take a close look at the Soviet Union's economy in light of its weak performance. We also examine the various problems of economically underdeveloped countries.

### The Complete *Economics* Package

- The *Test Bank* to accompany *Economics* includes a broad assortment of items to test students in recall, inference, calculation, and graph interpretation.
- All test items are available on microcomputer software.
- The student *Study Guide* to accompany *Economics* helps students review and relearn key concepts and provides a broad assortment of exercises and

applications to test and reinforce student understanding. Each chapter in the guide is divided into six parts: "Chapter in Perspective," "Learning Objectives," "Review of Key Concepts," "Helpful Hints," "Self-test," and "Something to Think About."
- The *Instructor's Manual* to accompany *Economics* includes brief introductions to the rationale and scope of each chapter, a complete list of suggested readings for each chapter, and additional teaching resources.
- Two-color overhead transparencies.

## Acknowledgments

No book of this scope can be written without a great deal of help and advice. Our book is certainly no exception. A large number of "official" reviewers—listed at the end of this section—were invaluable in improving the quality of our work, and we express our deep gratitude to them. In the category "friendly unofficial critics," we gratefully acknowledge the advice of the following: Richard Ault, Don Bellante, Andy Barnett, Raymond Battalio, Steve Caudill, Charles DeLorme, Roger Garrison, Randy Holcombe, George Horton, John Jackson, Mark Jackson, Charles Maurice, François Melese, Steve Morrell, Richard Saba, David Saurman, and David Whitten. For special help and assistance we wish to thank Gary Anderson, Bob Hebert, Bill Shughart, and Mark Thornton. To our friend Keith Watson, who authored the Study Guide and the Instructor's Manual to this book and who provided expert assistance throughout the entire project, we owe a very special debt. In this same category we wish to thank Greg Tobin, our developmental editor at Little, Brown, whose impact appears on every page of this book. His sound determination to make improvements from concept to the final product is the most valuable kind of assistance that authors could obtain. We also wish to thank Al Hockwalt and Will Ethridge, former and present economics editors at Little, Brown, for their faith and support. We extend our gratitude to development editor Shelley Roth, production editor Sally Stickney, designer George McLean, copyeditor Barbara Flanagan, and editorial assistant Max Cavitch. We also want to thank graduate students Don Boudreaux, Brian Goff, Ladd Jones, Yvan Kelly, Karen Palasek, Kendall Somppi, Deborah Walker, and Biff Woodruff for their aid at various stages of the project. Secretaries Cynthia Spinks and, most especially, Pat Watson who typed the bulk of the manuscript several times, were very able help. Our official reviewers were:

Richard K. Anderson, Texas A & M University
Ian Bain, University of Wisconsin—Milwaukee
Robert Barry, College of William and Mary
W. Carl Biven, Georgia Institute of Technology
Ronald G. Brandolini, Valencia Community College
Jacquelene M. Browning, Texas A & M University
Bobby N. Corcoran, Middle Tennessee State University
Judith Cox, University of Washington
Larry Daellenbach, University of Wisconsin—La Crosse
Harold W. Elder, University of Alabama
Donald Ellickson, University of Wisconsin—Eau Claire
Keith D. Evans, California State University—Northridge
Susan Feiner, Virginia Commonwealth University
David Gay, Brigham Young University
Kathie Gilbert, Mississippi State University

Otis Gilley, University of Texas—Austin
William R. Hart, Miami University (Oxford)
J. Paul Jewell, Kansas City, Kansas Community College
Ki Hoon Kim, Central Connecticut State University
Patrick M. Lenihan, Eastern Illinois University
Herbert Milikien, American River College
Jim McKinsey, Northeastern University
Norman Obst, Michigan State University
Samuel Parigi, Lamar University
Glenn Perrone, Pace University
John Pisciotta, Baylor University
E. O. Price III, Oklahoma State University
John Price, San Francisco State University
Robert Pulsinelli, Western Kentucky University
Mark Rush, University of Florida
Don Tailby, University of New Mexico
Allan J. Taub, Cleveland State University
Chris Thomas, University of South Florida
Abdul M. Turay, Mississippi State University
Michael Watts, Purdue University
Donald A. Wells, University of Arizona
Walter J. Wessels, North Carolina State University
George Zodrow, Rice University
Armand J. Zottola, Central Connecticut State University

RBE and RDT

# A Note to Students

We have designed special features in *Microeconomics* that make it easier for you to preview, read, and review the contents of each chapter. Become acquainted with these features and you will give yourself a better chance to succeed on quizzes and exams.

Before you read each chapter, take a few minutes to preview its topics. Begin with the brief chapter overview in the first few paragraphs. Then glance over the section headings and subheadings and read over the chapter summary at the end. After reading the chapter through slowly for the first time, check your comprehension by using the review questions and the list of key terms at the end of each chapter. Reading the "Focus" boxes and the "Economics in Action" box is another good way to test your understanding of concepts. (For each chapter, the *Study Guide* to accompany *Economics* includes a section of additional "Helpful Hints.")

During your second reading, be sure to go over each graph and table in detail, reading the caption and rereading the text description slowly. Try to ask yourself questions as you read and review each graph, and try drawing the graph yourself. After reading the chapter a second time, test your comprehension by going over the end-of-chapter questions and working through the problems there. (The *Study Guide* contains additional short-answer problems.)

Before each quiz, use the key-term definitions printed in the margins to help you review chapter concepts. Also reread the chapter summary. (The *Study Guide* includes a sample selection of multiple-choice test items for each chapter.) With all of these tools at hand, you should find the study of economics a bit less of a chore!

# Brief Contents

## I The Power of Economic Thinking   1

1. Economics in Perspective   3
   APPENDIX: Working with Graphs   20
2. Economic Principles   28
3. Markets and the U.S. Economy   46
4. Markets and Prices: The Laws of Demand and Supply   68

## II The Microeconomic Behavior of Consumers, Firms, and Markets   97

5. Elasticity   99
6. The Logic of Consumer Choice   123
   APPENDIX: Indifference Curve Analysis   134
7. The Firm   142
8. Production Principles and Costs to the Firm   158
9. The Competitive Firm and Industry   177
10. Monopoly: The Firm as Industry   203
11. Monopolistic Competition, Oligopoly, and Cartels   228

## III Microeconomic Principles of Input Demand and Supply   261

12. Marginal Productivity Theory and Wages   263
13. Labor Unions   286
14. Capital and Interest   306
15. Rents, Profits, and Entrepreneurship   321
    APPENDIX: Putting the Pieces Together: General Equilibrium in Competitive Markets   341
16. The Distribution of Income   344

## IV Microeconomics and Public Policy   373

17. Market Structure and Public Policy   375
18. Market Failure and Public Policy   401
19. Taxation   421
20. The Theory of Public Choice   440

## V International Trade and Economic Development 461

- 21 International Trade  *463*
- 22 The International Monetary System  *487*
- 23 Economic Systems and Economic Development  *514*

**Glossary**  *543*
**Index**  *555*

# Contents

## I  The Power of Economic Thinking  1

### 1  Economics in Perspective  3

**What Economics Is (and What It Isn't)**  3

Economics: A Working Definition  4   The Economic Condition: Scarcity  4   Scarce Resources and Economic Problems  5

**The Power of Economic Thinking**  7

Resources Cost More Than You Think  7   Economic Behavior Is Rational  8   Choices Are Made at the Margin  9

FOCUS   Grubbing the Rational Way  10

Prices Are the Signals to Produce More  11   Economists Won't Say Who Should Get What  12   We All Need Money, But Not Too Much  13   Voters Choose the Role of Government—The Economist Only Criticizes  13

**The Role of Theory in Economics**  14

The Need for Abstraction  14   The Usefulness and Limitations of Theory  15   Positive and Normative Economics  16   Microeconomics and Macroeconomics  16   Why Do Economists Disagree?  17

ECONOMICS IN ACTION   Marginal Analysis in Everyday Decisions  19

APPENDIX   Working with Graphs  20

**The Purpose of Graphs**  20

**How to Draw a Graph**  22

General Relations  22   Complex Relations  24

**Slope of the Curve**  25

### 2  Economic Principles  28

**Opportunity Cost: The Individual Must Choose**  28

**Opportunity Cost and Production Possibilities: Society Must Choose**  30

The Law of Increasing Costs  32   Unemployment of Resources  32   Choices Are Made at the Margin  32

**How the Production Possibilities Frontier Shifts**  33

Factors Causing a Shift  33   Economic Growth: How Economies Progress  34

Specialization and Trade: A Feature of All Societies  35
FOCUS   How Societies Can Regress: Armageddon Economics  36
FOCUS   Less-Developed Countries, Recent U.S. Growth, and Production Possibilities  37
Absolute Advantage  37   Comparative Advantage  39   Exchange Costs  41

ECONOMICS IN ACTION   Comparative Advantage: The Case of Bio Bio  44

## 3 Markets and the U.S. Economy  46

### The U.S. Market System in Perspective  46
Comparative Systems  46   Contemporary Economic Organizations  48   Money in Modern Economies  49   The Circular Flow in a Market Economy  50

### Institutions of American Capitalism  50
Property and the Law  50   Free Enterprise  52   Competitive Economic Markets  52

FOCUS   Adam Smith's "Invisible Hand"  53

The Limited Role of Government  53

### The Mixed System of American Market Capitalism  54
An Evolving Competitive Process  54   The Expanded Role of Government  54

### Growth of Government  56
The Size of the Federal Budget  56   Government Expenditures  58   Government Receipts: The U.S. Tax System in Brief  60   The United States and Other Mixed Economies  61   The Economic Effects of Increased Government  62

ECONOMICS IN ACTION   Exchange in the Bazaar Economy  64

ECONOMICS IN ACTION   The Mixed Economy of Sweden: How Much Government Is Too Much?  66

## 4 Markets and Prices: The Laws of Demand and Supply  68

### An Overview of the Price System  69

### The Law of Demand  71
The Individual's Demand Schedule and Demand Curve  72   Factors Affecting the Individual's Demand Curve  72   Reviewing the Law of Demand  75   From Individual to Market Demand  76

### Supply and Opportunity Cost  78

### The Law of Supply and Firm Supply  79
Changes in Quantity Supplied and Shifts in the Supply Curve *81*
Market Supply *82*

### Market Equilibrium Price and Output  82
The Mechanics of Price Determination *83*   Price Rationing *85*
Effects on Price and Quantity of Shifts in Supply or Demand *85*

### Simple Supply and Demand: Some Final Considerations  88
Price Controls *88*

FOCUS   Usury and Interest Rates: An Application of Price Controls  *90*

Static Versus Dynamic Analysis *90*   Full Versus Money Prices *91*
Relative Versus Absolute Prices *91*

FOCUS   Full Price and a New Orleans Restaurant  *92*

ECONOMICS IN ACTION   When Other Things Do Not Remain Equal: The Pope and the Price of Fish   93

POINT-COUNTERPOINT   Adam Smith and Karl Marx: Markets and Society   95

## II  The Microeconomic Behavior of Consumers, Firms, and Markets   97

### 5  Elasticity   99

#### Price Elasticity of Demand   100
Formulation of Price Elasticity of Demand *100*   Elastic, Inelastic, and Unit Elastic Demand *101*   Elasticity Along the Demand Curve *103*
Relation of Demand Elasticity to Expenditures and Receipts *106*

#### Determinants of Price Elasticity of Demand   109
Number and Availability of Substitutes *109*   The Importance of Being Unimportant *110*   Time and Elasticity of Demand *110*

#### Other Applications of Elasticity of Demand   111
Income Elasticity of Demand *111*

FOCUS   Time, Elasticity, and the U.S. Demand for OPEC Oil   *112*

Cross-Elasticity of Demand *113*

#### Elasticity of Supply   114
The Elasticity of Labor Supply *114*   Time and Elasticity of Supply: Maryland Crab Fishing *115*

FOCUS   Gasoline, Cigarettes, and Elasticity: The Effects of an Excise Tax   *117*

ECONOMICS IN ACTION   The U.S. Farm Problem   120

## 6 The Logic of Consumer Choice 123

**Utility and Marginal Utility** 123

**Consumer Equilibrium: Diminishing Marginal Utility Put to Work** 125

Basic Assumptions *125* Balancing Choices Among Goods *126*

FOCUS The Diamond-Water Paradox *128*

**From Diminishing Marginal Utility to the Law of Demand** 128

The Substitution Effect and Income Effect *129* Marginal Utility and the Law of Demand *129*

**Some Pitfalls to Avoid** 130

FOCUS Consumers' Surplus *131*

ECONOMICS IN ACTION: Is It Rational to Vote? 133

APPENDIX **Indifference Curve Analysis** 134

**Indifference Curves** 134

Characteristics of Indifference Curves *135* The Budget Constraint *138* Consumer Equilibrium with Indifference Curves *139*

**Indifference Curves and the Law of Demand** 140

## 7 The Firm 142

**Market and Firm Coordination** 142

Distinguishing Market and Firm Coordination *143* Least Cost and Most Efficient Size *144*

**Team Production** 145

The Problem of Shirking *146* Enter the Manager *146* Monitoring the Manager *147* Voluntary Acceptance of Managers' Commands *147*

**Size and Types of Business Organizations** 147

Scale of Production *148* Categories of Ownership *148* Other Types of Enterprises *151*

FOCUS Corporate Takeovers and "Greenmail" *152*

**The Balance Sheet of a Firm** 153

ECONOMICS IN ACTION The Economics of Law Firms 156

## 8 Production Principles and Costs to the Firm 158

**Types of Costs** 158

Explicit and Implicit Costs *159* Economic Profits Versus Accounting Profits *159*

FOCUS  The Opportunity Cost of Military Service  *160*
Private Costs and Social Costs  *160*

## Economic Time  161
## Short-Run Costs  161

Fixed Costs and Variable Costs  *161*   Diminishing Returns  *163*
Diminishing Marginal Returns and Short-Run Costs  *166*

## Long-Run Costs  170

Adjusting Plant Size  *170*   Economies and Diseconomies of Scale  *171*   Other Shapes of the Long-Run Average Total Cost Curve  *172*

## Shifts in Cost Curves  173
## The Nature of Economic Costs  173
## Costs and Supply Decisions  174

ECONOMICS IN ACTION   Costs to the Firm and Costs to Society: The Case of Pollution  *176*

# 9 The Competitive Firm and Industry  177

## The Process of Competition  177
## A Purely Competitive Market  178
## The Purely Competitive Firm and Industry in the Short Run  180

The Purely Competitive Firm as a Price-Taker  *180*

FOCUS  The Stock Market: An Example of Perfect Competition  *181*

The Demand Curve of the Competitive Firm  *182*   Short-Run Profit Maximization by the Purely Competitive Firm  *183*   A Numerical Illustration of Profit Maximization  *185*   Economic Losses and Shutdowns  *187*   Supply Curve of a Purely Competitive Firm  *188*   From Firm to Industry Supply  *189*

## The Purely Competitive Firm and Industry in the Long Run  191

Equilibrium in the Long Run  *191*   The Adjustment Process Establishing Equilibrium  *191*   The Long-Run Industry Supply Curve  *194*

FOCUS  The Effect of an Excise Tax on a Competitive Firm and Industry  *195*

## Pure Competition and Economic Efficiency  197

The Competitive Market and Resource Allocation  *198*   Reality Versus Pure Competition  *199*

ECONOMICS IN ACTION   Competitive Markets in Disequilibrium  *201*

## 10 Monopoly: The Firm as Industry 203

### What Is a Monopoly? 203
FOCUS Mercantilism and the Sale of Monopoly Rights 204
Legal Barriers to Entry 205 Economies of Scale 206

### Monopoly Price and Output in the Short Run 206
The Monopolist's Demand Curve 206 The Monopolist's Revenues 208 Total Revenue, Marginal Revenue, and Elasticity of Demand 210 Short-Run Price and Output of the Monopolist 211 Monopoly Profits 213 Differences from the Purely Competitive Model 215

### Pure Monopoly in the Long Run 215
Entry and Exit 215 Adjustments to Scale 215 Capitalization of Monopoly Profits 216

### Price Discrimination 217
When Can Price Discrimination Exist? 218 A Model of Price Discrimination 218
FOCUS Perfect Price Discrimination 220

### The Case Against Monopoly 220
The Welfare Loss Resulting from Monopoly Power 220 Rent Seeking: A Second Social Cost of Monopoly Power 222 Production Costs of a Monopoly 223 Monopoly and the Distribution of Income 224

### The Case for Monopoly 224

### ECONOMICS IN ACTION State Liquor Monopolies 227

## 11 Monopolistic Competition, Oligopoly, and Cartels 228

### Monopolistic Competition 229
Characteristics of Monopolistic Competition 229 Short-Run and Long-Run Equilibrium Under Monopolistic Competition 232
FOCUS The Pros and Cons of Advertising 233
Does Monopolistic Competition Cause Resource Waste? 237 How Useful Is the Theory of Monopolistic Competition? 239
FOCUS Why Is There a Funeral Home in Most Small Towns? 240

### Oligopoly Models 241
Characteristics of Oligopoly 241 Models of Oligopoly Pricing 243

### Cartel Models 248
Characteristics of Cartels 248 Cartel Formation 249 The Cartel in Action 250 Cartel Enforcement 250 Why Cartels May or May Not Succeed 252

### From Competition to Monopoly: The Spectrum Revisited 253

ECONOMICS IN ACTION  The Prisoner's Dilemma and Automobile Style Changes  256

ECONOMICS IN ACTION  All That Glitters: The De Beers Diamond Cartel  257

POINT-COUNTERPOINT  Alfred Marshall and Joan Robinson: Perfect Versus Imperfect Markets  259

# III Microeconomic Principles of Input Demand and Supply  261

## 12 Marginal Productivity Theory and Wages  263

### Factor Markets: An Overview  264
Derived Demand  264   Firms as Resource Price Takers  264

### Marginal Productivity Theory  266
The Profit-Maximizing Level of Employment  266   Profit-Maximizing Rule for All Inputs  269

### The Demand for Labor  269
The Short-Run Market Demand for Labor  269   The Firm's Long-Run Demand for Labor  271   Changes in the Demand for Labor  271   The Elasticity of Demand for Labor  273   Monopoly and the Demand for Labor  274

### The Supply of Labor  275
Individual Labor Supply  275
FOCUS  The Marginal Revenue Product of Professional Baseball Players  276
Human Capital  278   Other Equalizing Differences in Wages  280
FOCUS  Does It Pay to Go to College?  281

### Marginal Productivity Theory in Income Distribution  282

ECONOMICS IN ACTION  Regional Wage Differences  284

## 13 Labor Unions  286

### Types of Labor Unions  286
Craft Unions  287   Industrial Unions  288   Public Employees' Unions  288

### Union Activities  289
Union Goals  289   Union Objectives and the Elasticity of Demand  290

FOCUS   The Political Power of Unions   *292*
Increasing the Demand for Union Labor   *292*

## Monopsony: A Single Employer of Labor   293

Monopsony and the Minimum Wage   *295*   Bilateral Monopoly and the Need for Bargaining   *295*   Collective Bargaining   *296*   Strikes   *297*   Political Influence in Bargaining   *298*

## Union Power Over Wages: What Does the Evidence Show?   299

FOCUS   The Right-to-Work Controversy   *300*
The Impact of Unions on Labor's Share of Total Income   *301*

## Conclusion   303

ECONOMICS IN ACTION   Toward a New Theory of Labor Unions   *304*

# 14 Capital and Interest   306

## Roundabout Production and Capital Formation   306
## The Rate of Interest   308
## Demand and Supply of Loanable Funds   309

Variations Among Interest Rates   *310*

## Interest as the Return to Capital   311
## The Nature of Returns to Owners of Capital   312

Pure Interest   *312*   Risk   *313*   Profits and Losses   *313*

## The Present Value of Future Income   314
## Investment Decisions   315

FOCUS   Present Value and Calculating the Loss from Injury or Death   *316*

## The Benefits of Capital Formation   317
ECONOMICS IN ACTION   Stocks and Bonds   319

# 15 Rents, Profits, and Entrepreneurship   321

## Types of Rent   321
## Land Rents   322

Pure Economic Rents   *322*   The Supply of Land to Alternative Uses   *324*

FOCUS   Land Rent, Location, and the Price of Corn   *325*

Land and Marginal Revenue Product   *326*

Specialization of Resources and Inframarginal Rents  326
The Economics of Accountants and Superstars  329
Rents and Firms  331
    Quasi-Rents *331*   Monopoly Rents *331*
Rent Seeking  333
Profits and the Allocation of Resources  333
    Profit Data *334*   Rate of Return on Capital *335*   Accounting Versus Economic Profits *336*   Profits and Entrepreneurship *336*
General Equilibrium  337
    FOCUS   Venture Capitalism  *338*
ECONOMICS IN ACTION   Computer Entrepreneurs  339
APPENDIX   Putting the Pieces Together: General Equilibrium in Competitive Markets  341

## 16 The Distribution of Income  344

The Individual and Income Distribution  345
    Labor Income *345*   Income from Savings and Other Assets *346*   Taxes and Transfer Payments *346*

How Is Income Inequality Measured?  347
    Interpretation of Money Distribution Data *348*   Economists' Measures of Income Inequality *348*   Problems with Lorenz and Gini Measures *350*

Income Distribution in the United States  351
    Adjustments to Money Income *352*
    FOCUS   Do the Very, Very Rich Get Richer?  *353*
    Historical Changes in Adjusted Income *355*   Age Distribution and Income *356*   Rags to Riches Mobility *356*

Some Characteristics of Income Distribution  357
    Family Income Characteristics *357*   Poverty *359*   Poverty During Recession *360*   Race and Sex Discrimination in Wages *362*   Programs to Alleviate Poverty *363*

The Justice of Income Distribution  365
    FOCUS   Pareto Movements and Income Distribution  *366*

ECONOMICS IN ACTION   Who Are the Economic Minorities in the United States?  368

POINT-COUNTERPOINT   Thomas Malthus and Gary Becker: The Economics of Population Growth  370

# IV Microeconomics and Public Policy 373

## 17 Market Structure and Public Policy 375

### Measuring Industrial Concentration 376

Aggregate Concentration 376  Industry Concentration Ratios 376  Differences in the Number of Firms 378

FOCUS  New Merger Guidelines 381

Changes in the Number of Firms 381

### Monopoly Profit Seeking 382

Cartels 384  Trusts 385  Collusion 385  Mergers 385  Government-Managed Cartels 386

### Public Policy 386

Antitrust Law 386

FOCUS  Cutthroat Competition 388

The Value of Antitrust Policies 388  Enforcement of Antitrust Policies 389  An Economic View of Antitrust Cases 392  Regulation 392  Government Ownership 394  Laissez-Faire 395

FOCUS  What Happened to the IBM and GM Monopolies? 397

Conclusion 397

**ECONOMICS IN ACTION**  The Breakup of AT&T  399

## 18 Market Failure and Public Policy 401

### The Economics of Common Property 402

Fishing in International Waters 402  Drilling for Oil 403

### Externality Theory 404

Negative and Positive Externalities 404  Determining the Need for Government Intervention 404  Correcting Relevant Externalities 405

### Public Goods Theory 411

FOCUS  Saving Endangered Predators 412

Public Provision of Public Goods 412  Free Riders 413  Pricing Public Goods 414

### Cost-Benefit Analysis 416

### A Role for Government 418

**ECONOMICS IN ACTION**  Is Urban Mass Transit Worth the Cost? 419

## 19 Taxation 421

### Government Expenditures 421

### Tax Revenues 423

### Theories of Equitable Taxation  423
Benefit Principle  *424*
FOCUS  Tax Freedom Day  *425*
Ability-to-Pay Principle  *426*

### Types of Taxes  427
Personal Income Tax  *427*   Corporate Income Tax  *429*   Social Security Taxes  *429*   Property Taxes  *430*   Sales and Excise Taxes  *430*

### The Effects of Taxes  430
Basic Effects on Price and Quantity  *431*   Tax Incidence  *432*   The Allocative Effect of Taxes  *433*   The Aggregate Effect of Taxes  *433*

### Tax Reform  434
The Need for Tax Reform  *434*
FOCUS  The Laffer Curve  *435*
Suggested Reforms  *436*

**ECONOMICS IN ACTION**   If You Are Taxed More, Will You Work Less?  438

## 20  The Theory of Public Choice  440

### The Self-Interest Axiom  441

### Normative Public Choice Analysis  442
Majority Voting  *442*
FOCUS  A National Town Meeting  *444*
Constitutional Choice  *444*

### Positive Public Choice Analysis  446
Political Competition  *446*   Logrolling  *448*   Voting  *449*   Interest Groups  *450*   Bureaucracy  *452*
FOCUS  Interest Groups and the British Factory Acts  *453*

### The Growth of Government  454

**ECONOMICS IN ACTION**   Voter Preferences and Majority Voting  457

**POINT-COUNTERPOINT**   A. C. Pigou and Ronald Coase: Solving the Problem of Social Costs  459

# V  International Trade and Economic Development  461

## 21  International Trade  463

### The Importance of International Trade  464
National Involvement in International Trade  *464*   Why Trade Is Important: Comparative Advantage  *465*   Barriers to Trade  *471*

### The Effects of Artificial Trade Barriers  471
FOCUS  State Protectionism  472

Trade and Tariffs: Who Gains? Who Loses? 472  The Benefits of Free Trade 474  Welfare Loss from Tariffs 475  Why are Tariffs Imposed? 475

### The Case for Protection  477

National Interest Arguments for Protection 477  Industry Arguments for Protection 478

FOCUS  The Effects of Textile Quotas  479

### U.S. Tariff Policy  481

Early Tariff Policy 481  Modern Tariff Policies 482  Where Does the United States Stand in the Battle for Free Trade? 483

**ECONOMICS IN ACTION**  When Is a Rose Not a Rose? The Flower Industry Seeks Protection  485

## 22 The International Monetary System  487

### The Foreign Exchange Market  488

Rates of Exchange 488  How Are Exchange Rates Determined? 489

### Floating Exchange Rates  489

Exchange Rates Between Two Countries 489  Changes in Floating Exchange Rates 491

### Fixed Exchange Rates  494

The Operation of a Fixed Exchange Rate System 495  Changes in Fixed Exchange Rates 496

FOCUS  Macroeconomic Policy and the Foreign Exchange Market  497

### The Gold Standard  498

### The Balance of Payments  500

Exports and Imports: The Balance of Trade 501  Net Transfers Abroad 502  Current Account Balance 502  Net Capital Movements 502  Statistical Discrepancy 502  Transactions in Official Reserves 502  The Balance of Payments and the Value of the Dollar 503

### The Evolution of International Monetary Institutions  504

The Decline of the Gold Standard 504  Bretton Woods and the Postwar System 505

FOCUS  Should We Return to a Gold Standard?  506

The Role of the Dollar in the Bretton Woods System 507  The Current International Monetary System 507

FOCUS  World Debt Crisis  510

**ECONOMICS IN ACTION**  International Capital Movements and the Dollar  512

## 23 Economic Systems and Economic Development 514

### Comparative Economic Systems 514
### The Generality of Economic Analysis 516
FOCUS The Socialist Calculation Debate 517

The Law of Demand 518 Opportunity Cost 518 Diminishing Marginal Returns 518 Self-Interest 519

### The Soviet-Style Economy 519

A Command Economy 519 Central Planning 520 The Legal Private Sector in the Soviet Economy 522 Comparative Performance 523

FOCUS The Soviet Underground Economy 524

Income Distribution in the Soviet-Style Economy 527

### The Problem of Economic Development 527

The Characteristics of Less-Developed Countries 528 The Economic Gap Between Developed and Less-Developed Countries 529

### Causes and Cures for Underdevelopment 531

The Vicious Circle of Poverty 531 Population Growth and Economic Development 532 International Wealth Distribution and Economic Development 533 Property Rights Arrangements and Economic Development 534

### Conclusion: Explaining Economic Development 537
### ECONOMICS IN ACTION The Rising Japanese Economy 539
### POINT-COUNTERPOINT David Ricardo and Gunnar Myrdal: Does Free Trade Always Benefit the Traders? 541

### Glossary 543

### Index 555

# Microeconomics

# I

# The Power of Economic Thinking

# Economics in Perspective

Does economics matter? More to the point, why should you spend precious time and money learning economics when there are so many other activities, products, and services—not to mention other college courses—competing for your attention? The answer: Economics touches all facets of our lives as consumers of hamburgers and home computers, as voters for political candidates, and as workers and employers. Economics analyzes why we are poor or rich as individuals and extends its scope to government policies about inflation, unemployment, economic growth, and international trade. Close study of economics thus gives an entirely new perspective on a wide variety of human activities and institutions.

## What Economics Is (and What It Isn't)

Most people would say that economics deals with the stock market or with how to make money by buying and selling gold, land, or some other commodity. This common view contains a grain of truth but does not touch on the richness, depth, and breadth of the matter. Economics is a social science—the oldest and best developed of the social sciences. As such, it studies human behavior in relation to three basic questions: What **goods** and **services** are produced? How are goods and services produced? For whom are goods and services produced?

**Goods:** All tangible things that humans desire.

**Services:** All forms of work done for others—such as medical care and car washing—that do not result in production of tangible goods.

All societies and all individuals have faced these three questions. Since goods and the wherewithal to produce them have never existed in limitless amounts, the insistent questions—How? What? and For whom?—must be asked; for at least two hundred years, economists have tried to analyze how individuals and societies answer them. Consider some famous economists' definitions of economics:

*Adam Smith* (1776): Economics or political economy is "an inquiry into the nature and causes of the wealth of nations."[1]

*Nassau William Senior* (1836): Political economy is the science that treats the nature, the production, and the distribution of wealth.[2]

*Karl Marx* (1848): Economics is the science of production. Production is a social force insofar as it channels human activity into useful ends.[3]

*Alfred Marshall* (1890): "Political Economy or Economics is a study of mankind in the ordinary business of life . . . it is on the one side a study of wealth; and on the other, and more important side, a part of the study of man."[4]

*Lionel Robbins* (1935): "Economics is the science which studies human behavior as a relationship between ends and some means which have alternative uses."[5]

*Ludwig von Mises* (1949): Economics "is a science of the means to be applied for the attainment of ends chosen, not . . . a science of the choosing of ends."[6]

*Jacob Viner* (1958): "Economics is what economists do."[7]

*Milton Friedman* (1962): "Economics is the science of how a particular society solves its economic problems."[8]

### Economics: A Working Definition

There is merit in each of the preceding definitions; economists like them all. Economics *is* the study of how nations produce and increase wealth. Economics also studies the activities of people in producing, distributing, and consuming wealth. It analyzes how people and particular societies choose among competing goals or alternatives, but not how or why the goals are chosen. All this is "what economists do."

A common thread runs through all definitions of economics. However narrow or broad, each definition emphasizes the inescapable fact that **resources**—the wherewithal to produce goods and services—are not available in limitless quantities and that people and societies, with unlimited desires for goods and services, must make some hard choices about what to do with the resources that are available. Our working definition of economics includes these elements and may be expressed as follows:

> Economics is the study of how individuals and societies, experiencing limitless wants, choose to allocate scarce resources to satisfy their wants.

**Resources:** Inputs necessary to supply goods and services. Such inputs include land, minerals, machines, energy, and human labor and ingenuity (called the factors of production).

### The Economic Condition: Scarcity

What, exactly, is **scarcity**? More to the point, what are scarce resources? Dorothy Parker, an American humorist, once said, "If you can't get what

---

[1] Adam Smith, *An Inquiry into the Nature and Causes of the Wealth of Nations*, ed. Edwin Cannan (1776; reprint, New York: Modern Library, 1937).

[2] N. W. Senior, *An Outline of the Science of Political Economy* (1836; reprint, New York: A. M. Kelley, 1938).

[3] Karl Marx, *Capital*, trans. Ernest Untermann, ed. F. Engels, 3 vols. (Chicago: Charles Kerr, 1906–1909).

[4] Alfred Marshall, *Principles of Economics* (London: Macmillan, 1920), p. 1.

[5] Lionel Robbins, *The Nature and Significance of Economic Science*, 2nd ed. (London: Macmillan, 1935), p. 16.

[6] Ludwig von Mises, *Human Action* (New York: Regnery, 1966), p. 18.

[7] Jacob Viner in Kenneth Boulding, *The Skills of the Economist* (Cleveland: Howard Allen, 1958), p. 1.

[8] Milton Friedman, *Price Theory* (Chicago: Aldine, 1962), p. 1.

**Scarcity:** Limitation of the amount of resources available to individuals and societies relative to their desires for the products that resources produce.

you want, you'd better damn well settle for what you can get." The entire study of economics amplifies and expands on Parker's proposition. As individuals and as a society, we cannot get all of what we want because the amount of available resources is limited. Our ancestor *homo erectus* faced this problem, as do the bush people of the African Kalahari desert today; King Louis XIV of France and Howard Hughes could not get all of what they wanted; Americans and the people of underdeveloped Chad face limits. The role of the economist and of economics in general is to explain how we can make the most of this problem of scarcity—in Parker's terms, how to get as much as we can of what we want.

The most important problem in economics is that while the wants of individuals and societies must be satisfied by limited resources, the wants themselves are not limited; rather, they are endless. We are never satisfied with what we have. Individuals are forever lured by more tempting foods, more cleverly engineered computers, more up-to-date fashions. Societies continually desire safer highways, more accurate missiles, greater Social Security benefits, or more cancer research.

Scarcity is of course relative to time and fortune. Our generation has many more services and goods to choose from than our parents and grandparents did. The quantity and quality of goods and services may have grown from primitive to modern times, but the supply of resources needed to produce them is limited, and human wants are not.

**Free goods:** Things that are available in sufficient quantity to fill all desires.

You may feel that some things are not scarce and that the best things in life—such as love, sunshine, and water—are free. In economic terms, **free goods** are goods that are available in sufficient supply to satisfy all possible demands. But are many things truly free? Surface water is usually unfit for drinking except in areas far from human habitation. Water suitable for drinking must be raised to the surface from deep wells or piped from reservoirs and treatment plants, operations involving resources that are scarce even when water itself is not. Scarcity of winter sun in the North results in costly winter vacations in warm states. And if you think love is free . . . .

**Economic goods:** Scarce goods.

**Costs:** The value of opportunities forgone in making choices among scarce goods.

All scarce goods—from television to chlorinated water—are called **economic goods.** Their scarcity leads to **costs.** While it is customary to associate cost with the money price of goods, economists define *cost* as the value of what individuals have to forgo to acquire a scarce good. Since all unlimited wants cannot be met with scarce resources, individuals have to make choices—between, for instance, more steaks and more computer games; societies may have to choose between safer highways and more accurate missiles. Cost is therefore the direct result of scarcity of resources. Scarcity of resources means that both individuals and societies must endure the costs of acquiring more of any good or service. That cost is the value of the good given up in place of the good chosen.

### Scarce Resources and Economic Problems

There are basically two categories of scarce resources: human resources and nonhuman resources (see Table 1–1). **Human resources** encompass all types of labor, including specialized forms of labor such as management or entrepreneurship. **Nonhuman resources** include the rest of the bounty: land, natural resources such as minerals and water, capital, and still other resources such as technology and time.

**Human resources:** All forms of labor used to produce goods and services.

**Nonhuman resources:** Inputs other than human labor involved in producing goods and services.

Examples of human resources abound. By definition, all human resources apply talent and energy to produce goods and services. The cook at the Chicken Shack, the hair stylist at the Mad Hacker, the chief executive of a

**TABLE 1–1  Economic Resources**

Economic resources include all human and nonhuman resources that are scarce in supply. Technology and time are scarce and can be categorized as resources.

| Human Resources | Nonhuman and Other Resources |
| --- | --- |
| Labor, including entrepreneurship and management | Land |
|  | Natural resources, including minerals and water |
|  | Capital |
|  | Technology |
|  | Time |

computer firm, and the assembly-line worker at a General Motors plant all represent human resources. Obviously, labor includes a huge variety of skills, both general and precise. Economists are interested not only in the scarcity of labor but in its equally scarce quality. The quality of human resources can be enhanced through investments in education and training.

Economists view *entrepreneurship* as a special form of labor. An entrepreneur is a person who perceives profitable opportunities and who combines resources to produce salable goods or services. Entrepreneurs attempt to move resources from lower- to higher-valued uses in the economy and take the risk that, by so doing, they can make profits. Lemonade-stand entrepreneurs, for example, see an opportunity to make a profit by combining lemons, ice, and cups and by selling the final product. *Management*, a second special form of labor, guides and oversees the process by which separate resources are turned into goods or services. The successful lemonade entrepreneur could hire a manager to oversee the opening of new stands around the neighborhood.

Human resources utilize nonhuman resources such as land, minerals, and natural resources to produce goods and services. A plot in Manhattan, an acre in Iowa, a coal deposit in Pennsylvania, a uranium mine in South Dakota, and a timber stand in Oregon are all scarce nonhuman resources. New deposits of minerals can be discovered, forests can be replanted, and agricultural land can be reclaimed from swamps. But at any one time, the available supply of nonhuman resources is limited.

*Capital*, a second category of nonhuman resources, comprises all machines, implements, and buildings used to produce goods and services either directly or indirectly. A surgeon's scalpel, a factory, an electric generator, and an artist's brush are all used to produce goods and services and thus are considered capital.

Many different forms of capital may be needed to produce a single economic good. With a wheat harvesting machine, a South Dakota farmer can reap a huge crop. But the wheat must also be milled into flour and transported from South Dakota to bakeries in California. Once the wheat has arrived, bakeries must utilize brick or convection ovens to produce bread. The harvesting and milling machines, the railroad, and the baker's ovens are all capital goods, created to increase the amount of final production.

Capital—and the resources used to produce it—is scarce. To create capital, we must sacrifice consumer goods and services because the production of capital takes time away from the production of goods that can be consumed in the present. Societies and individuals must therefore choose between immediate consumption and future consumption. That choice is crucial to growth and ultimate economic well-being. We return to this important issue in Chapter 2.

*Technology* and *time* can also be regarded as resources. Technology, in general, is a resource composed of all "know-how," inventions, and innovations that help us get more from scarce resources. Finer distinctions can be made. Technology is knowledge of production methods. An improvement in technology implies that we produce more with a certain amount of inputs. Existing technology is the outcome of many inventions, some of which were the invention of new resources—such as aluminum, radium, hybrid plants. All inventions that increase the productivity of labor and capital can be thought of as improvements in technology. Innovation is the application of technology to the production of goods and services.

Information is a scarce and costly resource. The acquisition of information on which to base purchasing or managerial decisions has never been free. In the nineteenth century and well into the twentieth, businesses hired armies of bookkeepers to provide sufficient information for managers to use to make decisions. The digital computer, developed in the mid-twentieth century, made information storage and retrieval far less costly. In fact, technology has progressed so rapidly that the quantity of information that could be stored in a warehouse-sized computer in 1950 can now be placed on a chip the size of a fingernail with a lot of room to spare! Technology, then, is a resource that helps make other resources less scarce.

Time is another scarce resource because while its quantity is fixed, the things we could do with it proliferate. We must therefore constantly make choices. We may choose to spend an evening listening to a New Wave rock band instead of studying accounting or hearing a Bach organ recital. We may choose a career in acting or dance instead of in law or computer programming. Retirees may choose to return to the classroom rather than spend time fishing or playing bridge. Like all human and nonhuman resources, time is scarce and is a cost that we inevitably incur when we make a choice.

Scarcity of resources is at the core of all economic problems. Resources can be augmented over time; indeed, we are much better off materially than our grandparents, and our grandparents were better off than their grandparents. At any one time, however, individuals and societies cannot get all of what they want. Given scarcity, individuals and societies must make choices, and a primary role of economists is to analyze scarcity and the process of choosing.

## The Power of Economic Thinking

All economists of all political stripes have common fundamental ways of thinking. These perspectives, at the core of the science of economics, appear many times throughout this book. When properly and consistently used, they brand a person as adept in the economic way of thinking. A look at these economic perspectives in simple, commonsense language should convince you that economics and economic reasoning are closely related to decisions you make every day.

### Resources Cost More Than You Think

What does it cost you to take a skiing weekend in the mountains or to make a trip to the beach during spring break? Your instant reply might include costs of gasoline, auto depreciation, air fare, lift tickets, food, drink

**Accounting costs:** Direct costs of an activity measured in dollar terms; out-of-pocket costs.

tainment, and a motel room. These money expenditures are called **accounting costs**.

Economists do not look upon costs in the traditional manner of the accountant. Rather, economists look at what must be given up—in the case of the ski trip, what must be given up to make the trip. Economists ask what opportunities must be forgone when we choose anything, and they call the forgone opportunities **opportunity costs**.

**Opportunity costs:** The value placed on opportunities forgone in choosing to produce or consume scarce goods.

Mary makes a beach trip for three days. She is a college student and has a part-time job at a local restaurant earning $30 per night. Mary could have worked and earned $90 rather than take the trip. Economists would say $90 is one of the opportunity costs of making the trip.

Such opportunity costs also exist for society. Use of government lands in Wyoming and Montana as recreation areas or national forests entails a cost. Through lease or purchase, these lands could be used as a source of oil, minerals, and timber. Such use would contribute to society's well-being, but park land serves the recreational needs of society as well. Whatever choice society makes for the use of the land will incur a cost in economic terms. An opportunity must be forgone when the land is used in either manner.

A favorite saying of economists is "There is no such thing as a free lunch." The first fundamental principle of economics is that most things in life come at the opportunity cost of something forgone. They are never free.

### Economic Behavior Is Rational

The second fundamental principle of economics is that people behave according to **rational self-interest**. Economists selected from a number of alternative views of human behavior and settled on a very simple one—that of *homo economicus* (economic man or woman). Rather than viewing humans as inconsistent, incompetent, selfish, or altruistic, economists argue that human behavior is predictably based on a person's weighing the costs and benefits of decisions. This choice making is influenced by constraints such as the availability of options, personal tastes, values, and social philosophy. A student will choose to eat lunch at the local health food restaurant rather than the fast-food cafeteria on campus if the personal benefits of doing so, say eating nutritious food in a pleasant atmosphere, outweigh the costs, such as longer lines and greater distance to be walked.

**Rational self-interest:** The view of human behavior espoused by economists. People will act to maximize the difference between benefits and costs as determined by their circumstances and their personal preferences.

When costs or benefits change, behavior may change. Consider Jack, a poor but honest man. When he finds $100 cash in a phone booth, he turns it in to authorities. Had Jack found $1000, he might have acted similarly, but discovery of $100,000 might have caused Jack to pause. It may be more costly to be honest under some circumstances.

The view of the individual espoused by economists has always been subject to misinterpretation. When economists say that humans are self-interested, they do not mean that other views of human behavior and motivation are unimportant or irrelevant. Altruistic or charitable behavior, for example, is perfectly compatible with *homo economicus*. If a person values altruism, then an act that benefits others will carry emotional benefits and will therefore be in that person's rational self-interest. Economists simply maintain that people calculate the costs and benefits of their decisions in acting in their own self-interest.

If the IRS were to quadruple the income tax deduction for charitable contributions, economists would predict an increase in charity. No economist would deny that love, charity, and justice are important aspects of human behavior. Indeed, an economist has argued that "that scarce resource Love" is in fact "the most precious thing in the world" and that humans are therefore compelled to conserve it in ordering the affairs of the world.[9] Economists have merely advanced the simple but powerful proposition that, given personal tastes, values, and social philosophy, rational self-interest is a better guide to predicting behavior than any other assumption about why people act as they do.

The economists' view of the rational self-interested individual applies not only to what is usually termed "economic behavior" but also to other realms of behavior. Animals other than humans, such as birds or rats, can be shown to be self-interested in that they act to maximize their own well-being given their constraints (see Focus, "Grubbing the Rational Way"). Politicians maximize their self-interest by wooing voters to elect and reelect them. Since politicians face periodic elections, political behavior such as garnering support from particular voters can be predicted using the economic self-interest assumption. Even matters such as dating, marriage, and divorce have been subjected to analysis using the self-interest assumption! After more than two hundred years of economic theorizing, rational self-interest remains the most powerful predictor of most behavior.

### Choices Are Made at the Margin

Economists do not ordinarily analyze all-or-nothing decisions but are concerned with decisions made at the **margin**—the additional costs or benefits of a specific change in the current situation. An individual consumer, for example, does not decide to spend his or her entire budget solely on food, cassette tapes, or weekends at the beach. Consumers purchase hundreds of goods and services. Their choice to purchase or not purchase additional units of any one good is based on the additional (or *marginal*) satisfaction that that single unit would bring to them.

**Marginal analysis** is a method of finding the optimal, or most desirable, level of any activity—how much coffee to drink, how much bread to produce, how many store detectives to hire, and so on. In an economic sense, every activity we undertake involves *both* benefits and costs, so an optimal level of an activity is the point at which the activity's benefits outweigh its costs by the greatest amount. For example, the optimal level of coffee drinking is reached when coffee's total benefits (its taste and stimulating effects, perhaps) exceed its costs (the expense, the health risks, the acid indigestion) by the greatest degree.

You might find it easy to decide how much coffee to drink on a particular evening, but what of other, more complicated decisions? How, for example, would a baker decide how much bread to produce? He or she might try to add up the total costs and benefits of baking each loaf, but this could be a costly if not impossible task. Through marginal analysis, the baker would look at his activity in small steps. If the net benefit (total benefit − total costs) increases every time a loaf goes in the oven, then the baking will continue. If the net benefit decreases, then the baking will decrease or stop as well.

Or suppose you are the manager of a large department store and want to reduce the amount of shoplifting that takes place. By hiring additional store

**Margin:** The difference in costs or benefits between the existing situation and a proposed change.

**Marginal analysis:** Study of the difference in costs and benefits between the status quo and the production or consumption of an additional unit of a specific good or service. This, not the average cost of all goods produced or consumed, is the actual basis for rational economic choices.

---

[9]D. M. Robertson, "What Does the Economist Economize?" in *Economic Commentaries* (London: Staples, 1956), p. 154.

## FOCUS  Grubbing the Rational Way

The economists' view of rational behavior—that humans react, predictably, to their perceptions of costs and benefits—has been recognized in many forms for some time. Rational behavior under scarcity was certainly recognized by Charles Darwin and other early natural scientists. Is the economists' view time-bound and applicable only to twentieth-century industrialized countries and only to humans, as some critics argue, or does it apply in all times and cultures and among all other species as well?

The coal tit is a common forest-dwelling English bird that resembles the American chickadee. This small bird evidently hasn't paid any attention to claims that economic behavior is not predictable. Surprisingly, the behavior of the coal tit provides an excellent example of economic principles in operation. This bird responds "rationally" to "price" changes and even searches for the "cheapest market" to shop in!

The coal tit survives through the winter by eating the larvae of the eucosmid moth, which it finds lying dormant in small cavities just under the surface of pine cones. The bird finds the moths by tapping on the outside of the pine cones. Naturalists can distinguish in the spring between moth larvae that were eaten by the coal tit, were destroyed by some other cause, or survived to become moths. An ecologist and naturalist named Gibb determined the number of moth larvae in pine cones in different areas of the forest in the fall. The next spring, Gibb returned to see how many had fallen prey to the coal tit over the winter.

Using the data collected by the naturalist, economist Gordon Tullock[a] derived the graph shown in Figure 1–1. (The appendix to this chapter contains a general explanation of graphs.) The horizontal axis represents the number of larvae in different localities. The vertical axis signifies the number of larvae per 100 pine cones in each locality. Curve A plots the number of larvae found in the fall, before coal tit predation began; Curve B shows the number of larvae that survived predation, counted in the following spring. Hence the space between curve A and curve B repre-

[a] Gordon Tullock, "The Coal Tit as a Careful Shopper," *American Naturalist* (November 1969), pp. 77–80.

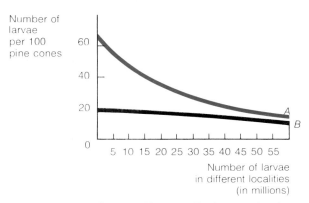

**FIGURE 1–1  Larvae Density Before and After the Coal Tits' Predation**

sents the number of larvae consumed during the winter by the birds in different localities.

An interesting relation is immediately apparent: The coal tits have obviously concentrated their predation in localities where the density of larvae is over 60 per 100 cones. As density of available larvae decreases, so does predation. Tullock explains that the amount of energy spent in seeking out each larva is the "price" the bird must pay to consume the larva. Where the density of larvae is high, this price is low, and vice versa. In other words, the coal tit consumes more larvae when the price is low and less larvae when the price is high.

But this is not all. The fact that the coal tits systematically seek out those areas of the forest with higher density of larvae and avoid the low-density areas indicates that the birds are behaving much like we would expect a careful shopper to behave. They are searching for the cheapest market and then shopping there.

This apparently rational behavior of the coal tit does not imply that the tiny bird engages in conscious economic calculation. It is more likely that searching for the higher-density areas is an instinctive response that developed in the birds as the result of natural selection. Its value in increasing the coal tits' chances of survival over a long, cold winter when food is scarce is obvious. But the economic efficiency of the behavior remains striking regardless of the coal tits' inspiration.

---

detectives, you will benefit from fewer thefts. But the detectives must be paid, so the net benefit is the amount you save minus the costs of the detectives. If store detectives cost $200 per week, you can perform an experiment. Hire an additional detective and see whether losses from shoplifting exceed $200 per week. If additional losses are more than $200, hire additional detectives, if less, let detectives go. Stop hiring when losses per week are exactly equal to the weekly wage of detectives. (*Note that the optimal level of shoplifting you will permit is not zero!*)

For another example, suppose a shipbuilding company that produces four ships at a time consistently receives more orders than it can fill. The land on which the company is located allows room for expansion. In deciding whether to enlarge the operation, the rational manager must weigh the marginal costs and benefits of producing more ships. Will it be profitable to produce one more ship? That is, will the net benefits increase? To accommodate work on five ships at one time, the company would have to enlarge its boat yard, construct storage space for more shipbuilding materials, and pay its current workers overtime or train new workers. Will the extra costs of building one more ship be warranted, given the expected benefit, the current selling price of a ship? Can the company better maximize profits by building ten ships at a time? By making no change unless ship prices rise? Rational self-interest requires isolating and considering the specific effects of specific changes—of choices made at the margin, in the area of change. (Economics in Action at the end of this chapter provides further illustrations of marginal analysis.)

### Prices Are the Signals to Produce More

What products and services are produced in our economic system and how are they produced? In a market economy **prices** are the essential signals that tell producers and resource suppliers what and how to produce. Take a common product such as apple jelly. Apple jelly is not free at the supermarket. Since it bears a price, you can be sure that scarce resources—such as apples, sugar, and jars—are being used to produce it and that someone finds it profitable to do so.

**Prices:** The opportunity costs established in markets for scarce goods, services, and resources.

*Buyers and sellers in action at a commodity exchange*

What happens if consumers suddenly develop a craving for apple jelly, preferring it to all other kinds? Consumers express this craving by buying as much apple jelly as they can, leaving such products as grape jam and apricot preserves gathering dust on the shelves. Meanwhile, the relative unavailability of apple jelly causes the price to rise because there are insufficient resources devoted to apple jelly production to meet the new demand for it. Manufacturers, noting the increased sales and rising prices, recognize that additional production of apple jelly will be profitable. Their reaction is to produce additional jelly by ordering more jars, buying more apples from apple farmers, and hiring more labor and machinery. Again, higher prices of apples, jars, and other resources are the signal to the apple farmers and the other resource suppliers to sell more of their inputs and services. Why do they do it? Because higher prices mean greater profitability to them.

All goods, including apple jelly, are produced in this manner. Consumers are sovereign in that they decide what is to be produced—TVs, automobiles, beets—with the scarce resources in our society. Economists say that consumers transmit their desires through **markets,** which are simply arrangements whereby buyers and sellers exchange goods or services for money or some other medium. Buyers and sellers can be physically together in a market, as in a livestock market, or they can be separated by geographic distance or by wholesalers and other intermediaries, as when the buying and selling of a company's stock is carried out through stockbrokers. A market may even be an abstraction, as is the labor market, in which the supply of and demand for jobs and workers are juggled on a broad scale. The prices established in these markets (through mechanics to be explored in Chapter 4) are the key to what gets produced and how much is produced in our economic system. Economists therefore give a great deal of attention to the role and function of prices and markets in explaining how things get done.

**Market:** An arrangement that brings together buyers and sellers of products and resources.

## Economists Won't Say Who Should Get What

Economists study poverty and wealth but they will not, as economists, say how rich people should be or how much poverty is tolerable. Economists do not decide whether a college football coach deserves $16 million on a five-year contract or whether a Nobel Prize–winning physicist deserves more. Economists simply explain that the value of all individuals—expressed as their income—is determined by the relative desires for their services on the one hand and by the relative scarcity of their talents and abilities on the other. Perhaps good football coaches are relatively more scarce and more highly desired than good physicists.

Most products and services—such as food, vacations, and domestic help—are allocated to those with the greatest desire and ability to pay for them. Our ability to pay is determined by the value of our own particular services or resources in the marketplace. This value depends on our education, on-the-job training, health, luck, inheritance, and a host of other factors.

The distribution of "public" goods and services—such as highways and bridges, MX missiles, and "free" movies on campus—is conducted by national, local, and student governments through the filter of politics and voting. A decision whether to add seating capacity to the football stadium or to double the size of the English department at a public university may be made by a university committee appointed by the president and board of trustees, who may be appointed by the governor, who is elected by the citizens of the state. Ultimately, therefore, decisions on how resources are to be distributed are made by voter choices. If voters do not like the decisions made in the political process, they have a periodic right to change the decision makers. Chosen by voters, politicians make economic decisions based on their perceptions of costs and benefits to the public—as well as to their own chances of reelection. In government as in the workings of the market, the chief role of economists is not to say who should get what but to describe and analyze the process—the costs, benefits, and incentives—through which the distribution of all products and services takes place.

## We All Need Money, But Not Too Much

Economists also place a special significance on the role of money in our lives. The trade of goods for goods, called **barter,** will not work in any economy that completes several million transactions every day. Barter may have been sufficient in primitive societies in which few things were produced and traded, but as individuals became specialized, using more and more of their talents and skills, the number and kind of traded commodities and services grew. As the number of commodities grew, the cost of transacting by barter grew enormously. A common denominator, acceptable to all, developed within economic societies to reduce the costs and inconveniences of barter. That common denominator is **money,** which serves as a medium of exchange.

Money is not limited to coins and paper dollars. Almost anything—shells, feathers, paper, gold, and cattle, for instance—can serve as money. The important point to economists is that money, whatever it is, reduces the costs of barter and increases specialization and trade. At the same time, economists are concerned that money should not be available in such quantities as to become unacceptable and valueless to people who produce, buy, and sell. For society, at least, huge increases in the money supply may be too much of a good thing.

When the quantity of money increases beyond its use as a means of making transactions, confidence in the medium of exchange and its value (in terms of the quantity of goods and services a unit of it can purchase) deteriorates. Economists call this phenomenon **inflation.** Inflation was an especially sad fact of American life during the 1970s, and it is a persistent element in less-developed nations around the world. Inflation, if left untreated, can result in sheer collapse of entire economies with reversion to primitive barter conditions. Economists therefore are very concerned with the relation between the production and trade of goods and the quantity of money available to facilitate these crucial activities. Breakdowns must be prevented, and economists have valuable insights on preventive measures.

## Voters Choose the Role of Government— The Economist Only Criticizes

In a functioning democratic system, society at large chooses an economic role for government by electing politicians. In general, that role has included such activities as taxation for and provision of collective goods such as national defense, highways, and education; the regulation of monopoly; pollution control; control of the money supply; and alleviation of poverty through welfare programs. In its economic role of taxing and spending, moreover, government can affect economic factors such as inflation, the degree of unemployment, economic growth, and international trade.

Economists have always been concerned with the effects of government

---

**Barter:** Direct exchange of goods or services.

**Money:** A generally accepted medium of exchange.

**Inflation:** A sustained increase in prices; a reduction in the purchasing power of money.

activity in these areas. In the United States, economists are in the thick of government. Since 1946 an official Council of Economic Advisers has been appointed by the president to aid in the formulation and implementation of economic policy. Almost every agency of government employs teams of economists to give advice in their areas of expertise.

In their role as advisers, economists primarily evaluate, from the perspective of economic theory or analysis, the effects of proposals or, more correctly, of marginal changes in economic policies. Economists do not argue, for example, that additional funds should be allocated to Social Security and taken away from water reclamation projects. They simply evaluate the costs and benefits of such proposals.

Economists are not confined to giving advice on specific issues such as the Social Security program or the effects of advertising regulation by the Federal Trade Commission. Larger issues are within economists' purview as well. Predicting the effects of government taxing and spending policies on employment, interest rates, and inflation is a very large part of economists' role. Economists might predict, for example, the impact of a tax cut on private spending, interest rates, the private production of goods and services, and inflation. As such, economists are concerned with what is termed *economic stabilization*.

Despite economists' influence, however, the economic role of government is ultimately decided by voters in periodic elections. Specific economic functions are not decided in elections, but politicians run on platforms pledging to "increase social welfare spending," "increase defense spending," "reduce the debt," "balance the budget," "increase environmental regulation," and so on. In a loose sense, then, voters decide economic policies. Economists' role, although critical, is simply to criticize and evaluate the implications of government's economic policies. Economists advise and implement. They do not make ultimate decisions about what the government should do.

## The Role of Theory in Economics

Economists must be able to discern fundamental regularities about human behavior. Only if behavior is regular and consistent, on average, among individuals or from generation to generation will economists be able to predict the results of behavior. The means through which economists organize their thoughts about human behavior and its results are called **models**. Economists, like all scientists, construct models to isolate specific phenomena for study. Constructing models requires assumptions and abstractions from the real world.

**Model:** An abstraction from real-world phenomena that approximates reality and makes it easier to deal with; a theory.

### The Need for Abstraction

The economist, the biochemist, or the physicist could never handle all the details surrounding any event. Thousands of details are involved in even the seemingly simple act of buying a loaf of bread. The friendliness of the clerk, the way the buyer is dressed, and the freshness of the bread are all details surrounding the purchase, and these details may be different each time a buyer makes a purchase. Likewise many details surround a particular chemical reaction or an astronomical event like the appearance of a comet. Economists, however, like chemists or astronomers, are interested in particular details about an event, such as how much the bread cost or how many loaves the buyer purchased.

Economists must *abstract* from extraneous factors to isolate and understand some other factor because, as one economist put it, people's minds are limited and nature's riddles are complex. Economists must assume that the extraneous factors are constant—that "other things are equal"[10]—or that, if altered, they would have no effect on the relations or model under consideration. Humanity has never progressed very far in understanding anything—be it chemistry, astronomy, or economics—without abstracting from the multiplicity of irrelevant facts. *Thus, all economic models are of necessity abstractions. Good models, those that perform and predict well, use relevant factors; poor models do not.*

An economic model (or, essentially the same thing, a theory) can be expressed in verbal, graphical, or mathematical form. Sometimes verbal explanation is sufficient, but graphs or algebra often serve as convenient shorthand means of expressing models. Economics is like other sciences in this regard. A complicated chemical or physical process can be described in words, but it is much more convenient to use mathematics. All methods of expressing models will be used in this book, but we rely primarily on words and graphs. The appendix to this chapter explains how to read and interpret graphs.

## The Usefulness and Limitations of Theory

In recognizing the regularities of human behavior and in committing them to a model, economists or scientists do not have to rethink every event as it occurs. Theory saves time and permits extremely useful predictions about future events based upon regularities of human behavior.

Economic theory has its limits, though. Since a complex world is the economist's workshop, economic theory cannot be tested in the way chemical or physical theories can. Economists can seldom find naturally existing conditions that provide a good test of a model, let alone a sequence of such conditions, and it is difficult to conduct controlled laboratory experiments on human beings to determine their economic behavior. Because of the difficulties of testing, economists are further from certainty than, say, geneticists or biologists. The accuracy of economics is somewhat more like that of meteorology. Total accuracy is not within the meteorologist's abilities. At least, total accuracy is not worth what it would cost to obtain in either weather forecasting or economics. The very bizarre weather of 1983—including a great drought in Australia, coastal storms in California, and floods along the lower Mississippi River—were caused by a complicated host of factors related to the jet steam, volcanic eruptions, and equatorial currents. It took meteorologists a good deal of time to sort out the relevant causes by testing and applying their models.

In economics, the problem of imperfect conditions for testing is a limitation to theory, but theory is essential nonetheless. Nothing can substitute for the usefulness of theory in organizing our thoughts about the real world and in describing the regularities of economic behavior.

In constructing models, moreover, economists and all other scientists must be careful to follow the rules of logic and to avoid common fallacies. Two of these are the **fallacy of composition** and the ***post hoc* fallacy.** The fallacy of composition involves generalizing based only on a particular experience. Suppose that you had never seen a fox, and then the first one you saw was white. Should you conclude that all foxes are white? If you did, you would

**Fallacy of composition:** Generalization that what is true for a part is also true for the whole.

---

[10]The Latin phrase *ceteris paribus*, meaning "other things being equal," is common economic shorthand for an assumption about human behavior.

commit the fallacy of composition. The assumption that if everyone stands up at a crowded football game each will get a better view founders on the same fallacy.

The *post hoc* fallacy (from *post hoc, ergo propter hoc*, "after this, therefore because of this") is the false assumption of cause-and-effect relations between events. The ancient Aztec Indians of Mexico believed that the spring ritual of sacrificing children and virgins appeased the gods of agriculture, bringing a bountiful harvest. When crops appeared in the fall, high priests insisted that the good crop was caused by the sacrifice. If the crop was insufficient, more children and virgins had to be wary the following spring. In economic analysis, the *post hoc* fallacy may be involved in statements such as "high interest rates cause inflation" or "after hoarding sugar, the price of sugar rises" or "the government subsidy to Chrysler Motor Company saved the business." Economists and all who are interested in sound thinking must constantly guard against such fallacies.

**Post hoc fallacy:** From *post hoc, ergo propter hoc*, "after this, therefore because of this." The inaccurate linking of unrelated events as causes and effects.

### Positive and Normative Economics

Economics is usually a positive rather than a normative science. Positive statements describe what is or predict what will be under certain circumstances, whereas normative prescriptions entail value judgments about what should be. For the most part, economists confine themselves to positive statements. Individual values, concepts of justice, or tastes are normative matters and generally do not enter economists' analyses of issues. These matters are not testable in any accepted scientific sense and are therefore excluded from economics.

Consider U.S. welfare programs. To say that additional tax dollars should be spent on welfare would involve a normative judgment on the part of economists. Once society has decided to spend more on welfare, however, economists can make positive statements about the effects of such spending. Or economists may observe the welfare system as it stands and analyze how the system could work more efficiently in achieving its goals. In other words, economists must show, once society has decided on a welfare system and on some dollar amount to be devoted to welfare, how alternative systems of welfare distribution would alter the effectiveness of the program. Economists make positive statements in presenting alternatives and their effects so that society might get the most for the dollars spent.

On a smaller scale, imagine a student facing an afternoon choice between watching soap operas and studying for an economics quiz. In positive terms, economists can only describe the alternatives open to the student. An observer would be on normative grounds if he or she insisted that the student should avoid TV all afternoon and spend the time preparing for the quiz. Free choice is, in this as in all cases, the prerogative of the economic actor. Economists can only array the alternatives. On occasion, normative statements slip into economic discussion, as they undoubtedly do in this book. Beware of them and learn to distinguish them from positive economics.

**Positive economics:** Observations or predictions of the facts of economic life.

**Normative economics:** Value judgments about how economics should operate, based on certain moral principles or preferences.

### Microeconomics and Macroeconomics

Economics is divided into two main parts: microeconomics and macroeconomics. **Microeconomics,** like microbiology, concerns the components of a system. Just as a frog is made up of individual cells of various kinds, individual sales and purchases of commodities from potatoes to health care are the stuff of which the whole economy is composed. Microeconomists are thus concerned with individual markets and with the determination of relative

**Microeconomics:** Analysis of the behavior of individual decision-making units within an economic system, from specific households to specific business firms.

prices within those markets—the price of potatoes versus the price of hot dogs. Supply and demand for all goods, services, and factors of production are the subject matter of microeconomics. Microeconomists address questions such as the following: Will the use of larger quantities of solar energy reduce the total energy bill of Americans? Will an increased tax on cigarettes increase or decrease federal revenues from the tax? Will the quality and quantity of nursing services in the state of Wyoming be changed by the licensing of nursing in that state?

**Macroeconomics** is the study of the economy as a whole. The overall price level, inflation rate, international exchange rate, unemployment rate, economic growth rate, and interest rate are some issues of concern to macroeconomists. Macroeconomists analyze how these crucial quantities are determined and how and why they change.

Some issues contain elements of both microeconomics and macroeconomics; for instance, a proposed U.S. tariff on Japanese auto imports. Microeconomists would be interested in the effects of the tarriff on auto prices, on the American auto industry, and on the quantity of cars bought by U.S. consumers. Macroeconomists would be primarily interested in the effects of the tariff on international trade and on total spending. They might address such questions as: Would the tariff increase or decrease economic growth in the United States and in Japan? Would total employment in the United States or Japan change because of the tax?

Daily television reports and newspapers are filled with discussions of both macroeconomic and microeconomic issues: Do tax cuts to business spur investment and economic growth? Will an OPEC price decrease or an alteration in supply affect oil and gas prices in America? Is the economy headed for more inflation, greater unemployment, or both? Will a tightening of the money supply cause an increase or decrease in interest rates? Does foreign competition mean fewer jobs for domestic autoworkers? Will price competition cause some airlines or trucking firms to shut down? Should the federal government balance the budget or not? A mastery of the fundamental and general principles of economics will give you a good foundation for formulating intelligent and reasoned answers to important questions such as these.

## Why Do Economists Disagree?

Government or academic economists are forever predicting economic doom or economic prosperity for the same future time period. If economics contains fundamental principles agreed on by all, why do economists always seem to be fighting over predictions or over causes and effects? Why do economists disagree? As we noted earlier, one of the limitations of economic science is its inability to test theory. This imperfection means that alternative theories may be offered to explain the same events, such as the effects of deficits or of the Great Depression of the 1930s. Alternative theories mean divergent policy prescriptions and different perspectives on the effects of policy prescriptions.

Even when economists agree on theoretical apparatus, they may differ on the magnitude of the effects suggested by the theory. And secondary effects—effects not considered in a theory—can occur after the initial impact of a policy change has taken place. Economists often disagree on the nature and importance of these secondary effects. Would the addition of new monetary incentives for an all-volunteer army increase or decrease our military effectiveness? Economists of good faith and common fundamentals might disagree. We do not wish to overemphasize economists' disagreements or the

---

**Macroeconomics:** Analysis of aspects of the economy as a whole.

"All the experts say the market is going to turn around, but I say maybe."

Drawing by Saxon; © 1984 The New Yorker Magazine, Inc.

"iffy" or unfinished nature of economics. Meteorologists of equal training and ability might well disagree over whether it will rain tomorrow.

Economics, like meteorology, is an inexact science, and it is likely to remain so. But economic theory and prediction are accurate and testable enough to provide extremely important insights into numerous issues that touch our lives. And the science of choice and scarcity is (like meteorology) constantly being improved from both theoretical and empirical perspectives. The well-established power of economic thinking, as revealed in the following chapters, is the result of two hundred years of such continuing improvements. Sound understanding of the principles of economic thinking will improve your understanding of the world.

## Summary

1. Economics is basically concerned with three questions: What goods and services are produced? How are goods and services produced? For whom are goods and services produced?
2. Economics is the study of how individuals and societies choose to allocate scarce resources given unlimited wants.
3. Scarce resources include human and nonhuman resources as well as time and technology. Human resources consist of labor, including entrepreneurship, and management. Nonhuman resources include land, all natural resources such as minerals, and capital.
4. All scarce resources bear an opportunity cost. This means that individuals or societies forgo opportunities whenever scarce resources are used to produce anything.
5. Economic behavior is rational: Human beings are assumed to behave predictably by weighing the costs and benefits of their decisions and their potential actions.
6. Economic choices are made at the margin, that is, the additional costs or benefits of a change in the current situation. Prices are the signal that indicates individual and collective choices for goods and services as well as the relative scarcity of the resources necessary to produce the goods and services.
7. Economic thinking also concerns the role of money, inflation, unemployment, and stabilization of the overall economy. The study of prices of individual products and inputs is called microeconomics, and the study of inflation, unemployment, and related problems is called macroeconomics.
8. Economics is a more inexact and imprecise science than chemistry or physics. It is accurate enough to predict much economic behavior, however, and to provide insight into a large number of important problems.

## Key Terms

| | | | | |
|---|---|---|---|---|
| goods | costs | margin | money | microeconomics |
| services | human resources | marginal analysis | model | macroeconomics |
| resources | nonhuman resources | prices | fallacy of composition | |
| scarcity | accounting costs | market | *post hoc* fallacy | |
| free goods | opportunity costs | barter | positive economics | |
| economic goods | rational self-interest | inflation | normative economics | |

## Questions for Review and Discussion

1. What problem creates the foundation of economic analysis? Is this problem restricted to the poor? Is it relevant to the animal kingdom?
2. Can wants be satisfied with existing resources?
3. What are resources? How is the resource capital different from the resources land and labor?
4. What is the difference between behaving in one's self-interest and behaving selfishly?
5. What functions do prices have in the economy? Why were things rationed during World War II?
6. What is the basic function of money? What happens when there is too much money?
7. Do economists run the economy?
8. Why do economists abstract from reality when formulating theories? Does this imply that their theories have no relevance in the real world?
9. "The stock market crash of 1929 preceded the Great Depression. Therefore, the fall in stock prices caused the Great Depression." "Periods of inflation are frequently followed by increases in the stock of money. Therefore, inflation causes increases in the money supply." "A occurs after B. Therefore, B causes A." These statements violate which fallacy?
10. "The distribution of income in the United States is not fair." What type of statement is this? Can "fair" be used to describe something without being normative?

## ECONOMICS IN ACTION

### Marginal Analysis in Everyday Decisions

Far from being an abstract, theoretical idea, marginal analysis—making decisions at the margin, or letting "bygones be bygones"—is central to everyday life. Individuals, governments, and businesses are always making decisions based on considerations of marginal benefits and marginal costs. Here are some examples.

You, as a student, may face the following types of decisions on any given day: (1) whether to change majors; (2) whether to take an optional quiz in your economics course. You will make such decisions at the margin—that is, on the basis of marginal (not total) costs and benefits. Contemplating a change in majors would involve both costs and benefits, some monetary, others nonmonetary. How many more credit hours, for example, are required before your current degree program is finished? How does this figure compare to the number of hours required in the alternative degree program? If the latter number is greater, how much income would you forgo by spending more time in school? Clearly, your decision hinges on all of the additional (or marginal) costs and benefits from changing majors. The same principle holds in even less important decisions. Should you take an optional economics quiz to improve your grade? The marginal costs of the quiz include your time spent studying (you could be studying for some other exam), and the marginal benefits depend on whether and how much an excellent quiz grade will improve your average in economics.

Government, through the political process, also makes decisions at the margin. The completion of the MX missile program or the decision to finish a nuclear power plant are examples. These decisions will not be made on the basis of how much has already been spent on the projects but on all of the additional costs and benefits attached to completion. Political opposition, for example, will increase the marginal costs to those who vote to complete the MX project, whereas political (voter) support will increase their marginal benefits. If the costs of completing a nuclear power plant no longer outweigh the benefits of ample electrical power, then it is likely the plant will sit half-finished.

Businesses also make decisions at the margin. Suppose you are the manager of a large grocery store located close to campus. On what basis would you decide to stay open all night rather than close at midnight and open at 6 A.M.? Clearly, the additional costs would include the additional electricity required plus the additional labor cost of cashiers, stock persons, and so on. The additional benefit is the sales revenue the store would make from staying open six more hours each day. In other words, the decision would not be based on the costs of building the grocery store, installing shelves, and so on. The decision, like all decisions, will be based on marginal considerations.

## Questions

1. On what will you base your decision whether to eat another piece of pizza at a pizza parlor that advertises "all you can eat" for $5?
2. Apply marginal analysis to a decision you are facing right now. What are the marginal costs and benefits? How will you make your decision?

# APPENDIX
# Working with Graphs

Economists frequently use graphs to demonstrate economic theories or models. This appendix explains how graphs are constructed and how they can illustrate economic relations in simplified form. By understanding the mechanics and usefulness of graphs, you will find it much easier to grasp the economic concepts presented in this book.

## The Purpose of Graphs

Most graphs in this book are simply pictures showing the relation between economic variables, such as the price of a good and the quantity of the good that people are willing to purchase. There are many such pairs of variables in economics: the costs of production and the level of output, the rate of inflation and the level of unemployment, the interest rate and the supply of capital goods, for example. Graphs are the most concise means of expressing the variety of relations that exist between such variables.

Figure 1–2 illustrates these ideas. It is a bar graph that shows the relation between the U.S. gross national product (GNP), a measure of the value of all goods and services produced annually, and time, in this case the period between 1960 and 1980. The GNP variable is plotted along the vertical line on the left side of the graph. The time variable is plotted along the horizontal line at the bottom of the graph. From this graph you can roughly estimate the level of GNP for each year between 1960 and 1980. For instance, GNP in 1967 was somewhere around $800 billion. You can also see that GNP tended to increase over the years 1960–1980. By working with the graph, you could estimate the rate of increase of GNP from year to year or the percentage increase over a particular period, such as 1965–1970.

This one concise picture—the graph—contains a great deal of information. Aside from economy of expression, graphs have a great deal of cognitive appeal: People can more easily grasp concepts demonstrated through pictures than concepts demonstrated through words.

Some of the graphs in this book are bar graphs; many others use the Cartesian coordinate system, which consists of points plotted on a grid formed by the intersection of two perpendicular lines. See Figure 1–3. The horizontal line is the **x-axis,** and the vertical line is the **y-axis.** The intersec-

**x-axis, y-axis:** Perpendicular lines in a coordinate grid system for mapping variables on a two-dimensional graph. The x-axis is the horizontal line; the y-axis is the vertical line. The intersection of the x- and y- axes is the origin.

**FIGURE 1-2**

**A Bar Graph**

This graph shows the levels of U.S. GNP between the years 1960 and 1980 at five-year intervals.

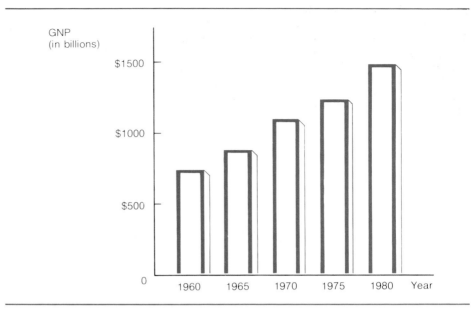

**FIGURE 1-3**

**A Linear Graph**

Any pair of numerical values can be plotted on this grid. The upper-right quadrant, shaded here, is the portion of the graph most often used in economics.

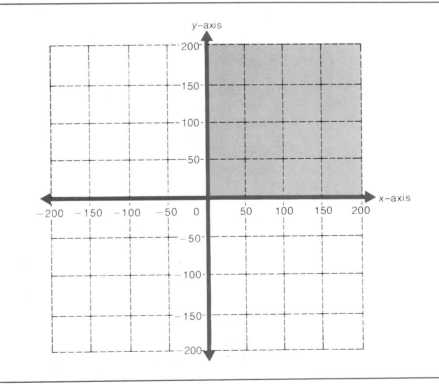

tion of the two lines is the point 0, called the *origin*. Above the origin on the vertical axis, all values are positive. Below the origin on the vertical axis, all values are negative. Values to the right of the origin are positive; values to the left are negative. The economic variables we use in this book are usually positive; that is, to plot economic relations we will usually use the upper-right portion, or quadrant, of a graph such as the one in Figure 1–3.

## How to Draw a Graph

Suppose you wish to graph the relation between two variables. Variable $y$ is the number of words memorized, and variable $x$ is the number of minutes spent memorizing words. Table 1–2 presents a set, or schedule, of hypothetical data for these two variables.

We can plot the data from Table 1–2 on a graph using points in the upper-right quadrant formed by the $x$-axis and the $y$-axis. See Figure 1–4. The variable number of words memorized is measured along the $y$-axis and is considered the $y$ value. The variable number of minutes spent memorizing is measured along the $x$-axis and is the $x$ value. From the data in Table 1–2 we see that each increase of one minute of memorizing resulted in four more memorized words. The graph in Figure 1–4 illustrates this relation.

The points marked as large dots in Figure 1–4 represent pairs of variables. Point A represents the pair on the first line of Table 1–2: 4 words and 1 minute. Point B represents 8 words and 2 minutes, and so on. When we connect the points we have a straight line running upward and to the right of the origin. Lines showing the intersection of $x$- and $y$-values on a graph are referred to as **curves** in this book, whether they are straight or curved lines.

Curves show two types of relations between variables: positive and negative. The relation between the $x$ and $y$ values in Figure 1–4 is positive: as the $x$ value increases, so does the $y$ value. (Or, as the $x$ value decreases, the $y$ value decreases. Either way, the relation is positive.) On a graph, a **positive relation** is shown by a curve that slopes upward and to the right of the origin. In a **negative, or inverse relation,** the two variables change in opposite directions. An increase in $y$ is paired with a decrease in $x$. Or a decrease in $x$ is paired with an increase in $y$. Figure 1–5 shows negative, or inverse, relations on a graph. The curve for a negative relation slopes downward from left to right—the opposite of a positive relation.

## General Relations

Throughout this book, some graphs display general relations rather than specific relations. A general relation does not depend on particular numerical values of variables, as was the case in Figure 1–4. For example, the relation between the number of calories a person ingests per week and that person's weight (other things being equal) is positive. This suggests an upward-sloping line, as shown in Figure 1–6a, with calories on the $x$-axis and weight on the

**Curve:** Any line, straight or curved, showing the correlation between two variables on a graph.

**Positive relation:** A direct relation between variables in which the variables change in the same direction. A positive relation has an upward-sloping curve.

**Negative, or inverse, relation:** A relation between variables in which the variables change in opposite directions. A negative relation has a downward-sloping curve.

TABLE 1–2  Schedule of Hypothetical Data

| Number of Words Memorized ($y$) | Number of Minutes Spent Memorizing ($x$) |
|---|---|
| 4 | 1 |
| 8 | 2 |
| 12 | 3 |
| 16 | 4 |
| 20 | 5 |

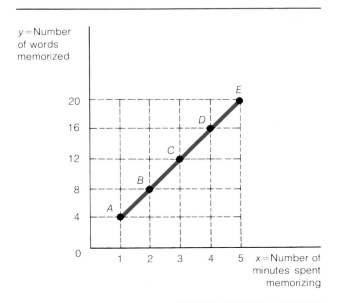

**FIGURE 1–4  A Simple Line Graph**

Curve AE shows the essential relation between two sets of variables: the number of words memorized and the number of minutes spent memorizing. The relation is positive; as the variable on the x-axis increases or decreases, the variable on the y-axis increases or decreases, respectively. A positive relation is shown by an upward-sloping line tracing the intersection of each pair of variables.

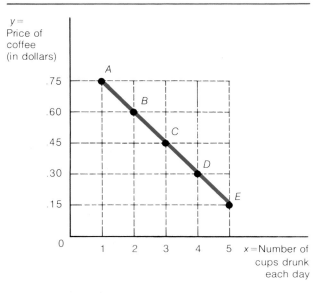

**FIGURE 1–5  A Negative Relation Between Two Variables**

As the price of a cup of coffee increases, the number of cups drunk each day decreases. When variables move in opposite directions, their relation is negative, or inverse. Negative relations are shown by a curve that slopes downward and to the right of the origin.

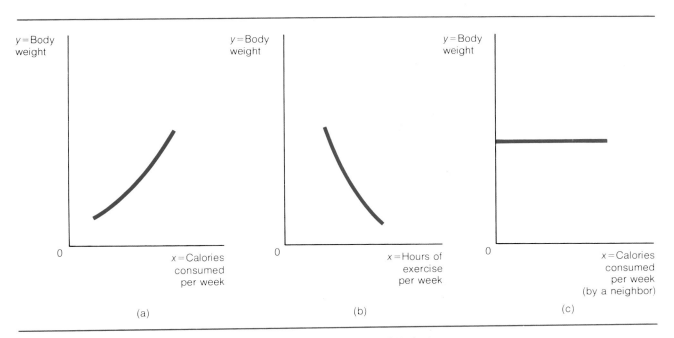

**FIGURE 1–6  General Graphical Relations**

General relations are (a) positive, (b) inverse, or (c) independent. General relations are not measured numerically on the y- or x-axis.

y-axis. Figure 1–6 represents a general relation, so it does not need numerical values on either axis. It does not specify the number of calories required to maintain a particular body weight; it simply shows that an increase in the number of calories consumed increases body weight. Figure 1–6b shows an inverse relation between body weight and the amount of exercise per week, again a general relation. Sometimes two variables are not related. Figure 1–6c shows that a person's weight is independent of a neighbor's caloric intake. Thankfully, no matter how much your neighbor eats, it has no effect on your weight.

At this point you should be able to construct a graph showing simple relations. For example, graph the relation between the weight of a car and the miles per gallon it achieves or between the length of the line at the school cafeteria and the time of day.

## Complex Relations

Occasionally relations are more complex and do not fall into the simple positive or negative category. Figure 1–7 shows that for some relatively low $x$ values, the variables on the graph are positively related, but as the $x$ value increases, a point is reached where the two variables become inversely related.

From the origin to $x^*$ the curve slopes positively. But for values greater than $x^*$, the curve slopes negatively. To illustrate, we let the $x$-axis show the number of pieces of pizza that a friend consumes in an evening and the $y$-axis show the total amount of pleasure she receives from each additional

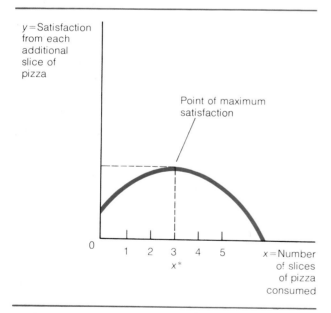

**FIGURE 1–7  Variables Both Positively and Negatively Related**

For lower values of $x$—in this example, the number of pizza slices eaten—the $y$ value—the amount of satisfaction—increases. After some point, say 3 slices, the $y$ value decreases with increasing values of $x$.

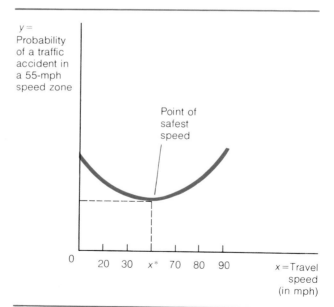

**FIGURE 1–8  A U-shaped Curve**

Some relations yield a U-shape. Variables along the $x$- and $y$-axes are inversely related at first, but after some critical value, here shown as $x^*$, they become positively related.

slice. At first, the more she eats the happier she becomes, but eventually a critical point is reached, after which the more she eats the less pleasure she receives from each additional slice.

Many other relations can exist. A U-shaped curve, as graphed in Figure 1–8, shows that the $x$ value is at first inversely related to the $y$ value and then positively related. After some critical value, $x^*$, the $y$ value begins to rise as the $x$ value increases. The $y$ axis in this example might represent the probability of having an accident on an interstate highway and the $x$-axis might show the miles per hour at which a car travels. To avoid accidents, $x^*$ is the optimal speed. Driving at speeds lower than $x^*$ increases the probability of an accident, but rates faster than $x^*$ mph also increase the chances of an accident.

## Slope of the Curve

**Slope:** The ratio of the change in ($\Delta$) the $x$ value to the change in ($\Delta$) the $y$ value; $\Delta x/\Delta y$.

In graphical analysis the amount by which the $y$ value increases or decreases as the result of an increase or decrease in the $x$ value is the **slope** of the curve. The slope of the curve is an important concept in economics. Much economic analysis studies the margin of change in a variable or in a relation between variables, and the slope of a curve measures the marginal rate of change. In Figure 1–4, for example, every change in the $x$ value—every increase of 1 minute—is associated with an increase of 4 words memorized. The slope of the curve in Figure 1–4 is the rate of change. The concept "change in" is expressed by the symbol $\Delta$. So in Figure 1–4 the slope of the line $AE$ is $\Delta y/\Delta x$, or $4/1 = 4$.

For the straight-line curves in Figures 1–4 and 1–5, the slope is constant. Along most curves, however, the slope is not constant. In Figure 1–8, an increase in speed from 55 mph to 60 mph may increase the probability of an accident only slightly, but an increase from 60 mph to 65 mph may increase the probability by a greater amount. Not only does the probability increase for each mile-per-hour increase but it increases by a greater amount. This indicates that the slope not only is positive but also is increasing.

The slope of a curved line is different at every point along the line. To find the slope of a curved line at a particular point, draw a straight-line tangent to the curve. (A tangent touches the curve at one point without crossing the curve.) Consider point $A$ in Figure 1–9. The slope at $A$ on the curved line is equal to the slope of the straight-line tangent to the curved line. Dividing the change in the $y$ variable ($\Delta y$) by the change in the $x$ variable ($\Delta x$) yields the slope at $A$. In Figure 1–9, every change in $x$ from 0 to 1 and 1 to 2, and so on, results in a change in $y$ along the curve. At point $A$, the change is $y = -4$. So $\Delta y/\Delta x = -4/1 = -4$. At lower points along the curve, such as $B$, the curve is flatter, that is, less steep. The line tangent at $B$ has a slope of $-1$. As we move from point $A$ to point $B$, equal increases in $x$ result in smaller and smaller decreases in $y$. In other words, the *rate* at which $y$ falls decreases as $x$ increases along this particular curve.

Other relations of course lead to different slopes and different changes in the slope along the curve. Along the curve in Figure 1–10, not only does the $y$ value increase as the $x$ value increases but the slope increases as the $x$ value increases. For equal increases in the $x$ value, the incremental changes in the $y$ value become larger.

Interpreting the essential concepts illustrated in graphs is a necessary part

**FIGURE 1–9**

**Slope Along a Curve**

The slope of a curve at a particular point is the slope of a straight-line tangent to the curve at that point.

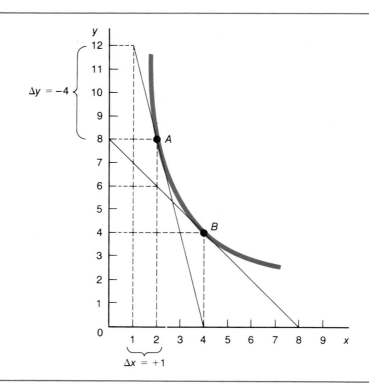

**FIGURE 1–10**

**A Curve with Increasing Slope**

As the x variable increases, the change in the y variable increases.

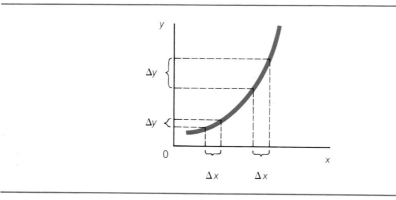

of learning economic principles. Most of the graphs used in this book are really quite simple and require only a few minutes of examination to grasp the relation between the two variables. With a certain amount of practice you can easily understand any graph in the book.

## Summary

1. Graphs are a concise expression of economic models, or theories. Graphs usually show the relation between two variables, such as price and quantity.
2. Linear graphs are drawn with two perpendicular lines, called axes. The x-axis is a horizontal line. Variables measured along the x-axis are called x values. The y-axis is a vertical line. Variables along it are called y values.

3. Lines showing the correlation between x and y values are called curves. An upward-sloping curve indicates a positive relation between variables: As the x value increases or decreases, the y value does likewise. A downward-sloping curve indicates a negative, or inverse, relation between variables: As the x value increases, the y value decreases and vice versa.
4. Relations between variables may be specific or general. Specific relations are based on numerical quantities measured on either the x- or y-axis or both. General relations are not based on specific numerical quantities.
5. Variables can be both positively and negatively related. In such complex relations, the curve is either bow-shaped or U-shaped.
6. The slope of a curve measures the ratio of change in the x-value to change in the y value, expressed as $\Delta y/\Delta x$. Along a straight-line curve, the slope is constant. Along a curved line, the slope is different at every point.

## Key Terms

x-axis          curve                negative, or inverse, relation
y-axis          positive relation    slope

## Questions for Review and Discussion

1. Explain how the slope of a line tells you whether the relation between the two variables illustrated is direct or inverse. If you drew a line illustrating your score on a history test as you study more hours, what would its slope be?
2. Plot the following points on a graph with X on the horizontal axis and Y on the vertical. Connect the dots to show the relation between X and Y.

|   | X | Y |
|---|---|---|
| A | 0 | 2 |
| B | 2 | 4 |
| C | 4 | 8 |
| D | 6 | 14 |
| E | 8 | 22 |

What is the slope of the line between the points A and B? What is the slope between C and D? Is this a straight line?

3. Curves can be represented in algebraic form by an equation. For example, the points along a certain line can be demonstrated by the following equation:

$$Y = 10 - 2X$$

The value of Y can be found by inserting different values of X and solving the equation for Y. Some of the values of Y for different values of X are shown in the following table. Fill in the blanks by solving the equation for the values of Y for each value of X given:

| X  | Y  |
|----|----|
| 0  | 10 |
| 2  | 6  |
| 4  | —  |
| 6  | —  |
| 8  | —  |
| 10 | —  |
| 12 | —  |

Draw this set of points on a graph. Does this show a direct or inverse relation between X and Y? What is the slope of this line between the first and second points? Between the second and third? Is this a straight line?

4. If you measured body weight on the horizontal axis and the probability of dying of heart disease on the vertical axis, what would be the slope of a curve that showed the relation between these two variables?

# 2

# Economic Principles

The essence of the economic problem is how to get as much value as we can from limited and costly resources. To do so, we constantly make choices. A college student may choose to spend a weekend at the beach rather than buy a new sweater and skirt. Another may choose to sleep late rather than attend an eight o'clock chemistry class. Every choice must be made at some cost.

Economists emphasize that decisions to use scarce resources bear an opportunity cost. The **opportunity cost** of a decision is the next most preferred or next-best alternative to a good or activity (the new clothes or the class instruction) that one chose to forgo to obtain some other good (a beach weekend or a late sleep). Economists have developed tools for expressing the choice and opportunity cost that are central to all decisions.

**Opportunity cost:** The next-best alternative that is lost when undertaking any activity.

## Opportunity Cost: The Individual Must Choose

Suppose that a student is trying to decide between typing an overdue paper and studying for an economics quiz and that these two activities are the most desired of all activities on a particular afternoon. The student chooses among five combinations of hours spent typing and numerical grades on the economics quiz (see Table 2–1). Choice (1), no typing, results in a perfect grade on the quiz. Choice (2), 1 hour of typing, gives the student 4 hours of studying and a grade of 90 on the quiz. The third, fourth, and fifth alternatives mean still more typing finished but lower grades on the quiz. In fact, in the fifth and final choice, the student does not study for the quiz at all—yielding a failing grade—but does manage to type the entire paper.

These alternatives are graphed in Figure 2–1. The vertical axis plots the possible hours of typing, and the horizontal axis presents the possible quiz

TABLE 2-1  The Individual Chooses: Typing or Grades

Individuals must make choices from many alternatives. Most frequently these choices are based on evaluations of the combined costs and benefits of the alternatives. The benefit of typing the entire paper would be chosen at the cost of earning a failing grade on the quiz. The choice is not just all or nothing, however. The individual has intermediate options involving varying combinations of some studying and some typing.

| Choice | Pages Typed | Grade Earned |
|---|---|---|
| (1) 0 hours typing, 4 hours studying | 0 | 100 |
| (2) 1 hour typing, 3 hours studying | 4 | 90 |
| (3) 2 hours typing, 2 hours studying | 8 | 75 |
| (4) 3 hours typing, 1 hour studying | 12 | 55 |
| (5) 4 hours typing, 0 hours studying | 16 | 0 |

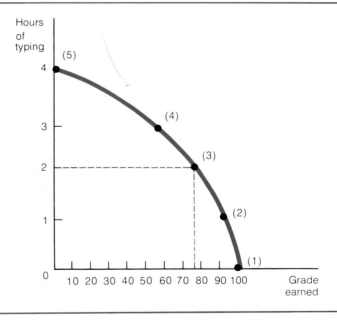

FIGURE 2-1  The Individual Chooses: Typing or Grades

As the hours of typing increase, the score on the economics quiz falls.

grades. The points representing the combinations of grade and pages typed, as given in Table 2-1, can be connected as a curve. Choice (3), 2 hours of typing and 2 hours of studying, is shown as point (3). With that choice, the student types 8 pages and gets a 75 on the quiz.

Several aspects of Figure 2-1 are important. *All possible combinations* of typing time and study time are represented by the curved line, meaning that choices between typing and studying are *continuous*. Since they are continuous, the student could choose any combination of hours of typing and quiz grade—say a half-hour of typing and a grade of 95. If the choices were *discrete*, the student would be limited to the five choices labeled (1) to (5). For convenience, economists ordinarily assume that economic relations are continuous.

A second crucial point is that Figure 2–1 and Table 2–1 constitute what economists call a *model*. Economists construct models to isolate specific phenomena for study. The model summarized in Figure 2–1 and Table 2–1 assumes a number of things about the student in the example. First, the choices are continuous rather than discrete. Second, we assume as given the amount of time the student puts into the course before the afternoon in question. Also assumed is the student's IQ. Does the student have the flu, a factor that would detract from performance on the quiz? Is the overdue paper more important to the student's academic standing than the economics quiz? Are there other alternatives such as washing a car or spending time in language lab? Is the fact that one alternative has a future payoff, such as a high-paying economics-related job, relevant to the choice? Certainly matters such as these are significant, but we assume them as given with respect to the choice at hand.

Economists also emphasize that the decision to study or to type is made at the margin. Practically, making decisions at the margin means that we are faced not only with all-or-nothing choices but also with degrees of balancing units of one choice against another. Each choice would be made in terms of the costs of an additional hour of studying or typing forgone. Satisfactions from both activities are thus balanced **at the margin**.

**Choices at the margin:** Decisions made by examining the benefits and costs of small, or one-unit, changes in a particular activity.

In making decisions at the margin we assume that all other factors are equal, or fixed. Each time we make a decision it is based on some fixed or given factors resulting from previous decisions. In our example, the student would weigh, for instance, the amount of previous study, a fixed factor in choosing typing or studying on the afternoon in question.

Previous decisions cannot be changed and are gone forever, but that does not mean that they have no impact on current choices. It simply means that we make choices in the present with unchangeable past choices as given and unalterable. This is another aspect of making choices at the margin.

## Opportunity Cost and Production Possibilities: Society Must Choose

So far we have focused on individual choices, but the same principles apply to society's choices. Just as an individual is limited by the scarcity of time in choosing among options for an afternoon, society is constrained by scarce resources in producing and consuming alternative goods. Let us examine an abstract example. Suppose a society produces only two goods—oranges and peanut butter—and that all its resources and technology are fixed and constant in amount and level.

Given the state of technology and assuming that the society uses all its resources, the society could produce either 100 million jars of peanut butter or 16 million tons of oranges (see Table 2–2). Or the society may choose to devote some of its resources to producing peanut butter and some to producing oranges. These choices, called society's production possibilities, are numbered A–E in Table 2–2. From this information a production possibilities

TABLE 2–2  Society Decides: Oranges or Peanut Butter

Society must choose among many alternative goods to produce. The cost of one good is the lost production of other goods.

| Choice | Peanut Butter (millions of jars) | Oranges (millions of tons) |
|---|---|---|
| A | 100 | 0 |
| B | 90 | 4 |
| C | 75 | 8 |
| D | 50 | 12 |
| E | 0 | 16 |

**Production possibilities frontier:** The situation represented by a curve that shows all of the possible combinations of two goods that a country or an economic entity can produce when all resources and technology are fully utilized and fixed in supply.

curve can be constructed (see Figure 2–2). A production possibilities curve depicts the alternatives open to society for the production of two goods, given all existing resources, human and nonhuman, and an existing state of technology. When resources are fully employed, moreover, society is said to be located on its **production possibilities frontier.**

At choice A in Figure 2–2, society produces 100 million jars of peanut butter and no oranges, while at choice E society uses all of its resources in orange production. Choices in between, with some of both commodities produced, are shown at points B, C, and D. Continuous choices of all other possible combinations are shown along the curve; resources are assumed to

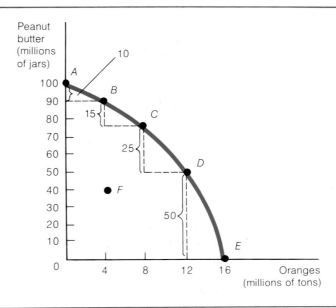

FIGURE 2–2  Production Possibilities Curve

As more and more oranges are produced, the opportunity cost per orange increases. The opportunity cost of producing 4 million tons of oranges is 10 million jars of peanut butter. The opportunity cost of producing 8 million tons of oranges is an additional 15 million jars of peanut butter, and so on.

be fully employed at all points on the curve. The curve therefore shows all of the possibilities for the production of oranges and peanut butter.

### The Law of Increasing Costs

Just as there was an opportunity cost to the student in choosing to study for the quiz or to type the paper, there is an opportunity cost for society in any choice between producing oranges and producing peanut butter.

Consider a society that chooses nothing but peanut butter, choice A in Figure 2–2. The opportunity cost is that no oranges can be produced. A move to choice B, representing a shift of some resources from peanut butter production to orange production, means a sacrifice of 10 million jars of peanut butter, but society gains 4 million tons of oranges. Thus the opportunity cost of the first four million tons of oranges is 10 million jars of peanut butter. To get the second four million tons of oranges, however, a larger quantity of peanut butter must be given up—15 million jars. Thus as society becomes more and more specialized in production of either oranges or peanut butter, the opportunity cost per unit rises. This effect is called the **law of increasing costs**.

Why does the law of increasing costs hold? As we know from experience, resources, training, and talents are not all alike and are not perfectly adaptable to alternative uses. Land suited to orange growing is not equally suited to peanut farming and vice versa. Orange pickers are not equally adept at the manufacture of metal lids for peanut butter jars or even peanut farming without additional training. The most-suited human and nonhuman resources are moved into production first, but as production of a good increases and becomes more specialized, less-adaptable resources must be used to produce it. The costs of the good produced increasingly rise because greater quantities of another good must be given up to get additional quantities of the good produced. A society of peanut butter lovers might be growing peanuts in greenhouses in North Dakota, and concert pianists and nuclear technicians might be operating the machinery in peanut butter processing plants.

### Unemployment of Resources

The law of increasing costs always applies when resources are fully employed. When there is a degree of **unemployment of resources,** however—such as a certain amount of land lying idle—society finds itself at a point *within the* production possibilities curve. Point F in Figure 2–2 represents such a situation. From point F to the curve, additional oranges or peanut butter can be produced without any increase in opportunity costs until resources become fully employed. The production possibilities curve is thus referred to as a *frontier* when it represents full employment of resources. Full employment means that society is realizing its maximum output potential. Many economic problems are the result of unemployment, a subject that gets a good deal of attention in the study of economics.

### Choices Are Made at the Margin

As with individuals, society's choices are made at the margin. Suppose, for example, that a society has for years been one of peanut butter lovers and orange haters. Suppose further that long-time devotion of society's resources to peanut butter production "warped" resources—made them more adaptable over time—to peanut butter production. Past decisions to warp the resources would make current additional orange production more costly. A sudden decision by society to produce more oranges would have to be made at the

---

**Law of increasing costs:** As more scarce resources are devoted to producing one good, the opportunity costs per unit of the good tend to rise.

**Unemployment of resources:** A situation in which human or nonhuman resources that can be used in production are not so used.

margin. Thus, past decisions obviously affect current costs, but past decisions cannot be changed.

## How the Production Possibilities Frontier Shifts

If we keep in mind the meaning of the production possibilities frontier, we will also understand that an increase or decrease in society's human or nonhuman resources can shift society's output potential. An understanding of such changes is central to the understanding of economic growth—or the lack of it.

### Factors Causing a Shift

The production possibility frontier shifts inward or outward in response to any change in the quantity of human resources, any change in the quantity of nonhuman resources, or any change in technology. It makes sense that changes in these factors shift the production possibilities frontier because we previously assumed they were constant.

Suppose a technological improvement occurs in peanut butter production that increases the productivity of peanut harvesters—perhaps a fertilizer or new crop techniques. In such a case, society's production possibilities relating to peanut butter are increased without a similar decrease in orange production. The result is a shift outward and upward on the peanut butter end of the production possibilities curve. Such a change is shown in Figure 2–3a as a shift from the $PP_1$ curve to the $PP_2$ curve.

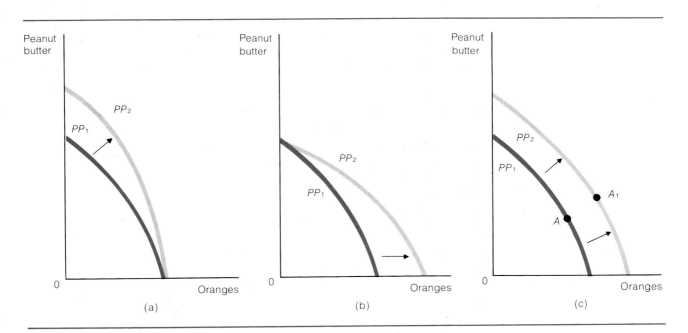

**FIGURE 2–3   Shifting the Production Possibilities Frontier**

(a) A technological improvement in peanut butter production shifts the production possibilities curve outward and upward at the peanut butter end. (b) A technological improvement in orange production shifts the production possibilities curve outward and rightward at the orange end. (c) A general increase in technology or resource supply shifts the entire production possibilities curve outward.

In Figure 2–3b, a productivity change brought about by technological improvement occurs in orange production. Here the production possibilities curve shifts outward and rightward on the orange end; if society prefers, additional oranges can be produced without any sacrifice of peanut butter.

In Figure 2–3c, a general change in productivity, technology, or resource supply is assumed. An increase in the quantity of labor (through a simple growth in population), for example, shifts the entire production possibilities frontier outward from $PP_1$ to $PP_2$. Suppose society now chooses the peanut butter/orange combination $A_1$. More of both goods are produced and consumed as a result of the technological improvement. But, society faces new trade-offs once it is producing on its new frontier. Opportunity costs arise again in resource utilization, just as they did at combination A on production possibility curve $PP_1$. Society cannot escape scarcity, yet it can improve its production possibilities through growth in technology, quantity of human or nonhuman resources, or resource productivity.

### Economic Growth: How Economies Progress

We can understand the nature of economic growth with the aid of the production possibilities curve model. The key to economic growth in society is related to the growth or change in capital stock and other crucial resources.

Before considering the concept of capital stock, think about how life must have been for primitive peoples. Humans, having just barely earned the designation *homo sapiens* (wise man), initially had to use most of their time and resources to survive. Later they began to use tools to hunt and kill wild animals. It is not enough to say that primitive peoples found better ways of hunting. In addition to developing the technology of spear-making, primitives had to refrain from some consumption of wild game to have more in their future.

The creation of such **capital stock** (a supply of items used to produce other items) takes time and bears an opportunity cost in present consumption. The benefits of creating capital, however, are in increased amounts of future consumption. In other words, forming new capital stock requires that people save—that is, abstain from consuming in the present. When investment in capital stock results from this saving, capital formation and growth occur. The formation of capital goods—spears, lathes, tractors, and so on—requires a redirection of resources from consumption of goods to production of capital goods. The opportunity cost of acquiring capital is thus consumption goods, but the reward for society is an increase in future productive capacity in consumption goods, capital goods, or both.

We can illustrate this phenomenon with the production possibilities curve. Figure 2–4 shows two production possibilities frontiers that contrast capital goods production (the vertical axis) with consumption goods production (the horizontal axis) for two hypothetical societies with initially similar production possibilities curves. Consider society A's choice of consumption and capital goods in 1985, represented by point $K_A$ in Figure 2–4a. Clearly, society A devotes a larger proportion of its resources to capital goods than to consumption goods. Saving (abstaining from current consumption), investing, and capital formation are all high in that society. The payoff for society A will come later, say in 1995, with vastly enlarged productive capacity, designated by the new production possibility curve, $PP_{1995}$, in Figure 2–4a.

Society B chooses to use almost all of its resources in 1985 for consumption goods, with only a small quantity of resources devoted to capital goods, a combination represented by point $K_B$ in Figure 2–4b. The consequence of

**Capital stock:** Supply of items used in the production of goods and services; these items include tools, machinery, plant and equipment, and so on.

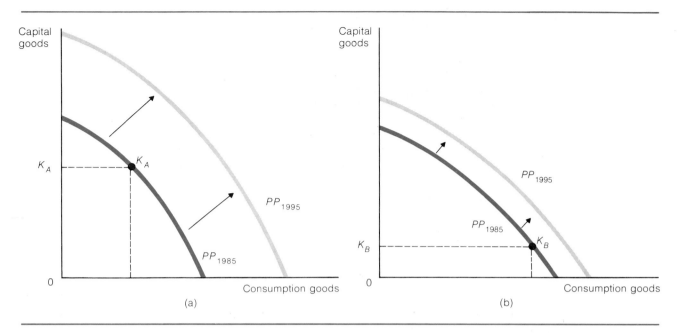

**FIGURE 2-4** Growth Choices on the Production Possibilities Curve
(a) By producing more capital goods, $K_A$, in 1985, society A can produce more consumption goods in the future than (b) society B, which produces only $K_B$ capital goods in 1985.

**Economic growth:** An increase in the sustainable productive capacity of a society.

this choice is low economic growth over the following decade. By 1995, the production possibilities of that society have grown by only a small amount, as curve $PP_{1995}$ in Figure 2-4b indicates. Thus economic growth in the future is largely determined by current decisions about production. Technological changes are important factors in the progress of society, but to take advantage of technology, society must sacrifice some present consumption. (See Focus, "How Societies Can Regress: Armageddon Economics," and Focus, "Less-Developed Countries, Recent U.S. Growth, and Production Possibilities.")

**Specialization:** Performance of a single task in the production of a good or service to increase productivity.

In addition to choosing to put some resources into the creation of capital stock, societies can increase their output by specializing in and trading, or exchanging, goods and services. **Specialization** simply means that the tasks associated with the production of a product or service are divided—and sometimes subdivided—and performed by many different individuals to increase the total production of the good or service. The principles of specialization are therefore related to how all societies and economic organizations try to overcome the basic problem of scarcity.

## Specialization and Trade: A Feature of All Societies

Virtually all known peoples have engaged in specialization and trade, and theories and principles related to specialization and trade have been part of economists' tool kits for more than two hundred years. Adam Smith (1723–1790), the Scottish philosopher and recognized founder of economics, pub

## FOCUS: How Societies Can Regress: Armageddon Economics

As we know from the ancient Greek and Roman experiences, civilization and its production possibilities can decline or be lost. An example from recent history also illustrates this point. In 1939 and 1940 the productive capabilities of Nazi Germany were huge and were growing larger. Massive quantities of resources were feeding a war machine for conquering the world.

Just as in the case of orange and peanut butter production, there was a trade-off, here between military goods and private goods (sometimes called a choice between guns and butter). The opportunity cost of the huge arms buildup in Germany was the resources that could have been devoted to other production. We can conceptualize this position as point $A_{1940}$ in Figure 2–5.

The result of Germany's militarism is well known. By 1945, the country lay in ruins. More than half of its prewar population and 85 percent of its prewar productive capacity were destroyed. We might depict the production possibilities curve as having shrunk to the one labeled 1945 in the figure. On this curve, at choice $B_{1945}$, society must devote all of its resources to private consumption goods (such as food) just to survive.

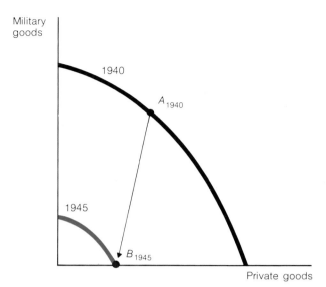

**FIGURE 2–5  Armageddon Economics**
The loss of resources shifts the production possibilities curve toward the origin, indicating that society must use all of its resources to produce private consumption goods just to survive.

---

lished his great work *An Inquiry into the Nature and Causes of the Wealth of Nations* in 1776 (a year of declarations). It formally established the science of what we now call economics.

At the very core of economics, according to Smith, is the ability of individuals and societies to deal with the facts of scarcity through specialization and trade. In the *Wealth of Nations*, Smith used the example of a pin factory, a seemingly trivial manufacturing activity he had observed directly, to evaluate specialization:

> . . . a workman not educated to this business . . . nor acquainted with the use of the machinery employed in it . . . could scarce, perhaps, with his utmost industry, make one pin in a day, and certainly could not make twenty. But in the way in which this business is now carried on, not only the whole work is a peculiar trade, but it is divided into a number of branches, of which the greater part are likewise peculiar trades. One man draws the wire, another straights it, a third cuts it, a fourth points it, a fifth grinds it at the top for receiving the head; to make the head requires two or three distinct operations; to put it on, is a peculiar business, to whiten the pins is another; it is even a trade by itself to put them into the paper; and the important business of making a pin is, in this manner, divided into about eighteen distinct operations, which, in some manufactories, are all performed by distinct hands, though in others the same man will sometimes perform two or three of them.[1]

---

[1] Adam Smith, *An Inquiry into the Nature and Causes of the Wealth of Nations*, ed. Edwin Cannan (1776; reprint, New York: Modern Library, 1937), pp. 4–5.

## FOCUS  Less-Developed Countries, Recent U.S. Growth, and Production Possibilities

The production possibilities curve offers insights into many important matters. The economic growth rates of many less-developed countries, for example, are quite low compared with those of the United States and other, more-developed countries. Yet many poor developing countries teem with resources, and modern technology is available to them. Because they must devote a huge quantity of resources to present consumption to meet the demands of high population growth, their future economic growth is severely limited. Some economists point not only to high population growth rates but also to low levels of education as probable causes of the failure of some countries to grow economically.

The production possibilities curve can also help explain the slowed rate of U.S. economic growth in the 1970s and early 1980s. The problems have been a lagging growth in labor productivity, sharply curtailed access to vital natural resources such as oil and other essential inputs, and a slower growth rate of saving and investment, with consequent reductions in the rate of production of capital goods. The U.S. production possibilities frontier expanded more slowly over this period than it did, on average, over the past one hundred years.

---

Smith's point in describing the pin factory is that when tasks are divided, permitting each individual to concentrate on a single element in the production of a good, output increases over what it would be if each individual produced the entire good. Specialization and the **division of labor** led to increased output. As a result, people became more dependent on one another for all goods. We may help produce autos, for example, but we still want eggs for breakfast. Smith recognized that mutual dependence could be a problem but thought that the potential for increased output with given resources was well worth the cost.

**Division of labor:** Individual specialization in separate tasks involved in production of a good or service; increases overall productivity and economic efficiency.

If Smith's pin factory does not stir your imagination, think of any organization of which you have ever been a part and try to list the number of distinct divisions of labor in it. Think of a larger factory, a steel mill, a football team, a fraternity, a church. Specialization exists in almost everything we do.

In the modern world, nations, states, firms, and individuals specialize and trade according to two principles—absolute advantage and comparative advantage.

### Absolute Advantage

**Absolute advantage:** Ability of a nation or an economic entity to produce a good with fewer resources than some other entity requires.

**Absolute advantage** simply means that one production unit (individual, firm, state, or nation), given an equal expenditure of resources, is more efficient than another at producing some good or service. In other words, that production unit can produce more from the same resources.

For example, Sally, the best physician in town, meets Sam, the best auto mechanic in town. Given equal expenditures of resources, Sally is more efficient than Sam at health care, while Sam is more efficient than Sally at auto repair. That is, with an equal expenditure of all resources, including time, Sally can produce more medical services than Sam and Sam more auto repair services than Sally. Therefore, it is more efficient for Sally to specialize in medicine and to trade her services for Sam's.

Similarly, states and nations may find mutual benefit in specializing and trading when each has an absolute advantage in producing different goods. Assume that the states of Texas and Idaho are isolated and that both are

**TABLE 2–3  Absolute Advantage: Output Before Specialization**

Texas has an absolute advantage in the production of beef: Texas can produce one pound of beef with fewer resources than Idaho needs. Idaho has an absolute advantage in the production of potatoes.

|  | Texas | | Idaho | | |
| --- | --- | --- | --- | --- | --- |
|  | Yearly Production (millions of pounds) | Resources Spent (percent) | Yearly Production (millions of pounds) | Resources Spent (percent) | Total Output |
| Beef | 40 | 25 | 25 | 25 | 65 |
| Potatoes | 100 | 75 | 300 | 75 | 400 |

**TABLE 2–4  Absolute Advantage: Output After Specialization**

Through specialization, Idaho's output of potatoes increases and Texas's output of beef increases. The increase in total output allows both states to gain from trade.

|  | Texas | | Idaho | | |
| --- | --- | --- | --- | --- | --- |
|  | Yearly Production (millions of pounds) | Resources Spent (percent) | Yearly Production (millions of pounds) | Resources Spent (percent) | Total Output |
| Beef | 160 | 100 | 0 | 0 | 160 |
| Potatoes | 0 | 0 | 400 | 100 | 400 |

able, with the same resources, to produce the quantities of beef and potatoes reported in Table 2–3. Further assume that the quantities are actually demanded and consumed within the two states. From Table 2–3, it is clear that Texas is more efficient at producing beef than is Idaho. It is 1.6 times more efficient (40/25), to be exact. Idaho, on the other hand, is absolutely more efficient at producing potatoes—in fact, 3 times (300/100) more efficient. The reasons for these absolute advantages, though somewhat irrelevant for the working of the principle, may include climatic differences, different endowments or qualities of resources including labor or land, or differing levels of technology. Whatever the underlying reasons, Table 2–3 shows the respective outputs and consumptions before specialization.

Suppose that Texas and Idaho agree to specialize: Idaho will produce only potatoes, Texas will produce only beef, and they will trade to help meet each other's needs. If we assume that output proportions remain the same—that Idaho, for example, will be able to produce 400 million pounds of potatoes with 100 percent of its resources since it could produce 300 million pounds with 75 percent—the respective outputs after specialization appear as in Table 2–4. Texas now produces 160 million pounds of beef and no potatoes, while Idaho produces 400 million pounds of potatoes and no beef. Note the result: There is a clear gain from specialization—an increase of 95 million pounds of beef.

On what terms will the parties trade? How will they determine a fair trade ratio between beef and potatoes? Though we can discern the limits to the trading terms, it is impossible to determine what the exact trade will be without additional information. The actual **terms of trade** would be determined by the relative bargaining strength of Texans and Idahoans, which is itself determined by the demands for the two goods within the two states. But we do know the limits. Idaho would be willing to trade 100 million pounds of potatoes for a quantity of beef greater than 25 million pounds (the quantity of beef Idaho produced and consumed before specialization). If Idaho accepted any less beef, the state would be worse off than before and would not have agreed to specialize. At the same time, for 100 million pounds of potatoes (Texas's prespecialization production and consumption), Texas would not be willing to part with more than 120 million pounds of beef—a quantity that would leave Texas only as well off as before specialization.

If the two states split the gain, 100 million pounds of potatoes would trade for 60 million pounds of beef. Texas would then be able to consume 100 million pounds of both potatoes and beef, and Idaho would have 300 million pounds of potatoes and 60 million of beef. Both states would be better off.

To summarize, our example shows that Texas and Idaho have specialized, traded, and gained according to the principle of absolute advantage. Specialization permitted Texas and Idaho to devote all of their respective resources to the production to which they were best suited. Trade ensued, after which both states are better off.

### Comparative Advantage

Adam Smith explained the principle of absolute advantage in his *Wealth of Nations* in 1776. It was left to his English followers, economists David Ricardo (1772–1823) and Robert Torrens (1780–1864) to develop a somewhat less obvious but critically important principle of specialization—the theory of **comparative advantage.** In a situation of comparative advantage, nations or economic entities can benefit from trade of two goods even though one has an absolute advantage in the production of both. This is because of a difference in opportunity costs for each nation or entity, giving both the less productive and the more productive entity a relative advantage in specializing in production of one of the goods. Comparative, rather than absolute, advantage is the modern principle most closely associated with the basis for trade between any two parties.

Ricardo developed the principle of comparative advantage by observing and analyzing Britain's international trade position early in the nineteenth century, especially during the Napoleonic Wars with France. As in Ricardo's original examples, international specialization can illustrate the principle in modern terms. Look at Table 2–5, where the hypothetical outputs of oil and cosmetics are given for the United States and West Germany. While the United States is able to produce more (in absolute terms) of both goods, the United States is 1.5 times (90/60) more efficient at producing oil than cosmetics, whereas West Germany is twice as efficient (20/10) at producing cosmetics than oil. If these production rates are also the rates at which the two commodities can be exchanged within the two countries before specialization, oil is cheaper relative to cosmetics in the United States, and cosmetics are cheaper than oil in West Germany (half as cheap, to be exact).

After specialization (see Table 2–6), the same total quantity of cosmetics is produced, but a net increase of 20 million barrels of oil is obtained. The 20 million barrels of oil is the net gain from specialization.

---

**Terms of trade:** The price ratio or range of price ratios at which two entities are likely to trade.

**Comparative advantage:** Ability of a nation or an economic entity to produce a good at a lower opportunity cost than some other entity.

**TABLE 2–5 Comparative Advantage: Output Before Specialization**

Oil has a lower opportunity cost in the United States than in West Germany. West Germany has a comparative advantage in cosmetics, for it is relatively more efficient at producing cosmetics than at producing oil.

|  | United States | | West Germany | | |
|---|---|---|---|---|---|
|  | Yearly Production (millions of barrels) | Resources Spent (percent) | Yearly Production (millions of barrels) | Resources Spent (percent) | Total Output |
| Oil | 90 | 75 | 10 | 75 | 100 |
| Cosmetics | 60 | 25 | 20 | 25 | 80 |

**TABLE 2–6 Comparative Advantage: Output After Specialization**

Specialization according to comparative advantage results in a net increase of 20 million barrels of oil.

|  | United States | | West Germany | | |
|---|---|---|---|---|---|
|  | Yearly Production (millions of barrels) | Resources Spent (percent) | Yearly Production (millions of barrels) | Resources Spent (percent) | Total Output |
| Oil | 120 | 100 | 0 | 0 | 120 |
| Cosmetics | 0 | 0 | 80 | 100 | 80 |

How can this gain be divided between the United States and West Germany? As stated earlier, we cannot determine that division without knowing the relative bargaining strength of the two nations. But the limits to the bargain can be determined.

In exchange for 60 units of cosmetics (the United States' prespecialization consumption), West Germany will not be willing to take less than 10 million barrels of oil. If West Germany took less than 10 million barrels of oil, it would be worse off than before specialization; in that case West Germany would not specialize. The United States would not give more than 30 million barrels of oil for 60 million barrels of cosmetics. To do so would make the United States worse off than before. Thus the trade must be something greater than 10 million barrels of oil for 60 million barrels of cosmetics but something less than 30 million barrels of oil for 60 million barrels of cosmetics. If they split the net gain, the United States gives up 20 million barrels of oil, retaining 100 million barrels of oil and 60 million barrels of cosmetics, and West Germany retains 20 million barrels of cosmetics and acquires 20 million barrels of oil through trade. Both nations are better off after specialization and trade, although the United States is absolutely more efficient in producing both of the traded goods. For both nations, specialization and trade, according to the principle of comparative advantage, expanded the consumption possibilities in oil.

## Exchange Costs

The models of absolute and comparative advantage—along with all other economic models—are simplifications of reality. A number of important assumptions hide behind our simple discussions, some of which we have already mentioned. But there are also costs to the process of the exchange that must be accounted for when calculating the benefits of specialization and trade. We classify these **exchange costs** as transaction costs, transportation costs, and artificial barriers to trade.

**Transaction Costs.** Transaction costs are all the resource costs (including time-associated costs) that are incurred because of exchange. Transaction costs occur every time goods and services are traded, whether exchanges are simple (purchase of a pack of gum) or complex (a long-term negotiated contract with many contingencies).

Here's a simple example: Gwen goes to the supermarket to purchase a pound of coffee. What are the costs of the transaction? Gasoline and auto depreciation must of course be considered as resource costs, but the principal cost is the opportunity cost of Gwen's time. Gwen might have spent this time working or playing tennis instead of grocery shopping. Gwen's wage rate might then serve as her opportunity cost. If the shopping trip takes 30 minutes and if Gwen's wage rate is $15 an hour, the time part of her transaction costs is $7.50.

In this simple case the contracting and negotiating are instantaneous—Gwen simply gives the money to the checkout person, takes her coffee, and the transaction is complete. In other, more complex exchanges—such as the purchase of a house, a car, or a major appliance or the negotiation of a long-term labor contract—contracting and negotiating costs can be substantial. Think of the time and other resource costs associated with long-term supply contracts such as U.S. arms deals with allies or negotiations for ammonia plants in China. Every detail must be studied, formulated, and then spelled out.

The important point is that transaction costs include resources used by each party to the exchange. The higher the resource costs, the lower the benefit that comes from specialization and trade. Institutions—new legal arrangements, new marketing techniques, new methods of selling—have emerged and are continuously emerging, to reduce all forms of transaction costs. Gwen could have purchased coffee at a convenience store nearer to her house than the supermarket is. The price of the coffee might have been somewhat higher, but Gwen's time costs, and therefore her total transaction costs, would have been lower. In a broader sense, laws about contracts, and law itself, are means by which transaction costs are reduced. The invention of money and the development of various forms of money and financial instruments are responses to transaction costs associated with barter and more primitive means of exchange. We will return to these issues in the next chapter.

In addition to money, intermediaries such as wholesalers and advertisers developed over the ages to facilitate exchange. The creative marketing of goods and services from bazaars to discount stores to media advertising has increased consumer information and thus reduced the costs of making transactions. Middlemen have lowered transaction costs by decreasing the risk of exchange. The production of some goods entails some risk and uncertainty on the part of buyers and sellers. Planting wheat in the spring for sale in the

---

**Exchange costs:** The value of resources used to make a trade; includes transportation costs, transaction costs, and artificial barriers to trade.

**Transaction costs:** The value of resources used to make a purchase, including time, broker's fees, contract fees, and so on.

fall obviously entails some uncertainty about what prices will be at harvest. Middlemen-speculators who deal in futures provide sellers and buyers assurance of prices in the future. This type of middleman makes profit on the miscalculations of buyers and sellers but performs the service of reducing uncertainty and thereby increasing trade and specialization.

**Transportation Costs.** A second impediment to trade is transportation costs, resource costs associated with the physical transport of products from place to place. The higher these resource costs, the lower the benefits from specialization and trade.

We did not include transportation costs in our initial examples of absolute and comparative advantage. Suppose that transport of beef to Idaho and potatoes to Texas was costly because of the rugged terrain between the two states. High transportation costs would reduce the gains that Texans and Idahoans might enjoy from specialization and trade. If transportation costs were high enough, possible advantages to specialization and trade could be wiped out completely. Cheaper transportation costs, such as those created by new trade routes in the Renaissance or by the invention and spread of the railroad in nineteenth-century America, permit more trade and open up opportunities for new forms of and increases in specialization. The Pony Express, the invention of the automobile and truck, and the dawning of air freight transport were all boons to specialization and increased output.

**Artificial Barriers to Trade.** The final impediments to specialization and trade are government-imposed restrictions such as tariffs, quotas, and outright prohibitions on the import or export of goods. More localized restrictions include minimum-wage laws and specific restrictions on an industry or in an area. Such impositions either reduce or eliminate the benefits of specialization and trade. Governments always have reasons for these restrictions, but the reasons must be closely scrutinized because the benefits from international and domestic specialization and trade are potentially huge for consumers in all nations. Artificial barriers have the power to reduce economic welfare by reducing or eliminating the benefits of specialization according to the law of comparative advantage (see Economics in Action, "Comparative Advantage: The Case of Bio Bio," at the end of this chapter).

The possible effects of restrictions on trade can be seen in the large and growing volume of trade between Japan and the United States in commodities such as TVs, automobiles, stereo equipment, and musical instruments. Special-interest groups—such as American autoworkers and manufacturers—have lobbied for import tariffs and other trade restrictions to protect their own interests, which include increased demand for American-made products and therefore increased domestic production. Government-enforced tariffs make Japanese goods more expensive, however, causing a reduction in the general well-being of Americans. Artificial restrictions on trade, whatever their purpose, reduce the advantages of specialization and trade. As a matter of economic principle, therefore, most economists are generally advocates of the advantages of free trade over any type of trade restriction. Further, any institution or mechanism that reduces the costs of exchange usually gets the support of economists because greater specialization permits better utilization of scarce resources. Specialization, in other words, helps us get more of what we want, given a limited amount of resources.

---

**Transportation costs:** The value of resources used in the transportation of goods.

**Artificial barriers to trade:** Any restrictions created by government that inhibit trade, including quotas and tariffs.

## Summary

This chapter described several principles that provide a foundation to economic thinking:

1. Economic choices, both for the individual and for society, always involve an opportunity cost.
2. Opportunity cost includes all costs or opportunities forgone in the decision to engage in a particular activity.
3. The law of increasing costs means that as more of one good is consumed by a society, the opportunity costs of obtaining additional units of that commodity rise. The cost increase results because resources become less adaptable as production becomes more specialized.
4. Choices are ordinarily made not in all-or-nothing fashion but at the margin. Both individuals and societies, therefore, calculate the cost of consuming additional units of some good or service.
5. The production possibilities curve shows the possible quantities that could be produced of any two goods given the state of technology and society's scarce resources.
6. Changes in technology or increases (or decreases) in the amount of resources cause outward or inward movements in the production possibilities frontier.
7. Greater quantities of output can be obtained with society's scarce resources when people specialize and trade. Trade takes place according to one of two principles: absolute advantage or comparative advantage.
8. Trade can take place between two individuals or economic entities even if one of the entities is more efficient at producing all goods. All that is required is that each entity be relatively more efficient than the other in some production.
9. Transaction costs, transportation costs, and artificial trade barriers such as tariffs and quotas reduce the benefits obtainable from specialization and trade.

## Key Terms

opportunity cost
choices at the margin
production possibilities frontier
law of increasing costs
unemployment of resources
capital stock

economic growth
specialization
division of labor
absolute advantage
terms of trade
comparative advantage

exchange costs
transaction costs
transportation costs
artificial barriers to trade

## Questions for Review and Discussion

1. What did reading this chapter cost you? Did you include the price of the book? What will reading the next chapter cost? Does that include the price of the book?
2. Do government-sponsored financial aid programs for college students influence the amount of education produced? Do these programs shift the production possibilities curve?
3. What does a movement along the production possibilities frontier suggest? What does a point inside the curve suggest?
4. A subsidy to farmers who purchase tractors and combines increases the production of this farm machinery. Does this cause an increase in the production possibilities curve or just a movement along the curve? Can subsidies cause economic growth?
5. Why would a country with an absolute advantage in the production of all goods be willing to trade with other countries?
6. Alpha can produce 60 bottles of wine or 40 pounds of cheese. Beta can produce 90 bottles of wine or 30 pounds of cheese. Both have constant costs of production. Draw their production possibilities curves. What is Alpha's cost of one bottle of wine? What is Beta's cost of one pound of cheese? If they trade, who should specialize in cheese?
7. What are the costs of going to college? Does the marginal benefit outweigh the marginal cost?
8. Is the lost present consumption associated with the production of capital goods worth the benefit of the new capital?
9. Does Japan have an absolute advantage over the United States in the production of televisions and stereo equipment or is it just a comparative advantage?
10. Who is hurt by and who benefits from an import quota on foreign beef?
11. How does the cost of purchasing a loaf of bread at a supermarket compare with the cost of purchasing a loaf of bread at a convenience store?

## ECONOMICS IN ACTION

### Comparative Advantage: The Case of Bio Bio

There are terrible economic problems in Bio Bio, a coastal region of western Chile with a land mass about the size of Massachusetts and Connecticut combined. In 1983, 33 percent of its 1.4 million inhabitants were unemployed. Bio Bio lawyers, business people, and even economists blame the region's current economic problems on the theory of comparative advantage. Is the situation in Bio Bio a case of comparative advantages gone wrong?

Bio Bio's comparative advantage, prior to the opening of the Panama Canal in 1914, was in shipping. The region's coastal cities enjoyed prosperity by offering refueling and resupplying services to ships that had to pass through the Strait of Magellan. After the Panama Canal opened, the Chilean government sought to prop up its fragile economy by enforcing protective tariffs on textiles and by offering direct subsidies to coal miners. The government poured money into state-owned enterprises such as steel mills and petrochemical plants. For years, these actions created jobs and forestalled economic chaos.

The region's present economic problems emerged, it is argued, after the overthrow of the Marxist government of Salvador Allende by the Chilean military in 1973. Economic reforms, at the suggestion of University of Chicago-educated free-market economic advisers, included the elimination of all protective subsidies and tariffs to industries across Chile and, in general, the institution of free-market production on the basis of comparative advantage.

Free-market economic advisers of the new Pinochet government argued that Bio Bio had a comparative advantage in fish and timber production. Other industries that were rapidly developing new markets included ceramics and the production of rose hips, an ingredient in natural foods and vitamins. These industries have grown substantially since free-market policies were instituted but not enough to take up the slack in employment. The region suffered bitterly during the worldwide recession of the early 1980s. Are the government's free-trade policies a failure? Is comparative advantage a cruel hoax to these people?

The answer is no: The Bio Bio problem is essentially an "adjustment problem." Comparative advantage in ship service and related trade worked in favor of residents until the Panama Canal was opened. Afterward the government located industries in Bio Bio through subsidies and tariffs. Subsidized workers did well in every industry. But when subsidies and tariffs ended, Bio Bio found its real comparative advantage in fish, lumber, and rose hips. Of course, capital and labor resources of the area were concentrated in the previously subsidized industries.

The cost of retraining textile workers to be fishermen is not zero. The quality and skills of a large part of the labor force cannot be changed quickly. Hence, problems arise when the government encourages people, for extended periods of time, to invest capital and skills in activities where no comparative advantage exists.

It is not clear, in other words, that the previous Bio Bio "prosperity" was genuine. Consumers in Bio Bio paid higher prices for both types of goods produced—those that were subsidized through the tariff protection and those for which a comparative advantage existed but which were not produced. Taxpayers in all regions

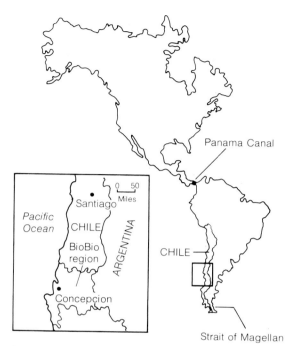

*Prior to 1914 and the opening of the Panama Canal, the Bio Bio region of Chile was a popular refueling stop for ships en route between the Atlantic and Pacific oceans. Since 1914, the region has suffered severe unemployment problems.*

*Source:* Everett G. Martin and Fernando Paulson, "Victims of a Theory: Chilean Region, Competitive in Few Products, Hits Hard Times Under Rule of 'Chicago Boys,'" *Wall Street Journal*, May 5, 1983, p. 60.

of Chile subsidized the owners and workers in government-favored industries. The many consumers and taxpayers paid for the prosperity of the few.

Artificial barriers to trade are costly to any society because resources are not directed to their most valued uses by comparative advantage. Comparative advantage will work in Bio Bio, as elsewhere, once resources adapt to new, more suitable production.

## Question

Assume that the United States decides to become self-sufficient in energy production and reduces imported energy sources to zero by means of a prohibitive tariff. What effects would the oil tariff have on the benefits due to specialization and exchange in the short term and over a longer period of time?

# 3

# Markets and the U.S. Economy

The vehicles for economic growth, as we have seen, are technology and resource development, specialization, and trade. But what makes these vehicles possible? Why are they more apparent in some societies than in others? To better understand the answers to these questions and the workings of the American market economy—the system most closely analyzed in this book—we present an overview of how the American economy works, and we contrast it with the ways in which other societies have chosen to deal with production and distribution in the face of scarcity.

## The U.S. Market System in Perspective

Rice paddies in China, nut-and-fruit-gathering societies in Africa, government enterprise in Russia, free-trade zones in Hong Kong, computer hardware development in California's Silicon Valley—all these arrangements represent different answers to the "what," "how," and "for whom" questions under different economic systems. An **economic system** is the particular form of social arrangements through which the three fundamental questions are answered.

**Economic system:** A means of determining what, how, and for whom goods and services are produced.

### Comparative Systems

Three basic types of economic systems have emerged in response to scarcity of resources: traditional, command, and market societies. But all societies, including the United States, actually contain some elements of each type. It is very difficult to identify a pure type of economic organization in contemporary societies.

**Traditional Societies.** A **traditional society** is characterized by subsistence food gathering, primitive agriculture, or nomadic herding. Such societies

> **Traditional society:** An economic system in which the "what," "how," and "for whom" questions are determined by customs and habits handed down from generation to generation.

typically suffer acute scarcity. Specialization and trade, crucial factors in expanded production, are minimal. The answer to "what is produced" is limited to fulfillment of the basic needs of the community (food, shelter, clothing). Most activities are geared to basic survival. Little or no time is left for innovation or for the development of technology after basic needs have been met.

The "how" of production—means such as hunting tools, agricultural implements, and the construction of shelter or storage—change very little through time as skills and roles are passed on from father to son, mother to daughter. Specialization exists in the traditional society, but it is narrowly restricted by the social arrangements surrounding production.

Distribution—the answer to the "for whom" question—is based solely on social arrangements rooted in what the society regards as "superior qualities" of its members—the strongest and most successful hunters, best firekeepers, most talented rugmakers, the oldest or wisest people. Members of society with such superior qualities receive the best of the traditional society's output.

> **Command society:** An economic system in which the questions of "what," "how," and "for whom" are determined by a central authority.

**Command Societies.** In a **command society**, "what," "how," and "for whom" questions are answered by some central authority—a single individual (king, dictator, pharaoh) or a group of individuals (Communist party, military junta). The central authority determines what is going to be produced. In ancient Egypt, pharaohs, the personifications of the gods, were able to direct massive quantities of resources into the construction of pyramids and other monuments through slave labor. These monuments were constructed at the opportunity cost of consumption goods and services for the Egyptian population.

Expressions of command are found in contemporary societies as well. Perhaps the most familiar contemporary example of a command society is the Soviet Union. Over the last fifty years, the members of the ruling Politburo in the Soviet Union have ordered that a vast quantity of resources be poured into military goods at the expense of consumer goods and capital goods used to produce consumer goods. Other advanced nations, including the United States, invest large quantities of resources in national defense, although decisions to do so are ultimately directed through a democratic voting process in many of the nations.

Most production in a command society takes place in government-owned or -sponsored enterprises. Invention and innovation can be prominent in command societies, but their development and the use of their results are ordinarily controlled by authority. Free thought, the handmaiden of innovation, is not encouraged, and particular sciences and arts (for example, space technology and weaponry in the Soviet Union) tend to thrive while others (in the Soviet Union, those necessary for production of consumer goods, such as TVs and refrigerators) languish.

A large degree of discretionary power exists in a command society, and this power leads to a selective distribution of wealth. Those highly placed or those in the favor of the highly placed have first choice of products and services produced. In the Soviet Union rewards are based on membership and power within the Communist party bureaucracy. Opulent resort homes in the Soviet Union, for example, belong mostly to well-placed government and military bureaucrats. (We discuss the economics of the command society in more detail in Chapter 23.)

Historically, traditional and command societies have been most common.

Egyptian, Greek, Roman, and medieval societies (dictatorships, monarchies, and aristocracies) were composed of both traditional and command characteristics. In the seventeenth and eighteenth centuries, however, England and other European countries underwent a significant decline of centrally controlled and regulated economic life. These developments ushered in the market society as a new form of economic organization.

**Market Societies.** In a **market society,** impersonal forces lead consumers and producers to answer the three fundamental questions of production. Consumers answer the "what" question. Production of goods and services such as personal computers and auto repair is determined by "dollar votes." Just as political votes elect a president, dollar votes—or money spent on products and services—express consumers' demands. Suppliers of products and services, interested in making profits, respond to these votes by providing the products and services in just the right quantities. (Chapter 4 explains exactly how this occurs, through the laws of supply and demand.)

The "how" question in the market society is, at any one time, answered by available technology and by suppliers' profit-motivated desire to produce goods most cheaply given the price of resources. Most goods can be produced by a variety of methods. Avocadoes, for example, can be picked by hand or by harvesting machines. The avocado grower will choose whatever method minimizes the costs of production. In a market economy prices are the signals not only for what to produce and how to produce in the present but also for how new technologies and new resources can be brought into production over time.

When one highly demanded resource becomes scarce and therefore high-priced, producers will attempt to substitute other resources for it. If possible substitutions are limited, alternative technologies and new types of resources may be developed. In mid-nineteenth-century America, for example, whale oil—used primarily for lighting—became scarce and high-priced as the supply of whales depleted. The high price, however, provided the incentive to develop alternative lighting fuel such as fossil fuels. Some historians believe that the depletion of the whales encouraged the discovery of petroleum and the use of oil derived from it.

Prices also answer the "for whom" question in a market society. Just as the prices and quantities of all goods and services available in a market system are determined by the demands of buyers and sellers, so are the "prices" (such as wage rates, rental rates, and interest rates) of resources used to produce them. The value of particular resources in general depends on the demand for the product that the resource helps produce. For example, the owner of property bordering Central Park in New York City may expect to receive a higher rental than one who owns desert property in Nevada. Luciano Pavarotti, a world-renowned operatic tenor, may be expected to receive a higher wage than a college president. The money rewards to these resources are determined not only by the demand for them but also by their relative scarcity. First-class tenors are scarcer than college presidents.

## Contemporary Economic Organizations

There are no pure economic systems—all societies contain mixed elements of traditional, command, and market systems. In most contemporary societies, however, one type of economic organization predominates. Within

---

**Market society:** An economic system in which individuals acting in their self-interest determine what, how, and for whom goods and services are produced, with little or no government intervention.

some Third World countries, especially in tribal cultures, tradition predominates. But more advanced countries also include some elements of tradition. For instance, generations of families of doctors, lawyers, and crafts people are common in American life. Commandlike elements also exist in the American economy. Taxes in general require societal consent, but the power to print money is a command function given to government through the Constitution until changed by the will of the people. Indeed, we may think of all of the functions of local, state, and federal governments in the United States as temporary command functions that may be changed periodically through the elective process.

The United States is in general a market economy with elements of command and tradition, while the Soviet Union and China are primarily command societies with some traditional and market characteristics. The Russians have at times used prices to allocate resources, but within an overall command scheme. Chinese society, since the death of Mao Tse-tung in 1976, has begun using market incentive in the production and sale of goods, especially agricultural products. These economies are primarily command-oriented, however. Basic decisions to allocate most available resources are left to central authority in the Soviet Union and China. Individual citizens cannot decide, either through dollar votes or through the political process, to change the direction of major resource allocations. In the United States, on the other hand, consumers and voters are sovereign decision makers with regard to how resources are allocated and to how the three fundamental questions are answered.

The important question is which of these economic systems expands or retards specialization, production, and trade. Of market economies and command economies—the two leading systems in the modern world—which is better at furthering economic growth and well-being? The question is loaded because the answer in part depends on the value one places on political and economic freedom. But most Western economists see the market system as the greatest force for economic growth within the context of political and economic freedom. The U.S. market economy—its functioning and its problems—therefore receives the lion's share of attention in this book. We show that the American economy is a one-of-a-kind economy but that it shares some important features with other, similar market systems. One shared feature is the use of money and a circular flow between money, goods, and resources.

## Money in Modern Economies

**Barter:** The trading of goods for goods with no medium of exchange such as money.

Despite their differences, a feature of all developed economic systems is the use of money. **Barter**—the exchange of goods for goods—is cumbersome and costly. For a system to take advantage of specialization, output must grow steadily. For output to expand, trade must take place at a low cost. But the costs of bartering goods are numerous, for the value of each good must be recorded in terms of each other good traded. As the number of goods traded increases, calculations become overwhelming. Money was invented as a medium of exchange and as a unit of accounting. It is a good in terms of which the value of all other goods can be calculated. Money lowers the cost of transacting and is therefore essential to any modern society.

We can think of money as a substitute for real things traded. Wages, for example, represent a trade of the output of a worker for the goods and services consumed by the worker. We may earn wages as a bricklayer and purchase fried chicken at a fast-food establishment. In a real sense, we are bartering bricklayer labor for fried chicken. Money simply substitutes for real trades.

### The Circular Flow in a Market Economy

The exchange of money for goods and services takes place throughout the United States and other developed economies. One can conceptualize this exchange as a **circular flow of income** (Figure 3–1), a cyclical pattern of money payments and services and goods produced.

In Figure 3–1, the arrows on the outer circle represent the flow of real goods and services between business firms and households, while the arrows on the inner circle represent the reciprocal flow of money payments. In the circular flow of products, business firms produce products and services with resources (land, labor, and capital) from resource suppliers. To complete the circle, households that purchase products and services are virtually all resource suppliers—of labor, land, or capital. The flow of products has a corresponding and counterbalancing money flow in the opposite direction. For goods and services, firms receive money payments (business receipts) from households, and for resources (labor, land, and capital) resource owners receive money income in the form of rents, wages, and interest from firms. The quantities and the mix of resources used and products produced are determined by price signals. These signals, as we will show more clearly in Chapter 4, are the result of the impersonal forces of supply and demand.

Figure 3–1 is not meant to depict a functioning economy in detail but to serve as a model for real and money flows. Missing are the complexities of flows by which producers supply goods to other producers and of the tax and expenditure flows by which government provides public goods and services. These complexities are discussed in the study of macroeconomics.

**Circular flow of income:** The flow of real goods and services, payments, and receipts between producers and suppliers.

## Institutions of American Capitalism

While money and real goods circulate through all modern economic systems like blood through the veins and arteries of the body, societies vary in the institutions they create for ownership and use of goods and money. In a **socialist economy,** goods and services and resources are owned by the society as a whole and distributed, in command fashion, by the government. A **capitalist economy** relies primarily on market forces and the profit motive for production, distribution, and consumption of goods and services. Capitalism is characterized by individual ownership of property, free enterprise, open competition, and a minimal role for government.

**Socialist economy:** An economic system in which the means of production are owned and controlled by the government.

**Capitalist economy:** An economic system in which the means of production are privately owned.

### Property and the Law

In the American economy the individual's right to own and dispose of property is regarded as basic. Property includes both physical property (such as houses and automobiles) and intellectual or intangible property. For example, the ownership of poems, songs, and books by their authors is protected by copyright laws; inventions are protected by patents. The legal apparatus set up to protect such rights may even make a distinction between property and property rights. Rental of a carpet cleaner from U-Rent-Um gives the renter certain property rights over the cleaner but not ownership of the property itself. In all societies, rights to use property are limited. For example, the Environmental Protection Agency has used its legislatively derived power to limit the rights of businesses to pollute air and water and has established

Institutions of American Capitalism 51

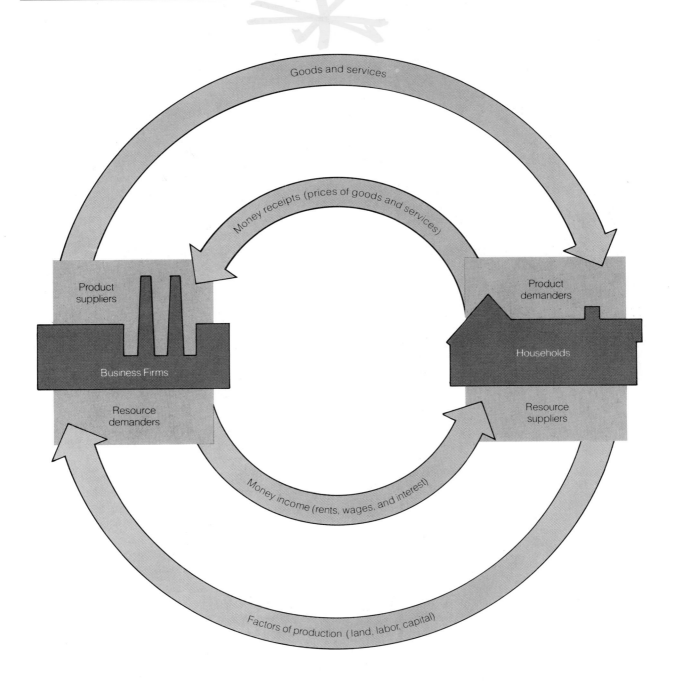

**FIGURE 3-1  The Circular Flow of Income**

Economic activity between business firms (producers) and households (consumers) takes place in a cyclical pattern. The arrows on the outer circle represent the flow of real goods and services (consumer products and production factors such as labor); the arrows on the inner circle represent the reciprocal flow of money payments for real goods and services.

worker protection standards (such as those limiting the use of asbestos) that restrict businesspeople's free use of property.

In a capitalist society, the law also protects property rights. Without some guarantee that property rights will be protected, there would be little incentive to accumulate capital stock and therefore to grow economically. Without state guarantees of rights to property, individuals would have to protect their own property at high cost.

Legal protection of property rights often emerges as a response to market activities. Consider an example from the American frontier west. In the early days of silver and gold prospecting, individuals were forced to protect their own mining claims. As their claims grew, individual protection entailed higher opportunity costs; that is, higher costs in time spent away from mining or prospecting and in potential loss of equipment or mined ore. Thus, the collective benefits of enforcement through police, courts, and prisons outweighed the costs of such enforcement to individual prospectors—paying taxes, serving on posses, and the like. Property rights and the laws that protect them emerge when benefits to the parties involved begin to exceed the costs of acquiring and enforcing the rights. Such rights to private property, established by law, have been a part of American capitalism since the foundation of the nation.

### Free Enterprise

**Free enterprise,** the freedom to pursue one's economic self-interest, is an intrinsic part of the capitalist system. Men and women are free to choose their line of work with few or no governmental restraints or subsidies, and businesspeople are free to combine any resources at their command to produce products and services for profit. Laborer-consumers are free to produce, purchase, and exchange any good or service so long as their activity does not infringe on others' rights. Free enterprise, in sum, means that

1. laborers are free to work at any job for which they are qualified;
2. business firms and entrepreneurs can freely combine resources, at competitive market prices, to take advantage of profit opportunities; and
3. consumers can decide what products and services will be produced.

**Free enterprise:** Economic freedom to produce and sell or purchase and consume goods without government intervention.

### Competitive Economic Markets

The American economic system is also characterized by free competitive markets. **Competition** entails two important conditions: a large number of buyers and sellers and free entry and exit in the market. When these two conditions are met, the self-interested actions of buyers and sellers tend to keep prices of goods and services at a reasonable level, usually the costs of production plus a normal profit for the sellers.

When the number of buyers and sellers is large, no individual buyer or seller can affect the market price of a product or service. Many millions of individuals purchase canned soup, for example, but no one buyer purchases enough to affect the market price of the soup. Likewise, the existence of competing suppliers means that no individual seller can acquire enough power to alter the market for his or her gain. Sellers of canned soup are numerous enough so that no one seller can affect the price of soup by increasing or decreasing output.

Crucial to the effects of large numbers is the condition that firms be free to enter and leave markets in response to profit opportunities or actual losses. New firms entering particular lines of business, bankruptcies, and business

**Competition:** A market situation satisfying two conditions—a large number of buyers and sellers and free entry and exit in the market—and resulting in prices equal to the costs of production plus a normal profit for the sellers.

## FOCUS   Adam Smith's "Invisible Hand"

Economists tend to praise a free and unfettered competitive market system because the rational and self-interested forces that characterize economic behavior lead not to a permanent state of chaos but to a harmony of interests. Adam Smith had great insight into the matter more than two hundred years ago in his "invisible hand" passage in the Wealth of Nations:

> Every individual necessarily labours to render the annual revenue of the society as great as he can. He generally, indeed, neither intends to promote the public interest, nor knows how much he is promoting it. By preferring the support of domestic to that of foreign industry, he intends only his own security; and by directing that industry in such a manner as its produce may be of the greatest value, he intends only his own gain, and he is in this, as in many other cases, led by an invisible hand to promote an end which was not part of his intention. Nor is it always the worse for the society that it was no part of it. By pursuing his own interest he frequently promotes that of the society more effectively than when he really intends to promote it. I have never known much good done by those who affected to trade for the public good. It is an affectation, indeed, not very common among merchants, and very few words need be employed in dissuading them from it.[a]

Smith felt that individuals' tendency to act in their own self-interest is a natural law and a natural right that precedes the existence of government. The exertion of these individual rights in a competitive market setting, furthermore, creates the greatest good for the greatest number in society. Smith's view, although a mainstream perspective in American capitalism, has been amended to accommodate government provisions of goods when the market fails to provide them in sufficient quantities.

[a]Adam Smith, An Inquiry into the Nature and Causes of the Wealth of Nations, ed. Edwin Cannan (1776; reprint, New York: Modern Library, 1937), p. 423.

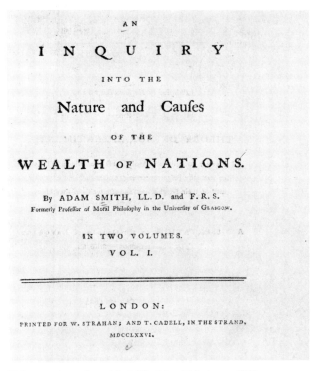

Title page from the original Wealth of Nations, 1776

---

failures are expected consequences of a competitive system. New fast-food restaurants open every day in anticipation of profits. Airlines declare bankruptcy and leave the industry—a sure sign of losses. Competition requires that entry and exit into business be free and unregulated. The Focus, "Adam Smith's 'Invisible Hand,'" discusses some aspects of the competitive system.

Coordinating the billions of individual decisions involved in competition is an interconnected system of prices for inputs and outputs that is so complex that no individual or computer can fully comprehend it. We begin to study the intricacies of the price system in Chapter 4.

### The Limited Role of Government

American capitalism as an economic system requires an attitude of laissez-faire (from the French, meaning roughly "to let do"). Laissez-faire has come to mean minimum government interference and regulation in private and

**Laissez-faire economy:** A market economy that is allowed to operate according to competitive forces with little or no government intervention.

economic lives. In a pure **laissez-faire economy,** government has a role limited to setting the rules—a system of law establishing and defining contract and property rights, ensuring national defense, and providing certain goods that the private sector cannot or would not provide. The last category includes roads, canals, and a banking system. Another important feature of laissez-faire society is the government's legally limited or sharply curtailed access to the taxing power.

## The Mixed System of American Market Capitalism

**Mixed capitalism:** An economy in which both market forces and government forces determine the allocation of resources.

In truth, no society ever conforms totally to the laissez-faire ideal. The ideal is modified in two ways: in an altered notion of the competitive process and in an expanded role for government. Such a modification is called **mixed capitalism.**

### An Evolving Competitive Process

Was any country ever composed of so many competing buyers and sellers to become, in Adam Smith's phrase, a "nation of shopkeepers"? Although historical data are less than perfect, we are fairly certain that purely competitive market structures did not exist even in Adam Smith's time. With the Industrial Revolution, capital requirements of firms were such that the most efficient firms—those producing goods at lowest cost—became larger. Certain industries and markets no longer had large numbers of competing sellers. Economists call such markets imperfectly competitive. They are also called oligopolies (characterized by a few competitors) or monopolies (having a single producer).

Some economists have argued that the decline in the number of competitors in some markets has led to concentrations of economic power in the hands of a few and to the demise of the laissez-faire competitive system. Modern economic research into the competitive process disputes this position, however. In the new view, competition is not to be described by a given number of sellers and buyers but rather by a rivalry for profits—that is, a process. Such rivalry—or even the potential for it, as long as individuals and businesses are free to enter and exit the market—produces results similar to competition among many buyers and sellers. One or two sellers in an industry can be competitive as long as entry and exit in the market is possible.[1]

### The Expanded Role of Government

The most important modification in the traditional conception of laissez-faire capitalism is an expanded social and economic role of government. Since the turn of the century, and especially since the 1930s, the relative size of government in the United States has grown dramatically in both social and economic spheres. In the 1960s and 1970s we saw large increases in government payments to individuals through Social Security, Aid to Families with Dependent Children, Medicare and Medicaid, unemployment compensation, and other welfare programs. The direct economic activity of government has grown apace.

---

[1] See Isreal M. Kirzner, *Competition and Entrepreneurship* (Chicago: University of Chicago Press, 1973), for more details on rivalrous competition.

# The Mixed System of American Market Capitalism

**Public goods:** Goods that no individual can be excluded from consuming, once it has been provided to another.

**Public Goods and Externalities.** Theoretically, underlying the government's role in economic life is the failure of a free-market society to satisfy all of its members' needs. The market society can fail in its ability to provide **public goods** such as national defense. Since national defense protects all citizens regardless of whether they pay for it, no one is likely to contribute to defense voluntarily. The private market fails in the sense that public goods such as defense would not be provided (or provided in sufficient quantity) unless government assumed responsibility.

Another cause for government intervention in a free-market economy is what economists call an *externality*. An externality is an unintended by-product of some activity, and it often involves environmental protection. A beautiful garden creates a **positive externality** in that it confers benefits to neighbors for which they do not pay. A **negative externality** might arise from a factory belching smoke or a firm dumping chemical wastes into a stream. In such a case, costs are imposed on members or segments of society rather than limited to the perpetrators of the externality. Negative externalities have led to various government interventions when the market has failed to limit the cost to the perpetrator. Taxes, subsidies, quotas, prohibitions, and assignment of legal liability are examples of government intervention (see Chapter 18).

**Positive externality:** A benefit of producing or consuming a good that does not accrue to the sellers or buyers but can be realized by a larger segment of society.

**Negative externality:** A cost of producing or consuming a good that is not paid entirely by the sellers or buyers but is imposed on a larger segment of society.

**Industry regulation:** Government rules to control the behavior of firms, particularly regarding prices and production techniques.

**Antitrust and Monopoly Regulation.** Another broad area of increased government participation in the U.S. free-market economy is **industry regulation**. In the early decades of this century, antitrust laws prohibiting price discrimination, collusion among producers, and deceptive advertising practices were passed in an attempt to restore competition where it no longer existed. Such laws continue to be enforced today.

Even earlier, however, economists and politicians believed that government had a role whenever competition could not exist, perhaps because of economies of large-scale production, or what economists call *natural monopolies*. Such monopolies are created when each seller can produce more and more output at lower and lower costs. Eventually it becomes profitable for only one seller to supply the *total* quantity demanded of a good, thereby creating a monopoly. Federal, state, and local governments undertook the regulation—not the ownership or operation—of transportation, communications, energy, and many other industries that were regarded as natural monopolies. Government regulation, in this view, was regarded as a substitute for competition where viable competition could not exist because of industry production and cost conditions.

Some economists (see Chapter 17) have strongly disputed this view and question the existence of large-scale economies (natural monopoly) in many of these regulated industries. Some contemporary economists believe that regulation of prices and profits must fail, either because regulation has been ineffective or because these industries are more competitive than previously thought. Questions have been raised about the self-interested supply of regulation by politicians combined with the self-interested demand for regulation by firms and industries. Do industries and other interest groups use the government regulatory apparatus for their own benefit? Should broad areas of regulation of industry, such as regulation of transportation and trucking, be eliminated? These and many other issues concerning the expanded role of government are in hot debate. A firm foundation in economic theory is required to answer these important questions.

**Economic stabilization:** A situation in which aggregate economic variables such as inflation rate, unemployment rate, interest rate, and growth rate are fairly constant over time.

**Economic Stabilization.** A final, but crucially important, part of the expanded role of government is in macroeconomic **stabilization** of the economy—that is, the government's efforts to promote full employment of resources without creating increases in the price level (or inflation). Taxation, expenditure policies, and the money supply can be intentionally changed by the federal government to help maintain full employment and promote economic growth at noninflationary levels.

Whether the government is or is not capable of achieving macroeconomic goals of full employment without inflation is a subject of debate among economists. Indeed, there are several alternative views on the role of government in economic stabilization. For the present, however, we should note that the mission to stabilize business cycles of inflation and unemployment has been a generally accepted role of government since the 1930s—the years of massive unemployment of resources known as the Great Depression. Much of the impetus to assign this new function to government came from the writings of the British economist John Maynard Keynes (1883–1946). Keynes believed that government policy makers could actually influence employment and inflation through spending and taxing policies and could thereby prevent depressions or severe reductions in economic activity.

The role of government in American economic life is much larger than it was fifty years ago. The American market system is modified laissez-faire. Government has provided a large number of social and economic goods, regulated markets and externalities such as pollution, and attempted to establish a high rate of economic growth through full employment without inflation. The microeconomic effects of government policy receive a great deal of attention later in this book. A brief look now at the relative size of government's role in the private economy will provide some perspective on the modified system of laissez-faire that constitutes the American market system.

## Growth of Government

**Private sector:** All parts of the economy and activities that are not part of government.

How big is government in our mixed American economy? Should the economic role of government be larger or smaller in relation to that of the **private sector**—all nongovernment activities? This critical matter is the subject of an ongoing debate among political candidates, members of Congress, journalists, local, state, and federal voters, intellectuals, academics, and private citizens. Rather than developing value judgments about how big government should be, economists analyze the probable effects of political decisions about the role of government in the present and in the past. To provide background information for such an economic analysis, we examine the size of the U.S. government's economic role by looking at its expenditures and the taxes it levies to support them, contrast the level of U.S. government taxation and spending with that of some other developed nations, and then relate this information to patterns of economic growth.

### The Size of the Federal Budget

We can get an overview of the role of the federal government in the U.S. economy by looking at a gross measure of government expenditures and receipts over the past three and a half decades. Figure 3–2 depicts the federal budget (expenditures and tax receipts) for the years 1950–1984. After a slow

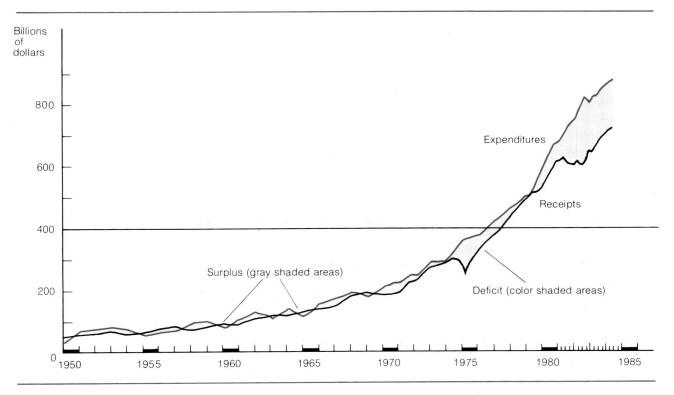

FIGURE 3–2  **Growth of the Federal Budget, 1950–1984**
Since 1950 the federal budget (the total of expenditures and receipts) has grown both absolutely, in total dollars, and as a percentage of the gross national product. Federal deficits have also grown dramatically in recent years. In 1984, the federal deficit was over $175 billion.
*Source:* Board of Governors of the Federal Reserve System, *Historical Chart Book* (Washington, D.C.: Government Printing Office, 1984), p. 51.

rate of growth through the mid-1960s, expenditures and tax receipts began to increase at a faster pace.

The Great Society welfare programs begun by President Lyndon B. Johnson in the mid-1960s and the defense expenditures of the Vietnam War were partly responsible for the absolute increases in government spending and taxation. Especially during the 1970s, deficits appeared when expenditures exceeded tax revenues. (The economic effects of these deficits are quite widespread.) The government's economic growth is not only of recent origin, however. The Franklin D. Roosevelt administration's social and economic programs in the 1930s—a response to the largest worldwide depression in modern history—were an initial and important force for the expansion of government into a mixed economy. The participation of government at all levels—federal, state, and local—has increased dramatically since 1930.

Since prices of goods and services have risen considerably during the twentieth century, the absolute dollar increase in government expenditures does not necessarily indicate whether government has grown bigger relative to the private sector. For this information, economists often look at government expenditures as a percentage of the **gross national product,** or GNP. The GNP is the aggregate value of all goods and services produced in the country over some period, usually a year. Using this measure, econo-

**Gross national product:** The dollar value in terms of market prices of all final goods and services produced in an economy in one year.

mists have found that while the government accounted for less than 10 percent of all purchases of goods and services as a percent of gross national product in 1929, by 1984 government purchases of goods and services were responsible for more than 20 percent of GNP. Since 1960, government's percentage of purchases of goods and services has remained fairly constant at 20 percent, but this constancy understates the growth of the government's role in the economy. The government's tax receipts at all levels in 1984 accounted for more than 30 percent of GNP. We can understand the discrepancy between expenditures and receipts and the expanding role of government by examining the kinds or distributions of expenditures at the various levels of government and then at the ways the government collects revenues.

## Government Expenditures

There are two kinds of government expenditures: direct purchases of goods and services and transfer payments. **Direct purchases** of newly produced goods and services include such items as missiles, highway construction, police and fire stations, consulting services, and the like. In other words, the government purchases real goods and services. **Transfer payments** are the transfers of income from some citizens (via taxation) to other citizens; these are sometimes called *income security transfers* or payments. Examples of transfer payments are Social Security contributions and payments, Aid to Families with Dependent Children, food stamp programs, and other welfare payments. These transfers do not represent direct purchase by the government of new goods and services, but they influence purchases of goods and services in the private sector. They are a growing part of government's role in the mixed economy.

> **Direct government purchases:** Real goods and services such as equipment, buildings, and consulting services purchased by the government.
>
> **Government transfer payments:** Money transferred by government through taxes from one group to another, either directly or indirectly; also called income security transfers.

**The Distribution of Federal Expenditures.** Out of the thousands of items in the federal budget, we can use six major categories to compare expenditures as a percentage of the total federal budget in 1960 and 1984 (see Figure 3–3). Since providing national defense is one of the major functions of the federal government, we would expect defense to account for a large proportion of federal outlays, and it does. National defense expenditures represented about one-quarter of all federal outlays in 1984. The largest single item in the 1984 federal budget, however, was not defense expenditures but income security transfers, which made up about 33 percent of total outlays. Expenditures on interest service of the federal debt, education, and natural resources ranged from over 12 percent to about 1 percent, respectively. The remainder, accounting for only 17 percent of the federal budget, went to such activities and projects as the administration of justice, science and technology, transportation, agriculture, international affairs, energy, the environment, revenue sharing, and the running of general government.

A mere recital of the proportions of outlays in the 1984 budget is not as interesting as a more dynamic picture of how these outlays changed over the two and one-half decades preceding 1984. Using Figure 3–3 we can compare the distribution of expenditures in 1960 and 1984. In 1960, fully 50 percent of federal expenditures was for national defense, while only 19 percent went to income security. Health-related expenditures have also grown from less than 1 percent in 1960 to more than 10 percent in 1984.

Changes in defense spending and in transfer payments between 1960 and 1984 are part of a clear trend over the period, shown in Figure 3–4. Over most of the 1960s, transfer payments grew at a faster rate than defense pur-

## FIGURE 3–3
### Federal Expenditures, 1960 and 1984

These two illustrations show percentage shares of federal spending. Expenditures at the federal level between 1960 and 1984 underwent dramatic changes. The percentage spent on defense was almost cut in half and the percentage of the budget spent for income security transfers increased by 14 percent. Education and health also received increased shares of expenditures. (Data for 1983 and 1984 are estimated from budget revisions in April 1983.)

Source: Facts and Figures on Government Finance (Washington, D.C.: The Tax Foundation, 1983), p. 108.

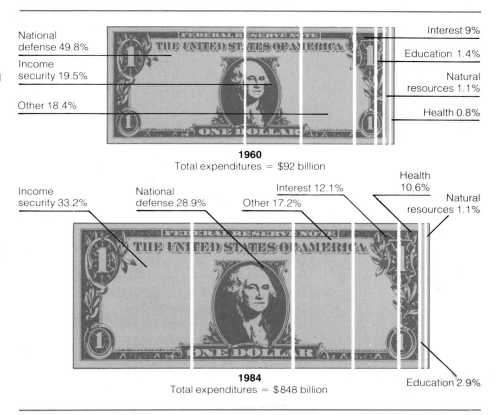

## FIGURE 3–4
### Growth in Government Expenditures, 1950–1984

The dramatic growth rate in transfer payments over the 1960s and 1970s is shown in the figure. Absolute amounts spent on transfer payments overtook defense expenditures in about 1970.

Source: Board of Governors of the Federal Reserve System, Historical Chart Book (Washington, D.C.: Government Printing Office, 1984), p. 53.

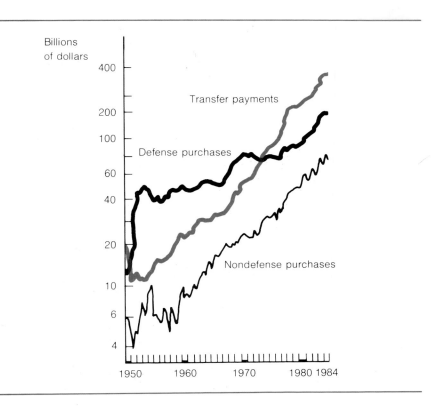

chases, reflecting the decisions of President Johnson and Congress to attack poverty and social imbalance. In spite of the fiscal pressures of the Vietnam conflict, transfer payments overtook defense purchases in absolute amounts—that is, in actual billions of dollars spent—in about 1970 and have exceeded them every year since.

Will this trend continue? Ronald Reagan campaigned for and won the presidency in 1980 and 1984 partly on this issue, promising to slow the growth rate in income security expenditures and to raise it on defense expenditures. The final outcome is unclear.

**State and Local Expenditures.** Figure 3–5 shows that the primary public goods provided by state and local government are education, highways, public welfare, hospitals and police, fire, and correctional institutions. Economists, voters, and other observers tend to view the federal government as the principal economic agent in the mixed economy. The truth is, however, that state and local governments combined are larger purchasers of goods and services (as a percentage of GNP) than the federal government. The big difference between the economic impact of the federal government and state and local governments is the huge federal redistribution of funds through the tax system from some citizens to other citizens. When income security transfers are included, the economic impact of the federal budget is larger than that of state and local governments.

### Government Receipts: The U.S. Tax System in Brief

Goods and social transfers provided at all levels of government are paid for out of taxation. The type of taxes levied at federal, state, and local levels varies a great deal.

**Federal Taxation.** The principal source of federal revenues, as shown in Figure 3–6, is the individual income tax. In 1984, the income tax accounted for 45 percent of total federal receipts.

Second in order of importance at the federal level, representing 36.7 per-

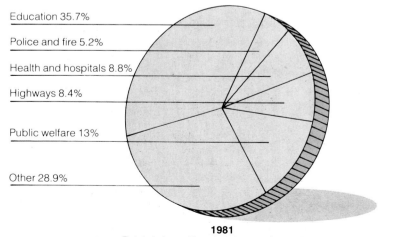

**FIGURE 3–5**
**State and Local Expenditures by Category**

Principal expenditures at the state and local levels of government are on education, highways, public welfare, and health and hospitals.

Source: *Facts and Figures on Government Finance* (Washington, D.C.: The Tax Foundation, 1983), p. 178.

**FIGURE 3–6**

**Distribution of Federal Tax Receipts, 1959 and 1984**

Social insurance tax receipts have more than doubled as a percentage of total receipts between 1959 and 1984, while the relative contribution of the individual income tax has remained almost constant. (Data for 1983 and 1984 are estimated from budget revisions in April 1983.)

Source: *Facts and Figures on Government Finance* (Washington, D.C.: The Tax Foundation, 1983) pp. 26, 101.

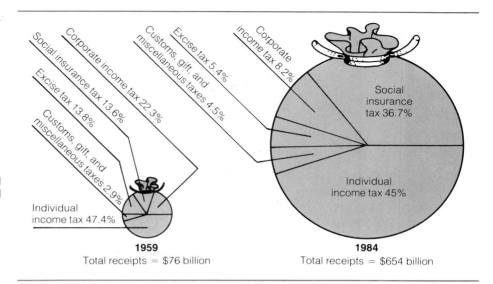

cent of receipts in 1984, are receipts from social insurance taxes and other contributions. These taxes are, principally, the payroll taxes paid jointly by employees and employers that are used to finance Social Security, disability compensation, and other payments. Receipts from these taxes have grown, dramatically, from less than 10 billion in 1950 to more than 200 billion in 1984.

Taxes on corporate income accounted for about 8 percent of revenues in 1984 and have generally declined since 1970 as a percentage of federal revenue. Other sources of federal revenues include federal excise taxes on goods such as liquor, tobacco, and gasoline, customs deposits paid on imports and exports, and estate and gift taxes.

**State and Local Receipts.** State and local governments rely primarily on property taxes and sales taxes for revenue. An additional revenue source, of varying importance from state to state, is the state income tax. Only about 10 percent of state receipts were from state income tax in 1984. Transfers of revenue from the federal to state and local governments, called grants-in-aid, have assumed increasing importance over the past twenty years. In 1984, for example, federal grants accounted for the highest single percentage of revenue for states and municipalities.

### The United States and Other Mixed Economies

The preceding discussions give some indication of the kinds of activities pursued by government in a mixed economy as well as the kinds of taxes the government relies on for revenue. The relative size of government is only hinted at by the breakdowns of outlays and receipts, however. To understand how mixed the American economy is, we can consider some international comparisons.

Figure 3–7 shows the growth of the government's public expenditures as a percentage of gross domestic product for five Western industrialized nations between 1965 and 1977. **Gross domestic product** (GDP), like GNP, is a measure of a country's production of goods and services, but GDP measures final goods and services produced with resources located within the country.

**Gross domestic product:** A measure of the final goods and services produced by a country with resources located within that country.

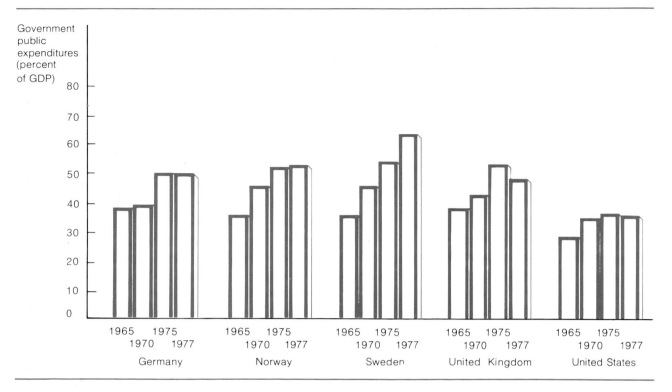

FIGURE 3–7  Relative Growth Rates of Government Public Expenditures in Five Western Democracies, 1965–1977

The bars represent government expenditures as a percentage of gross domestic product (GDP). By this measure, the U.S. economy is the least mixed of five Western democracies.

Source: Organization for Economic Cooperation and Development, *Main Economic Indicators*, various issues.

(GDP is determined by subtracting income earned by foreign investments from the GNP.)

Public expenditures as a percentage of GDP grew between 1965 and 1977 in all of the countries shown in Figure 3–7. The percentage of government expenditures in Sweden almost doubled between 1965 and 1977. By 1977, more than 60 percent of spending in Sweden was directed by the government. Germany, Norway, and the United Kingdom were about evenly divided in their mix between public and private participation in the economy by 1977. The bottom line: Of the leading Western industrialized nations, the U.S. economy is the least mixed in that the government directs only one-third of the country's spending. The trend over the past fifty years, however, has been toward more government participation in Western democracies.

### The Economic Effects of Increased Government

Do economists care about how mixed economies are? As noted in Chapter 1, most economists believe that rational self-interested forces and private enterprise are the key factors leading to maximum output, growth, and efficiency. In this view, governmental-directed enterprises—or self-perpetuating bureaucracies—do not normally provide maximum incentives for work, creativity, technology, and economic progress.

High income taxation needed to finance growing government budgets tends to reduce private savings and work effort, forcing taxpayers to engage in more leisure activity and other nonmarket activity. High corporate taxation discourages business investment in capital goods, technology, and innovation, resulting in low rates of economic growth. Most economists would predict that countries with low rates of government participation will have higher rates of capital formation over time, and vice versa. Japan, for example, has only about one-third of its spending directed by the government, but it has one of the highest rates of private investment in the world. In 1980, investment as a percentage of GDP was over 31 percent in Japan; that is, 31 percent of the country's resources were devoted to the production of capital goods for *future* production. The United Kingdom, on the other hand, where government spending accounted for more than 43 percent of GDP in 1978, had the lowest rate of capital formation of Western democracies (17.8 percent of GDP in 1980).

While these relations are merely suggestive, many economists are concerned about the economic impact of larger and larger governments on growth and economic progress (see Economics in Action, "The Mixed Economy of Sweden," at the end of this chapter). Government, as we have seen, has important and legitimate functions. Few economists, moreover, would question some redistribution of income based on a concept of justice since a market system does not automatically produce a just distribution of income or wealth. The question plaguing economists is, as usual, a marginal one: Will added government control over a given amount of private resources increase or decrease satisfaction and economic incentives to work, produce, and invest? Is the relative size of government a tonic to economic society or a sedative? These are extraordinarily difficult questions to answer. Much of this book is devoted to both microeconomic and macroeconomic analyses of these critical questions since American capitalism is, increasingly, a blend of larger government combined with free-market forces.

## Summary

This chapter discussed alternative economic systems and their characteristics, with particular emphasis on the American economy.

1. Specialization and trade, which are responses to the problem of scarcity, take place within economic systems. The three main types of economic systems are traditional, command, and market systems.
2. The primitive, traditional society handles the fundamental "what," "how," and "for whom" questions by employing skills and customs passed down from generation to generation. Such societies are economically static because the level of output is strictly limited.
3. Command societies answer the three fundamental questions through a central authority that directs the allocation and distribution of scarce resources. Such societies can be very specialized and technologically advanced.
4. Market societies answer the fundamental questions through the interplay between consumers' dollar votes and the free and unregulated decisions of producers and resource suppliers.
5. All real-world systems are combinations of traditional, command, and market characteristics. Further, modern economic systems are mixed economies because they contain elements of both free-market forces and government provision of goods, services, and income security transfers.
6. Some of the important features of American capitalism include the individual's right to own and dispose of property, a legal system protecting property and contracts, free enterprise, competitive economic markets, and a traditionally limited economic role for government.
7. An expanded economic role for government characterizes contemporary American capitalism. The basis for this role is the provision of goods such as

national defense that would not be produced in sufficient quantity by the private sector. Government also intervenes in the market to block the effects of negative externalities, to regulate industry competition, and to help achieve economic stabilization.
8. Local, state, and federal governments directed about one-third of America's resources in 1984 through purchases of goods and services and through the redistribution of income. Income security transfers are, in absolute terms and as a percentage of total spending, the largest item in the federal budget, whereas expenditures on defense are half the percentage today that they were in 1960.
9. Economists study mixed economies and the role of government to determine whether an enlarged public sector increases economic well-being or reduces incentives to work and to invest capital.

## Key Terms

| | | | |
|---|---|---|---|
| economic system | socialist economy | public goods | gross national product |
| traditional society | capitalist economy | positive externality | direct government purchases |
| command society | free enterprise | negative externality | government transfer payments |
| market society | competition | industry regulation | gross domestic product |
| barter | laissez-faire economy | economic stabilization | |
| circular flow of income | mixed capitalism | private sector | |

## Questions for Review and Discussion

1. How are the "what," "how," and "for whom" questions determined in a market society and in a command society?
2. "They don't build cars like they used to. These days cars wear out before they are paid for." This type of statement is heard frequently. Who determines the quality and durability of products?
3. What is capitalism and what does it have to do with property rights and economic freedom?
4. What are some of the roles of government in the U.S. economy? Has the role of government increased in size and scope? Does this hinder or help the rate of economic growth?
5. What are the primary activities of state and local governments?
6. Can free enterprise exist in a country where a dictator or king has absolute power?
7. "The local cable TV company provides slow services, and it doesn't have many channels." If this cable company has a monopoly granted by the government, is Adam Smith's invisible hand at work?
8. The Oakland city government was disappointed when the Raiders, a professional football team, moved to Los Angeles. Is a professional football team a public good?
9. Does individual self-interest hinder economic growth and well-being if there is competition?
10. What can government do to improve the general economic welfare? Has it done such things?

## ECONOMICS IN ACTION

### Exchange in the Bazaar Economy

The main features of a market economy—a well-defined division of labor, the determination of production and distribution by impersonal forces, and specialized institutions serving to minimize transaction costs—are ordinarily associated by economists with industrialized, highly technological societies such as the United States. But this association is apparently not a necessary relation. Complex institutions of market exchange can emerge in a setting that, at least in technological terms, seems quite backward.

An interesting case in point is the bazaar economy of Morocco, which has been studied in detail by Clifford Geertz.[a] Bazaars exist in a society that in most respects has changed little in hundreds of years; by far the most important economic activity is subsistence agriculture.

Bazaars (in Arabic, *suqs*) are quite literally market-

[a] Clifford Geertz, "Suq: The Bazaar Economy in Sefrou," in *Meaning and Order in Moroccan Society: Three Essays in Cultural Analysis*, ed. L. Rosen (Cambridge: Cambridge University Press, 1979); "The Bazaar Economy: Information and Search in Peasant Marketing," *American Economic Review* 28 (May 1978), pp. 28–32.

# Exchange in the Bazaar Economy

*Scene from a Moroccan bazaar*

places. They represent either permanent or in some cases periodic centers in which almost all trade occurs. Their principal features have remained virtually the same for hundreds of years. To the eyes of the typical tourist, a bazaar may appear to be chaos incarnate—a blur of shouting, wildly gesticulating men and women in colorful costumes hawking a confusing array of goods, from bolts of cloth to dates, from cattle to utensils. Goods do not have fixed prices—every trade is assumed to be entirely negotiable. To make matters seem even more baffling, the bazaar is only partially "monetized" in its exchange; much trade occurs as barter.

However, the bazaar is anything but chaotic. In fact, the *suq* represents a complex market structure that resembles the market economies of modern industrialized countries to a remarkable degree.

Despite the apparent technological backwardness of Moroccan society, participants in the bazaar economy invariably occupy specialized niches in a well-defined division of labor. For example, virtually no artisans market their own wares. Shopkeepers and traders in general tend to specialize in one good or in a narrow range of goods. Geertz identified at least 110 different specialized occupational categories in the bazaars around Segrou, a town at the foot of the Middle Atlas Mountains, adding that many more occupations exist but are hard to categorize. This degree of specialization alone suggests a complex and highly organized economic system. Moreover, movement from one occupation to another is common; according to Geertz, "no one in the bazaar can afford to remain immobile; it's a scrambler's life" (1979, p. 185).

Specialized institutions have emerged that minimize the costs of exchange. There is a group of professional auctioneers; another of specialized brokers who act as agents for bazaar traders; and another of *arbitrateurs*, who travel from one bazaar where goods are relatively low-priced to sell them at another bazaar where they can be relatively high-priced. As Geertz explains, in a bazaar—as in any modern market economy—"search is the paramount economic activity, the one upon which virtually everything else turns, and much of the apparatus . . . is concerned with rendering it practicable" (1978, p. 216).

But perhaps the most striking feature of the bazaar is its impersonality. Prices are determined by a process of impersonal competitive bidding in which any buyers and sellers can participate. This is not to imply that bazaar participants (*bazaaris*) are total strangers to one another—usually they are not—but rather that in the bazaar all participants are equals, regardless of their cultural background. Groups who might never associate with each other under any other circumstances (and who may even dislike each other) trade with one another in the bazaar. Jews and Arabs trade freely, although outside of the bazaar they have little or nothing to do with one another. The bazaar does not depend on any particular group for its characteristics but is an impersonal mechanism of exchange. As Geertz notes, "A suq is a suq, in Fez or in the Atlas, in cloth or in camels. The players differ (and the stakes), but not the shape of the game" (1979, p. 175).

Finally, the bazaar economy is completely unregulated. There is no government intervention of any kind. There is not even a government court system—disputes in the bazaar are settled by private arbitrators (specialists in the business of settling disputes). But disputes are relatively uncommon because in the highly competitive bazaar economy, one of the most valuable capital assets a trader can have is a reputation for honesty. Overall, the system seems to function smoothly—if noisily—without government regulation.

## Question

Would one expect to find much race or sex discrimination in highly organized markets—such as in the U.S. stock or commodity exchanges—characterized by large numbers of buyers and sellers? Why or why not? Do you think that the possibilities for discrimination are greater when the numbers of buyers and/or sellers are small? Why?

## ECONOMICS IN ACTION

### The Mixed Economy of Sweden: How Much Government Is Too Much?

Economists have long defended the important economic roles of government in Western democracies. There is little debate about government's role in providing public goods, handling externalities such as pollution, and dealing with economic stabilization. But can the size of government get out of control relative to the size of the private sector? Some economists think so, and many point to Sweden as an example.

In 1977, nearly two-thirds of the gross domestic product of Sweden passed through the hands of either the national or local governments. In Sweden, the public sector, as in other Western democratic governments, has assumed responsibility for many services, such as education, employment, care of the sick and aged, and protection of the environment. But the Swedish government's role as redistributor of income, when coupled with its function as purchaser of goods and services, has made it a far more pervasive economic force than the country's private sector. What have been the economic consequences of the growth of government in Sweden?

Public goods and income transfers must be paid for by citizens, and Sweden, a nation of some 8.3 million people, has the highest tax rates of all Western democracies. Consider the national income tax and the local income tax rates for 1977 presented in Table 3–1. Taxes are translated to U.S. dollars in the table.

For single people earning $40,000 per year in Sweden, the total tax payment is more than 60 percent of $40,000, and, further, the rate on additional income (the marginal percentage rate) is 80 percent! Married people earning $100,000 per year would keep only about $25,000.

Take-home income is of course not the only income a person receives. All citizens are provided with free public services such as medical treatment. Unemployment characterized only 2.5 percent of the labor force in 1981. All individuals have a right to employment in the public sector. Certainly the Swedish welfare state produces welfare for some people, and the economist would count these welfare goods and services as benefits. But what are the costs?

Many economists claim that the high tax rates on Swedish individuals and corporations inhibit work effort, private investment, and formation of capital. If the emergence of the welfare state has in fact had these effects, they should show up in the distribution of employment between the public and the private sector and in the growth record of the Swedish economy. Consider in Table 3–2 how the composition of the public and private labor force changed in percentage terms between 1970 and 1981. Over this decade employment in all private-sector activities either failed to grow or declined. But employment in national and, especially, local government mushroomed. More than 30 percent of all people employed in Sweden now work for the national or local government.

Raw employment percentages in private and public enterprise do not reveal possible growth in the productivity of labor because technological changes may have

**TABLE 3–1  Swedish Individual Income Tax Rates, 1977**

| Pretax Income (U.S. dollars) | Single Persons | | | Married Couples (one wage-earner) | | |
|---|---|---|---|---|---|---|
| | Tax (U.S. dollars) | Percent of income | Marginal rate (%) | Tax (U.S. dollars) | Percent of Income | Marginal rate (%) |
| 5,000 | 1,129 | 22.6 | 32 | 775 | 15.5 | 32 |
| 10,000 | 2,844 | 28.4 | 39 | 2,490 | 24.9 | 39 |
| 15,000 | 5,292 | 35.3 | 56 | 4,938 | 32.9 | 56 |
| 20,000 | 8,344 | 41.7 | 68 | 7,990 | 39.9 | 68 |
| 25,000 | 12,109 | 48.4 | 78 | 11,755 | 47.0 | 78 |
| 40,000 | 24,119 | 60.3 | 80 | 23,765 | 59.4 | 80 |
| 50,000 | 32,577 | 65.2 | 85 | 32,223 | 64.4 | 85 |
| 75,000 | 53,832 | 71.8 | 85 | 53,477 | 71.3 | 85 |
| 100,000 | 75,086 | 75.1 | 85[a] | 74,732 | 74.7 | 85 |

[a]Maximum rate

Source: Svenska Handelsbanken, *Sweden's Economy in Figures* (1982), courtesy of the Swedish Embassy, Washington, D.C.

TABLE 3–2  Swedish Employment by Sector, as Percentage of Labor Force

|  | Occupation | 1970 (%) | 1981 (%) |
|---|---|---|---|
| Private Sector | Agriculture, fishing, forestry | 8.2 | 5.3 |
|  | Mining, manufacturing | 26.6 | 22.0 |
|  | Electricity, gas, water works | 0.7 | 0.7 |
|  | Construction | 9.3 | 7.2 |
|  | Wholesaling, retailing, restaurants | 14.5 | 13.3 |
|  | Other private and business services[a] | 18.9 | 18.4 |
|  | Total | 78.2 | 66.9 |
| Public Sector | National government | 6.2 | 7.3 |
|  | Local governments | 14.1 | 23.3 |
|  | Total | 20.3 | 30.6 |
|  | Unemployed | 1.5 | 2.5 |
|  | Total labor force (thousands) | 3,910 | 4,330 |

[a]Transport, communications, banking, insurance, property management, and so on.
*Source:* "The Swedish Economy" (Stockholm: The Swedish Institute, 1982).

made labor more productive in both private and public activity. If such were the case in Sweden, we would expect the results to show up in economic growth rates in gross domestic product. Consider in Figure 3–8 the annual percentage average growth in the volume of GDP between 1960 and 1981. By 1982, Sweden had the lowest growth rate of twenty-four Western member nations of the Organization for Economic Cooperation and Development, an international trade and finance organization.

Many economists argue that these results are not surprising. Incentives to work and invest have been progressively reduced with high individual and corporate tax rates and with promises of public employment and guaranteed bureaucratic income. Low and even negative growth rates are the result of reduced private capital investment. This is not to say that the emergence of the welfare state in Sweden has not produced benefits. Economists, however, note that these benefits have come at the cost of work effort, investment, and economic growth. Sweden may find itself with less and less to redistribute as its welfare state grows larger.

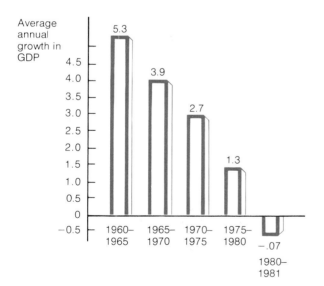

FIGURE 3–8  Average Annual Percentage Growth of Swedish Gross Domestic Product
*Source:* Adapted from "The Swedish Economy" (Stockholm: The Swedish Institute, 1982).

## Question

The country of Atlantis provides welfare for all of its unemployed citizens. The law presently states that if welfare recipients find work they will lose an equal amount of welfare benefits. How does this affect the desire to find work? What would happen if the law were changed so that all welfare benefits would not be removed with increases in earned income?

# 4

# Markets and Prices: The Laws of Demand and Supply

Simple specialization began to take place in primitive cultures as individuals recognized unique abilities in themselves and in others. As we have seen, increasing specialization led to organized markets where people bought and sold goods, to the use of money, and to increasingly large groups of buyers and sellers. Economists define these organized markets—such as bazaars, the stock exchange, or Saks Fifth Avenue—as places or circumstances that bring together demanders (buyers) and suppliers (sellers) of any goods or services.

What are the motivations of demanders and suppliers in the exchange of goods and services in these organized markets? How and why do some things get produced and sold at certain prices while other items are not produced or sold at all? Products such as tape cassettes were unknown to our grandparents. How and why do they get produced and sold to us for a certain price at a local music store? What happens to demand and supply when governments or other agencies attempt to intervene in organized markets through policies such as agricultural price supports and rent controls? Can the appearance of surpluses or shortages be predicted? These questions can all be approached through the single theory of supply and demand—the economist's basic tool.

Every day the news media provide dramatic evidence of the workings of the market system. The price of silver rises by 45 percent in one day as massive investments in silver futures by two of the world's richest oil magnates trigger a bandwagon effect. Leak of a technological breakthrough in a certain computer firm creates a frenzy of stock buying, driving up the price of the firm's stock overnight. Crude oil prices fall sharply as a result of the dissolution of the OPEC cartel.

In the familiar economic transactions of everyday life, we too enter markets where buyers and sellers congregate to buy and sell a great variety of products and services. The typical American supermarket sells thousands of products, and as we wander through the store we can view a price system in action. In the produce section, for instance, quantities and prices of fruits and vegetables depend on the quantities consumers want and on the season. Early crops usually bring in the highest prices. It is not uncommon for watermelon to sell for more than two dollars a pound in March but only fifteen cents a pound by the Fourth of July.

What determines who will get the early melons or how they will be rationed among those who want them? Why are prices and quantities constantly rising and falling for millions of goods and services in our economy? How do new products find their way to places where buyers and sellers congregate? The answers are simple. In a market society, the self-interest of consumers and producers, of households and businesses, determines who gets what and how much. To paraphrase Adam Smith, it is not to the benevolence of the butcher and the baker that we owe our dinner but to their self-interest. The primary way that consumers and producers express their self-interest is through the economic laws of supply and demand. Sticking a price tag on a product does not imply price-setting power, as anyone who has run a garage sale knows. In a market system, demand and supply determine prices, and prices are the essential pieces of information on which consumers, households, businesses, and resource suppliers make decisions. High melon prices in March will encourage suppliers and discourage demanders, whereas low prices in July will encourage demanders and discourage suppliers. Before investigating the mechanics of these laws of supply and demand, we consider a simple overview of the price system.

## An Overview of the Price System

As market participants, households and businesses play dual roles. Businesses supply final output of products and services—rock concerts, bananas, hair stylings—but also must hire or demand resources to produce the outputs. Households demand rock concerts, bananas, and hair stylings for final consumption but also supply labor and entrepreneurial ability as well as quantities of land and capital to earn income for the purchase of products and services.

**Products market:** The forces created by buyers and sellers that establish the prices and quantities of goods and services.

**Resources market:** The forces created by buyers and sellers that establish the prices and quantities of resources such as land, labor, and capital.

As Figure 4–1 shows, businesses and households are interconnected by the **products (outputs) market** and by the **resources (inputs) market**. Each market depends on the other; they are linked by the prices of outputs and inputs. The particular mix of goods and services exchanged in the products market depends on consumer demands in that market plus the cost and availability of necessary resources. For example, the groups that are featured at a rock concert will depend on what the targeted audience wants to hear plus the ability of that audience to pay the price demanded by the groups and the groups' availability on the chosen date. Similarly, the particular mix of resources available at any one time or through time is determined by what households are demanding—subject also to the availability of the resources. If land suitable for banana growing is available, it is most likely to be sold for banana plantations if households are demanding a lot of bananas, thereby making it possible for banana growers to pay landowners handsomely for their land.

Prices are the impulses of information that make the entire system of input and output markets operate. Take the prices of fad goods: At times certain

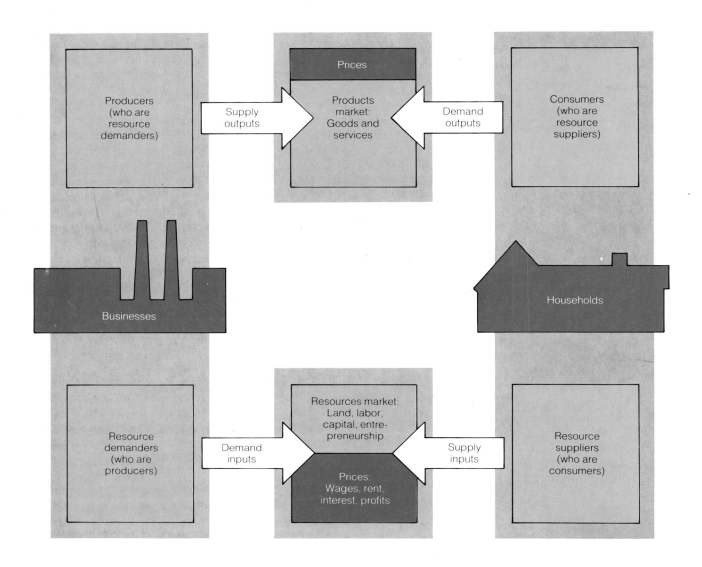

FIGURE 4–1  The Product and Resource Markets: A Circular Flow

Businesses play a dual role in the market economy: They are the suppliers of goods and services as well as the demanders of resources. Households also have a dual role: They are both demanders of goods and services and suppliers of resources.

goods or services—such as Rubik's cubes and hula hoops—have quickly appeared and then disappeared. When a fad begins to catch on, prices tend to be high because there is a high and rapidly growing demand and because resources necessary to produce the good may be scarce and command a high price. It takes time to adapt resources to the production of the fad good.

Machinery must be developed to mold plastic into Rubik's cubes. Labor must be drawn from other uses into the new production. Marketing channels must be established so that sellers and buyers can converge at convenient points of trade. Demand for the fad good is transmitted through businesses to the factors of production through a system of prices. Initially high prices signal the scarcity of goods and resources. As more entrepreneurs and businesses perceive the profit opportunities associated with the fad and as more resources are discovered or developed to produce it, prices of both the inputs and the outputs change accordingly.

The price system reacts similarly with goods that remain in the market for a longer time. Consider the development of computers and computer technology in the 1960s. Initial investments by producers were substantial as businesses rushed to introduce computer systems. In particular, the wages of computer programmers and technicians were high because there was a relative scarcity of workers possessing the skills necessary to use and produce computers. High demand and relative scarcity mean high wages. However, high wages are also an excellent piece of information that may encourage changes that ultimately lower wages. From the 1960s to the present, many schools teaching computer technology have emerged, reducing the scarcity of this resource. Demand for computer technology, however, has grown over time. The wages for people supplying services essential to production depend on consumers' demand for computers and on price (wage)-signaled supply conditions in the market for these services.

The informative signals of a price system work whether goods have short lives (Rubik's cubes) or long lives (computers). The prices formed in both product and resource markets reflect the relative desires of consumer-demanders for particular goods and services as well as the relative scarcity of the resources required to produce them. The very fact that a product or service bears a price means that scarcity exists. Supply and demand in all markets is at the core of scarcity and, therefore, of economics. These critical notions must be understood with the greatest possible clarity. We begin with demand.

## The Law of Demand

**Law of demand:** The price of a product and the amount purchased are inversely related. If price rises, the quantity demanded falls; if price falls, the quantity demanded increases.

**Quantity demanded:** The amount of any good or service consumers are willing to purchase at some specific price.

**Demand curve:** A graphic representation of the quantities of a product that people are willing and able to purchase at all of the various prices.

What determines how much of any good—Rubik's cubes, cassette tapes, or hair styling—consumers will purchase over some time period? Economists have answered that question for hundreds of years in the same manner—by formulating a general rule, or law of demand. The **law of demand** states that, other things being equal or constant, the quantity demanded of any good or service increases as the price of the good or service declines. In other words, **quantity demanded** is inversely related to the price of the good or service in question.

The relation between price and quantity demanded is a fact of everyday experience. The reaction of individuals and groups to two-for-the-price-of-one sales, cut-rate airline tickets, and other bargains is common proof that quantities demanded increase with decreases in price. Likewise, gas price hikes will lower the quantities of gasoline demanded. The formalization of this inverse relation between price and quantity demanded is called a law because economists believe it is a general rule for all consumers in all markets. Imagine a graphic representation of the law of demand—called a **demand curve**—for two hypothetical consumers.

### The Individual's Demand Schedule and Demand Curve

Suppose that we observe the behavior of Dave and Marcia over one month. These two music lovers own tape players and thus are willing to purchase, or demand, cassette tapes. To determine Dave's and Marcia's demand for tapes, we need only to vary the price of tapes over the month, assuming that all other factors affecting their decisions remain constant, and observe the quantities of tapes they would demand at those prices. This information is summarized in Table 4–1.

Table 4–1 shows a range of tape prices available to Marcia and Dave over the one-month period and the quantities (numbers) of tapes that each would purchase, all other things being equal. Given factors such as their income and the availability of other forms of entertainment, neither person would choose to purchase even a single tape at $10 per tape. Dave, however, would buy one tape at $9 and two at $8. Marcia would not buy her first tape until the price was $7. Each would purchase more tapes as the price falls. Thus, Dave's and Marcia's tape-buying habits conform to the law of demand.

We obtain the individuals' demand curves by simply plotting or transferring the information from Table 4–1 to the graphs in Figures 4–2a and 4–2b. The prices of tapes are given on the vertical axis of each graph, and the quantities of tapes demanded per month are given on the horizontal axis. The various combinations of price and quantity from Table 4–1 are plotted on the graphs. Each demand curve is then drawn as the line connecting those combinations of price and quantity. For both Dave and Marcia, the demand curve slopes downward and to the right (a negative slope), indicating an increase in quantity demanded as the price declines and a decrease in quantity demanded as the price rises.

### Factors Affecting the Individual's Demand Curve

**Factors affecting demand:** Anything other than price that determines the amount of a product that people are willing and able to purchase.

In addition to the price of a good or service, there are dozens, perhaps hundreds, of other factors and circumstances affecting a person's decision to buy or not to buy. These **factors affecting demand** include income, the price of related goods, price expectations, income expectations, tastes, the number

**TABLE 4–1  Two Consumers' Demand Schedules**
While individuals' demand schedules may differ, they do not violate the law of demand. For both people, the quantity demanded increases as the price falls.

| Price of Cassettes (dollars) | Quantity Demanded (one month) Dave | Marcia |
|---|---|---|
| 10 | 0 | 0 |
| 9 | 1 | 0 |
| 8 | 2 | 0 |
| 7 | 3 | 1 |
| 6 | 4 | 2 |
| 5 | 5 | 3 |
| 4 | 6 | 4 |
| 3 | 7 | 5 |
| 2 | 8 | 6 |
| 1 | 9 | 7 |
| 0 | 10 | 8 |

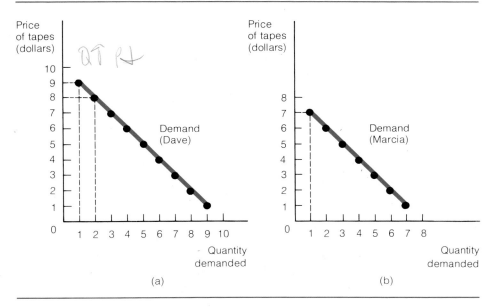

**FIGURE 4–2  Demand Curves for Two Consumers**

A consumer's demand for a product is the quantity that he or she is willing to purchase at each price. The demand curve is downward sloping for both Dave (a) and Marcia (b): As the price falls the quantity demanded increases, and as the price rises the quantity demanded decreases.

of consumers, and time. Even though this list of other factors is limited, it is too difficult to handle simultaneous variations among all factors in expressing a demand curve or schedule.

**Holding Factors Other Than Price Constant.** To isolate the effect of price on demand, the nonessential factors must be stripped away. We want to know what quantity of tapes Dave and Marcia would choose to purchase in a month at various possible prices, given that other factors affecting their decision do not change. This condition is called **ceteris paribus** ("other things being equal") by economists. It is essential to the development of any economic theory of model dealing with real-world events since all events cannot be controlled. Economists hold factors such as income and the price of related goods constant when constructing a demand schedule or curve.

*Ceteris paribus:* All other things held constant.

This does not mean that these factors do not change, but if they do change, the demand schedule or curve must be adjusted to account for them. Laboratory scientists are in a better position than economists to hold conditions constant. Chemists can perform controlled experiments, but economists, like weather forecasters, deal with a subject matter that can rarely be controlled. But like their scientific counterparts, economists must use scientific methods to organize real-world events into theories of how things work. The economists can use these theories to predict some of the effects on the demand schedule of changes in factors other than price.

**Changes in Demand.** A simple but crucial distinction exists between a change in Dave's or Marcia's *demand* for tapes and a change in their *quantity demanded* of tapes. Other things being equal, a change in the price of tapes will change the quantity demanded of tapes, as we have seen. A change in

**Change in demand:** A shift of the demand curve or a situation in which different quantities are purchased at all previous prices.

any factor other than the price of tapes will shift the entire demand curve to the right or left. Economists call this a **change in demand.** We consider some possible changes, why they take place, and how they affect the demand curve.

*Change in Income.* Marcia's or Dave's income may change, and such a change would necessitate a redrawing of the entire demand curve for cassettes. For most goods, a rise in income means a rise in demand. For instance, if Marcia's income increases from $500 to $800 a month, she will demand more tapes at every price because she can afford more. Figure 4–3 shows that, given a new, higher income, Marcia's demand curve shifts to the right for every price of tapes. When demand increases, quantity demanded is increased at every price.

Although the theory that rising income means greater demand for goods holds true for most goods, it does not apply to all goods. Economists make a distinction between normal goods and inferior goods. **Normal goods** are those products and services for which demand increases (decreases) with increases (decreases) in income; the demand for **inferior goods** actually decreases (increases) with increases (decreases) in consumers' income. Joe's demand for Honda automobiles may decrease as his income increases; a Honda automobile is an inferior good to him. Beth purchases less Häagen-Dazs vanilla ice cream as her income falls, indicating that Häagen-Dazs ice cream is a normal good for her.

**Normal good:** A product that an individual chooses to purchase in larger amounts as income rises or smaller amounts as income falls.

**Inferior good:** A product that an individual chooses to purchase in smaller amounts as income rises or larger amounts if income falls.

The terms *normal* and *inferior* contain no implications about quality or about absolute standards of goodness or badness. Indeed, a good or service that is normal for one consumer in a given income range may be inferior for another consumer in the same income range. It is even possible for a good to be normal for an individual consumer at certain levels of income and inferior at other levels. As one's income rises, for example, hamburger or compact cars may change from normal to inferior. This distinction should not detain us here (we will discuss it in more detail later), but it is important to note that a change in income will produce a shift in the demand curve.

*Price of Related Goods.* Suppose that the price of a good closely related to tapes—such as records or cassette tape players—changes during the month for which Dave's and Marcia's demand curves are drawn. What happens? Clearly, one of the assumptions about other things being equal has changed, and the demand curve will shift right or left depending on the direction of the price change and on whether the closely related good is a **substitute** for or a **complement** to the product under consideration.

**Substitutes:** Products that have a relation such that an increase in the price of one will increase the demand for the other or a decrease in the price of one will decrease the demand for the other.

Suppose that other forms of entertainment can substitute for tapes in Dave's or Marcia's budgets. If the ticket price of local movies or record prices decline during the month, the demand curve for tapes for both consumers would shift to the left, that is, the demand for tapes would decline. For every price of tapes, the quantity demanded would be lower. This shift is represented in Figure 4–4. If the price of a substitute good increased during the month, the demand for tapes would increase.

**Complements:** Products that have a relation such that an increase in the price of one will decrease the demand for the other or a decrease in price of one will increase the demand for the other.

If the price of a good or service complementary to tapes rises or falls, the demand curve for tapes would shift. Such a complementary good or service might be cassette tape players for home or auto, cassette tape carriers, or a new music store opening near Dave or Marcia. If the price of the complement increases, the demand for tapes would decrease (shift left). If the price of the complement decreases, demand for tapes would increase (shift right).

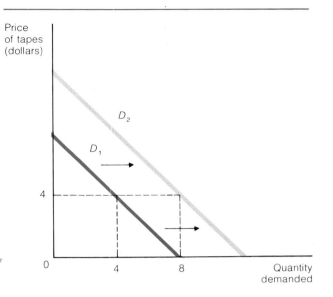

**FIGURE 4-3    Increase in Demand**
A change in any factors affecting demand causes a shift to the right or left in the demand curve. In this case, Marcia's income increases, causing an increase in her demand for tapes at every price. The demand curve shifts to the right. At a price of $4 per tape, Marcia previously would purchase 4, but given her increased income, she would now purchase 8 tapes. In this representation, D = demand.

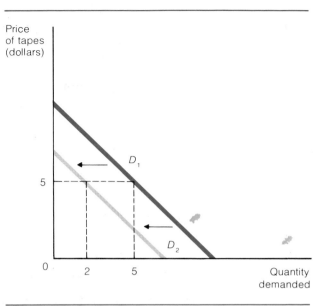

**FIGURE 4-4    Decrease in Demand**
A leftward shift of the demand curve indicates a decrease in demand. In this case, the price of a substitute good decreased, causing Dave to demand fewer tapes at every price. The demand curve shifts to the left. Before the price change in the substitute good, Dave would purchase 5 tapes at $5 each, but given the change in the substitute good, he would purchase only 2 tapes at $5 each.

*Other Factors Shifting the Demand Curve.* A number of factors other than income and the price of related goods can cause a shift in the demand curve. Among these are consumers' price and income expectations, consumers' tastes, and the time period. If the price of tapes or the income of consumers is expected to change in the near future, the demand for tapes during the month will be altered. Marcia may discover during the month that she will receive an inheritance sometime in the future. The basis under which her original demand curve was derived changes because she anticipates a change in income. Or back-to-school tape sales might be announced in the middle of July, causing an increase in demand during the time period. Likewise, any alteration in the time period under examination—changing from a month to a day, week, or year—will alter the construction of the curve. A purchase of four pizzas per month at $6.95 would be represented differently than a weekly consumption of one pizza. A change in the time period requires a redrawing of the demand curve.

## Reviewing the Law of Demand

The demand curve expresses an inverse relation between the price of a good and the quantity of the good demanded, assuming a number of constant factors affecting demand. As price rises, quantity demanded falls; as price falls, quantity demanded rises. When any one of the non-price factors affecting demand changes, we must reevaluate the demand schedule and curve. In general, we identify only the most important factors affecting demand curves. If we have missed some important factor affecting demand—

Dave's carburetor unexpectedly burns out in the middle of the month, for example—that factor must be accounted for in analyzing demand.

Economists predict that individuals (and collections of individuals), all things being equal, will purchase more of any commodity or service as its price falls. To verify this prediction, the individuals in question do not even have to be fully aware of their behavior; they need only act in the predicted manner. Individuals' response to sales of any kind—they buy more when price declines—is evidence that a general and predictable law of demand exists. That law is a fundamental tool of economists' analyses of real-world events.

### From Individual to Market Demand

While an individual's demand curve is sometimes of interest, economists most often focus on the **market demand** for some product, service, or input such as automobiles, intercontinental transport, or farm labor. Market demand schedules are simply the summation of all individual demand schedules at alternative prices for any good or service. An increase in the number of consumers increases the market demand curve and a reduction decreases market demand. The key is to add up the quantities demanded by all consumers at alternative prices of the good or service in question.

We can use the tape demand example to understand market demand. We constructed individual demand schedules for Dave and Marcia by varying the price and observing the quantities of tapes that they would buy at those prices, other things being equal. To determine the market demand schedule we simply observe the behavior of all other consumers in the same market situations.

Table 4–2 begins with the data on Dave's and Marcia's demand from Table 4–1. The table also contains a summary of quantity demanded for all other consumers at every price and, finally, the total quantity demanded for

> **Market demand:** The total demand for a product at each of various prices, obtained by summing all of the quantities demanded at each price for all buyers.

### TABLE 4–2  Market Demand Schedule

The total market demand for a product is found by summing the quantities demanded by all consumers at every price.

| Price of Tapes (dollars) | Quantity Demanded (one month) | | | |
|---|---|---|---|---|
| | Dave | Marcia | All Other Consumers | Total Market Demand |
| 10 | 0 | 0 | 0 | 0 |
| 9 | 1 | 0 | 39 | 40 |
| 8 | 2 | 0 | 78 | 80 |
| 7 | 3 | 1 | 116 | 120 |
| 6 | 4 | 2 | 154 | 160 |
| 5 | 5 | 3 | 192 | 200 |
| 4 | 6 | 4 | 230 | 240 |
| 3 | 7 | 5 | 268 | 280 |
| 2 | 8 | 6 | 306 | 320 |
| 1 | 9 | 7 | 344 | 360 |
| 0 | 10 | 8 | 382 | 400 |

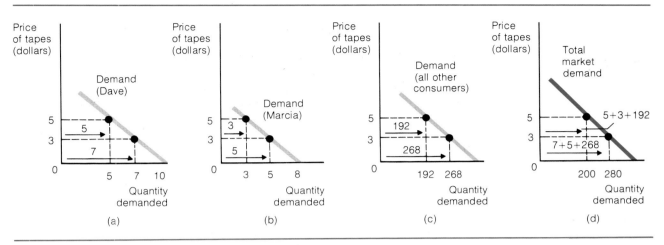

**FIGURE 4–5  Market Demand Curve**

The total market demand curve for a product is obtained by summing the points on all the individual demand curves horizontally. This is accomplished by selecting prices and summing the quantities demanded by all individuals to obtain the total quantity demanded at each price.

all consumers, or the total market demand. At a price of $10 no one, including Dave and Marcia, wants to buy tapes. At a slightly lower price, $9 per tape, Marcia does not choose to buy, but 40 tapes are sold. The total market demand is, therefore, 40 tapes at a price of $9. Note that actual numbers of tapes sold to all consumers would likely be much higher in any given market. We use low numbers for simplicity. The important point is that the market demand schedule for tapes or any other privately produced product or service is constructed in precisely this manner.

The market demand schedules can be represented graphically as market demand curves (see Figure 4–5). Dave's and Marcia's demand curves are repeated from Figure 4–2. The demand of all other consumers, taken from Table 4–2, is plotted in Figure 4–5c. The total market demand, shown in Figure 4–5d, is simply the horizontal addition of the demand curves of Figures 4–5a, b, and c.

As in the case of individual demand curves, the market demand curve is downward sloping (negatively sloped) and drawn under the assumption that all factors other than the price of tapes remain constant. If the incomes of consumers change, or the price of goods or services closely related to tapes is altered, the market demand would shift right or left, as in the individuals' demand curves. Economists must focus closely on these related factors in any real-world application. Changes in the demand for any product—compact cars, energy, crude oil—will be closely related to factors such as income changes and the price of substitutes and complements.

The important concept of market demand summarizes only half of the factors creating and affecting prices. Like the cutting blades of a pair of scissors, two sets of factors simultaneously determine price. Demand is the first factor. Supply is the second, and not necessarily in that order of importance.

## Supply and Opportunity Cost

Indirectly, we have already encountered a supply concept—that of opportunity cost along a production possibilities frontier, which we discussed in Chapter 2. We will briefly review those concepts before shifting attention from the behavior of the consumer to that of the producer.

An opportunity cost—the value of alternative products forgone—is incurred whenever any good or service is produced. Consider again the trade-off between producing oranges and peanut butter in a two-good world, as we did in Chapter 2. See Figure 4–6(a). A given stock of resources is available for the production of these two goods. Society, of course, will choose some combination of the two, and there is an opportunity cost in terms of resources in making a choice of more peanut butter and fewer oranges or more oranges and less peanut butter.

Consider what the real **marginal** (or additional) **opportunity cost** of producing more oranges is in terms of peanut butter forgone. The marginal cost (short for marginal opportunity cost) of more oranges is simply the amount of peanut butter sacrificed to produce the oranges. This is plotted in Figure 4–6b. From Figure 4–6a we see that society's choice of 4 million tons of oranges is made at a cost of 10 million tons of peanut butter (point G in Figure 4–6b). An additional, or marginal, 4 million tons of oranges can be produced at an additional cost of 15 million tons of peanut butter (point H). Additional oranges will be produced at higher and higher opportunity costs of peanut butter. Figure 4–6b gives the marginal cost curve in terms of real

**Marginal opportunity costs:** The extra costs associated with the production of an additional unit of a product; those costs are the lost amounts of an alternative product.

### FIGURE 4–6
### Production Possibilities and Marginal Opportunity Cost

The opportunity cost of a product rises as more and more of that product is produced. This is shown in the production possibilities curve as well as in the marginal opportunity cost curve.

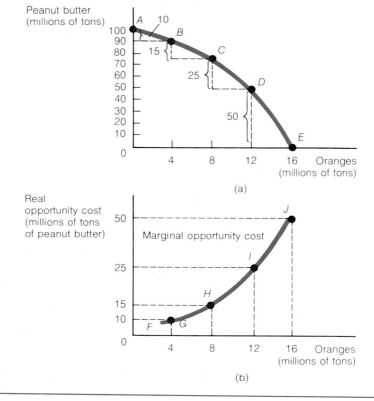

opportunity cost. Note that marginal cost increases as more oranges are produced. Why? Common sense suggests that there are increasing costs of transforming resources from peanut butter production to orange production. At first the resources drawn from peanut butter are easily adapted to producing oranges (that is, there is a low opportunity cost); as more and more oranges are produced, less-adaptable and less-talented resources must be used. The concept of the law of supply, discussed below, is totally analogous to this real opportunity cost except that prices (or opportunity cost in money terms) serve as the proxy for the real output of some good given up (peanut butter) to produce another good (oranges).

Thus, any production involves a real opportunity cost to society. The increasing money costs of producing additional tapes, for example, is merely a reflection of the higher real opportunity costs of drawing resources from other productions. It is crucial to remember that money prices, which represent real opportunity costs, are merely symptoms of the real factors underlying the economy. Scarcity and opportunity cost do not vanish, in other words, in a highly developed economy. Relative resource prices and, consequently, costs of production are but relative signals of scarcity.

## The Law of Supply and Firm Supply

**Law of supply:** The price of a product and the amount that producers are willing and able to offer are directly related. If price rises, then quantity supplied rises; if price falls, then quantity supplied falls.

**Quantity supplied:** The amount of any good or service that producers are willing to produce at some specific price.

**Supply schedule:** A schedule or curve that shows the quantities of a product that producers are willing and able to offer at all prices.

The **law of supply** states that, other things being equal, firms and industries will produce and offer to sell greater quantities of a product or service as the price of that product or service rises. There is a direct relation between price and quantity supplied: As price rises, **quantity supplied** increases; as price falls, quantity supplied decreases. The assumption of other things being equal is invoked, as in the case of demand, so that the important relation between price and quantity supplied may be specified exactly. This relation, called a **supply schedule,** shows the quantities of any good that firms would be willing to supply at alternative prices over a specified time period.

The method for constructing an individual firm's (and the market's) supply schedule is identical to the method we used for individual and market demand. All factors affecting supply except the price of the good or service are held constant. The price of the good or service is varied and the quantities that the firm or the industry will supply are specified.

To see how the supply curve is drawn, we turn to the supply side of cassette tapes. Suppose that there are a number of firms supplying tapes in a given geographic area and that the output per month of two typical firms (Grooves Inc. and Joe's Tapes) and all other firms combined is as shown in Table 4–3. The supply schedules of Grooves Inc. and Joe's Tapes are given in the table by the combination of price and quantity supplied. That is, given alternative prices of cassettes and the assumption that all other things are equal, Grooves Inc. and Joe's Tapes specify the quantity of tapes that they would be willing to supply during a one-month period. (Again, numbers are kept arbitrarily low for simplicity.) At a price of $10 per tape, Grooves Inc. would be willing to supply 24 tapes, but if the price falls to $3 per tape, Grooves Inc. will supply only 3 tapes.

As in the case of demand schedules, the information from supply schedules can be graphically expressed as supply curves (see Figure 4–7). The individual supply curves for Grooves Inc. and Joe's Tapes conform to the law of supply—other things being equal, as price rises, the quantity supplied increases and as price falls, quantity supplied decreases. Note that the supply curve for the individual firm is sloped upward and to the right. This is be-

## TABLE 4–3  Individual Firm and Market Supply Schedules

Individual firms' supply schedules follow the law of supply: As the price of the product increases, the quantity supplied increases. The total market supply is obtained by summing the quantities supplied by all firms at every price.

| Price of Tapes (dollars) | Quantity Supplied (one month) | | | |
|---|---|---|---|---|
| | Grooves Inc. | Joe's Tapes | All Other Firms | Total Market Supply |
| 10 | 24 | 16 | 410 | 450 |
| 9  | 21 | 14 | 365 | 400 |
| 8  | 18 | 12 | 320 | 350 |
| 7  | 15 | 10 | 275 | 300 |
| 6  | 12 | 8  | 230 | 250 |
| 5  | 9  | 6  | 185 | 200 |
| 4  | 6  | 4  | 140 | 150 |
| 3  | 3  | 2  | 95  | 100 |
| 2  | 0  | 0  | 50  | 50  |
| 1  | 0  | 0  | 0   | 0   |
| 0  | 0  | 0  |     |     |

cause the marginal opportunity cost of resources used for increased tape production rises as more tapes are produced. As more tapes are produced, less-adaptable resources are drawn into tape production, just as in the case of peanut butter and orange production. To increase the quantity of tapes supplied, a firm may have to enlarge its quarters by buying and converting buildings formerly used for other purposes, incur the costs of hiring and training workers who have never made tapes before, and perhaps redesign its product to use alternative materials as original materials become more scarce. These

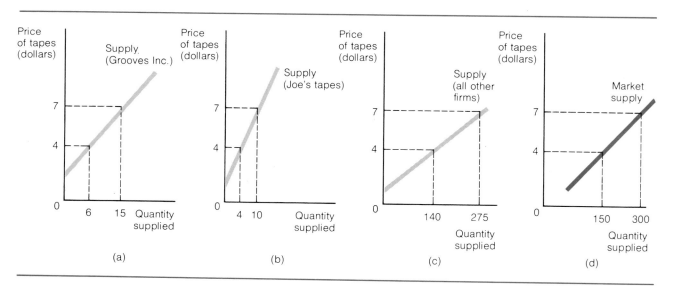

FIGURE 4–7  Market Supply Curves

The market supply, or total quantity supplied, of a product is obtained by summing the quantities that will be supplied by the two individual firms and by all other firms at every price. For example, at a price of $7 the total quantity supplied is 300 (15 + 10 + 275).

increases in marginal costs may make it unprofitable for the firm to supply more tapes unless the price of tapes rises.

## Changes in Quantity Supplied and Shifts in the Supply Curve

As in the case of demand, a change in price will alter the quantity that producers are willing to supply, and a change in any other factor will cause a **change in supply,** indicated by a shift in the supply curve either right or left. An increase in price will increase the quantity supplied, but a decrease in price will reduce the quantity supplied. The supply curve is positively sloped—upward and to the right—and, as we saw, the demand curve is negatively sloped—downward and to the right. When **factors affecting supply**—that is, factors other than price—change, the whole curve shifts. Some examples follow.

**Changes in Cost of Production.** The most important influence on the position of the supply curve is the cost of producing a good or service. The price of resources—labor, land, capital, managerial skills—may change, as may technology or production or marketing techniques peculiar to the product. Any improvement in technology or any reduction in input prices would increase supply, that is, it would shift the supply curve to the right.

Suppose that the price of plastic materials used in cassette tape construction falls or that the wages of salespeople available to tape stores decline. As the production or sales costs to firms producing and selling tapes decline, the quantity supplied increases and the supply curve shifts to the right for every price of tapes. An increase in supply is shown in Figure 4–8a. At price $P_0$, the firm was willing to supply quantity $Q_0$ of tapes, and the supply curve was $S_0$. After the firm (and all other firms) experiences a reduction in costs, the supply curve shifts rightward to $S_1$, indicating a willingness to supply a quantity $Q_1$ at price $P_0$. An improvement in production or sales techniques or a reduction in the price of some resource shifts the supply curve to the right.

**Change in supply:** A shift in the supply curve or a situation in which different quantities are offered at all of the previous prices.

**Factors affecting supply:** Anything other than price that determines the amount of a product that producers are willing and able to offer.

### FIGURE 4–8
**A Shift in the Supply Curve**

As factors other than price change, the supply curve shifts. When input costs fall, the quantity supplied increases from $Q_0$ to $Q_1$ at the price $P_0$, shifting the supply curve to the right from $S_0$ to $S_1$. An increase in input costs causes a decrease in quantity supplied from $Q_0$ to $Q_2$, and the supply curve shifts to the left, from $S_0$ to $S_2$.

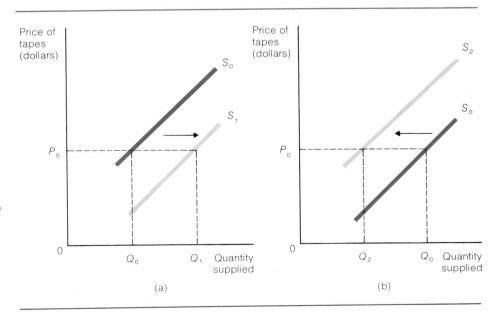

Any decline in production or sales techniques or an increase in any input cost shifts the supply curve to the left, as shown in Figure 4–8b.

**Other Determinants of Supply.** Factors other than cost of production changes can affect the location of the supply curve. One such factor is changes in producer-seller price expectations. The supply curve, like the demand curve, is drawn for a certain time period. If expectations of future prices change drastically in a market—for example, prices of the good or service are expected to rise suddenly—suppliers would withhold current production from the market in anticipation of higher prices. The current supply would be reduced, causing the supply curve to shift to the left, as in Figure 4–8b.

One other factor deserves mention. The supply curve is drawn for a given time period. A change in the time period (from one month to one week or one year) will alter the dimensions of the supply curve; it must be redrawn if the time period changes.

### Market Supply

**Market supply:** The total supply of a product, obtained by summing the amounts that firms offer at each of the various prices.

**Market supply** is simply the addition of all firms' quantities supplied for every price. If the number of firms in a given market increases or decreases, the supply curve would increase or decrease accordingly because the market supply curve is constructed by adding all the supply curves of individual firms. The market supply of tapes represented by the price-quantity combinations of the first and last columns of Table 4–3 is plotted in Figure 4–7d. The market supply curve is obtained by plotting the total quantities of tapes that would be produced and sold at every price during the time period.

The market supply curve is positively sloped—that is, total quantity supplied increases with increases in price—because the real marginal opportunity cost of tape production rises as more tapes are produced. Resources become more costly as more and more inputs are diverted from other activities into tape production. Remember that money cost is simply a proxy for real marginal opportunity cost.

## Market Equilibrium Price and Output

**Market:** Any area in which prices of products or services tend toward equality through the continuous negotiations of buyers and sellers.

**Perfect market:** A market in which there are enough buyers and sellers so that no single buyer or seller can influence price.

We now put the concepts of supply and demand together to understand how market forces work to establish a particular price and output. Before proceeding, we must set forth a few terms. A **market** is simply any area in which prices of products or services tend toward equality through the continuous negotiations of buyers and sellers. Competitive (self-interested) forces of both buyers and sellers guarantee this result. All other things being equal, a buyer of dog food will always choose the seller with the lowest price, whereas a seller will choose if possible to sell at higher prices. Buyers will not pay more than price plus transportation costs, and sellers will not take less. Only one price is possible.

In a **perfect market,** both buyers and sellers are numerous enough that no single buyer or seller can influence price. In addition, buyers and sellers are free to enter or exit the market at any time. In this case of perfect competition, no single seller sells enough of the commodity and no buyer buys enough of the product or service to influence price or quantity. In a perfect market the **law of one price** holds: After the market forces of supply and demand, of buyers and sellers, are at rest or in equilibrium, a single price for a commodity (accounting, of course, for transportation and other costs) will

## Market Equilibrium Price and Output

**Law of one price:** Exists in a perfect market. After the market forces of supply and demand reach equilibrium, a single price for a commodity prevails.

prevail. If a single price did not prevail, anyone could get rich by buying low and selling at a higher price, thereby driving prices to equality. The self-interested, competitive forces of buyers and sellers acting through supply and demand guarantee this important result.

### The Mechanics of Price Determination

Price and output are determined in a market from the simple combination of the concepts of supply and demand already developed in this chapter.

**Tabular Analysis of Supply and Demand.** Table 4–4 combines the data on the market supply and demand for cassette tapes and contrasts the quantities supplied and quantities demanded at various prices. The numbers used in Table 4–4 come from Table 4–2 (market demand for tapes) and 4–3 (market supply of tapes). The principles discussed here apply to supply and demand functions in any market.

Consider a price of $10 for tapes in Table 4–4. At the relatively high price of $10, the quantity of tapes supplied would be 450 while quantity demanded would be zero. That is, suppliers would be encouraged to supply a large number of tapes at $10, but consumers would be discouraged from buying tapes at that high price. If a price of $10 prevailed in this market, even momentarily, a **surplus** of 450 tapes would exist, that is, would remain unsold on the sellers' shelves.

**Surplus:** The amount by which quantity supplied exceeds quantity demanded when the price in a market is too high.

These unsold inventories of tapes are the key to what happens if the price of tapes rises above $5 in this market. Such inventories would create a competition among sellers to rid themselves of the unsold tapes. In this competition sellers would progressively lower the price. Consider Table 4–4 and assume that the price is lowered to $7. At $7 the quantity supplied of tapes is 300, while the quantity demanded is 120—a surplus of 180 tapes. Only when price falls to $5 per tape is there no surplus in the tape market.

**Shortage:** The amount by which quantity demanded exceeds quantity supplied when the price in a market is too low.

Now consider a relatively low price: $3 per tape. As the hypothetical data of Table 4–4 tell us, the quantity demanded of tapes at $3 would far exceed the quantity that sellers would be willing to sell or produce. There would be a **shortage** of 180 tapes; that is, 280 minus 100. Clearly some potential buy-

**TABLE 4–4  Market Supply and Demand**

The equilibrium price is established when quantity supplied and quantity demanded are equal. Prices above equilibrium result in surpluses; prices below equilibrium result in shortages.

| Price (dollars) | Quantity of Tapes Supplied | Quantity of Tapes Demanded | Surplus (+) or Shortage (−) |
|---|---|---|---|
| 10 | 450 | 0 | 450 (+) |
| 9 | 400 | 40 | 360 (+) |
| 8 | 350 | 80 | 270 (+) |
| 7 | 300 | 120 | 180 (+) |
| 6 | 250 | 160 | 90 (+) |
| 5 | 200 | 200 | 0 |
| 4 | 150 | 240 | 90 (−) |
| 3 | 100 | 280 | 180 (−) |
| 2 | 50 | 320 | 270 (−) |
| 1 | 0 | 360 | 360 (−) |
| 0 | 0 | 400 | 400 (−) |

**Equilibrium price:** The price at which quantity demanded is equal to quantity supplied; when this price occurs there will be no tendency for it to change, other things being equal.

ers would be unable to buy tapes if tape prices remain at $3. In fact, some buyers would be willing to pay more than $3 per tape rather than go without music. These buyers would bid tape prices up—offer to pay higher prices—in an attempt to obtain the product.

As the price bid by buyers rises toward $5, sellers will be encouraged to offer more tapes for sale; simultaneously, some tape buyers will be discouraged (will buy fewer tapes) or drop out of the market. For instance, at a price of $4 per tape, sellers would sell 150 tapes while buyers would demand 240, creating a shortage of 90 tapes. The shortage would not be eliminated until the price reached $5.

**Equilibrium price** in this market is $5; equilibrium quantity is 200 tapes supplied and demanded. At this price there is no shortage or surplus. A price of $5 and a quantity of 200 are the only price-output combination that can prevail when this market is in equilibrium, that is, where quantity demanded equals quantity supplied. The very existence of shortages or surpluses in markets means that prices have not adjusted to the self-interest of buyers and sellers. *Equilibrium* means "at rest." More precisely, in economic terms, equilibrium is that price-output combination in a market from which there is no tendency on the part of buyers and sellers to change. The free competition of buyers and sellers leads to this result.

**Graphic Representation of Supply and Demand.** The most common and useful method of analyzing the interaction of supply and demand is with graphs. Figure 4–9 displays the information of Table 4–4 on a graph that combines the market demand curve of Figure 4–5d with the market supply curve of Figure 4–7d.

The interpretation of Figure 4–9 is identical to the interpretation of Table

**FIGURE 4–9**
**Equilibrium Price and Quantity**

The equilibrium price is established at the point where quantity demanded and quantity supplied meet. At prices below equilibrium, the quantity demanded exceeds the quantity supplied and the price is bid upward. At prices above equilibrium, the quantity supplied exceeds the quantity demanded and the price is bid downward.

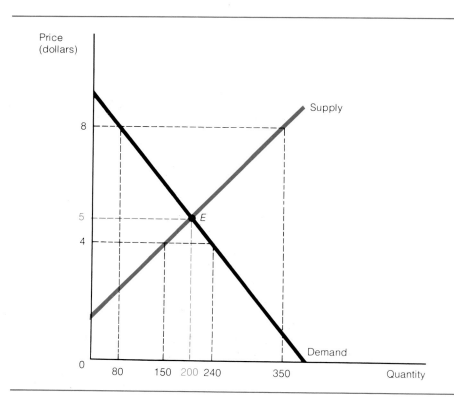

4–4, but the point of equilibrium is pictured graphically. Equilibrium price and quantity for tapes is established at the intersection of the market supply and demand curves. The point where they cross is labeled *E*, for equilibrium. At equilibrium, both demanders and suppliers of tapes are mutually satisfied. Any price higher than $5 causes a surplus of tapes; that is, a higher price eliminates some demanders and includes more suppliers. Any price below $5 eliminates some suppliers and includes more demanders.

The theory of supply and demand is one of the most useful abstractions from the world of events that is available to economists. The following three parts of this text will amplify this statement, but here we will discuss a few of the implications of supply and demand for society and for public policy.

### Price Rationing

Prices, which are formed through the interaction of supply and demand, are **rationing** devices. This means that scarce resources are channeled to those who can produce a desired product in the least costly fashion for demanders who most desire the product. Another way of saying this in economic terms is that resources flow to their most highly valued uses. Consider our hypothetical market for tapes again. Suppliers who are able and willing to produce or sell tapes at a cost below $5 (including a profit) are "successful" in that their tapes will be purchased. Demanders who are willing and able to purchase tapes at a price at or above $5 are the successful buyers of tapes. Only the most able sellers and buyers of tapes are successful in this market. High-cost producers (above $5) and buyers with a low preference for tapes (below $5) are eliminated from the market. Tapes and all other goods are rationed by a price system—by the free interplay of supply and demand. In such a system no conscious attempt is made by any organization (such as government) to allocate scarce resources on the basis of factors such as presumed need, eye color, morals, skin color, or ideas of justice. As such, the market system plays no favorites. A price rationing system, in other words, ensures that only the most able suppliers and demanders participate in markets.

### Effects on Price and Quantity of Shifts in Supply or Demand

Both individual and market supply and demand functions are constructed by assuming that other things are equal. What happens if other things do not remain equal, if factors other than price change? We can summarize these other factors and indicate their influence on equilibrium price and quantity in any market obeying the law of one price.

**Demand Shifts.** Any change in a factor other than the price of a good will alter the basis on which a demand curve is drawn—that is, it will shift the curve left or right. (Remember the difference between a change in demand, which is caused by a change in a non-price factor, and a change in quantity demanded, which is caused by a change in price.) A number of shifting factors are summarized in Table 4–5, which indicates the nature of the change, the direction in which demand will shift, and the effects on equilibrium price and quantity. We discussed all of these possible changes earlier in the chapter. Note that when demand changes, price and quantity move in the same direction.

These facts can be seen graphically in Figure 4–10. Factors causing an increase in demand from Table 4–5 will have the effects on price and quantity shown in Figure 4–10a. An increase in demand shifts the whole demand

---

**Rationing:** Prices are rationing devices; the equilibrium price rations out the limited amount of a product produced by the most willing and able suppliers, or sellers, to the most willing and able demanders, or buyers.

## 86   4 / Markets and Prices: The Laws of Demand and Supply

**TABLE 4–5   Factors Shifting the Demand Curve**

Changes in factors other than the price of the product will shift the demand curve either to the right or to the left, changing equilibrium price and quantity.

| Factors Changing Demand | Effect on Demand | Direction of Shift in Demand Curve | Effect on Equilibrium Price | Effect on Equilibrium Quantity |
|---|---|---|---|---|
| Increase in income (normal good) | Increase | Rightward | Increase | Increase |
| Decrease in income (normal good) | Decrease | Leftward | Decrease | Decrease |
| Increase in income (inferior good) | Decrease | Leftward | Decrease | Decrease |
| Decrease in income (inferior good) | Increase | Rightward | Increase | Increase |
| Increase in price of substitute | Increase | Rightward | Increase | Increase |
| Decrease in price of substitute | Decrease | Leftward | Decrease | Decrease |
| Increase in price of complement | Decrease | Leftward | Decrease | Decrease |
| Decrease in price of complement | Increase | Rightward | Increase | Increase |
| Increase in tastes and preferences for good | Increase | Rightward | Increase | Increase |
| Decrease in tastes and preferences for good | Decrease | Leftward | Decrease | Decrease |
| Increase in number of consumers of good | Increase | Rightward | Increase | Increase |
| Decrease in number of consumers of good | Decrease | Leftward | Decrease | Decrease |

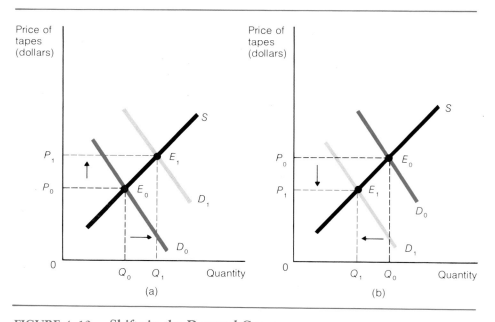

**FIGURE 4–10   Shifts in the Demand Curve**

(a) An increase in demand from $D_0$ to $D_1$ will increase the equilibrium price and the equilibrium quantity. (b) A decrease in demand from $D_0$ to $D_1$ will decrease both price and quantity.

curve rightward from $D_0$ to $D_1$. An excess demand—a shortage—opens up at price $P_0$. All the self-interested market forces discussed earlier now come into play. Demanders bid prices up to $P_1$, where additional quantities are supplied by firms and where a new equilibrium, $E_1$, is established. Thus, an increase in demand has the effect of increasing the equilibrium price *and* the quantity of the product demanded and supplied. A decrease in demand causes the demand curve to shift leftward (shown in Figure 4–10b). This has the effect of reducing prices and quantities demanded.

**Supply Shifts.** The effects of supply changes are summarized in Table 4–6. Clearly, any increase or decrease in a factor such as resource prices or price expectations will cause increases or decreases in the whole supply schedule. Effects of such changes are shown graphically in Figure 4–11.

Figure 4–11a shows the effects on price and quantity of an increase (rightward shift) in the supply schedule. At price $P_0$, excess supply—a surplus—opens up, and self-interested competitive firms and buyers bid prices down to $P_1$. Note, however, that while equilibrium price decreases, equilibrium quantity *increases* from $Q_0$ to $Q_1$. A price decline from $P_0$ to $P_1$ means that an additional quantity of the product or service will be demanded by consumers. Figure 4–11b shows the effects on price and quantity of a factor that decreases the supply curve (shifts it leftward). A decrease in supply has quantity-decreasing (from $Q_0$ to $Q_1$) but price-increasing ($P_0$ to $P_1$) effects, as Figure 4–11b shows. In supply shifts, unlike demand shifts, price and quantity change in opposite directions.

**Shifts in Both Supply and Demand.** Multiple shifts in both supply and demand are also possible, but the results are less predictable. Suppose in both supply and demand in a particular market that the incomes of consumers rise (and the product is a normal good) and that, simultaneously, resource input prices rise. In this case, as a look at Tables 4–5 and 4–6 will verify, demand increases and supply decreases. What will happen to equilibrium price and quantity? An increase in demand and a decrease in supply will clearly have

TABLE 4–6  Factors Shifting the Supply Curve

Factors changing the supply schedule will change equilibrium price and quantity in opposite directions.

| Factors Changing Supply | Effect on Supply | Direction of Shift in Supply Curve | Effect on Equilibrium Price | Effect on Equilibrium Quantity |
|---|---|---|---|---|
| Increase in resource price | Decrease | Leftward | Increase | Decrease |
| Decrease in resource price | Increase | Rightward | Decrease | Increase |
| Improvement in technology | Increase | Rightward | Decrease | Increase |
| Decline in technology | Decrease | Leftward | Increase | Decrease |
| Expect a price increase | Decrease | Leftward | Increase | Decrease |
| Expect a price decrease | Increase | Rightward | Decrease | Increase |
| Increase in number of suppliers | Increase | Rightward | Decrease | Increase |
| Decrease in number of suppliers | Decrease | Leftward | Increase | Decrease |

## FIGURE 4–11
### Shifts in the Supply Curve

(a) An increase in supply from $S_0$ to $S_1$ will lower the equilibrium price and increase the equilibrium quantity. (b) A decrease in supply from $S_0$ to $S_1$ will increase the equilibrium price and lower the equilibrium quantity.

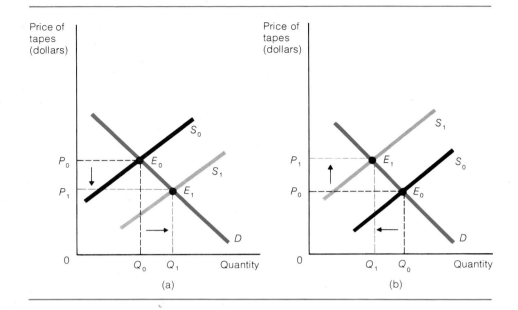

price-increasing effects (see Figures 4–10a and 4–11b), but equilibrium quantity is not predictable. Why? Because an increase in demand has a quantity-*increasing* effect, while a decrease in supply has a quantity-*decreasing* effect. Thus, whether equilibrium quantity increases or decreases depends on the magnitude of the relative shifts in the supply and demand curves. More information is needed before the economist can predict an outcome about quantity in such a case.

## Simple Supply and Demand: Some Final Considerations

Many real-world events—new entrants into markets, oil price increases, new computer technology—are analyzed using the simple laws of supply and demand. The theory of supply and demand is one of the most powerful tools the economist has to analyze the real world, as you will see throughout the rest of this book. At the outset, however, it is important to understand some limitations and some possible applications and extensions of the simple mechanics described in this chapter.

### Price Controls

As we have seen, a major effect of the free interplay of supply and demand is the rationing of scarce goods by a system of prices. One of the best ways to understand price rationing and its usefulness is to examine what happens when government intervenes in freely functioning markets. Such interventions into the natural functioning of supply and demand are called **price controls**. Rent controls, price controls, agricultural price supports, usury laws, and numerous other policies are examples of such "tinkering" in free markets. We deal with many of these matters elsewhere in the book, but simple supply and demand provide the basis for an initial discussion of the rationing of goods and services by nonprice means.

Take the hypothetical market for tape cassettes. The market demand and supply curves are reproduced in Figure 4–12 from the data in Table 4–4.

**Price control:** Government intervention in the natural functioning of supply and demand.

# Simple Supply and Demand: Some Final Considerations

**Price ceiling:** A maximum legal price established by government to protect buyers.

Suppose the government decides that the price of tapes is too high—"unjust" to consumers. Such is the logic of most state usury laws—"interest rates are too high"—or price controls—"prices are too high." By decree, the government orders the price of tapes to be $4 *or less*, thus establishing a **price ceiling**, a maximum legal price. At a price of $4, quantity demanded (240) exceeds quantity supplied (150), creating a shortage of 90 tapes. Clearly, the desires of buyers and sellers are not synchronized at a price of $4. What will happen? The government will have to police the market, incurring what economists call an enforcement cost. Otherwise, some profit-hungry tape sellers will charge black market prices (above $4) for tapes. Such behavior is perfectly possible since there are buyers willing to pay prices higher than $7 for tapes. Owing to these market forces, such price ceilings seldom, if ever, work. Buyers and sellers find profitable ways to evade them. In most cases consumers would prefer higher prices to non-price rationing which entails waiting lines, the purchase of coupons, or illegal purchases.

**Price floor:** A minimum legal price established by government.

Alternatively, suppose the government decides that tape prices are too low and that suppliers must be protected. Many agricultural suppliers have been protected in this manner. In this case a **price floor**—a minimum legal price—is instituted, say at $7 per tape (see Figure 4–12). At a price of $7 the quantity supplied of tapes is 300, while the quantity demanded is only 120, a surplus of 180 tapes. In this case a different (and familiar) problem arises. Assuming that the market is adequately policed—that is, that tape suppliers are disallowed from selling tapes at a price lower than $7—surpluses of tapes build up. The government must stand ready with tax dollars or some other device to buy up and store the surplus tapes to support the price floor. Such policies can become quite costly, as price support programs for butter, milk, cheese, peanuts, and grain have demonstrated. An understanding of

**FIGURE 4–12**
**Price Controls in the Market for Tapes**

A price that is fixed either above or below equilibrium will create a surplus or a shortage of a product. Prices set above equilibrium are called *floors* and prices fixed below equilibrium are called *ceilings;* a price floor is a minimum legal price and a price ceiling is a maximum legal price.

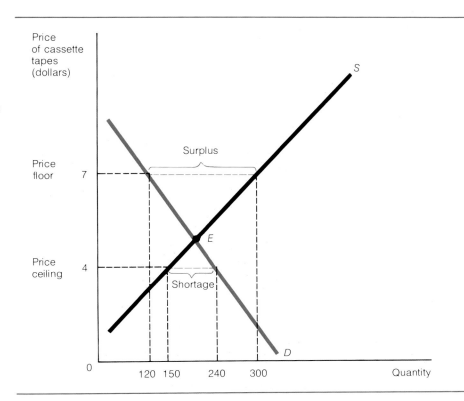

## FOCUS  Usury and Interest Rates: An Application of Price Controls

Price controls have been known to all ages and all civilizations. The Roman Emperor Diocletian invoked general price and wage controls prior to the fall of the Roman Empire. During the Middle Ages, church doctrine combined with economic policy to create civil laws related to interest-taking. The laws, called *usury laws*, originated in the belief that an uncontrolled market produced interest rates that were too high. The laws consisted of a legal limit on the amount of interest that lenders could charge or borrowers pay. Such laws survive today in the enactments of certain U.S. state legislatures that wish to protect borrowers and in federal regulations establishing the maximum interest rate allowable on small savings deposits.

The mechanics of the usury laws are simple; we demonstrate them in Figure 4–13. The figure shows the free-market supply and demand for loanable funds. As market interest rates increase, the quantity supplied of loanable funds rises and the quantity demanded declines. As market rates fall, the quantity of loanable funds supplied falls while quantity demanded rises. Market equilibrium occurs at point E, where quantity demanded equals quantity supplied ($Q_E$) and where the equilibrium price—the market interest rate—is $i_E$.

What happens when the government (Diocletian or the Ohio state legislature) declares that market rate $i_E$ is too high for the poor and that henceforth the maximum rate will be $i_C$, a ceiling rate below $i_E$? At rate $i_C$ the quantity of loanable funds demanded, $Q_D$ (point C in Figure 4–13), exceeds the quantity supplied, $Q_S$ (point B). A shortage of funds, BC, develops at the ceiling interest rate because banks are unwilling to lend the quantity of funds demanded at rate $i_C$. The available funds, $Q_S$, must be rationed because of the shortage. In the absence of price rationing, the process of rationing can take other forms. Unscrupulous lenders—"loan sharks"—may charge black market rates for the limited funds. Lenders may begin demanding additional collateral or better credit standing for loans at the ceil-

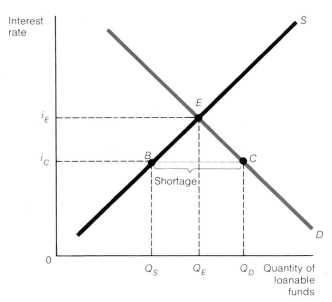

**FIGURE 4–13  Interest Rate Ceiling**

When a ceiling is placed on interest rates ($i_C$), a shortage (BC) develops, the difference between the quantity of funds that lending institutions are willing and able to supply ($Q_S$) and the quantity of funds that borrowers demand ($Q_D$). Without the ceiling, the forces of supply and demand interact to create equilibrium (E) between interest price and the quantity of loanable funds made available.

ing rate, reducing their risk and making it more difficult for less creditworthy borrowers to obtain funds. Such borrowers are often the poor whom the laws were designed to protect. The poor are often compelled to acquire funds from loan sharks in such circumstances. Usury laws, like all other forms of price controls, have built-in effects that often hurt those who are supposed to be the beneficiaries.

---

simple supply and demand points out the problems and costs associated with government (or other) interventions in a market system. See Focus, "Usury and Interest Rates," for a further discussion of price controls.

### Static Versus Dynamic Analysis

The analysis discussed in this chapter is called *static equilibrium analysis*. This means that (1) time does not enter into the discussion—it is not dynamic; and (2) price-quantity combinations move from one equilibrium to another ($E_0$ to $E_1$, for instance, in Figure 4–11) with no indication of the process by which firms and consumers actually move. The *dynamic analysis* of market supply and demand considers the time and process involved in moving from one equilibrium situation to another. How long, for example, does it take for a world oil price increase to influence prices in the various markets using oil as an input? A static analysis cannot tell us. It can indicate only the probable direction of price and quantities. What is the role of the entrepreneur in seeking profit opportunities when demand or supply schedules shift and markets are in disequilibrium, that is, experience shortage or surplus? Again, static analysis provides no method for answering questions about disequilibrium. A number of these issues will be discussed and supply and demand will be analyzed more thoroughly in a real-world context in the following chapters on microeconomics.

### Full Versus Money Prices

Another distinction must be made about prices resulting from the forces of supply and demand: There is a difference between the **money price** of a product or service and the **full price.** Consider the money price of a haircut for which the customer pays $12.50 in cash. Is this the full price of a haircut? Do consumers react to the money price of products and services or to what economists call the full price?

Money price is often not the only cost to consumers. In the case of a haircut, we must account for the time spent traveling to and from the salon, the time spent waiting, and the time spent with the hairdresser. The economist accepts the truth of the adage "time is money." Time, like diamonds, is a scarce resource, and it bears an opportunity cost. Time costs are often estimated in terms of the wage rate forgone, the leisure time forgone, or, generally, as the opportunity cost of the consumer's next-best alternative. Thus the full price of a haircut or any other good or service includes the price in money terms plus any other resource costs required in the purchase of the commodity. (Focus, "Full Price and a New Orleans Restaurant," gives another example.) Prices as interpreted in the simple model of demand and supply of this chapter are to be regarded as full prices. Consumers react to full prices, not money prices. No one, for example, should take an advertisement for "free puppies" or "free kittens" literally. There is no such thing as a free puppy!

**Money price:** The dollar price that sellers charge buyers.

**Full price:** The total cost to an individual of obtaining a product, including money price and other costs such as transportation or waiting time.

### Relative Versus Absolute Prices

In addition to the distinction between money prices and full prices, we make the distinction that consumers react to relative prices, not to absolute prices. **Relative price** is the price ratio in consumption of one product relative to another product or all other products. **Absolute price** is the price of a product measured in terms of money.

Many of us come from states that produce some agricultural commodity that varies in quality—citrus fruits from Florida or California, lobsters from Maine, apples from North Carolina or Washington. Residents of such states often complain, "They are shipping the good apples out," or ask, "Why do the high quality apples go to New York?" The theory of supply and demand tells us why.

Suppose that the absolute price of a good apple in Washington State is 10 cents and the absolute price of a bad apple is 5 cents. In this case, the relative price of good versus bad apples in Washington is two to one: A good apple is worth two bad apples. Now assume that the cost of transportation of each apple to New York is 5 cents. In New York, a good apple therefore costs 15 cents and a bad apple costs 10 cents. The relative price of good to

**Relative price:** The price of a product related in terms of other goods that could be purchased rather than in money terms.

**Absolute price:** The price of a product measured in terms of money.

## FOCUS: Full Price and a New Orleans Restaurant

New Orleans is considered a restaurant town by gourmets. Restaurants such as Le Ruth's, Antoine's, Brennan's, and Commander's Palace consistently offer some of the best cuisine to be found anywhere. One of our favorite French Quarter restaurants, however, is called Galatoire's. Galatoire's has justifiably earned a reputation for producing some of the highest-quality meals in the city. Menu prices, however, have remained low relative to other famous establishments and, even more important, they have remained low over the past several decades.

The question is, Have both menu prices and full prices remained low at Galatoire's? If not, why not? Finally, what evidence might be offered for a real price increase? How would an economist view the matter?

The economist would focus on elements in the full price of a meal at Galatoire's. It is notable that the restaurant has not enlarged its classic physical plant over the years but that, over time, longer and longer lines form outside the door at many hours during the day and evening. Galatoire's, unlike most of its competition, does not take reservations or accept credit cards—only cash will do. Moreover, Galatoire's enforces a dress code—tie or coat for men (evenings and all day on Sundays) and, until a few years ago, dresses (no pants) for women.

All these factors would tend to increase the full price of dining at Galatoire's. There are opportunity costs to waiting for a table (time is a scarce resource); the time and resources spent dressing for dinner are also applicable to the full price. Unpredictability of the length of the line on arriving at the restaurant can also make the real price higher or lower than anticipated, although diners can form hunches about when the line is apt to be shortest or nonexistent.

These additional costs mean that the full price exceeds the money price printed on the menu. However, the economist must calculate all benefits as well as all costs associated with purchasing products or services. Consider the possibility that some customers get positive benefits from dressing for dinner and from being surrounded by those similarly attired. Others may place a high value on the prestige or satisfaction of dining at the legendary restaurant. Such factors would tend to increase the benefits to these consumers. The full price paid for a meal at Galatoire's varies among consumers, depending on opportunity cost. As we will see in Chapter 6, however, consumers always will marginally balance the perceived costs of buying products and services with the perceived benefits associated with consumption. When price or price formation is discussed in this book, it is therefore the full price, not the nominal or money price, that is being considered.

---

bad apples is therefore different: Good apples are only one and one-half times more expensive than bad apples in New York. Because the relative price of good apples is lower in New York than in Washington, a relatively greater quantity of good apples are demanded in New York. Therefore, the good apples are shipped out of Washington to New York.

The theory of supply and demand discussed in this chapter has not explicitly stressed the distinction between real prices and relative prices. But it is important to interpret even basic supply and demand theory as a theory of relative, not absolute, prices and full, not money, prices. With these distinctions taken into account, the theory of supply and demand forms the foundation for the understanding of how markets function and therefore for the whole science of economics.

## Summary

1. The extension of specialization from earlier societies is the modern market society, where individuals and collections of individuals buy and sell—demand and supply—millions of products and services.
2. Demand can be expressed as a schedule or curve showing the quantities of goods or services that individuals want to purchase at various prices over some period of time, all other factors remaining constant.
3. Supply can be expressed as a schedule or curve of the quantities that individuals or businesses are willing to sell at different prices over a period of time, other factors remaining constant.
4. Equilibrium prices and quantities are established for

any good or service when quantity supplied equals quantity demanded.
5. A change in the price of a good or service changes the quantity demanded or quantity supplied. A change in demand or supply occurs when some factor other than price is altered. When these factors change, the demand or the supply curve shifts to the right or left, either raising or lowering equilibrium price and quantity.
6. The market system, through supply and demand, rations scarce resources and limited quantities of goods and services among those most willing and able to pay for them. Products and services, moreover, appear and disappear in response to the market system of supply and demand.
7. Price controls instituted by governments tend to create shortages or surpluses of products. Market forces usually result in some form of rationing other than price rationing under such circumstances.

## Key Terms

products market
resources market
law of demand
quantity demanded
demand curve
factors affecting demand
*ceteris paribus*
change in demand
normal good
inferior good
substitutes
complements
market demand
marginal opportunity costs
law of supply
quantity supplied
supply schedule
change in supply
factors affecting supply
market supply
market
perfect market
law of one price
surplus
shortage
equilibrium price
rationing
price control
price ceiling
price floor
money price
full price
relative price
absolute price

## Questions for Review and Discussion

1. What happens to the demand for a product if the price of that product falls? What happens to the quantity demanded?
2. What happens to the supply of coal if the wages of coal miners increase?
3. If income falls, what happens to the demand for potatoes? Are potatoes inferior goods?
4. If price is above equilibrium, what forces it down? If it is below equilibrium, what forces it up?
5. What is a shortage? What causes a shortage? How is a shortage eliminated?
6. If the demand for cassette tape players increases, explain the process by which the market increases the production of tape players. What are the costs of tape players?
7. A price ceiling on crude oil has an effect on the amount of crude oil produced. With this in mind, explain what happens to the supply of gasoline if there is a price ceiling on crude oil.
8. What happens to the supply of hamburgers at fast-food restaurants if the minimum wage is increased?
9. What is the full price of seeing a movie? Is it the same for everyone?
10. "Lately the price of gold keeps going up and up, and people keep buying more and more. The demand for gold must be upward sloping." Does this statement contain an analytical error?

## ECONOMICS IN ACTION

### When Other Things Do Not Remain Equal: The Pope and the Price of Fish

Perhaps the major lesson about the laws of demand and supply is the importance of holding other things constant. Economists have singled out consumers' income, the price of related goods, and other factors as being of special importance in causing changes in demand (see Table 4–5). Similarly, factors such as the price of inputs or technology create shifts in supply (see Table 4–6). Yet the assumption of other things being equal includes all other factors in consumers' and producers' environment as well. Changes in this environment, including institutions and regulations, must sometimes be taken into account when analyzing price and quantity movements. Consider, for example, some effects of lifting the ban on eating meat on Fridays for Roman Catholics in the northeastern United States.

For years, Roman Catholics were required to abstain from eating meat on Fridays. This ban, which can be thought of as a constant over time, helped support the worldwide fishing industry (there are almost 600 million Catholics in the world), including the U.S. com-

TABLE 4–7   Decline in Demand for Fish After Papal Decree

Monthly prices of all seven species of fish declined in the percentages indicated, from a 2 percent reduction for scrod to a 21 percent decline for the price of large haddock. On average, the price of all species of fish declined by 12 1/2 percent.

| Species | Percentage Change in Price of Fish After Papal Decree (monthly) |
| --- | --- |
| Sea scallops | −17 |
| Yellowtail flounder | −14 |
| Large haddock | −21 |
| Small haddock (scrod) | −2 |
| Cod | −10 |
| Ocean perch | −10 |
| Whiting | −20 |
| All species (average) | −12 1/2 |

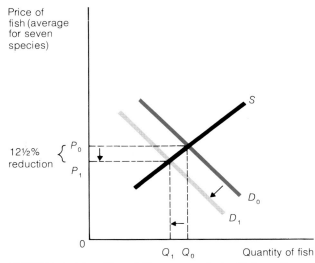

FIGURE 4–14   Papal Decree and the Price of Fish

Following the papal decree lifting the Friday abstinence law, the price of fish declined from $P_0$ to $P_1$ and the quantity also declined, from $Q_0$ to $Q_1$. The demand curve shifted to the left, from $D_0$ to $D_1$.

mercial fishing industry. In February 1966, however, Pope Paul VI issued a decree that allowed local Catholic bishops to end, at their discretion, the rule requiring abstinence from meat on the approximately forty-six Fridays that do not occur during Lent. In December 1966, U.S. Catholics were no longer bound to the rule of abstinence on non-Lenten Fridays.

Economist Frederick W. Bell statistically estimated the impact of the pope's decree on the markets for seven kinds of fish produced and consumed in the northeastern United States.[a] The northeastern United States was chosen because 45.1 percent of the population there is Roman Catholic, more than twice the number in any other area of the United States. To isolate the effects of the decree and estimate a demand function, Bell had to assess the relative importance of all factors affecting fish prices. These factors included many of those discussed in this chapter: the quantity supplied of the particular species of fish; the aggregate personal income for people living in the northeastern United States; cold storage holdings of the particular species of fish; importation of fish into New England; the price of related goods (meat and poultry) and competing fish species; a factor to capture the effects of Lenten and non-Lenten months; and a factor to capture the effects of the papal decree.

In spite of some statistical difficulties, Bell discovered some interesting facts using monthly data. In comparing a period after the decree, December 1966 to August 1967 (excluding Lenten months), with a ten-year period from January 1957 to November 1966, Bell found reductions in demand and price after the decree for all seven species of fish, given in Table 4–7.

Figure 4–14 shows graphically the major economic effects of the pope's decree. Demand curve $D_0$ and supply curve $S$ represent the initial situation. Market forces underlying demand and supply create a price $P_0$ for all species of fish. After the papal decree, the demand curve shifts to the left at every price, creating an average price decline, from $P_0$ to $P_1$, of 12½ percent and a reduction in quantity from $Q_0$ to $Q_1$. An exit of labor and capital from the fishing industry could be expected if the price- and revenue-depressing effects of the papal decree persisted. Careful analysis of all factors affecting supply and demand, even those that seem remote, must be made when dealing with changing market prices and quantities.

## Question

A coffee bug infests the Latin American coffee producing region. The bug destroys a substantial portion of the coffee crop. What would happen to the supply and demand, the equilibrium price and quantity of coffee? What factors would influence the new levels of output and price? If some consumers chose a substitute such as tea during the crisis, what would you predict about the coffee market when the coffee bug and its effects are totally eliminated?

[a] Frederick W. Bell, "The Pope and the Price of Fish," American Economic Review 58 (December 1968), pp. 1346–1350.

## POINT-COUNTERPOINT

## Adam Smith and Karl Marx: Markets and Society

Adam Smith

Karl Marx

ADAM SMITH, one of the most important figures in the history of economics, actually began his career as a lecturer in moral philosophy at Scotland's Glasgow College in 1751. Moral philosophy in Smith's time encompassed a wide range of topics, including natural theology, ethics, jurisprudence, and the field we now know as economics. In 1776, coincident with the Declaration of Independence, Smith published his second book (his first was a treatise on moral behavior), *An Inquiry into the Nature and Causes of the Wealth of Nations*, known usually by the shorter title *Wealth of Nations*. The book, which won Smith much attention from scholars of the day, brought together most of what was then known about the workings of the market system. Smith's insights are still being taught today, more than two hundred years later.

Smith was born in Kirkcaldy on the east coast of Scotland in 1723 and lived most of his life in his native country. Although known for his brilliant lectures (and his many eccentricities), Smith did not devote his entire career to teaching. In 1778 he accepted a well-paying job as commissioner of Scottish customs, a post he kept until his death in 1790.

KARL MARX "*looked* like a revolutionary," writes Robert Heilbroner in *The Worldly Philosophers*. "He was stocky and powerfully built and rather glowering in expression with a formidable beard. He was not an orderly man; his home was a dusty mass of papers piled in careless disarray in the midst of which Marx himself, slovenly dressed, padded about in an eye-stinging haze of tobacco smoke."[a]

Coauthor with Friedrich Engels of the *Communist Manifesto*, which predicted the inevitable downfall of capitalism and the triumph of communism, Marx spent most of his life in difficult circumstances. His activities as a radical in the communist movement caused his exile from his native Germany as well as from Belgium and France. In 1849, a year after the publication of the *Manifesto*, he settled in London, where he and his family survived through the benevolence of Engels and where Marx researched and wrote *Das Kapital*, a theory and history of capitalism and its ills. Marx died in 1883 in London at the age of sixty-five.

As economic theorists, Smith and Marx represent opposite views. Smith was one of the most eloquent defenders of free markets and the promise of capitalism. Marx, who wrote in response to the miseries of the European working class during the Industrial Revolution of the late eighteenth and nineteenth centuries, argued for a new social order and for the overthrow by the working class of capitalists.

### The "Invisible Hand"

Smith's views of the free-market system are summarized in a passage from the *Wealth of Nations* in which he writes that individuals pursuing their own self-interest are "led by an invisible hand to promote an end which was no part of [their] intention."[b] Smith believed that by freely exchanging goods and services across markets, individuals contribute to the public good—the aggregate wealth of society—even though they act from purely self-interested motives. In other words, markets cause individuals to benefit others even though they intend only to benefit themselves.

To Smith, voluntary market exchange coordinated the decisions of consumers and producers and generated economic progress. Producers strive in competition

---

[a]Robert L. Heilbroner, *The Worldly Philosophers* (New York: Simon and Schuster, 1953), p. 131.

[b]Adam Smith, *An Inquiry into the Nature and Causes of the Wealth of Nations*, ed. Edwin Cannan (1776; reprint, New York: Modern Library, 1937), p. 423.

with one another to satisfy consumers with the most appropriate and cheapest goods and services, not out of the goodness of their hearts, not because government planners instruct them to do so, but simply because they maximize the profits of their enterprises by doing so. Markets coordinate supply and demand by way of the price system. Consumers express their relative preferences in their decisions about what to buy; producers attract customers by producing goods at the least cost. In this system of coordination without command, individuals pursuing their own interests are led "as if by an invisible hand" to mesh their interests with those of other individuals trading across markets.

Smith was not opposed to government but argued that its proper role in society was to provide a legal framework—police and courts—within which the market could operate as well as to provide certain other services (including national defense, highways, and education) that he felt the market itself would not supply or would tend to supply in inadequate amounts. Smith also felt that government should provide welfare services for the poor. But he strongly believed that government could best assist the market economy achieve growth by stepping out of the way, that is, by not engaging in most forms of regulation and by ending grants that gave monopoly privileges to favored groups and individuals.

In Smith's view, income was distributed in a market economy by the production of wealth. An individual's income was a strict function of the value of his or her output. Smith did not feel that inequality in income by itself was unfair because the invisible hand ensured that individuals' wealth (or lack of it) was a measure of how much their efforts benefited society as a whole. Anyone can increase his or her income in a free market by serving the consumers in a new, better, or faster way.

### The "Anarchy of Production"

Karl Marx rejected Smith's view of the market process. In *Das Kapital* he argued that Smith's writing represented merely the interests of the ruling capitalist class. To Marx, the market process was a system of exploitation by which owners of capital robbed their employees by paying them wages less than the worth of their labor (a situation he termed the "alienation of labor").

The alternative social system that Marx thought would eliminate this exploitation and at the same time greatly increase the efficiency of production was a "general organization of the labor of society . . . [that] would turn all society into one immense factory."[c] He viewed the market economy as one of general disorganization. Its main feature was "the anarchy of production" where producers overproduced and consumers were forced to accept goods they neither wanted nor needed. He claimed that the market process, left to itself, could not coordinate diverse individual plans.

In Marx's view, Smith's "invisible hand" was a euphemism for describing the economic system in which "chance and caprice have full play in distributing the producers and their means of production among the various branches of industry. . . . the division of labour within the society [the theme of much of Smith's *Wealth of Nations*] brings into contact independent commodity-producers, who acknowledge no other authority but that of competition, of the coercion exerted by the pressure of their mutual interests. . . . the same bourgeois mind which praises division of labour in the workshop [as a conscious organization that increases productivity] denounces with equal vigour every conscious attempt to socially control and regulate the process of production."[d] Smith's "invisible hand" was not only invisible but also unbelievable. Only the central planning of economic activity by society (government), which owned all means of production, could coordinate the needs of consumers and producers and eliminate the wastefulness of capitalism. To Marx it was nonsense to describe the market as organized economic activity because there was no "organizer"—coordination of economic activity requires the conscious, centralized control of the economy.

While Adam Smith described the emerging market economy of his day and offered reforms (most of which could be summarized by the phrase "less government"), Marx offered a vision of economic organization that did not exist at the time he wrote but that he maintained was the inevitable wave of the future. In a sense he was proven correct. Followers of the teachings of Marx and his admirer Lenin (who filled in many of the details of what a central planning system would look like in practice) imposed avowedly Marxist-socialist, centrally planned economies on Russia (1917), China (1949), most of the countries of Eastern Europe, and some African and Latin American nations. Today about one-third of the world's population lives in economies organized in accordance with the ideas of Marx, each economy intended to resemble an "immense factory."

In another sense Marx's teachings appear to have failed. The centrally planned economies seem to function poorly—providing low per capita income and poor rates of economic growth—relative to the modern versions of the capitalist economies whose central principle of market organization was so clearly seen by Smith.

---

[c]Karl Marx, *Das Kapital*, ed. Max Eastman (1867; reprint, New York: Modern Library, 1932), p. 83.

[d]Marx, *Das Kapital*, p. 83.

# II

# The Microeconomic Behavior of Consumers, Firms, and Markets

# 5

# Elasticity

The laws of supply and demand help economists organize their thoughts about real-world problems. But it is not enough to understand that an inverse relation exists between price and quantity demanded or that a decrease in supply causes price to rise in a market. For supply and demand theory to explain and predict economic events, economists must be able to say how much the quantity demanded or supplied of a product will change after a price change. The laws of supply and demand tell us nothing about how responsive quantity demanded or quantity supplied is to a price change. But if, for example, you owned a fast-food restaurant, you would be very interested to know how responsive your customers would be to a hamburger price discount. The law of demand tells you that you would sell more hamburgers, but it does not tell you how many more.

The deteriorating conditions of U.S. highways and bridges led the federal government in 1983 to levy an additional five-cent-per-gallon tax on gasoline. The Reagan administration intended that the tax would generate sufficient revenues to allow the government to make the necessary repairs. But will the tax have the hoped-for effect? The answer depends on the responsiveness of gasoline consumers to the price increase. If they are *very* responsive—if they reduce gasoline consumption a great deal in response to the price rise—tax revenues may be insufficient to finance the repair project.

The Drug Enforcement Administration initiates a crackdown on the import of illegal drugs such as cocaine and marijuana. With stepped-up enforcement, a smaller quantity of drugs is smuggled into the country by organized crime. As a result, the street price of these drugs goes up steeply. Will the revenues of organized crime increase or decrease? There will be some reduction in quantity demanded—that is, fewer drugs will be sold at higher prices. The important question is, how much of a reduction? Clearly, the answer will depend on the responsiveness of drug users to the increase in price.

**Elasticity:** A measure of the responsiveness of one variable caused by a change in another variable; the percent change in a dependent variable divided by the percent change in an independent variable.

You are contemplating two business alternatives: buying a gourmet delicatessen or opening a new travel agency. Best estimates tell you that consumers' incomes are expected to rise by 5 percent per year over the next eight years. On these grounds, which business should you enter? How much will increases in consumers' incomes drive up the demand for gourmet foods relative to the demand for travel and leisure?

Elasticity, the topic of this chapter, helps us answer such important, practical questions. The responsiveness of consumers to changes in price (elasticity of demand) is the best-known application of elasticity, but **elasticity** is a far more general concept. It refers to the responsiveness of any "effect" to any change in "cause."

Elasticity is always calculated in percentage terms, as a ratio of the percent change in the effect (quantity demanded or quantity supplied of gasoline, drugs, TV sets, and so on) to the percent change in the cause (income change, price change, or any other change). Any calculation of elasticity is therefore independent of absolute numbers such as quantity, income, or prices.

Consider the relation between egg production and the weather. Assume that egg production falls with increasing henhouse temperatures. The elasticity of egg production with respect to changes in temperature is calculated by dividing the percentage change in production by the percentage change in temperature. If, according to a farmer's records, a 10 percent increase in henhouse temperature resulted in a 30 percent reduction in egg production, we say that the chickens are very responsive to temperature changes; if a 10 percent increase in temperature reduced egg production by only 2 percent, we say that the chickens are not very responsive to temperature changes.

All kinds of elasticities are measured in the same way. The five-cent tax on gasoline will increase the price of gasoline by some percentage and it will reduce the amount sold by some other percentage, large or small, according to gas buyers' response to the price change. The degree of responsiveness is measured by simple division of the two percentages; the amount of tax revenues will depend on this response. More drug enforcement reduces supply and increases prices by some percentage. Drug purchases are reduced by some percentage depending on buyers' responsiveness to the price change, but, given the effects of addiction on drug users, the decrease in purchases will probably be small. Percentage increases in income will, similarly, cause percentage increases in gourmet food and travel consumption. The percentages of these changes can be estimated from similar past experiences, and businesses can use the information to make decisions.

## Price Elasticity of Demand

Elasticity, as these examples suggest, is a general and wide-ranging concept with many applications. The most common and important applications relate to demand.

**Price elasticity of demand:** A measurement of buyers' responsiveness to a price change; the percent change in quantity demanded divided by the percent change in price.

### Formulation of Price Elasticity of Demand

The formal measurement of demand elasticity, like the generalized concept itself, is simple and straightforward. **Price elasticity of demand** is the percent change in quantity demanded divided by the percent change in price:

$$\epsilon_d = \text{Price elasticity of demand coefficient} = \frac{\%\text{ change in quantity demanded}}{\%\text{ change in price}}.$$

# Price Elasticity of Demand

**Demand elasticity coefficient:** The numerical representation of price elasticity of demand: $(\Delta Q/Q) \div (\Delta P/P)$.

Price elasticity of demand is expressed as a number. If, for example, a 10 percent reduction in the price of jogging shoes causes a 15 percent increase in the quantity of jogging shoes demanded, then the ratio, called the **demand elasticity coefficient**, $\epsilon_d$, is

$$\frac{15\%}{10\%} = 1.5.$$

This elasticity coefficient or ratio, 1.5, and the percent changes in price and quantity demanded from which it was calculated, are independent of the absolute prices and quantities of jogging shoes. If a 10 percent rise in the price of jogging shoes causes a 2 percent decline in sales, the price elasticity coefficient is calculated in the same way:

$$\frac{2\%}{10\%} = 0.2.$$

The price elasticity of demand coefficient (or, simply, the demand elasticity coefficient) always measures consumers' responsiveness, in terms of purchases, to a percent change in price.[1] To determine the effect of price changes alone, all other changes—such as differences in the quality of goods—must be held constant.

## Elastic, Inelastic, and Unit Elastic Demand

The size of the elasticity coefficient is important because it measures the relative consumer responsiveness to price changes. If the number obtained from the elasticity calculation is greater than 1.00, we say that demand is elastic over the price and quantity range; the percent change in quantity demanded is greater than the percent change in price. Above a coefficient of 1.00, degrees of elasticity vary. A demand elasticity coefficient of 1.5 for jogging shoes means that consumers are somewhat responsive to a price change. A coefficient of 6.0 for pizza means that buyers of pizza are much more responsive to a change in price (four times more responsive, in fact). The larger the demand elasticity coefficient is above 1.0, the more elastic demand is said to be.

**Elastic demand:** A situation in which buyers are very responsive to price changes; the percent change in quantity demanded is greater than the percent change in price; $\epsilon_d > 1$.

**Unit elasticity of demand:** A situation in which the percent change in quantity demanded is equal to the percent change in price: $\epsilon_d = 1$.

**Unit elasticity of demand** means that the elasticity coefficient equals 1. In this case, a given percent change in price is exactly matched by the percent change in quantity demanded. If, for example, a 2 percent increase in the price of candy bars causes a 2 percent reduction in purchases, demand would be of unit elasticity. The same would be said if an 8 percent decrease in the price of Volkswagens caused an 8 percent increase in quantity demanded.

**Inelastic demand:** A situation in which buyers are not very responsive to changes in price; the percent change in quantity demanded is less than the percent change in price; $\epsilon_d < 1$.

An **inelastic demand** coefficient is a number less than 1, meaning that a percent change in quantity demanded is less than the percent change in price that caused the change in quantity. If the price of salt increases by 5 percent and the quantity demanded decreases by 2.5 percent, the demand elasticity coefficient would be 0.5, placing it in the inelastic category. An elasticity number lower than 0.5 for salt would mean that demand is relatively more inelastic.

---

[1] Notice that in actual calculation, the demand elasticity coefficient will always be negative owing to the inverse relation between price and quantity demanded (the law of demand). If price goes up, quantity demanded for a good or service goes down, and vice versa. Unless otherwise noted, this point is irrelevant to the interpretation of elasticity. We will accordingly eliminate use of a negative sign before the demand elasticity coefficient.

The various categories of price elasticity of demand can be shown with demand curves. Figure 5–1 depicts three responses to a 20 percent increase in pizza prices over a given time period in a small college town. In Figure 5–1a, a 20 percent increase in price causes a 5 percent reduction in quantity demanded. This means that pizza consumers are not very responsive to a change in price: $\epsilon_d$ is inelastic—it equals 0.25, which is less than 1. In Figure 5–1b, a 20 percent increase in pizza price reduces pizza consumption by exactly 20 percent, meaning that demand is of unit elasticity: $\epsilon_d = 1$. Figure 5–1c shows an elastic demand—a 20 percent price increase causes a 50 percent reduction in the quantity of pizzas demanded: $\epsilon_d$ is 2.5 (greater than 1).

The elasticity of two special forms of the demand curve are also of interest in analyzing economic problems. Figure 5–2 shows a completely inelastic and a completely elastic demand curve. Total or complete price inelasticity means that consumers are not responsive at all to price changes. Increases or decreases in price leave quantity demanded unchanged—the demand curve is vertical, as in Figure 5–2a. One might think of the demand for heroin or other addictive drugs—at least for certain price ranges and certain levels of use—as being completely inelastic. The price elasticity of demand coefficient is zero along such a curve: $\epsilon_d = 0$.

A completely elastic demand curve—along which an infinitely large or small quantity is demanded at a given price—is shown in Figure 5–2b. Such a demand curve occurs in a competitive market situation, discussed fully in Chapter 9. In a competitive market, producers are so numerous that they are unable to affect prices. The existence of a huge number of wheat farmers, for example, means that a single small producer is unable to affect the mar-

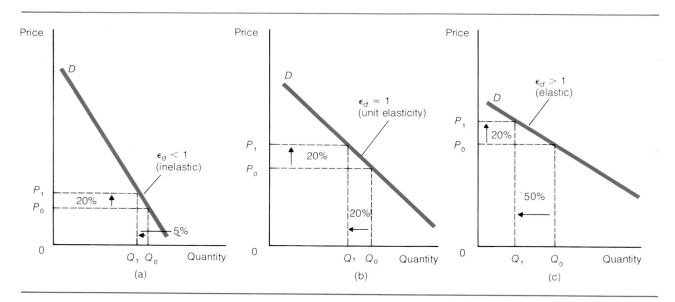

FIGURE 5–1  Price Elasticity of Demand
(a) The percent reduction in quantity demanded (from $Q_0$ to $Q_1$, 5%) is less than the percent increase in price (from $P_0$ to $P_1$, 20%). Demand is inelastic, $\epsilon_d < 1$. (b) The percent reduction in quantity demanded equals the percent increase in price. Demand is unit elastic, $\epsilon_d = 1$. (c) The percent reduction in quantity demanded is larger than the percent increase in price. Demand is elastic, $\epsilon_d > 1$. The same relations hold for reductions in price or increases in quantity demanded.

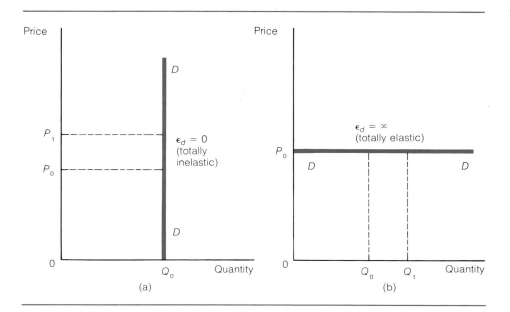

**FIGURE 5-2**

**Demand Curves May Be Totally Inelastic or Totally Elastic**

(a) Quantity demanded is unresponsive to price changes (totally inelastic). (b) Price is unaffected by smaller or larger quantities demanded (totally elastic). Such a situation occurs in a competitive market.

ket price by altering production. If the wheat farmer raises his or her price by even a small amount, buyers will not purchase any of the farmer's production. Buyers can get all the wheat they want at the (lower) prevailing market price. In these competitive circumstances, a single wheat farmer faces an infinitely elastic demand curve ($\epsilon_d = \infty$). Quantity demanded is supersensitive to price increases. The farmer can sell as much or as little as he desires at the market price.

To summarize, an elasticity coefficient less than 1.0 means that demand is inelastic, and a coefficient greater than 1.0 means that demand is elastic over a given price range. A coefficient equal to 1.0 indicates unit elasticity of demand. The demand curve can, under certain circumstances, be completely elastic or completely inelastic.

### Elasticity Along the Demand Curve

Our discussion of demand elasticity has contained a crucial qualification: that the coefficient (whether elastic, inelastic, or unitary) has meaning only over certain (or relevant) price ranges. We now investigate exactly what this means. To do so requires that we be even more specific about the basic elasticity concept and its algebraic formulation.

Table 5-1 and Figure 5-3 show the market demand function—or price-quantity pairs—that constitute the demand schedule for tapes. (Table 5-1 and Figure 5-3 reproduce some of the information presented in Table 4-2 and Figure 4-5d relating to the market demand curve.)

Is it ever correct or meaningful to ask, "What is the elasticity of this or any other demand curve?" Some simple calculations from the market demand function of Table 5-1 or Figure 5-3 will tell us immediately that the answer is no. Remember that demand curves are downward sloping because quantity demanded drops as price rises. Every negatively sloped demand curve will, in general, contain portions that are elastic, unit elastic, and inelastic. The elasticity coefficient will vary along any straight (linear) or curving (nonlinear) demand curve in all but a few special cases. To verify

## TABLE 5–1  Market Demand Schedule for Tapes

As always, the price of tapes is inversely related to the quantity of tapes demanded.

| Price of Tapes | Quantity Demanded |
|---|---|
| 10 | 0 |
| 9 | 40 |
| 8 | 80 |
| 7 | 120 |
| 6 | 160 |
| 5 | 200 |
| 4 | 240 |
| 3 | 280 |
| 2 | 320 |
| 1 | 360 |

this fact consider the simple elasticity expression and its algebraic counterpart once more:

$$\epsilon_d = \frac{\%\text{ change in quantity demanded}}{\%\text{ change in price}} = \frac{\Delta Q/Q}{\Delta P/P}.$$

In the algebraic expression, $Q$ simply means quantity, $\Delta$ means "a change in," and $P$ means price. In the numerator, percent change in quantity demanded is determined by dividing the change in quantity demanded by the initial quantity demanded ($\Delta Q \div Q$). The same is done in the denominator for initial price and change in price to determine the percent change in

### FIGURE 5–3
**Differing Elasticities Along a Market Demand Curve**

The elasticity of demand is different between various points along a downward-sloping demand curve. If prices fall by $1 between A and B, demand is elastic. If prices fall by $1 between C and D, demand is inelastic.

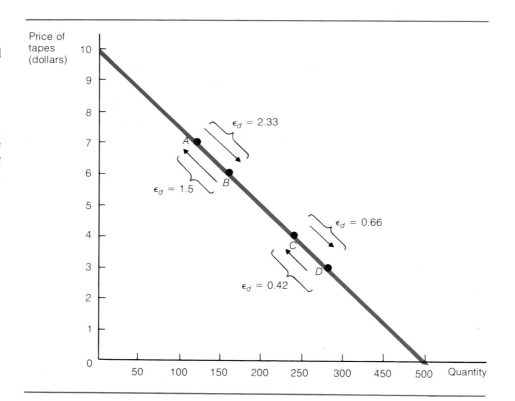

price. Then elasticity can be expressed as the percent change in quantity demanded ($\Delta Q/Q$) divided by the percent change in price ($\Delta P/P$). This simple formulation can be rearranged thus:

$$\frac{\Delta Q/Q}{\Delta P/P} = \frac{\Delta Q}{Q} \times \frac{P}{\Delta P}.$$

We will use this expression to make some calculations.[2]

Suppose from Table 5–1 and Figure 5–3 that the price of tapes declines from $7 to $6. Quantity demanded would increase from 120 to 160 tapes. How would elasticity be calculated? Returning to our simple formula, we can calculate the elasticity across this price range as follows:

$$\epsilon_d = \frac{\Delta Q/Q}{\Delta P/P} = \frac{\Delta Q}{Q} \times \frac{P}{\Delta P} = \frac{40}{120} \times \frac{7.00}{1.00} = 2.33.$$

The number of tapes sold increased by 40 over the original quantity of tapes, those purchased at $7 (120). Thus, $40/120$ is multiplied by the ratio of the original price, $7, to the change in price, $1. The elasticity coefficient is 2.33, which means that demand is elastic over this range of prices. Consumers are responsive to a price reduction from $7 to $6.

Now assume that tapes are selling at $4 and that the price is reduced to $3. Using the same method, we see that

$$\epsilon_d = \frac{\Delta Q/Q}{\Delta P/P} = \frac{\Delta Q}{Q} \times \frac{P}{\Delta P} = \frac{40}{240} \times \frac{4.00}{1.00} = 0.66.$$

The coefficient indicates that for a decline in price over the price range from $4 to $3 demand is inelastic. Consumers are far more responsive to a one-dollar price reduction from $7 to $6 than they are to a one-dollar price reduction from $4 to $3.

This example illustrates that elasticity varies all along a demand curve. It is not correct to say that the demand for anything—salt, cigarettes, tapes—is elastic or inelastic without identifying some specific price range.

The knowledge that elasticity changes along demand curves also points out another difficulty in accurately calculating elasticity. Since elasticity varies at all points along the demand curve, it also varies along the arc, or length, A to B or C to D. This is called **arc elasticity**. In other words, there will be different elasticity coefficients for price changes *between* $4 and $3, such as a price increase from $3.25 to $3.75. Indeed, we will even get a different elasticity coefficient for a price *increase* from $3 to $4, as we verify in the following calculation (also see Figure 5–3):

**Arc elasticity:** A measure of average elasticity across all intermediate points between two points along a demand curve.

$$\epsilon_d = \frac{\Delta Q/Q}{\Delta P/P} = \frac{\Delta Q}{Q} \times \frac{P}{\Delta P} = \frac{40}{280} \times \frac{3.00}{1.00} = 0.42.$$

Whereas a price decrease from $4 to $3 yielded an elasticity of 0.66, a price increase from $3 to $4 yields an elasticity of 0.42, clearly a more inelastic consumer response. Why the difference? The size of the price change means that widely different initial prices and quantities are being expressed in the simple formula.

[2] Note that this expression for elasticity can be reorganized again to equal ($\Delta Q/\Delta P$) × ($P/Q$). This expression tells us immediately that elasticity and slope are different concepts. The slope of the demand curve is ($\Delta P/\Delta Q$); it shows how price changes with unit or other changes in quantity. Elasticity is the inverse of the slope ($\Delta Q/\Delta P$), that is, the slope "turned upside down," multiplied by the ratio of some specific price and quantity ($P/Q$) along the demand curve.

One compromise solution to the problem of determining the exact elasticity of demand over such a price range is to take the average elasticity within the arc between the two price-quantity combinations.[3] There are other methods of calculation as well. Naturally, the smaller the price change, the more precise is the simple formula $\epsilon_d = (\Delta Q/Q) \div (\Delta P/P)$ at estimating the true elasticity between two points.

While there are several means of calculating elasticity and while these calculations (as with many other representations of actual economic data) are approximations, elasticity is often very useful in assessing real-world problems and policies. We will continue to use the simple formula as our approximation.

### Relation of Demand Elasticity to Expenditures and Receipts

As indicated at the beginning of this chapter, it is often necessary to estimate the effect of a proposed increase or decrease in price on total revenue—how much the government will make (or lose) by raising gas taxes or a fast-food chain will make (or lose) by offering a half-price sale on hamburgers. Estimates of elasticity from previous experience can make such projections possible. Or, if the price change and consumer response have already occurred, we can look at what happened to determine the elasticity of demand by examining either what customers have spent for a good or what businesses have received for it.

Common sense tells us that an industry's revenues are the same as consumers' expenditures for the industry's product. The number of items bought by all consumers multiplied by the price paid for each item is the same as the number of items sold by all producers multiplied by the average price charged for the items. Elasticity, therefore, is related both to **total expenditures** of consumers and to **total revenues** or receipts of businesses.

**Total revenue:** Price times quantity sold; total expenditures.

To understand how elasticity is related to consumer expenditures, we return to the example from Chapter 4 of a single consumer purchasing cassette tapes. Table 5–2 reproduces Dave's demand and total expenditures (quantity times price) for tapes, and Figure 5–4 reproduces his demand curve with the associated elasticities. Dave's total expenditures on tapes begin to rise as the price of tapes falls below $10. As the price falls to $9, $7, and $5, Dave's quantity demanded *and* his total expenditures on tapes rise. Notice that, for decreases in price, if elasticity is greater than 1, total expenditures will increase. Why? When demand is elastic, there is an inverse relation between price and total expenditures because the percent change in quantity demanded dominates (is larger than) the percent change in price. But as price is reduced below $5, Dave's total expenditures begin to decline, even though his consumption of tapes continues to increase. In the price range in which

---

[3] To calculate the average elasticity in the $3 to $4 price range for tapes, the prices $3 and $4 and the quantities 240 and 280 are given equal weight. To express the formula algebraically, we can call one quantity $Q_1$ and the other $Q_0$. The price at $Q_1$ is $P_1$; the price at $Q_0$ is $P_0$. The average elasticity of their relation can then be calculated as follows:

$$\epsilon_d = \frac{\frac{Q_1 - Q_0}{(Q_1 + Q_0)/2}}{\frac{P_1 - P_0}{(P_1 + P_0)/2}} = \frac{Q_1 - Q_0}{Q_1 + Q_0} \times \frac{P_1 + P_0}{P_1 - P_0} = \frac{280 - 240}{280 + 240} \times \frac{3 + 4}{3 - 4} = 0.53.$$

The average elasticity coefficient is thus between those calculated for a price increase and for a price decrease. This is one method of handling the difference in elasticity between two points.

# Price Elasticity of Demand

**TABLE 5–2  Elasticity and Consumer Expenditures**

The average elasticity of demand is greater at higher prices and falls as price falls. At the midpoint (near $5), the elasticity is equal to 1 and total expenditures are at a maximum.

| Price of Tapes (dollars) | Quantity Demanded | Total Expenditure | Average Elasticity of Demand |
|---|---|---|---|
| 10 | 0 | 0 | |
| | | | 19 |
| 9 | 1 | 9 | |
| | | | 5.66 |
| 8 | 2 | 16 | |
| | | | 3 ? |
| 7 | 3 | 21 | |
| | | | 1.85 |
| 6 | 4 | 24 | |
| | | | 1.22 |
| 5 | 5 | 25 | |
| | | | 0.81 |
| 4 | 6 | 24 | |
| | | | 0.53 |
| 3 | 7 | 21 | |
| | | | 0.333 |
| 2 | 8 | 16 | |
| | | | 0.176 |
| 1 | 9 | 9 | |
| | | | 0.05 |
| 0 | 10 | 0 | |

Handwritten annotations: $\left(\begin{array}{c}8\\7\end{array}\right)$ $\frac{1}{3} \cdot 7$ $\left(\begin{array}{c}2\\3\end{array}\right)$ $\frac{1}{2} \cdot \frac{8}{1} = 4$ $\left(\begin{array}{c}16\\21\end{array}\right)$

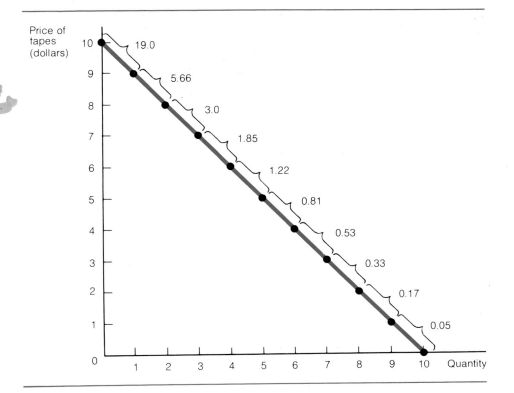

**FIGURE 5–4**
**An Individual's Elasticity of Demand**

The elasticity of demand varies along a downward-sloping demand curve. As price falls, the elasticity falls.

demand is inelastic (a coefficient of less than 1) total expenditures decrease as price falls. Here the percent decline in price dominates the percent increase in quantity, and total expenditures fall. In other words, prices and total expenditures change in the same direction if demand is inelastic.

The same relation between price and total expenditures is found when prices increase. If demand is inelastic, expenditures rise as prices increase from some low level, say $2, to a new level or over a range. If demand is elastic for that price range, total expenditures actually fall as prices increase. At some price-quantity combination, demand is unit elastic, $\epsilon_d = 1$. For Dave in Figure 5–4, unit elasticity of demand occurs around $5, when Dave's total expenditures for tapes will remain constant whether price is increasing or decreasing by some minuscule amount. It is at this point that Dave's expenditures are at a maximum.

Being able to estimate the ways that price elasticity of demand affects the total expenditures of consumers is of obvious value to businesses seeking to maximize revenues.

If total expenditures are known, the relations between the direction of the price changes and total expenditures can be used as a "back-door" method for determining elasticity, though not of calculating elasticity coefficients. The relations between the direction of the price change, the elasticity, and total expenditures (or revenues) are summarized in Figure 5–5. As the price falls from $P_1$, where consumption of the good is zero, to $P_2$, the midpoint on the demand curve, demand is elastic, and total expenditures (TE) and total receipts (TR) are rising. At $P_2$, price elasticity of demand is unitary, and total receipts and expenditures remain constant as the price rises or falls around this price. As the price falls below price $P_2$ (where quantity $Q_2$ is demanded), the demand elasticity coefficient falls below 1. Below $P_2$, total

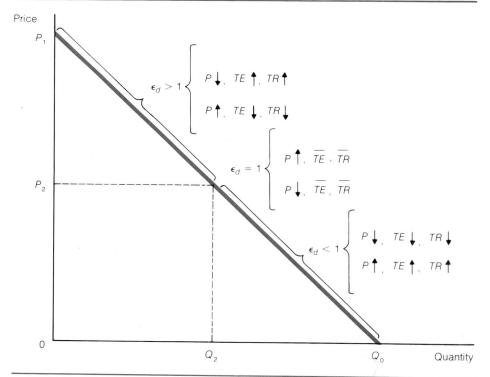

**FIGURE 5–5**

**Total Expenditures and Total Receipts Along a Demand Curve**

As the price falls ($P \downarrow$) along the demand curve, both total receipts (TR) and total expenditures (TE) rise, indicating that elasticity is greater than 1. At the point of unit elasticity, TE and TR remain constant. As the price falls from the midpoint, TE and TR decline owing to inelasticity of demand. Rising price ($P \uparrow$) has the opposite effects on expenditures and receipts.

expenditures and receipts fall, and demand is inelastic, becoming more inelastic as the price approaches zero.

The major point to remember is that elasticity varies along any ordinary demand curve. It makes no sense to say, for instance, that salt is an inelastically demanded commodity without reference to some price range. The demand for salt may be elastic over some price range and inelastic over another. In the next section we consider what makes the demand elasticity coefficient greater than, equal to, or less than 1 and what factors determine consumer responses to price changes.

## Determinants of Price Elasticity of Demand

There are three major determinants of price elasticity: (1) the number and availability of substitutes, (2) the size and importance of the item in the consumer's budget, and (3) the time period involved. Since all three factors interact, the condition of other things being equal must be invoked to determine the specific effect of any one factor.

### Number and Availability of Substitutes

By far the most important predictor of demand elasticity is the ability of consumers to find good substitute products. If the price of one brand of toothpaste rises, many consumers may respond by switching to a different brand. But this substitution will happen only when alternatives are available.

Food is a vital commodity. Everyone must eat; elasticity of demand for nutrients in general is therefore approximately zero over relevant price ranges—completely inelastic. For most of us, there is no substitute for food. But it is possible to substitute between kinds of food. While the demand for food as a whole tends to be inelastic, the consumer may substitute between broad food groups such as meat and seafood and also between foods within the broad groups. Consumers of meat, for instance, have a wide choice of substitutes, such as chicken, beef, pork, duck, possum, or alligator. Consumers are far more sensitive to changes in the price of beef when other substitutes are available. Elasticity of demand for meat itself is lower than that for kinds of meat.

What about the elasticity of demand for beef versus beef products such as hamburger, sirloin steak, beef tails, and so on? The consumer can substitute among beef products. When the price of sirloin rises relative to hamburger, the consumer can substitute hamburger for sirloin. The elasticity of demand for hamburger is therefore higher than the elasticity of demand for meat. By now, the general rule must be apparent: The broader the product or service group, the lower the elasticity because there are fewer possibilities for substitution. We expect the elasticity of demand for Budweiser beer to be more elastic than the demand for beer, just as the demand for a Ford or Mercedes-Benz is more elastic than the demand for automobiles. Ordinarily, the wider the selection and substitutability of similar products and services, the larger the elasticity of demand for some product.

In Chapter 9 we will see that elasticity is a crucial factor in determining how competitive markets are structured. One view of competition is that it exists when products are perfectly substitutable—one seller's wheat is identical to another seller's, for example. A number of closely substitutable products—such as hair stylists in a large city—would also indicate a very competitive market. When goods are demanded and produced with few or no close substitutes, the market changes from competition to monopoly.

### The Importance of Being Unimportant

The size of the total expenditure within the consumer's budget is another determinant of elasticity. Ordinarily, the smaller the item in the consumer's budget, the less elastic the consumer's demand for the item will be over some price ranges. A salt user may be insensitive to price increases in salt simply because expenses for salt are a small part of his or her budget. If the price of salt were to rise too high, the user might substitute alternatives, such as artificial salt or lemon juice. Thus, in calculating elasticity it is important, at some point, to analyze both substitutability and the size of the item in the consumer's budget.

On the other hand, the effects of size in one's budget and substitutability may offset each other in determining elasticity of demand. Take, for instance, the demand for electricity by the poor and the aged living on Social Security or other transfer payments. Inflation and rising energy costs have caused electricity rates to soar over the last decade. How would such consumers respond to rising electricity prices, which obviously take up a large portion of their total budget? Using the "importance of being unimportant" criterion, we might be tempted to say that their demand for electricity is highly elastic. But economic theory and common sense tell us that electricity consumers will not be very responsive to price increases because there are few if any viable substitutes for electricity (except perhaps for intolerable house temperatures or highly expensive conversion to oil heat, when electricity is used for heating and cooling). Thus, substitutability—the second determinant of elasticity—outweighs the importance of being unimportant.

The condition of other things being equal clears up our understanding of the elasticity determinants. We can correctly state that, given some constant degree of substitutability, the more unimportant the commodity is to a consumer, the lower demand elasticity will be. This simply indicates that tastes, substitutability, and importance of the commodity in the consumer's budget must all be examined in gauging the demand elasticity of products over given price ranges.

### Time and Elasticity of Demand

Time is the final factor affecting the demand curve. We alluded to these effects at the beginning of this chapter, but we can now understand them explicitly.

As a simple example, a local market for tennis rackets is represented in Figure 5–6. (For simplicity we may neglect the supply function.) Assume that the initial price is $P_0$ and the initial quantity demanded is $Q_0$. What happens to elasticity of demand if tennis rackets go on sale; that is, if that price falls to $P_1$ per unit, a 15 percent decline?

The answer depends on how long it takes for consumers and potential consumers to adjust to the new price. If the sale is totally unpublicized, there may be no immediate reaction to the new price. As consumers gain information, however, increases in demand ensue. Thus we may think of the demand curve as rotating around a point ($A$ in Figure 5–6) as news of the sale becomes more widespread. Greater quantities ($Q_1$, $Q_2$, and so on) will be sold *through time*. For the given price change, the elasticity of demand will be different at different times—that is, it will depend on whether it is calculated on the first day of the sale ($\epsilon_d = 0.33$) or one week ($\epsilon_d = 0.66$) or one month later ($\epsilon_d = 1.33$). Percent changes in quantity demanded are greater as time passes, meaning that the demand elasticity coefficient is

**FIGURE 5–6**
**Elasticity of Demand Over Time**

As time goes by, a given price change ($\Delta P$) may be associated with larger and larger changes in quantity demanded ($\Delta Q$). Since the percent change in quantity demanded grows over time, the elasticity of demand grows as well.

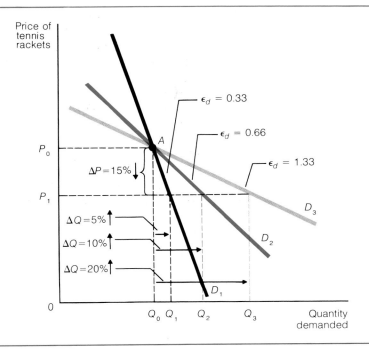

larger and larger (up to some limit, of course). The general rule holds: Elasticity of demand increases the longer any given price change is in effect.

The time period of adjustment is a crucial factor in calculating or estimating all types of actual elasticities. Not only do tastes, substitutability, and size within the consumer's budget affect elasticity, but the ability of consumers to recognize and adjust to changes also plays a part. (Focus, "Time, Elasticity, and the U.S. Demand for OPEC Oil," is an example of the effect of time on elasticity of demand.)

## Other Applications of Elasticity of Demand

The concept of elasticity is not restricted to percent changes in quantity demanded to percent changes in price, that is, to price elasticity of demand. In general, we can calculate an elasticity of any dependent variable (effect) to a change in any independent variable (cause). We will consider two other important applications of this versatile economic concept: income elasticity of demand and cross elasticity of demand.

### Income Elasticity of Demand

**Income elasticity of demand:** A measure of buyers' response to a change in income in terms of the change in quantity demanded; the percent change in quantity demanded divided by the percent change in income.

As you will recall from Chapter 4, a consumer's income is an important determinant (independent variable) of the demand for goods and services. It is often very informative to inquire about consumers' **income elasticity of demand.** Producers of all kinds are interested in the magnitude of consumption changes as incomes rise or in consumption habits within various income groups. Budget data compiled by the government and other sources can be used to calculate recent and historical trends in changing consumption patterns as incomes change.

The mechanics of income elasticity of demand are identical to those in-

## FOCUS  Time, Elasticity, and the U.S. Demand for OPEC Oil

For a fast-driving, freewheeling, high energy-using country like the United States, the successful embargo of oil supplies to this country by the OPEC oil cartel in 1973–1974 was a sobering experience. Since OPEC began restricting the supply of oil, the effects on prices of oil and oil products, including gasoline, have been dramatic, especially between 1973 and 1978 (see Figure 5–7).

As the figure shows, OPEC's control over supply had the effect of shifting the supply curve to the left. Initially the price of oil rose from $P_0$ to $P_1$ along the demand curve (assumed constant) for oil (and implicitly for oil products). Short-run elasticity for oil was quite low, indicating that Americans did not instantaneously or even rapidly adjust to the reduced supply.

The effects of time on elasticity of demand were very different, however, as the ability to find substitute sources of energy grew. Americans, both consumers and producers, began to find ways to economize on high-priced oil. Consumers reduced auto travel, participated in car pooling, and began buying smaller cars. Substitutes for private auto travel, such as urban transit systems, began to gain support in the cities. All in all, these effects produced a 7 to 10 percent reduction in gasoline consumption by 1983. Consumers and producers also economized on home and industrial uses of oil-based energy. Improved home insulation and use of alternative fuel sources, such as solar energy, were common substitutions. Producers substituted alternative resources, including other forms of energy, for high-priced oil.

What were the effects of such substitution over time on the demand for oil and oil products? By the early 1980s elasticity of demand had risen significantly, opening up an excess supply of oil (AB) at price $P_1$. After time adjustments, the demand curve for oil became more elastic ($D_2$). The prices of oil and oil products fell to lower levels, such as $P_2$. Along with and, indeed, partially because of the rising elasticity of demand for oil through time, the OPEC cartel lost its

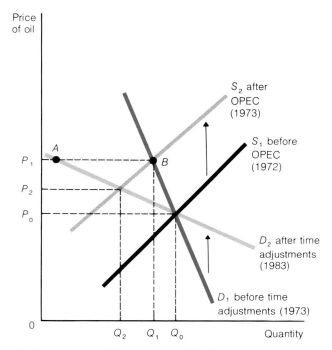

**FIGURE 5–7  Effect of the OPEC Cartel on Demand Elasticity over Time 1973–1983**

The OPEC cartel's restriction of output effectively shifted the whole supply curve to the left, resulting in higher prices for American consumers. As consumers reduced quantity demanded for OPEC products, however, the demand elasticity for oil rose substantially, creating a reduced quantity demanded ($Q_2$) and a lower price ($P_2$).

punch. Disputes among OPEC nations became not so much political as economic in differences over elasticity projections. The bottom line is that elasticity of demand is profoundly affected by time and the ability to substitute.

---

volved in the calculation of price elasticity; only the independent variable changes. With price held constant, income elasticity is the percent change in quantity demanded resulting from (divided by) a given percent change in income. It is expressed as $\epsilon_y$, with Y representing income:

$$\epsilon_y = \frac{\% \text{ change in consumption of a good}}{\% \text{ change in income}} = \frac{\Delta Q}{Q} \div \frac{\Delta Y}{Y}.$$

The income elasticity coefficient, $\epsilon_y$, may be positive or negative depending on whether the good is normal or inferior. (Recall from Chapter 4 that a good is normal if an increase in income results in greater consumption and inferior if an increase in income results in a reduction in consumption of the product.) When applying the income elasticity formula, if $\epsilon_y$ is greater than zero, the good is normal; if $\epsilon_y$ is less than zero, the good is inferior. For the moment we discuss only normal goods.

If the elasticity coefficient $\epsilon_y$ is greater than 1, demand (or consumption) is said to be income elastic; if $\epsilon_y$ is less than 1, the product is income inelastic; and if $\epsilon_y$ equals 1, income is unit elastic. Suppose that income in the United States rises by 10 percent in 1986 and that the quantity of new automobiles consumed over the year increases by 8 percent. Clearly, new automobiles are a normal good (since their consumption increases along with the increase in income). But what is the income elasticity? The simple computation is the following:

$$\epsilon_y = \frac{8\%}{10\%} = 0.8.$$

The income elasticity of demand for automobiles in this hypothetical calculation is less than 1 but greater than zero. What does this mean for the auto industry? Other things being equal (such as price and tastes), the demand for automobiles will rise but at a slower pace than income. This fact is obviously important to groups in society such as auto manufacturers, investors, boat dealers, and airlines.

Income elasticity for a particular good may be determined for any individual consumer or for consumers as a group. The practical importance of this calculation is undeniable. An individual deciding between opening a gourmet food shop and a travel agency during a period of rapidly rising income would be very interested to know, for example, that income elasticity for gourmet foods is perhaps 0.2, while the same coefficient for Mediterranean vacations is 6.2. However, to achieve accuracy in actual use, all other factors, such as substitutability and the price of the product, must be kept constant. If these factors vary, as they often do in the real world, their impact on consumption must be determined and integrated into the analysis.

## Cross Elasticity of Demand

**Cross elasticity of demand:** A measure of buyers' responsiveness to a change in the price of one good in terms of the change in quantity demanded of another good. The percent change in the quantity demanded of one good divided by the percent change in the price of another good.

**Substitutes:** Two goods whose cross elasticity of demand is positive.

**Cross elasticity of demand** simply reveals the responsiveness of the quantity demanded of one good to a change in the price of another good. As such, a cross elasticity coefficient can define either substitute or complementary products or services. In more general terms, cross elasticity is an extremely useful economic tool for identifying groups of products whose demand functions are related.

First there are substitute products. If the price of one good rises and, other things being equal, the quantity demanded of another good increases, those products are **substitutes**. If Dorothy's demand for Bayer aspirin rises 85 percent following a 10 percent rise in the price of Bufferin, her cross elasticity of demand for aspirin is +8.5. The coefficient of cross elasticity of demand is calculated thus:

$$\epsilon_c = \frac{\%\text{ change in quantity demanded of one good}}{\%\text{ change in price of another good}} = \frac{85}{10} = +8.5.$$

For substitute commodities, then, the cross elasticity coefficient is positive because an increase in the price of one good causes an increase in the quantity demanded of the other. The larger the elasticity coefficient (in absolute terms), the more substitutable the products or services are.

**Complements:** Two goods whose cross elasticity of demand is negative.

Next there are items that are complements in consumption, such as bacon and eggs; left shoes and right shoes; gasoline and automobiles; light bulbs, lamps, and electricity. Goods are **complements** when an increase in the price of one good results in a decrease in the quantity demanded of the other. Gin and vermouth are the two essential ingredients (besides olives) in a martini. If there is a 4 percent increase in the price of vermouth and a 16 percent reduction in the quantity consumed of gin, other things being equal, we can call gin and vermouth complementary products. Note that the cross elasticity coefficient is negative:

$$\epsilon_c = \frac{\% \text{ decrease in quantity demanded of one good}}{\% \text{ increase in price of another good}} = \frac{-16}{4} = -4.0.$$

A negative number is obtained for the cross elasticity coefficient for complementary goods since the increase in the price of one is always associated with a decrease in the demand for the other.

In calculating cross elasticity, therefore, a positive sign indicates that goods are substitutes and a negative sign indicates that they are complements. The absolute size of the coefficient, moreover, tells us the degree of substitutability or complementarity. A coefficient of $-28.0$ indicates a greater degree of complementarity than one of $-4.0$, for example.

## Elasticity of Supply

The versatile concept of elasticity which we have so far related to demand can also be applied to problems related to supply. We will encounter many of these important concepts throughout the book, but the main theme of elasticity of supply and two variations are introduced here. **Elasticity of supply is the degree of responsiveness of a supplier of goods or services to changes in price or some other variable.**

**Elasticity of supply:** A measure of producers' or workers' responsiveness to price or wage changes; price elasticity of supply is the percent change in quantity supplied divided by the percent change in price.

A price elasticity of supply coefficient can be mechanically calculated in the same manner as all other elasticities. To determine the relation between a change in quantity supplied and a change in price, the simple formula can be applied:

$$\epsilon_s = \frac{\% \text{ change in quantity supplied}}{\% \text{ change in price}} = \frac{\Delta Q_s}{Q_s} \div \frac{\Delta P}{P} = \frac{\Delta Q_s}{Q_s} \times \frac{P}{\Delta P}.$$

Such a simple coefficient or some more elaborate "average elasticity" can be calculated for any supply curve at any instant in time just as for the elasticity of demand described earlier. Elasticity of supply can also be applied to any kind of supply curve, from beets to ball bearings. We turn to an input supply curve—a labor supply curve.

### The Elasticity of Labor Supply

A labor supply curve, such as the one shown in Figure 5–8, simply shows the amount of work (measured in number of hours) that a laborer, whom we will call Sam, would be willing to supply at alternative wage rates. Factors such as the worker's wealth, tastes, preferences, and other factors affecting work decisions are held constant. Each person's labor supply function differs because of these factors.

Ordinarily the labor supply curve is positively sloped, like the supply curve for any commodity or service. An increase in the wage rate from $W_0$ to $W_1$ in Figure 5–8 causes Sam to increase the number of hours he is willing to work from $L_0$ to $L_1$; likewise, a wage increase from $W_2$ to $W_3$ increases Sam's

## FIGURE 5–8
### Elasticity of Labor Supply

Starting from a relatively low wage rate ($W_0$), a 10 percent increase in Sam's wage leads him to supply 25 percent more work, an elastic response. A 10 percent rise at a higher wage ($W_2$) increases Sam's hours worked by only 3 percent, an inelastic response.

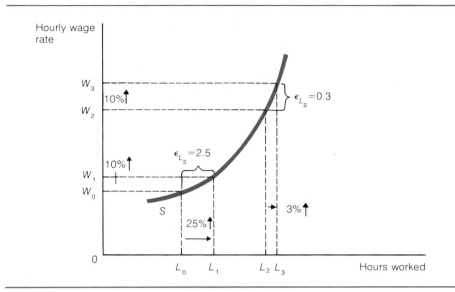

supply of work from $L_2$ to $L_3$. The question is, What is the elasticity of the labor supply schedule and what does it tell us?

The elasticity of labor supply shows the responsiveness of labor supply to a percentage change in the wage rate. It is calculated as follows:

$$\epsilon_{L_s} = \frac{\% \text{ change in quantity of labor supplied}}{\% \text{ change in the wage rate}} = \frac{\Delta L}{L} \div \frac{\Delta W}{W} = \frac{\Delta L}{L} \times \frac{W}{\Delta W}.$$

We can make an elasticity calculation for two different segments of Sam's supply curve. Starting from a relatively low wage rate, $W_0$, an increase in wages of 10 percent will produce a 25 percent increase in the number of hours worked. The coefficient $\epsilon_{L_s}$ is 2.5, which indicates an elastic supply of work effort (because it is greater than 1).

But see what happens to an identical wage rate increase (10 percent) starting from a higher wage, $W_2$. Sam increases the number of hours worked by 3 percent, producing an elasticity coefficient, $\epsilon_{L_s}$, of 0.3—clearly an inelastic response. Why the drastic change in Sam's response to wage increases?

Some personal reflection will give us clues to Sam's work behavior. Work and leisure are two ways of using the scarce resource of time. The more hours we work, the less time we have for leisure. As we work more hours per day in response to wage rate increases, the relative opportunity cost of working (leisure forgone) begins to increase. At some point, laborers start to substitute leisure for work and income. Elasticity of labor supply is one method of calculating this trade-off (although other factors such as income level also shape labor supply). Other things being equal, an inelastic labor supply coefficient for a given wage rate change means that the laborer demonstrates a preference for leisure over work (nonmarket time over market time).

### Time and Elasticity of Supply: Maryland Crab Fishing

An extremely important issue, which we have already related to demand, concerns the time dimension over which the economist calculates supply elasticities. A time dimension exists in all facets of life and human activity, and it is important in the economic activities of suppliers and producers.

A supply elasticity coefficient would not ordinarily capture the full response of suppliers over time to a given price change. To illustrate, a crab fisherman from Chesapeake Bay daily brings in a catch and offers it for sale. The supply curve for crabs on any given day would be totally inelastic. On any given day, the quantity supplied of crabs would be completely unresponsive to price changes.

How then does the market establish a price? Price is determined by the interaction of supply and demand. If the demand for crabs on some particular day happens to be $D_0$ in Figure 5–9, price will settle at $P_c$ and the entire quantity of crabs (a commodity that we assume, unrealistically, is not storable) will be sold.

Now suppose that owing to a change in consumers' taste, Maryland crabs become more desirable, and the demand curve shifts permanently to $D_1$. The fisherman's good fortune is revealed to him when he brings in his usual catch of $Q_0$ crabs. All of a sudden he finds that his catch brings a higher price $P_0$. How will the fisherman respond? If the price $P_0$ is higher than his average production costs, meaning a higher-than-normal profit, the fisherman will adjust by shifting available resources into crab fishing as soon as possible. If he has idle boats or nets and if there is plenty of labor available, all will be put to use. The act of producing more crabs takes time because resources are not instantly adaptable to crab fishing.

During ensuing days, weeks, or months, more crabs will be offered for sale, resulting in a more elastic supply curve for crabs. Such a supply curve over the initial adjustment period may look like $S_1$ in Figure 5–9. The price may temporarily fall to $P_1$ given that the demand for crabs remains stationary at $D_1$. Comparing the quantity $Q_1$ sold at price $P_1$ with the previous quantity of crabs sold, $Q_0$, shows us that over some adjustment period the supply curve for crabs is not completely inelastic but that quantity supplied is responsive to the initial price change. The response of the fisherman to price changes on any given day will still be nil; but over time, he will adjust resources so as to increase the amount of crabs he offers for sale.

If price $P_1$ is still abnormally profitable, the fisherman will continue to

### FIGURE 5–9
### Time and Elasticity of Supply

If the demand for crabs shifts from $D_0$ to $D_1$, then price rises quickly to $P_0$. As time goes by and crab fishermen are able to adjust inputs, the price begins to fall to $P_1$. After a long period of time and all adjustments are made to the increased demand, price falls to $P_2$. Over time, elasticity of supply increases.

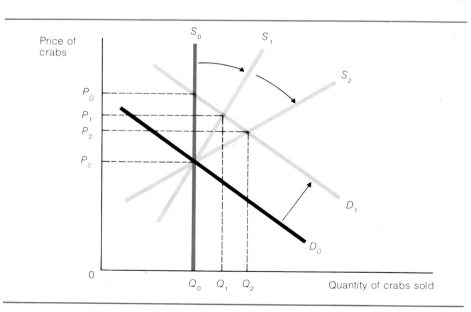

## FOCUS  Gasoline, Cigarettes, and Elasticity: The Effects of an Excise Tax

In 1983 the Reagan administration levied a 5-cent-per-gallon excise tax on gasoline in an effort to acquire tax revenues for highway and bridge repairs. The effects of such a tax are of interest not only to consumers and producers of gasoline but also to legislators. By how much will the price of gasoline increase? By how much will the amount of gasoline purchased decrease? Will tax revenues be as great as expected? The elasticities of the demand for and supply of gasoline are the keys to answering these and other questions.

An excise tax is a simple per unit tax on the sale of a particular item. Tax collectors determine the amount of the taxable good sold by a retail or wholesale firm and require that the firm pay the amount of the tax times the quantity sold.

To gauge what may happen under the gasoline excise tax, consider the mechanism of what happens under a similar situation: the levying of an excise tax on cigarettes. The equilibrium effects of an excise tax on cigarettes are shown in Figure 5–10. The hypothetical supply and demand for cigarettes before the excise tax are points $S_1$ and $D$. The equilibrium price, $E_1$, is 70 cents per pack and the quantity is 50 million packs per day. The effect of a 30-cent-per-pack excise tax can be shown by shifting the supply curve vertically upward by 30 cents to $S_2$. For each quantity along $S_2$ the price at which producers were willing to offer that quantity has now increased by 30 cents. Producers were previously willing to offer 50 million packs for 70 cents each, but now they must receive $1 per pack to offer 50 million packs.

The new equilibrium quantity, $E_2$, is now lower, at 40 million. This change occurs because buyers respond to price changes; their elasticity of demand for cigarettes is greater than zero. As the excise tax puts upward pressure on the price, a smaller quantity is demanded. Producers are willing to offer 40 million packs at a price of 55 cents before the tax. The posttax equilibrium price to consumers, $P_c$, is 85 cents. From this price, producers must pay the 30-cent excise tax and receive the net price, $P_p$, of 55 cents per pack.

The total revenues received by the government from the tax may not be as great as some politicians expected. Before the tax was instituted, the quantity bought was 50 million. A simple multiplication of 30 cents times 50 million would yield an overestimate of tax revenues. The actual tax revenues are 30 cents times 40 million, the new equilibrium quantity. For any excise tax—including that levied on gasoline—the elasticity of supply and demand must be taken into account before a projection is made for tax revenues.

The elasticities of supply and demand also determine the relative burden of the excise tax. Who actually pays the tax—producers or consumers—is shown by the

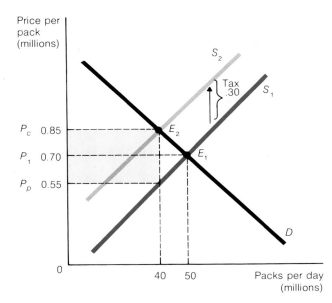

**FIGURE 5–10  Effect of an Excise Tax on Cigarettes**

When a 30-cent excise tax is added to the price of a pack of cigarettes, the entire supply curve for cigarettes shifts from $S_1$ to $S_2$. At the new equilibrium price, 85 cents, the elasticity of demand is greater than zero; specifically, consumers will reduce their consumption of cigarettes by 10 million packs and total tax revenues will equal 40 million times the tax. The shaded area represents government tax revenues, the 30-cent excise tax per pack multiplied by the number of packs of cigarettes sold after the tax is imposed.

change in price to the buyers and sellers. In this example, the price to consumers increased 15 cents and the net price to producers fell 15 cents. Here the burden is shared equally by consumers and producers, but this is not necessarily the case.

If demand had been relatively less elastic, then price would have increased more to consumers than it fell to producers. Figure 5–11a shows that with less elastic demand more of the burden is shifted to consumers. Also, Figure 5–11b shows that if supply had been less elastic more of the burden would have been shifted back to the producers. Figure 5–11c shows that a more elastic demand also shifts more of the burden to producers.

To summarize the effects of an excise tax: The lower the elasticity of demand, the greater the price increase or tax burden to consumers. The lower the elasticity of supply, the greater the net price decrease or burden to the producers. And the greater the elasticity of supply or demand, the lower the total tax revenues.

What then are the effects of a 5-cent excise tax on

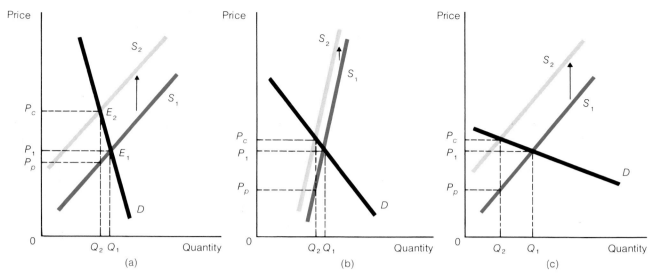

FIGURE 5–11  Elasticity and the Relative Burden of a Tax
(a) With less elastic demand, more of the burden of an excise tax is shifted to consumers: at $Q_1$ consumers pay the equilibrium price $P_1$, but at $Q_2$ following a tax, consumers must pay $P_c$. (b) With less elastic supply, more of the burden of a tax is shifted to producers. (c) More elastic demand also shifts more of the burden to the producers.

gasoline? The price rises, but not by the full 5 cents. The producers and consumers share the burden of the tax. However, the burden may not be shared equally. The tax revenues are the amount of the tax times the quantity sold *after* the tax is instituted. The change in price, the burden of the tax, and the level of tax revenues are all determined by the elasticity of supply and demand.

shift resources into crab production by purchasing new boats and equipment and by training new workers. Again, such activity takes time, but the effect will be to increase crab supply (perhaps to $S_2$). The end result in the crab market will depend on the adaptability and availability of resources for crab fishing and on the cost of producing new inputs. But the general rule is that elasticity of supply will tend to increase over time. The longer the time period of adjustment to an initial change in price—whether price is rising or falling—the more elastic supply schedules will be. This principle applies to all supply curves, including market or industry supply curves. In the case of market supply curves, the adjustment to a change in demand and price will depend on how long it takes to draw on unused or idle capacity or on resources from other industries; in the case of a permanent price drop, the adjustment will depend on the time it takes to decrease the use of resources. The full impact of economic policy changes—levying taxes on industries, for example—often hinges on the elasticity of supply over time.

Real-world events are seldom so isolated and data so accurate as to provide the economist with means to calculate elasticities precisely. It is nonetheless important to a great number of economic and practical problems, some discussed in this chapter, that elasticities can be estimated. To estimate elasticities, the economist must apply precise analytical tools to often complicated policy situations. In Focus, "Gasoline, Cigarettes, and Elasticity: The Effects

## Summary

1. Elasticity is the ratio of the percent change in effect to the percent change in some cause. If changes in quantity demanded (the effect) are caused by changes in price, elasticity of demand is the percent change in quantity divided by the percent change in price. This ratio is called the demand elasticity coefficient ($\epsilon_d$).
2. A demand elasticity coefficient greater than 1 means that consumers are responsive to price changes; demand is elastic. When $\epsilon_d = 1$, demand is unit elastic. When $\epsilon_d$ is less than 1, consumers are not very responsive to price changes; demand is said to be inelastic.
3. Elasticity can also be derived (in a general, shorthand manner) by examining total expenditures (or total revenues) as price rises or falls. If total expenditures rise (fall) as price rises (falls), $\epsilon_d$ is less than 1. If total expenditures remain constant as price rises or falls, $\epsilon_d = 1$. If total expenditures fall (rise) as price rises (falls), the demand elasticity coefficient is greater than 1, that is, it is elastic.
4. There are three major determinants of demand elasticity: (1) the number and availability of substitutes; (2) the size and importance of the item in the consumer's budget; and (3) the time period over which the coefficient is calculated.
5. Other applications of the elasticity concept related to demand are the relation between income changes and changes in quantity demanded (income elasticity of demand) and between price changes for one good and changes in quantity demanded for another complementary or substitute good (cross elasticity of demand between two goods).
6. An elasticity coefficient can be calculated for all kinds of supply curves. A supply elasticity coefficient is simply the percent change in quantity supplied divided by the percent change in price. The concept can be applied to all input and output supply curves, including labor supply.
7. Time is at the center of most important economic calculations, including elasticity. A general rule regarding time and elasticity is that the elasticity of demand and supply increases with the time that a change (in price, for instance) is in effect.

## Key Terms

elasticity
price elasticity of demand
demand elasticity coefficient
elastic demand
unit elasticity of demand
inelastic demand
arc elasticity
total revenue
income elasticity of demand
cross elasticity of demand
substitutes
complements
elasticity of supply

## Questions for Review and Discussion

1. The formula for elasticity is
$$\frac{\%\Delta Q_d}{\%\Delta P} = \frac{\Delta Q_d/Q}{\Delta P/P} = \frac{\Delta Q_d}{Q} \times \frac{P}{\Delta P}.$$
Using this formula derive the elasticity of demand for a product when the price changes from $1 to $1.50, $2 to $2.50, $3 to $3.75, and the change in quantity demanded is 15 to 10, 25 to 5, 50 to 30, respectively.
2. What are the three determinants of consumers' sensitivity to a change in price? Do these always work in the same direction?
3. Is it feasible to talk about *an* elasticity all along a single demand or supply curve? Why or why not?
4. Given that a price change remains in effect over a period of time, will elasticity increase or decrease? Why?
5. What is meant by cross elasticity? How is it algebraically different from elasticity of supply and elasticity of demand, which we calculated earlier?
6. What do cross elasticities indicate about relations between two goods?
7. What might reports about the link between salt consumption and high blood pressure do to the elasticity of demand for salt?
8. Is the demand elasticity coefficient for large industrial consumers of electricity larger or smaller than that for residential use? If electric companies lower the price to both groups, would total revenues from each group change in the same direction?
9. Suppose a friend has an allowance of $10 per week. She spends all of her weekly income on banana splits. What is her elasticity of demand for banana splits?

10. A college town pizza parlor decides to offer a back-to-school two-for-one special, in effect cutting the price of a pizza in half. More pizzas will be sold according to the law of demand. Will the total receipts of the parlor increase, decrease, or remain the same if pizza consumption increases by 30 percent? By 80 percent?

## ECONOMICS IN ACTION   The U.S. Farm Problem

Contemporary American farmers are experiencing severe problems related to elasticity and to long-run trends in technology. The annual income of farmers fluctuates greatly, and the relative price of agricultural products has been steadily declining while farmers' costs have risen. In spite of government programs to alleviate the problem, many farms go out of business every year. Elasticity and the longer-run economic environment under which these businesses operate help explain their current plight.

Both the demand for food and the supply of food are relatively inelastic over short time periods. Price inelastic demand means that small changes in farmers' supply of food result in large changes in price. If weather conditions are particularly favorable, a small increase in production (supply) can result in a large decrease in price. Also with inelastic demand, total revenue falls as price falls. On the other hand, a natural disaster (a late or early ice storm, a hurricane, or an insect plague) causes a decrease in farm output and raises prices considerably, causing total revenue to rise. These situations are presented graphically in Figure 5–12. The production and demand conditions of individual crops—oranges, wheat, avocados—change continually. Frequent booms and busts caused by changing market conditions create short-run income-maintenance problems for individual farmers.

A problem of greater concern is the long-run trend. Every year more and more farm land is being converted to housing, highways, and shopping malls. Since 1880 the number of farms has decreased from about 4 million to 2½ million. Does this mean that agricultural products are becoming more scarce?

In the early 1800s Thomas Malthus earned economics a reputation as "the dismal science" by suggesting that the human population is doomed to a subsistence wage. This would result from a rate of population expansion greater than the rate of growth in agricultural

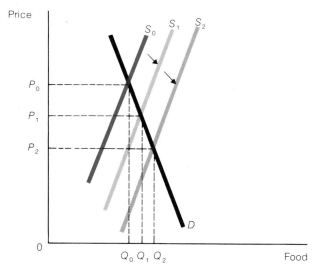

FIGURE 5–12   **Short-Run Fluctuations of Food Prices**

The demand for agricultural goods is relatively inelastic, and short-run supply fluctuations (shifts to $S_0$, $S_1$, or $S_2$) affect price and revenues dramatically. If farmers' output rises from $Q_1$ to $Q_2$, then price falls substantially along with total revenue. If output falls from $Q_1$ to $Q_0$, price and total revenue rise substantially.

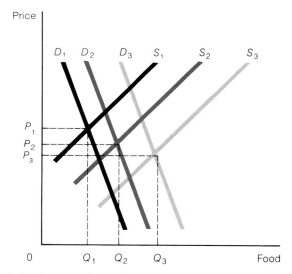

FIGURE 5–13   **Long-Run Changes in Food Prices**

In recent years the relative price of agricultural products has been falling. Supply has been shifting to the right owing to tremendous strides in food production technology ($S_1$ to $S_2$ to $S_3$). These supply shifts have outstripped increases in the demand for agricultural goods ($D_1$ to $D_2$ to $D_3$) caused by growth in population.

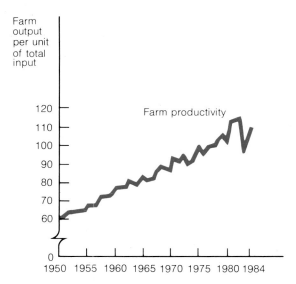

**FIGURE 5-14   Farm Productivity, 1950–1984**

Productivity—measured as farm output per unit of farm input—has doubled since 1950 on U.S. farms. Rapid technological advance has made this dramatic growth in productivity possible.

Source: U.S. Government Printing Office, *Economic Report of the President* (1985), p. 339.

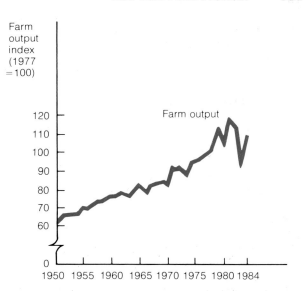

**FIGURE 5-15   U.S. Farm Output, 1950–1984**

American farm output has grown steadily over the past thirty-five years. With the exception of small and temporary downturns in selected years, farm output has doubled over the period.

Source: U.S. Government Printing Office, *Economic Report of the President* (1985), p. 339.

products. His prediction was that the demand for food would increase faster than the supply. With the increase in food prices, a larger and larger percentage of people's incomes would be spent on food. People's urge to procreate would force population increases, but starvation, disease, and wars would ultimately limit population growth to the growth in the food supply.

Malthus's grim prediction has not panned out for the United States. Even though there are fewer farms and a smaller percentage of U.S. land devoted to crop production, agricultural products have not become more scarce. Instead, the supply of food has been increasing faster than the demand for food. In fact, the price of food relative to other goods and the percentage of income spent on food have fallen steadily through the years. See Figure 5-13.

Tremendous improvements in technology are the reasons for these trends. The U.S. Department of Agriculture and many universities have developed new production techniques, disease and insect control, and high-yield hybrid plants.

Figures 5-14 and 5-15 show farm productivity and the actual growth in total output between 1950 and 1984. A major measure of productivity change is farm output per unit of farm input. Figure 5-14 shows that in spite of a declining number of farms and farmers, output per unit has almost doubled between 1950 and 1984 from an index of 60 in 1950 to an index of 109 in 1984. This statistic indicates that farm technology has been advancing at an extremely rapid pace in the last thirty-five years. The net result of these factors is faster growth in total output which is seen to be the case in Figure 5-15.

This increase in technology has allowed output prices to fall while input prices have increased, and farmers' income has fallen as a result. As Figure 5-16 shows, current dollar net farm income (in terms of 1967 dollars) has fallen from about 19 billion dollars in 1950 to 8 billion dollars in 1984. This decline is the result of a number of long-run forces discussed above—farm failures due to falling farm prices caused by vastly improved technology and lower growth in the demand for farm products.

Under normal unrestricted competitive conditions, gains in productivity would create only short-run problems for farmers. As fewer farms were needed, competitive forces would drive some farmers out of business. In the long run, the surviving farms would earn a normal profit. Several government programs have been implemented to help support farm incomes and to encourage farmers to remain in the agricultural industry in spite of economic losses. These programs alleviate the risks of farming and ensure an overabundance of agricultural products for the U.S. consumer, but they also contribute to long-run problems for farmers. Rather than encouraging farmers to stay in an unprofitable industry, government programs could be instituted to encourage farmers to leave the industry. This would possibly eliminate both the losses in the industry and the need for government income support programs.

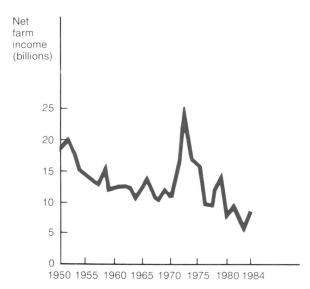

**FIGURE 5-16    Net Farm Income, 1950-1984 (in 1967 dollars)**

With the exception of a few years, farm income has tended to decline dramatically between 1950 and 1984.

Source: U.S. Government Printing Office, *Economic Report of the President* (1985), p. 338.

In summary, the low income of farmers is the result of improved productivity, low short-run elasticities of supply and demand, and government price-income support programs. These effects have resulted in too many farmers and an unnecessary burden on taxpayers in the form of government support to farmers. All of us who consume agricultural products have benefited from falling relative prices. In addition, the gain in comparative advantage has allowed a substantial increase in exports of agricultural products to foreign countries. The industry's output and efficiency are growing at a much faster rate than in most other industries, but many individual farmers have problems. Low elasticities of demand and supply over short time periods are a large part of the problem.

In 1985, an increasing number of U.S. farmers were experiencing business failures. The strong dollar in international exchange markets encouraged imports of farm products from abroad and discouraged exports of U.S. farm commodities. The problem of the typical farm family has also been exacerbated by falling land prices. In the early 1980s, farmers borrowed heavily to purchase land, only to see the value of that land diminish rapidly. In the first half of 1985 many small farmers threatened with bankruptcy petitioned Congress and President Reagan for additional aid. But the push for a balanced budget has made large increases in support unlikely. The family farm may be a vanishing part of the American experience.

## Question

In addition to oil, Iran was a major supplier of pistachio nuts to the United States. In the Khomeni era, U.S. imports of the nuts fell to zero (in 1980). Analyze the possible short-run and long-run effects of demand and supply elasticity on pistachio prices and quantities traded in the United States. What are the predicted effects upon California pistachio nut growers?

# 6

# The Logic of Consumer Choice

We have seen in Chapters 4 and 5 that the law of demand is an accurate and useful generalization about human behavior. Other things being equal, individuals will consume more of a good when its price is lower. Knowing that the law of demand works, however, does not tell us why it works. In this chapter we seek to understand why individuals increase their purchases of a good or service when its price falls. The heart of the analysis involves how consumers deal with the problem of scarcity. In other words, given that a consumer has a fixed budget or income, how does he or she allocate the budget among goods and services to obtain the most satisfaction from consumption?

This chapter's analysis of consumer choice is presented in two alternative forms. In the body of the chapter, individual choice is analyzed in terms of marginal utility; in an appendix at the end of the chapter, the same analysis is cast in terms of indifference curves. Both approaches to the analysis of individual choice-making behavior are useful and introduce new tools and insights with which to study economic behavior in general.

## Utility and Marginal Utility

**Utility:** The ability of a good to satisfy wants; the satisfaction obtained from the consumption of goods.

Why do we demand anything at all? The obvious answer is that we get satisfaction or pleasure from consuming goods and services. A vacation at the beach, a ticket to the big game, dinner at a fine restaurant all give us pleasure. Economists call this pleasure or satisfaction **utility.** Economics, unlike psychology, does not provide any fundamental answers to questions such as why some people prefer red rather than blue shirts or why so many people like chocolate. Economics analyzes the economic results of people's preferences—the observations that people will demand or pay for those things that give them utility.

**Marginal utility:** The change in total utility that results from the consumption of one more unit of a good; the change in total utility divided by the change in quantity consumed.

**Principle of diminishing marginal utility:** As more and more of a good is consumed, eventually its marginal utility to the consumer will fall, all things being equal.

**Total utility:** The total amount of satisfaction obtained from the consumption of a particular quantity of a good; a summation of the marginal utility obtained from consuming each unit of a good.

Although economics cannot offer an answer to questions such as why people like chocolate, economics is interested in the intensity of consumer desires for goods. Why will individuals pay $1.50 for a magazine and $500 for a vacation at the beach? In other words, why is the intensity of demand, as reflected in the prices people are willing to pay, greater for some goods than for others? Economists address questions like these with the aid of the **principle of diminishing marginal utility.** This principle is a simple proposition: As people consume a good in greater and greater quantities, eventually they get less and less extra utility from further increases in consumption.

Although the utility derived by a consumer from some good—shoes, books, raspberries—depends on many factors such as past experience, education, and psychological traits that we may not be able to explain, the principle of diminishing marginal utility allows us to predict certain bounds for consumption behavior. Other things being equal, an individual will not pay more for additional units of a good. And with increasing consumption, sooner or later the consumer will begin to pay less because the good has diminishing marginal utility.

The principle of diminishing marginal utility is based on common sense. At lunch you might be ravenously hungry from skipping breakfast. The satisfaction, or utility, you get from eating the first hamburger in the cafeteria would likely be great. But after one giant burger, or certainly after two or three, the additional satisfaction you experience from one additional hamburger must decline. The amount of money that you would be willing to pay for additional burgers—amounts that reflect their marginal utility for you—would also decline.

The principle of diminishing marginal utility can be numerically and graphically illustrated, as in Table 6–1 and Figure 6–1. Table 6–1 indicates the relation between the quantity of ice cream consumed, the **total utility** for this consumer of all the ice cream eaten, and the marginal utility of each additional scoop for the ice cream eater. For convenience, we construct imaginary units termed "utils" with which to measure quantities of utility going to the ice cream consumer. Note that as the number of scoops consumed increases, total utility—a measure of the total satisfaction gained from the entire amount consumed—increases, but at a decreasing rate. Marginal utility—the satisfaction gained from each additional scoop—declines as additional scoops are consumed within a limited time framework: one visit to the ice cream parlor.

#### TABLE 6–1  Total Utility and Marginal Utility

In one sitting, an ice cream lover can consume anywhere from one to five scoops of ice cream. The satisfaction the consumer gains from ice cream is measured in imaginary units called utils. Total utility increases as the consumer devours more and more ice cream, but marginal utility—the satisfaction associated with each additional scoop—declines.

| Scoops of Ice Cream | Total Utility (utils) | Marginal Utility (extra utils per scoop) |
|---|---|---|
| 1 | 6  | 6 |
| 2 | 11 | 5 |
| 3 | 15 | 4 |
| 4 | 18 | 3 |
| 5 | 20 | 2 |

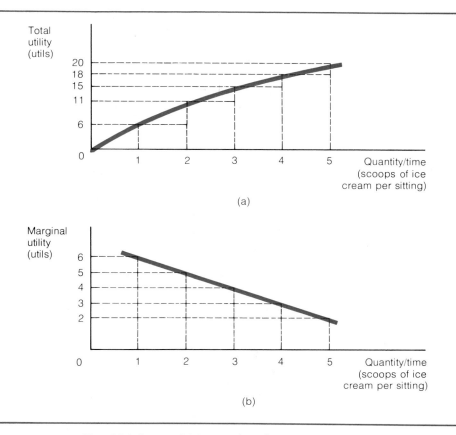

**FIGURE 6–1  Total Utility and Marginal Utility**
(a) Total utility increases at a decreasing rate as more and more ice cream is consumed in one sitting. (b) Marginal utility declines with each additional scoop of ice cream at one sitting.

In Figure 6–1a, the step line represents the increase in total utility generated by the increased consumption of ice cream. A smooth curve has been drawn through the steps to show the relation between units of consumption and total utility. Notice that total utility increases at a decreasing rate. Figure 6–1b shows the corresponding change in marginal utility. As more ice cream is consumed, the marginal utility of each additional scoop declines.

## Consumer Equilibrium: Diminishing Marginal Utility Put to Work

If the principle of diminishing marginal utility applies only to a consumer's behavior in an ice cream store, then it is not very useful. To be useful, the concept of diminishing marginal utility must help us explain behavior in a world in which many different goods are consumed.

### Basic Assumptions

To use diminishing marginal utility to understand consumer behavior, four postulates of consumer behavior need to hold.

1. Each consumer desires a multitude of goods, and no one good is so precious that it will be consumed to the exclusion of other goods. Moreover,

goods can be substituted for one another as alternative means of yielding satisfaction. For example, the consumer good of exercise can be satisfied by jogging, playing basketball, hiking, swimming, or a number of other activities. Dinner offers a variety of choices—steak, chicken, pizza, and so on.

2. Consumers must pay prices for the things they want. This seems obvious, but for the purposes of the following analysis, it is important to assume that consumers face fixed prices for the things they consume.
3. Consumers cannot afford everything they want. They have a budget or income constraint that forces them to limit their consumption and to make choices about what they will consume.
4. Consumers seek the most satisfaction they can get from spending their limited funds for consumption. Consumers are not irrational. They make conscious, purposeful choices designed to increase their well-being. This does not mean that consumers do not make mistakes or sometimes make impulsive purchases that they later regret. But gaining experience over time as they deal in goods and services, consumers try to get the most possible satisfaction from their limited budgets given their past experience.

### Balancing Choices Among Goods

Armed with these postulates, we can better understand the behavior of consumers in the real world. To see how diminishing marginal utility works in a practical setting, assume that you—a consumer—are deciding how to choose between two goods: sweatpants and socks. For each additional dollar you spend on sweatpants, you experience an increase in utility. Simultaneously, you forgo the utility you could have experienced by spending the dollar on socks. Because you seek to maximize the satisfaction you get from your consumption expenditures (postulate number 4), you will spend your additional dollar on the good that yields you the largest increase in utility, that is, the good with the greatest marginal utility for you.

According to the principle of diminishing marginal utility, whether the utility of one good is greater than the other is a function of the amount of each good consumed. If you already have a relatively large sock collection but relatively few pairs of sweatpants, the marginal utility of socks will tend to be low and that of sweatpants high for you. You are therefore more likely to spend an additional dollar on sweatpants than on socks.

In this simple world of two goods, you will continue to spend additional dollars on sweatpants until the marginal utility you receive from an additional dollar spent on sweatpants is equal to the marginal utility of an additional dollar spent on socks. This process of reaching equality in the two marginal utilities per dollar occurs naturally because the marginal utility of sweatpants declines relative to the marginal utility of socks as you purchase more sweatpants. When the marginal utility per dollar of consuming the two goods, or whatever number of goods are available to you, is equal, you have maximized the total utility of your purchases, subject to the resources you have to spend. In other words, at the point of equality of marginal utility per last dollar spent, you will have no incentive to alter the pattern of your choices between socks and sweatpants. This condition of balance in consumer purchases, from which there is no tendency to change, is called **consumer equilibrium.**

How do economists take the differing goods—as well as their differing marginal utilities—into account? Note that consumer equilibrium occurs

**Consumer equilibrium:** A situation in which a consumer maximizes total utility within a budget constraint; equilibrium implies that the marginal utility obtained from the last dollar spent on each good is the same.

where the marginal utility of a dollar's worth of socks is equal to the marginal utility of a dollar's worth of sweatpants. To accommodate the differing goods, the following equation is used to describe the condition of consumer equilibrium:

$$\frac{\text{Marginal utility of socks}}{\text{Price of socks}} = \frac{\text{Marginal utility of sweatpants}}{\text{Price of sweatpants}}.$$

Equilibrium is reached when the marginal utility of the last pair of socks purchased divided by the price of a pair of socks is equal to the marginal utility of the last pair of sweatpants purchased divided by the price of a pair of sweatpants.

This equation is just a representation of consumer equilibrium. To see what it means, assume that a pair of socks costs $1 and a pair of sweatpants costs $10, and that you, the consumer, are initially not in equilibrium. This means that the above equation does not hold for you, that is,

$$\frac{\text{Marginal utility of socks}}{\$1} \neq \frac{\text{Marginal utility of sweatpants}}{\$10}.$$

Suppose that you begin by buying a pair of sweatpants because their marginal utility to you is initially higher than that of socks:

$$\frac{\text{Marginal utility of socks}}{\$1} < \frac{\text{Marginal utility of sweatpants}}{\$10}.$$

As you purchase more sweatpants, the marginal utility of an additional pair declines relative to the marginal utility of an additional pair of socks. At each point you are trying to get the most possible satisfaction from your expenditures on the two goods—you are balancing the marginal utility of spending a dollar toward an additional pair of sweatpants against the marginal utility of spending a dollar on another pair of socks.

At the end of this process, as the ratio of the marginal utility of sweatpants to the price of sweatpants declines relative to the ratio of the marginal utility of socks to the price of socks, equality between the two ratios is reached. Consumer equilibrium is restored at the point where an additional pair of sweatpants adds 10 times as much utility as a pair of socks. Put another way, you will adjust your consumption of the two goods until the marginal utility of a dollar's worth of sweatpants is equal to the marginal utility of a dollar's worth of socks. This is what the equation of consumer equilibrium means and how it is reached by a consumer.

The two-good case of gym clothes is a simplification of the consumer's actual choices among thousands of goods. To express the general condition of consumer equilibrium, the equation given above is written with $MU_x$, $MU_y$, and so forth standing for the marginal utilities of different goods and $P_x$, $P_y$, and so forth representing the corresponding prices of the goods:

$$\frac{MU_x}{P_x} = \frac{MU_y}{P_y} = \frac{MU_z}{P_z} = \ldots.$$

The simplicity of this neat equation is deceptive. The only numbers we have to put in the equation are the prices of the goods. We do not know—or need to know—what the subjective marginal utilities of the consumer are. The important point about the equation is that it is a proposition about individual behavior. It outlines how consumers will achieve balance or equilibrium in allocating their incomes among goods and services. In a sense,

## FOCUS: The Diamond-Water Paradox

Adam Smith formulated the diamond-water paradox when he observed:

> Things which have the greatest value in use frequently have little or no value in exchange; and on the contrary, those which have the greatest value in exchange have frequently little or no value in use. Nothing is more useful than water; but it will scarce purchase anything, scarce anything can be had in exchange for it. A diamond, on the contrary, has scarce any value in use; but a very great quantity of other goods may frequently be had in exchange for it.[a]

Water is useful but cheap; diamonds are not useful but expensive. This seems paradoxical. But using the principle of diminishing marginal utility, what errors can we find in Smith's reasoning about the paradox?

First, Smith failed to grasp the importance of the relative scarcity of a commodity in determining its value in use, or marginal utility. He compared a single diamond with the total supply of water. Had he compared the marginal utility of a single diamond with a single gallon of water, no other water being available, the paradox would have disappeared, for the scarce water would be considered quite valuable. As economic theorists later discovered, water commands little in exchange because its supply is so abundant relative to the intensity of consumer desire for it. Diamonds are scarce relative to consumer desires and therefore command much in exchange. There is no diamond-water paradox when one focuses on the value of the marginal unit of supply to the consumer. Water is plentiful and cheap; diamonds are scarce and expensive.

Second, Smith makes a personal judgment of utility when he suggests that diamonds have no value in use. Many wearers and investors in diamonds would disagree on this point. Modern utility theory does not allow judgments to be made that some preferences are good and others bad.

Of course, Adam Smith did not have modern utility theory available to him when he wrote his statement. Indeed, it was in trying to resolve this simple paradox that modern utility theory was developed by economists in the late nineteenth century in Europe and England.

[a] Adam Smith, *An Inquiry into the Nature and Causes of the Wealth of Nations*, ed. Edwin Cannan (1776; reprint, New York: Modern Library, 1937), p. 28.

---

consumers assign their own utility numbers and arrange their pattern of consumption to achieve equilibrium. Consumers solve the equation for themselves as an expression of their rational choices.

Is consumer choice always rational? In describing the famous diamond-water paradox (see Focus, "The Diamond-Water Paradox"), Adam Smith pondered what seemed to be an irrational willingness of consumers to spend vast sums of money on "useless" goods like diamonds. The principle of diminishing marginal utility, developed many years after Smith's writings, helps us understand the rationality behind such choices.

## From Diminishing Marginal Utility to the Law of Demand

The marginal utility equation helps us understand the relation between diminishing marginal utility and the law of demand. Assume that you, the consumer, are in equilibrium with socks priced at $1 and sweatpants priced at $10. The equation of consumer equilibrium then looks like this:

$$\frac{MU_{socks}}{\$1} = \frac{MU_{sweatpants}}{\$10}.$$

If the price of sweatpants falls to $5, this equality will be upset. The resulting disequilibrium will be temporary, however, for you will act to restore equality by consuming more sweatpants. As you consume more pairs of sweatpants, the principle of diminishing marginal utility tells us that the marginal utility of sweatpants for you will fall in proportion to their lower price. When the marginal utility of sweatpants has fallen by the same proportion that their price has fallen, consumer equilibrium will be restored.

Note the relation of this behavior to the law of demand. Because the price has been reduced, you will consume more sweatpants until a dollar's worth of sweatpants generates no more utility for you than a dollar's worth of socks—or anything else. The law of demand says that consumers will respond to a fall in the relative price of a good by purchasing more of that good. Diminishing marginal utility thus provides a behavioral basis for the law of demand. When price falls, consumption increases until consumer equilibrium is again established.

### The Substitution Effect and Income Effect

Consumers' tendency to buy more of a good when its price drops in relation to the price of other goods is called the **substitution effect.** The relatively cheaper good is substituted for other now relatively more expensive goods. Sweatpants are substituted for socks when the price of sweatpants drops. Technically, the substitution effect refers to that portion of the increase or decrease in quantity demanded of a good that is the direct result of its change in price relative to the price of other goods.

Price, however, is not the only relevant factor affecting the quantity demanded of a good. Changes in wealth or income will also tend to shift the demand curve for goods, as we saw in Chapter 4. The larger an individual's income, the more he or she will demand of most goods.[1] Other things being equal (primarily the prices of other goods remaining the same), the fall in price of a particular good will raise the **real income,** or buying power, of the individual consumer. This change in real income means that the consumer can buy more goods under the same budget constraint as before. This **income effect** can be stated more technically. It is that portion of the change in the quantity demanded of a good that is the direct result of the change in the individual's real income that resulted from a price change.

### Marginal Utility and the Law of Demand

The principle of diminishing marginal utility and the law of demand are thus closely related. A price change leads to a change in quantity demanded, which is composed of a substitution effect and an income effect. When the price of a good falls relative to the prices of other goods, the principle of diminishing marginal utility, acting by means of the substitution effect, tends to increase the quantity of the good that the consumer demands. The income effect is a separate and distinct influence that generally reinforces the substitution effect; that is, the income effect tends to increase the quantity demanded of the good whose price has fallen. Though there are cases where the income effect can pull in the opposite direction from the substitution effect, as in the case of an inferior good, economists generally predict that the strength of the substitution effect in such cases will outweigh the strength of the income effect, leading to an increase in the quantity demanded of a good whose relative price has fallen.

---

[1] Increases or decreases in quantity demanded as a result of an increase in income will depend upon whether the good in question is normal or inferior. Goods for which consumption increases as the consumer's income rises, such as medical care, are called normal goods. When the consumption of a good falls with increases in income, we say that such goods are inferior goods. Used cars may be an example of an inferior good. See Chapter 4 for a discussion of these terms.

---

**Substitution effect:** The change in the quantity demanded of a particular good that results from a change in its price relative to other goods.

**Real income:** The buying power of money income; the quantities of goods and services that may be purchased with a given amount of dollar income.

**Income effect:** The change in quantity demanded of a particular good that results from a change in real income, which has resulted in turn from a change in price.

Sometimes people claim that they have found exceptions to the law of demand, but the exceptions turn out, on close examination, to be based on simple mistakes in reasoning. For example, it sometimes happens that a fall in price of a good seems to result in a decline in the quantity demanded of that good. A department store may actually sell fewer record albums after it lowers record prices. If we look closer at such cases, we may find that, in the calculations of consumers, the relative price of records has not fallen. Consumers may judge that the store is about to go out of business and, as a result, expect that record prices will soon be even lower. What is crucial to consumers is that, relative to the expected future price, the present "sale" price of records has actually risen. This is the law of demand in action—relative and not absolute prices matter to consumers.

One more of the many possible examples of an apparent, but not real, exception to the law of demand is the case of what is sometimes referred to as a "prestige" good. The demand for caviar, for example, is sometimes said to be much higher as an expensive item than it would be if its price per pound were comparable to, say, tuna fish. The high price of caviar is said to actually increase the demand for caviar, contrary to the law of demand. But this notion is obviously confused. Would the demand for caviar continue to increase as the price rose to $1000, $10,000, or $1,000,000 per pound? Of course it would not. Although prestige goods are usually expensive goods, they do not defy the law of demand. If they did, we would see their prices rise without limit, which does not happen.

Although there may be exceptions of the law of demand, they are extremely hard to find. You might at first think of exceptions that are probably not really exceptions after all. The law of demand seems to describe the way people behave in virtually all cases.

Although the law of demand does seem to explain consumer behavior, what might not be so clear at this point is the potential usefulness of utility analysis. How can subjective magnitudes such as marginal utilities be of any practical relevance? To help answer this question, Focus, "Consumers' Surplus," shows the relation between the important economic concept of **consumers' surplus** and utility analysis. In addition, Economics in Action at the end of the chapter applies marginal utility analysis to the act of voting. These two examples illustrate some ways that utility analysis can be applied to real-world problems.

**Consumers' surplus:** The benefits that consumers receive from purchasing a particular quantity of a good at a particular price, measured by the area under the demand curve from the origin to the quantity purchased, minus price times quantity.

## Some Pitfalls to Avoid

The principles of diminishing marginal utility and marginal utility analysis are useful constructs for analyzing individual behavior. As we have seen, they help explain individual choice behavior and they offer a richer understanding of the law of demand. In using this analysis, however, some fairly common misunderstandings must be avoided:

1. The individual wishing to maximize utility is motivated by personal self-interests, but the principle says nothing about what those interests may be. They may be based on purely selfish, greedy goals or on humanitarian concern for others. Marginal utility analysis does not specify the goals or desires of individuals; it is useful only in understanding and analyzing individual behavior once those goals have been specified.

2. It is important to remember that only individuals make economic decisions. When we refer to a government agency, a corporation, or a snor-

## FOCUS  Consumers' Surplus

One stock item in the economist's analytical tool kit is an application of utility theory—the concept of consumers' surplus. Consumers' (or consumer's) surplus is the difference between the amount of money that consumers (or a consumer) would be willing to pay for a quantity of a good and the amount they actually pay for that quantity.

Figure 6–2 shows a market demand schedule for tomatoes. If the price of tomatoes is $P_0$, a quantity of $Q_0$ of tomatoes would be purchased, and consumers would spend a total of $P_0$ times $Q_0$, or the area $P_0BQ_00$, on tomatoes.[a] But consumers would be *willing* to spend a lot more money for $Q_0$ of tomatoes. For some lower quantity, $Q_1$, tomato consumers would be willing to pay price $P_1$, higher than $P_0$. For still lower quantities, some consumers would pay still higher prices. Thus, we may identify the whole shaded area under the demand curve in Figure 6–2 as a surplus of utility that consumers receive when the price of tomatoes is $P_0$ per pound. They would be willing to pay higher prices, but they only have to pay $P_0$.

If the surplus utility of individual consumers and of consumers in general can be measured and compared in terms of dollars, the consumers' surplus is a way of measuring the net utility or welfare produced by the consumption of goods. Such an application of utility analysis clearly can be useful. When government decides to produce goods and services that will not flow through a market, such as a nuclear-powered submarine, some means must be found to evaluate the economic welfare created by such production. Cost-benefit

[a] Throughout this book, whenever we identify a geometric area in a graph, we proceed clockwise, beginning from the northwest corner.

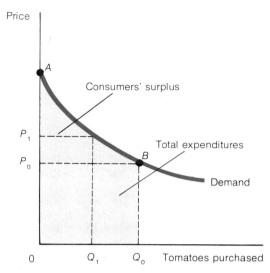

**FIGURE 6–2   Marginal Utility and Consumers' Surplus**

At a price of $P_0$, consumers buy $Q_0$ tomatoes and spend a total of $P_0BQ_00$ in the process. They would have been willing, however, to spend up to $ABQ_00$ for the tomatoes. This means that the consumer earns a surplus, in this case equal to $(ABQ_00 - P_0BQ_00) = ABP_0$. The triangle above the market price and below the demand curve is called consumers' surplus. Under certain conditions it measures the net benefit to the economy of the production of a good.

analysis, discussed in detail in Chapter 18, is often used in an attempt to apply the concept of consumers' surplus to evaluating the effects of government programs on the economy.

---

keling club as making a decision, we are only using a figure of speech. Take away the individual members who compose these organizations, and nothing is left to make a decision. This reminder is a principle of positive economics, necessary to avoid confusion.

3. Economics is concerned with the effects of scarcity on the lives of people. It does not address the issue of human needs. A need is not an object that can be measured in a way that everyone can agree on (like the size of a desk or the height of a building). Needs are subjective, just like love, justice, and honor. Economics is neutral with respect to subjective judgments. Marginal utility analysis and the law of demand allow us to discuss resource allocation without having to argue about what "real needs" are.

4. Marginal utility analysis is an explanation, not a description, of individual choice behavior. Economists do not claim that individual consumers actually calculate marginal utility trade-offs before they go shopping. In-

deed, most consumers, if asked, would probably deny that they behave in the way that marginal utility analysis suggests they do. The proof is obviously in the pudding. Individuals behave so as to generate the same outcomes they would generate if they actually did calculate and equate marginal utilities. Marginal utility analysis explains the outcomes we observe rather than describing the mental process involved.

## Summary

1. As individuals, we demand goods and services because the consumption of material things gives us utility, or satisfaction.
2. The principle of diminishing marginal utility says that the more we consume of a good, the less utility we get from consuming additional amounts of that good.
3. Consumer equilibrium occurs when individuals obtain the most possible utility from their limited budgets. Their consumption pattern is in balance, and they cannot increase their total utility by altering the way they allocate their given budget among goods.
4. In equation form, consumer equilibrium is

$$\frac{MU_x}{P_x} = \frac{MU_y}{P_y} = \frac{MU_z}{P_z} = \ldots$$

In other words, the ratio of marginal utility to price is equal among all the goods a consumer consumes.
5. The principle of diminishing marginal utility underlies the law of demand. In the above equation, if the price of x falls, the equality of the marginal utility ratios is upset. To restore equality, the consumer will consume more of the good that has fallen in price, driving its marginal utility down proportionate to its fall in price. This is equivalent to the type of behavior predicted by the law of demand. Other things being equal, more of a good is consumed when its price falls.
6. The increase in consumption of a good when its price falls is caused by two effects. More is consumed because the price of the good has fallen relative to the prices of other goods. This is called the substitution effect. In addition, more is consumed because the consumer has more real income owing to the fall in price. This is called the income effect.
7. Generally, the substitution and income effects taken together lead to increased consumption of a good when its price falls.
8. The law of demand is a widespread empirical regularity. We observe its action everywhere. Supposed exceptions to the law are usually just misapplications or misunderstandings of basic economic principles.
9. Marginal utility analysis does not specify or judge goals, address issues of social needs, refer to group behavior, or describe the mental process by which consumers make choices; it merely explains the observed patterns of consumers in making those choices.

## Key Terms

utility
marginal utility
principle of diminishing marginal utility
total utility
consumer equilibrium
substitution effect
real income
income effect
consumers' surplus

## Questions for Review and Discussion

1. What is meant by *utility*? Is it calculable? Why do economists not speak of *economic needs*?
2. Define *marginal utility*.
3. What are the four postulates of consumer preference?
4. Express consumer equilibrium algebraically in terms of marginal utility. Explain what your expression means in words.
5. "I love seafood so much that I could never get enough of it." Why would an economist disagree with this statement? Could there ever be a time when the person making this statement might actually receive negative utility from consuming more seafood? Give your answer in a graph and in words.
6. Explain the difference between substitution effects and income effects. Which would we expect to dominate during a period of changing demand? During economic downturn and deflation? During economic prosperity and inflation?
7. Oil is essential to the economic life of the nation. Our machines and homes cannot run without it. Yet a quart of oil is cheaper than a ticket to the Super Bowl. How can oil be so precious and yet so cheap? Resolve this oil–Super Bowl paradox.

## ECONOMICS IN ACTION: Is It Rational to Vote?

You decide to vote in a presidential election. Typically, more than 60 million people cast their votes for their favorite candidate—Democratic, Republican, Socialist, Libertarian, or other. Your vote therefore has only $1/60{,}000{,}000$ worth of influence on the outcome of the election.

We also know that voting is not costless. To make a special trip to the polling place, voters must pay for transportation, take time off from work or leisure, and so forth. Moreover, prior registration often means another trip to the voter registration office. Thus, voting is not free—it clearly places opportunity costs on the voter.

Suppose, in a radical simplification, that all these opportunity costs equal $1; that is, it costs you $1 to vote. Given that you are 1 out of 60 million voters who vote and that it costs you $1 to vote, what must your vote be worth to you to make it rational for you to go to the polls on election day? Since you have only a tiny effect on the election, the outcome would have to have an enormous impact on you to make it worth your while to vote. This concept can be represented algebraically by the formula

$$\frac{1}{60{,}000{,}000} \times \text{Benefits} = \$1,$$

which gives the large value of $60 million necessary in benefits to you to justify the cost of voting. In terms of this analysis, it is hardly reasonable to expect the outcome of any given election to be worth $60 million to the average voter. Therefore, no one should vote, and democracy would fall on its face.

Yet herein lies a paradox. A great number of people do vote, and the question is, Why? Economics provides a simple explanation for the conditions under which it is rational to vote. Obviously, voters are smarter than to view themselves as only 1 vote out of 60 million. They have identified—before the election and during the campaign—with parties and individual candidates, and they will make their own estimates of the prospects of the candidates in the election.

Economics says that voters will behave in terms of the marginal utility and marginal cost of voting. Notice that the above example was stated in terms of the *average* influence of a voter on the election. The fact that the average influence of any one voter is low does not tell us anything about his or her marginal influence on the election. (As we stress repeatedly, the distinction between average and marginal is a crucial one in economics.) While we cannot directly measure the marginal influence of a voter in an election, we can indirectly gauge the marginal benefit from voting. The marginal influence of a vote will be greater the closer the election is predicted to be. In other words, your vote counts for more in a close election than in a lopsided election. Therefore, economic theory leads to the following prediction: Voting turnout will be positively related to expected closeness of the election. In other words, voter turnout will be heavier the closer the election is expected to be.

This is precisely what Barzel and Silverberg found in studying gubernatorial elections in the United States. The closer an election was predicted to be, the larger the voter turnout was.[a] So the simple economic proposition that individuals behave according to marginal cost and marginal benefit provides an explanation for the conditions under which voting is rational. This is not to argue that the only reason people vote is to obtain narrow personal benefits. Benefits can be construed in a wide variety of ways, such as patriotic duty, citizenship, support for an ideological cause, voting for a friend, and so forth. The point is not what the benefits are but that, depending on the closeness of the election, these benefits are likely to be higher or lower relative to costs. Thus, if your friend is running for Congress, you are more likely to vote if he or she is in a tight contest than if your friend is a shoo-in.

### Question

Logically, how should each of the following situations affect voter turnout on election day?
a. Rain
b. An international crisis
c. A close race
d. Television coverage

---

[a]Yoram Barzel and Eugene Silverberg, "Is the Act of Voting Rational?" *Public Choice* (Fall 1973), pp. 51–58.

# APPENDIX
## Indifference Curve Analysis[2]

Suppose that we wanted to know how a consumer felt about consuming various combinations of two goods, such as vanilla and strawberry ice cream. One approach would be to ask the person how much utility she gets from consuming four scoops of vanilla and nine scoops of strawberry per week. She would perhaps answer, "50 utils." This would not be a very helpful answer because we do not know how much 50 utils represent to this person.

A second approach would be to observe how the individual chooses among various combinations of vanilla and strawberry over a specified time. We could then see, through her **revealed preferences,** how she ranks the various combinations. For example, does she prefer a combination of four vanilla and nine strawberry scoops to a combination of nine vanilla and four strawberry? That is, when offered a choice of the two combinations, does she choose one over the other? Proceeding in this way, we are more likely to derive useful information about preferences. This type of approach yields the concept of an indifference curve.

**Revealed preference:** A consumer's ordering of combinations of goods demonstrated through observations of the consumer's actions.

## Indifference Curves

**Indifference set:** A group of combinations of two goods that yield the same total utility to a consumer.

Indifference curves are based on the concept of **indifference sets.** Indifference sets can be easily illustrated. Suppose that four possible combinations of two goods confront an individual—say four possible combinations of quantities per week of vanilla and strawberry ice cream. Four combinations are given in Table 6–2. Assume that the consumer, through her revealed behavior, obtains the same total utility from consuming 10 scoops of vanilla and 3 scoops of strawberry as she would from consuming 7 scoops of vanilla and 5

[2] We present indifference curve analysis in an appendix because it is not necessary to know the technique to understand any of the basic points about economics presented in this book. Indifference curve analysis is, however, an integral part of more advanced courses in economics, and many students who plan to do more work in economics will want to learn the technique.

TABLE 6–2  Combinations of Ice Cream Yielding Equal Total Utility

The data represent the consumer's observed behavior. The consumer derives as much satisfaction from consuming combination A, 10 scoops of vanilla and 3 scoops of strawberry ice cream per week, as she does from consuming combination D, 4 scoops of vanilla and 9 scoops of strawberry. The total utility of each bundle, A, B, C, and D, is equivalent.

| | Scoops Per Week | |
|---|---|---|
| Combination | Vanilla | Strawberry |
| A | 10 | 3 |
| B | 7 | 5 |
| C | 5 | 7 |
| D | 4 | 9 |

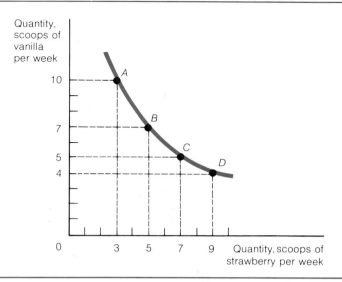

**FIGURE 6-3   An Indifference Curve**

This indifference curve is based on the data in Table 6-2. The curve shows the various combinations of vanilla and strawberry among which the consumer is indifferent. The marginal rate of substitution of one good for the other is given by the slope of the indifference curve. The indifference curve is convex, which means that its slope decreases from left to right along the curve. This behavior of the marginal rate of substitution is due to the principle of diminishing marginal utility.

scoops of strawberry, and so on through the table. Each combination is equivalent to the others in terms of the total utility yielded to the consumer. For example, if we compare combinations C and D, the extra scoop of vanilla in combination C exactly makes up for the utility lost by consuming 2 fewer scoops of strawberry. The consumer reveals herself to be indifferent among these various combinations of the two goods; each yields the same level of utility to her. An indifference set, then, is all combinations of the two goods among which the consumer is indifferent.

With this information about the individual's evaluation of the combinations, we can draw an indifference curve, as in Figure 6-3. Each point on the graph represents a combination of scoops of vanilla and strawberry ice cream per week. The points representing combinations A through D are shown on the graph. These points and points in between them are joined by a smooth curve because they are members of an indifference set. The curve is the **indifference curve,** a curve made up of points that are all members of a particular indifference set. Any point on the indifference curve is equally preferred to any other by the consumer; all yield the same level of total utility.

**Indifference curve:** A curve that shows all the possible combinations of two goods that yield the same total utility for a consumer.

### Characteristics of Indifference Curves

Indifference curves have certain characteristics designed to show established regularities in the patterns of consumer preferences. Five of these characteristics are of interest to us.

1. Indifference curves slope downward from left to right. This negative slope is the only one possible if the principles of consumer choice are not to

be violated. An upward-sloping curve would imply that the consumer was indifferent over the choice between a combination with less of both goods and another with more of both goods. The assumption that consumers always prefer more of a good to less requires that indifference curves slope downward from left to right.

2. The absolute value of the slope of the indifference curve at any point is equal to the ratio of the marginal utility of the good on the horizontal axis to the marginal utility of the good on the vertical axis. In Figure 6–3, the slope of the indifference curve between A and B is about $-1.5$, or simply 1.5 as in absolute value. This absolute value tells us that in the area of combinations A and B, the marginal utility of strawberry is approximately one and one half times that of vanilla. In this region, about three scoops of vanilla can be substituted for two scoops of strawberry without lowering the consumer's total utility. For this reason the slope of the indifference curve is called the **marginal rate of substitution,** in this case substitution of strawberry for vanilla ice cream. The marginal rate of substitution tells us the rate at which one good can be substituted for another without gain or loss in utility.

**Marginal rate of substitution:** The amount of one good that an individual is willing to give up to obtain one more unit of another good.

3. Indifference curves are drawn to be convex: The slope of an indifference curve decreases as one moves downward to the right along the curve. This convexity reflects diminishing marginal utility. As the quantity consumed of one of the goods increases, the marginal utility of that good declines. Hence, the ratio of the marginal utilities—the slope of the indifference curve—cannot be the same all along the curve. The slope must decrease as we move farther down and to the right because the quantity of strawberry ice cream consumed is increasing and the marginal utility of strawberry is therefore decreasing, while the quantity of vanilla ice cream is decreasing and the marginal utility of vanilla is therefore increasing.

In Figure 6–3, the marginal rate of substitution of strawberry for vanilla falls from 1.5 between points A and B to 0.5 between points C and D. As more strawberry is consumed, its marginal utility (measured by how much vanilla the consumer is willing to give up to get strawberry while remaining on the same indifference curve) falls. Less vanilla is consumed, and its marginal utility rises. Between A and B, the consumer is indifferent between three scoops of vanilla and two of strawberry. Between C and D, fortified with more strawberry, the consumer is indifferent between one scoop of vanilla and two scoops of strawberry. The marginal utility of vanilla has risen, and the marginal utility of strawberry has declined.

4. A point representing any assortment of consumption alternatives will always be on some indifference curve. Figure 6–4 presents a graph of combinations of bacon and eggs. We can select a point on the graph—such as A, representing 10 eggs and 1 bacon strip, or D, representing 9 bacon strips and 1 egg—and that point will have a corresponding indifference curve through it showing other bacon and egg combinations that generate the same level of total utility for the consumer. We can see from the graph that this particular consumer is indifferent among 10 eggs and 1 bacon strip (A), 5 eggs and 4 bacon strips (B), and 3 eggs and 8 bacon strips (C) per week.

Moreover, we know that the consumer will prefer any point on the ABC indifference curve, $I_1$, to any point on the DEF indifference curve, $I_2$, because points on $I_1$ contain more of both goods than do points on $I_2$. The farther from the origin the indifference curve lies, the higher the level of utility the individual will experience.

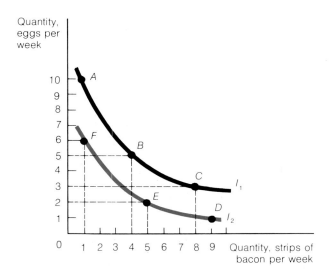

**FIGURE 6–4  An Indifference Map**

This consumer indifference map illustrates combinations of bacon and eggs. Any point on the graph will be associated with an indifference curve. Since a large number of indifference curves can be placed on the graph, it is called an indifference map. Indifference curves that are farther from the origin, such as $I_1$, represent higher levels of utility for the consumer because greater quantities of both goods are consumed.

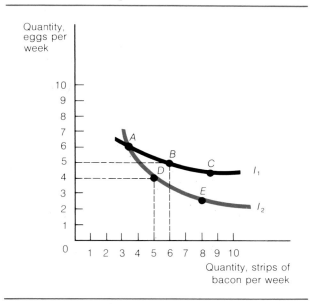

**FIGURE 6–5  Indifference Curves Cannot Cross**

Because preferences are transitive, point A cannot lie on two different indifference curves. If it did, the consumer would be indifferent among the combinations of bacon and eggs in point A, point B with higher quantities of both, and point D with lower quantities of both.

**Indifference map:** A graph that shows two or more indifference curves for a consumer.

**Transitivity of preferences:** A rational characteristic of consumers that suggests that if A is preferred to B and B is preferred to C, then A is preferred to C.

We can draw as many indifference curves on the graph as there are indifference sets confronting the consumer. However many we draw—two or two hundred—the resulting graph is termed an **indifference map.**

5. Indifference curves, which are always drawn for only one individual over a given time period, never cross. The reason for this fact is called **transitivity of preferences,** which simply means that if an individual prefers carrots to squash and squash to artichokes, he or she will also prefer carrots to artichokes. Indifference curves that cross would violate this assumption. Figure 6–5 illustrates this point.

In this graph two indifference curves *do* cross. The point where they cross (point A) lies on both indifference curves, $I_1$ and $I_2$. Since, by definition, all points along one indifference curve are equally preferred by the individual, the consumer will be indifferent among choices A, B, and C on $I_1$ and A, D, and E on $I_2$. This implies that the consumer is indifferent among all these points. But we can see that this is impossible—if the consumer were indifferent between point B on $I_1$ and point D on $I_2$, this would mean that having more of both bacon and eggs (point B) made him or her no better off than having less of both (point D). Other things being equal, consumers always prefer more to less, so indifference curves cannot intersect without implying a type of irrationality that we do not observe in the real world.

**Budget constraint:** A line that shows all the possible combinations of two goods that an individual is able to purchase given a particular money income and price level for the two goods; budget line or consumption opportunity line.

### The Budget Constraint

Although we know that an individual consumer prefers a point on an indifference curve that is farther from the origin to a point closer to the origin, this knowledge does not enable us to establish which indifference curve represents the best an individual can achieve with a limited budget. We can solve this problem by introducing a **budget constraint,** represented on the graph by a budget line.

In Figure 6–6a, we have assumed that the prices of two goods—apples and oranges—are the same, $1 per pound. If the consumer has a weekly budget of $10, the budget line will be a straight line running from the $10 level of apples (on the vertical axis) to the $10 level of oranges (on the horizontal axis). This line simply reflects the fact that the consumer can allocate his budget between apples and oranges in any way he sees fit—$10 on apples and zero on oranges, $10 on oranges and zero on apples, or any other combination that adds up to $10, such as point A: $5 for each.

How does a change in price affect the budget line? Figure 6–6b shows that the individual's budget remains the same ($10) but that the relative prices of the two goods have changed. While the price of apples remains $1 a pound, the price of oranges has risen to $2 a pound. Although the consumer can still allocate his entire income to the purchase of either apples or oranges, $10 spent on apples would purchase 10 pounds of apples, while the same amount spent on oranges would yield only 5 pounds of oranges.

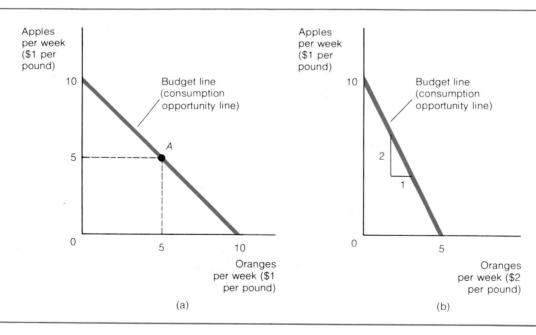

**FIGURE 6–6  Budget Constraint**

The budget line, or consumption opportunity line, depends on the relative prices of two goods. The absolute value of the slope of the budget line is the ratio of the price of the good on the horizontal axis to the price of the good on the vertical axis. (a) The budget line represents all possible combinations of apples and oranges at $1 per pound each within a budget constraint of $10 per week. The absolute value of the slope of the line is 1. (b) The budget line, still within the budget constraint of $10 per week, has changed to reflect an increase in the price of oranges to $2 per pound. The absolute value of the slope of the line is 2.

## FIGURE 6–7
**Determining Consumer Equilibrium with Indifference Curves**

The consumer attains maximum satisfaction at point A, where the budget, or consumption opportunity, line is tangent to the highest possible indifference curve. At A the slope of the indifference curve, or the ratio of the marginal utilities of the two goods, is equal to the slope of the budget line, or the ratio of the prices of the two goods. This relation reflects the same condition for consumer equilibrium as that suggested by marginal utility analysis.

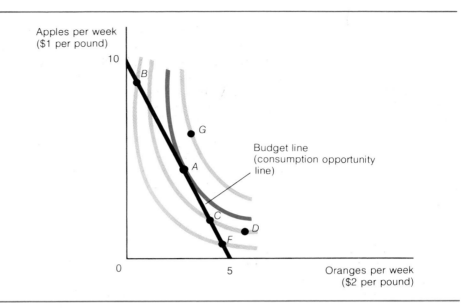

The budget line is thus drawn to reflect any combination of prices for the two goods. The budget line is also called the consumption opportunity line because it shows the various combinations of goods that can be purchased at given prices with a given budget. The absolute value of the slope of the budget line is equal to the ratio of the price of the good on the horizontal axis to the price of the good on the vertical axis. As illustrated in Figure 6–6b, where the price of oranges is $2 a pound and the price of apples $1 a pound, the absolute value of the slope is 2.

### Consumer Equilibrium with Indifference Curves

The combination of an individual's indifference curves and budget line allows us to represent consumer equilibrium in a way that is equivalent to the method of marginal utility analysis. This alternative method is illustrated in Figure 6–7, which combines the budget line from Figure 6–6b with a set of indifference curves.

We can see that the consumer represented here will prefer point C to point F because C lies on a higher indifference curve. She will also prefer point A to point C. All three of these points are actual opportunities confronting the consumer; she can afford them, given her budget constraint. Other things being equal, she would prefer point G to point A. However, point G is beyond the limit of her budget line. The best point she can achieve—the point on the highest indifference curve that can be reached within her budget constraint—is point A. This point represents consumer equilibrium—the combination yielding maximum utility from a given budget.

At A, the relevant indifference curve is exactly tangent to the budget line. This means that the slope of the budget line is equal to the slope of the indifference curve. We therefore know that in equilibrium the ratio of the marginal utility of the good on the horizontal axis (oranges) to the marginal utility of the good on the vertical axis (apples), indicated by the slope of the indifference curve, is equal to the ratio of the price of oranges to the price of apples, indicated by the slope of the budget line. Or, expressed somewhat differently:

$$\frac{\text{Marginal utility of oranges}}{\text{Marginal utility of apples}} = \frac{\text{Price of oranges}}{\text{Price of apples}}.$$

With terms rearranged, this formula is exactly the same result we arrived at earlier in the chapter when we discussed consumer equilibrium in marginal utility terms without the aid of indifference curves. That is, the above expression is equivalent to

$$\frac{\text{Marginal utility of oranges}}{\text{Price of oranges}} = \frac{\text{Marginal utility of apples}}{\text{Price of apples}}.$$

The two approaches thus yield similar predictions about consumer behavior.

## Indifference Curves and the Law of Demand

Indifference curve analysis can also be used to demonstrate the law of demand. Demand curves are obtained by allowing the price of one of two goods to change and finding the new equilibrium quantities demanded of the two goods. If we allow the price of oranges to fall from $2 a pound to $1 a pound, the budget line shifts outward as in Figure 6–8. With the new budget line, a new equilibrium will occur. The new indifference curve tangent to the new budget line will be farther from the origin, indicating that the lower price of one good increases the level of total utility. As shown in the figure, the lower price of oranges also results in a larger quantity purchased. The consumer equilibrium obtained through indifference curve analysis thus yields downward-sloping demand curves, as predicted by the law of demand.

The increase in the quantity demanded of oranges from $O_1$ to $O_2$ is composed of a substitution effect and an income effect. The substitution effect is reflected in the lower relative price of oranges as the budget line shifts

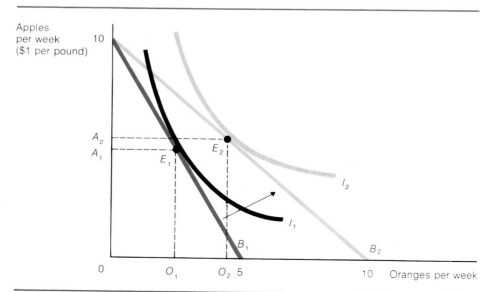

**FIGURE 6–8**

**Indifference Curves and the Law of Demand**

As the price of oranges falls from $2 a pound to $1 a pound, the budget line shifts to the right from $B_1$ to $B_2$. The consumer reaches a higher level of utility and maximizes satisfaction by obtaining the highest indifference curve, $I_2$ rather than $I_1$, possible. As the price of oranges falls, the consumer purchases a larger quantity, from $O_1$ to $O_2$. This is precisely what the law of demand predicts—that, other things being equal, quantity demanded varies inversely with price.

outward. This lower relative price leads to increased orange consumption, just as the law of demand predicts. The income effect measures the degree to which the consumer's real income has increased owing to the fall in the price of oranges. In other words, the shift of the budget line outward means that the consumer is now wealthier. Part of this increased wealth is spent on oranges and part on apples, the consumption of which rises from $A_1$ to $A_2$.

## Summary

1. The method of indifference curves is an alternative way to study the process of individual choice and consumer equilibrium. It is based on the idea of observing how an individual chooses among consumption alternatives.
2. An indifference set is a group of consumption alternatives that yield the same total utility to the consumer.
3. An indifference curve is a graphical representation of an indifference set. Each point on the curve represents the same total utility to the consumer.
4. Indifference curves have negative slopes, are convex, and do not intersect.
5. The slope of an indifference curve is called the marginal rate of substitution of one good for another.
6. An indifference curve is associated with any combination of goods selected by the consumer. An indifference map consists of a set of possible indifference curves of the consumer.
7. The budget line shows the relative prices of the goods an individual consumes and the amounts of the goods that he or she can consume without spending beyond the budget constraint.
8. In the indifference curve approach, consumer equilibrium occurs at the point where the consumer's budget line is tangent to the highest possible indifference curve. This condition is the same as that derived for consumer equilibrium using marginal utility analysis, namely, $MU_x/P_x = MU_y/P_y$.

## Key Terms

revealed preference
indifference set
indifference curve
marginal rate of substitution
indifference map
transitivity of preferences
budget constraint

## Questions for Review and Discussion

1. Explain why indifference curves cannot intersect.
2. What is the relation between the principle of diminishing marginal utility and the marginal rate of substitution?
3. Suppose that the consumer feels that each combination of x and y in the following table yields the same total utility. What comment would you make about the individual's choice process?

| Combinations | Goods | |
|---|---|---|
| | x | y |
| A | 10 | 7 |
| B | 15 | 8 |
| C | 20 | 9 |
| D | 25 | 10 |
| E | 30 | 11 |

4. What are the characteristics of indifference curves? What is the rationale for each characteristic? What would be the consequences of violating each of these characteristics?

# 7

# The Firm

We are now at the point in our study of microeconomics where the concept of the business firm comes into play. If you look back briefly at the circular flow of income diagram in Chapter 3, you will see that firms occupy a critical role in a market economy. In fact, there are over 16,000,000 business firms in the United States, ranging in size from the small corner grocery store owned by one individual to the immense multinational corporations owned by millions of stockholders.

What do all of these firms have in common? Why are they organized in different ways? Who controls these firms and to what end? This chapter will explore each of these important questions.

In broad terms, the economic function of a business **firm** is to combine scarce resources (factors of production) to produce goods and services demanded by consumers. Firms that perform this function well make profits and survive; firms that perform this function poorly experience losses and fail. This brief definition of a firm, however, touches only the surface of the matter. To understand the role of firms in a market economy, we need to explore how a firm organizes resources.

**Firm:** An economic institution that purchases and organizes resources to produce desired goods and services.

## Market and Firm Coordination

Production in a market economy is based on the principle of the **division of labor:** Individuals specialize in different productive activities, all of which are coordinated by the price system. The division of labor in a market economy is a vast network of activities based on such factors as the different levels of skill possessed by members of the labor force. Some individuals find their calling in farming, others in medicine. But what coordinates the divi-

# Market and Firm Coordination

sion of labor so that we can be assured of a supply of food and a supply of medical services? What is the mechanism by which the production of goods and services by millions of individuals is coordinated so that we can get what we want when we want it?

Basically, there are two methods of economic coordination in a market economy: market coordination and firm coordination.

## Distinguishing Market and Firm Coordination

**Division of labor:** An economic principle whereby individuals specialize in the production of a single good or service, increasing overall productivity and economic efficiency.

**Market coordination:** The process that directs the flow of resources into the production of desired goods and services through the forces of the price mechanism.

**Market coordination** is what we have studied so far in this book. It is the use of the price system to provide incentives to suppliers and demanders to produce and consume goods and services in the appropriate amounts. Market coordination refers to the myriad daily economic activities that are guided by the "invisible hand" of the market. Our study of demand and supply is the study of how the market coordinates activities.

Not all activities in a modern economy are coordinated by the market, however. Part of the division of labor is carried out within firms. Automobile assembly-line workers do not sell their output directly in a market. Rather, they supply labor to the auto company, which employs managers to direct and coordinate the uses of their labor within the firm. **Firm coordination** is a productive process that depends on managerial rather than market direction.

**Firm coordination:** The process that directs the flow of resources into the production of a particular good or service through the forces of management organization within a firm.

**Manager:** An individual or group of individuals that organize and monitor resources within a firm to produce a good or service.

In some respects a business firm is like a command economy within a market economy. Decision making in the firm is centralized and performed by **managers**. Managers decide what and how to produce, just as central planners do in a command economy such as that of Russia or Cuba. Within the firm, resources are not bought and sold but are transferred at the command of a manager. The firm, in short, seems to be a contradiction in the marketplace; it appears to conduct its production activities outside the price system.

The analogy of the firm and a command economy is somewhat misleading. A firm is a voluntary institution in which individuals cooperate through free contractual relationships. The manager acts as a specialist in overseeing production and makes decisions that can only figuratively be called commands. Employees agree voluntarily to follow the manager's directions when they join the firm.

The distinction between market and firm coordination is not absolute either. Resource allocation within the firm is not really outside the price system although it might superficially look that way. Managers transfer and allocate resources within the firm in ways that are efficient given the prices for equivalent resources on the market outside the firm. The manager's decision to transfer a quantity of a particular resource within the firm is based on prices, even though resources are not actually bought and sold within the firm.

To understand this point, consider the example of a hypothetical gourmet restaurant, the Greenhouse. It has a luncheon and dinner trade five days a week, it serves brunch on Sundays, and it runs a catering business. After two years of operation, luncheons and Sunday brunch are very crowded, while there are few customers in the evening. In response, the manager of the restaurant will probably transfer resources such as labor, linens, and food from nighttime to daytime use. The transfer is within the firm, but the manager will make the decision by judging the marginal cost of transferring resources—linens, buspersons, waiters, and so on—as if each were being hired or purchased afresh outside the firm. Transfer of resources within the firm is therefore subject to the laws of supply and demand just as if the resources were newly contracted from outside the firm.

If market and firm coordination are so similar, what accounts for the existence of firms? In actuality, much production is undertaken by independent individuals, and it is common for individual demanders to contract with individual suppliers for particular goods and services. For example, Saks Fifth Avenue, Tiffany's, and Neiman-Marcus contract with individual artists and artisans for their wares. Little, Brown and Company—the publisher of this book—hires some work out to freelance editors and art directors rather than having all editorial services performed by in-house staff. Economic coordination between contracting individuals in such cases is achieved directly through the operation of the price system. So an interesting question arises: Why is all production not undertaken by freelance individual operators? Why will an owner of capital goods hire employees for long periods of time instead of contracting out for the performance of specific tasks as they arise?

### Least Cost and Most Efficient Size

Ronald H. Coase was the first economist to pose and answer this important question.[1] His answer was simple and profound: Firms exist when they are the least costly form of economic coordination; their size is determined by what is most efficient (least costly) for production. We do not use the market to organize all production because it is not costless to use markets. Using the market necessitates finding out what prices are, negotiating and enforcing contractual agreements with suppliers, going to court when promises are not kept, paying transactions costs, and engaging in various other costly and resource-using activities. For these reasons, not all production in a market economy is coordinated through market exchange. It costs less in many cases to organize production in a firm where resource allocation is directed and coordinated by managers.

Recall the Greenhouse restaurant and some of the reasons that it organized as a firm. The production of food services—luncheons and dinners—requires labor and many other inputs such as refrigeration, stoves, and silverware. It would be too costly to hire cooks or waiters sporadically for specific tasks—catering a wedding or simply serving the regular dinner hours. Using the market afresh to contract for every single input or task required would be costly and inefficient. Instead, a firm is organized to manage and coordinate a variety of inputs. Waiters and cooks are hired on a long-run basis, and some (but possibly not all) capital is leased or purchased through similar long-run arrangements.

In firm production, the inside of the firm will be run by managerial coordination; the firm will deal with the outside world through market coordination. Which activities the firm chooses to organize internally will be related to the cost of organizing the same activities through market exchange. Automobile manufacturers can buy tires in a contractual arrangement with tire companies, or they can make their own tires. The relative costs of these alternatives will determine how the auto company gets its tires. The restaurant may go into the direct market to obtain bookkeeping services or lease, in the short term, specialized capital such as silver serving trays for special catering jobs. It may be too costly to purchase silver trays for one or two occasions a year or to hire a permanent bookkeeper or accountant.

Firms exist in a market economy because they represent a least-cost means

---

[1] Ronald H. Coase, "The Nature of the Firm," *Economica* (November 1937), pp. 386–405.

of organizing production. However, firms will not grow without limit because at some point the marginal cost of organizing a task within the firm will exceed the marginal cost of organizing the same task through market contracting. This limit on firm size will evolve naturally because the managerial cost of organizing and keeping track of inside production rises along with the number of tasks the firm undertakes. There will be a natural division of production in an economy between firms and markets, determined by the relative costs of producing goods and services in each way.

Suppose the automobile company in its early years produces many of its car parts, such as the seat covers and the spark plugs, because buying these inputs from independent suppliers entails too many risks. This future supply may be uncertain, perhaps because the courts do not strongly enforce supplier contracts for failure to supply these inputs. If, over time, the courts become more concerned about this type of problem and supply contracts come to be more strongly enforced in the event of a failure to supply inputs, the automobile firm can begin buying the auto parts it needs from independent suppliers rather than making them in-house. In the new legal environment, the future supply of seat covers is more certain. In principle, this is how Coase's theory works over time.

## Team Production

It will further our understanding of why firms exist and how they are organized if we look at the internal operations of a firm in more detail. A fundamental principle involved in the operations of many firms is **team production,** by which several people work together to accomplish a task. None of the individual team members produces a separate product. The classic example is the assembly line invented by Henry Ford.

Production by means of a team is sometimes more efficient than production by individuals working separately. For example, two workers making

**Team production:** An economic activity in which workers must cooperate, as team members, to accomplish a task.

"We see ourselves as a team. Wes may discover you have a radiator that won't make it through the summer, while Smitty may decide your transmission needs work, and at the same time Jamie just may come to the conclusion that your whole front end has been twisted out of shape."

Drawing by Booth; © 1984 The New Yorker Magazine, Inc.

jogging shoes—one stamping out the material and the other stitching it together—may be able to produce more pairs of shoes than the two could produce if each were independently assembling whole shoes. Or it may be possible to repair more automobile engines per day if two or three mechanics work on one engine together than if each mechanic concentrates on repairing a separate engine.

Management of the firm can be seen as a device that organizes and **monitors** team production. Management as such functions as a substitute for the discipline of market competition. Management is the firm's solution to a problem that would otherwise limit productive efficiency in group efforts: shirking.

> **Monitor:** An individual who coordinates team production and discourages shirking.

### The Problem of Shirking

In individual production, a person's output can easily be measured, but in team production, only the total output of the team as a whole is observable. The contribution of each team member to the total output is harder to detect. Since only the total output of a team is observable, each team member has an obvious incentive to shift the burden of work onto the rest of the team. Such behavior is called **shirking.** This term is not used in a negative sense. We are not saying that workers are naturally lazy; we are simply saying that shirking can be a rational form of behavior when the worker is part of a team production process. Shirking reduces the total output and productivity of the team, but the individual worker who shirks does not bear the full consequences of his or her reduced effort; part of the consequences are shifted to the other team members. Anyone who has ever played basketball, worked in an office or a fast-food restaurant, or participated in any team production process is aware of this problem.

> **Shirking:** A sometimes rational behavior of members of a team production process in which the individual exerts less than the normal productive effort.

Where team production does not exist, individual producers are disciplined by competition from other producers and other products. The individual producer can dawdle or take frequent work breaks if he or she chooses, but the individual bears the cost of reduced production directly in lower income.

In a team setting, it is costly to monitor the performance of each team member. When the increased productive efficiency of team production exceeds this cost of monitoring, team production will replace independent production by individuals. In other words, when teams can produce products at a lower cost than individuals can, firms will exist.

### Enter the Manager

The manager is that individual (or group) assigned by the firm to monitor team members and direct them to perform their tasks in a manner that enhances team productivity. The manager is more than just a specialist relieving individual team members of the responsibility for monitoring each other's performance. He or she coordinates the production process in an effort to minimize the costs of production. Managers have the power to hire or fire team members and to renegotiate wage contracts and redirect work assignments. As a result, they are able to discipline shirking and to reward superior productive contributions by team members more effectively than would be possible in a system in which every member monitors every other.

Return momentarily to the Greenhouse restaurant example and suppose that an elaborate banquet is planned for sixty people. The chef, kitchen crew, waiters, and buspeople are alerted. Without management, all members of the team would be responsible for monitoring each other's performance; all would have an incentive to shirk work, placing responsibility on other members of the team. If the food does not appear on time or if it is cold and unpalatable, the chef may blame the kitchen help, the kitchen personnel may blame the waiters, and the waiters may blame the buspeople. It might be more efficient for a manager to step in to organize team activity and to monitor the effort of each member of the food-producing and food-serving team. The manager-monitor would discipline shirking and reward outstanding performance by individual members of the team. The manager, in other words, would attempt to minimize the costs of producing the banquet.

### Monitoring the Manager

Of course, within any firm there remains another monitoring problem: Who will monitor the monitor? That is, what is to prevent the manager from shirking? Here competition serves to discipline shirking. More specifically, the threat of losing his or her job to a competing manager will force a manager to avoid shirking. Inefficient managers can and will be fired when more efficient managers are available.

Another incentive to avoid shirking is positive rather than negative: making the managers **residual claimants** who share in the profits of team production. Extra profits that result from more efficient management can be used as a reward to managers. It is common for managers to be given a share of the profits so that they will manage the company effectively, thereby increasing their own wealth. For example, in 1982 the president of Federal Express, a company that provides overnight mail delivery, received a salary of $413,590 and stock gains of $51,129,126 (*Forbes Magazine*, June 6, 1983). This is an astounding amount of profits for a corporate president to receive. Further down the scale was the president of Revco Drugs, who received $30,000 in stock gains.

### Voluntary Acceptance of Managers' Commands

What incentive is there for team members to accept the manager's dictates? The manager's power is entirely contractual; team members have voluntarily agreed to accept the manager's commands within the firm. Team members will enter into such contracts because they are better off with a monitor than they would be without one. At first this appears paradoxical. How can employees be better off if they are made to work harder or more efficiently? Why would they voluntarily contract to be disciplined? Even though each team member might be better off if he or she alone could shirk, if every team member shirks, each will bear the full burden of lowered production in reduced income. Hence, every team member will prefer that the entire team be monitored, even if this reduces his or her own opportunities for shirking.

**Residual claimant:** The individual or group of individuals that share in the excess of revenues over costs, that is, profits.

## Size and Types of Business Organizations

As we have seen, firms evolve when team effort can produce goods and services at a lower cost than individual effort. This occurs because of the increased efficiency that may be gained when labor and capital are coordinated through managerial talents. In the following sections, we consider the varying sizes of firms and the advantages and disadvantages of various ways that firms can be organized.

## Scale of Production

The most efficient size for firms (the size that performs at the lowest cost of production) varies greatly. The most efficient **scale of production** can be as small as a hot dog stand on a street corner or as large as a General Motors assembly plant.

A firm's scale of production affects the amount of capital equipment it needs. Owners of firms that require large-scale production are forced to make large expenditures on buildings, inventories, machines, and other tools of production. These purchases can amount to millions of dollars. On the other hand, owners of firms with a small scale of production may spend only a few thousand dollars on capital equipment.

When an individual decides to start a business, the amount of money required is determined by the desired scale of production. The larger the scale of production, the larger the amount of funds needed to purchase capital equipment. Often an individual is not able or willing to provide all of the necessary capital. Fortunately, there is more than one way to finance a new firm. An individual may seek other people with complementary talents, ambitions, and funds to start a new business enterprise, sharing the responsibilities, risks, and profits. The way in which a firm's ownership is organized depends on many things—the scale of production, the required capital funds, and the individual talents of owners.

## Categories of Ownership

Generally, a firm is organized to fit one of three legal categories: a proprietorship, a partnership, or a corporation. The nature of ownership is different in each category.

**Proprietorships.** A **proprietorship** is a firm that has a single owner who is liable—or legally responsible—for all the debts of the firm, a condition termed **unlimited liability**. The sole proprietor has unlimited liability in the legal sense that if the firm goes bankrupt, the proprietor's personal as well as business property can be used to settle the firm's outstanding debts.

More often than not, the sole proprietor also works in the firm as a manager and a laborer. Obviously, most single-owner firms are small. Many small retail establishments are organized as proprietorships.

The primary advantage of the proprietorship is that it allows the small businessperson direct control of the firm and its activities. The owner, who is the residual claimant of profits over and above all wage payments and other expenses, monitors his or her own performance. The sole proprietor faces a market price directly. It is up to the sole proprietor to decide how much effort to expend in producing output. In other words, the sole proprietor can be his or her own boss.

The primary disadvantage of the proprietorship is that the welfare of the firm largely rests on one person. The typical sole proprietor is the chief stockholder, chief executive officer, and chief bottle washer for the firm. Since there are only twenty-four hours in a day, the sole proprietor faces problems in attending to the various aspects of the business.

Nevertheless, as Table 7–1 shows, proprietorships are the dominant form of business organization in the American economy. In 1979, they constituted 76 percent of all firms. Though plentiful, they generated only 8 percent of total business revenue. The small business sector is composed of many small firms.

## Size and Types of Business Organizations 149

**TABLE 7–1  Types of Firms and Total Receipts, 1979**

While the absolute number of sole proprietorships is greater than the number of corporations, their volume of business (total receipts) is much lower.

|  | Number of Firms | Percent of Total Firms | Total Receipts (millions) | Percent of Total Receipts |
|---|---|---|---|---|
| Proprietorships | 12,330,000 | 76 | $   487.8 | 8 |
| Partnerships | 1,300,000 | 8 | $   253 | 4 |
| Corporations | 2,557,000 | 16 | $5,136 | 88 |
| Total | 16,187,000 | 100 | $5,876.8 | 100 |

*Source: Statistical Abstract of the United States,* 1982–1983, Table 878, p. 529.

**Partnership:** A firm that has two or more owners who have unlimited liability for the firm's debts and who are residual claimants.

**Partnerships.** A **partnership** is an extended form of the proprietorship. Rather than one owner, a partnership has two or more co-owners. These partners—who are team members—share financing of capital investments and, in return, the firm's residual claims to profits. Jointly they perform the managerial function within the firm, organizing team production and monitoring one another's behavior to control shirking. A partnership is a form well suited to lines of team production that involve creative or intellectual skills, areas in which monitoring is difficult. Imagine trying to direct a commerical artist's work in detail or monitoring a lawyer's preparation for a case. If the lawyer is looking out the window, is he or she mentally analyzing a case or daydreaming? By making lawyers residual claimants in their joint efforts, the partnership ensures self-monitoring. (Economics in Action at the end of this chapter analyzes the law firm partnership in more detail.)

The partnership also has certain limitations. Individual partners cannot sell their share of the partnership without the approval of the other partners. The partnership is terminated each time a partner dies or sells out, resulting in costly reorganization. And each partner is considered legally liable for all the debts incurred by the partnership up to the full extent of the individual partner's wealth, a condition called **joint unlimited liability.** Because of these limitations, partnerships are usually small and are found in businesses where monitoring of production by a manager is difficult.

**Joint unlimited liability:** The unlimited liability condition in a partnership that is shared by all partners.

As Table 7–1 shows, partnerships constitute a small percentage of firms and business revenues in the U.S. economy. One should not underestimate the significance of partnership firms, however, since many important services, such as law, accounting, medicine, and architecture, are organized in the partnership form.

**Corporation:** A firm that is owned by one or more individuals who hold shares of stock that indicate ownership and rights to residuals but who have limited liability.

**Corporations.** While **corporations** are not the most numerous form of business organization in the United States, they conduct most of the business. In 1979, corporations accounted for just 16 percent of the total number of firms in the economy but 88 percent of total business revenues. This means that many large firms are corporations and that this form of business organization must possess certain advantages over the proprietorship and the partnership in conducting large-scale production and marketing.

**Share:** The equal portions into which the ownership of a corporation is divided.

In a corporation, ownership is divided into equal parts called **shares** of stock. If any stockholder dies or sells out to a new owner, the existence of the business organization is not terminated or endangered as it is in a proprietorship or partnership. For this reason the corporation is said to possess the feature of continuity.

**Share transferability:** The power of an individual shareholder to sell his or her portion of ownership without the approval of other shareholders.

*Share Transferability.* Another feature that makes the corporation radically different from other forms of business organization is **share transferability**—the right of owners to transfer their shares by sale or gift without having to obtain the permission of other shareholders. For many large organizations, shares of stock are traded on a stock market such as the New York Stock Exchange. Most corporations, however, are smaller, and their shares are traded so seldom that they are not even listed on formal stock exchanges. The shares of these firms are traded by independent stockbrokers on the over-the-counter market.

Share transferability is the most economically important feature of the corporation. It allows owners and managers to specialize, increasing efficiency and profitability in the firm. Owners of stock in a corporation do not need to be concerned with the day-to-day operations of the firm. All that owners need to do is observe the changing price of the firm's shares on the stock market to decide whether the company is being competently managed. If they are dissatisfied with the performance of the company, they can sell their stock. Managers, on the other hand, specialize in reviewing the day-to-day operations of the corporation.

*Accumulation of Capital.* Specialization in ownership means that relatively large amounts of capital can be accumulated because ready transferability of shares renders investment by individuals feasible. Because stock can be traded at the individual owners' discretion, the risk of owning it is reduced. If the firm's performance sags, dissatisfied investors can sell their shares with a minimum of loss.

Many people who have neither the time nor the inclination to bear the burden of management in a firm nevertheless want to invest in firms operated by skilled specialists and to share in the firms' profits. The resources provided by this investment enable the firms to operate on a larger scale. The stockholders primarily bear the changes in value for those resources—both increases and decreases—and thereby partially free the managers from bearing those risks.

**Limited liability:** The legal term indicating that owners of corporations are not responsible for the debts of the firm except for the amount they have invested in shares of ownership.

*Limited Liability.* Another feature of the corporation that distinguishes it from other forms of business organization is **limited liability.** This means that corporate shareholders are responsible for the debts or liabilities of the corporation only to the extent that they have invested in it. For example, if an investor is a millionaire and has invested $10 in one share of XYZ corporation, under no circumstances will he or she risk the loss of more than the invested $10, even if XYZ declares bankruptcy, leaving large unpaid debts. (By contrast, sole proprietorships and partnerships are characterized by unlimited liability; sole owners or partners are legally responsible for the firm's debts up to the amount of their entire personal wealth.) Many investors prefer investments in which their risk of personal loss is strictly limited; the amount of direct investment in corporations is therefore increased as a result of the limited liability involved.

*The Separation of Ownership and Control.* Not all observers see the corporation as the goose that lays the golden egg. Critics of the modern corporation often claim that it is inefficient because ownership and control are separated: Shareholders are the owners of the firm, but the control of the firm is vested in professional managers. Except for the cases in which managers

# Size and Types of Business Organizations    151

are also large stockholders in the firm, the problem of separating ownership and control is that management has different objectives for the firm than shareholders. Shareholders are interested in increasing profits and raising the value of their shares. Incumbent managers often have other objectives. They may want to be captains of industry and heads of large firms; they may seek job security or plush corporate headquarters; or they may want to work with friendly rather than competent colleagues. To the extent that managers can pursue such goals at the expense of firm profitability, the separation of ownership and control is a real problem, and shareholder wealth is reduced as a consequence.

This problem may be more apparent than real, however. The corporate shareholder is not held in bondage by the corporation's management—he or she can sell shares at any time. Because the shareholder has this salable right in the capital value of the firm, changes in the behavior of managers that produce losses will provide the inducement to replace inefficient managers with more efficient ones. Such a change may come through a corporate takeover in which another firm or group of investors buys controlling interest in the firm, a stockholder rebellion, or some other means. The point is simply that mechanisms exist to pull errant managers back into line. A subsequent rise in stock price would provide shareholders a gain from the managerial replacement. This potential gain provides a powerful incentive for stockholders to ride herd on management by monitoring the value of their shares (but not by monitoring managerial activities in detail). (See Focus, "Corporate Takeovers and 'Greenmail,'" for a recent and well-publicized method of monitoring corporate managers.)

## Other Types of Enterprises

The proprietorship, the partnership, and the corporation are not the only possible types of firms in the economy. Other types of firms are not generally numerous, nor do they account for a large portion of economic activity in the United States, but we mention some briefly.

**Labor-Managed Firms.** Most firms in Yugoslavia are owned and managed by the employees. The employees share in the profits of the enterprise in addition to earning their normal wage. This form of profit-sharing enterprise is gaining popularity in West Germany and exists to a limited degree in the United States. Profit-sharing programs in U.S. firms allow laborers to share in the performance of the firms. While this is not strictly labor management of the firm, since the workers do not make managerial decisions, it is perhaps a step in the direction of labor management. Other examples include the phenomena of workers buying a plant when management decides to shut it down, as happened recently at a Bethlehem Steel plant in Johnstown, Pennsylvania, and the addition of union leaders to the boards of directors of corporations. The most notable example of this is in the automobile industry, where the president of the United Auto Workers sits on the board of the Chrysler Corporation. Also in this vein are numerous cooperatives around the country in which individuals band together to produce simple clothing or craft goods such as quilts, furniture, and Christmas tree ornaments.

**Labor-managed firms** are very attractive to employees. Any increase in profits resulting from increased productivity accrue to the workers, which adds an incentive to be efficient and avoid shirking. Increases in wages and strikes only serve to decrease profits and thus occur less often. Also, since

**Labor-managed firm:** A firm that is owned and thus managed by the employees of the firm, who have the right to claim residuals.

## FOCUS  Corporate Takeovers and "Greenmail"

Corporations account for the largest share of total receipts among the various forms of the firm. The principal manner in which large corporations acquire capital to begin business or to make large capital expansions is the issuing of stock shares that are tradable on stock exchanges such as the New York Stock Exchange or the American Stock Exchange. Stocks are supplied and demanded just as any other commodity or service (computers or automotive oil changes). The price of stocks reflects the forces of supply and demand, the demand being determined primarily by the stock purchasers' assessments of the current and future value of the corporation.

Occasionally, the value of a corporation's assets is not accurately represented by the price of stock shares. This sometimes happens when managers are incompetent—that is, when they fail to maximize the wealth of the shareholders. Some investors may come to realize that the stock prices of the firm are too low and do not reflect the value of the corporation's assets. These investor-entrepreneurs may engage in what is called a *takeover bid strategy*: An entrepreneur (usually one already possessing and in control of large-scale assets) finds a corporation whose stock prices are undervalued with respect to the value of the assets of the corporation they represent. He or she then offers to buy up stock in that corporation, thereby driving the value of the shares upward. The entrepreneur is usually followed by other investors in these purchases. Existing shareholders are then faced with a choice: either to sell their stock at the higher prices or to stick with existing management. When sufficient shares are acquired to gain voting control of the corporation, the entrepreneur may attempt to take over the corporation entirely. The takeover entrepreneur in effect gives management of the corporation a choice: Buy back the stock at the new higher price or submit to new corporate control and possible firing. Management may try to prevent stockholders from selling by using all kinds of legal maneuvers, or it may attempt to stave off the "raiders" by selling assets to buy up stock.

Such takeover strategy has become common. In 1984 and 1985, T. Boone Pickens twice attempted to take over the Phillips Petroleum Company of Bartlesville, Oklahoma. The Phillips Company was able to buy out his bid by purchasing Pickens' stock. The al-

*T. Boone Pickens*

ternative was to submit to new corporate control. Pickens made $89 million on the deal! Such "blackmail" of the corporation is appropriately called "greenmail."

What are the economic implications of takeover bids? While such entrepreneurial activities may appear to carry a "robber baron" flavor, they merely reflect the free market at work. In order for investors to be able to make rational choices between stocks and other assets, shares of stock must reflect the real values of the assets underlying the prices of all assets, including stocks. If a company is being poorly or improperly managed, society is not obtaining maximum output and investors are not receiving maximum returns from their investments. In this sense, the takeover entrepreneur who observes such disparities and acts on them is merely bringing the value of the tradable stock assets to par with the value of the corporation. If the entrepreneur makes a mistake and acquires a corporation whose value is below the price he or she paid for it, the entrepreneur will suffer the consequences.

---

the workers manage the firm, they can determine their own working conditions, safety standards, and fringe benefits.

This type of firm faces one important constraint, as do all other kinds of firms. For a firm to survive in the long run it must be efficient with respect to the costs of production. To compete with other firms, the labor-managed firm must keep its costs as low as possible. Excessive wages, fringe benefits, or costly programs such as office beautification could force firms out of business. Any management decisions by labor must be as efficient as decisions made by traditional managers of other firms. For this reason the working conditions and incomes of employees of successful labor-managed firms do not and cannot differ significantly from those in other firms in a market economy.

In addition, management by labor does not completely eliminate incentives to shirk. When an individual worker's extra effort or lack of it are shared equally with a large number of other employees, then the rewards or costs of these activities to the individual are severely diminished.

In the long run the survival of firms is in the hands of owners and managers whose incomes depend on their special talents and efficiency. Labor-owned and -managed firms must provide this talent and efficiency as well as the labor.

**Nonprofit firm:** A firm in which the costs of production and revenues must be equal and which does not have a residual claimant.

**Nonprofit Firms.** Nonprofit firms—such as churches, country clubs, colleges, cooperatives, and mutual insurance companies—do not have a group of owners to whom a residual return accrues. No one owns a college in the sense of making a profit from its operation. Generally, funds raised by such organizations must be spent to further the purposes of the organization. Since the reward for good performance is spread among members of the organization, the individual incentive to do a good job is much reduced. Theoretically, this means that workers in nonprofit institutions may have greater incentives to shirk and to pursue their own interests on the job than do workers in profit-making institutions.

**Publicly owned firm:** A firm owned and operated by government.

**Publicly Owned Firms.** Governments sometimes own and operate basic public services, such as railroad, airline, and water companies. One issue in these cases is whether government is a more efficient provider of these services than privately owned enterprises. The answer seems to be no. Note the distinction between publicly owned firms and private firms that are subject to public regulation. (Publicly regulated firms are treated in detail in Chapter 17.) Government water costs more than private water; government garbage collection costs more than private garbage collection; government airlines are less efficient than private airlines. Why? The answer appears to be that operators of government enterprises do not face the test of profit or loss directly as do their private counterparts. Losses in a publicly owned firm can be made up from general tax revenues, for example. Without the discipline of profit and loss, costs can be higher in public than in private enterprises. (We will take up the issues of profits and losses in succeeding chapters.)

## The Balance Sheet of a Firm

**Balance sheet:** An accounting representation of the assets and liabilities of a firm.

**Asset:** Anything of value owned by the firm that adds to the firm's net worth.

The financial status of a business enterprise is represented by a **balance sheet,** a statement in which the dollar value of all the firm's **assets** corresponds to an equal total value of **ownership claims.** This correspondence is necessary and exact. An asset is a resource owned by the firm; an ownership claim identifies each party who has a property right or a claim to the firm's assets or wealth. A **liability** is simply a debt of the company—what the company owes to various creditors.

### TABLE 7–2  Balance Sheet of Joan Robinson, a Sole Proprietor

The net worth or equity of the firm to Joan Robinson is equal to total assets minus total debts. In a sole proprietorship the single owner has complete claim to all of the equity.

| Assets | | Debts | |
|---|---|---|---|
| Cash holdings | $ 4,000 | Accounts payable | $ 8,000 |
| Equipment | 20,000 | Mortgage payable | 20,000 |
| Warehouse inventory | 14,000 | Total debts | 28,000 |
| Land and building | 100,000 | Joan Robinson's equity | 110,000 |
| Total assets | $138,000 | Total debts plus equity | $138,000 |

### TABLE 7–3  Balance Sheet of Smith and Ricardo, a Partnership

The balance sheet of a firm that is a partnership is very much like that of a sole proprietorship. In this case, the firm's equity is divided equally between the partners.

| Assets | | Debts | |
|---|---|---|---|
| Cash holdings | $ 60,000 | Accounts payable | $ 11,000 |
| Equipment | 10,000 | Mortgage payable | 40,000 |
| Warehouse inventory | 52,000 | Total debts | 51,000 |
| Land and building | 120,000 | Adam Smith's equity | 95,500 |
|  |  | David Ricardo's equity | 95,500 |
| Total assets | $242,000 | Total debts plus equity | $242,000 |

**Ownership claims:** The legal titles that identify who owns the assets of a firm.

**Liability:** Anything that is owed as a debt by a firm and therefore takes away from the net worth of the firm.

**Net worth:** The value of a firm to the owners, determined by subtracting liabilities from assets; also called *equity*. For corporations, net worth is termed *capital stock*.

Underlying the balance sheet is a fundamental identity:

Value of assets − Value of total claims to ownership = Value of liabilities (the amount owed) + Value of owned property in the firm.

Another familiar way of expressing this identity is

$$\text{Assets} = \text{Liabilities} + \text{Net worth}.$$

Net worth is

$$\text{Net worth} = \text{Assets} - \text{Liabilities}.$$

**Net worth,** also called equity, is the amount of a firm's wealth left over for the owner(s) after all liabilities are met from the firm's assets. If liabilities exceed assets, the firm's net worth is negative.

A simplified hypothetical balance sheet of a sole proprietorship is presented in Table 7–2. In the sole proprietorship, all of the firm's assets are owned by a single individual who is personally liable for all of the firm's debts. The proprietor's own stake in the business is his or her net worth, or equity; this is the difference between the firm's assets and the firm's debts. In this case, Joan Robinson's net worth is $110,000.

Table 7–3 reproduces a simplified balance sheet for a hypothetical partnership. Note that it is identical to the sole proprietorship's balance sheet except that the net worth of the firm is divided equally between partner Smith and partner Ricardo. Equity in partnerships need not always be divided equally between the two (or more) partners. However they divide the equity, each of the partners is individually liable for the firm's entire debt.

TABLE 7–4  Balance Sheet of Robert Malthus, Inc., a Corporation

The capital stock of a corporation is the firm's net worth. This equity is divided among the shareholders, the owners of the firm.

| Assets | | Debts | |
|---|---|---|---|
| Cash holdings | $ 60,000 | Accounts payable | $ 13,000 |
| Equipment | 10,000 | Mortgage payable | 40,000 |
| Warehouse inventory | 52,000 | Total debts | 53,000 |
| Land and buildings | 120,000 | Capital stock | 189,000 |
| Total assets | $242,000 | Total debts and capital stock | $242,000 |

In Table 7–4 we present a simplified balance sheet for a hypothetical corporation. The corporation is legally defined as an individual separate and distinct from its shareholders. Hence, all of its assets are held in the name of the firm. The firm's assets minus its liabilities equals its net worth, termed the *capital stock*. Shareholders hold claims to this capital stock, which represents their equity in the firm. This equity may be paid out to shareholders in the form of dividends or plowed back into investments in the firm in the form of retained earnings. As we have seen, shareholders are legally liable for the firm's debts up to, but not exceeding, the extent to which they have invested in the firm.

The balance sheet for a firm can be likened to a snapshot of the firm's operations taken at an instant in time. The balance sheet may look one way today and radically different next week or next year.

The firm is a fascinating and complex institution. It arises and continues because of its efficiencies relative to simple exchange within markets. Some people see the modern corporation as an unchained beast that seeks to rule the world. The analysis in this chapter leads to a more benign view of the corporate form. It is basically a powerful instrument for economic good, which is not to say that corporations should be completely unfettered from public control. We explore such matters in Chapter 17 on economic regulation. For the present we turn our attention in the next chapter to the conditions of real production and costs in the business firm.

## Summary

1. The basic function of a business firm is to combine inputs to produce outputs.
2. Economic coordination in a market economy can take place through prices and markets or within firms. Market coordination relies on price incentives; firm coordination relies on managerial directives.
3. Firm coordination of economic activities exists because there are costs to market exchange and sometimes efficiencies in large-scale production.
4. Team production takes place in firms. In team production, group output can be observed but individual output cannot.
5. Team production makes it natural for individual team members to shirk, that is, to reduce individual effort in achieving the team goal. It is the function of managers in the firm to control shirking.
6. A proprietorship is a single-owner firm whose owner faces unlimited liability for the contractual obligations of the firm. With unlimited liability, the personal wealth of the owner, above and beyond what he or she has invested in the firm, can be drawn on to settle the firm's obligations in the event of bankruptcy.
7. A partnership is owned by two or more individuals who face shared unlimited liability for the contractual obligations of the firm.
8. A corporation is owned by shareholders who have limited liability. Shareholders are liable only for the amount they have invested in the corporation.

9. Corporations are owned by shareholders and typically controlled by professional managers. This potential division of interests has been a source of criticism of the modern corporation.
10. A balance sheet is a financial statement of the economic well-being of a firm. Net worth, which can be positive or negative, is a measure of the wealth of a firm's owner(s). It can be derived from a balance sheet and is defined as assets minus liabilities.

## Key Terms

firm
division of labor
market coordination
firm coordination
manager
team production
monitor
shirking
residual claimant
scale of production
proprietorship
unlimited liability
partnership
joint unlimited liability
corporation
share
share transferability
limited liability
labor-managed firm
non-profit firm
publicly owned firm
balance sheet
asset
ownership claims
liability
net worth

## Questions for Review and Discussion

1. Describe the workings of a balance sheet as presented in this chapter; be sure to include definitions for assets, liabilities, and net worth.
2. Compare and contrast limited liability and unlimited liability. Which of them characterizes (a) a sole proprietorship, (b) a partnership, (c) a corporation?
3. What are the major advantages and disadvantages of the corporation as a form of business organization?
4. What is the concept of team production? Does shirking mean that workers are naturally lazy? Give some examples of team production.
5. What type of firm is your college or university? How would you expect it to operate compared to a private corporation?
6. Is it inefficient for the control and ownership functions of large corporations to be separated?
7. What is the single most efficient way to extract effective performance from managers?
8. Since it would be easy for members of a team to shirk on their work, why does team production exist, from the firm owners' point of view?
9. What types of business activity should be organized as partnerships? Why?
10. A football team is by definition a team production process. Determine three ways that the coaches at your school work to monitor and control individual players' performances on the football team. Do they, for example, film practices and games or do they give differential rewards for superior performance?

## ECONOMICS IN ACTION   The Economics of Law Firms

A law firm is an example of team production in action. Most law firms are organized as legal partnerships; partners are mutually liable for commitments made on behalf of the firm by any partner. This liability does not end with the size of the partner's investment in the firm; it extends to the limit of each partner's personal assets (including his or her house). In the event that the partnership goes into bankruptcy, creditors can attach the personal assets of the partners to settle claims against the firm. That is what unlimited liability means legally; what does it mean economically?

As a result of unlimited liability, partners will be very careful about how they conduct their business. In particular, they will screen candidates for partners in the firm very carefully. A young lawyer is first given an associate status in the firm and must learn and compete with other young associates to become a partner after five or six years. Only the brightest and most trustworthy young lawyers rise through the ranks to become partners in the firm. This screening and training process is clearly motivated by the unlimited liability condition—the cost of making a mistake in the selection of a partner can be very high.

Unlimited liability also means that each partner has to keep track of what the other partners are doing. The natural effect of this consideration is that most law firms are small. It is more costly to keep track of the activities of more partners. Costs are also easier to hold

*Source:* Arleen Leibowitz and Robert Tollison, "Free Riding, Shirking, and Team Production in Legal Partnerships," *Economic Inquiry* 18 (July 1980), pp. 380–394.

down in a smaller partnership. The data on law firms tend to support these arguments. In the mid 1970s, for example, the average law partnership contained only 3.4 partners. Indeed, most of the firms (52 percent) providing legal services consisted of single practitioners.

Although the typical law firm is small, partnerships seem to be growing more popular in the legal services industry than single practitioners. One possible explanation for this trend is that lawyers are now allowed to advertise. Previously, lawyers in small towns were isolated from competition because advertising about price was not allowed. Now that lawyers can advertise, larger firms and legal clinics in urban areas can put competitive pressure on small-town lawyers who provide standard legal services such as wills, divorces, and house sales.

When organized as a partnership rather than a proprietorship, a law firm is a little society of profit sharers. This seems necessary because it is difficult to conceive of paying a lawyer a wage and monitoring his input activities. A lawyer is engaged in what might be loosely called artistic production. How would a monitor know if a lawyer was working on a case when he or she was staring out the window or walking down the street? An external observer cannot tell whether the lawyer is goofing off or considering a new legal angle on a case. Making the lawyer a partner with shared responsibilities in the firm and a share of the firm's profits is a way to make sure that the lawyer devotes his or her time to cases and not to shirking. Thus, one rationale for the partnership is to promote efficient behavior by artistic producers such as lawyers, accountants, architects, and commercial artists.

## Question

Although legal advertising has helped many law firms grow, a number of lawyers are vehemently opposed to advertising legal services. What economic motives might these lawyers have? Does this opposition to legal advertising have anything to do with the partnership structure of law firms?

# 8

# Production Principles and Costs to the Firm

The typical business firm buys inputs at fixed prices, produces a product or service, and sells the product or service at the prevailing market price. The firm's use of inputs leads to costs, and its sale of outputs leads to revenues. The firm has a clear incentive to make revenues exceed costs by as much as possible. When it is successful, the firm earns a profit.

In this chapter we develop concepts related to the firm's costs. Economists do not define costs as the explicit outlays for inputs. Instead, they stress the idea of opportunity costs, or what must be paid to or for a resource to keep it employed in its present use. This amount depends on the value of the next-best alternative use for the resource. As we build on this concept, we will introduce many new terms that economists use in isolating specific costs and analyzing their relations to outputs produced. These ways of pinpointing costs lead to ways of predicting whether it would cost more or less per unit for a firm to increase or decrease its amount of capital or plant size. The tools we develop here are therefore important for understanding how firms maximize profits.

## Types of Costs

As we have seen, consumers have virtually unlimited wants, but resources to satisfy their wants are limited. Costs of production arise from the fact that resources have alternative uses. The same resources that are used to produce a good to satisfy consumers' demands can also be used to produce other goods. For resources to be drawn into the production of a particular good, they must be bid away from their present uses. Production of a newspaper might require, among other things, that wood pulp be turned into newsprint rather than notebook paper, that photographers be enticed to work as pho-

# Types of Costs

tojournalists rather than as portrait photographers, and that youngsters be hired to distribute newspapers rather than earn money by baby-sitting or mowing lawns. The expenditures necessary to do all of these things are called **costs of production.** Costs of production are therefore the value of resources in their next-best uses. Focus, "The Opportunity Cost of Military Service," illustrates this point in more detail.

**Costs of production:** Payments made to the owners of resources to ensure a continued supply of resources for production.

## Explicit and Implicit Costs

The opportunity costs of production include both explicit and implicit costs. Take, for example, a small, owner-operated dry cleaning firm organized as a sole proprietorship. The dry cleaner has to purchase labor, cleaning fluids, hat blocks, bagging machines, insurance, accounting and legal services, advertising, and so on, to produce dry cleaning services. The costs of these resources, entered onto the firm's balance sheet, are called **accounting**, or explicit, **costs.** They are so named because they are quite visible; they are the wages and bills the sole proprietor must pay to conduct business.

**Accounting costs:** Payments that a firm actually makes, in the form of bills or invoices; explicit costs.

The firm does not make explicit payments for all the resources it uses to produce dry cleaning services. The owner of the firm may not enter a wage for his own services on the balance sheet. Omitting his salary from the balance sheet, however, does not mean that the owner's services are free. They carry an **implicit cost,** valued by what the owner could have done with his time instead of working in his firm. Implicit costs are the opportunity costs of resources owned by the firm. They are called implicit costs because they do not involve contractual payments that are entered on the firm's balance sheet.

**Implicit costs:** The value of resources used in production for which no explicit payments are made; opportunity costs of resources owned by the firm.

The main implicit cost that is not entered on the firm's balance sheet is the **opportunity cost of capital.** Individuals who invest in firms expect to earn a normal rate of return on their investment. For instance, they could have placed their capital in money market certificates and earned at least a 10 percent rate of return. In this case, the 10 percent rate of return represents the opportunity cost of capital invested in business ventures. Unless investors earn at least the opportunity cost of capital, they will not continue to invest in a business.

**Opportunity cost of capital:** The value of the payments that could be received from the next-best alternative investment; the normal rate of return.

The **total cost of production** is the sum of explicit and implicit costs. Total cost is the value of all the alternative opportunities forgone as a result of production of a particular good or service.

**Total cost of production:** The value of all resources used in production; explicit plus implicit costs.

## Economic Profits Versus Accounting Profits

Businesspeople must compare their total costs with their total revenues (all the money they take in) to determine whether they are making a profit. **Economic profit** is the difference between total revenue and total cost. It exists when the revenue of the firm more than covers all of its costs, explicit and implicit. Economic loss results when revenue does not cover the total cost. A firm's economic profit is said to be zero when its total revenue equals its total cost. Zero economic profit does not mean that the firm is not viable or about to cease operating. It means that the firm is covering all the costs of its operations and that the owners and investors are making a normal rate of return on their investment.

**Economic profit:** The amount by which total revenues exceed total costs.

Since accounting costs do not usually include implicit costs—such as the value of the sole proprietor's time or the opportunity cost of capital invested in the firm—accounting costs are an understatement of the opportunity costs of production. This means that **accounting profits** will generally be higher

**Accounting profit:** The amount by which total revenues exceed accounting costs.

## FOCUS: The Opportunity Cost of Military Service

Recent attempts to establish an all-volunteer army have been based on the concept of opportunity costs. That is, to induce people to volunteer for military service, the U.S. government offers them benefits comparable to those they would forsake in civilian careers.

Assume that potential recruits have civilian jobs and that economics is the only consideration in their decision to join the army. If government can offer recruits a package including salary, room and board, clothing, education, health care, and travel opportunities that is equal to or greater than their opportunity cost—that is, what they would earn as civilians—then perhaps people will sign up in numbers considered sufficient for national defense.

If the opportunity cost to new recruits is not balanced by the army's offer, people will not volunteer in sufficient numbers. In this case, the government may return to the draft to meet its quota for national defense. When this happens, forced conscription can be seen as a tax—"the draft tax." The draft tax is equal to what an individual could have earned as a civilian minus any compensation he or she gets from serving in the military. If a man was earning $10,000 as a civilian but the military pays him only $5,000 (including the value of benefits such as food and shelter), his draft tax is $5,000.

This example assumes that only money wages matter to the individual. If the individual is not indifferent about the choice between civilian and military work—if he or she strongly prefers to work as a civilian no matter what the pay for military service is—the financial cost of $5,000 understates the real opportunity cost of the draft to the individual.

---

than economic profits because economic profits are calculated after taking both explicit and implicit costs into account.

### Private Costs and Social Costs

We have so far discussed a world of **private costs,** in which the firm pays a market price for the resources it uses in production. The dry cleaner's costs of production were private costs, equal to the sum of explicit and implicit costs. The key aspect of the concept of private costs is that someone is responsible for them; that is, someone pays for the use of the resources. The owner of the dry cleaning firm incurred both explicit and implicit costs in his business, and he paid these costs to engage in business.

But what if the dry cleaner dumps his used cleaning fluid into the town lake adjacent to his plant? In this case he is using the lake as a resource in his dry cleaning business—as a place to dispose of a by-product of his production process—but he pays no charge for the use of the lake as a dumping site. No one owns the lake for purposes of dumping, and it is therefore unpriced for this use. The private costs of the dry cleaner therefore do not reflect the full costs of his operation. The cost of the use of the lake is not reflected in the dry cleaner's costs of production. Such a cost is called an **external cost** because the cost is borne by others.

When external costs arise from a firm's activities—such as polluting the air or water or contributing to erosion or flooding—the firm imposes **social costs.** These are the total of both private and external costs; the total cost to society of the firm's production includes the costs of using environmental assets such as the town lake.

Economics in Action at the end of this chapter discusses the concept of social cost in more detail. In Chapter 18, we further analyze external costs, including ways in which society can correct problems arising from external costs. The important point here is that private costs of production do not always reflect the full opportunity costs of production; the use of unpriced resources, such as the environment, may also carry a hidden price to society.

---

**Private costs:** The total opportunity costs of production for which the owner of a firm is liable.

**External costs:** The costs of a firm's operation that it does not pay for.

**Social costs:** The total value of all resources used in the production of goods, including those used but not paid for by the firm.

## Economic Time

Time is fundamental to the theory of costs. Economic time is not the same thing as calendar time. To the economist analyzing costs, the firm's time constraints are defined by its ability to adjust its operations in light of a changing marketplace. If the demand for the firm's product decreases, for example, the firm must adjust production to the lower level of demand. Perhaps some of its resources can be immediately and easily varied to adapt to the changed circumstances. In the face of declining or increasing sales, an automobile firm can reduce or increase its orders of steel and fabric. Inputs whose purchase and use can be altered quickly are called **variable inputs**. The auto firm with declining sales cannot immediately sell its excess plant and equipment or even reduce its labor force, however. Such resources are called **fixed inputs**; a reduction or increase in their use takes considerably more time to arrange.

There are two primary categories of economic time: the short run and the long run. The **short run** is a period in which some inputs remain fixed. In the **long run** all inputs can be varied. In the long run, the auto firm can make fewer cars not only by buying less steel but also by reducing the number of plants and the amount of equipment and labor it uses for car production. The time it takes to vary all inputs is different for different industries. In some cases, it may take only a month, in others a year, and in still others ten years. In August 1983, for example, the Chrysler Corporation announced that it planned to be able to respond to an increase in sales by boosting its output by 50 percent within twelve months. The relative brevity of this period needed to increase output was made possible by its purchase of a former Volkswagen assembly plant, an increase in work shifts, the return to production of resources left idle during earlier cutbacks, and a sharp increase in purchases of new capital stock.

**Variable input:** Factors of production whose quantity may be changed as output changes in the short run.

**Fixed input:** Factors of production whose quantity cannot be changed as output changes in the short run.

**Short run:** An amount of time that is not sufficient to allow all inputs to vary as the level of output varies.

**Long run:** An amount of time that is sufficient to allow all inputs to vary as the level of output varies.

## Short-Run Costs

In the short run, some of the firm's inputs are fixed and others are variable. Building on this distinction, the short-run theory of costs emphasizes two categories of costs—fixed and variable.

### Fixed Costs and Variable Costs

A **fixed cost** is the cost of a fixed input. In the short run, the fixed cost does not change as the firm's output level changes. Whether the firm produces more or less, it will pay the same for such things as fire insurance, local property taxes, and other costs independent of its output level. Fixed costs exist even at a zero rate of output. Perhaps the most important fixed cost is the opportunity cost of the firm's capital equipment and plant; that is, the next-best alternative use of these resources. This cost exists even when the plant is idle. The only way to avoid fixed costs is to shut down and go out of business, an event that can occur only in the long run.

**Variable costs** are costs that change as the output level of a firm changes. When the firm reduces its level of production, it will use fewer raw materials and perhaps will lay off workers. Consequently, its variable costs of production will decline. Variable costs are expenditures on inputs that can be varied in short-run use. A third concept, **total cost,** is the sum of fixed and variable costs.

It is useful to determine the cost of producing each unit of output (each

**Fixed costs:** Payments made to fixed inputs.

**Variable costs:** Payments made to variable inputs that necessarily change as output changes.

**Total cost:** All the costs of a firm's operations, including fixed and variable costs.

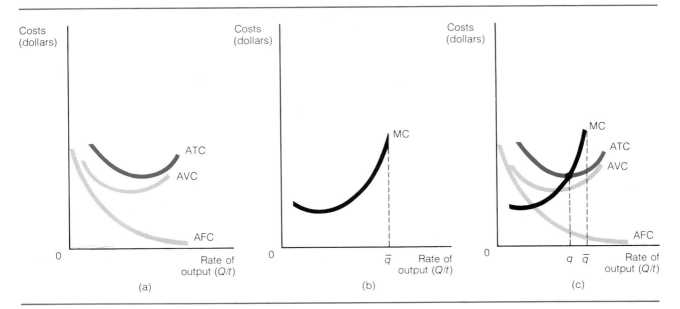

**FIGURE 8–1  Short-Run Cost Curves of the Firm**

(a) The behavior of average fixed cost (AFC), average variable cost (AVC), and average total cost (ATC) as output changes. ATC is the sum of AVC and AFC. (b) The behavior of the firm's marginal costs (MC) as output increases. MC ultimately rises as the short-run capacity limit of the firm is approached at $\bar{q}$. (c) All the short-run cost curves. Notice in particular the U shape of the ATC curve. ATC is high either where the firm's plant is underutilized at low rates of output or overutilized at high rates of output.

**Average fixed cost:** Fixed cost divided by the level of output.

**Average variable cost:** Variable costs divided by the level of output.

**Average total cost:** Total costs divided by the level of output, or average fixed cost plus average variable cost; unit cost.

**Marginal cost:** The extra costs of producing one more unit of output; the change in total costs divided by the change in output.

car, backpack, or ton of soybeans) in the short run as the level of output is increased or decreased. The fixed cost per unit is called **average fixed cost** (AFC) and is found by dividing the total fixed cost by the firm's total output of the good (referred to as $q$ throughout this chapter). Since fixed costs are the same at all levels of output, AFC will decline continuously as output increases. This relation is shown graphically in Figure 8–1a.

**Average variable cost** (AVC) is found by dividing total variable cost by the firm's output. It is also drawn in Figure 8–1a; it ordinarily has a U shape.

**Average total cost** (ATC)—the actual short-run cost to the firm of producing each unit of output—is total cost divided by total output. Average total cost can also be found by adding average fixed costs and average variable costs. Thus, in Figure 8–1a, the ATC curve is shown as the sum of the AFC and AVC curves. Average total cost is sometimes referred to as *unit cost*.

Although the average total cost curve provides useful information, economic decisions are actually made at the margin. In considering a change in the level of production, firms must determine what the marginal cost of the change will be.

**Marginal cost** (MC) is the cost of producing each additional unit of output; it is found by dividing the change in total costs by the change in output. The result of such calculations is a marginal cost curve, shown in Figure 8–1b. Marginal cost first declines as output rises, reaches a minimum, but then rises because it becomes increasingly hard to produce additional output with a **fixed plant.** Extra workers needed to produce additional output begin to

**Fixed plant:** A situation in which the firm has a given size of plant and equipment to which it adds workers.

get in each other's way; inefficiencies proliferate as workers must take turns using machines; storage room for inputs and outputs becomes filled. At some point the firm's maximum output from its fixed plant will be reached (shown at $\bar{q}$).

Figure 8–1c brings all the short-run cost curves together. Notice that the ATC curve is U shaped. At low levels of output a firm does not utilize its fixed plant and equipment very effectively. The firm does not produce enough relative to the fixed costs for its plant, causing the average total cost to be high. At high levels of output the firm approaches its capacity limit ($\bar{q}$). The inefficiencies of overloading the existing operation cause marginal costs to rise, raising the average total cost. These two effects combine to yield a U-shaped average total cost curve. Either low or high utilization of a fixed plant leads to high average total cost. The minimum point on the ATC curve, $q$, represents the lowest average total cost—the lowest cost per unit for producing the good. This represents the best short-run utilization rate for the firm.

Table 8–1 provides a shorthand reference to all these concepts, and later in the chapter we will discuss them in more detail with a numerical example and more graphical analyses.

### Diminishing Returns

The behavior and shape of the short-run marginal cost and average total cost curves can be understood more completely with the **law of diminishing marginal returns.** The law of diminishing marginal returns states that as additional units of a variable input are combined with a fixed amount of other resources, the amount of additional output produced will start to decline beyond some point. The returns—additional output—that result from adding the variable input will ultimately diminish.

**Law of diminishing marginal returns:** A relation that suggests that as more and more of a variable input is added to a fixed input, the resulting extra output eventually decreases to zero.

Though the language sounds formidable, the law of diminishing marginal returns is little more than formalized common sense. Suppose that the dry cleaner in our previous example has a plant of a given size that we will view as his fixed input. He begins operations by adding workers one at a time and observes what happens to the resulting output of dry cleaning. The first few additional workers are very useful. One specializes in cleaning, another in pressing, another in bagging, and so on. The output of dry cleaning increases

**TABLE 8–1  Short-Run Cost Relations**

| Terms | Symbols | Definition |
| --- | --- | --- |
| Average fixed cost = Fixed cost ÷ Total output | AFC = FC/q | A fixed cost does not vary with output, but the average fixed cost per unit declines as output rises. |
| Average variable cost = Variable cost ÷ Total output | AVC = VC/q | A variable cost changes as output changes. |
| Average total cost = Total cost ÷ Total output<br>= Average fixed cost + Average variable cost | ATC = TC/q<br>ATC = AFC + AVC | Total cost is the sum of fixed and variable costs. Average total cost per unit forms a U-shaped curve when graphed from very low to very high output levels. |
| Marginal cost = Change in total cost ÷ Change in output | MC = $\Delta$TC/$\Delta$q | Marginal cost rises as the short-run capacity of a firm's fixed plant is approached. |

**TABLE 8–2  Law of Diminishing Marginal Returns**

If units of a variable input (labor) are added one at a time to production with a fixed plant (growing soybeans with no change in technology on a fixed plot of land), the total product, marginal product, and average product all increase at first. But eventually, as more workers are added, they begin to get in each other's way. Total output gained by adding more workers drops, as predicted by the law of diminishing marginal returns.

| Variable Input (units of labor) | Total Product (tons) | Marginal Product (tons) | Average Product (tons) |
|---|---|---|---|
| 0 | 0 |  | 0 |
|  |  | 10 |  |
| 1 | 10 |  | 10.00 |
|  |  | 12 |  |
| 2 | 22 |  | 11.00 |
|  |  | 14 |  |
| 3 | 36 |  | 12.00 |
|  |  | 13 |  |
| 4 | 49 |  | 12.25 |
|  |  | 11 |  |
| 5 | 60 |  | 12.00 |
|  |  | 10 |  |
| 6 | 70 |  | 11.67 |
|  |  | 5 |  |
| 7 | 75 |  | 10.71 |
|  |  | 3 |  |
| 8 | 78 |  | 9.75 |
|  |  | −2 |  |
| 9 | 76 |  | 8.44 |
|  |  | −6 |  |
| 10 | 70 |  | 7.00 |

as the laborers are added. But this process cannot go on forever. At some point the output provided by one additional laborer will begin to diminish because capacity of the existing physical plant will be reached.

The law of diminishing marginal returns is a fact of nature. Imagine how the world would work if it were not true: All of the world's dry cleaning could be done in a single plant. And if diminishing returns did not exist in agriculture, all of the world's food could be grown on a fixed plot of land, say an acre.

The law of diminishing marginal returns can be illustrated numerically and graphically. A standard example comes from agriculture; we experiment by adding laborers to a fixed plot of land and observe what happens to output. To isolate the effect of adding additional workers (the variable input), we keep all inputs except labor fixed and assume that the technology does not change.

Table 8–2 is a numerical analysis of the results of such an experiment. The first column shows the amount of the variable input—labor—that is used in combination with the fixed input—land. The simplified numbers for **total product** and **marginal product** represent real measures of output such as tons of soybeans grown.

Marginal product is the change in total product caused by the addition of each additional worker. At first, as workers are added, total product expands rapidly. The first three workers show increasing marginal products (10, 12, and 14 extra tons of soybeans with the addition of each worker). Diminish-

**Total product:** The total amount of output that results from a specific amount of input.

**Marginal product:** The extra output that results from employing one more unit of a variable input.

ing marginal returns set in, however, with the addition of the fourth worker. That is, the marginal product of the fourth worker is 13 tons, down from the 14-ton marginal product gained by adding the third worker. Marginal product continues to decline as additional workers are added to the plot of land. This squares with our earlier definition of the law of diminishing marginal returns: As additional units of a variable input are added to a fixed amount of other resources, beyond some point the amount of additional output produced will start to decline. In fact, the addition of the ninth and tenth workers leads to reductions in total product. Marginal product in these cases is negative, meaning that the addition of these workers actually reduces output. It becomes more and more difficult to obtain increases in output from the fixed plot of land by adding workers. At some point the workers simply get in each other's way.

The last column in the table introduces the concept of **average product**, which is total product divided by the number of units of the variable input. It is the average output per worker. Average product increases as long as marginal product is greater than average product. Thus, average product increases through the addition of the fourth worker. The marginal product of the fifth worker is 11, which is less than the average product for five workers, and beyond this point the average product falls.

Figure 8–2 graphically illustrates the law of diminishing marginal returns

**Average product:** The output per unit of a variable input; total output divided by the amount of variable input.

**FIGURE 8–2**
**Total, Average, and Marginal Product Curves**
(a) Total product curve, plotting data from Table 8–2. (b) Average product and marginal product curves. Marginal product is the rate of change of the total product curve. The area of diminishing marginal returns begins when marginal product starts to decline. Average product rises when marginal product is greater than average product, and it falls when marginal product is less than average product.

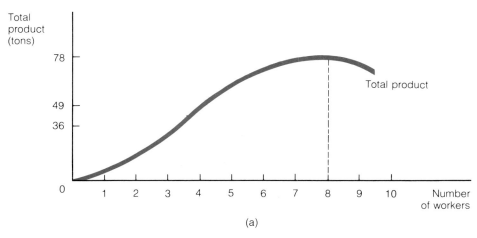

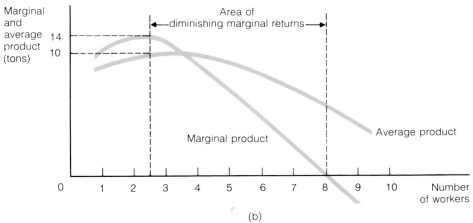

with the data from Table 8–2. Figure 8–2a illustrates the total product curve. Total product increases rapidly at first as the marginal product of the first three workers increases. Diminishing returns set in with the fourth worker, and thereafter the total product increases less rapidly. Beyond eight workers, the total product curve turns down and starts to decline. The maximum total product is reached with eight workers. The marginal product and average product curves are shown in Figure 8–2b. The marginal product curve is simply the slope of the total product curve. It rises to a maximum at three workers and then declines as diminishing returns set in. Eventually, it becomes negative at eight workers, indicating that total product is decreasing. The average product curve rises when marginal product is greater than average product, and it declines when marginal product is less than average product.

How can the law of diminishing marginal returns be applied to economic decision making? Consider the thinking of the farmer who owns the plot of land in this experiment. How many workers will he or she choose to hire? Before the area of diminishing marginal returns is reached, additional workers have an increasing marginal product. The farmer will add these workers and will go no further in hiring than the point where the marginal product falls to zero. Even if labor were free, the farmer would not go beyond this point because there would be so many workers on the land that the marginal product of an additional worker would be negative and the total product would decline. This implies that it is rational for the farmer to operate somewhere in the area of diminishing marginal returns.

Precisely how many workers will maximize profits? The answer depends on the cost of the workers. Assume that each worker is paid a wage rate equivalent to 4 units of output. The farmer will compare the marginal product of each worker with the wage rate. The fourth worker adds 13 units of output and costs the farmer the equivalent of only 4 units. In fact, the marginal product of the first seven workers exceeds their wage rate. However, the marginal product of the eighth worker is 3 units of output, which is less than the wage rate. The rational farmer will therefore hire seven workers. This is the law of diminishing marginal returns in action. It points the farmer toward the rational utilization of factors of production.

### Diminishing Marginal Returns and Short-Run Costs

What does the law of diminishing marginal returns imply about the behavior of the firm's short-run cost curves? Once diminishing marginal returns set in, more and more of the variable input—in our example, labor—is needed to expand output by an additional unit. If the price of the variable input is fixed, the firm's marginal costs will rise as a reflection of diminishing returns. Adding more workers to the plot of land at a fixed wage rate eventually leads to both diminishing returns and rising marginal costs. These are two ways of looking at the same thing.

To see this point more clearly, recall the definition of marginal cost from Table 8–1:

$$MC = \frac{\text{Increase in total cost}}{\text{Increase in output}}.$$

Since the increase in total cost is equal to the increase in the number of units of the variable input times the price of the variable input (which we assume is constant), the expression can be rewritten as

$$MC = \frac{\text{Increase in quantity of variable input}}{\text{Increase in output}} \times \text{Price of variable input.}$$

We know that marginal product is defined as the increase in output associated with an increase in the quantity of the variable input. Thus the first term in the MC expression is the reciprocal of marginal product. MC can further be rewritten as

$$= \frac{1}{\text{Marginal product of variable input}} \times \text{Price of variable input}$$

$$= \frac{\text{Price of variable input}}{\text{Marginal product of variable input}}.$$

This means that marginal cost is inversely related to marginal product. That is, if the price of the variable input is constant, then increases in marginal cost are associated with decreases in marginal product. The law of diminishing marginal returns is equivalent to increasing marginal cost.

Table 8–3 presents a numerical illustration of how the law of diminishing marginal returns affects a firm's short-run costs. Columns (2), (3), and (4) show how total costs behave as output increases in the short run, and columns (5), (6), and (7) show the behavior of the corresponding average cost concepts.

In this example, we keep the numbers low for simplicity. Total fixed cost is constant at $10 per day; average fixed cost per unit is the total fixed cost divided by output. Fixed costs must be paid in the short run regardless of whether the firm operates. The level of average fixed cost falls continuously as output is increased.

Total variable cost is the sum of the firm's expenditures on variable inputs. Average variable cost is the total variable cost divided by output. Notice that total variable cost first increases at a decreasing rate and that after nine units of output it increases at an increasing rate.

Total cost is the sum of total fixed and total variable costs. Average total cost is the sum of average variable and average fixed costs, or simply total cost divided by output.

Marginal cost, shown in column (8), is the change in total cost that results from producing one additional unit of output. The behavior of marginal cost reflects the law of diminishing marginal returns. In this example, marginal cost falls through the production of the ninth unit of output and rises thereafter. The rising portion of the marginal cost schedule reflects diminishing marginal returns for additions of the variable input.

The data in Table 8–3 are plotted in Figure 8–3, which graphically demonstrates the behavior of short-run cost curves. Figure 8–3a illustrates the concepts of total cost, and Figure 8–3b presents the corresponding average and marginal cost concepts. Notice in part b that MC intersects AVC and ATC at their minimum points. This is because marginal cost bears a definite relation to average variable cost and average total cost. In the case of average total cost, marginal cost lies below average total cost when average total cost is falling and above average total cost when average total cost is rising. Marginal cost is equal to average total cost at the point where average total cost is at a minimum. In Table 8–3, this occurs between the thirteenth and fourteenth units of output. At lower levels of output, the MC values in column (8) are less than the ATC values in column (7). ATC therefore declines. At higher levels of output, the MC values are greater than the ATC values,

## 8 / Production Principles and Costs to the Firm

### TABLE 8-3  Short-Run Cost Data for a Firm

The various concepts of short-run costs are expressed in both total and average terms. Columns (1) through (4) present the firm's total cost data; columns (5) through (8) present the cost data in average terms. Notice especially that short-run marginal cost ultimately rises because of the law of diminishing marginal returns.

| | Total Cost Data (per day) | | | Average and Marginal Cost Data (per day) | | | |
|---|---|---|---|---|---|---|---|
| (1) | (2) | (3) | (4) | (5) | (6) | (7) | (8) |
| Output (units) | Total Fixed Cost | Total Variable Cost | Total Cost (2) + (3) | Average Fixed Cost (2) ÷ (1) | Average Variable Cost (3) ÷ (1) | Average Total Cost (4) ÷ (1) | Marginal Cost Δ(4) ÷ Δ(1)[a] |
| 0 | $10.00 | $ 0 | $10.00 | $ 0 | $0 | $ 0 | |
| 1 | 10.00 | 1.60 | 11.60 | 10.00 | 1.60 | 11.60 | $1.60 |
| 2 | 10.00 | 3.00 | 13.00 | 5.00 | 1.50 | 6.50 | 1.40 |
| 3 | 10.00 | 4.35 | 14.35 | 3.34 | 1.45 | 4.78 | 1.35 |
| 4 | 10.00 | 5.70 | 15.70 | 2.50 | 1.43 | 3.93 | 1.30 |
| 5 | 10.00 | 6.90 | 16.90 | 2.00 | 1.38 | 3.33 | 1.20 |
| 6 | 10.00 | 8.00 | 18.00 | 1.67 | 1.33 | 3.00 | 1.10 |
| 7 | 10.00 | 9.00 | 19.00 | 1.43 | 1.29 | 2.71 | 1.00 |
| 8 | 10.00 | 9.90 | 19.90 | 1.25 | 1.24 | 2.49 | .90 |
| 9 | 10.00 | 10.20 | 20.20 | 1.11 | 1.13 | 2.24 | .30 |
| 10 | 10.00 | 11.85 | 21.85 | 1.00 | 1.19 | 2.19 | .65 |
| 11 | 10.00 | 13.15 | 23.15 | 0.91 | 1.20 | 2.11 | 1.30 |
| 12 | 10.00 | 14.65 | 24.65 | 0.83 | 1.22 | 2.05 | 1.50 |
| 13 | 10.00 | 16.40 | 26.40 | 0.77 | 1.26 | 2.03 | 1.75 |
| 14 | 10.00 | 18.45 | 28.45 | 0.71 | 1.32 | 2.03 | 2.05 |
| 15 | 10.00 | 20.90 | 30.90 | 0.67 | 1.39 | 2.06 | 2.45 |
| 16 | 10.00 | 23.80 | 33.80 | 0.62 | 1.49 | 2.11 | 2.90 |
| 17 | 10.00 | 27.20 | 37.20 | 0.59 | 1.60 | 2.19 | 3.40 |
| 18 | 10.00 | 31.15 | 41.15 | 0.56 | 1.73 | 2.29 | 3.95 |
| 19 | 10.00 | 35.70 | 45.70 | 0.53 | 1.88 | 2.41 | 4.55 |
| 20 | 10.00 | 40.95 | 50.95 | 0.50 | 2.05 | 2.55 | 5.25 |

[a] Δ = change in.

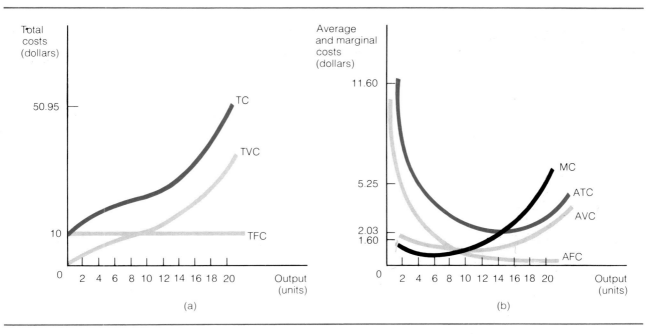

**FIGURE 8–3  Short-Run Cost Curves**

(a) Total cost data plotted from Table 8–3. (b) Average cost and marginal cost data. Notice the U-shaped average total cost curve. At low levels of output, ATC is high because AFC is high, and the firm does not utilize its fixed plant efficiently. At high levels of output, ATC is high because MC is high as the firm approaches the capacity limit of its plant. These effects give ATC a U shape.

and ATC rises. The same analysis holds for the relationship between MC and AVC.

This type of relation holds for marginal–average series in general. If a baseball player is batting .400 and goes 3 for 4 today (.750), his batting average rises because the marginal figure (.750) lies above the average (.400). If he goes 1 for 4 today (.250), his average falls because the marginal figure (.250) lies below the average (.400). An example can be found closer to home. Suppose that your average grade in economics to date is 85. If you score 90 on your next exam, your average will rise. If you score 80, it will fall. When the marginal figure is above the average, it pulls the average figure up, and vice versa. This same relation holds for average total cost and marginal cost, which both Table 8–3 and Figure 8–3 verify.

In sum, the firm's short-run cost curves show the influence of the law of diminishing marginal returns. First, the firm has certain fixed costs independent of level of output. These costs correspond to what the firm must pay for fixed inputs, and they must be paid whether or not the firm operates. Second, assuming the price of variable inputs is constant, marginal costs reflect the behavior of the marginal product of the variable input. As output rises, marginal costs first decline because marginal product is increasing and it requires less and less of the variable input to produce additional units of output. At some point, however, diminishing marginal returns set in, and it takes more and more of the variable input to produce additional units of output. When this happens, marginal costs start to rise. Third, marginal cost will eventually rise above average variable cost and average total cost, causing these costs to rise as well, which results in a U-shaped average total cost curve.

## Long-Run Costs

In the short run, some inputs are fixed and cannot be varied. These fixed inputs generally are characterized as a physical plant that cannot be altered in size in the short run. Short-run costs therefore show the relation of costs to output for a given plant. In the long run, however, all inputs are variable, including plant size. As economic time lengthens and the contracts that define fixed inputs can be renegotiated, a firm owner can adjust all parts of his or her operation. In fact, the scale of a firm's operations can be adjusted to best fit the economic circumstances that prevail in the long run; an owner can choose to arrange the organization in the best possible way to do business. Stated in terms of plant size, the owner will seek the plant size that minimizes long-run costs of producing the profit-maximizing output.

### Adjusting Plant Size

The choice of plant size affects production costs. An owner might have the choice of four plants, such as those depicted in Figure 8–4. The ATC curves are short-run average total cost curves that correspond to four different-sized plants. Which plant is best from the owner's point of view? The answer depends on the profit-maximizing level of output. If the firm wants to produce less than $q_1$, the plant represented by $ATC_1$ is the lowest-cost plant for those levels of output. For outputs between $q_1$ and $q_2$, $ATC_2$ is the lowest-cost plant. For outputs between $q_2$ and $q_3$, $ATC_3$ is the lowest-cost plant. For even larger outputs—beyond $q_3$—the plant represented by $ATC_4$ is the lowest-cost plant.

In effect, the owner's best course of action is depicted by *PLAN* in Figure 8–4. Given that inputs can be varied to build any of these plants, the owner will move along a path such as *PLAN*, gradually expanding output by expanding plant size, seeking the lowest-cost way to produce the profit-maximizing output in the long run.

To determine the optimum long-run plant size, we introduce a new concept: **long-run average total cost** (LRATC). This measure shows the lowest-

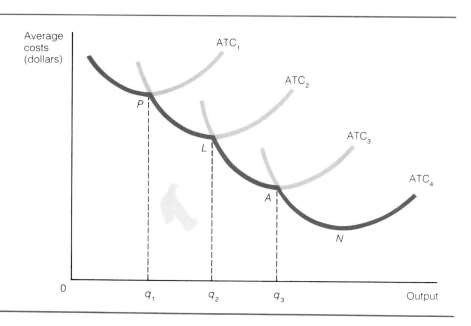

**FIGURE 8–4**

**Adjusting Plant Size to Minimize Cost**

The figure shows short-run average total cost curves for four plant sizes. If these are the only alternative plant sizes available, the long-run average total cost (or planning) curve will be given by *PLAN*. That is, starting with plant size *P*, the rational owner expands plant size to *L, A,* and then *N* in the long run. Each point represents movement to a larger and lower-cost plant size.

**FIGURE 8–5**

**Long-Run Average Total Cost Curve**

LRATC is the planning curve of the firm owner. It shows how plant size can be adjusted in the long run when all inputs are variable. The plant size shown at $q$ represents the lowest possible unit cost of production in the long run. Economies of scale prevail before $q$ on the curve, and diseconomies of scale prevail past $q$.

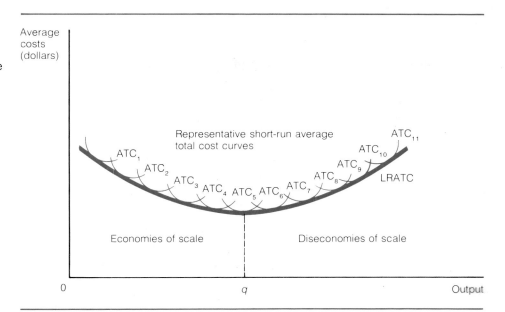

**Long-run average total cost:** The lowest possible cost per unit of producing any level of output when all inputs can be varied.

cost plant for producing each level of output when the firm can choose among all possible plant sizes. In Figure 8–4, PLAN is the long-run average total cost curve when only four plants are possible. In the long run, however, more than four plant sizes are possible. Indeed, the number of possible plant sizes is unlimited. In effect, the owner sees the world of possibilities in terms of a long-run average total cost (or planning) curve such as that drawn in Figure 8–5. The long-run average total cost curve is smoothly continuous and allows us to see the full sweep of a firm owner's imagination. On its downward course, the long-run average total cost curve is tangent to each short-run average total cost curve *before* the point of minimum cost for each given plant size. On its upward course it touches each short-run average total cost curve *past* the point of minimum cost. Only at the bottom of the long-run U shape do the minimum points of the long-run and short-run average total cost curves coincide. In effect, the long-run average total cost curve is an envelope of short-run average total cost curves.

### Economies and Diseconomies of Scale

Why does long-run average total cost have a U shape, falling to $q$ and rising thereafter? The answer involves two new concepts: economies and diseconomies of scale.

**Economies of scale:** The relation between long-run average total cost and plant size that suggests that as plant size increases, the average cost of production decreases.

The initial falling portion of the LRATC curve is due to **economies of scale.** To a certain point, long-run unit costs of production fall as output increases and the firm gets larger. There are a number of reasons why a larger firm might have lower unit costs: (1) A larger operation means that more specialized processes are possible in the firm. Individual workers can concentrate and become more proficient at more narrowly defined tasks, and machines can be specially tailored to individual processes. (2) As the firm grows larger and produces more, workers and managers gain valuable experience in production processes, learning by doing. Since workers and managers of a larger firm produce more output, they acquire more experience. Such experience can lead to lower unit costs. (3) Large firms can take advantage of mass production techniques, which require large setup costs. Setup costs are

**FIGURE 8-6**
**Alternative LRATC Curves**

Not all LRATC curves are U-shaped. (a) Constant returns to scale. The lowest possible production costs per unit exist within a large range between output levels $q_1$ and $q_2$. Both small and large firms can have the same long-run unit costs between $q_1$ and $q_2$. (b) Increasing returns to scale. The more output the firm produces, the lower its long-run average total cost.

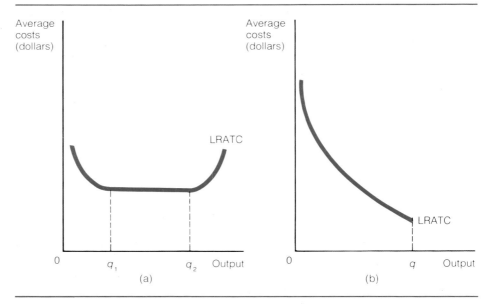

**Diseconomies of scale:** The relation between long-run average total cost and plant size that suggests that as plant size increases, the long-run average total cost curve increases.

most economical when they are spread over a large amount of output. Production techniques such as the assembly line used by large automobile manufacturers would result in very high unit costs if used by a small producer of specialized cars.

The fact that the LRATC curve rises after $q$ is due to **diseconomies of scale**. In this range of outputs, the firm has become too large for its owner to control effectively. Managers do not have the monitoring technology to hold costs down in a very large firm, and bureaucratic inefficiencies creep in. If such bureaucratic problems did not exist, firms would be much larger. Indeed, a single firm could produce all the world's output of a particular good or service. However, diseconomies may be hard to observe in the real world because it is in the firm's self-interest to correct its operations to keep costs down. Large firms will reorganize, spin off component parts, hire new managers, and in general seek ways to avoid diseconomies of scale.

### Other Shapes of the Long-Run Average Total Cost Curve

The LRATC curve in Figure 8-5 shows a unique ideal plant size at $q$. There is only one minimum point on this U-shaped curve. This means that there is a small range of plant sizes that are efficient in a particular industry, and plant sizes in the industry will tend to cluster at the level of $q$. The fact that most discount department stores are approximately the same size illustrates this point.

**Constant returns to scale:** The relation that suggests that as plant size changes, the long-run average total cost does not change.

Figure 8-6 shows two other possible shapes the LRATC curve can take. Figure 8-6a illustrates **constant returns to scale.** In certain industries an initial range of economies of scale prevails up to a minimum efficient size of $q_1$. Beyond $q_1$, a wide variation in firm size is possible without a discernible difference in unit cost. Small firms and large firms can operate with the same unit costs over this range of outputs. This flat portion of the LRATC curve shows constant returns, or the same unit costs, for a range of output levels from $q_1$ to $q_2$. Beyond $q_2$, diseconomies of scale begin. This is apparently a very common LRATC curve in the real world because we observe both small and large firms prospering and surviving side by side in many industries, such as publishing and textiles.

**Increasing returns to scale:** The relation that suggests that the larger a firm becomes, the lower its long-run average total costs are.

Figure 8–6b shows an LRATC curve that exhibits economies of scale over its whole range; this is called **increasing returns to scale.** In such cases, the larger the firm, the lower its costs. This type of LRATC curve is representative of such industries as utilities and telephone service.

## Shifts in Cost Curves

Our analysis of cost curves has been based on the familiar *ceteris paribus,* or other things constant, assumption. In other words, we held certain factors constant in the discussion of short-run and long-run costs. What factors did we hold constant, and how do they affect cost curves?

*Resource prices* have been held constant. If resource prices rise or fall, the firm's cost curves will rise or fall by a corresponding amount. If the price of gasoline falls, the cost curves of a trucking firm fall; this is illustrated by the fall from $ATC_1$ and $MC_1$ to $ATC_2$ and $MC_2$ in Figure 8–7.

*Taxes and government regulation* have been held constant. If government increases the excise tax on gasoline, the trucking firm's cost curves will rise. In fact, the average and marginal costs of the trucking firm will rise by the amount of the tax. Similarly, if government imposes more stringent highway weight limits for trucks, the costs of trucking firms will increase.

*Technological change* has been held constant. Advances in technology make it possible to produce goods and services at lower costs. The invention of the diesel engine, which is more durable and less expensive to operate than the conventional gasoline engine, shifted the cost curves of trucking firms downward. Trucking services can now be produced with fewer resources because of this technological improvement.

## The Nature of Economic Costs

Opportunity cost is a forward-looking concept. Opportunity costs are incurred when the decision maker decides about the future use of resources; they are the expected costs of possible alternatives available to an owner.

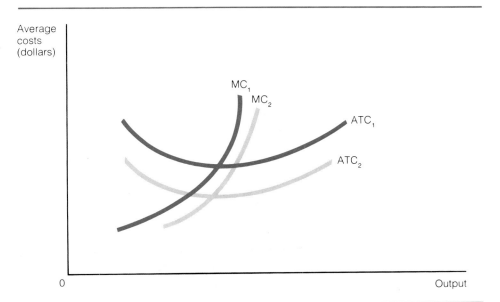

**FIGURE 8–7**
**The Effect of a Decrease in Resource Prices on Costs**

As resource prices decrease, the cost curves of the firm fall from $ATC_1$ and $MC_1$ to $ATC_2$ and $MC_2$. If resource prices increased, the cost curves of the firm would rise.

In our discussion of the long-run average total cost curve, we saw that the owner's choice was among potential plants of different sizes. The plants do not yet exist—they are only in the planning or blueprint stage of development. The owner, of course, will base his or her estimate of what the LRATC curve looks like on past experience in the industry or in other industries, so the past is of some help in estimating what will happen in the future. The choice of plant size, however, involves looking into the future and making a decision under uncertainty. To select a plant size means forgoing certain future alternative plant sizes.

In contrast, accounting costs are **historical costs,** or costs that have been incurred in the past. While historical costs can be useful in estimating economic costs such as short-run marginal cost and long-run average total cost, historical costs look backward; opportunity costs look forward. Historical costs are often poor guides to opportunity costs. Consider the case of sunk costs. **Sunk costs** are incurred as a result of an earlier decision; they are historical costs. For better or worse, earlier decisions cannot be reversed. For this reason economists argue that sunk costs are irrelevant to current decisions. The following examples exemplify this principle and will convince you that it is true.

A jeweler purchases some diamond rings and necklaces. The price of diamonds doubles after her purchase. At what price will she sell the diamonds—the old price of diamonds plus her usual markup or the new price? The amount she paid for the diamonds is a sunk cost. The current price of diamonds determines their worth in the market. She will sell the diamonds at their current price plus her normal markup.

Manhattan, proverbially, was purchased for $24. Suppose you own an acre of Manhattan today. Would your selling price have anything to do with the original purchase price?

You buy a car for $8000. After driving it for a day and imposing only negligible wear and tear, you decide that you do not like it. You put it on the market, and your best offer is $6000. Should you refuse to sell the car for $6000 because you paid $8000 for it?

Each of these examples involves the fallacy of sunk costs. The point is that, viewed rationally, past costs exert no influence on current decisions. This does not mean that we do not learn from experience; it simply means that current decisions are based on current or expected costs and benefits, not on past costs and benefits.

**Historical costs:** Costs of production from the past.

**Sunk costs:** Past payment for a presently owned resource.

## Costs and Supply Decisions

Costs help explain the supply or output decisions of firms. Firms compare their expected costs and expected revenues in deciding how much output to produce. In the short run, the relevant comparison is between marginal cost and expected revenue. If the latter exceeds the former, the firm will supply additional units of output. In the long run the owner of a firm decides whether to enter an industry or to expand output within an industry by comparing long-run average total cost with expected revenue. Again, if expected revenue exceeds marginal cost, the firm enters or increases production, and industry output is expanded. There is a critical link between firm costs and the industry supply curve; we examine this link in detail in the next chapter.

## Summary

1. Costs result from the fact that resources have alternative uses. An opportunity cost is the value of resources in their next-best uses.
2. Explicit costs are like accounting costs; they are the bills that the firm must pay for the use of inputs. Implicit costs are the opportunity costs of resources owned by the firm. The total cost of production is the sum of explicit and implicit costs.
3. Private costs are payments for the use of inputs by the firm. External costs arise when the firm uses an input without paying for its services. The sum of private and external costs is social cost.
4. The short run is a period of economic time when some of the firm's inputs are fixed. The long run is a period over which all inputs, including the size of the firm's plant, can be varied.
5. The law of diminishing marginal returns states that when adding units of a variable input to a fixed amount of other resources, beyond some point the resulting additions to output will start to decline.
6. The law of diminishing marginal returns implies that the marginal cost curve of the firm will ultimately rise. As the firm adds more variable inputs to its fixed plant, diminishing marginal returns set in at some point, and marginal cost will start to rise as the firm approaches its short-run capacity.
7. Short-run costs are (a) fixed costs—costs that do not vary with the firm's output; (b) variable costs—the costs of purchasing variable inputs; (c) total costs—the sum of fixed and variable costs; and (d) marginal costs—the change in total cost with respect to a change in output. These costs can be expressed in total or average (unit) terms. The short-run average total cost curve is U shaped. When marginal cost is below average total cost, the latter falls. When it is above average total cost, the latter rises.
8. Long-run average total cost shows the lowest-cost plant for producing output when the firm can choose among all possible plant sizes. It is the planning curve of the firm; it helps the firm pick the right-sized plant for long-run production.
9. A U-shaped long-run average total cost curve results from economies and diseconomies of scale. Economies of scale cause long-run unit costs to fall and diseconomies of scale cause long-run unit costs to rise as output is expanded.
10. Economic or opportunity costs are the expected future costs of forgoing alternative uses of resources. Accounting or historical costs are costs that have been incurred in the past. Sunk costs are historical costs and not relevant to present decisions.

## Key Terms

costs of production
accounting costs
implicit costs
opportunity cost of capital
total cost of production
economic profit
accounting profit
private costs
external costs
social costs
variable input
fixed input
short run
long run
fixed costs
variable costs
total cost
average fixed cost
average variable cost
average total cost
marginal cost
fixed plant
law of diminishing marginal returns
total product
marginal product
average product
long-run average total cost
economies of scale
diseconomies of scale
constant returns to scale
increasing returns to scale
historical costs
sunk costs

## Questions for Review and Discussion

1. Explain the difference between explicit and implicit costs.
2. State and explain the law of diminishing marginal returns.
3. What is the definition of marginal cost? How is it linked to the law of diminishing marginal returns?
4. Define economies and diseconomies of scale and give an example of each.
5. How long is the long run? How does it compare to the short run?
6. When is the marginal cost curve below the average total cost curve? When is it above the average total cost curve?
7. An engineering student invests in obtaining an engineering degree. Let us say that his degree cost him $50,000 in opportunity costs. Five years later, he decides he wants to be an artist. Is the $50,000 investment in the engineering degree relevant to his decision to become an artist?
8. Derive the long-run cost model discussed in this chapter using the short-run model. Be sure to include short-run average total cost (ATC), marginal cost (MC), and long-run average total cost (LRATC). Where is the single most preferred operating position in the long-run model?

## ECONOMICS IN ACTION

### Costs to the Firm and Costs to Society: The Case of Pollution

Costs to a private firm can sometimes differ from full social costs when the firm's production activities create unwanted by-products—external costs. Suppose an iron-smelting firm belches foul-smelling smoke into the air and dumps toxic wastes into streams. The same process that produces goods and services demanded by society also imposes costs on society in health hazards and pollution. What, if anything, is to be done about the costs to society of reducing or eliminating the costly pollution?

It will help to visualize the difference between firm costs and social costs, which we present in Figure 8–8.

Economists have proposed a number of solutions to the externality problem. Cambridge economist A. C. Pigou, in his 1920 work *The Economics of Welfare*, argued that taxes could be levied upon the perpetrators of negative externalities.[a] In a theoretical sense at least, such taxes would have the effect of making the costs faced by the firm include all social costs. Implementation of Pigou's general proposal could take a number of forms: (1) The public could build a treatment plant. (2) Pollution standards could be (and have been) imposed. (3) A tax could be levied on units of waste discharged. All of these proposals to alleviate the effects of externalities have advantages and disadvantages, but all contemplate government intervention into private markets.

In 1960, Ronald Coase challenged Pigou's analysis of externalities.[b] He argued that externality relations can be thought of as bilateral. Stream polluters would not create an externality if there were no downstream population. Cigarette smokers would not impose costs on nonsmokers if the latter did not position themselves in the way of smoke. It takes two to make an externality. Coase also argued that society itself might "internalize" many externalities. Restaurants, for example, can offer smoking and nonsmoking sections. In these cases, the market would adjust to the externality without governmental interference. Coase argued further that if the judicial system assigns liability to the party in the externality relation who would incur the least cost in correcting the situation, market forces may be sufficient to generate efficient solutions to these problems. Perhaps, for example, a scheme that assigns the responsibility for workplace safety to either employers or employees rather than a government regulatory pro-

[a] A. C. Pigou, *The Economics of Welfare* (London: Macmillan, 1920).
[b] Ronald Coase, "The Problem of Social Cost," *Journal of Law and Economics* 3 (October 1960), pp. 1–44.

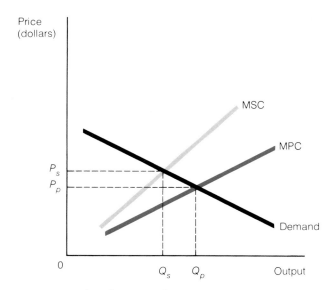

**FIGURE 8–8  Costs to the Firm, Costs to Society**

In the case of an iron-smelting firm, the marginal social costs (MSC) exceed the marginal private costs (MPC) of production. The marginal social cost curve includes all of the firm's private costs of production plus the costs to society of the firm's pollution. In the case of the polluting firm, MSC is greater than MPC, and "too much," $Q_p$, of the good is produced at "too low" a price, $P_p$. If all costs are accounted for, including external or social costs, a lower output, $Q_s$, would be produced at a higher price, $P_s$.

gram—such as that administered by the Occupational Safety and Health Administration—is the most efficient way to promote on-the-job safety. These are clearly important matters for debate and analysis, and we return to them in more detail in Chapter 18.

### Questions

1. Assume that shrimping firms who fish the Gulf of Mexico "overfish," depleting the stock of shrimp. Such overfishing causes the quantity of shrimp to be lower in the future. This means that the marginal social costs of shrimping are greater than the marginal private costs facing the shrimping firms. How might this problem be reduced or eliminated?
2. Suppose that your neighbor creates a beautiful flower garden. Your neighbor pays all of the costs of the garden. Does he or she receive all of the benefits?

# 9

# The Competitive Firm and Industry

Competition exists in virtually every aspect of life. Students compete for grades, animals compete for habitat, college football teams compete for national championships, government agencies compete for budget appropriations, firms compete for customers. In each case, scarcity causes the competition. Funds available to support government agencies are limited; defense and social agencies therefore compete for scarce budget dollars. Likewise, there can be only one national champion in college football. College teams therefore compete for this scarce distinction. If resources were not scarce but freely available, there would be no need for competition. Everyone could have all they wanted of whatever they wanted.

This chapter introduces the economist's model of how firms and industries behave under circumstances of pure competition. It is the first of three chapters devoted to analyzing the impact of industry structure on price and output. In these chapters we look at four models of industry behavior: pure competition, monopoly, monopolistic competition, and oligopoly. In this chapter we discuss the usefulness of the abstract model of pure competition and its relevance for real-world problems. We also use new analytical techniques relating the costs and output decisions of individual firms to the output of an industry as a whole.

## The Process of Competition

Most people naturally think of competition as the process of competition or the conduct of competitors. Seen in this way, the important question in determining whether firms are competing with one another is whether they are exhibiting rivalrous, competitive behavior. Do they compete hard for customers? Do they try to outperform one another? Do they use a variety of methods—persuasive advertising, a carefully chosen location, an attractive price—to win and keep customers? Do they seek the best managerial talent available? Are they forward-looking and innovative? Do they seek to eliminate waste and inefficiency in their company? The list easily could go on, but the point is already clear: Competition is normally thought of as a process of rivalry among firms.

This process of rivalry leads to better and cheaper products for consumers. Business firms compete to make profits, but the competitive process actually forces firms to meet consumer demands at the lowest possible level of profit. As Adam Smith observed more than two hundred years ago, firms' self-interest is harnessed by the competitive process to promote the general well-being of society:

> It is not from the benevolence of the butcher, the brewer, or the baker, that we expect our dinner, but from their regard for their own interest. We address ourselves, not to their humanity but to their self-love, and never talk to them of our own necessities but of their advantages.[1]

The process of competition channels the pursuit of individual self-interest to socially beneficial outcomes, and for this reason economists put the study of the competitive process at the heart of their science.

An important point about the process of competition is that it takes place in an uncertain world. Individuals have to make conjectures as they make resource commitments for the future. Some conjectures will turn out to be correct, and individuals who forecast correctly will survive and prosper. Individuals whose conjectures turn out not to be correct will not fare so well in the future. The concept of competition as a process of rivalry under uncertain conditions is especially important for understanding entrepreneurial behavior in the economy, a topic we return to in Chapter 15.

Although competition generally benefits consumers by harnessing the self-interest of producers, producers have incentives to subvert the competitive process to increase their profits. Adam Smith also recognized this tendency:

> People of the same trade seldom meet together, even for merriment and diversion, but the conversation ends in a conspiracy against the publick, or in some contrivance to raise prices.[2]

Independent action among competitors is by no means guaranteed. Indeed, firms may have incentives to avoid competition and independent action in favor of collusion and higher profits. This tendency leads to monopolies of one sort or another, which we discuss in more detail in Chapter 11. Here we simply note that society can influence the rules under which firms compete. Price fixing may or may not be allowed; property rights may or may not be defended; business taxes may be high or low. The process of competition is affected by government regulations, which may be more or less restrictive. Indeed, as we will see, this is a two-way street. Government affects business, and business affects government. Chapter 17 is devoted to an analysis of government regulation of business.

## A Purely Competitive Market

In contrast to the concept of the competitive process just described, pure competition is an abstract model of competitive behavior that emphasizes the importance of industry structure. In particular, it stresses the number of

---

[1] Adam Smith, *An Inquiry into the Nature and Causes of the Wealth of Nations*, ed. Edwin Cannan (1776; reprint, New York: Modern Library, 1937), p. 14.
[2] Smith, *Wealth of Nations*, p. 128.

independent producers in an industry. The difference between the competitive process model and the pure competition model can be compared with the behavior of runners in a race. To determine how close a race will be, the competitive process approach would ask how intensely the runners are competing. Are they striving hard to win? The pure competition approach would ask how many equally qualified runners are in the race. The process view is clearly a richer, more natural interpretation of competition. Nonetheless, the model of pure competition is a useful abstraction that, interpreted carefully, can help us understand the competitive behavior of producers in real-world industries.

**Market:** A coming together of buyers and sellers for the purpose of making transactions.

The model of pure competition is based on the concept of a **market**. As you recall from Chapter 4, market prices are subject to the laws of demand and supply, and within a market prices of a good or service will tend toward equality. Market forces even transcend geographical boundaries, given free flow of information. Buyers at point $x$ will not pay more for a commodity than the price at point $y$ plus transportation costs. Buyers at point $y$ will not pay more for a commodity than the price at point $x$ plus transportation costs. Price deviations will quickly be spotted by buyers, restoring the market's tendency to one price. Suppose that the price at point $y$ fell below the price at point $x$ plus transportation costs. What would happen? Buyers would shift their purchases from point $x$ to point $y$, decreasing demand and lowering price at point $x$ and increasing demand and raising price at point $y$. This shift of buyers would restore the tendency to equality of price.

The behavior of sellers in a market is equally predictable. In fact one way to determine if two commodities are in the same market is to check whether price reductions or increases for the commodity in one area are matched by price reductions or increases in other areas, accounting also for transportation costs. If the price of gasoline goes down in Atlanta, does it also go down in Birmingham? In Charlotte? In Washington, D.C.? In New York? In Chicago? In Seattle? In other words, do sellers in other areas of the country respond with competitive price reductions? To the extent that they do, we can delineate the market for gasoline from the tendency of all sellers to adjust prices in the same direction.

Building on this concept of a market, we can define four conditions that characterize a **purely competitive market**.

**Purely competitive market:** A coming together of a large number of buyers and sellers in a situation where entry is not restricted.

**Homogeneous product:** A good or service produced by many firms such that each firm's output is a perfect substitute for the other firms' output, with the result that buyers do not prefer one firm's product to another firm's.

1. Firms in a purely competitive market sell a **homogeneous product.** That is, they sell identical or nearly identical products that are perfectly substitutable for one another. This means that advertising does not exist in a purely competitive market. Why would one firm want to advertise the advantages of its product if it is perfectly substitutable for the product of competitors?
2. A large number of independent buyers and sellers exist in a purely competitive market. This assumption rules out the possibility of collusion among buyers or sellers to affect price and output in the industry. Moreover, the large number of buyers and sellers ensures that the purchases or sales of any one buyer or seller will not affect the market price. Each buyer and seller is small relative to the total market for the commodity and exerts no perceptible influence on the market.
3. There are no barriers to entry or exit in a purely competitive market. Features of economic life such as control of an essential raw material by one or a few firms or government regulation of firms' behavior in the market do not exist under pure competition.

**Perfect information:** A condition in which information about prices and products is free to market participants; combined with conditions for pure competition, perfect information leads to perfect competition.

4. A perfectly competitive market also offers **perfect information** to buyers and sellers. Everybody in the market has equal, free access to information about the location and price of the product.

These four conditions are the assumptions on which the theory of pure or perfect competition is based. This theory is designed to explain the behavior of many independent buyers and sellers, none of whom has any perceptible influence on the market. The assumptions may seem unrealistic. Products, for example, are rarely homogeneous, and industries with a large number of sellers are rare in a modern economy. But in economics, the realism of the assumptions is not the point of this or any other model. The point is the empirical relevance of the model. Does it explain behavior in real-world markets? As we will see, the purely competitive model helps us analyze important actual markets and industries such as the stock market and agriculture. (See Focus, "The Stock Market," for one application of the concept of pure competition.) Moreover, the purely competitive model gives us an analytical framework for what might be loosely described as the "ideal" working of an economy. In this sense the model provides a benchmark against which other industry models, such as pure monopoly, can be measured.

## The Purely Competitive Firm and Industry in the Short Run

How then do firms behave within a purely competitive industry? We focus first on the purely competitive firm's decisions about how much output to supply in the short run; that is, over a period of time when it cannot adjust its plant size.

### The Purely Competitive Firm as a Price Taker

In a purely competitive market an individual firm cannot influence the market price for its good or service by increasing or decreasing its output. Because each seller is only a small part of the total market, its actions have no perceptible influence on the market. The competitive firm is called a **price taker**: It must accept the going market price for its product.

**Price taker:** An individual buyer or seller who faces a single market price and is able to buy or sell as much as desired at that price.

As an example of price-taking behavior, consider the cotton market in 1978–1979. World cotton output was estimated to be 65,337,000 bales in 1978. Of this total, assume that the Commonwealth of Virginia produced a minuscule amount—100 bales in 1978 and 200 in 1979. Further assume that, over a given price range, the price elasticity of demand for cotton worldwide ($e_m$) is 0.25, which means that the market demand curve for cotton is inelastic.

First consider what happened to the world price of cotton as a result of the 100-bale increase in Virginia's cotton production between 1978 and 1979. Other things being equal, the percent change, in this case a decline, in world price is given by

$$\frac{\text{Percent change in world output}}{e_m} = \frac{100 \text{ bales}/65{,}337{,}000 \text{ bales}}{0.25}$$

$$= 0.0000061 = 0.00061\%.$$

(Remember that $e_m$ equals percent change in world output divided by percent change in world price, so percent change in world output divided by $e_m$ equals percent change in world price.)

## FOCUS: The Stock Market: An Example of Perfect Competition

The stock market—the market in which shares of ownership of firms are traded—might at first seem like an implausible example of a competitive market. The popular press typically insinuates that the stock market is basically little more than a gambling casino, with ticker-tape machines replacing one-armed bandits. Alternatively, the stock market is portrayed as a kind of free-for-all of combat among an elite few investment tycoons.

But closer observation dispels these popular misconceptions. The stock market is a highly competitive market. Many thousands of firms engaged in hundreds of distinct lines of business offer their shares to many thousands of potential buyers every day. Entry into the market on both sides of potential transactions is practically unrestricted. The shares firms offer for sale are highly substitutable as potential investments to buyers, making them essentially homogeneous goods.

The highly competitive nature of the stock market is obscured by the fact that only a few hundred of the largest corporations are traded in the organized stock exchanges (in the United States, these exchanges are collectively referred to as "Wall Street"). Since the shares of large corporations are traded frequently, it is efficient for trade to occur among exchanges organized for this purpose. The organized stock exchanges radically reduce transaction costs, benefiting both buyers and sellers. But it is misleading to confuse the overall stock market with the organized stock exchanges. Many thousands of other, smaller firms offer stock shares for sale in the so-called over-the-counter market, outside of the larger organized exchanges. All these firms—those participating in the organized exchanges and those trading outside—are competing for the funds of potential investors. Entry into the market is open even to the tiniest firm that wants to offer its shares to the public.

On the buyers' side, although large institutional investors such as well-known mutual funds like the Dreyfus Fund receive much publicity, there are many small investors because the costs of entry into the market—the transaction costs—are very low, both in the organized exchanges and in the market generally. In 1982, for example, General Motors had 1,122,000 common stockholders, General Electric had 502,000 shareholders, Eastman Kodak had 221,000, and Gulf Oil 302,000. In each case, the overwhelming majority of these shareholders were individuals with relatively small investments.

The price of the firm's shares reflects, at any given moment, the sum of all economically relevant information that the market possesses with respect to that firm's past, present, and expected future performance. That price is the result of bidding in the market, which in turn reflects the judgment and relevant knowledge of all interested parties. Hence, stock prices are extremely sensitive to information about any change that affects the firm and its future prospects. An accidental oil field explosion in Sumatra will probably be reflected in the price of oil company stock in seconds, as will be the influence of a flash flood in Alaska on the price of shares in a civil engineering firm. The stock market's first and foremost function is as a communications mechanism: It maintains a flow of such information among buyers, sellers, and firms.

In short, the stock market, rather than representing a kind of floating financial dice game, is probably one of the most efficient and competitive markets imaginable.

---

Now we can derive the elasticity of demand facing Virginia producers of cotton ($e_v$). Since we know the 1978–1979 percent change in Virginia output (100 to 200 bales, or 100%) and since we have just computed the percent change in world price caused by the increase in Virginia production, we have all the information we need to calculate the price elasticity of demand facing Virginia producers:

$$e_v = \frac{\text{Percent change in Virginia output}}{\text{Percent change in world price}}$$

$$= \frac{100\%}{0.00061\%}$$

$$= 163{,}934.$$

The Virginia producers face an extremely elastic demand curve for their output of cotton. If we took only one Virginia producer of cotton as a representative purely competitive firm, the elasticity result would be dramatically larger.

### The Demand Curve of the Competitive Firm

From this example it is easy to see why the competitive firm is called a price taker. Expansions or contractions in an individual firm's output within a competitive market have virtually no effect on the market price. To capture this condition analytically, we draw the firm's demand curve as a straight line at the level of market price in Figure 9–1.

Figure 9–1a shows the demand curve (d) facing Virginia cotton producers, and Figure 9–1b shows the world market demand (D) and supply (S) curves for cotton. The demand curve of the Virginia producers is perfectly elastic with respect to the price of cotton; it is drawn as a flat line at the level of market price (P). A perfectly elastic demand curve means that the firm can sell all it wants to sell at the prevailing market price. If it tried to sell its output at a slightly higher price, demand for its product would vanish because buyers can purchase cotton at the lower market price in whatever quantities they choose. If the firm is trying to maximize its profits, it has no reason to sell its product for less than the market price.

The scale for price is the same in both parts of Figure 9–1. They obviously are not the same for quantity because of Virginia's minuscule proportion of world output. Keep this in mind as you interpret the diagrams in this chap-

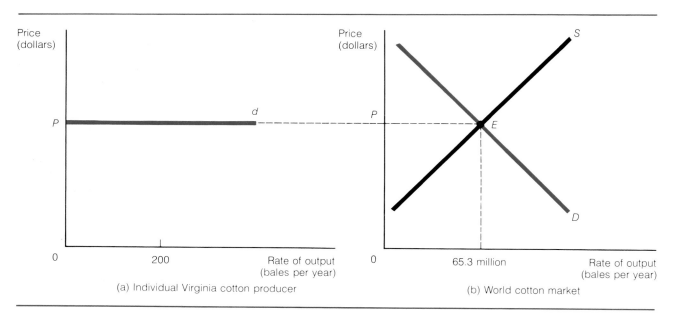

**FIGURE 9–1   Firm and Industry Demand in a Purely Competitive Market**

In a purely competitive industry, demand for an individual firm's product is perfectly elastic. The perfectly elastic demand curve facing one individual Virginia cotton producer is shown in (a) at the level of P, which is also the equilibrium price for the world market in (b) at an output of 65.3 million bales per year. Note that the scale for price is the same for both the individual producer and the world market but that the scale for quantity is much different for the two.

ter. Always check to see if the horizontal axis represents *firm* or *industry* quantity.

To summarize: Every firm in a purely competitive industry is a price taker and faces a demand curve such as that of the Virginia cotton producers in Figure 9–1. And every purely competitive firm produces a tiny proportion of industry output.

### Short-Run Profit Maximization by the Purely Competitive Firm

The purely competitive firm is a price taker and faces a given price represented by a perfectly elastic demand curve. Under these circumstances how does the firm decide how much of its product to produce? The simple answer is that the firm compares the costs and benefits of producing additional units of output. As long as the added revenues from producing another unit of output exceed the added costs, the firm will expand its output. By following this rule, the firm is led to maximize its **profits** in the short run.

**Profits:** The amount by which total revenue exceeds total cost.

The added cost of producing an additional unit of output is the marginal cost. We know from our analysis in Chapter 8 that the short-run marginal cost curve of a firm will eventually rise. This is due to the law of diminishing marginal returns, which comes into play as the firm uses its fixed plant more intensively in the short run.

The extra revenue from producing an additional unit of output is **marginal revenue** (MR). Simply, marginal revenue is the addition to total revenue from the production of one more unit of output:

**Marginal revenue:** The change in total revenue that results from selling one additional unit of output; the change in total revenue divided by the change in amount sold.

$$\text{MR} = \frac{\text{Change in total revenue}}{\text{Change in output}}.$$

In this expression, the denominator is simply 1, a one-unit increase in output, and the numerator is always the market price (P) because the demand curve facing the competitive firm is perfectly elastic. Therefore, marginal revenue equals market price, MR = P, for the purely competitive firm. In fall 1984, the market price for used newspaper was $10 per ton, so paper recyclers could assume that their marginal revenue for each additional ton of paper they collected and sold would be $10.

The purely competitive firm will decide how much to produce in the short run by comparing marginal cost with marginal revenue. To maximize its profits, the firm will produce additional units of output until marginal cost and marginal revenue are equal. Figure 9–2 illustrates this process.

The firm's perfectly elastic demand curve (d) in Figure 9–2a is drawn at the level of market price (P) from part b. As we have just seen, d is also the marginal revenue curve for the firm. Revenue will increase by the market price of one unit each time output increases by one unit. In other words, MR = P. We have also drawn the marginal cost (MC) and average total cost (ATC) curves of the firm.

The owners of the firm want to make as much money as possible, so their problem is to find the level of production or output that yields the largest profit. Suppose that the firm is producing hot pepper sauce. It presently produces a certain quantity $q_1$. Should the owners expand their level of production? At this point, the answer is yes because additional units of pepper sauce add more to revenue than to costs. In other words, marginal revenue exceeds marginal cost at $q_1$, as you can see by comparing points *mr* and *mc* at quan-

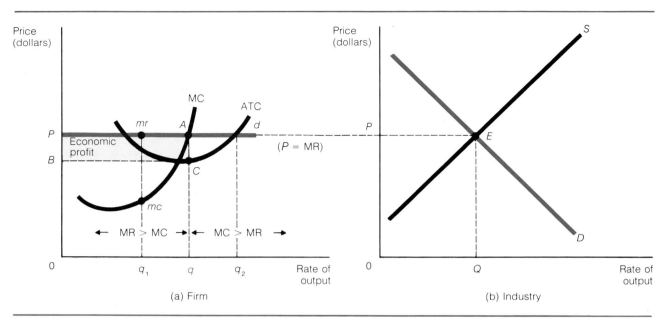

**FIGURE 9–2  Short-Run Output Choice and Profit Maximization**
(a) The short-run choice of output confronting the individual firm. Before the rate of output $q$, additional production adds more to revenue than to cost. Beyond $q$, additional production adds more to cost than to revenue. Therefore, the purely competitive firm will produce at $q$. At this rate of output the firm earns an economic profit equal to the difference between total revenues and total costs: $PAq0 - BCq0 = PACB$. (b) Prevailing market conditions in the industry.

tity $q_1$. Since the owners want to maximize their profits, they will therefore expand output beyond $q_1$ as far as $q$.

Suppose instead that the owners are operating at an output such as $q_2$. What will their profit-maximizing reaction be in this case? At this level of production, an additional unit of output adds more to cost than it adds to revenue. That is, marginal cost exceeds marginal revenue. The rational profit-maximizing response of the owners will be to lower production. Through a process of trial and error, they will find that their best level of output is $q$. At a market price of $P$, they can do no better for profits than to produce $q$ jars of hot pepper sauce. The short-run equilibrium level of production for the purely competitive firm is therefore defined by the condition $P = MC$. Since we know that price and marginal revenue are the same for the purely competitive firm, we can also write $P = MR = MC$.

Using the average total cost curve in Figure 9–2, we can see exactly how well the hot pepper sauce firm fares by producing at level $q$. The total revenue of the firm equals its sales, which are the level of output $q$ times the price $P$ for the output. **Total revenue** at $q$ therefore equals $Pq$, represented by the area $PAq0$ in Figure 9–2. Total cost is the level of average total cost (C) times the level of output $q$. Total cost at output $q$ thus equals $Cq$, represented by $BCq0$. Total revenue exceeds total cost in this case: $PAq0 - BCq0 = PACB$. This firm's short-run economic profit is shown by $PACB$, the return in excess of the total cost of production.

In the real world, businesspeople do not spend a lot of time trying to draw marginal cost and marginal revenue curves for their firms. Moreover, they

**Total revenue:** The total amount of money received by a firm from selling its output in a given time period; price times quantity sold.

operate in an uncertain environment where future costs and prices cannot be known with certainty. Despite these considerations, however, the P = MR = MC rule for profit maximization may have predictive power. It is a rule based on common sense. If additional production promises to add more to revenue than to cost, most businesses will try to increase their production. If additional units will probably add more to cost than to revenue, most businesses will cut back production. Such behavior leads to the P = MR = MC result even when businesspeople know nothing about the economic rule involved.

## A Numerical Illustration of Profit Maximization

Another way to understand how the choice of output levels allows a competitive firm to maximize profits is to examine the specific costs and revenues at each level of production. Table 9–1 presents a numerical schedule for the hot pepper sauce firm. The firm's output level is given in column (1). Column (2) shows that the firm confronts an unvarying market price of $2.40 per jar for its output. Price equals marginal revenue (P = MR) since this is a purely competitive firm. The marginal cost and total cost data in columns (3) and (5) are taken from Table 8–3 (p. 168), where these cost concepts were introduced and discussed. Total revenue in column (4) is the sales of the firm, or simply price times output, column (1) × column (2). Profit in

### TABLE 9–1  Profit Maximization for a Purely Competitive Firm

This numerical schedule of the costs and revenues the hot pepper sauce firm faces at each level of output provides data for determining the maximum profit level in two ways. One way is to find the greatest positive difference between total revenue and total cost, reflected as profit in column (6). The other way is to compare the marginal revenue and marginal cost columns, (2) and (3). The highest profit occurs at the output level where the two columns coincide—between 14 and 15 jars.

| (1) Rate of Output (jars per day) | (2) Price (= Marginal Revenue) | (3) Marginal Cost | (4) Total Revenue (1) × (2) | (5) Total Cost | (6) Profit (4) − (5) |
|---|---|---|---|---|---|
| 1 | $2.40 | $1.60 | $ 2.40 | $11.60 | − $9.20 |
| 2 | 2.40 | 1.40 | 4.80 | 13.00 | − 8.20 |
| 3 | 2.40 | 1.35 | 7.20 | 14.35 | − 7.15 |
| 4 | 2.40 | 1.30 | 9.60 | 15.70 | − 6.10 |
| 5 | 2.40 | 1.20 | 12.00 | 16.90 | − 4.90 |
| 6 | 2.40 | 1.10 | 14.40 | 18.00 | − 3.60 |
| 7 | 2.40 | 1.00 | 16.80 | 19.00 | − 2.20 |
| 8 | 2.40 | 0.90 | 19.20 | 19.90 | − 0.70 |
| 9 | 2.40 | 0.30 | 21.60 | 20.20 | 1.40 |
| 10 | 2.40 | 0.65 | 24.00 | 21.85 | 2.15 |
| 11 | 2.40 | 1.30 | 26.40 | 23.15 | 3.25 |
| 12 | 2.40 | 1.50 | 28.80 | 24.65 | 4.15 |
| 13 | 2.40 | 1.75 | 31.20 | 26.40 | 4.80 |
| 14 | 2.40 | 2.05 | 33.60 | 28.45 | 5.15 |
| 15 | 2.40 | 2.45 | 36.00 | 30.90 | 5.10 |
| 16 | 2.40 | 2.90 | 38.40 | 33.80 | 4.60 |
| 17 | 2.40 | 3.40 | 40.80 | 37.20 | 3.60 |
| 18 | 2.40 | 3.95 | 43.20 | 41.15 | 2.05 |
| 19 | 2.40 | 4.55 | 45.60 | 45.70 | − 0.10 |
| 20 | 2.40 | 5.25 | 45.00 | 50.95 | − 2.95 |

column (6) is the difference between total revenue and total cost, or (4) − (5).

There are two ways to find the profit-maximizing rate of output in Table 9–1. First, we can examine the difference between total revenue and total cost, given in column (6). Notice that at low and high rates of output the firm's economic profits are negative, ranging from a loss of $9.20 per day because of the high cost of producing only 1 jar of sauce to a loss of $2.95 for trying to produce 20 jars a day with an overcrowded fixed plant. The maximum profits the firm can earn in the short run occur at a rate of output of 14 jars of sauce per day and are equal to $5.15 per day, the largest dollar amount in column (6).

Figure 9–3a illustrates the total revenue–total cost approach to maximizing profits. The figure graphs columns (4) and (5) with respect to the output of the firm. The maximum profits possible in the short run occur where the total revenue line (TR) exceeds the total cost curve (TC) by the largest vertical difference. As Table 9–1 indicates, this takes place between 14 and 15 jars of sauce.

The second way to find the profit-maximizing rate of output using the numerical schedule in Table 9–1 is to compare marginal cost and marginal revenue, columns (2) and (3). As long as marginal revenue is greater than marginal cost, it pays to produce additional units of output. Clearly, marginal revenue ($2.40) is greater than marginal cost up to and including an output of 14 jars of sauce. This means that the point of maximum profit occurs between 14 and 15 jars per day. If the firm can turn out jars only one

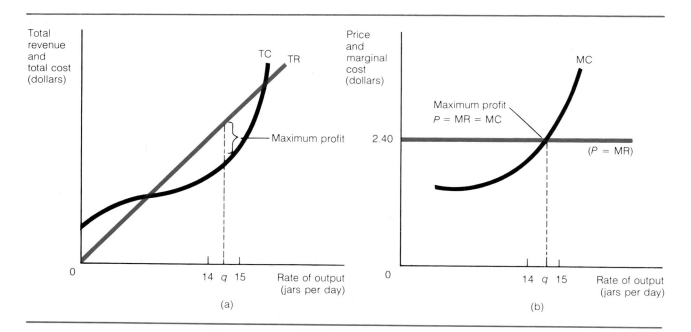

**FIGURE 9–3  Profit-Maximizing Output for a Purely Competitive Firm**
(a) Total revenue–total cost approach to profit maximization. The profit-maximizing rate of output occurs where the vertical distance between total revenue and total cost is greatest. (b) Marginal revenue–marginal cost approach. Profit maximization occurs where price, or marginal revenue (which are identical for the purely competitive firm), equals marginal cost ($P = MR = MC$). Both approaches result in the same profit-maximizing rate of output.

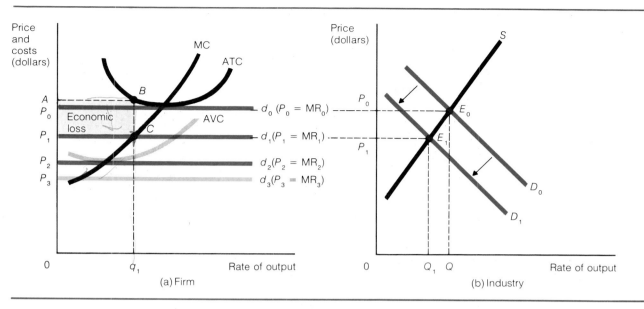

**FIGURE 9-4  Short-Run Loss Minimization**

(a) Firm's behavior in the face of economic losses caused by (b) decreased industry demand from $D_0$ to $D_1$ and a price drop from $P_0$ to $P_1$. At $P_1$, the firm will minimize losses by continuing to operate in the short run because it covers all of its variable costs and some of its fixed costs by operating. At $P_2$, the firm may either operate, covering its variable costs, or shut down, eliminating its variable costs. At $P_3$, the firm will shut down in the short run because the price is too low to cover even its variable costs. It minimizes losses by shutting down and paying only its fixed costs, which must be paid whether the firm operates or not.

at a time, the firm will cease producing additional sauce after 14 jars, since the marginal cost of producing the next jar is $2.90 and the marginal revenue is only $2.40. Units of output beyond 14 add more to costs than to revenue; that is, they cause profits to decline. Notice, then, that both the total revenue–total cost and the marginal revenue–marginal cost approaches yield the same answer for the profit-maximizing rate of output.

Figure 9–3b illustrates the marginal revenue–marginal cost approach. Again, note the equivalence of the two approaches by comparing the optimal rate of output, between 14 and 15 jars per day in both graphs.

### Economic Losses and Shutdowns

What will the firm do if the short-run situation changes for the worse? What if, for example, demand for hot pepper sauce declines because the government publishes an adverse report on the health consequences of eating too much hot pepper sauce? In this case, the market demand curve for pepper sauce will decrease, as shown in Figure 9–4b, shifting to the left from $D_0$ to $D_1$. This results in a price reduction in the hot pepper sauce market, causing the market price confronting the firm to fall from $P_0$ to $P_1$. The firm's situation is drawn in Figure 9–4a. Note that the average variable cost curve (AVC) has been added becaue it is now an important consideration.

We see in Figure 9–4a that the firm incurs an economic loss because the new market price, $P_1$, is below its average total cost curve. The firm's revenues are not sufficient to cover its total costs, and it therefore loses money on its operations. What will the firm do in the face of an economic loss? The firm has two options in the short run.

If it expects the adverse effect on sales to be short-lived, the firm can continue to operate in the face of short-run losses. While the price remains depressed at $P_1$, the firm can minimize losses by following the same rule it followed to maximize profits: It determines the output level at which price is equal to marginal cost and lowers production to that level, $q_1$ in the figure. Total cost at an output of $q_1$ is $ABq_10$, and total revenue is $P_1Cq_10$. The firm therefore incurs an economic loss equal to the difference between the two: $ABCP_1$. Why might the firm continue to operate in this case while making a loss? It covers all of its variable costs and some of its fixed costs at a price of $P_1$. If the firm stopped operating, it would still have to pay its fixed costs. These payments must be made whether or not the firm operates. By continuing to operate at $P_1$, the firm earns *something* toward the payment of its fixed costs.

Suppose, however, that the market price for hot pepper sauce continues to fall in the industry, say to $P_3$. At $P_3$, the firm does not take in enough revenue to cover even its variable costs of production. The loss-minimizing policy for the firm at this point is to cease operations. By shutting down, the firm is ensured of having to pay only its fixed costs. If it tried to operate at a price below its average variable cost, such as $P_3$, it would not only have to pay its fixed costs but also would incur a deficit in its average variable cost account. The loss-minimizing policy for the firm is to **shut down** when the price falls below its AVC curve.

Suppose that the market price falls to point $P_2$ where it is just equal to the minimum AVC? What can the firm do in this case? Either operating or shutting down is a reasonable option. If it operates, it can cover its variable costs but none of its fixed costs. If it shuts down, it still must pay its fixed costs. Other things being equal, the firm should be indifferent about whether it operates in the short run under these conditions.

Keep in mind that we are discussing *short-run* policies for the firm. The short run is a period of time in which fixed inputs cannot be varied. Given that fixed costs must be paid, the firm's objective in the short run, when price falls below ATC and losses begin, is to minimize its losses. The firm's long-run adjustment will depend on what it expects to happen to prices and costs in the industry. If it expects prices to rise, the firm may shut down temporarily, keep its plant intact, and plan to reopen at some time in the future. Indeed, the firm may use the shutdown period to reorganize in an effort to have lower costs when production begins again. If the firm expects price to remain so depressed in the industry that losses are likely to continue into the future, it may act to sell off its plant and equipment and go out of business.

### Supply Curve of a Purely Competitive Firm

The previous considerations help us understand a competitive firm's willingness to supply products as the market price changes, even when it declines. We saw in Chapter 4 that a supply curve shows the relation between the quantity supplied of a good and its price, other things being equal. What is the supply curve of a purely competitive firm in the short run? We answer the question in Figure 9–5.

We have drawn the firm's marginal cost curve and average variable cost curve and a series of different price levels that the firm faces from a compet-

---

**Shutdown:** A loss-minimizing option of a firm in which it halts production in the short run to eliminate its variable costs, although it must still pay its fixed costs.

**FIGURE 9–5**

**Short-Run Supply Curve of the Purely Competitive Firm**

The short-run supply curve of the purely competitive firm is its marginal cost curve (MC) above the point of minimum average variable cost. The firm will not operate below $P_1$ because it cannot cover its variable costs of production. Above $P_1$, the firm produces where $P = MR = MC$.

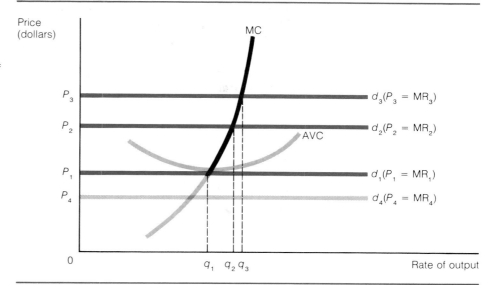

itive market. We know that the general profit-maximizing rule for the competitive firm is to operate at the level of output at which $P = MR = MC$. We also know that the loss-minimizing rule for the firm is to shut down when price falls below the minimum average variable cost. With these two facts, we can derive a supply curve for the competitive firm.

We know that the firm will not produce at a price such as $P_4$ in Figure 9–5. Since $P_4$ is below the minimum AVC, the firm will shut down and produce no output at that price. (Notice that no output corresponding to $P_4$ is given along the horizontal axis.) This decision of the firm applies at any price below $P_1$, which is the point of minimum AVC. At $P_1$, the firm will just cover its variable costs and will be economically indifferent about whether or not it operates. We have drawn the firm as producing $q_1$ units of output at the price $P_1$. At prices above $P_1$, the firm will operate in the short run and produce at the output level at which $P = MR = MC$. We have shown two such cases, at $P_2$ and $P_3$. In other words, the firm responds to changes in price above $P_1$ by producing output along its marginal cost curve. The marginal cost curve of the purely competitive firm above the point of minimum average variable cost is the **short-run firm supply** curve; it shows the relation between price and quantity supplied in a period of time when the firm's plant size is fixed.

**Short-run firm supply:** The portion of the marginal cost curve above the minimum average variable cost.

### From Firm to Industry Supply

It is a simple step from the competitive firm's short-run supply curve to the short-run industry supply curve. The **short-run industry supply** curve is the horizontal sum of the individual firms' marginal cost schedules above their points of minimum average variable cost. This process, which is analogous to the way in which we derived market demand curves from individual demand curves in Chapter 4, is illustrated in Figure 9–6.

**Short-run industry supply:** A summation of all the existing firms' short-run supply curves.

For simplicity, assume that the industry consists of two firms producing hot pepper sauce. At a price of $P_0$, firm A produces 3 jars of hot pepper sauce per day and firm B produces 2 jars; at these outputs, $P = MR = MC$ for

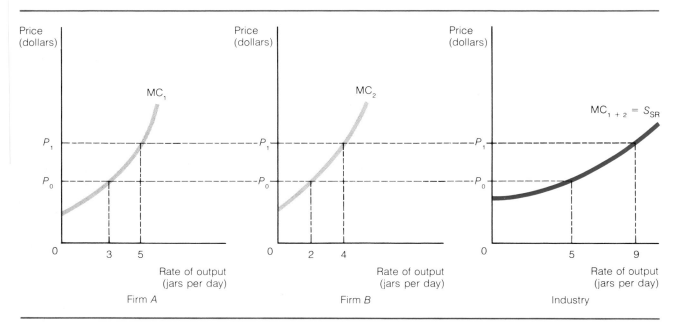

**FIGURE 9–6  From Firm to Industry Supply Curves**

The short-run supply curve of a competitive industry is the horizontal sum of individual firms' marginal cost curves above their respective points of minimum average variable cost. At price $P_0$, firm A supplies 3 units of output, and firm B supplies 2 units. Market supply is 5 units at $P_0$. Other points on the industry supply curve, such as $5 + 4 = 9$ units at $P_1$, are derived in the same way.

each firm. In each case $P_0$ is above the firm's minimum AVC. This point on the industry supply curve is 5 jars per day at a price of $P_0$. That is, 5 jars is the horizontal sum of 3 and 2 jars produced by the two firms. If price rises to $P_1$, the firms expand production, to 5 and 4 jars, respectively, along their MC curves. The corresponding point for the industry is at 9 jars of pepper sauce. Other points on the short-run industry supply curve ($S_{SR}$) are obtained in the same manner. The short-run supply curve of a competitive industry is the horizontal sum of all the individual firms' marginal cost curves above their respective points of minimum average variable cost.

Three points should be kept in mind about this discussion. First, a purely competitive industry encompasses many independent producers; the two-firm model used to illustrate the derivation of the industry supply schedule is therefore an abstraction. In a real case of pure competition, thousands of marginal cost curves would have to be summed horizontally. Second, our analysis is conducted in the economic time frame of the short run, in which firms adjust to price changes within the limits imposed by their fixed plant sizes. In the next section we will see what happens to a competitive industry in the long run. Third, the intersection of the industry demand curve and the short-run market supply curve determines the market price for the industry. Both firm and industry are in short-run equilibrium. The firm produces at a level where $P = MR = MC$ for its fixed plant, and industry demand equals industry supply. These two conditions define a short-run equilibrium for the purely competitive model.

# The Purely Competitive Firm and Industry in the Long Run

The long run is a period of economic time during which firms can select the lowest-cost plant to produce their output and firms can enter and exit the industry. The result of these adjustments by firms in the face of economic profits or losses is to move the industry toward equilibrium.

## Equilibrium in the Long Run

**Long-run competitive equilibrium** occurs when two conditions are met: (1) quantity demanded equals quantity supplied in the market and (2) firms in the industry are making a normal rate of return on their investments, a situation economists call **zero economic profits.**

Economic profits are returns above and beyond the total (explicit plus implicit) costs to the owner of or investor in a firm. They are returns above the opportunity cost of the owner's capital investment in the firm; that is, they are above the normal return that an owner could expect to make on a capital investment of some other form such as a money market certificate. Economic profits therefore attract the notice of other investors. They are a signal that prods others to try to capture above-normal returns by entering the industry. Economic profits also lead firms already in the industry to seek to expand their scale of operations. Both cases will result in an increase in output in the industry. This causes the short-run industry supply curve to increase and the market price to fall, erasing economic profits and returning the rate of return in the industry to a normal level. This level is referred to as zero economic profits; investors do make a profit, but no more than what their money would have earned through the prevailing rate of return on any other investment.

If firms in the industry are making economic losses, the opposite situation holds. Firms will cut back operations, and some firms will leave the industry. The short-run industry supply curve decreases, causing the market price to rise and restoring a normal rate of return to surviving investors.

A long-run equilibrium state for a purely competitive industry is depicted in Figure 9–7. The two conditions for this equilibrium are illustrated in the diagram: (1) demand equals supply at the industry level of output, and (2) price just equals the minimum average total cost of the firm. In other words, each firm earns a normal rate of return on its investment in the industry.

## The Adjustment Process Establishing Equilibrium

Establishment of equilibrium is a continual process in the purely competitive industry. To see how the position of long-run equilibrium is reached, consider a condition of long-run equilibrium in the hot pepper sauce industry, which is depicted in Figure 9–7.

**Rightward Shift in Demand.** Now imagine that hot pepper sauce not only is given a clean bill of health by government researchers but also is praised as a cure for the common cold by a prominent scientist. The industry demand curve for hot pepper sauce suddenly shifts to the right, as shown in Figure 9–8b, raising the market price to $P_1$ from the previous equilibrium price $P_0$.

At first, individual firms in the industry (Figure 9–8a) adjust to the higher price $P_1$ by expanding output from $q_0$ to $q_1$ within their fixed plant sizes, that is, along their MC curves. This increase in firms' output is reflected at the

---

**Long-run competitive equilibrium:** A situation in an industry in which economic profits are zero and each of the many firms is operating at minimum average total cost.

**Zero economic profits:** The condition that faces the purely competitive firm in the long run; long-run equilibrium in a competitive industry leads to a condition where $P = MR = MC = LRATC$, which means that firms in the industry earn just a normal rate of return on their investment, or a zero economic profit.

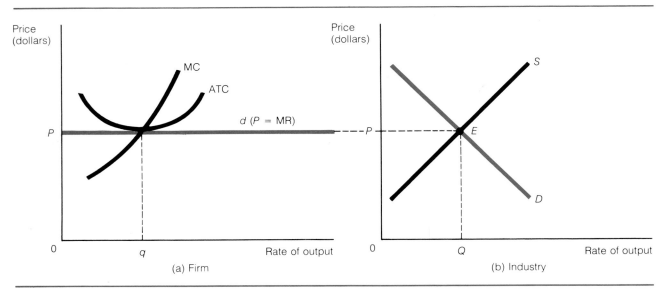

**FIGURE 9-7  The Purely Competitive Industry in Long-Run Equilibrium**

In a long-run equilibrium state, two conditions are met: (a) the competitive firm makes a zero economic profit, or normal rate of return; (b) industry demand equals industry supply. Price, and therefore revenue, is exactly equal to the minimum total cost, including a normal rate of return for investors.

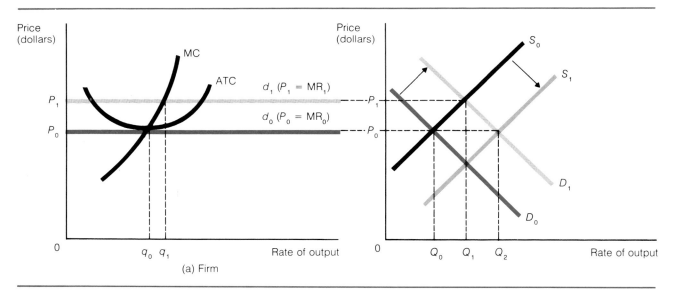

**FIGURE 9-8  Industry Adjustment to Long-Run Equilibrium: An Increase in Market Demand**

Demand increases from $D_0$ to $D_1$ in the industry, and price rises from $P_0$ to $P_1$ (b). As demand and price rise, firms now earn profits in excess of their average total costs (a). The opportunity for higher-than-normal profits induces expansion of output by firms in the industry and entry by new firms. As this expansion of output takes place, the industry supply curve shifts to the right until a normal rate of return and the original price $P_0$ are restored in the industry.

industry level by the movement along $S_0$ as price rises from $P_0$ to $P_1$. Industry output thus initially rises from $Q_0$ to $Q_1$. However, $P_1$ is above the firm's ATC, so firms in the industry are earning economic profits. Other firms will therefore be attracted to the industry, and existing firms will have an incentive to expand their scale of operations to capture more profits. The effect of both adjustments will be to shift the market supply curve to the right, from $S_0$ to $S_1$. Eventually long-run equilibrium is restored at the original price $P_0$, where firms again earn a normal rate of return. In the process, industry output has further expanded, from $Q_1$ to $Q_2$. The economy now produces and consumes more hot pepper sauce.

Two points should be noted about this process. First, the long-run adjustment process in this example returned price to the original level of minimum average total cost, that is, to the previously prevailing price $P_0$. This is a special case of long-run adjustment in an industry in which the prices of resources do not change when industry output is expanded. We discuss this and other long-run adjustment possibilities in more detail later in the chapter.

Second, entry can take place from both without and within the industry. Entry from without is the entry of new firms. Entry from within is the expansion of old firms. Entry from within can take place only if there is a range of long-run firm sizes consistent with minimum-cost production. If the long-run average total cost curve of firms in the industry were U shaped, only one firm size would offer lowest cost in the long run. When this is the case, entry takes place entirely by new firms. The existing firms will not expand their scale of operations because they are already operating with the lowest-cost plant size. An existing firm can expand its scale of operations only if its LRATC curve exhibits a range of constant returns to scale (see Figure 8–6a, p. 172, and its accompanying discussion).

**Leftward Shift in Demand.** Whereas economic profits lead to the expansion of output in an industry, economic losses lead to a contraction of industry output. Figure 9–9b illustrates a reduction in industry demand from $D_0$ to $D_1$. At the resulting lower price of $P_1$, firms incur economic losses because $P_1$ is less than firms' average total cost (Figure 9–9a). In the short run, firms' output is reduced, depicted by movement down the MC curve, or firms shut down if $P_1$ is less than their minimum average variable cost. The short-run drop in production by individual firms causes quantity supplied in the industry to fall to $Q_1$. In the long run some firms will leave the industry by going out of business and others will reduce their scale of operations. The drop in production (to $Q_2$) causes the industry supply curve to shift to the left, from $S_0$ to $S_1$, putting pressure on the industry price to rise. Ultimately, price will rise enough to restore the original price $P_0$ and a normal rate of return to firms in the industry.

What are some of the uses of these analyses? First, they depict the desirable effects of competition on the use and allocation of resources. Signals to producers about what consumers want are sent through the market system, and producers respond by adjusting and supplying what consumers want in a way that minimizes the costs of production. Second, the model generates testable propositions, such as the assertion that excess returns should promote entry in a competitive market, from both without and within. An indirect way to test for the presence of a competitive market is to see if excess returns or the presence of economic profits leads to an expansion of output in the industry, or if a decline in output follows economic losses.

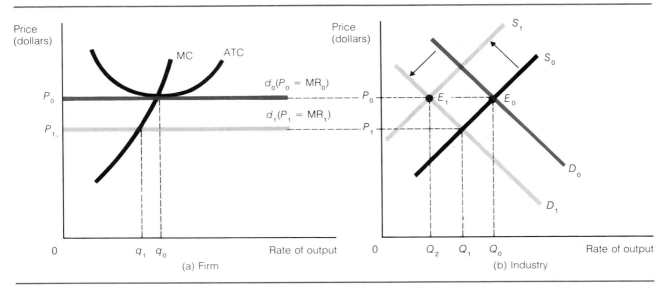

**FIGURE 9–9  Industry Adjustment to Long-Run Equilibrium: A Decrease in Market Demand**

Industry demand declines from $D_0$ to $D_1$, and market price falls from $P_0$ to $P_1$ (part b). At $P_1$ firms incur economic losses and cut back production to $q_1$. Industry output falls to $Q_1$. Losses cause some firms to leave the industry and others to reduce their scale of operations. The overall decline in production shifts the industry supply curve to the left, from $S_0$ to $S_1$, thus restoring the original price $P_0$ and a normal rate of return to firms that remain in the industry.

Such an application of this model could be useful to antitrust analysts, for example, whose job is to detect the presence of monopolistic behavior in the economy. Focus, "The Effect of an Excise Tax," describes an application of the competitive model.

### The Long-Run Industry Supply Curve

**Long-run industry supply:** The quantities of a product that all firms are willing and able to offer at all the various prices when the number of firms and scales of operation of each firm are allowed to adjust to the equilibrium level.

The **long-run industry supply** curve represents the minimum price at which firms will supply output to the market over the long run—that is, during a period long enough for entry into and exit out of the industry to occur and for firms to adjust their plant sizes to the lowest-cost level. The long-run supply curve reflects what happens to the prices of inputs as the output of an industry is increased or decreased. In our example of long-run equilibrium, depicted in Figures 9–8 and 9–9, price returned to its original level after first rising or falling with rise or fall in demand. As we noted, this is a special case that does not always apply because the costs of inputs may change as industry output changes. There are three possibilities related to input costs: they may remain constant, increase, or decrease, depending on the nature of the industry.

**Constant Cost.** In Figures 9–8 and 9–9, long-run adjustments led to the restoration of the original price in the industry ($P_0$). In the first case the size of the industry expanded, and in the second case the size of the industry contracted. In both cases, price returned to the level of $P_0$ as entry and exit and adjustments by firms in the industry took place.

# FOCUS: The Effect of an Excise Tax on a Competitive Firm and Industry

Although the model of a purely competitive market is based on abstract assumptions, it is sometimes useful for making predictions about economic events. As we have pointed out, the mark of a good theory is its ability to explain and predict, not the realism of its assumptions. The model of the purely competitive firm and industry can be used to make predictions about the impact of taxation on firms.

Suppose the government decides to place an excise tax of 10 cents per pound on peanut production. Figure 9–10 illustrates the effects of such a tax. The immediate effect of the tax is to raise each peanut farmer's marginal cost curve ($MC_0$) vertically by the amount of the tax to $MC_1$. Higher marginal costs lead peanut farmers to cut their output from $q_0$ to $q_1$. Output thus falls in the industry to $Q_1$, reflected in the shift in the industry supply curve from $S_0$ to $S_1$. The decrease in industry supply causes the market price of peanuts to rise from $P_0$ to $P_1$. As price rises, peanut farmers respond by producing more output, represented along the $MC_1$ curve. Farmers expand their output to $q_2$, and industry output correspondingly expands to $Q_2$. As the market price for peanuts rises, farmers act to restore their output to its original level.

The economic effects of the tax are clear. An excise tax on competitive firms leads to higher costs, lower production, and higher prices.

Notice, however, that the price of peanuts does not rise by the full amount of the tax. The tax was 10 cents per pound, and the price rises by only 5 cents per pound. This means that consumers pay part of the excise tax in the form of higher prices. Who pays the other part? Peanut producers avoid paying part of the tax by reducing their output. This reduction in peanut production leads to a reduction in the demand for factors of production. The other part of the tax is actually paid by the owners of resources, the values of which have fallen because fewer peanuts are now produced. Excise tax payments (pounds of peanuts sold times 10 cents = $P_1abP_2$) are split between peanut consumers and resource owners in the peanut industry. Excise tax payments in this case are calculated by the amount of peanuts sold at the higher price ($Q_2$) times the tax (10 cents). In Figure 9–10, this equals $Q_2 \times ab$, represented by $P_1abP_2$. These payments are split between peanut consumers and resource owners, shown by the shaded areas in the figure.

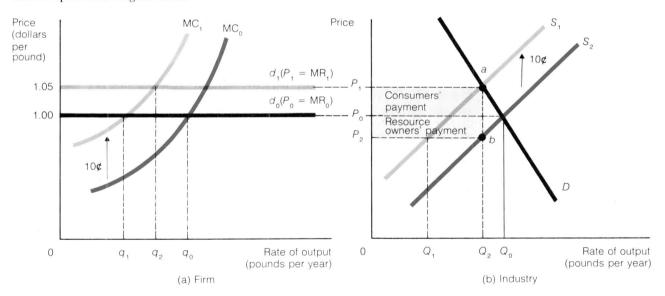

(a) Firm

(b) Industry

**FIGURE 9–10  Effect of an Excise Tax on a Competitive Market**

In a competitive market, an excise tax will raise industry price and reduce industry output. The tax is paid partly by consumers in higher prices and partly by resource owners in the industry who experience a decreased demand for their resources. The amount of the peanut tax is $P_1abP_2$ in part (b) calculated by multiplying the tax rate (10 cents) by the amount of peanuts sold that are subject to the tax ($Q_2$). Payment of the tax is split between consumers and owners of resources, shown in the shaded areas of part (b).

**FIGURE 9–11**

**Long-Run Price in a Constant-Cost Industry**

The long-run industry supply curve (LRS) for a constant-cost industry is a perfectly elastic long-run supply at the level of market price. Expansions in industry output do not cause changes in resource costs to individual firms.

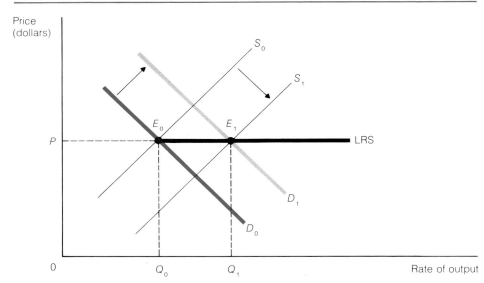

Constant-cost industry: An industry in which the minimum average cost of producing a good or service does not change as the number of firms in the industry changes; an industry for which the supply of resources is perfectly elastic, resulting in a perfectly elastic industry supply.

Figures 9–8 and 9–9 are examples of a **constant-cost industry,** an industry in which expansions or contractions of industry output have no impact on input prices or costs of production. The long-run industry supply (LRS) curve in this case is perfectly elastic—a flat, straight line at the level of the long-run equilibrium price in the industry, as shown in Figure 9–11. Notice that the curve is derived by connecting points $E_0$ and $E_1$, the points where long-run industry demand equals supply, that is, in long-run equilibrium.

Constant cost is most likely to occur in industries where the resources used in industry production are a small proportion of the total demand for the resources in the economy. Take, for example, the toothpick industry. A major expansion, say a tripling, of the output of toothpicks would probably not have much impact on the price of the wood used in the industry because far more wood is consumed in other uses, such as building materials, paper, and firewood. The cost of wood to toothpick producers would therefore remain the same as the industry expanded.

Increasing-cost industry: An industry in which the minimum average cost of producing a good or service increases as the number of firms in the industry increases; such an industry has an upward-sloping long-run supply curve.

**Increasing Cost.** In an **increasing-cost industry,** expansions of industry output lead to higher input prices and therefore higher costs of production for individual firms. This type of industry exhibits the common slope of the long-run industry supply curve because the expansion of most industries puts upward pressure on costs in the industry. An increase in the demand for chicken meat will cause the prices of chicken feed, farmland, and chicken coops to rise. An increase in the demand for newspapers will lead to higher prices for paper, printers, reporters, and so on. These increases in costs lead to a higher long-run price for the goods produced. To obtain more of these goods, consumers must pay higher prices in the long run.

The effects of industry expansion on price and supply in an increasing-cost industry are shown in Figure 9–12. Initially, the firm and industry are in equilibrium at $E_0$, with a market price of $P_0$. Demand increases, the market demand curve shifts to the right from $D_0$ to $D_1$, and price rises initially to $P_1$. Entry into the industry and increased production are encouraged by the higher price, and the additional output is represented by a shift in the

## FIGURE 9–12
### Long-Run Price Changes in an Increasing-Cost Industry

Expansions of industry output in response to higher demand $D_1$ and resulting higher price $P_1$ lead to higher resource costs facing individual producers in the industry. Equilibrium is reached at $P_2$ as costs to producers rise and price falls from $P_1$ to $P_2$. The long-run industry supply curve (LRS) for an increasing-cost industry, determined by connecting equilibrium points $E_0$ and $E_1$, slopes upward to the right.

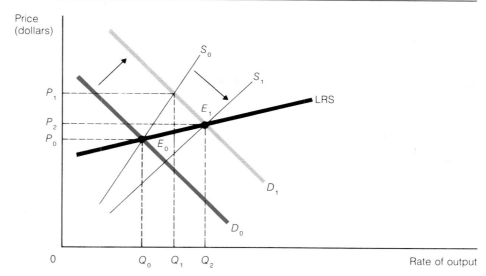

supply curve from $S_0$ to $S_1$. As production expands, however, resource costs to producers in the industry rise, causing the individual firms' cost curves to shift upward. Thus, as entry takes place, price falls from its height at $P_1$, and costs to producers rise until a new equilibrium is reached at $E_1$. This new equilibrium price is higher than the initial equilibrium price of $P_0$. Connecting the two equilibrium points at $E_0$ and $E_1$ again yields the long-run industry supply curve. The long-run industry supply curve in an increasing-cost industry slopes upward to the right.

**Decreasing-cost industry:** An industry in which the minimum average cost of producing a good or service decreases as the number of firms in the industry increases; such an industry has a downward-sloping long-run supply curve.

**Decreasing Cost.** A less typical type of long-run industry supply is demonstrated by a **decreasing-cost industry**. In this industry, expansion of industry output leads to lower input prices and lower costs for individual firms and hence to a lower long-run market price for the product. This case is depicted graphically in Figure 9–13. When industry demand increases and the market demand curve shifts from $D_0$ to $D_1$, price first rises from $P_0$ to $P_1$. In response, the industry supply curve shifts from $S_0$ to $S_1$, and price to consumers falls to $P_2$. Connecting the two points of industry equilibrium, $E_0$ and $E_1$, yields a downward-sloping long-run industry supply curve.

This is an unusual type of long-run industry supply, but a decreasing-cost industry is logically possible. As the clothing industry expands, for example, the costs of certain inputs may fall. Producers of cutting and sewing machines may experience economies of scale, leading to lower prices for their machines. Hence, it is possible for the long-run supply curve of clothing producers to exhibit the decreasing-cost phenomenon. We stress, however, that this is more a logical prospect than a reality.

## Pure Competition and Economic Efficiency

The model of a purely competitive market is often used by economists as a benchmark or ideal against which other models of market structure are compared. This point will be more obvious in Chapter 10, where we discuss the market structure of pure monopoly. But we first need to understand in what

## FIGURE 9–13
**Long-Run Price Changes in a Decreasing-Cost Industry**

Though an unusual case, it is logically possible that industry expansion can lead to lower input costs. As demand increases from $D_0$ to $D_1$ and the supply curve shifts from $S_0$ to $S_1$, prices ultimately settle at $P_2$, lower than both the initial market price $P_0$ and the short-run price $P_1$.

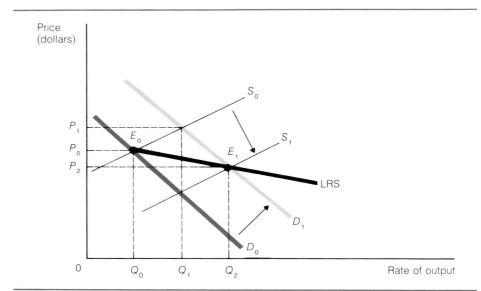

**Productive efficiency:** A situation in which the total output of an industry is obtained at the lowest possible cost for resources.

**Allocative efficiency:** A situation in which the socially optimal amount of a good or service is produced in an industry, given the tastes and preferences of society and the opportunity cost of production.

sense the operation of the purely competitive market represents an ideal outcome in the economy and in what sense the model is only an ideal and not a reality.

### The Competitive Market and Resource Allocation

Pure competition produces what most economists would agree is an ideal or optimal resource allocation. Purely competitive allocation of resources is optimal or best in two fundamental senses.

First, a competitive industry minimizes the resource costs of producing output. In effect, competition forces firms to produce their output at the minimum long-run average total cost and at a price just sufficient to cover that cost. This **productive efficiency** can be seen in long-run equilibrium in competitive firms, which produce at the point where $P = ATC = MC$. Consumers benefit because they get the goods they desire at the lowest possible cost. Inefficient, high-cost firms will incur economic losses and will be driven from competitive markets. Surviving will be firms who can provide industry output at the lowest cost. The 1978 lifting of government regulations on the airline industry created a highly competitive situation in which a number of airlines with excess costs suffered severe losses. Some were forced out of business. By contrast, Northwest Airlines thrived because of the leanness of its operations. By avoiding frills such as lavish company headquarters and by promoting values such as employee productivity, Northwest has kept its cost per passenger seat mile 13 percent below the industry average. Such productive efficiency characterizes firms that survive in highly competitive markets.

Second, a competitive market also creates **allocative efficiency**, which simply means that competition causes resources to flow to their most highly valued uses. In a competitive market, consumer demands are met as long as consumers are willing to pay for the production of additional output at a price that is higher than the cost of the additional resources required to produce the output.

This situation is expressed in the now familiar condition $P = MC$. Price reflects the desire of consumers for additional units of a good. Marginal cost represents the opportunity cost of the resources necessary to produce an additional unit of the good. If $P$ is greater than MC, consumers will pay more for additional units of production than the cost to produce them. When this is the case, resources in the economy will be reallocated from other uses considered less valuable by consumers. If a chicken farmer, for instance, finds that consumers will now pay more for the relatively scarce rooster hackle feathers prized for fly-fishing than they will pay for relatively abundant chicken meat, the farmer will shift the use of some chicken coops, chicken feed, and farmhands from production of meat birds to production of roosters bred for hackle feathers. Chicken lovers might find fewer chickens for sale.

If $P$ is less than MC, consumers will pay less for additional units of production than the cost to produce them. Resources will be reallocated out of such production. If the price for hackle feathers is less than the cost, then fewer hackle feathers will be available for fly-fishing. If $P$ equals MC, resource allocation is ideal in the sense that the things consumers want are produced by competitive firms in the exact quantities and combination (of feathers and food, for instance) that consumers desire and, as we saw above, at the lowest cost.

The competitive model is really Adam Smith's concept of the "invisible hand" at work. Acting in their self-interest, consumers and producers create a mutually beneficial outcome. Looking out for their own interest, producers seek to maximize profits, and yet in a competitive market the result is that consumers' desires are met in the most efficient way possible. Consumers selfishly demand the goods and services they want, but competitive producers balance those demands against other demands for other goods and services. An incredibly complex process of consumer demand and producer response is put in motion by the behavior of each individual in heeding his or her own interest. No central planner is required, and yet a result emerges that is ideal.

Perfect competition is not perfect in every sense, however. While a purely competitive world leads to an ideal resource allocation, it might not lead to an ideal distribution of income among people. Also, while pure competition accommodates consumers' preferences, these preferences themselves may not conform to anyone's image of the ideal state of human existence. Individuals are free to buy and sell as they see fit. If consumers want gidgets rather than great books, they are free to make this choice.

### Reality Versus Pure Competition

In addition to not being ideal in every sense, the model of pure competition is an abstraction rather than a situation that actually occurs in its perfect form. As we turn to the analysis of other models of market structure in the next two chapters, it is useful to point out some of the important ways in which the model of pure competition diverges from real-world markets.

1. In some industries, large firms experience economies of scale and can produce at lower unit costs than small firms. In the long run, therefore, many sellers cannot exist in such industries, and the market structure becomes less than purely competitive.

2. Competition in the real world usually involves more than simply price changes for a homogeneous product. More often than not, competition for consumer purchases involves such factors as location, product design, advertising, and many other aspects of nonprice competition.

3. The model of pure competition emphasizes the establishment of indus-

try equilibrium. But as we all know, the world is generally a place of disequilibrium or change. Technology changes, consumer demands change, owners and investors have new ideas, and so on. The world is rarely at rest. In a world of change, it is the process of adapting to change and moving from one equilibrium state to another that is the most interesting and important way to study competitive behavior.

4. Finally, consumers actually do not desire homogeneous products. Consumer preferences vary widely, encompassing many aspects of products such as their quality, location, color, and design.

To address some of these aspects of more realistic behavior, we turn to the discussion of other models of market behavior and structure. We begin with the model of pure monopoly, which is at the other extreme of the competitive continuum from pure competition. In fact, the model of pure monopoly describes the behavior of industries consisting of a single producer rather than many producers.

## Summary

1. Economists study competition in two basic senses: as a rivalrous, natural process among competitors and as an abstract concept described by the model of a purely competitive market.
2. A purely competitive market is characterized by many buyers and many sellers, a homogeneous product, no barriers to entry or exit by firms, and free information. Examples include the stock market and agricultural markets.
3. The purely competitive firm is a price taker: The firm is so small in relation to the total market for its product that its output has no influence on the prevailing market price. The demand curve facing the purely competitive firm is perfectly elastic at the level of the prevailing market price.
4. Marginal revenue is the change in total revenue caused by an increase in output. Since the purely competitive firm faces a perfectly elastic demand curve, each unit is sold at the prevailing market price. Thus, marginal revenue is equal to market price for the pure competitor.
5. In the short run, the pure competitor decides how much output to produce by setting marginal cost equal to market price. If price is above the average total cost, the firm earns an economic profit in the short run. If price is below the average total cost, the firm has an economic loss. If price is below the minimum average variable cost, the firm will shut down in the short run to minimize its losses.
6. The supply curve of a purely competitive firm is its marginal cost curve above the point of minimum average variable cost. The supply curve of a purely competitive industry in the short run is the horizontal sum of individual firms' marginal cost curves above their respective points of minimum average variable cost.
7. Short-run equilibrium for the purely competitive firm and industry occurs when quantity supplied and quantity demanded are equal in the market and each firm is producing at the level where $P = MR = MC$.
8. In the long run, firms in a purely competitive industry can enter and exit the industry and seek the optimal plant size in which to produce their output. The presence of economic profits leads to entry and expansion in the industry, and economic losses lead to the opposite.
9. Long-run equilibrium in a purely competitive industry results when firms in the industry earn zero economic profits and industry demand and supply are equal. In the long run, the purely competitive firm's position is such that price equals marginal cost equals long-run average total cost: $P = MC = LRATC$.
10. The long-run industry supply curve in a purely competitive industry reflects what happens to firm costs as industry output expands or contracts. In a constant-cost industry, resource prices and firm costs are unchanged by industry expansion, and the long-run industry supply curve is perfectly elastic. An increasing-cost industry experiences rising costs as industry output expands; the long-run industry supply curve slopes upward to the right. In a decreasing-cost industry, resource prices fall as industry output expands, and the long-run industry supply curve slopes downward to the right.
11. The working of the purely competitive model represents a benchmark against which the working of the economy can be measured. In this model, resources flow to their most highly valued uses, and output is produced at the lowest cost in terms of resources used. The model of pure competition is useful in some applications but is an abstraction and does not describe most real-world markets.

## Key Terms

| | | | | |
|---|---|---|---|---|
| market | price taker | short-run industry supply | long-run industry supply | decreasing-cost industry |
| purely competitive market | profits | long-run competitive equilibrium | constant-cost industry | productive efficiency |
| homogeneous product | marginal revenue | zero economic profits | increasing-cost industry | allocative efficiency |
| perfect information | total revenue shutdown short-run firm supply | | | |

## Questions for Review and Discussion

1. What is a market? What is a purely competitive market?
2. Why are price and marginal revenue the same thing in a purely competitive firm?
3. What is the supply curve of a purely competitive firm? Of a purely competitive industry?
4. Describe the adjustment process between points of long-run equilibrium in an increasing-cost industry.
5. In the peanut excise tax example in the second Focus essay (p. 195), what would have happened if the tax had been applied to one firm in the industry rather than to the whole industry?
6. Suppose that a competitive firm is earning economic profits because its owner figured out a way to lower the costs of production. Since the source of the reduced costs is known only to the owner, how can entry take place? Should such information be proprietary (belong to the owner), or should the owner be required to tell potential competitors the reason for the lower costs?
7. Which of the following markets could be analyzed with the competitive model: (a) automobiles; (b) Swiss cheese; (c) blue jeans; (d) cheeseburgers; (e) trash collection; (f) television news; (g) janitorial services?
8. How long will a firm in a competitive industry endure economic losses before it leaves the industry? In your answer assume that price is above minimum AVC but below ATC.

## ECONOMICS IN ACTION

### Competitive Markets in Disequilibrium

Changes in demand and supply in competitive markets typically lead to a smooth adjustment to a new equilibrium as consumers and producers take appropriate actions in the marketplace. It is possible, nonetheless, that changes in some competitive markets may not follow such a smooth pattern. Consider the case of the *cobweb effect* in agricultural markets.

The cobweb effect arises from the uncertainty of agricultural production and from the fact that farmers do not know when they plant a crop what price the crop will bring after it is harvested. It takes a year or more to grow and harvest most crops, and farmers must decide how much of each crop to plant before they know what next year's price for the crops will be. This process works fine as long as agricultural markets are in equilibrium, but what happens when something knocks a market out of equilibrium?

Figure 9–14 graphically depicts the cobweb effect in terms of the market for Halloween pumpkins. The demand curve, $D$, represents the relation between this year's price and the quantity demanded in the same year. The supply curve for pumpkins, $S$, shows the relation between the price of pumpkins this year and the quantity supplied of pumpkins next year. Pumpkin growers decide how many hills of pumpkins to plant as a function of the level of price this year. The market is presently in equilibrium at a price of $P_0$. At this price, $Q_0$ pumpkins are bought and sold.

If the market is in equilibrium, pumpkin growers will bring $Q_0$ pumpkins to the market each year, and no problem of disequilibrium arises. Suppose, however, that drought destroys a large part of the pumpkin crop one year. In this case, the quantity of pumpkins brought to the market falls to $Q_1$.

The shortage of pumpkins at $Q_1$ causes market price to rise to $P_1$ this year. Farmers respond by planting pumpkins in the higher amount of $Q_2$ to be brought to the market next year. Next year's *supply* is thus a function of this year's *price*. But next year, when $Q_2$ pumpkins are brought to the market, they will fetch a price of only $P_2$ because the higher price of $P_1$ reflected a shortage that no longer exists. From here, the cycle starts over again. At a price of $P_2$, farmers will plan to bring $Q_3$ pumpkins to the market. These pumpkins will sell for $P_3$, and so on.

The market is in disequilibrium, and it is not returning to equilibrium instantaneously. It is proceeding in cobweb-like fashion, bouncing around the equilibrium

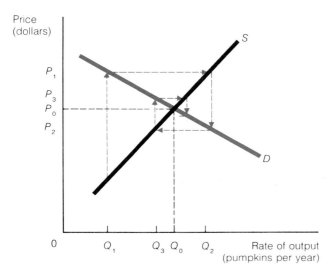

**FIGURE 9-14  A Cobweb Process in the Market for Halloween Pumpkins**

values of $P_0$ and $Q_0$ and approaching these values more closely each year. If there are no further disturbances in the market, equilibrium eventually will be restored. The cobweb process in Figure 9-14 is stabilizing because the old equilibrium values of $P_0$ and $Q_0$ will ultimately be reached if the market is not further disturbed.

The point of the cobweb model is that competitive markets, such as those in agriculture, can exhibit periods of instability in which the adjustment to equilibrium is not smooth. This instability is costly to the economy. In agriculture, farmers may alternately plant too much and then too little of crops.

Of course, we have analyzed the farmer's planting decisions in starkly simple terms. Farmers certainly take more information into account than this year's price for a crop when deciding how much to plant. Regardless of the data they use, however, the fact that they have to plant a crop before they know what price it will bring leads to the potential for cobweb-like behavior in agricultural markets. Indeed, even more complicated and less stable cobweb models—such as the corn-hog cycle in which this year's corn planting is a function of this year's hog prices—can be identified and analyzed.

## Question

Is a cobweb process more likely to be at work in the automobile industry or in asparagus production? (Asparagus plants take three years to mature). Is one more likely in turkey production and consumption than in the Christmas tree ornament industry?

# 10

# Monopoly: The Firm as Industry

*I*n a purely competitive market, each firm's production is such a small proportion of industry output that a single firm has no influence on the market price. As we have seen, the competitive firm is a price taker. The opposite extreme to a purely competitive firm is a pure monopoly: a single seller in an industry. The pure monopoly, in fact, *is* an industry. It alone faces the industry demand curve for its output, and it can affect market price by changing the amount of output that it produces. For this reason, we refer to the pure monopoly firm as a *price searcher:* It must seek the price that maximizes its profits. The ability of the pure monopoly firm to affect market price is not absolute, however. Even a single seller in an industry is subject to competitive pressures from the makers of substitute products.

## What Is a Monopoly?

**Pure monopoly:** An industry in which a single firm produces a product that has no close substitutes and in which entry of new firms cannot take place.

A **pure monopoly** is an industry composed of a single seller of a product with no close substitutes and with high barriers to entry. This definition seems clear, but we must use some care in applying it.

Examples of pure monopoly in the sense of a single seller in an industry are rare. In sixteenth- through eighteenth-century Europe, monarchs granted monopoly rights to individuals for a variety of productive undertakings. These were pure monopolies because each was given the exclusive right to run an entire industry. For example, the king of England created only one English East India Company, which was empowered with a monopoly over the trade with India. This monopoly power was manifested in the high prices the East India Company charged for the goods it brought back from India and sold in England. (See Focus, "Mercantilism and the Sale of Monopoly Rights," for a more detailed discussion of such historical monopolies.)

## FOCUS: Mercantilism and the Sale of Monopoly Rights

Mercantilism refers to the form of economic organization that dominated Western Europe from roughly 1500 to 1776. The English and French economies of the time were typical mercantile economies. Both were characterized by monarchies in which a king or queen represented the central government. As practiced by these monarchs, mercantilism involved widespread and detailed regulation of the economy. More often than not, this regulation took the form of creating monopoly rights for favored individuals. The mercantile economies therefore came to be characterized by the existence of pure monopolies in such diverse areas as brewing, mining, trading, playing cards, and so on, endlessly.

The creation and protection of these monopoly rights had a clear purpose—to raise revenue for the "needs" of the sovereign. Such needs included the expenses of the king's court and the resources needed to fight foreign wars. In other words, the monarchies sold monopoly rights to raise revenue.

Why did sovereigns use monopolies rather than taxes for revenue? Fundamentally, tax collection was a relatively inefficient means to raise revenue for the central state because the costs of monitoring and controlling tax evasion were high. Barter and nonmarket production were widespread in the agricultural economy of the times; moreover, commercial record keeping was not highly developed. Tax collection was therefore a difficult and unattractive alternative as a source of revenue.

The granting of monopoly rights as a means to raise revenue did not have the same deficiencies as taxation. Most important, competition among potential monopolists revealed to the state authorities the worth of such privileges. There were no problems of evasion or guessing at taxable values in this case. Those who wanted the monopoly right would come to the king or queen and make an offer for the right. The potential monopolists were buying the agreement of the ruler to protect and enforce the monopoly privileges that he or she granted.

Thus, pure monopoly had a prominent place in economic history because of its effectiveness at raising revenue for the state. As the state's power to tax has increased in modern times, this revenue-raising role of monopolies has diminished.

*The coat of arms of the East India Company. The throne of England granted the East India Company monopoly rights over most trade with India and the Far East.*

---

Even in these historical cases of monopoly, the monopoly was usually granted to a group of merchants. These monopolies were therefore not technically a single seller; rather, they were a group of sellers acting as a single seller. This is a broader meaning of monopoly, and, in fact, we typically observe monopoly in the real world in groups of sellers acting together. Such associations of sellers are called cartels. The analysis of pure monopoly presented in this chapter is relevant to cartels, but pure monopolies and cartels are not the same. A group of sellers must somehow organize to form a cartel agreement; the single pure monopolist faces no such organizational problem. In this chapter, we discuss pure monopoly in the sense of a single seller facing the industry demand curve. In Chapter 11, we return to the cartel problem and see what extra elements have to be added to monopoly theory to account for cartel behavior.

A second problem in applying the definition of monopoly concerns the criterion that there be no close substitutes. All products exhibit some degree of substitutability. The monopolist's isolation from competition from substitute products is therefore a matter of degree. Consider electricity, usually sold by firms holding local monopolies on its production and distribution. Are there good substitutes for electricity? The answer depends on what the electricity is used for. If it is used for lighting services in residential homes, the few possible alternatives—such as candles and oil lamps—are not very good substitutes for the convenience of electric lighting. If it is used for heating, however, the range of substitute products is greater. The wood stove, oil, and gas industries are competitors with the electric utility company in the market for home heating. The degree of substitutability for the monopolist's product is an important determinant of monopoly behavior. Where there are close substitutes, the monopolist cannot substantially raise prices without losing sales. Where there are no close substitutes, the monopolist has more power to raise prices.

A third consideration in defining monopolies is high barriers to new entry by potential competitors, the basic source of a pure monopoly. There are several types of barriers to new competition: legal barriers, economies of scale, and control of an essential resource.

## Legal Barriers to Entry

Sometimes the power of government is used to determine which industry or firm is to produce certain goods and services. Such legal barriers to entry take several forms.

First, it may be necessary to obtain a **public franchise** to operate in an industry. As we noted in the Focus essay, the monarch granted a franchise determining who could sell Asian goods in England in the seventeenth and eighteenth centuries. In modern times, franchises are granted by government for a variety of undertakings. The U.S. Postal Service, for example, has an exclusive franchise to deliver first-class mail. Many universities offer exclusive franchises to firms that provide food service on campus. Similar arrangements for food and gas service are made along toll roads such as the New Jersey Turnpike. The essence of an exclusive public franchise is that a monopoly is created; competitors are legally prohibited from entering franchised markets.

Second, in many industries and occupations a **government license** is required to operate. In most states a license is required to enter occupations such as architecture, dentistry, embalming, law, professional nursing, pharmacy, schoolteaching, medical practice, and veterinary practice. At the federal level, an operating license is required from the Federal Communications Commission to open a radio or television station. If you want to operate a trucking firm that carries goods across state lines, you must obtain a license from the Interstate Commerce Commission. Licensing therefore creates a type of monopoly right by restricting the ability of firms to enter certain industries and occupations.

Third, a **patent** grants an inventor a monopoly over a product or process for seventeen years in the United States. The patent prohibits others from producing the patented product and thereby confers a limited-term monopoly on the inventor. The purpose of a patent is to encourage innovation by allowing inventors to reap the exclusive fruits of their inventions for a period

---

**Legal barriers to entry:** A legal franchise, license, or patent granted by government that prohibits other firms or individuals from producing particular products or entering particular occupations or industries.

**Public franchise:** A right granted to a firm or industry allowing it to provide a good or service and excluding competitors from providing that good or service.

**Government license:** A right granted by state or federal government to enter certain occupations or industries.

**Patent:** A monopoly granted by government to an inventor for a product or process, valid for seventeen years (in the United States).

of time. Yet a patent also establishes a legal monopoly right. In effect, the social benefit of innovation is traded off against the possible social costs of monopoly.

### Economies of Scale

In some industries, low unit costs may be achieved only through large-scale production. Such economies of scale put potential entrants at a disadvantage. To be able to compete effectively in the industry, a new firm has to enter on a large scale, which can be costly and risky. The effect is to deter entry. Only rarely do new firms attempt to enter the automobile manufacturing industry, for example, because it uses highly automated production techniques on a large scale to keep costs down.

In a **natural monopoly,** economies of scale are so pronounced that only a single firm can survive in the industry. In such a case competition will not work, and government enacts some sort of regulatory scheme to control the natural monopoly. Public utilities such as natural gas, water, and electricity distribution are examples of natural monopolies that are regulated. The regulation of natural monopoly is discussed in more detail in Chapter 17.

A firm may own all of an essential resource in an industry. In this case new entry is barred because potential entrants cannot gain access to the essential resource. The De Beers Company of South Africa, for example, controls 80 percent of the world's known diamond mines. This makes the company a virtual monopolist in the diamond market.

**Natural monopoly:** A monopoly that occurs because of a particular relation between industry demand and the firm's average total costs that makes it possible for only one firm to survive in the industry.

## Monopoly Price and Output in the Short Run

Even though pure monopoly is rare, the theory of monopoly can be applied to a large number of situations in the economy. These range from the behavior of government-created monopolies to natural monopolies, patent rights, and diamond companies. Moreover, as we will see in the next chapter, virtually every firm has some control over the prices it charges. The degree of control varies with the type of market in which the firm sells, but the fact that the firm can choose its price makes the concept of the monopoly firm as price searcher relevant to understanding firm behavior in general. In other words, the analysis of pure monopoly is useful in helping us understand the pricing behavior of all firms, from the corner grocer to IBM.

The monopoly firm, like the competitive firm, faces both long- and short-run production and pricing decisions. In this section we define more carefully what the concept of price searching means, and we examine how the monopoly firm decides how much output to produce and what price to charge in the short run.

### The Monopolist's Demand Curve

The demand curve of the pure monopolist is fundamentally different from that of the pure competitor. The purely competitive firm, as we saw in Chapter 9, is a price taker; it accepts market price as given but can sell as much as it wants at the prevailing market price because demand for its product is perfectly elastic. The monopoly firm faces a downward-sloping market demand curve and must find the price that maximizes its profits. This means that to sell more, the monopolist must lower price. The monopoly firm therefore confronts the problem of finding the best price-output combination. For this reason, we say that the monopolist is a **price searcher.** The process of price searching occurs in any firm that faces a downward-sloping

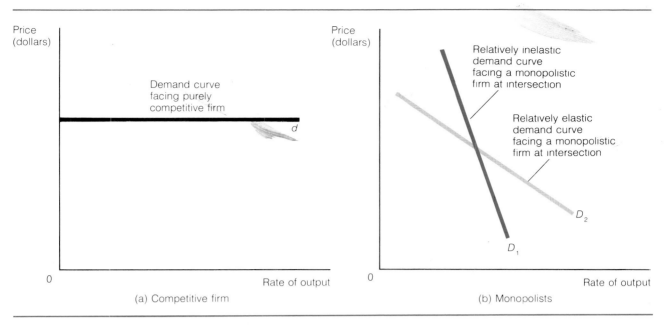

**FIGURE 10–1   Demand Curves for Competitive and Monopoly Firms**
(a) A perfectly elastic demand curve d facing a purely competitive firm. This firm is a price taker: It can sell all of its product that it wants at the prevailing market price. (b) Two market demand curves facing different monopolists. The firm facing demand curve $D_1$ has more leeway than the firm facing curve $D_2$ in choosing price-output combinations because consumers cannot easily find substitutes for the former's product. But both monopolies are price searchers. They face a downward-sloping demand curve: to sell more, they must lower their price.

**Price searcher:** A firm that must choose a price from a range of prices rather than have a single price imposed on it; such a firm has a downward-sloping demand curve for its product.

demand curve, however slight the downward slope may be. A local dry cleaning firm, for example, has some control over its price and must therefore search for the best price for its service.

Figure 10–1 illustrates the difference in the demand curves facing price takers and price searchers: The former is perfectly elastic, while the latter slopes downward. How elastic or inelastic is the price searcher's demand curve? Economic theory does not give us a single answer to this question. Recall from Chapter 5 that the degree of elasticity along a demand curve depends on a number of variables, including the price and the number of close substitutes for the monopolist's product. We noted earlier that even though monopolists are synonymous with the industries they occupy, they are not immune to competition from producers of substitute products. The cheaper and more numerous these substitutes are, the more elastic the monopolist's demand curve will be. The more elastic a monopolist's demand curve, the less valuable is its monopoly position in setting prices. Figure 10–1b represents the demand curves for two different monopoly industries. The demand curves for these two industries, $D_1$ and $D_2$, both slope downward, but $D_1$ is more inelastic than $D_2$ because the latter monopoly faces stiffer competition from substitutes.

For this reason a monopoly in tuna fishing, for example, would not be a tremendously valuable monopoly right. While tuna is a unique product, significant price increases for tuna would cause consumers to switch to other protein sources, such as chicken salad rather than tuna salad for sandwiches.

A monopoly over crude oil production, however, where consumers have few viable consumption alternatives in the short run, has proved to be tremendously valuable, as the price-raising success of OPEC testifies.

The basic point to remember about the demand curve facing the monopolist is that it is downward sloping. This means that the monopoly firm must lower price to sell more and must search for the best, or profit-maximizing, level of output. In other words, the monopolist must be able to answer the question, Does the revenue gained from lowering price exceed the revenue lost from lowering price? To answer this question, the monopolist must employ the concept of marginal revenue.

### The Monopolist's Revenues

To understand the usefulness of marginal revenue figures in price searching, consider yourself the owner of a monopoly. You have discovered a special kind of rock on your land that, when split, reveals a natural hologram. The mysterious beauty of these rocks is highly appealing to rock collectors and gift purchasers. If your aim is to earn the maximum profit from the production and sale of the rocks, how do you decide what price to charge for them?

By trial and error—for you have no previous experience on which to base your pricing decision—you find that the revenues, costs, and profits you encounter at different prices vary considerably, as illustrated in Table 10–1. Column (1) shows the decreasing prices you must charge to sell increasing numbers of rocks. Together, columns (1) and (2) represent the market de-

**TABLE 10–1  Revenues, Costs, and Profits for the Pure Monopolist**

These data are for the hologram rock monopolist. Columns (1) and (2) show the components of the demand curve facing the firm. Column (3) gives total revenue, or sales. Column (4) is the important concept of marginal revenue, which is the change in total revenue divided by the change in output. Columns (5) and (6) are the cost data. Maximum profits in column (7) occur where marginal cost equals marginal revenue, at 6 units of output per day.

| (1) Rate of Output (units per day) | (2) Price (per unit) | (3) Total Revenue (per day) (1) × (2) | (4) Marginal Revenue $\Delta TR \div \Delta Q$ | (5) Total Cost (per day) | (6) Marginal Cost | (7) Profit (per day) (3) − (5) |
|---|---|---|---|---|---|---|
| 0  | —       | —       |         | $ 40.00 |         | −$40.00 |
| 1  | $24.00  | $ 24.00 | $24.00  | 50.00   | $10.00  | − 26.00 |
| 2  | 23.00   | 46.00   | 22.00   | 58.00   | 8.00    | − 12.00 |
| 3  | 22.00   | 66.00   | 20.00   | 65.00   | 7.00    | 1.00    |
| 4  | 21.00   | 84.00   | 18.00   | 75.00   | 10.00   | 9.00    |
| 5  | 20.00   | 100.00  | 16.00   | 87.00   | 12.00   | 13.00   |
| 6  | 19.00   | 114.00  | 14.00   | 100.00  | 13.00   | 14.00   |
| 7  | 18.00   | 126.00  | 12.00   | 115.00  | 15.00   | 11.00   |
| 8  | 17.00   | 136.00  | 10.00   | 135.00  | 20.00   | 1.00    |
| 9  | 15.50   | 139.50  | 3.50    | 160.00  | 25.00   | − 20.50 |
| 10 | 14.00   | 140.00  | .50     | 190.00  | 30.00   | − 50.00 |

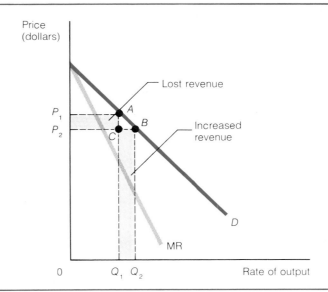

**FIGURE 10–2**

**The Dual Effects of a Price Reduction on Total Revenues**

When the monopoly firm lowers price, it gains revenue from the additional output sold, $Q_1$ to $Q_2$, and loses revenue on the output previously sold at price $P_1$ and now sold at a lower price, $P_2$. The net effect is a marginal revenue curve that lies below and inside the demand curve.

mand curve for hologram rocks, of which you are the sole seller. Your total revenue per day (also known as *sales*) is column (1) × column (2), or price × quantity.

Marginal revenue (MR) is the change in total revenue brought about by a change in output, or $\Delta TR/\Delta Q$. In Table 10–1, column (4) reflects the change in column (3) each time you lower price and produce an additional rock. In Chapter 9, we saw that for the purely competitive firm, MR was equal to price because the firm did not have to lower its price to sell additional units of output (see Table 9–1, p. 185). However, as Table 10–1 shows, MR is not equal to price for you as a monopoly firm, except for the very first unit you sell, because you face a downward-sloping demand curve. When you cut your price, as for example from $22 to $21, two conflicting influences affect your total revenues. You gain additional revenue from the additional units you sell at the lower price. But since the price reduction also applies to output that you were previously able to sell at a higher price, in effect you *lose* revenue on these units. In other words, the three rocks per day you could previously sell at $22 are now priced at $21, a reduction of $3 from your total revenues. Your marginal revenue is therefore $18 in this case.

When demand is downward sloping, MR ($18) is less than price ($21) by the amount of the revenue lost on units that would have sold at the higher price. For price searchers, marginal revenue is less than price (P > MR) because of the need to lower price to sell additional units of output.

Figure 10–2 provides another way of looking at the fact that marginal revenue is always lower than price for the monopoly firm. At a price of $P_1$, total revenue is $P_1 \times Q_1$. At a price of $P_2$, total revenue is $P_2 \times Q_2$. Marginal revenue is the difference between these two total revenue rectangles. To see its components more clearly, consider what happens when price is cut from $P_1$ to $P_2$. First, additional revenue is generated by the sale of extra units at the lower price. This additional revenue is indicated in the shaded rectangular area $CBQ_2Q_1$ above increased sales. Second, there is a loss of revenue on units previously sold for $P_1$ but now sold for $P_2$. This loss is represented by the shaded area $P_1ACP_2$. Thus, the marginal revenue de-

## FIGURE 10–3
### Changes in Elasticity of Demand and Total Revenue as Price Changes

The elasticity of the monopolist's demand curve is linked to the behavior of total revenue as price changes. In the elastic ($\epsilon_d > 1$) portion of the demand curve, price reductions cause total revenue to rise. When demand is unit elastic ($\epsilon_d = 1$), price reductions leave total revenue unchanged. When demand is inelastic ($\epsilon_d < 1$), price reductions lead to a fall in total revenue. The point of unit elasticity in part a, beyond which marginal revenue becomes negative—30 cans per day—is also the point in part b at which total revenue stops rising and starts declining.

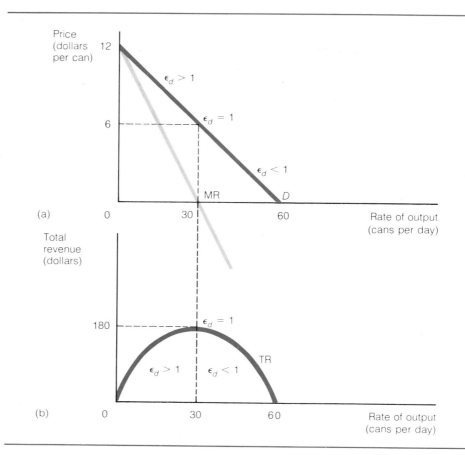

rived from a price reduction by the monopolist is less than the new price charged. Except for the first unit of output, the MR curve of the monopolist therefore lies below or inside the demand curve.

Even though the monopoly firm faces no direct competition, it still must search for its best price. This search for the most profitable price-output combination explains why the monopoly firm lowers price even though it has no direct competitors.

### Total Revenue, Marginal Revenue, and Elasticity of Demand

The demand curve facing the monopoly firm shows the amount of output that can be sold at different prices. It also shows how the firm's revenues vary as price and output are changed. By observing this, we can learn something about the elasticity of the monopolist's demand curve over various ranges. Figure 10–3 illustrates the relation between total revenue, marginal revenue, and the elasticity of the monopolist's demand curve at various points. Suppose that we are joint owners of a firm with a monopoly on Gas-Saver, a unique additive that doubles gas mileage if one can of it is added to each tankful of gas. If we conduct a pricing experiment along our demand curve, we can observe its effects on our revenues. We start with a high price of $12 per can and lower it gradually to $6. At $12, we will not sell any of the stuff, for consumers know that a can of it will cost them more than the gas they save. As we lower the price toward $6, we sell more and more.

Figure 10–3b shows what happens to our total revenue as the price we charge falls. Over the range of sales up to 30 cans per day, our total revenue rises. In Figure 10–3a, this rise is reflected by the positive marginal revenue. As we saw in Chapter 5, a demand curve is elastic when price reductions cause total revenue to rise. Thus, our demand curve is elastic up to the level of 30 units of output and a price of $6. Beyond 30 units of output, however, we observe that as we further lower our price, our total revenue falls, and our marginal revenue is negative. This drop means that we have reached the inelastic portion of our monopoly demand curve. Here, price reductions cause our sales to decline.

The regions of our Gas-Saver demand curve that are elastic and inelastic are labeled as $\epsilon_d > 1$ and $\epsilon_d < 1$, respectively, in Figure 10–3. The point labeled $\epsilon_d = 1$ is where the demand curve is unit elastic. This point occurs where marginal revenue equals zero and where total revenue is at a maximum. It is the dividing line between the elastic and inelastic portions of the monopolist's demand curve. The link between MR and elasticity of demand can be stated thus: MR goes from positive to negative as demand goes from elastic to inelastic.

With this simple model, we can begin to discuss the rational pricing strategy of the monopolist. Suppose that our Gas-Saver monopoly firm has no costs of production. What price should we set to maximize our revenue? Figure 10–3b clearly illustrates that we can do no better than to be at the top of the total revenue curve. At this point, we reach our total maximum revenue: $180 per day. Reading vertically up to part a, you can see that to earn this maximum total revenue, we should produce 30 cans of Gas-Saver per day and sell them for $6 each. The profit-maximizing monopolist with no costs of production will set price along the demand curve where demand is unit elastic, or equal to 1. A monopoly firm will never operate along the inelastic portion of its demand curve because marginal revenue is negative along this portion. Increases in output and reductions in price actually cause total revenue to fall.

This simple analysis helps us understand the pricing behavior of any firm facing a downward-sloping demand curve. Such firms will not set price in the inelastic part of their demand curve because they can earn more revenue by raising price. Soon we will introduce the cost curves of the monopoly firm, which further strengthen this result. For the present we note that this result and the discussion of this section are based on the properties of a straight-line or linear demand curve.[1]

## Short-Run Price and Output of the Monopolist

Although for simplicity we did not consider input costs in the previous discussion, we introduce them now, because all firms incur costs of production. The fact that the monopolist possesses a monopoly in the output market says nothing about its position in input markets. In this respect the monopoly firm is like the competitive firm; it is one of many buyers of inputs. The cost curves of the monopolist therefore resemble those of the competitive firm.

By combining the cost and revenue concepts of monopoly, we can analyze the choice of the best, or profit-maximizing, rate of output and price for the monopoly firm in the short run. To maximize profits, the monopolist follows

---

[1] With a linear demand curve, the marginal revenue curve will bisect the horizontal axis and any line parallel to the horizontal axis. In Figure 10–3, for example, MR bisects the horizontal axis at 30 units of output.

the same rule that the competitive firm follows—it sets marginal cost equal to marginal revenue. The logic is identical in both cases. If MR is greater than MC, it pays to produce additional units of output because they add more to revenues than to costs. If MR is less than MC, it pays to reduce production because extra output adds more to costs than to revenues. The monopolist will therefore do best by producing at the output level where MC equals MR.

Look back at the cost and profit data in Table 10–1 for the hologram rock monopoly to check the viability of this rule for profit maximization. Comparing marginal revenue in column (4) with marginal cost in column (6), we see that the first 6 rocks produced per day add more to revenue than to costs; that is, MR > MC over this range of outputs. The sixth unit of output adds $14 to revenues and $13 to costs. It pays to produce that sixth unit. Producing the seventh rock, however, has a marginal cost of $15 and a marginal revenue of $12. The firm would be losing money if it increased operations to 7 rocks a day. The maximum profit occurs at a rate of output of 6 rocks per day and a price of $19 for each. The profitability of this decision is recorded in column (7), which is the difference in total revenue from column (3) and total cost from column (5). The maximum daily profit of $14 occurs when 6 rocks are produced, the point at which MR equals MC.

The same point is illustrated graphically in Figure 10–4, where a monopoly's marginal cost curve is combined with its demand and marginal revenue curves. To determine the profit-maximizing rate of output, the monopolist would find the point along the horizontal axis where MR = MC, in this case $Q_m$. For outputs less than $Q_m$, MR > MC, and it pays the monopolist to expand production. In this area, extra units of output add more to revenue than to cost. For outputs greater than $Q_m$, it pays the firm to reduce output. At $Q_m$, MR = MC, and the firm can do no better than produce this output. Profits are maximized at $Q_m$.

The monopolist sets a price for $Q_m$ units of output by reading the market value of this output off the demand curve at a point directly above the point where MR = MC. In other words, the price for this output is read off the

### FIGURE 10–4
**Profit-Maximizing Price for a Monopoly**

The monopoly firm maximizes profits where MR = MC. This means that $Q_m$ is the profit-maximizing output and $P_m$, read off the demand curve, is the profit-maximizing price for the monopoly firm.

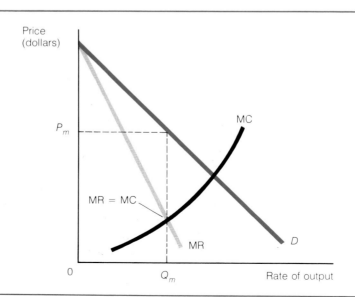

**FIGURE 10–5**
**Monopoly Profits**
At point B, the monopoly firm would just cover its average total cost. But since it holds a monopoly, it can charge price $P_m$, creating excess profits equal to $AB \times Q_m$, represented by the area $P_mABC$. In other words, $P_mABC$ is equal to total revenue, $P_mAQ_m0$, minus total costs, $CBQ_m0$.

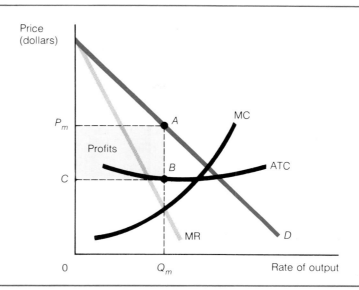

**Profit-maximizing price:** The price at which the difference between total revenue and total cost is greatest; the price at which marginal cost equals marginal revenue.

demand curve by drawing a straight line from the intersection of MR and MC to the demand curve and over to the price axis, as shown by the dashed lines in the figure. Price is not set where MR = MC on the marginal revenue curve; it is set by reference to the demand curve. In Figure 10–4, $P_m$ is the **profit-maximizing price.**

Figure 10–4 illustrates a principle based on economic behavior, but in real-world firms executives do not calculate MR and MC and set them equal to one another to determine their best price. Business decisions are more complicated, and more often than not businesspeople will be observed using rough and ready approaches to pricing and production decisions. However, through a process of trial and error, price-searching firms will grope toward the MR = MC solution. It may not seem as if they are applying the MR = MC rule, but the practical effect of their decision making ultimately works itself out in these terms.

Note also that although both the purely competitive and the monopolist firm follow the MR = MC rule, the result is fundamentally different in the two cases. For the competitive firm, marginal revenue is the same as price. The competitive firm therefore maximizes profits by producing where P = MC. Marginal revenue is not the same as price for the monopolist; as we have seen, the monopolist produces where P > MC. We will elaborate further on this important difference between competition and monopoly later in the chapter.

### Monopoly Profits

Figure 10–4 shows how the monopolist applies the MR = MC rule, but it does not tell us anything about the level of profits the monopoly earns. To discuss this concept we add the monopolist's average total cost to the analysis in Figure 10–5.

Nothing is changed from Figure 10–4. The monopolist continues to set MR = MC, selling $Q_m$ units at a price of $P_m$. Just as in the case of the purely competitive firm, the monopolist's profits are determined with respect to the average total cost curve. At point C, the monopolist would just cover its average total cost. But its monopoly position and the demand for its product allow it to set price higher, at $P_m$, on its demand curve, creating profits in excess of its costs. In Figure 10–5, these excess profits are given by the shaded area $P_mABC$, or the amount by which total revenue, the area $P_mAQ_m0$, exceeds the total cost of producing $Q_m$, the area $CBQ_m0$. Another way to look at monopoly profits in such a graph is to say that the monopolist makes $AB$ profits per unit of output, where $AB \times Q_m = P_mABC$.

This is a very profitable monopoly. These returns are in excess of the total costs of the firm, which means that they are above the opportunity cost of the monopolist's capital investment. Were this a purely competitive firm, these excess returns would stimulate new entry and expansion in the industry. Remember, however, that this is a pure monopoly. Entry cannot occur here, by definition. So, unlike the profits of the competitive firm, which ultimately are restored to a normal rate of return, the profits of the monopolist can persist in the long run.

Not all monopolies make monopoly profits. Some lose money. Such a case is graphed in Figure 10–6. In this case demand is not sufficient to cover the average total cost of the monopolist at the level of output where MR = MC. Total revenue at $Q_m$ is represented by $P_mCQ_m0$; total cost is represented by the larger area $ABQ_m0$. Total cost exceeds total revenue by the amount of the shaded area, $ABCP_m$. This monopolist makes an economic loss in the short run. As long as $P_m$ is high enough to cover the monopolist's variable costs, the monopoly will continue to operate in the short run if it expects market conditions for its product to improve. Over time, however, if the losses depicted in Figure 10–6 persist, the monopolist will cease operations and go out of business. The investment of the monopoly firm is not earning a competitive rate of return, and, by closing down, the firm is free to move

**FIGURE 10–6**

**An Unprofitable Monopoly**

Not all monopolies make profits. This monopoly makes $BC$ losses per unit of output because demand for its product is not sufficient to cover its average total costs. Its total revenue at $Q_m$ is represented by $P_mCQ_m0$, and its total cost by $ABQ_m0$. Therefore, its total losses are represented by the shaded area, $ABCP_m$. If the situation does not improve in the long run, this monopoly will go out of business.

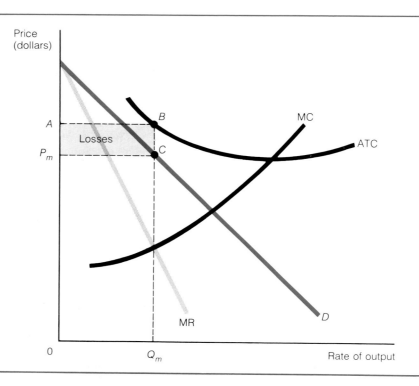

its investment elsewhere in the economy where it can earn higher rates of return.

It may seem curious that monopolists should make losses, but this is a very real phenomenon. Many holders of patents, for example, never market their products because there is no demand for them. The fact that one has a monopoly does not automatically make it a valuable monopoly.

### Differences from the Purely Competitive Model

Before we move on to long-run considerations for the monopolist, we note two important differences between the monopoly firm and the purely competitive firm. First, in a pure monopoly no distinction is made between the firm and industry. The pure monopoly firm *is* the industry, so there is no room for a separate theory of industry versus firm behavior, such as the one we presented for pure competition.

Second, recall that the marginal cost curve of a competitive firm (above the point of minimum average variable cost) is its supply curve because the firm sets marginal cost equal to price to maximize profits. Its pricing behavior traces out a relation between quantity supplied and price. No such relation exists for pure monopoly. The monopolist controls the quantity of output produced and sets MR = MC, but, as we have seen, MR does not equal price for the monopolist. Because the monopolist does not set MC equal to price, no unique relation exists between MC and price in the case of pure monopoly. To know what the monopolist will produce at a given price, we need to know more than the firm's MC; we need to know the shape and position of its demand and marginal revenue curves as well. The lesson is simply that the monopolist does not have a supply curve, that is, a unique relation between the amount it produces and the price.

## Pure Monopoly in the Long Run

Like all firms, the pure monopolist must face the long run, the period of economic time over which the firm can vary all of its inputs and enter or exit the industry. As we will see, the long run for the pure monopolist is not analogous to the long run for a purely competitive firm and industry. Certain adjustments take place, such as the capitalization of monopoly profits, but entry and other efficiency-enhancing adjustments do not take place. Monopoly and its effects persist in long-run equilibrium.

### Entry and Exit

A pure monopoly may not be profitable, as we saw above, and hence may leave the industry. Exit is a distinct possibility. Entry is not. By definition, high and prohibitive barriers to entry exist with monopoly; entry by new firms is barred. Long-run adjustments in the form of competitive entry do not take place in the case of the monopoly.

### Adjustments to Scale

The monopoly firm can adjust its scale of operations in the long run. That is, given that it makes a profit, the monopolist can seek to produce the profit-maximizing rate of output in the most efficient plant. Figure 10–7 illustrates the normal plant size result of long-run adjustment for the monopolist.

The monopoly firm produces where MR = MC at $Q_m$ and operates to the left of the point of minimum long-run average total cost. In contrast, the

**FIGURE 10–7**

**Monopoly Plant Size in the Long Run**

The monopoly firm can adjust its plant size in the long run along its LRATC curve. Typically, this adjustment results in the selection of an efficient plant size to the left of minimum long-run average total cost, at $Q_c$, the point at which competitive firms must operate.

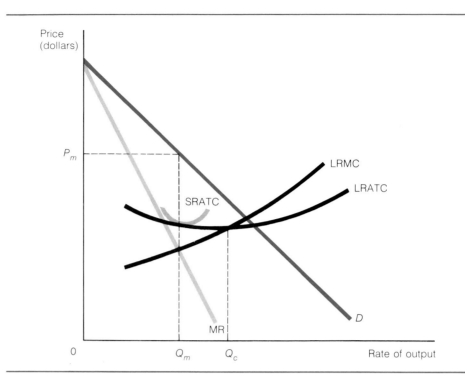

purely competitive firm was forced by competition to produce at the level where long-run average total cost was at a minimum, at $Q_c$. The monopolist produces $Q_m$ in the most efficient plant that it can, given by short-run average total cost (SRATC) for the lowest-cost plant in which to produce $Q_m$. But compared with the purely competitive firm, it is operating inefficiently, that is, in a plant size with unit costs greater than minimum long-run average total cost. We discuss this potential efficiency cost of monopoly later in this chapter. This result is by no means unique. It is the standard depiction of the monopolist's selection of a long-run plant size. Other results are possible if the demand and marginal revenue curves are in different positions.

### Capitalization of Monopoly Profits

The final long-run adjustment by the monopoly firm concerns what happens to its profits in the long run. We have said that entry is not possible with pure monopoly. But there is no reason that the monopoly cannot be sold to aspiring investors. In this respect, monopoly rights are like any other resource: Where competition is free, resources will flow to their most highly valued uses. Figure 10–8 shows the outcome of this process.

The monopoly is earning profits represented by $P_mABC$ in long-run equilibrium. Suppose that its owners decide to retire and sell the firm. What would they ask for the firm? Rationally, they would include in their asking price the value of the monopoly profits. They would not give these profits away to a buyer. As a result, the average total cost to the buyer would rise to reflect the value of the monopoly profits, from $LRATC_1$ to $LRATC_2$ in Figure 10–8. The price of the monopoly firm is given by the expected present value of $P_mABC$, which includes the value of the monopoly profits.

A useful principle has emerged. When monopoly profits are capitalized in this manner, the new owners of the monopoly firm may not earn any mo-

## FIGURE 10–8
### Capitalization of Monopoly Profits

If the monopoly is sold, the asking price will include the value of the monopoly profits, represented by $P_mABC$. The average total cost to the buyers is depicted by a shift in the LRATC curve from $LRATC_1$ to $LRATC_2$. The new owners therefore face an LRATC curve of $LRATC_2$; they will not make monopoly profits but will make only a normal rate of return on their investment.

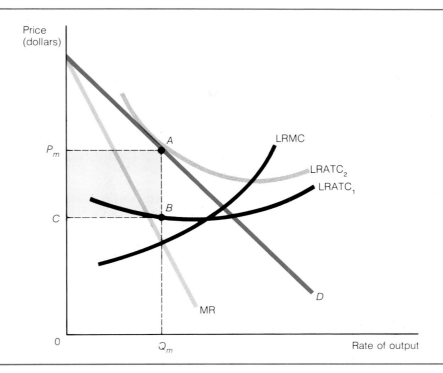

nopoly profits; they will most likely earn a normal rate of return on their investment in the monopoly right. Remember that a normal rate of return is embodied in the long-run average total cost curve. The LRATC curve faced by the new owners of the monopoly is $LRATC_2$, which is just tangent to the demand curve at $P_m$. At $P_m$, the new owners just cover their total costs, including a normal rate of return on their investment. The monopoly profits are taken by the original owners, who captured the value of $P_mABC$ when they sold the firm. This does not mean, of course, that the monopoly must cease to exist because it is not profitable. $P_m$ and $Q_m$ continue to prevail in the marketplace, but the current owners make only a normal rate of return on their investment. Why, then, did the new owners buy the firm? Obviously, they think they can run the firm at lower cost than the old owner and make profits through more efficient management of the monopoly.

## Price Discrimination

In our discussion of monopoly theory so far, the profit-maximizing monopoly has charged a single price for its product. Each customer who buys the product pays the same price. Under certain conditions, however, a monopoly can make more money by charging different customers different prices for its product. For example, movie theaters and airlines often charge lower prices to children and senior citizens than to others, and utilities charge different rates for electricity to businesses and residences. This method of pricing is called **price discrimination.**

Two points should be kept in mind as we discuss price discrimination. First, price differences that reflect cost differences do *not* constitute price discrimination. For example, large buyers (those purchasing more goods) are often charged less per unit for some goods than are small buyers. The general

**Price discrimination:** The practice of charging one buyer or group a different price than another group for the same product. The difference in price is not the result of differences in the costs of supplying the two groups.

reason for this difference is that selling costs per unit are lower when dealing with the large buyer. Each sale may require the same paperwork, but the paperwork cost is lower per unit of sales for the large buyer. The lower price charged to the large buyer is not the result of price discrimination; it reflects the seller's different costs of serving the two customers. Second, equality of prices across buyers does not necessarily imply the absence of price discrimination because the costs of supplying the buyers may be different. For these reasons, price discrimination is a tricky concept, and care must be used in asserting its presence in the economy. Some forms of price discrimination are illegal under the Robinson-Patman Act (1936), which is enforced by the Federal Trade Commission.

### When Can Price Discrimination Exist?

Price discrimination can occur only under certain conditions. First, the firm must have monopoly power and face a downward-sloping demand curve for its output. In other words, the firm must be a price searcher, not a price taker. A price taker obviously cannot charge different prices to different customers because the price it charges is determined by the prevailing market price, over which it has no influence. Second, the monopolist's buyers must fall into at least two clearly and easily identifiable groups of customers who have different elasticities of demand for the product. Third, the separation of buyers is crucial. Without the ability to separate buyers, the seller will not be able to keep buyers who are charged a low price from reselling the product to buyers who are charged a high price. Such behavior would undermine a price discrimination scheme and lead to a single price for the monopolist's product. The identification and separation of groups of consumers must be possible at low cost to make price discrimination worthwhile for the firm.

### A Model of Price Discrimination

Figure 10–9 shows how a monopolist can gain from price discrimination. There are two groups of customers for the monopolist's product. The demand of buyers in market A is relatively more elastic than the demand of buyers in market B.

Two points are crucial in understanding the analysis in Figure 10–9. First, the marginal cost to the monopolist of supplying both markets is the same. The best way to think about this condition is to assume that the output sold to each market is produced in the same plant. MC is thus at the same level in both market A and market B. Second, to maximize profits, the monopolist sets marginal cost equal to marginal revenue.

Following the MC = MR rule in this case, the monopolist sets MC = MR in both markets. In market A this leads to a price of $P_1$; in market B it leads to a price of $P_2$. The buyers in market B with the less elastic demand are charged a higher price than the buyers in market A ($P_2 > P_1$). These are the profit-maximizing prices in the two markets. The monopolist benefits from price discrimination to the degree that its profits go up relative to what they would be if it sold its output at a single price to all buyers.

The monopolist must be able to keep the two markets separate to sustain a price discrimination scheme. If it does not, buyers in market A could profitably resell the product to buyers in market B and break down the price discrimination system. In 1984 Apple Computer Inc. offered to sell all students at twenty-four universities its then-new Macintosh computer for less than half of the full price being charged to other customers. Students and profiteers immediately saw the potential for profit in reselling the computers

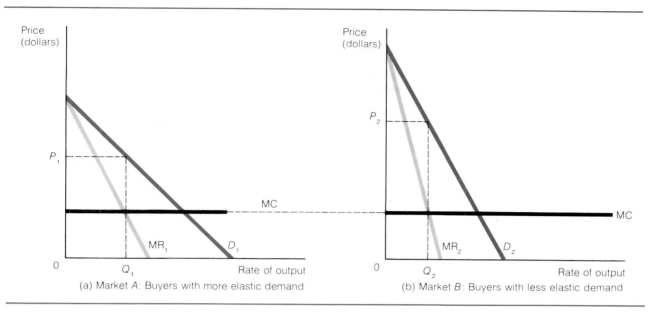

**FIGURE 10–9  Price Discrimination**

The figures demonstrate a case of price discrimination for the same product in two markets. Consumers in market A have a relatively more elastic demand than those in market B. The profit-maximizing monopolist confronts market A consumers with a lower price. That is, $P_1 < P_2$. For a price discrimination scheme to be viable, the monopolist must prevent buyers in market A from reselling to buyers in market B.

to nonstudents at marked-up but still below retail prices. Communication between students with the cut-rate Macintoshes and nonstudents desiring Macintoshes quickly sprang up through newspaper ads and black marketeers.

The seller who wants to practice price discrimination must therefore be able to distinguish and separate customers on the basis of their elasticity of demand. Moreover, the seller must be able to perform this feat without a large expenditure of resources. Thus, most real-world price discrimination schemes are based on general characteristics of customers, such as age, education, income, and sex. For example, children and older people are often charged lower prices for certain products because their age is thought to mean that they have more elastic demand for these products.

Another subtle factor that provides a basis for price discrimination is differences in the opportunity cost of time among customers. The practice of giving discount coupons for grocery purchases, for example, allows sellers to separate buyers into elastic and inelastic demand categories. Buyers in the inelastic demand category will not take the time and trouble to collect coupons and redeem them. These are customers who place a relatively high value on their time. They value other uses of their time and do not want to take the effort to clip coupons. Buyers in the elastic demand group place a lower value on their time and will take the time to clip coupons and convert them into lower prices at the store. Sellers, by using coupons, are therefore able to separate buyers into classes and charge them different prices based on their different elasticities of demand. Focus, "Perfect Price Discrimination," gives another example of price discrimination: the pricing of medical services according to patients' income.

## FOCUS   Perfect Price Discrimination

In the example graphed in Figure 10–9, the price-discriminating seller separated buyers into two groups and charged each group a different price. Perfect price discrimination occurs when the seller is able to charge each buyer a different price that reflects just what he or she is willing to pay for the product or service.

Perfect price discrimination is difficult to practice because it is obviously hard to separate buyers so that each buyer's willingness to pay can be determined. A possible use is the fee schedule of a plastic surgeon. The surgeon can estimate patients' willingness to pay by such factors as their income or insurance coverage and then charge different prices to each customer for similar services. Resale is out of the question with medical services. Thus, the conditions for perfect price discrimination can be met.

Figure 10–10 illustrates one plastic surgeon's pricing scheme. He charges each patient along the demand curve for plastic surgery just what that patient is willing to pay for the operation. For example, patient 1 is charged $P_1$, patient 2 is charged $P_2$, and so on. The cost of performing the surgery, indicated by MC, is assumed to be the same for all patients. All told, the surgeon extracts the maximum possible revenue from patients, represented by the total area $ACQ_00$, and a large profit in excess of cost, represented by $ACB$.

Because it is costly to separate buyers into separate categories and to prevent resale among categories of buyers, perfect price discrimination is rarely found in the real world. But pricing of medical services sometimes fulfills the conditions for perfect price discrimination.

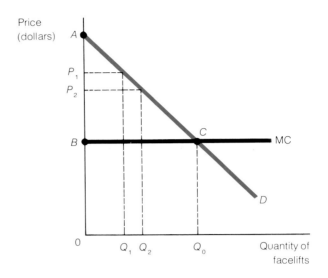

**FIGURE 10–10   Perfect Price Discrimination**
In this case of perfect price discrimination, the plastic surgeon charges each patient just what he or she is willing to pay for his services. Patient 1 pays $P_1$, patient 2 pays $P_2$, and so on. This surgeon's perfect price discrimination can capture the area $ACB$ as a return in excess of costs.

## The Case Against Monopoly

Monopoly is generally viewed as "bad" for the economy in comparison to the "good" expected from a competitive organization of markets. The indictment of monopoly by economists is based on arguments that monopolies are lacking in both efficiency and equity or fairness. After considering arguments against monopoly, we will examine the theory that monopolies provide certain social benefits that may offset their social costs.

### The Welfare Loss Resulting from Monopoly Power

Contrived scarcity: The action of a monopoly that reduces output and increases price and profits above the competitive level.

The main efficiency argument against monopoly is that this market structure leads to **contrived scarcity,** that is, that a monopoly withholds output from the market to maximize its profits. Remember that the monopolist sets price where MC = MR but where P > MR. At the profit-maximizing output for the monopolist, price therefore exceeds marginal cost. In the case of pure competition, price reflects the marginal benefit that consumers place on additional production, and marginal cost reflects the economic cost of the resources necessary for additional production. When price is greater than marginal cost, this is a signal to producers to increase output. The monopoly firm, of course, will not do this because this approach is not consistent with maximizing its profits. We thus say that the monopolist causes a contrived scarcity, a condition in which the monopoly's product is short in supply and high in price.

This contrived scarcity is a social cost to the economy. Its nature and magnitude can be estimated with an economic model that contrasts the effects of monopoly and competition in a given market. We give such a model in Figure 10–11.

Figure 10–11 shows the industry demand and supply functions, $D$ and $S$, for some commodity, say cranberry juice. There are no economies of scale in this industry. The average total cost of producing cranberry juice is the same at all levels of output. As defined in Chapter 9, this is a constant-cost industry with a flat long-run supply curve along which MC = ATC.

Suppose that the industry is organized competitively, with many producers of cranberry juice. Long-run industry equilibrium is established at $E$, where industry demand and supply are equal. The juice output of the various firms is priced according to the marginal cost of producers along the industry supply curve and according to the wishes of the consumers along the demand curve. In other words, price is equal to the long-run marginal cost of production.

Now we ask a special question: At the competitive price-output combination $P_c$ and $Q_c$, how much surplus do consumers receive? They are willing

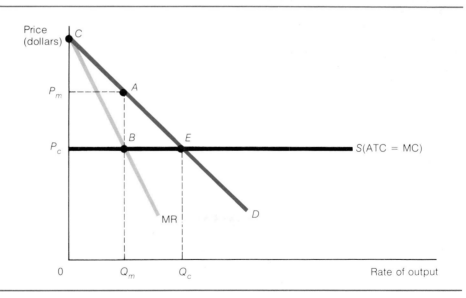

**FIGURE 10–11   Welfare Loss Due to Monopoly Power**

This graph illustrates the hypothetical case of converting a competitive industry into a monopoly. Consumer surplus at price $P_c$ and output $Q_c$ in a constant-cost industry is represented by $CEP_c$. When the industry becomes a monopoly, it cuts back output to $Q_m$, resulting in a decreased consumer surplus of $CAP_m$ and monopoly profits of $P_mABP_c$. The welfare loss to the economy, that is, the amount of real income that does not reappear for anyone after the monopoly is formed, is represented by $AEB$. The area of monopoly profits can be considered a transfer from consumers to the monopoly or an additional social cost of the monopoly.

to pay the amounts given by the demand curve, but they pay only $P_c$. They thus receive a surplus of real income in this case represented by the large area above the industry supply curve and below the industry demand curve, or $CEP_c$. This triangle represents the concept of consumer surplus discussed in Chapter 6: the amount that consumers would be willing to pay over what they have to pay for a commodity.

Now suppose that this cranberry juice industry is turned over to a single seller, a monopolist. As a result, the monopolist restricts juice output to $Q_m$ (contrived scarcity) and raises price to $P_m$ to maximize its monopoly profits. What has happened to the level of consumer surplus in this market? The consumer surplus before monopoly is given by $CEP_c$, but consumer surplus after monopoly is only $CAP_m$. The difference in these two areas is the trapezoid $P_mAEP_c$.

The area $P_mAEP_c$ is made up of two components. First we consider the rectangle $P_mABP_c$, depicting monopoly profits, to represent a transfer from consumers to the monopolist. In other words, these dollars do not leave the economy: They are taken out of the pockets of consumers and put into the pocket of the monopolist. (We later discuss both the efficiency and equity aspects of this transfer of wealth.) The triangle $AEB$ is left over. Who gets the real income represented by this triangle? The answer is no one. It simply vanishes when the monopoly is formed. This triangle is therefore called the **welfare loss due to monopoly.** It is a loss because it vanishes when the monopoly is formed. It is a cost to the economy because it does not reappear as income to someone.

Arnold Harberger was the first scholar to attempt to state the total amount of this welfare loss in the manufacturing sector of the U.S. economy.[2] Using some complicated economic formulas, he found that the losses in the U.S. manufacturing sector for 1929 were on the order of 0.1 percent of GNP. In other words, the simple welfare loss to monopoly did not loom large in the economy. Subsequent scholars have used variations of Harberger's technique and have generally found the welfare losses due to monopoly to be low. Thus, even though the welfare loss due to a single monopoly looks large in Figure 10–11, it does not appear to be large as a factual matter in the context of the whole economy.

### Rent Seeking: A Second Social Cost of Monopoly Power

The results of Harberger and other scholars suggest that in terms of the welfare loss it imposes, monopoly is not a big problem in our economy. This finding seems paradoxical in an economy such as ours that is dominated by very large and powerful firms. How could the social cost of these giant, powerful concerns be so small?

Gordon Tullock has suggested a solution to this apparent paradox.[3] He considered the area $P_mABP_c$ in Figure 10–11—previously defined as monopoly profits, a simple transfer from consumers to the monopolist—and asked the question, What happens if people compete for these transfers? That is, suppose the monopoly procures its price by hiring lawyers to lobby the government for a monopoly right. The expenditure of resources on lawyers to win the transfer is socially wasteful because these expenditures add nothing

**Welfare loss due to monopoly:** The lost consumers' surplus resulting from the restricted output of a monopoly firm.

---

[2]Arnold Harberger, "Monopoly and Resource Allocation," *American Economic Review* 44 (May 1954), pp. 77–87.
[3]Gordon Tullock, "The Welfare Costs of Tariffs, Monopolies, and Theft," *Western Economic Journal* 5 (June 1967), pp. 224–32.

to the social product. The opportunity cost of hiring lawyers consists of the productive activities they would otherwise perform. To pay lawyers to take a dollar out of the pockets of consumers and put it into the pocket of the monopolist reduces GNP by the amount of what the lawyers would have produced if engaged in productive pursuits. Not only lawyers but also lobbyists, accountants, executives, and secretaries may be involved in the effort to secure a monopoly transfer. If the competition to capture these transfers is sufficiently strong, resources will be wasted in the process up to the amount of the transfer. This amount, when added to the welfare loss, is the amount represented by $P_mAEP_c$. When the loss is equal to the trapezoid $P_mAEP_c$ and not the triangle $AEB$, the social cost of monopolies is no longer small.

**Rent seeking:** The activity of individuals who spend resources in the pursuit of monopoly rights granted by government.

This process of competing for transfers is called **rent seeking,** and it is normally associated with the process of seeking monopoly privileges from government. The central proposition of rent-seeking theory is that the expenditure of resources to gain a pure monopoly transfer is a social cost. (We discuss these ideas in more detail in Chapter 15.)

### Production Costs of a Monopoly

Recall from our discussion of Figure 10–7 that the typical monopoly firm, after long-run adjustment, does not produce at the point of minimum long-run average total cost. In other words, the monopoly firm does not produce its output in the most efficient plant size in the long run. We learned in Chapter 9 that the competitive firm was compelled by competition to produce at the point of minimum long-run average total cost. By comparison, the monopoly produces its output inefficiently, resulting in another social cost of the monopoly relative to the competitive outcome.

**X-inefficiency:** The increase in costs of a monopoly resulting from the lack of competitive pressure to force costs to the minimum possible level.

This result assumes that the monopoly and the competitive firm operate along the same long-run average total cost curve. If they do not, a further social cost of monopoly power is possible. This cost was proposed by Harvey Leibenstein, who called it **X-inefficiency.**[4] His reasoning is straightforward. Purely competitive firms must minimize costs of production to survive in the long run. Are monopolists similarly constrained? Leibenstein's answer is no. He argues that the monopolist can afford to live a quiet life; although the monopoly firm gains from reducing costs, there is no competitive pressure to force it to do so. The monopoly firm will not be driven out of business if it fails to minimize costs of production. So the monopolist can take part of its income in terms of a looser, less-efficient organization. The higher production costs caused by this X-inefficiency are also a social cost of monopoly power.

Estimates of the costs of X-inefficiency are hard to derive, so its possible magnitude in the economy is subject to speculation. Moreover, there are forces working to mitigate X-inefficiency costs. As we saw in our discussion of monopoly in the long run, the right to operate a monopoly can be bought and sold like any other asset in the economy. If the monopoly firm is being operated inefficiently, investors can take it over, operate it more efficiently, and make a profit. The competitive pressure of such takeovers may or may not be sufficient to erase X-inefficiency costs, but takeovers surely play a role in keeping such costs to a minimum.

---

[4]Harvey Leibenstein, "Allocative Efficiency vs. X-Inefficiency," *American Economic Review* 56 (June 1966), pp. 392–415.

## Monopoly and the Distribution of Income

In addition to the potential efficiency cost arguments against monopoly—welfare loss, rent seeking, and inefficient levels of production—there is an equity argument. In simple terms, the creation of a monopoly transfers wealth from consumers to the monopolist. In Figure 10–11, this transfer is represented by the area of monopoly profits, $P_mABP_c$. The average consumer gets poorer; the monopolist gets richer. Most people would regard this transfer to the monopolist as unfair, but here we are clearly facing an ethical comparison. What is fair? The truth of the matter is that the economist is no better than anyone else at answering such a question.

We can make some economic points about the transfer of monopoly profits, however. As we saw in the discussion of the capitalization of monopoly profits, it is the original monopolist who receives the transfer. The current owners of a monopoly may be making only a normal rate of return on their investment in the monopoly right. Thus, if the government sets out to break up monopolies in the economy on the grounds that they affect the distribution of income in an undesirable way, government would have to take care to distinguish between current and original owners of monopolies. Also, the concept of rent seeking states that monopoly profits are competed away in wasteful competition to obtain the monopoly right from government. In this sense, the monopoly profits are not transferred to anyone; they are wasted.

# The Case for Monopoly

**Static inefficiency:** A condition, related to the concept of welfare loss due to monopoly power, which is summarized as the production of too little output at too high a price.

**Dynamic efficiency:** A firm may at first glance impose welfare costs on the economy due to its monopoly power, but may on a closer look be a progressive, innovative firm. In other words, there may be a trade-off in analyzing real firms between static inefficiency and dynamic efficiency, between monopoly power and innovation, and so on.

We have seen that monopoly leads to economic inefficiency compared with a purely competitive organization of the economy. The monopolist essentially produces too little output at too high a price ($P > MC$). This type of inefficiency, called **static inefficiency,** can be seen as a social cost of the condition that other things are equal.

What if other things are not equal? Specifically, what if the monopoly produces more new ideas and products than the competitive firm? If this is the case, there may be an offsetting **dynamic efficiency** that makes monopoly a desirable form of market organization. One way to look at these counterforces is that the static inefficiency caused by monopoly pushes the economy inside its production possibilities frontier. This loss of production is the cost of monopoly power. But the dynamic efficiency of monopoly pushes the production possibilities curve outward over time, potentially swamping the short-run costs of monopoly power. In other words, monopoly has costs and benefits for the economy, and the benefits (dynamic efficiency) may outweigh the costs (static inefficiency).

This view of the potential efficiency of large-scale, monopolistic enterprise was put forth by Joseph Schumpeter.[5] He stressed the advantages of the ability of large firms to finance large research laboratories and hire thousands of scientists. While these firms with well-known, productive research labs—such as Du Pont, AT&T, and IBM—are not typically pure monopolies, they are usually large firms with dominant positions within their industries.

To Schumpeter, the traditional theory that competition spurred innovation therefore seemed wrong. Small competitive firms were at a disadvantage in the innovation process because they could not afford large-scale research. Large firms were the key to innovation and success in the industrial order.

In Schumpeter's theory, the innovative monopolist did not possess a pe-

---

[5] Joseph A. Schumpeter, *Capitalism, Socialism, and Democracy* (New York: Harper, 1942).

rennial advantage in the marketplace. He believed that innovation would proceed apace over time, and no one large firm would have more than a transitory monopoly in the face of a constant supply of new ideas and innovations by other large firms. The monopoly that any one large firm achieved by being creative was short-lived.

Other famous economists continue to advance Schumpeter's vision of the industrial economy. John Kenneth Galbraith, for example, has argued forcibly in his many works on the economy that large firms are the source of innovation and economic advance.[6] Is the Schumpeter-Galbraith vision of the economy correct? That is, what are the sources of major innovations in the modern economy? Large firms? Small firms? Independent entrepreneurs and inventors?

Numerous studies have been conducted on this issue. Studies of the source of major inventions indicate that a surprisingly large number of innovations have come from the backyard shops of independent inventors. John Jewkes, David Sawers, and Richard Stillerman found, for example, that more than half of sixty-nine important twentieth-century inventions were produced by individual academics or by individuals unaffiliated with any research organization.[7] Among these inventions were air conditioning, the Polaroid Land camera, the helicopter, and xerography. Much invention is individually inspired and created in small-scale, personal labs and workshops.

Other studies have examined the relation between patents as a measure of innovation and firm size and have concluded that in most industries the middle-sized firms have an advantage over both the very large and the very small firms in the innovation process.[8] Although the reasons for this finding are not entirely clear, more patents are issued to medium-sized firms in most industries. Industrial creativity thus seems to stem from something other than size.

Thus, the mass of the evidence suggests that the Schumpeter-Galbraith vision of the economy is not exactly accurate. But, then, neither is the vision that pure competition is best for the economy. It seems to be the case that innovation springs from many sources, from the backyard or basement lab of the inventor to the lab of the large industrial company. Diversity produces innovation, or, better yet, *individuals* produce innovation, whether they are working in their own workshop or a large firm's lab. Perhaps the most typical scenario is that an individual or small firm comes up with a new idea, and a large firm provides the vehicle by which the idea is put into practice.

Finally, where is the balance struck between the costs and benefits of monopoly? It is hard to generalize on this issue. In filling out a report card on firms, the best approach is to be pragmatic. Does the firm possess (monopoly) power? If so, can we judge how large this power is and how much it costs the economy? Does this (monopoly) power carry with it some redeeming merit such as increased innovation? These are not easy questions to answer or to weigh in the balance, but they are the type of issues that must be confronted if the costs and benefits of monopoly are to be understood and its model applied intelligently.

---

[6]See, for example, John K. Galbraith, *American Capitalism* (Boston: Houghton Mifflin, 1956) and *The New Industrial State* (Boston: Houghton Mifflin, 1967).

[7]John Jewkes, David Sawers, and Richard Stillerman, *The Sources of Invention* (New York: St. Martin's, 1959).

[8]See Fredric M. Scherer, *Industrial Market Structure and Economic Performance* (Chicago: Rand McNally, 1970), Ch. 18, for a survey of such studies.

## Summary

1. A pure monopoly is a single seller of a product for which there are no close substitutes in a market characterized by high barriers to new entry.
2. The monopoly firm faces the industry demand curve. When it lowers price, it must accept a lower price for all units previously sold for a higher price. The new revenue at the lower price minus the old revenue at the higher price is equal to the marginal revenue of the monopoly firm. For the monopoly, marginal revenue is less than price.
3. The monopoly selects an output to produce by setting marginal cost equal to marginal revenue. Since marginal revenue is less than price, the monopoly firm sets prices where price is greater than marginal cost.
4. Not all monopolies make profits. In the short run it is possible for the monopoly to incur a loss. If the loss persists in the long run, the monopoly will go out of business.
5. In the long run the monopoly firm can adjust its plant size. Normally, however, the monopoly does not operate at the point of minimum long-run average total cost. Monopolies can be bought and sold in the long run.
6. Price discrimination requires that a seller be a monopoly firm, be able to separate buyers on the basis of their elasticities of demand, and be able to prevent buyers in the low-price market from reselling to buyers in the high-price market.
7. Where buyers can be separated into two classes, the monopoly firm will set marginal cost, which is the same in both markets, equal to marginal revenue in each market. This is the definition of price discrimination—price differences that do not reflect cost differences.
8. There are three potential efficiency costs of monopoly: the welfare cost due to monopoly power, which is the lost consumer surplus caused by monopoly; the rent-seeking cost of monopoly, which is the value of the resources used to capture monopoly profits; and the degree to which the monopoly firm has higher costs of production than a competitive firm.
9. Monopoly affects the distribution of income. Where there is no rent seeking, monopoly transfers wealth from consumers to the monopolist.
10. A potential advantage of monopoly is that it allows the large scale necessary to foster innovation. The monopoly or large firm might thus possess a dynamic advantage over a small, competitive firm. The evidence on this issue is mixed, but it appears that the most progressive firms in most industries are middle sized.

## Key Terms

pure monopoly
legal barriers to entry
public franchise
government license
patent
natural monopoly
price searcher
profit-maximizing price
price discrimination
contrived scarcity
welfare loss due to monopoly
rent seeking
X-inefficiency
static inefficiency
dynamic efficiency

## Questions for Review and Discussion

1. What is a barrier to entry? Why is it important in the definition of monopoly?
2. Depict the short-run equilibrium of a pure monopolist. Can the monopoly firm make a loss? If so, diagram what the loss situation for the monopoly might look like.
3. What are the conditions that need to be met before a monopoly can price discriminate?
4. What are three efficiency costs of monopoly? What is a potential advantage of a monopoly firm?
5. Insurance companies have started to offer lower rates for life insurance to individuals who do not smoke and who jog. Is this an example of price discrimination? Why or why not?
6. Why is lobbying for an income transfer such as monopoly profits a social cost?
7. Entry is not possible with pure monopoly, but the monopoly can sell its monopoly right. What is the difference?
8. Monopolies and large firms are sometimes seen as the culprits in causing prices in the economy to rise. Does the theory of monopoly presented in this chapter imply that prices under monopoly will rise quickly or will be high? Explain.

## ECONOMICS IN ACTION

### State Liquor Monopolies

The wholesale and retail distribution of liquor in the United States takes place in two basic ways. First, some state governments monopolize the sale of alcohol; these states are called control states. In 1980, there were eighteen control states, including Alabama, New Hampshire, North Carolina, and Virginia. In these states, consumers must purchase liquor in the state store. Second, the remaining thirty-two states and the District of Columbia are license states; individual firms can purchase licenses to conduct all phases of the wholesale and retail sale of alcohol. While entry is not completely free and the price of licenses varies across these states, entry is easier than in the monopoly states, where it is impossible. We therefore have a comparative case: states where liquor is sold by state monopolies versus states where liquor is sold by licensed competitors. The theory of monopoly suggests that price should be higher and output lower in the monopoly states. Examine the following 1980 data:

| Type of State Control | Nine-Brand Average Price[a] |
|---|---|
| License | $7.57 |
| Monopoly | 7.36 |

[a]These prices are inclusive of the federal excise tax on distilled spirits, which applies in all states, and the varying excise levies imposed by state and local governments.

The state monopolies, on average, sell liquor at a lower in-store price than the states with competitive suppliers! It would seem that the theory of monopoly has fallen flat on its face.

The key to understanding this case is understanding the nature of "price." We said above that the in-store price was lower in monopoly states. The in-store price is not the real price of a purchase of liquor. The full price of liquor includes such factors as the time it takes to find a store, the hours that the store is open, and the service provided by the store (for example, does it take credit cards and checks or provide help in carrying purchases to the car?).

Now consider the following data:

| Type of State Control | Number of Persons per Outlet |
|---|---|
| License | 2,961 |
| Monopoly | 29,139 |

One reason for the great difference in the numbers of potential customers served by each store is that there are simply many more suppliers in the license states, by about 10 to 1. The real price of liquor is therefore lower in these states, for stores are easier to find and are closer on average to the people who use them. In addition, license stores generally are open longer, offer better service, carry a larger selection of brands, accept checks and credit cards, and so on. For such reasons the real price of liquor is lower in competitive than in monopoly states.

Thus, monopoly theory does help us understand real-world monopolies. It predicts that price should be higher and output lower when a market is monopolized, and this is what we find in the case of the state liquor monopolies after taking account of the real price of liquor.

### Question

In various cities, states, and nations of the world, such services as airline transportation, water distribution, and garbage collection are provided by government and in private markets. Drawing on the above example of state versus private liquor sales, what might you observe about real price and quality of service with respect to airline transportation, water supply and garbage collection?

*Source:* Distilled Spirits Council of the United States, Inc., *Annual Statistical Review 1980: Distilled Spirits* (Washington, D.C., 1981).

# 11

# Monopolistic Competition, Oligopoly, and Cartels

Until the 1930s, microeconomic theory analyzed only two basic market structures: pure competition and pure monopoly. These structures are, of course, opposite extremes. In a purely competitive market, prices tend to equal costs in the long run and consumers' economic welfare is theoretically maximized, whereas in a monopoly market the single monopolist may restrict output, raise prices, and capture economic profits. In the 1930s, two economists, E. H. Chamberlin of Harvard University and Joan Robinson of Cambridge University, wrote books developing models of market structures that did not fit the mold of pure competition or pure monopoly. Both Robinson and Chamberlin emphasized that the real world was characterized by market structures that did not easily fit into existing economic theory.

All models of firms and markets other than the extremes of pure competition and pure monopoly are called theories of **imperfect competition.** The word *imperfect* does not imply a value judgment. It simply means that not all the conditions for pure competition are met.

Some imperfectly competitive markets contain large numbers of sellers (as in pure competition) selling slightly different products (unlike in pure competition). These are called monopolistically competitive markets.

Other structures are composed of small numbers of competing sellers who recognize that their actions have an impact on one another's sales, prices, and profits. Under these circumstances, firms may engage or try to engage in various forms of combination or collusion. Such associations are called oligopoly or cartel market structures. They are closer in economic results to pure monopoly than to pure competition. The spectrum of market structures, therefore, runs from pure competition to monopolistic competition to oligopoly and cartels to pure monopoly.

This chapter focuses on monopolistic competition, oligopolies, and car-

**Imperfect competition:** A market model in which there is more than one firm but the necessary conditions for a purely competitive solution (homogeneous product, large number of firms, free entry) do not exist.

tels. In assessing market structures, we pay particular attention to the contrasts between each of these models and the purely competitive model. Governmental attempts to steer certain industries toward market structures considered more socially beneficial are covered in depth in Chapter 17.

Why is it important to categorize and classify intermediate or imperfect structures, those between competition and monopoly? No real-world market is characterized by the exact characteristics of pure competition or pure monopoly. We may think of the market for ice cream as competitive with many sellers, but in reality every brand or flavor is slightly different. Or, at the other end of the spectrum, we may view our local water company as a monopoly, but there are a few water substitutes available, such as digging a well.

Because real-world markets seldom fit into the neat categories of competition or monopoly, it is very important to develop more realistic models. The economist, moreover, must assess the effects of these real-world market structures on consumers' and producers' welfare and on economic efficiency in general. We begin with monopolistic competition, the economic model closest to pure competition.

## Monopolistic Competition

In his search for more realistic models of firms and markets, E. H. Chamberlin developed a model close but not equivalent to pure competition. Since the model contained some monopolistic elements, Chamberlin labeled this theory "monopolistic competition."

### Characteristics of Monopolistic Competition

**Monopolistic competition:** A market model with freedom of entry and number of firms that produce similar but slightly differentiated products.

**Monopolistic competition** is a market structure in which a large number of sellers sell similar but slightly different products, free from entry and exit barriers. Advertising is the principal tool for differentiating products within monopolistic competition.

**Large Number of Sellers.** Monopolistic competition, like pure competition, is characterized by a large number of competing sellers. The effect of large numbers is the same in both cases: Collusion to fix prices or other kinds of cooperation is costly—so costly, in fact, that it ordinarily does not occur. The presence of many sellers is a competitive element in monopolistic competition. Unlike pure competitors, however, monopolistic competitors are not always price takers because of another characteristic—product differentiation.

**Differentiated products:** A group of products that are close substitutes, but each one has a feature that makes it unique and distinct from the others.

**Nonprice competition:** Any means that individual firms use to attract customers other than price cuts.

**Product Differentiation.** Competitive sellers sell a homogeneous or identical product, such as corn. Monopolistic competitors sell products that are highly similar but not identical. Products may be distinguished by brand names, location, services, even differences merely perceived by the consumer. Because the monopolistic competitor is selling a slightly **differentiated product,** the firm will have a degree of control over price, unlike the competitive firm. Product differentiation is an example of **nonprice competition,** a term commonly used to refer to any action other than price cuts taken by a competitor to increase demand for its product.

There are many varieties of nonprice competition. One is differences created in the products themselves. Consumers shopping for home computers can now choose among many firms' products, each differing slightly in hardware, software, styling, sturdiness, and user-friendly or state-of-the-art appeal. Sometimes distinctions between brands are more illusory than real. Although aspirin is essentially a homogeneous product, monopolistically competitive firms produce aspirin under many different brand names—Bayer, Excedrin, Cope, and so on—often at very different prices. The reason for the different prices for aspirin is that each brand is differentiated in consumers' minds, often by advertising or some other form of sales effort. Consumers who are convinced that brands of aspirin are actually different will have some allegiance to a particular brand. It is this "brand allegiance" that permits the seller to set price, within limits.

Location may also differentiate a seller's products. There may be ten Exxon gas stations within a city, alike in every respect, including the gasoline they sell. The gas stations each occupy a different location, however, and thus are free to set different prices. The Exxon station nearby the interstate might charge a higher price than the station downtown because travelers in a hurry are willing to pay more. In fact, many businesses have failed because of poor location. You may have noticed a fast-food restaurant go out of business while others selling identical products thrive in locations where parking and accessibility are better.

Service is another form of nonprice competition. Physically identical products may be offered for sale in supermarkets at identical nominal prices. However, one store may be untidy, another may employ rude checkers, and still another suffer slow-moving checkout lines. Markets with well-swept aisles and shelves, automated checkout, and friendly clerks will have differentiated their products by the services they offer.

Subtle product differentiation, whether by packaging, service location, or any other variable, is a common fact of life, but the degree of substitutability among such products is high. While demand is highly elastic under monopolistic competition, it is not infinitely elastic as in perfect competition. A sudden tenfold increase in the price of one brand of aspirin would send buyers scurrying for substitutes. The monopolistically competitive firm, in other words, like the firm in pure competition, is primarily a price taker. It has a small and limited degree of control over price, however, stemming from its ability to differentiate its individual product from those of its many competitors. This limited control over price is the monopoly element in these markets. Chamberlin's ideas are verified by common observation: *There exists practically no market that is not characterized by monopoly elements through some form of differentiation.*

**No Barriers to Entry.** Just as with competitive markets, entry into and exit from the monopolistically competitive market is free. This fact follows from the high cost of collusion in a market occupied by a large number of sellers. Although monopolistic competition shares the free-entry conditions of perfectly competitive markets, entry itself is not costless in either market structure. There are obvious resource costs and commitments in opening any business: hiring labor, investing capital in fixed items such as buildings, managing or hiring managers, and so forth. Free entry means that there is no government regulation over entry and that essential raw materials are not controlled by one or a few firms.

In addition to the start-up costs associated with entering any new business, entry into monopolistically competitive markets entails the cost of introducing a new product. New product development and market entry through advertising and other sales efforts constitute a cost that the competitive firm does not face. A farmer's entry into the white winter wheat market, for example, will not involve a sales or advertising campaign, whereas entry into the deodorant, paper towel, or restaurant market will. Product differentiation is simply part of the cost of entry under monopolistic competition. Entry is nonetheless assumed to be free, and in this characteristic, monopolistic competition is very much like perfect competition.

**The Importance of Advertising.** With the general characteristics of the monopolistically competitive structure in mind, we examine a typical firm's demand curve and assess the impact of **advertising** on it. The existence of nonprice competition coupled with a high degree of substitutability means that the firm's demand curve, while highly elastic, is downward sloping. Such a demand curve is drawn in Figure 11–1 for a hypothetical antacid product, Relief. The demand curve for Relief approximates that of a competitive firm, but it is downward sloping because Relief is not the same product as its competitors, Tums, Rolaids, Mylanta, Maalox, and so on.

As with the pure monopoly firm, the monopolistically competitive firm's marginal revenue curve lies below its demand curve, and for the same reason: Additional units can be sold only at a lower price. The main difference between a monopoly firm and a monopolistic competitor is that the monopoly firm sells a unique product—one with no close substitutes—while substitutes abound for the monopolistic competitor's product.

How elastic or inelastic will the demand curve be for Relief? The answer will depend on the size and intensity of consumer preferences for Relief, given the range and price of substitutes. Naturally, the firm will be interested in increasing the size of its market and in intensifying consumers' preferences for its product. The firm accomplishes these goals through product advertising. Advertising is essential to the introduction of products in monopolistic competition. It is also a crucial tool in the manipulation or management of demand.

Advertising is a variable under the monopolistically competitive firm's control, within the limits imposed by its advertising budget. The firm in monopolistic competition will seek, along with all firms who advertise, to equate the addition to revenue from the final dollar spent on advertising to

**Advertising:** Any communication that firms offer customers in an effort to increase demand for their product.

**FIGURE 11–1**
**Demand Curve Facing Monopolistically Competitive Firm**

The demand curve for the monopolistic competitor slopes downward because the firm sells a product differentiated from its competitors. The marginal revenue curve lies below the demand curve because additional units can be sold only at a lower price.

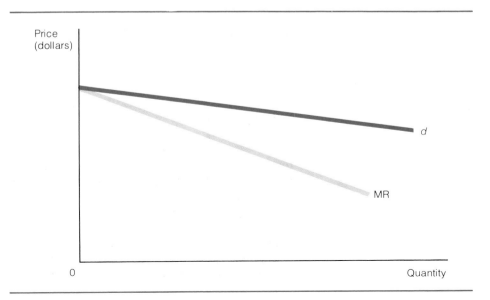

## FIGURE 11–2
### The Effects of Advertising on the Firm's Demand Curve

The firm advertising its product hopes to shift its demand curve to the right and to reduce its elasticity. In doing so, the firm tries to intensify demand for the product by distinguishing it in the minds of buyers.

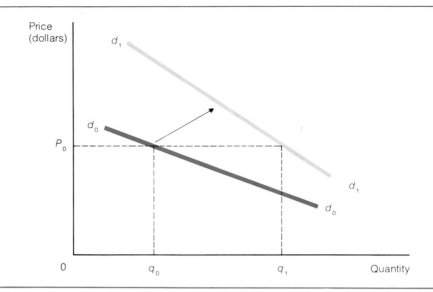

the marginal addition to cost. Naturally, firms seeking to enter markets will often spend large amounts of money to introduce their products. But whether the firm is an established competitor or a new entrant, the purpose of advertising is the same: to increase demand for the firm's product.

In Figure 11–2, demand curve $d_0d_0$ depicts a monopolistically competitive firm's demand curve. The hoped-for effect of increased advertising is a shift rightward in the position of the demand schedule to $d_1d_1$ *and* a reduction in the elasticity of the curve. Thus advertising is used not only to increase demand but to intensify demand by differentiating the good or service in consumers' minds so that the firm will be able to raise the price of the product. If the advertising of Relief is able to convince consumers that the product is "easier to swallow," "has no unwanted side effects," and the like, demand becomes more inelastic (less substitutable), and consumers are willing to pay more for it. Focus, "The Pros and Cons of Advertising," discusses advertising in more detail.

### Short-Run and Long-Run Equilibrium Under Monopolistic Competition

Now that we have defined the overall conditions for monopolistic competition, we can discuss how firms set prices. The firm in monopolistic competition has some control over price but, as we will see, its control differs in the short run and the long run.

**Short-Run Equilibrium.** Monopolistically competitive firms, like firms in all other market structures, always equate marginal cost to marginal revenue to maximize profits or minimize losses. In the short run, the firm may enjoy economic profits or endure losses.

Figure 11–3a gives a firm's short-run cost curves, demand curve, and marginal revenue curve. MC intersects, or is equal to, MR at quantity $q_0$. The firm produces that quantity and sets price $P_0$ for $q_0$ units of output. At this price and quantity, the firm will earn economic profits. These profits are

## FOCUS: The Pros and Cons of Advertising

Advertising and product differentiation are the hallmark of monopolistic competition. In contrast, competitive advertising is largely unnecessary under either pure competition (since products are homogeneous) or pure monopoly (because there is a single seller with no close substitutes).

No subject in the economics of market structure has been debated as vigorously as advertising. Some economists consider advertising wasteful and "self-canceling." Others judge it to be integral to the competitive process. We briefly consider three of the major arguments.

### Is Advertising Informative?

Critics of advertising argue that most ads are tasteless and wasteful assaults on consumers' senses. Since products and services in monopolistically competitive markets are, by definition, close substitutes, product characteristics are difficult to differentiate with *actual* differences. Rather, critics point out that advertising creates only imagined differences. A large number of competitive firms have the incentive to advertise in this manner, creating a confusing array of messages to consumers. In this view, advertising allocates demand among competitors without increasing the total demand for the product. To increase the total demand for a product, such as fast-food hamburgers, cat food, or breakfast cereal, advertising must reduce either consumers' savings or their expenditures on other goods. If advertising is unsuccessful at this task, the critics allege, it is unproductive because it merely allocates demand among competing firms producing goods that are fundamentally alike.

Defenders of advertising argue otherwise. In their view, nonfraudulent advertising offers real information about the existence of products and their characteristics. Such information thus lowers the consumers' cost of searching for goods. Further, actual brand comparisons made possible by a Federal Trade Commission ruling and price advertisement are by nature informative. Defenders of advertising argue that knowledge of any characteristic of a product or service produces information that permits consumers to make a rational choice among competing goods within their budgets. Unsatisfactory products are simply not repurchased. Shifts in consumers' brand allegiance resulting from advertising or any other cause are symptomatic of a working competitive system. According to this viewpoint, critics of advertising's alleged tastelessness are merely trying to substitute their own subjective judgments for consumer sovereignty.

### Does Advertising Lower Costs?

Critics argue that advertising expenditures, being largely duplicative and wasteful, increase the price of goods without producing corresponding consumer benefits. Defenders point out that, owing to economies of scale, increased production lowers per unit costs as output increases. Advertising that increases demand might actually lower the average costs of production, resulting in lower prices for consumers.

The argument is clearly empirical. If advertising can reduce unit costs with economies of scale, it could also increase unit costs if the firm is pushed to outputs characterized by increasing average costs. Unfortunately, the economies of scale argument is impossible to assess without some knowledge of the firm's cost curves.

In another variant of the cost argument, defenders of advertising argue that in monopolistically competitive markets advertising is a means of entry. It is part and parcel of the competitive process. Without advertising, existing firms would have greater market power—greater ability to charge higher prices and to exact economic profits. Defenders of advertising point to markets where advertising, especially price advertising, is restricted. For example, physicians, funeral directors, and dentists are all restricted in the amount and type of advertising legally permitted. When such nonprice competition is disallowed, defenders of unrestricted advertising argue, prices to consumers tend to be higher and available outputs tend to be lower than if advertising were free and unregulated.

### Does Advertising Create Social Waste?

A third, much-debated advertising issue, originated by economist John Kenneth Galbraith, relates to social waste. Galbraith argues that there is an imbalance between private goods and social goods—that too many automobiles, laundry detergents, and fashions (advertised private goods) are produced relative to dams, highways, and pollution-control facilities (nonadvertised social goods). Many private goods, moreover, are tasteless and ugly, while social goods produce societal welfare and utility.

Defenders of advertising argue that the social balance issue should not be settled by subjective aesthetic opinion but by consumer-voter-taxpayers. If voters are actually in control in a democracy, social wants are filtered through a political process, just as private wants are registered through the dollar votes of consumers. To say that advertising is useless and should be the object of regulation is to deny its potential benefits. Advertising, in the defenders' view, has permitted the rapid introduction of technological innovations with great consumer benefits. Advertising may make possible the sale of what some consider trash, but it also helps sell Mozart, Tolstoy, the ballet, and the New York Philharmonic. From the perspective of individualism and individual development, then, advertising has fostered a maximum of consumer choice and freedom.

Obviously, these arguments concerning the pros and cons of advertising cannot be settled here. Any stand on a particular issue depends on the facts of the particular case, one's philosophical preconceptions, and so on. However, it is important to understand at least some of the arguments within the context of monopolistic competition and all other market structures where advertising is observed.

calculated by multiplying price times quantity sold ($P_0 \times q_0$) to get the firm's total receipts, and then subtracting the total cost of $q_0$ units, $ATC_0 \times q_0$. Profits to the firm are shown in the shaded area of the figure. Notice that the firm's demand curve is highly elastic because of the large number of substitutes but that it is still negatively sloped because, under monopolistic competition, products are differentiated through advertising and other forms of nonprice competition.

Like firms in all other market structures, monopolistically competitive firms do not always make profits. Figure 11–3b depicts a loss-minimizing short-run equilibrium for a monopolistically competitive firm. In the case illustrated, the firm will sell $q_0$ units at price $P_0$—the price-quantity combination selected where MC equals MR. This time, however, total receipts, $P_0 \times q_0$, are exceeded by total costs, $ATC_0 \times q_0$. Losses (the shaded rectangle) are incurred. The firm, however, will remain in business in the short run because its total variable costs $AVC_0 \times q_0$ are covered by its total re-

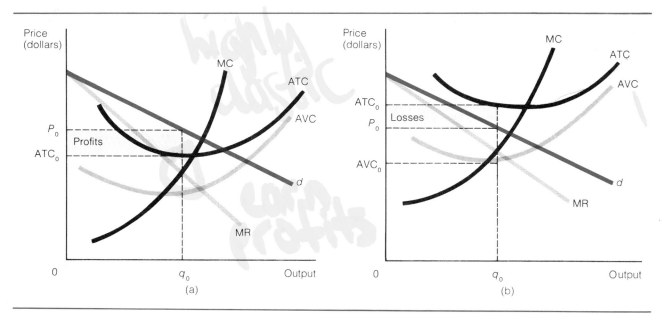

**FIGURE 11–3  Short-Run Profits and Losses for the Monopolistically Competitive Firm**

(a) The firm earns short-run profits. It sets price $P_0$ for $q_0$ units of output. Its profits are the difference between its revenue ($P_0 \times q_0$) minus its costs ($ATC_0 \times q_0$). (b) The firm suffers short-run losses. Costs are greater than revenue. The firm will continue operating, however, because its total variable costs ($AVC_0 \times q_0$) are less than its revenue.

ceipts, so there is some contribution to fixed costs. If these conditions prevailed in the long run, the firm (like the competitive or pure monopoly firm) would go out of business. In the long run, expenditures that are fixed in the short run become variable expenses and must be covered.

The monopolistic competitor does have one advantage over the pure competitor in the short run: the ability to manipulate demand through advertising and intensified product differentiation. The loss-minimizing firm in Figure 11–3b may increase short-term outlays on advertising or sales effort, thereby shifting demand and marginal revenue curves rightward. Such an increase may permit the firm to break even or to earn profits. Optimum advertising outlay for the firm occurs when one dollar of additional selling costs adds exactly one dollar to the firm's receipts. In this sense, the cost of advertising is treated like the cost of any other input, such as capital or labor.

**Long-Run Equilibrium: The Tangency Solution.** Long-run equilibrium for the firm in monopolistically competitive markets may take a number of forms. One of them, described by E. H. Chamberlin, is the famous **tangency solution** wherein the firm ends up making zero economic profits; that is, it simply breaks even. The assumptions of this model of the firm's economic behavior can be summarized as follows:

1. The firm operates in an industry composed of many sellers selling substitutable products, with the number of sellers constant throughout the period of adjustment.
2. Each seller assumes that its price actions will not provoke a response from rivals.
3. The degree of product differentiation (and thus advertising budgets) is constant and is determined by all competing firms; firms, therefore, manipulate prices to increase profits.
4. The firm represents all competitors in that their cost curves and demand curves (as well as their reactions) are identical.

A model conforming to these assumptions is shown in Figure 11–4. Two demand curves, $d_0d_0$ and $d_1d_1$, are reproduced along with a long-run average

**Tangency solution:** A long-run situation in which the firm's downward-sloping demand curve is just tangent to the average total cost curve, necessarily implying zero economic profits.

### FIGURE 11–4
### The Tangency Solution

Each firm believes it can increase profits by lowering price. When all firms lower price from $P_0$ to $P_1$, however, the demand curve shifts leftward for each firm. In final equilibrium, $E$, the point of tangency with the LRATC curve, all firms break even.

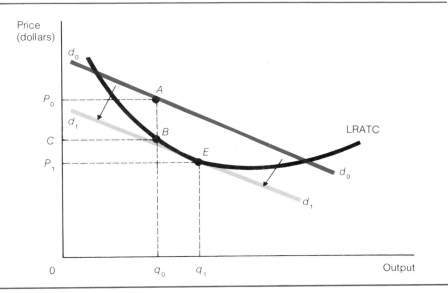

cost curve for the firm. Suppose that the representative firm (out of perhaps 100 firms producing similar products) initially finds itself at point A, charging price $P_0$ and selling quantity $q_0$. (The marginal revenue curves associated with $d_0 d_0$ and $d_1 d_1$ are omitted in Figure 11–4 for simplicity.) From the perspective of profitability, the firm, along with all other competitors, is earning economic profits in the amount represented by $P_0 ABC$, B being the long-run average cost of producing quantity $q_0$. Collusion to maintain profits at this level is impossible because of the large number of producers, so these economic profits cannot be sustained. The initial price and quantity combination selected, $P_0$ and $q_0$, will not be permanent in the face of Chamberlin's assumptions about firms' behavior.

Why? Because each firm believes that it can increase profits by lowering its price. In other words, each firm believes that its demand curve is more inelastic than it really is because it believes, erroneously, that rivals will not reduce prices when it does. In fact, rivals usually react to another's price drop by lowering their prices. This widespread lowering of price causes the demand curve for each seller to shift leftward until economic profits are zero, that is, until each firm breaks even. Finally, demand curve $d_1 d_1$ becomes tangent to the long-run average total cost function, LRATC (thus the term *tangency solution* to describe this case).

At point E in Figure 11–4, all firms will produce $q_1$ units of output and charge price $P_1$ per unit. Quantities above this amount and prices below $P_1$ would force firms out of business. Quantities less than $q_1$ and prices higher than $P_1$ would provoke the price adjustment responses described above. Under these circumstances, a stable equilibrium characterized by zero economic profits for all firms is generated.

A word of warning: While the tangency solution is perhaps the best-known analytical approach to monpolistic competition in the long run, it is only one of a large number of possibilities. A little reflection on the realism of the four assumptions in Chamberlin's model, listed earlier, will tell us why.

1. The initial existence of economic profits will certainly encourage the entry of new firms into the market, altering the effects of price responses.
2. We cannot assume that firms will continue to think that price reductions will not provoke similar responses from their competitors. Firm managers must have the limited vision of snails to remain unaware of other firms' repeated and identical responses to other firms' price reductions.
3. Monopolistic competitors can use product differentiation and advertising as well as price manipulation to maintain demand and profits. The degree of differentiation is artificially held constant in the model described in Figure 11–4.
4. It is obvious to the most casual observer that all firms are not alike with respect to demand and costs.

Thus the model described in Figure 11–4 reflects but one set of behavioral assumptions among many possibilities. Its value is considerable, however, because it describes a tendency to tangency. Real-world tangency solutions will likely not exist, but as long as the assumptions of Chamberlin's famous model *approximate* reality, an approach to a tangency solution may be observed and predicted in markets displaying the features of monopolistic competition. Additionally, the tangency model of monopolistic competition provides a convenient benchmark for a comparison of the attributes and economic effects of pure competition and those of monopolistic competition.

### Does Monopolistic Competition Cause Resource Waste?

A charge commonly leveled at monopolistic competition—unrelated to the alleged wastes of advertising—is that the tangency solution involves, and indeed requires, (1) underallocation of resources in production or (2) generation of excess capacity and prices higher than minimum long-run unit costs of production. This unflattering portrayal of monopolistic competition is possible when it is compared with the long-run competitive model.

**Monopolistic Competition versus Pure Competition.** Long-run equilibrium in the perfectly competitive model, discussed in Chapter 9, results in the following three conditions:

1. Price = Long-run marginal costs.
2. Price = Minimum long-run average total costs.
3. Total revenue = Total costs.

The first condition means that the consumer's marginal sacrifice (price) just equals the marginal opportunity cost of producing the last unit of a product or service consumed. The second condition means that consumers are getting the product at the lowest possible price at which the product can be produced. The third conclusion means that no competitive firm earns more than a normal profit in the long run. Firms are price takers under competition. Freedom of entry guarantees normal profits in the long run.

Consider, with the aid of Figure 11–5, how monopolistic competition compares to pure competition in these respects. Figure 11–5a shows the de-

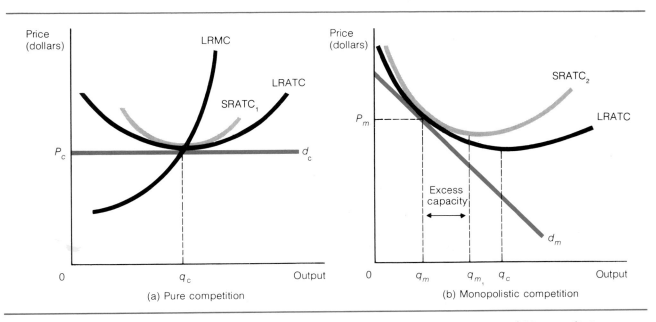

**FIGURE 11–5  A Comparison of Pure Competition and Monopolistic Competition**

In a technical sense, the monopolistically competitive firm (b) charges a higher price, $P_m$, than the purely competitive firm (a), $P_c$. The monopolistically competitive firm sets output at $q_m$ to maximize profits, but the most efficient use of its plant would be an output of $q_{m_1}$. The firm therefore produces at excess capacity, or underutilization of its scale of plant.

mand curve $P_c d_c$ and the short-run and long-run average cost curves for the typical firm in pure competition. Clearly, both long-run and short-run costs are tangent to the perfectly elastic demand curve for the firms in competition. Quantity $q_c$ is produced at price $P_c$, which equals long-run (and short-run) marginal cost. Figure 11–5b shows the downward-sloping demand curve $d_m$ of a monopolistically competitive firm in a tangency solution with the firm's long-run average cost curve. Given that the firms' cost curves are identical under both market structures, how do they compare in economic efficiency?

As Figure 11–5 shows, the following features result from monopolistic competition:

1. Price > Long-run marginal costs.
2. Price > Minimum long-run average costs.
3. Total revenue = Total costs.

The first conclusion means that the additional sacrifice (price) that consumers are willing to make for an additional unit of the good is greater than the marginal opportunity cost of producing that additional unit. In other words, resources are underallocated in the production of goods by monopolistically competitive firms, just as they usually are in production by pure monopolies.

The second conclusion is that consumers pay more for goods produced under conditions of monopolistic competition than they do when goods are supplied by perfect competitors. The fact that $P_m$ in Figure 11–5b is greater than $P_c$ in Figure 11–5a means that price is higher and output lower under monopolistic competition than under pure competition. Consumers are not paying the lowest possible unit cost. However, like perfect competitors, monopolistic competitors do not earn economic profits because a tangency solution means that total revenues equal total costs to the firm (or, alternatively, that the firm's average cost equals its average revenue).

Critics of monopolistic competition raise still another issue: that **excess capacity** is generated under the monopolistically competitive market structure. This criticism ordinarily holds for any less-than-perfectly competitive market. To understand what excess capacity means, consider Figure 11–5 again. The LRATC function is called an envelope curve because it is composed of a series of tangencies of points on the short-run average total cost curves (see Figure 8–5 on page 171). The short-run curves $SRATC_1$ and $SRATC_2$ drawn in Figure 11–5 are two such curves. To utilize fully any scale of plant or to produce at an optimum rate of output in the short run, the existing scale of plant or existing resources invested must be used at the lowest average cost of production. For the monopolistically competitive firm depicted in Figure 11–5b, the output corresponding to that lowest short-run average total cost of production is $q_{m_1}$, not $q_m$, the profit-maximizing quantity the firm has chosen. From society's point of view, the firm producing at $q_m$ is wasting, or underutilizing, resources. Excess capacity is the unused capacity of the scale of plant that the firm has built. From the firm's perspective, however, the scale of plant represented by $SRATC_2$ is perfect or optimal in that it produces the quantity $q_m$ more cheaply than any other scale of plant could. Thus critics argue that monopolistic competition creates excess capacity as well as the long-run resource inefficiencies outlined above.

**A Defense of Monopolistic Competition.** Is monopolistic competition as inefficient as the critics say? In assessing this issue we must be careful to distin-

---

**Excess capacity:** A situation in which industry output is not produced at the lowest possible average total cost, the result of underutilized plant size.

**Economic efficiency:** Proper allocation of resources from the firm's perspective.

**Economic welfare:** The situation in which products and services are offered to consumers at the minimum long-run average total cost of production.

guish between pure **economic efficiency** and **economic welfare.** Economic efficiency means that resources are properly allocated (at minimum cost for a given output) from the firm's perspective. Pure economic welfare means that consumers obtain all products and services at the minimum long-run average total cost of production. Chamberlin, for one, argued that the monopolistically competitive market structure is not wasteful in either sense because product differentiation creates variety and extends the array of consumers' choices.

If price is established close to minimum long-run average total cost, some pure inefficiency might exist with monopolistic competition, but consider what consumers get in return: varied products and a vastly expanded number of products and services from which to choose. If this variety is socially valued, advertising and product differentiation are not necessarily wasteful. Such differentiation, summed across all consumers, means that monopolistic competition may increase social welfare. A society that produces only white shirts may be efficient in a technical sense, but it is apparent that Americans have not chosen such a society. The repeated purchase of other kinds of shirts at higher prices means that consumers are willing to pay for differentiation and that their satisfaction is thereby heightened.

Some excess capacity may also be a benefit and not a cost in terms of consumer welfare. As indicated in Chapter 4, resource costs other than money price are involved in the purchase of goods. Time, for example, is an important cost in consuming goods. Excess capacity in the sale of some goods and services may simply be a means of lowering consumption costs. Taxi cabs typically line up and wait for fares in front of hotels and other busy establishments. This observed "excess capacity" may have a welfare benefit for consumers. A consumer's cost of transportation is certainly reduced by reduced waiting time. And as suggested in Focus, "Why Is There a Funeral Home in Most Small Towns?" even the excess capacity built into funeral homes is socially valued. The conclusion: Excess capacity itself may provide consumer welfare.

### How Useful Is the Theory of Monopolistic Competition?

When models of monopolistic competition and pure competition are subjected to real-world observation, is there a dime's worth of difference between them? How useful, in other words, is the model of monopolistic competition?

If the demand curve is very elastic when products are differentiated, the model of monopolistic competition is a close approximation of pure competition. As noted in Chapter 8, the perfectly competitive model is primarily an idealized benchmark with which real-world markets can be compared. Simple observation tells us that most actual markets conform to the set of characteristics we have defined as monopolistically competitive.

Thus, a study of markets with differentiated products and rapid and free entry of substitutes is a study of what is usually called "competition." With the ready emergence of substitutes, the "monopolistic" adjective in monopolistic competition may be downplayed. Firms' ability to set prices is critically restricted by such entry and even by potential entry. Viewed in this light, the model of monopolistic competition, with its emphasis on product differentiation and other forms of nonprice competition, is of great value in helping us understand how dynamic real-world markets function, change, or thrive.

## FOCUS: Why Is There a Funeral Home in Most Small Towns?

The funeral service business in the United States illustrates the workings of a monopolistically competitive market. Some observers and critics charge that funeral homes operating in small towns have a virtual monopoly on funeral services in their area. These firms are said to be able to price services above marginal costs by restricting output (monopoly power), to thus be able to take advantage of the extremely strained circumstances surrounding death, and to create excess capacity in the process. Is the fact that there is a funeral home in most small towns evidence of a degree of monopoly power and of excess capacity? Or is the market more competitive than commonly believed?

First we consider the information that consumers use in choosing funeral homes. Generally, more than one person goes along in viewing the home. This fact tends to reduce quick, emotional, or irrational choices. Past experience in dealing with firms and the recommendations of family and friends play a part in the selection of a funeral home. If the firm has a reputation of providing poor-quality services, even in small towns, we would expect new firms to enter the market. Indeed, one characteristic of the funeral industry is that of the long-standing firm, implying that quality of service has been at acceptable levels over the years.

But why are these firms in most small towns? One reason is that proximity is a consideration in making the choice. Most funeral directors in small towns also have other means of income because the "supply" of deaths in small towns is not as great as in large cities. Still, the home is there because people want quick service with proximity to where they live. People also like to deal with someone they know, someone who will provide speedy, high-quality, personal services. Waiting would itself be a cost to consumers if funeral service were not quickly available nearby.

Another issue related to monopolistic competition concerns excess capacity. The observed excess capacity in the funeral industry may not be excess at all in an economic sense. If the quality of the service depends on the funeral director's undivided attention, and if each funeral utilizes a major portion of the funeral home's physical plant, such as viewing rooms, then a funeral home may have to build in some excess capacity merely to avoid peak load problems. This is characteristic of many industries, particularly service industries, where the demand for output is unpredictable or variable and where output cannot be stored easily. Thus, these seemingly idle resources have economic value. Because they are idle some of the time, there will be sufficient resources at other times so that funerals can be performed with no delay.

All firms must have ways of dealing with variations

*In most small towns, funeral services are usually sold through independent, family-owned businesses. These small firms are able to operate profitably because of the special characteristics of the funeral industry.*

in demand, and there are several methods for balancing production and sales flows. Sometimes output can be produced and stored as inventories, which in turn fluctuate with demand and production conditions. Some firms allow queues to form when demand temporarily exceeds capacity. Others hold larger amounts of physical capacity and other resources—excess capacity—and thus have shorter (or no) queues, even during periods of high demand. The cost of these measures has its effect on the price of the product. Consumers, in their choice of firms, balance the price of the service or product against the time they would have to wait for delivery.

In the funeral industry, excess capacity can be explained by the nature of demand. Given the personal nature of funeral services, proximity as well as personal attention of the funeral director may be important qualities of service that the consumer purchases, along with the casket and other items. Even very small towns may therefore have funeral homes, each of which performs a relatively small number of funeral services per year. While this widespread existence of underused facilities appears to be inefficient, it occurs because people are willing to give up some of the benefits of econ-

## Oligopoly Models

In the remainder of this chapter we turn to the types of market structure between perfect competition and pure monopoly that we have designated as oligopolies and cartels. As we have seen so far, monopolistic competition lies closer to perfect competition than to the pure monopoly model. The demand curve facing the monopolistically competitive seller is downward sloping, but not by very much. The monopolistically competitive firm has some limited control over price, in contrast to the competitive firm. Oligopolies and cartels, however, lie closer to the pure monopoly model. Sellers in these markets face downward-sloping demand curves, but under certain circumstances they can raise price substantially above marginal cost without invoking the entry of new firms.

### Characteristics of Oligopoly

**Oligopoly:** A market model characterized by a few firms that produce either a homogeneous product or differentiated products and entry of new firms is very difficult or is blocked.

**Oligopoly** refers to a market dominated by a few sellers. Industries with a few sellers are quite characteristic of the modern marketplace: steel, automobiles, large mainframe computers, aircraft, breakfast cereal, and soft drinks are just a few of the products of industries with few sellers. Oligopolies may sell homogeneous products such as steel or differentiated products such as automobiles or soft drinks. In this sense, oligopolies may resemble either pure competitors or monopolistically competitive firms. Later we will consider the special problems of oligopolies selling differentiated products, but first we consider two important characteristics of markets that are dominated by a few sellers: mutual interdependence and barriers to entry.

**Mutual interdependence:** A relation between firms in which the actions of one firm have significant effects on the actions and profits of other firms.

**Mutual Interdependence.** The key to the existence of oligopoly is that the small number of sellers leads to interdependence among them. **Mutual interdependence** is a relation among firms such that what one firm does with its price, product, or advertising budget directly affects other firms in the market, and vice versa. If the Chrysler Corporation introduces two new types of automobiles or if the company increases its overall advertising budget by a significant percentage, sales of General Motors, Ford, and foreign import manufacturers will be directly affected. These oligopolistic competitors will know that their sales and profits are affected by Chrysler's actions and will react accordingly. Firms therefore invest resources in keeping track of their competitors' actions.

When mutual interdependence is recognized by competitors, each firm will do the best it can in the market (in price, product innovation, and advertising) depending on what it thinks its rivals will do. Obviously, mutual interdependence can lead to a variety of outcomes, and we will find that the theory of oligopoly consists of a number of plausible models. The variety of possible outcomes arises from the uncertainty facing sellers in this market; each model reflects how competitors expect rivals to react.

**Barriers to Entry.** Oligopoly firms are normally thought to possess a large degree of **market power.** Subject to the reactions of competitors and to the

**Market power:** A situation characterized by barriers to entry of rival firms, giving an established firm control over price and, therefore, profit levels.

ability of new competitors to enter the market, oligopoly firms are able to set prices and even make economic profits, much like pure monopolists. But how does a firm achieve market power in the oligopoly market structure?

The source of the market power of oligopolists, as in any market structure, is barriers to entry. The pure monopolist is the sole supplier of a product because of barriers to entry into its industry. In any locality, for instance, there is ordinarily no close substitute for cable TV or residential power service, and there are legal barriers to firms wishing to enter these markets. If entry were free in these markets, new competitors would be attracted by higher-than-competitive prices, and excess profits would be eroded. Entry barriers are likewise essential to the maintenance of market power by oligopolists.

For an oligopoly, an entry barrier is essentially a cost that confronts a potential entrant into the industry but does not affect the incumbent firms. There are both natural or artificial entry barriers in an oligopoly market.

*Natural Barriers.* Economies of scale are an example of a natural barrier to entry. The economies of large-scale production and distribution mean that long-run average costs decline as output grows. Minimum average cost is not reached until the firm is producing a huge proportion of the total industry output. Economies of scale in the U.S. automobile industry, for example, have promoted the presence of a few firms in the long run—far fewer than the hundreds of automobile manufacturers at the beginning of this century. Other natural entry barriers include high fixed costs, high risk, scarce managerial talent, personal technological knowledge, and high capital costs.

Great care must be used in interpreting natural entry barriers since most such barriers reflect the different efficiencies of firms in producing output. Society benefits when firms achieve economies of scale, produce with the best technology, and allocate scarce managerial talent carefully. The gross national product goes up as a result. If a natural force like economies of scale leads to the possibility of market power, then the firm may be subject to antitrust prosecution. This means that the firm may be subject to laws enforced by the Federal Trade Commission and the Justice Department, which limit or eliminate anticompetitive practices.

*Artificial Barriers.* Government restrictions—patents, government regulations, licensing, labor union legislation, tariffs, quotas, and other government actions—constitute artificial barriers to entry for ologopolists. Such restrictions are fairly self-evident, but consider some examples. Patent protection is the exclusive right to market an invention for a given number of years. Patent protection is said to promote inventions, but the patent law also creates and protects oligopolies or temporary monopolies. Patents on new drugs and such products as the Polaroid camera create temporary monopolies within oligopoly industries.

Government-sponsored import tariffs on foreign automobiles or steel are an example of government restrictions that encourage and protect oligopolies. Cartels or legal agreements to collude—in such industries as railroads, ocean transportation, communications, TV, and radio—are actually sponsored and enforced by government. Cartel arrangements legalized through regulation are probably the most important artificial barriers to entry established by government. The basic point is that oligopoly theory rests on the concept of barriers to entry, both natural and artificial. Without barriers to entry, oligopoly could not exist.

## Models of Oligopoly Pricing

Like other firms, oligopolies seek to maximize their profits. To do so, however, the oligopolist can pursue many different strategies because firms can compete on a number of bases—product development, prices, advertising budgets, and so on—and can form a whole spectrum of opinions about how rivals will react. Many theories of oligopoly based on price or nonprice competition have therefore developed. We present some of the models, representing the simplest to the more complex.

**Mutual Awareness.** A major category of oligopoly model involves the case where a few firms in an industry sell the same product. This is sometimes called *pure oligopoly*, reflecting the assumption of a homogeneous product. In the U.S. economy, the steel and aluminum industries are possible examples of pure oligopoly.

Suppose that there are only two firms in an industry (economists call such a situation a *duopoly*). Both are critically aware of the actions of the other, but they cannot communicate with one another because it is illegal to do so. What market outcome might result?

Obviously, a lot depends on the personalities of the two oligopoly owners. Suppose that a benign lack of competition—"live and let live"—is their characteristic attitude. Figure 11–6 illustrates the simple result.

The firms have the same MC curve and face the same industry demand curve, $D$. Mutual forbearance in this case will lead to equal shares of the market; $d$ represents one-half of industry demand. The $mr$ curve is drawn marginal to $d$, and hence $d$ and $mr$ represent the demand and marginal revenue conditions facing both firms. Price will be set at $P_1$, and each firm will produce $q_1$, where $2q_1 = Q_T$, or total industry output. This is not the same as the profit-maximizing situation for the pure monopolist. The pure monopolist would set MC = $d$ and would price and produce accordingly, whereas price is lower and quantity higher in a simple mutual awareness model.

This is an example of how two oligopolists might reach an implicit understanding with one another about market behavior. It is hard to say whether

**FIGURE 11–6**
**A Model of Mutual Awareness**

There are two firms in the industry with the same MC curve, and they tacitly agree to share the market equally. The $d$ represents one-half of industry demand, $D$. Each firm produces $q_1$, so the industry output $Q_T = 2q_1$. This simple model of oligopoly shows that price is lower and quantity greater than that which would exist under a pure monopoly model.

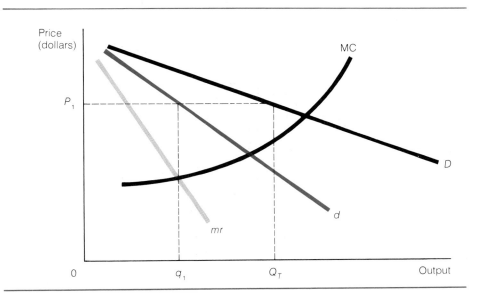

**FIGURE 11–7**
**The Effects of Differing Costs**

The duopolists in this case have different costs ($MC_A$ and $MC_B$), which lead them to prefer different prices ($P_A$ and $P_B$). Interdependence is harder to achieve, and firm B experiences great pressure to match the price of firm A.

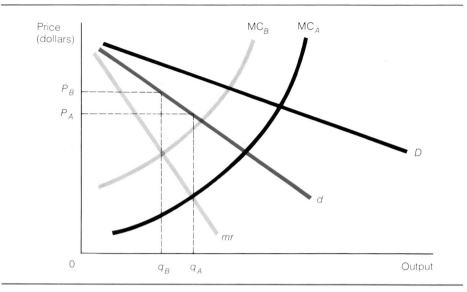

Kinked demand: A curve that has a discontinuous slope, the result of two distinct price reactions of competitors to changes in price.

this case represents a likely outcome, but it is surely based on simple assumptions.

Suppose, on the other hand, that the two firms had different costs or faced different demand curves. Figure 11–7 illustrates the case of different costs.

In this case firm A prefers a lower price and a larger output than firm B. One would guess that mutual forbearance would not work very well here. The firms are selling a homogeneous product, but one firm, A, is a more efficient producer than the other. Since the firms cannot explicitly communicate and set a common price-output policy to maximize profits, firm B is in a bad situation. It has little choice except to match firm A's price; it would otherwise lose its share of the market over time. Yet matching A's price may impose losses on B. Such considerations make mutual forbearance difficult when firms have different costs.

Virtually the same analysis holds if the firms have the same MC curve but face different demand schedules. Suppose that the industry is growing faster in the region of firm A than in that of firm B. Again, the pricing strategies of the duopolists will diverge, and it will be more difficult for the two firms to practice interdependent pricing. Indeed, it will be more difficult for the firms to avoid outright price competition. Since we generally expect firms to have different cost functions and to face different demand schedules, we expect interdependent pricing through mutual awareness and forbearance to be hard to implement in real-world markets.

**The Kinked Demand Model.** The Great Depression of the 1930s provided a setting for the development of a famous model of oligopolist competition. The **kinked demand** theory of oligopolist pricing was developed by Paul M. Sweezy in the 1930s to explain why prices in oligopolist industries with few sellers did not seem to change very much in the face of large-scale reductions in demand during the Depression. Prices were "sticky"—that is, they refused to fall or they fell slowly—over this period.

Figure 11–8 illustrates Sweezy's kinked demand model of oligopolist pricing. The graph looks complicated, but it is not if we understand Sweezy's

## FIGURE 11–8
### Sweezy's Kinked Demand Curve Model

Sweezy's basic assumption is that rivals will not follow price increases by a firm but will follow price decreases. This causes MR to be discontinuous over the range $BC$. If MC cuts MR in this segment, there will be no tendency for price or output to change when the level of marginal costs rises or falls between $B$ and $C$. Prices will be "sticky."

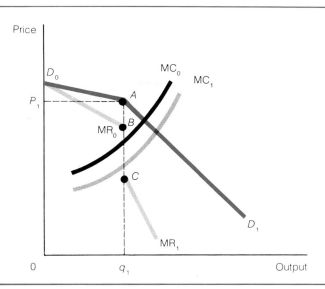

main assumption about oligopoly firm behavior: that rival firms will follow price decreases but will not follow price increases. In Figure 11–8, the price being charged is $P_1$, and output is $q_1$. As you can see, the market demand curve facing the firm has a "kink" in it. The slope of the upper portion, $D_0A$, reflects the assumption that if the firm raises its price, no rival will follow the price increase. $D_0A$ is an elastic portion of the kinked demand curve. The slope of the portion $AD_1$ is based on the assumption that rivals will match price decreases by the firm. Thus, the firm loses sales if it raises price above $P_1$ and does not gain sales if it lowers price below $P_1$. This market situation results in a kink in the demand curve at point $A$.

The two segments of the kinked demand curve are associated with two marginal revenue curves. $MR_0$ is the marginal revenue curve corresponding to $D_0A$, and $MR_1$ is the marginal revenue curve corresponding to $AD_1$. Connection of the two differently sloped MR curves results in a discontinuous marginal revenue curve. MR hits the dashed line in Figure 11–8 at $B$ and "jumps" to $C$, where it begins again as $MR_1$. The marginal revenue curve has a gap (or discontinuity) over the distance $BC$. It is useful to think of marginal revenue as consisting of $MR_0$, $BC$, and $MR_1$ in this case.

Now we can see why prices are "sticky" in this model. Applying the MC = MR rule, we see in Figure 11–8 that the MC curve cuts MR in the gap $BC$. Within this gap MC can rise as high as $B$ or fall as low as $C$, but applying the MC = MR rule will yield the same price for the firm. Thus, changes in marginal costs do not automatically lead to lower prices or higher quantities. In Figure 11–8, for example, when the firm's MC falls to $MC_1$, there is no change in its price.

Sweezy's assumption may or may not hold. In fact, the relation of the model to real-world pricing in concentrated markets seems open to question. Careful empirical studies tend to show a reasonable amount of upward and downward price flexibility in concentrated markets. Moreover, the kinked demand model is strictly a short-run theory of oligopoly pricing. It does not tell us how price comes to be set at $P_1$ or how firms adapt to situations where MC shifts outside the gap of MR.

**Price leadership:** A pricing behavior in which a single firm determines industry price.

**Dominant Firms.** Many industries consist of one or a few large firms and many smaller rivals. Examples are the airlines, steel manufacturers, and home computer industry. Pricing in such a market is often characterized by **price leadership,** which means that there is an unwritten agreement that the largest firm sets pricing policy for the industry. When the largest firm announces that it is raising its price by 5 percent, the other firms in the industry follow suit (and usually on the same day!). This unwritten code of industry conduct is one way to solve problems of interdependent pricing.

It is not always the largest firm that is the price leader; sometimes it is the most respected firm in the industry or the firm, called a "barometric" firm, that best reflects average cost, demand, and other conditions. And sometimes the role of price leader shifts among firms. Jones-Laughlin, though not the largest firm in the market, was for years the steel industry's price leader, whereas General Motors, clearly the dominant firm, provided a similar role for the automobile industry.

Figure 11–9 illustrates the dominant-firm model. The diagram looks more complicated than it is. $D$ is the industry demand curve; $\Sigma MC_f$ is the horizontal sum of the MC curves of the "fringe suppliers" in the industry. These are competitive firms that behave like price takers; they set MC equal to price. $D_d$ is the demand curve facing the dominant firm (or groups of firms), and $MR_d$ is the associated marginal revenue curve. The marginal cost curve of the dominant firm, $MC_d$, is lower than the marginal cost curves of the fringe suppliers. Its lower costs derive from economies of scale.

How does the model work? The dominant firm takes the existence of the

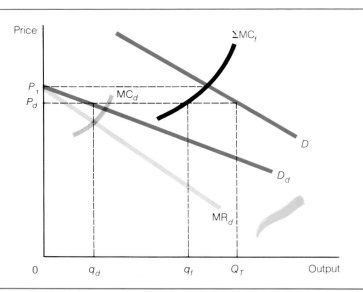

**FIGURE 11–9  A Dominant Firm Model**

$MC_d$ is the marginal cost curve of the dominant firm, and $\Sigma MC_f$ is the marginal cost curve of fringe suppliers. At price $P_1$, the competitive fringe supplies the whole market. At prices below $P_1$, a residual demand faces the dominant firm ($D_d = D - \Sigma MC_f$). The dominant firm, which has lower MC than the fringe suppliers, sets a monopoly price, $P_d$, over this portion of the market. The dominant firm produces $q_d$. The fringe firms accept $P_d$ as given and set $MC = P_d$, producing $q_f$; $q_d + q_f = Q_T$. The dominant firm is therefore the price leader in the industry, and the fringe suppliers are price takers.

fringe suppliers as a given, and it prices accordingly. At any given price, the dominant firm derives its demand curve by considering what the fringe firms will do. At price $P_1$, the fringe suppliers will supply the whole market ($\Sigma MC_f = D$). There is nothing left for the dominant firm to exercise its monopoly power over. At prices below $P_1$, fringe suppliers cannot supply the whole market; something is left over for the dominant firm to supply. This portion of the market is given by the horizontal difference between $\Sigma MC_f$ and $D$ at points below where $\Sigma MC_f = D$. In other words, at prices below $P_1$, the demand curve of the dominant firm is equal to the difference beteen $\Sigma MC_f$ and $D$.

Over this residual portion of the market, the dominant firm is a price setter. According to its own $MC_d$ and $MR_d$, it sets price at $P_d$ and produces $q_d$ units of output. The fringe firms accept $P_d$ as given and produce by equating $\Sigma MC_f$ to $P_d$. Fringe suppliers supply $q_f$. Since $q_d + q_f = Q_T$, market demand equals market supply at a price of $P_d$.

The dominant firm model mixes elements of monopoly and oligopoly theory. The dominant firm sets price for the industry, and other firms follow this pricing policy, but it sets price by acting like a conventional monopolist over the residual portion of the market not supplied by the fringe firms.

**Oligopolists with Differentiated Products.** Just as there are many cases of pure oligopoly, there are many cases in which few sellers supply differentiated products. Under such conditions, oligopolists may compete with each other and perhaps try to bar new entrants by advertising, quality variations, or pricing strategies.

*Advertising.* In oligopoly, as in monopolistic competition, advertising is used by rival firms to increase demand for their products (shift demand curves outward) through the creation of brand loyalty and product identity. Heavy advertising by a few large firms is often held to be an entry barrier to new competition. The argument is that new firms cannot enter because they cannot afford the heavy advertising expenses necessary to establish a presence in the market. But what is at stake in advertising? Basically, what advertising does is establish a firm's brand name, which represents the quality of the firm's products and services. A brand name reflecting reliable performance is quite valuable and reflects what might be called the advertising capital of the firm. However, if firms lie or produce shoddy products, their brand name capital value will fall.

If advertising creates a barrier to entry, we would expect to see firms earning excess profits (greater than a competitive rate of return) on their advertising capital investments in brand names. If they are earning only competitive returns on their advertising capital, advertising is not a barrier to entry. Incumbent firms and potential entrants would face an essentially competitive market in establishing brand names. There is some evidence that the rate of return on advertising investments across large U.S. firms is at the competitive level. This finding would undermine the argument that advertising is a barrier to entry into oligopoly industries.

*Quality Variations.* Quality variation and product development are other familiar methods of nonprice competition by oligopolists. Just as monopolistic competitors compete in this manner, so oligopolists will alter products to increase demand curve and to make demand more "intense" and inelastic in consumers' minds. Adding new colors to toothpaste, more or less chrome to automobiles, and more service to airline travel are familiar attempts to differentiate products and services.

Not all of this nonprice competition by oligopolists promotes efficiency in the economy. The first Economics in Action, "The Prisoner's Dilemma and Automobile Style Changes," at the end of the chapter illustrates how competition in styling between automobile producers can lead to wasteful results. In addition, this essay illustrates a particularly interesting form of oligopolist interaction called the prisoner's dilemma.

*Limit Pricing.* In oligopoly industries with a differentiated product, incumbent firms will seek to deter potential competition by limit pricing. **Limit pricing** is simply the practice of the incumbent to reduce price as much as necessary to limit entry.

Imagine a situation in which a new firm wants to enter an industry. Since other firms' products have already been differentiated, the new firm will have to incur heavy advertising expenses to become competitive with incumbent firms assumed to be earning economic profits. If a new entrant is successful in its bid for a share of the market, its sales come at the expense of the incumbent firms. To block this challenge, an incumbent firm may lower price and increase output in response to the threat of entry, voluntarily (if temporarily) reducing its profits. In general, the incumbent firm could set a price at which entry would not be worthwhile for any potential rival.

We are talking about oligopoly behavior. The potential entrant must make an estimate of what the incumbent firm will do in the face of entry. If the incumbent firm practices or even threatens limit pricing, entry may not be feasible. Indeed, the most rational thing for the incumbent firm to do is to keep price at profitable levels and at the same time make it clear to potential entrants that it will lower price and increase output if they enter.

> **Limit pricing:** The price behavior of an existing firm in which the firm charges a price lower than the current profit-maximizing price to discourage the entry of new firms and thus maximize its long-run profits.

## Cartel Models

In our discussion of oligopolies, we have assumed that competitor firms will not cooperate to set prices. But what if firms do cooperate? This question brings us to the topic of cartels.

Cartels and oligopolies are close cousins. The distinguishing feature of a **cartel** is an agreement among firms to restrict output in order to raise price and to achieve monopoly power over a market.

> **Cartel:** A formal alliance of firms that reduces output and increases price in an industry in an effort to increase profits.

### Characteristics of Cartels

Normally cartels are supported by agreements that limit entry and restrict and segment output among markets. Mutual interdependence—also a characteristic of oligopoly—exists prior to cartel agreements because the purpose of a cartel agreement is to limit or constrain rivals' actions that have effects on the prices and profits of the group of suppliers.

Cartel agreements may be legal or illegal. In countries like the United States with antitrust laws, cartels are illegal, but they sometimes operate covertly. An illegal cartel rests on private, collusive agreements that are out of sight of the antitrust authorities. In the 1960s certain firms made a secret agreement to fix the prices of electrical generators sold to municipalities; this illegal collusion was discovered and prosecuted under antitrust laws, with the guilty firms paying damages. Beyond the threat and reality of antitrust prosecution, such secret arrangements tend to fall apart on their own. Reasons for the fragile nature of cartels will be discussed later in the chapter, but most are related to the incentives for some members to cheat on the agreement.

A legal cartel is one supported by a law limiting entry and restricting competition among members. Such cartels may be made legitimate by the legal and political organization of an industry—communications, electrical distribution, and railroads, for example—under the umbrella of government regulation restricting competition and allowing price fixing. Or cartels may be made legitimate by legislation granting exceptions to antitrust statutes. Such is the case for exporters, labor unions, and farm organizations.

Legal status does not protect such cartels from breaking down in the long run. Technology and competition have all but destroyed the railroad cartel in the United States enforced by the Interstate Commerce Commission since the late nineteenth century. OPEC (the Organization of Petroleum Exporting Countries) was unable to maintain its full cartel status not because it was illegal within the participating countries but because of the lack of adequate enforcement and the unwillingness of certain nations, notably Saudi Arabia, to meet the terms demanded by other members. The Colombian coffee cartel of the 1960s broke down for similar reasons, and the De Beers diamond cartel (see the second Economics in Action at the end of the chapter), though legal, had some shaky times in the early 1980s. Both legal and illegal cartels tend to be fragile, therefore, though agreements enforced by law and the government are less so.

## Cartel Formation

Assume for the sake of simplicity that cartels do not violate any law and that we are analyzing a case in which the cartel is to be privately enforced by the cartel members and not by government. The first problem facing any potential cartel or association of sellers is how to get organized. In this respect the cartel is no different from any other group. The usual problem in getting any group to do anything can be roughly summarized as "passing the buck." This is quite a rational attitude for the individual cartel member to take. After all, if somebody else forms the cartel or takes the lead in restricting output, other members can generally enjoy the benefits of the cartel without bearing any of the costs of organization. This behavior is called "free riding."

Of course, if everybody in the cartel were a free rider, the cartel would never be formed. All potential members would hold back and expect to benefit from the organizational efforts of others. Economic theory does not have much to contribute to understanding how groups overcome these free-riding costs and get organized to do things.

Suffice it to say that a cartel has to get organized. Someone has to get the individual suppliers in the industry together and hammer out an agreement to restrict output in order to raise price. The details of this feat are not simple. Which firms will reduce output and by how much? How will the resulting cartel profits be shared? How will deviations from cartel policies be monitored and enforced? And so on.

None of these are easy questions for the cartel manager to answer or to reach agreement on. The problem becomes more pronounced the more firms there are in the industry. As a rule, smaller groups are easier to deal with than larger groups. This is the only general principle that we can enunciate about cartel formation.

**FIGURE 11–10**

**Organizing a Competitive Industry into a Cartel**

Price rises from $P$ to $P_c$ as the cartel is formed and output is cut back from $Q$ to $Q_c$. In the extreme case, the cartel price, $P_c$, is equivalent to the monopoly price.

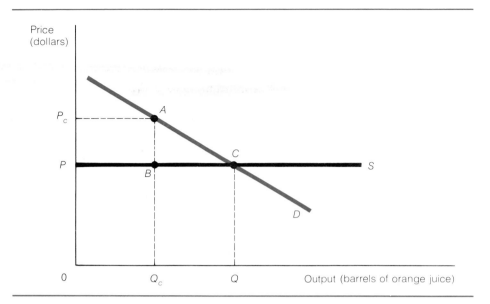

### The Cartel in Action

Figure 11–10 illustrates the basic sense of how a cartel operates. Imagine that orange juice producers try to form a cartel. Initially, the orange juice industry is competitively organized, with a number of firms. Orange juice industry demand and supply intersect to yield price $P$. Industry output is $Q$.

A cartel manager emerges and strikes an agreement among the competitive producers to cut back production from the industry's $Q$ to $Q_c$ and to raise price from $P$ to $P_c$. Cartel output $Q_c$ will ordinarily not be equivalent to the output a pure monopolist would select. Why? In an industry such as orange juice production, with a relatively large number of suppliers, the costs of eliminating all competitive behavior—the costs of enforcing the agreement—would be too high. Some competitive behavior will remain. Cartel price will therefore be lower than the profit-maximizing price of a pure monopolist.

The incentive to form cartels is nevertheless clear. Cartel members who were previously earning a competitive rate of return in a competitive industry now earn an agreed-upon portion of $P_cABP$—an equal share of cartel profits.

### Cartel Enforcement

**Cartel enforcement:** An effort by the administrators of a cartel to prevent its members from secretly cutting price below the cartel price.

Holding the cartel together is not easy. Once the cartel has been organized, has restricted output, and has raised price to $P_c$, what is the position of the individual orange juice firm in the cartel? If an individual firm in the cartel lowers its price slightly below $P_c$, it will face a very elastic demand curve for its output. Buyers will prefer the lower price and will switch their purchases to the lower-priced firm. The fact that individual firms in a cartel can significantly increase their sales and profits by secretly lowering their price puts tremendous pressure on the cartel to fall apart. Individual firms have a large incentive to cheat on the cartel agreement.

Of course, to expect such profits to hold up in the long run is unrealistic. Other firms would easily discover what was going on because they would suffer a drastic loss of sales. If they responded in kind with price reductions, the result would be that the secret price cutter would not experience an increase in sales and profits for long. Further spates of secret price cuts and price competition would lead to a return to competitive equilibrium.

Recognizing these possibilities, the individual firm in the cartel must form an estimate of what it can reasonably expect to gain from secret price reductions and compare this estimate to its share of cartel profits. A cartel will not be stable if the former is larger than the latter.

The general case in which a firm might expect to gain by secret price cutting is illustrated in Figure 11–11. Cartel equilibrium is at price $P_c$ and output $Q_c$, represented by $E_c$. $D/n$ represents the prorated firm demand curve, where $n$ is the number of firms in the industry, and $d_i$ represents the gains to an individual cartel member from secretly cutting price; $d_i$ is more elastic than $D/n$. With the more elastic $d_i$ curve, price increases would cause a sharp drop in sales, but price reductions cause a sharp gain in sales and profits at the expense of other cartel members. The gains of a secret price cut to $P_s$ are traced out by $d_i$ below $P_c$.

The key to cartel enforcement is now apparent. Means must be found to make $d_i$ approximate $D/n$. Perfect cartel enforcement exists when each cartel member knows that secret price cuts are not possible and will be matched by other members. In this case, the prorated demand curve, $D/n$, is the relevant demand curve. If a firm cuts price, it will immediately be matched by other cartel members. There is thus no gain to cutting price. Movements will be along the prorated firm demand curve, $D/n$. Where it is costly to detect secret price cuts, a demand curve like $d_i$ will confront each cartel member. Each firm must decide whether to cheat, depending on the expected profits from cheating relative to the expected profits from staying in the cartel. In general, we can gauge the effectiveness of cartel enforcement by the degree to which the cartel is able to make the cheating demand curve $d_i$ coincide with the prorated firm demand curve $D/n$. That is, how good is the cartel at detecting cheating?

Cartels will spend resources to control the incentive to cheat. Historically, many devices have been used. Perhaps the most effective device is a common sales agency for the cartel. All production is sold through the

**FIGURE 11–11**

**Cheating in a Cartel**

$P_c$ is the agreed-upon cartel price at cartel output $Q_c$. $E_c$ is therefore the cartel equilibrium. If secret price cuts cannot be detected, the cheating firm will face the more elastic demand curve, $d_i$, and can gain from cheating on the agreement. If such behavior can be detected, movements will be along $D/n$, the prorated firm demand curve. The degree to which $d_i$ diverges from $D/n$ reflects the efficiency of the cartel enforcement system.

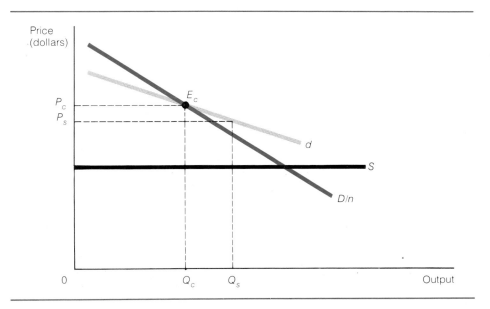

agency, and individual producers do not negotiate sales for themselves. Cheating is not a viable strategy under such a condition. Producers produce their allotted quotas, and the sales agency markets the agreed-upon cartel output ($Q_c$). Exclusive sales territories and customer lists are other ways to control incentives to cheat and to free ride on the cartel price. For example, exclusive territories are a way to divide markets, and intrusions by other cartel members are easy to spot.

### Why Cartels May or May Not Succeed

Collusive cartel agreements are more likely under some circumstances than others. We offer now a menu of factors thought to be conducive to cartel-like behavior. Our list is not meant to be exhaustive, but just illustrative of the types of economic conditions associated with a greater likelihood of cartel activity.

1. A cartel is easier to form among a smaller number of sellers than a larger number. The costs of any group decision-making process rise with the number in the group.
2. It is easier to reach a cartel agreement where members produce a homogeneous product. In such cases, nonprice competition is less of a problem. Sellers do not have to worry that their competitors will secure bigger market shares through advertising.
3. Collusion is easier to effect in growing and technologically progressive industries than in stagnant or decaying industries. In the latter case, excess capacity created by declining demand puts great pressure on individual firms to compete for customers and to cheat on the cartel agreement.
4. Many small buyers reduce the costs of collusion. Large buyers have a clear incentive to seek to undermine collusion among sellers. In effect, large buyers make cheating on the cartel price very worthwhile to individual sellers.
5. Low turnover among buyers makes collusion easier. If sellers cheat on the cartel agreement, frequent turnover of buyers makes it difficult to detect who is cheating.

Other factors could be added to this list of conditions favorable to the formation of private collusive cartels. Indeed, an easy way to find out what makes collusion attractive is to look at the areas of the economy in which the antitrust authorities most typically find such behavior (discussed further in Chapter 17).

At this point, one fact should be clear: Private collusive agreements are hard to sustain. The incentive to cheat on such agreements is very strong, and cartels therefore tend to be unstable. Moreover, where antitrust law prevails, cartels must operate on the sly to avoid law enforcement; this secrecy further reduces their ability to control cheating. Exclusive sales territories, for example, would likely be spotted by antitrust authorities, for customers would complain about high prices and the lack of competitive sales or service alternatives. It is little wonder, then, that cartels rarely last long unless they are supported by law.

Groups that have difficulty sustaining a private cartel agreement can petition government to pass a law making their cartel legal. More than this, the government can enforce the cartel by outlawing entry and price competition among members. So to say that private cartels are not stable and therefore unlikely to side-step the social benefits of competition in the long run is inaccurate if government steps in to reinforce such agreements.

# From Competition to Monopoly: The Spectrum Revisited

We are at last in a position to take an overview of the economic characteristics and welfare effects of market structures from pure competition (Chapter 8) to pure monopoly (Chapter 10) to those in between (the present chapter). Table 11-1 organizes the long-run economic tendencies and conclusions of the various structures we have examined.

The table distinguishes the five market structures by number of sellers, barriers to entry, product characteristics, and long-run market tendencies. The number of sellers ranges from many under competition and monopolistic competition to one in the case of pure monopoly. Significant entry barriers characterize oligopolies, cartels, and, of course, monopolies. These three structures are also characterized by the sale of *either* homogeneous or differentiated products, unlike the purely or monopolistically competitive structures.

Most important, Table 11-1 provides a category—long-run market tendencies—that gives us some insight into the consumer and economic welfare created under these alternative structures. As the table reveals, price tends to equal long-run average total cost in only two of the structures—pure and monopolistic competition. Price, moreover, tends to equal long-run marginal cost only in the purely competitive market structure. What does this mean in terms of consumers' economic welfare?

An important concept introduced in Chapter 6 aids us in interpreting the effects of market structure on social welfare. In Chapter 6, we introduced the concept of consumers' surplus, defined as the difference between the amount of money that consumers would be willing to pay for a quantity of a good and the amount that they actually pay for the good.

Figure 11-12 contrasts what happens to consumers' surplus for a given hypothetical market—personal computer software—if the market is organized

**TABLE 11-1  Characteristics and Consequences of Market Structures**

When static market structures are considered, only pure competition yields an efficient and welfare-maximizing allocation of resources. Resources are underallocated to the production of goods and services under all other structures from consumers' perspective.

| Type of Market Structure | Number of Sellers | Barriers to Entry | Type of Product | Market Tendencies |
|---|---|---|---|---|
| Pure competition | Many | No | Homogeneous | P = MC<br>P = ATC |
| Monopolistic competition | Many | No | Differentiated | P > MC<br>P = ATC |
| Oligopoly | Few | Yes | Homogeneous or differentiated | P > MC<br>P > ATC |
| Cartels | Relatively few, acting as one | Yes | Homogeneous or differentiated | P > MC<br>P > ATC |
| Pure monopoly | One | Yes | Unique | P > MC<br>P > ATC |

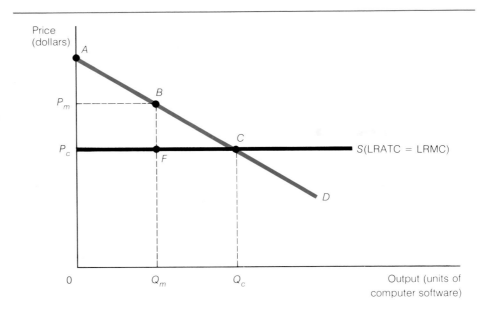

**FIGURE 11-12**

**The Effects of Competitive versus Monopolistic Markets**

In contrasting the higher price and lower quantity ($P_m$ and $Q_m$) produced under a monopolistic market structure with the lower price and higher quantity ($P_c$ and $Q_c$) theoretically possible in a purely competitive situation, $ACP_c$ represents the area of consumer surplus. This is real income available to anyone who can capture it. Under competition, it goes to consumers. Under monopoly, cartels, and oligopolies, part of it ($P_mBFP_c$) typically goes to producers and part (BCF) is lost to society.

along monopolistic rather than competitive lines. The industry demand curve for this market has the usual negative slope, but the industry has a flat supply curve in this simplified conception. In other words, this is a constant-cost industry, in which additional output may be produced at a constant average cost that is also equal to marginal cost. The supply curve is simply a flat line, as in Figure 11-12.

To verify what happens to consumers' surplus under the different market structures represented in Figure 11-12, look again at the long-run market tendencies of the various models in Table 11-1. If the computer software industry is purely competitive, it will theoretically be producing an output of $Q_c$ at price $P_c$, which is the lowest possible price because it is equal to both the average cost and the marginal cost of production. In this situation, consumers enjoy a surplus represented by $ACP_c$. In contrast, price exceeds both average cost and marginal cost in the market models of oligopoly, cartels, and pure monopoly. Economic profits are therefore captured at the expense of consumers in these markets. In Figure 11-12, $P_m$ and $Q_m$ represent the effects of a market structure on the monopoly end of the spectrum; the area $P_mBFP_c$ is redirected from consumers' surplus to producers' profits. Such a redistribution of real income would not worry economists so much were it not that society must also pay a deadweight loss (the area BCF in Figure 11-12) because price exceeds cost. This deadweight or social loss due to monopoly power or the semimonopoly power of cartels and oligopolies causes some economists to condemn such structures as reducers of consumer and social welfare.

As we have seen in this chapter, monopolistic competition is something of a special case. Under monopolistic competition, there is a long-run tendency for price to equal average total cost, although not marginal cost when economies of scale are considered. The benefits of differentiated products—a greater range of choices—also exist in this structure. But the only unambiguously welfare-maximizing structure is that of pure competition. As Table 11-1 and Figure 11-12 indicate, only competition provides the product in this example, computer software, at the lowest possible price (equal to both LRATC and LRMC) and in the greatest quantities possible, $Q_c$. Only pure competition maximizes society's welfare by maximizing consumers' surplus. Important deviations from this norm are prime subjects for society's economic policy concerns. The following two parts of this book, especially Part IV, discuss a number of important economic issues related to market structure.

## Summary

1. Monopolistic competition is a market structure that contains a large number of firms. Each firm sells a product that is similar to but slightly differentiated from the products of other firms in the industry. Freedom of entry and exit exist in this market structure.
2. Advertising is a means by which firms distinguish their products and thereby increase demand for them. Advertising and other forms of nonprice competition characterize monopolistic competition.
3. In the long run under monopolistic competition, a tangency between the firm's demand curve and its average cost curve can exist. If this tangency occurs, then the firm produces a rate of output less than the rate associated with minimum average total cost; that is, excess capacity exists.
4. The excess capacity of monopolistic competition may have some value. The larger number of firms may offer a larger variety of products or a lower real price to consumers than a competitive market with similar production costs.
5. Oligopoly is a market structure characterized by a few firms that sell either homogeneous or differentiated products. Entry into the industry is difficult because of either natural or artificial barriers.
6. In an oligopoly, the small number of sellers leads to mutual interdependence. The actions of one firm have significant effects on other firms in the industry. For this reason, the behavior of firms in oligopoly is difficult to predict or analyze.
7. The kinked demand curve theory of oligopoly encompasses two distinct reactions of firms to price changes. It is suggested that if one firm increases its price, no other firms will follow the price increase. On the other hand, if one firm lowers its price, all other firms will lower their prices as well. The resulting demand curve can result in a rigid price structure.
8. Cartels are formal agreements among firms within an industry to restrict output or to segment the market in an effort to increase profits. The result is similar to monopoly price and output in the industry.
9. There are potential profits for firms that cheat individually on a cartel. Cartels are therefore difficult to establish and maintain. The costs of enforcing the cartel agreement frequently prevent cartel's survival.
10. Cartels can be legal or illegal. Those created and enforced by the government or those allowed to exist by government sanctions are legal. Otherwise, cartels in the United States are breaches of antitrust legislation.

## Key Terms

imperfect competition
monopolistic competition
differentiated products
nonprice competition
advertising
tangency solution
excess capacity
economic efficiency
economic welfare
oligopoly
mutual interdependence
market power
kinked demand
price leadership
limit pricing
cartel
cartel enforcement

## Questions for Review and Discussion

1. Select two products or services that you regularly consume. Are these goods produced under monopolistically competitive conditions? What information do you have to have before giving a definite answer to this question?
2. What, exactly, is the role of product differentiation and advertising in monopolistic competition?
3. How many forms of competition can you name and analyze in addition to price competition? How is location a factor in competition?
4. Given that Coke is the number-one soft drink in the market, is it wasteful for the firm to continue to advertise on so large a scale? Give reasons for your answer.
5. What is the tangency solution to the market model of monopolistic competition? Are half-empty air-

planes on the New York–San Francisco route definite evidence of excess capacity and economic waste? Give reasons for your answer.
6. Compare the results of long-run equilibrium in pure competition and monopolistic competition from the standpoint of both efficiency and welfare.
7. What are the general characteristics of the oligopolist market model? Compare and contrast these to the characteristics of the monopolistic competition model.
8. Saudi Arabia is the dominant firm in the OPEC oil cartel. Describe how world oil prices might have been set in the 1970s.
9. What does the free-rider effect mean? Cite three personal experiences in which you have observed free-riding behavior.
10. Suppose that firms forming a cartel have different marginal costs. Describe, both in words and in a graph, how this situation complicates the various problems that a cartel must solve to be effective in raising prices.
11. Which of the following factors make collusion more or less likely: (a) the purchase of an input in the same market (such as pigs at the stockyard); (b) long-term contracts with buyers; (c) selling to governments; (d) salespeople staying in the same motel on the road; (e) an industry trade association and price list.

## ECONOMICS IN ACTION

### The Prisoner's Dilemma and Automobile Style Changes

A helpful analytical tool in analyzing the effects of nonprice competition among a small group of oligopolist producers of differentiated products is the concept of the prisoner's dilemma. The prisoner's dilemma originated in a story told by mathematician A. W. Tucker. Two people are caught in a serious crime, but the district attorney has evidence to convict them only for a lesser offense. The D.A. separates the prisoners and attempts to obtain a confession in this manner: Both prisoners are separately informed that (1) if one confesses, the confessor goes free and the other hangs; (2) if neither confesses, both will receive the modest penalty that goes with the lesser crime; (3) if both confess, both will receive a severe penalty short, of course, of hanging. Given the payoffs and the uncertainty, the expected result is a confession from both prisoners.

The prisoner's dilemma can be related to economic behavior. The automobile industry's annual change of style is an example of this type of oligopoly behavior.[a] A style or design change increases automobile production costs (retooling, refabrication, and so on) and at the same time tends to reduce profits. Yet it is consistently chosen by competing firms acting in their own self-interest. Their reasoning is illustrated in Figure 11-13, a hypothetical payoff table, in which two automobile firms are trying to maximize their profits, represented by the varying dollar figures within the boxes.

From the standpoint of the industry, joint profit maximization occurs in box A, where neither firm changes style. Industry profits are $120 here ($65 for Ford and $55 for GM), as opposed to $100 in boxes B and C and $90 in box D. Yet box A is not the profit-maximizing choice for either individual firm.

[a] Harold Bierman, Jr., and Robert D. Tollison, "Styling Changes and the 'Prisoner's Dilemma,'" *Antitrust Law and Economics Review* 4 (Fall 1970), pp. 95–100.

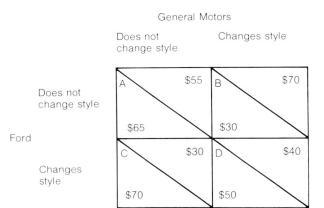

**FIGURE 11-13** The Prisoner's Dilemma and Automobile Style Changes

Various possible combinations exist when Ford and General Motors try to maximize profits by deciding whether to change styles. Profits for the firms' decisions are given in millions of dollars. The best choice for the industry is box A, where neither firm changes style, but the best profit-maximizing choice for each firm is to change style.

Take GM's decision. No matter what Ford does, GM's profits are higher if it changes its style. If it changes and Ford does not (box B), GM's profits are $70. If it changes and Ford does likewise, its profit will be only $40 (box D). But if it does not change style and Ford does, GM's profits will drop to the lowest point of all, $30 (box C). Ford management considers its own set of options and reaches a similar conclusion, namely, that Ford will always be better off by changing its style. If they both make decisions that tend to maximize their individual profits, both will introduce style changes and together earn a smaller amount of profit

($90) than they would have earned in the absence of a style change by either ($120).

This is the nature of a prisoner's dilemma. Both parties could be better off if they could communicate and stay in box A. Indeed, perhaps we would all be better off—pay less for cars—if automobile firms did not change styles every year or so. But since firms cannot legally communicate because we fear that they would talk about more than style changes, that is, would collude, we are stuck with the implications of the prisoner's dilemma. The general argument here is also applicable to other forms of nonprice competition among oligopolists, such as advertising.[b]

[b]The classic work applying game theory, of which the prisoner's dilemma is an example, to economic analysis is by mathematician John von Neumann and economist Oskar Morgenstern: *Theory of Games and Economic Behavior* (Princeton: Princeton University Press, 1944).

## Questions

1. What would be the pros and cons of allowing automakers to meet each year to decide collectively whether or not to change styles?
2. The home computer and computer software industries are both characterized by a rapid rate of technological advance. Assess the potential benefits and costs to consumers when there is (a) open competition in these areas, or (b) a cartel agreement which includes a slower introduction of new technology than would occur under competition.

---

## ECONOMICS IN ACTION

### All That Glitters: The De Beers Diamond Cartel

Recession and falling demand have been the ruination of many cartels. During periods of declining demand, the temptation to cheat is great. However, the De Beers diamond cartel—the strongest, longest-running, and most successful cartel of modern times—successfully warded off a threat of dissolution in the recession of the late 1970s and early 1980s.[a] De Beers' maneuvering during this period provides interesting insights into the enforcement powers of a cartel.

Founded by entrepreneur Cecil Rhodes in 1888, De Beers Consolidated Mines is a publicly held corporation that markets more than 80 percent of the uncut diamonds of both gem and industrial quality sold yearly in the world. De Beers owns some South African mines and acts as the marketing agent for virtually all other diamond-producing and -supplying nations, including Zaire and other African countries, Australia, and the Soviet Union. De Beers pays royalties out of its retail sales to these nations for exclusive rights to market their stones. Retailers have no choice but to purchase wholesale boxes of diamonds from De Beers; if they refuse, they are not necessarily invited to purchase diamonds again. Thus De Beers may be thought of as a marketing or middleman type of cartel.

The De Beers cartel system worked well until the

[a]John R. Emshwiller and Neil Behrmann, "Restored Luster: How De Beers Revived World Diamond Cartel After Zaire's Pullout," *Wall Street Journal*, August 7, 1983.

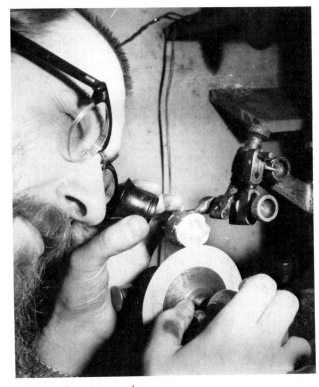

*A diamond cutter at work.*

late 1970s when several storm clouds appeared to threaten its market power. In 1979, an enormous load of diamonds called the Arggle mine was discovered in Australia. Even worse for the cartel, a worldwide recession created massive reductions in demand for diamonds. De Beers' earnings fell from $673 million in 1979 to $183 million in 1982. Rather than reduce price in the face of declining demand, De Beers' reaction was to remove massive supplies of diamonds from the market by cutting production at its own mines to shore up the cartel price. De Beers also reduced purchases from participating nations in return for lump-sum payments. In all, De Beers added more than one billion dollars' worth of diamonds to its inventory between 1979 and 1982.

Most troublesome of all, Zaire, the world's largest producer of diamonds, pulled out of the cartel in May 1981 and gave its diamond concession to three independent producers for five years. These independents marketed low-priced industrial-quality diamonds (called *boart*) from Zaire's Miba mine. De Beers' reaction was decisive. Though stockpiling diamonds in general, the company poured boart onto the market, causing the price of low-grade industrial diamonds to fall by two-thirds. De Beers was also charged with encouraging illegal diamond smuggling by Zairian citizens. The result: Zaire's revenues and profits fell precipitously. The country was even refused loans from international agencies to upgrade its mines. In March 1983 Zaire returned to the fold, dropping the three independent marketers in midcontract.

The icing on De Beers' cake came when Australia joined the cartel. The two majority stockholders of Australia's Arggle mine, which could produce a full 25 percent of current world production by 1985, had indirect ties to De Beers through a chain of stock ownership. Though there was political opposition and opposition from some of the partners, De Beers' Australian friends helped bring Arggle within the cartel, but with some concessions on the part of De Beers. Australia is now free to market 25 percent of its industrial stones and 5 percent of its gemstones outside the cartel.

Some observers believe that De Beers' concession sets a precedent that could weaken the cartel. Others believe this outcome unlikely. The Soviet Union, for example, could produce and sell enough diamonds to create chaos in world markets but chooses instead to play the cartel game. The Soviet Union apparently thinks that long-run price stability is preferable to possible short-run gains. Zaire, through costly misadventure, learned this lesson all too well.

### Question

The De Beers diamond cartel is largely a "private" cartel held together by private enforcement. Compare the enforcement problems that would likely be encountered in cartels such as De Beers to those found in government-supported ("legal") cartels.

## POINT-COUNTERPOINT

## Alfred Marshall and Joan Robinson: Perfect versus Imperfect Markets

Alfred Marshall

Joan Robinson

ALFRED MARSHALL (1842–1924), an English economist, was the strong-willed son of a harsh father, who early on tried to coerce young Alfred into the ministry. Much to his parent's dismay, Marshall rejected a theological scholarship to Oxford University and entered Cambridge, where he showed early brilliance as a mathematician and received his M.S. in mathematics in 1865. From mathematics, Marshall veered into metaphysics, joined a philosophy discussion group at Cambridge, and lectured for almost ten years in moral science at Cambridge. During this period Marshall became convinced of the overriding importance of economics to individual and social action. In 1877 Marshall married Mary Paley, a former student, and moved to Bristol, where they both lectured on political economy at University College. In 1885, they returned to Cambridge, where Marshall continued his long and illustrious career as lecturer in political economy.

The most important of Marshall's books was the modern bible of microeconomics *Principles of Economics* (1890),[a] which he revised through nine editions. Many of the concepts presented in today's introductory economics courses, including the principles of perfect competition among firms, were first set down systematically by Marshall. His mathematical precision and fondness for lucid example made the text and economics generally accessible and popular. In 1903, while his work on *Principles* continued, Marshall succeeded in establishing economics as a discipline separate from moral science. He is also credited with founding the neoclassical tradition in economics—the modern version of economic principles established by Adam Smith.

JOAN ROBINSON (1903–1983), like Alfred Marshall, spent most of her academic career at the University of Cambridge. She completed her B.A. in economics at Cambridge in 1925, the year after Marshall's death. She began a forty-year teaching career at Cambridge in 1931, at a time when Marshall's influence over the principles and methods of economics study was still great. Despite these close ties, Robinson is perhaps best known for her book *The Economics of Imperfect Competition* (1933),[b] which challenges many of Marshall's conclusions about business organization and his theory of perfect competition among firms.

Robinson's contributions to economics reach well beyond her theory of imperfect competition. In the mid-1930s she and a small group of Cambridge economists helped John Maynard Keynes write his monumental *General Theory of Employment, Interest, and Money,* which ushered in a new era of macroeconomic theory. She also wrote in the areas of economic development, international trade, capital theory, and Marxian economics.

Although Robinson later repudiated her own analysis of imperfect competition, her critique of Marshall's assumptions of how the economy operates presents an interesting chapter in the development of economic thought.

### The World of Perfect Competition

To Marshall, the microeconomic behavior of firms and individuals was kept in constant check by the competitive nature of markets. The dynamics of free enterprise, of limitless entry to and exit from markets enforced the norm of perfect competition among buyers and sellers. By his terms, each market is made up of sufficiently large numbers of competing firms buying and selling virtually identical products and services. No

---

[a] Alfred Marshall, *Principles of Economics*, 8th ed. (London: Macmillan, 1920).

[b] Joan Robinson, *The Economics of Imperfect Competition* (London: Macmillan, 1933).

single demander or supplier of goods can affect the market price because no single firm has a large enough stake in the market. Given these conditions, the market price for goods would always gravitate to a level equal to the firm's costs of production. As a result, economic profits could not persist. Under the rigors of perfect competition, no firm would be able to rise above its equals.

To demonstrate his model of perfect competition, Marshall chose a number of illustrations, including the market for fresh fish. Marshall noted that the ordinary daily price of fish varies according to factors such as the daily intensity of demand along with weather and luck, which affect the size of the catch. He then introduced into this illustration a cattle plague that permanently increased the demand for fresh fish over beef. How would the competitively organized fish market react? (Remember that no freezing or cold storage facilities were readily available in Marshall's day.)

Under competitive conditions, the price per pound of fish would immediately rise. Recognizing that the price rise was not a temporary phenomenon, fishermen would respond to the increased profits (since price would now be greater than costs of production) by using their boats, fishing crews, and nets more intensively. Supply would be increased by these activities as it would by the entry into the market of new fishing firms attracted by profits. Supply increases would ultimately bring price down to the average cost of catching fish.

Marshall's vision of perfect competition comports well with Adam Smith's theory of "the invisible hand." In fact, Marshall's theory of perfect competition describes the invisible hand at work. In his illustration, new fishing firms enter the market and old firms expand not for the benefit of customers but in the hopes of earning more profits. They act only in their own self-interest but in doing so promote the interest of society. The price of fish falls to the costs of production in the long run, and more fish are provided in the market.

Although Marshall did acknowledge the existence of monopolies and oligopolies, he believed that these market structures were a special case. Natural monopolies—those with large economies of scale in production (the phone company, the gas company, and so on)—could exist, as could monopolies due to artificial barriers to entry (government regulation, control of vital resources, and so on), but these were rare exceptions to competitive organization.

## Perfect Competition Turned on Its Head

"It is customary," Joan Robinson wrote in *The Economics of Imperfect Competition*, ". . . to open with the analysis of a perfectly competitive world, and to treat monopoly as a special case. . . . It is more proper to set out the analysis of monopoly, treating perfect competition as a special case." Beginning with these words, Robinson set out to turn Marshall's theory of perfect competition upside down.

Modern industrialized societies, she contended, were dominated by monopolies (a single firm in a market) or oligopolies (a few large firms in a market). To a greater or lesser degree, most markets were influenced by the power of monopoly control. In other words, individual firms in these markets were large enough to affect market prices on their own. Given these assumptions, Robinson pointed out, the prices of goods and services would not gravitate to a level equal to the firm's costs of production. Economic profits could persist, enabling the dominant firms to maintain or even increase their dominance.

Modern examples of imperfect competition include the automobile industry, dominated by industrial giants such as General Motors, or the fast-food industry, dominated by firms such as McDonald's. A single seller or a few sellers of running shoes and athletic equipment in a small college town is an example of monopoly on a smaller scale.

To further illustrate her point that imperfect, rather than perfect, competition was the norm, Robinson pointed to the pervasive evidence of price discrimination in markets. Whereas Marshall contended that only one market price could prevail under perfect competition, Robinson saw many instances in which a single good or service could command several different prices in the same market. Multiple prices for the same product could be possible, Robinson emphasized, only in a world of imperfect competition.

Robinson's theory of imperfect competition was not the only challenge made against Marshall's system of perfect competition. A similar theory was offered by American economist E. H. Chamberlin in his *Theory of Monopolistic Competition*,[c] also published in 1933.

Other important additions have been made to Marshall's theory of competition. Modern economists stress the manner in which information affects suppliers and demanders, the consequences of potential competition and rivalry on markets, and the impact on Marshall's model of interpreting price as "full price." The new theory of competitive markets has again come to dominate microeconomic analysis as the simple theory of competition did in Marshall's time.

[c]Edward H. Chamberlin, *The Theory of Monopolistic Competition: A Re-orientation of the Theory of Value*, 8th ed. (Cambridge: Harvard University Press, 1962), originally published in 1933.

# III

# Microeconomic Principles of Input Demand and Supply

# 12

# Marginal Productivity Theory and Wages

In Chapter 4, we presented an overview of the resource and products markets (see the circular flow model in Figure 4–1, p. 70). We showed business firms as both suppliers of goods and services and demanders of inputs such as land, labor, capital, and entrepreneurship. Households were depicted both as demanders of goods and services and as suppliers of inputs. Up to this point in our study of microeconomics, we have focused on the upper part of the circular flow model and studied how the prices of final products such as psychiatric services, home computers, or diapers are determined. We now turn to the lower part of the model to view ourselves as demanders or suppliers of inputs such as labor services, land, or capital. Specifically, we now inquire how all input prices—wages of day laborers or engineers, lumber, or machine prices—are determined. Once we know the economic process by which input values are determined, we will be able to understand better how and why people earn the incomes they do as owners of their own labor and other resources.

**Factor market:** The market in which the prices of resources (factors of production, or inputs) are determined by the actions of businesses as the buyers of resources and households as the suppliers of resources; also called resource market.

The markets in which input prices are determined are called **factor**, or **resource, markets.** The prices of factors such as wages, rent, and interest are determined through the interplay of supply and demand. Indeed, the theory of factor prices is just the familiar theory of demand and supply applied to factors of production.

We begin this chapter by presenting the marginal productivity theory of demand for one factor of production, labor. Although our example is labor, the marginal productivity theory is equally applicable to other factors of production, as we will see in Chapters 14 and 15, where we apply the theory to land and capital inputs. In the second part of this chapter, we present the theory of labor supply. Since wages are intimately linked to the level of income earned by labor, we devote the final section of the chapter to discussing the relevance of marginal productivity theory to income distribution.

## Factor Markets: An Overview

The price and quantity of inputs such as labor and land are determined by the same laws of supply and demand that we have discussed for outputs. To understand why this is so, consider the general characteristics of factor markets.

### Derived Demand

The demand for labor or any other factor of production is a **derived demand.** This term means that the demand for labor is directly related to the demand for the goods and services that the labor is used to produce. If the demand for computers increases, the derived demand for computer scientists and engineers will increase. If the demand for physical fitness increases, the derived demand for health foods will increase. If the demand for law and order increases, the derived demand for police officers will increase.

Derived demand also obeys the law of demand. All things being equal, if the price of health foods falls, more will be consumed; if the price of paper rises, fewer or shorter books will be published; if the rental prices on office space fall, more offices will be rented. So even though we call the demand for factors of production a derived demand, the law of demand applies to it in the same way that it applies to demand for final output.

**Derived demand:** The demand for factors of production that is a direct function of the demand for the product that the factors produce.

### Firms as Resource Price Takers

Just as there are different types of output market structures—such as competition, oligopoly, and monopoly—similar differences can occur in resource markets. Table 12–1 summarizes the characteristics of output and input market structures.

There can be perfectly competitive sellers and buyers of labor, for exam-

**TABLE 12–1  Characteristics of Output and Input Market Structures**

| Type of Market Structure | Output Markets | Input Markets |
| --- | --- | --- |
| Pure competition | Many sellers; no barriers to entry; homogeneous product | Many sellers; no barriers to entry; homogeneous resource |
| Monopolistic competition | Many sellers; no barriers to entry; differentiated product | Many sellers; no barriers to entry; differentiated resource |
| Oligopoly | Few sellers; barriers to entry; homogeneous or differentiated product | Few sellers; entry is difficult; homogeneous or differentiated resource |
| Cartels; Union | Relatively few sellers acting as one; barriers to entry; homogeneous or differentiated product | Relatively few sellers acting as one; barriers to entry; homogeneous or differentiated resource |
| Pure monopoly | One seller; barriers to entry; unique product | One seller; barriers to entry; unique resource |

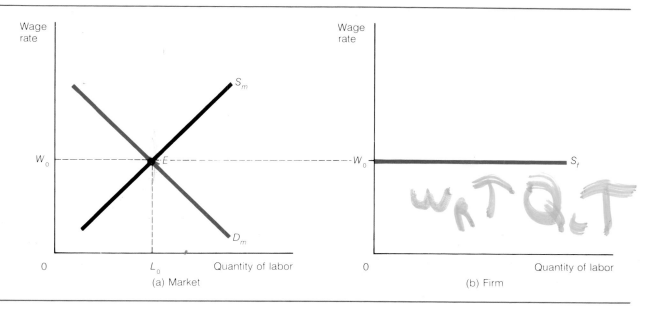

**FIGURE 12-1  The Firm as a Wage Price Taker**
(a) The equilibrium wage, $W_0$, is determined in the competitive market by the market supply, $S_m$, and demand, $D_m$, for a particular type of labor. (b) The firm must pay wage rate $W_0$ for workers but is able to hire all it wants at that wage. The supply of labor to the competitive firm, $S_f$, is infinitely elastic at the prevailing wage.

**Competitive labor market:** A labor market in which the wage rate of a particular type of labor is determined by the forces of supply by a large number of sellers of labor and demand by a large number of buyers of labor.

**Market demand for labor:** The sum of, or overall, demand for a particular type of labor by all firms employing that labor; the total level of employment of a particular type of labor at all the various wage rates.

**Market supply of labor:** The total amounts of labor that all individuals are willing to offer in a particular occupation at all the various wage rates.

ple. In such cases, there are large numbers of both buyers (employing firms) and sellers (workers) of a homogeneous type of labor. There can also be various types of imperfect competition in buying and selling labor, as we will learn in the next chapter. We limit this chapter to the competitive labor market. The following discussion of labor demand examines labor markets in which a large number of firms hire a single type of labor and a large number of people are willing to supply the labor.

In a **competitive labor market,** the wage rate is determined by the familiar forces of supply and demand. The overall or **market demand for labor** is the sum of all firms' demand for the type of labor in question. All of the firms may not be in the same industry—that is, they may not all produce the same product—but they all employ the same type of labor. For example, the demand curve in Figure 12-1a could represent the market demand for electricians. This overall demand would be found by summing the demand for electricians by all buyers of their labor, from AT&T and General Motors to construction companies, cable television companies, and even homeowners.

The overall or **market supply of labor** is obtained by summing the supply of individual workers. As indicated by the supply curve $S_m$ in Figure 12-1a, as the wage rate rises, a larger quantity of labor is supplied. The supply of labor to any particular occupation thus follows the familiar law of supply.

The interaction of supply and demand for labor results in an equilibrium wage $W_0$. Once this wage is established, the buyers and sellers of labor are price takers. That is, the firms must accept $W_0$ as the prevailing wage and may hire as many workers as they are willing and able to hire at that wage. Being perfectly competitive buyers of labor, employing firms face an infinitely elastic labor supply curve, as shown in Figure 12-1b. All buyers of any good or service who are price takers face an infinitely elastic supply curve.

# Marginal Productivity Theory

We cannot say exactly how many electricians or computer programmers will be demanded within a given time period, but we can present some general rules about how the numbers of people demanded in these occupations might rise or fall. In this analysis, we reencounter many principles from earlier chapters. All apply equally to the demand for resource inputs, including labor.

### The Profit-Maximizing Level of Employment

The firm hiring labor in a competitive market faces a market-determined wage. Since the equilibrium wage is established, the point of interest becomes the quantity of labor that an individual firm chooses to hire. If a firm is selling its product in a perfectly competitive product market, then how much labor will it hire at a particular wage rate?

To answer this question, let us say that the firm has many inputs, one of which is labor. Also, let the quantities of all inputs except labor be fixed. Labor, in other words, is the only variable input. (These circumstances, as you may recognize, describe a short-run situation for the firm.) To produce more output, the firm must hire more labor. The firm wishes to maximize profits and must therefore hire the profit-maximizing amount of labor.

Suppose the firm in this instance is a corn farmer who must decide whether to hire a worker to help pick and husk corn. How does the farmer make this decision? Hiring the worker will increase output, and the increase in production will be sold to obtain revenue. At the same time, hiring the worker will increase the cost of production because the farm worker must be paid. The farmer makes a profit-maximizing decision *at the margin*. If hiring the worker increases total revenue more than it increases total cost, then hiring the worker will increase profits. The farmer will always hire a worker if doing so increases profits.

Suppose the farmer is thinking about hiring a second worker. The same decision process is repeated. If the second worker adds more to revenue than to cost, then he or she will be hired. This process continues with every prospective worker. An important principle is in action: the *law of diminishing marginal returns*, which you may recall from Chapter 8. As more and more of a variable input is added to a fixed input, the marginal product of the variable input eventually declines. As more workers are hired, the extra output of each worker eventually falls. Since each additional worker adds less and less to total output, the additional revenues that each worker produces eventually fall below the additional cost of hiring. At some point, the farmer will stop hiring altogether. In short, the profit-maximizing farmer should hire all workers that add more to revenue than to cost but stop hiring at the point where the addition of revenue is just equal to the addition to cost.

The example of a corn farmer can also be expressed in economic terms. The extra output that each additional unit of labor adds to total output is called the **marginal product of labor**:

$$MP_L = \frac{\Delta TP}{\Delta L},$$

where $MP_L$ is the marginal product of labor and $\Delta TP$ is the change in total product brought about by adding one unit of labor, $\Delta L$. As we saw in Chapter 8, adding units of the variable input to other fixed inputs will eventually decrease the marginal product of the variable input in the short run.

> **Marginal product of labor:** The change in total output that results from employing one more unit of labor.

## TABLE 12–2  A Marginal Revenue Product Schedule for Labor

The marginal revenue product of labor in column (4) falls as more units of labor are employed because the marginal product of labor (column 2) falls. The firm will hire additional units of labor up to the point where the extra revenue these units generate just equals the extra cost of paying for them. At this point, 6 units of labor, MRP = MFC.

| (1) Units of Labor (worker-hours) | (2) Marginal Product of Labor $MP_L$ (bushels of corn) | (3) Marginal Revenue MR = P (dollars per bushel) | (4) Marginal Revenue Product MRP = MR × $MP_L$ = $\Delta TR/\Delta L$ | (5) Marginal Factor Cost (dollars per hour) MFC = Wage |
|---|---|---|---|---|
| 1 | 14 | $2 | $28 | 8 |
| 2 | 12 | 2 | 24 | 8 |
| 3 | 10 | 2 | 20 | 8 |
| 4 | 8  | 2 | 16 | 8 |
| 5 | 6  | 2 | 12 | 8 |
| 6 | 4  | 2 | 8  | 8 |
| 7 | 2  | 2 | 4  | 8 |
| 8 | 0  | 2 | 0  | 8 |

Table 12–2 summarizes the hypothetical choices available to the corn farmer. The first column shows the units of labor, in number of worker-hours. The second column shows the marginal product of the workers—the amount by which total output increases as one more unit, or hour, of labor is added to the production process. $MP_L$ decreases as the amount of labor increases, according to the law of diminishing marginal returns. The first unit of labor adds 14 units of output (bushels of corn), the second adds 12 units of output, and so forth.

When the extra corn produced by each additional unit of labor is sold, the resulting increase in total revenue is called **marginal revenue product** (MRP):

$$MRP_L = \Delta TR/\Delta L,$$

where $MRP_L$ is the marginal revenue product of labor, $\Delta TR$ is the change in total revenue, and $\Delta L$ is the change in the amount of labor hired.

Another method of expressing $MRP_L$ is

$$MRP_L = MR \times MP_L,$$

where MR is the marginal revenue (the increase in total revenue resulting from selling one more unit of output) and $MP_L$ is the marginal product of labor.

For the firm that sells its product in a perfectly competitive market, marginal revenue is equal to the price of the product. Thus, the marginal revenue product (MRP) is simply the price (P) of the product times the marginal product (MP) of each unit. As shown in Table 12–2, the marginal revenue product of the first unit of labor is found by multiplying its marginal product, 14, by the price of the product, $2, to obtain $28. The $MRP_L$ declines as more workers are hired because $MP_L$ declines.

Regardless of which economic shorthand is used, the profit-maximizing firm will continue adding units of labor as long as the additional revenues the labor produces are greater than the additional costs of the labor. The

**Marginal revenue product:** The change in total revenue that results from employing one more unit of a variable input.

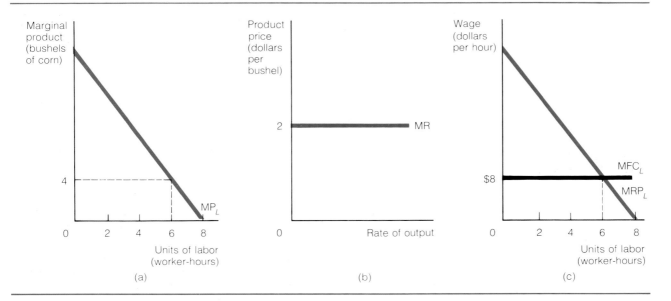

**FIGURE 12–2** **The Firm's Marginal Revenue Product Curve**

(a) The farmer's marginal product of labor ($MP_L$) curve is plotted from columns (1) and (2) in Table 12–2. (b) The demand curve for the farmer's product is determined by the product price ($P$ = MR in this perfectly competitive market). (c) The marginal revenue product ($MRP_L$) curve is found by multiplying $MP_L$ by MR. The profit-maximizing firm hires labor up to the point where marginal revenue product of labor equals marginal factor cost of labor ($MFC_L$). With a wage of $8 and a product price of $2, the firm will hire up to 6 units of labor.

**Marginal factor cost:** The change in total cost that results from employing one more unit of a variable input.

extra cost of hiring one more unit of labor is called the **marginal factor cost** of labor:

$$MFC_L = \Delta TC/\Delta L,$$

where $MFC_L$ is the marginal factor cost and $\Delta TC$ is the change in total cost. Under competitive conditions, the firm may purchase all of the labor it wants at the prevailing wage rate. Each additional unit of labor has a marginal factor cost equal to the wage rate (shown as $8 on the schedule).

The profit-maximizing firm will hire labor to the point where $MRP_L$ = $MFC_L$. With the numbers in Table 12–2, this equation indicates that it will be most profitable for the farmer to hire 6 units of labor. Each unit from the first to the sixth adds more to revenue than to cost. By hiring 6 units, the farmer adds as much as possible to profit. However, the seventh unit adds more to cost than to revenue and therefore will not be hired.

This process should sound familiar: It is the mirror image of how the competitive firm determines its profit-maximizing output. The firm follows the same process to equate its marginal cost to marginal revenue in hiring inputs as it does in selling its output. In fact, when the firm hires the profit-maximizing quantity of labor, that amount of labor will produce the profit-maximizing level of output.

The numbers used in Table 12–2 are presented in graph form in Figure 12–2. The $MP_L$ curve is drawn in part a by plotting the points from columns (1) and (2). The demand curve for the farmer's corn is shown in part b, where $P$ = $2. The $MRP_L$ is found by multiplying marginal revenue by marginal product; the result is the marginal revenue product curve in part c.

The prevailing wage is $8, stated as $MFC_L$. The profit-maximizing quantity of labor is the point where the $MFC_L$ curve crosses the $MRP_L$ curve. At this point, the marginal revenue product equals the marginal factor cost of labor.

### Profit-Maximizing Rule for All Inputs

Firms employ many inputs other than labor. The profit-maximizing rule for the level of employment is the same for each resource. The corn farmer may be planning to use fertilizer, for example. If he applies 1 unit of fertilizer—a 100-pound bag—the amount of corn produced will increase. The extra corn is treated as the marginal product of the first bag of fertilizer. The extra revenue that results is the marginal revenue product, and the extra cost of the bag is the marginal factor cost. If the MRP of the fertilizer is greater than the MFC, then the farmer will purchase and use the fertilizer. The profit-maximizing amount of fertilizer is found at the point where the MRP of fertilizer is just equal to the MFC.

This rule applies to all inputs, such as tractors, seed, land, tractor drivers, and water. The profit-maximizing quantity of all inputs is the amount that makes the MRP of each input equal to the MFC of each input:

$$\frac{MRP_1}{MFC_1} = \frac{MRP_2}{MFC_2} = \frac{MRP_3}{MFC_3} = \cdots = \frac{MRP_N}{MFC_N} = 1,$$

where the subscripts 1, 2, and 3, . . . indicate the MRP and MFC of the first, second, third, and so on, inputs, and $N$ indicates the total number of inputs used. Each ratio of MRP to MFC is equal to 1 because the MRP of each is equal to the MFC of each. Under this condition, all inputs are employed in the profit-maximizing amounts.

## The Demand for Labor

Marginal productivity theory gives insight into the purchasing behavior of firms. In the short run, when labor is the only variable input, the firm is willing to purchase or hire additional units up to the point at which MRP = MFC. The firm's marginal revenue product curve (Figure 12–2c) thus traces the relation between the price of labor (the wage rate) and the amount of labor a firm is willing to purchase. In other words, the marginal revenue product curve is also the firm's short-run demand curve for labor. Figure 12–3 illustrates this. At wage $W_0$ the firm chooses $L_0$ units of labor. A higher wage such as $W_1$ results in less labor hired. For each wage, the firm adjusts the level of employment to maintain the equation $MRP_L = MFC_L$.

In the following sections we examine the demand for labor from several perspectives.

### The Short-Run Market Demand for Labor

A single firm's demand for labor in the short run is equivalent to its marginal revenue product of labor. To obtain the overall market demand for labor in the short run, we must make some additional calculations.

Each firm will, of course, demand labor up to the point where $MRP_L = MFC_L$. Thus, at a particular wage, the market demand for labor may be shown as the sum of all the firms' $MRP_L$ curves. Figure 12–4 shows such a curve, $MRP_{L0}$. This curve represents simply the sum of all the firms' short-run demand curves for a particular type of labor. The curve labeled $MRP_{L0}$,

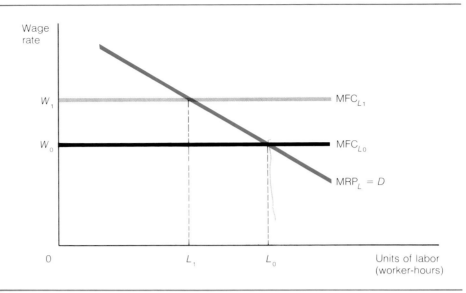

**FIGURE 12–3  The Firm's Short-Run Demand for Labor**

The firm hires labor up to the point where the wage is equal to the marginal revenue product of labor ($MRP_L$). The marginal revenue product curve shows the various quantities of labor the firm is willing to hire at all the different wage rates when labor is the only variable input.

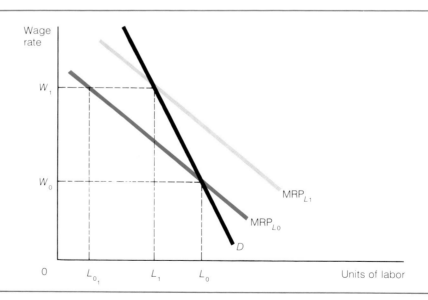

**FIGURE 12–4  The Short-Run Market Demand for Labor**

$MRP_{L0}$ represents the sum of all the firms' marginal revenue product curves for a particular occupation. When the wage for this occupation is $W_0$, the firms hire $L_0$ units of labor. If the wage rises to $W_1$, the firms initially hire the lower $L_{0_1}$ units of labor. However, the increase in wages causes an increase in the price of the product. The higher product price shifts the sum of the firms' $MRP_{L0}$ to $MRP_{L1}$. At $W_1$, when labor is the only variable input, the firms hire $L_1$ units of labor. $D$ represents the short-run market demand curve for labor because it allows for changes in the wage rate.

however, is not the short-run market demand curve for this type of labor, but it does establish one point on the short-run market demand curve, $D$.

To establish the whole demand curve, we must examine what happens when the wage rate changes. Imagine the overall demand for corn huskers. Many farmers employ huskers at wage $W_0$; the total number of corn huskers employed at $W_0$ is $L_0$. Recall that the $MRP_L$ curve is obtained by multiplying the price of corn by the marginal product of labor. If the wage rate rises to $W_1$, then each firm's costs of production will rise. The increased costs of production will decrease the supply of corn and increase the equilibrium price. To obtain the new $MRP_L$ curve ($MRP_{L1}$), we multiply the marginal product of labor by the new, higher price of corn, and we sum this quantity over all the firms. The new wage, $W_1$, and level of employment, $L_1$, establish a second point on the short-run market demand curve. The short-run market demand curve, $D$, differs from the sum of the $MRP_L$ curves because it allows for increases in the price of the product that result from wage increases.

### The Firm's Long-Run Demand for Labor

**Firm's long-run demand for labor:** The various quantities of labor that a firm is willing to hire when all inputs are variable.

We have so far concentrated on the short-run demand for labor, in which labor is the only variable input. In the long run, many adjustments occur when wages change. Allowed enough time, firms may vary all inputs. Also, at the industry level, not only does the price of products change directly with wages, but the number of firms changes as well.

Suppose that corn huskers demand a raise in wages. What happens in the long run? First, farmers will adjust to the higher price of labor by hiring fewer corn huskers; given enough time, they may turn to substitute inputs for this type of labor, perhaps by buying combines rather than hiring human huskers. Second, the increase in the price of labor increases the farmers' costs of production. With the increase in costs, the supply of the product decreases, and its price rises. Finally, the number of firms will adjust, establishing a new equilibrium. The number of firms, in the end, is likely to decrease. When wages rise, a larger number of workers lose their jobs in the long run than in the short run; or, if wages fall, a larger number of workers are hired. The long-run market demand is more elastic than the short-run market demand.

### Changes in the Demand for Labor

In applying the law of demand to labor, we have seen some adjustments at the firm level and at the industry level in response to wage rate changes. An increase in wage rate changed the price of the product, the employment of other inputs, and the number of firms. However, in establishing the demand curve for labor, many other things were held constant, such as the demand for the product, the prices of other inputs, and the level of technology. If these things were not held constant, then we would not have been able to establish the relation of wage rate to the amount of labor demanded.

What causes an increase or decrease in the demand for a particular type of labor? Why would the demand for chemical engineers suddenly increase? Or why would a restaurant suddenly hire more waiters and cooks? Such questions may be answered by examining the origins of the demand for labor and the behavior of profit-maximizing firms.

The demand for labor is derived basically from two things: the demand for the product the labor produces and the marginal product of the labor itself. If either of these increases or decreases, then the demand for labor will increase or decrease in the same direction. Understanding this relation simplifies things; however, further details are of value.

1. *If the demand for the product that labor produces changes, then the demand for labor will change in the same direction.* Remember that for a competitive firm, the marginal revenue product is equal to marginal revenue (the output price) times marginal product. If the demand for the good increases, then its price increases and the demand for labor (MRP) and all other inputs also increases. The demands for electronic technicians and computer components have therefore increased since the demand for personal computers has increased.

The opposite situation occurs when the demand for a product decreases. If the price of a product falls, then the demand for inputs used to produce that product falls. When the demand for American-made cars decreases, the demand for domestic autoworkers decreases (along with the demand for tires and steel).

2. *If the marginal product of labor changes, then the demand for that labor will change in the same direction.* The productivity of a particular type of labor depends to a very large extent on the firm's employment of other inputs in the production process. Additional expenditures on other inputs often increase the productivity of labor. For example, the marginal product of farm workers is influenced by the amounts of other inputs that the farmer buys, such as land, equipment, seed, fertilizer, and water. If such inputs increase the workers' productivity, their use will also increase demand for the workers' labor.

**Complementary inputs:** Two or more inputs with a relation such that increased employment of one increases the marginal product of the other.

**Substitute inputs:** Two or more inputs with a relation such that increasing the employment of one decreases the marginal product of the other.

The relation between the quantity of nonlabor inputs and the productivity of labor can be direct or inverse. In other words, increasing the amount of one input may either increase or decrease the productivity of labor. If increasing the quantity of one input increases the marginal product of labor, then this input and labor are **complementary inputs.** For example, if the farmer uses more fertilizer, then corn huskers can produce more output; their marginal product increases. Increases in inputs that complement labor will increase the demand for labor.

If increasing the amount of another input decreases the marginal product of labor, then that input is called a **substitute input** for labor. For corn pickers and huskers, a combine that picks and husks is a substitute input. If the farmer uses more combines, then the marginal product of human huskers and pickers will decrease, and so will the demand for them.

Why would a farmer suddenly choose to purchase more fertilizer or combines? A firm makes decisions on the basis of expected profits. For example, if the price of fertilizer falls, then it is in the farmer's interest to purchase more fertilizer. If the price of combines falls, then the farmer will purchase more combines. Thus, if the price of a complementary input falls, the demand for labor rises. If the price of a substitute input falls, the demand for labor decreases.

In addition to changes in the quantity of inputs, technological advances may shift the demand for labor. New discoveries about production techniques may increase or decrease the demand for labor in a particular occupation. In fact, this is generally what technological change does: It increases the marginal product of some workers and decreases that of others. When robots replace factory workers, the marginal product of the factory workers falls, and the demand for their labor decreases. At the same time, the marginal product of the workers who make robots goes up, and the demand for their labor increases.

## The Elasticity of Demand for Labor

We have shown how to find the firm's demand for labor, how to find the market demand for labor, and how the demand curve for labor can shift. We still need to say something about the **elasticity of demand for labor.** Remember that elasticity measures the responsiveness of buyers to changes in price. Specifically, the elasticity of demand for labor is equal to the percent change in the quantity of labor demanded divided by the percent change in the wage rate.

**Elasticity of demand for labor:** A measure of the responsiveness of employment to changes in the wage rate; the percent change in the level of labor employed divided by the percent change in the wage rate.

Keep in mind that we are not talking about labor in general. We are talking about the derived demand for a specific type of labor. For example, we want to understand why the level of employment of corn huskers decreases by 30 percent when the wage rate increases by 10 percent; or why the level of employment of nuclear physicists falls by only 10 percent when their wages increase by 30 percent. There are three useful rules to apply when analyzing the elasticity of demand for labor.

1. *The elasticity of demand for labor is directly related to the elasticity of demand for the product that labor produces.* If the demand for a product is highly elastic, then a small percent increase in price will decrease the quantity demanded by a large proportion. If the increase in price is caused by an increase in wages, the resulting large decrease in production will cause a large decrease in the quantity of labor demanded. Suppose that the demand for houses is very elastic, and the wage of carpenters increases. The increase in the price of houses that results from the higher wages will decrease the quantity of houses purchased by a relatively large amount. The decreased production of houses will then decrease the quantity of carpenters demanded by a relatively large amount.

The greater the elasticity of demand for corn, the greater the elasticity of demand for corn huskers. The less elastic the demand for nuclear reactors, the less elastic the demand for nuclear physicists. The elasticity of demand for labor is in part derived from the elasticity of demand for the product that labor produces.

2. *The elasticity of demand for labor is directly related to the proportion of total production costs accounted for by labor.* The total amount of money that a firm pays for labor is a percentage of its total costs. The larger the percentage of total costs accounted for by labor, the greater the elasticity of demand for labor. This relation occurs because of the effect of wage increases on the price of the product. If wages rise by $2 an hour and labor represents a large percentage of total costs, then total costs and thus product price will increase by a large amount. The large price increase will decrease the amount of the product purchased and will decrease greatly the quantity of labor demanded. On the other hand, if labor accounts for a very small percentage of total costs, the same $2-an-hour wage increase will have little effect on overall cost and price. The decrease in the amount of the product demanded will be very small, making the decrease in labor demanded also small.

If the percentage of corn huskers' wages in the farmer's total costs is very large, then an increase in the huskers' wages will result in a relatively large decrease in the number of workers that the farmer hires. Conversely, if the percentage of huskers' wages in the farmer's total costs is very small, then the same wage increase will result in a relatively small decrease in the number of huskers hired.

3. *The elasticity of the demand for labor is directly related to the number and*

*availability of substitute inputs.* The demand for labor is highly elastic if there are good, relatively low-priced substitutes for labor. If the wage rate rises and there is another comparably priced input that can do the job of labor, then the relative amount of labor hired will decrease significantly. If there are no close substitutes for labor, or the substitutes are of a much higher price, then a wage increase will reduce employment by a lesser amount.

Once again, we see that the behavior of firms is similar to the behavior of consumers. The determinants of the elasticity of demand for labor and other inputs parallel the determinants of the elasticity of demand for goods by consumers. A larger percentage of total expenditures for purchases of some item—labor by firms or a good by consumers—creates greater elasticity of demand. A large number of substitutes for an item creates a highly elastic demand.

### Monopoly and the Demand for Labor

We have spoken so far of demand for labor only under conditions of competition. What if the firm demanding labor enjoyed a monopoly in its output market but was still a purely competitive buyer of labor? For example, what would the farmer's demand for labor be if he were the only producer of corn? The farmer's decision rule for hiring labor would include a minor twist. Recall from Chapter 9 that the demand curve for the monopolist's product is downward sloping. This suggests that as the firm hires more workers and increases output, the price of the product must fall for the firm to sell the extra output. As the firm hires more labor, not only is the marginal product of labor falling but the product price is falling as well.

Even though the price of the product is falling as more output is produced, marginal revenue product, MRP, still describes the change in total revenue that results from hiring one more unit of labor. As before, MRP is equal to marginal revenue times marginal product. The difference is that the marginal revenue of the competitive seller is equal to the market price, whereas the monopolist's marginal revenue is a downward-sloping curve, derived from its demand curve.

A graphical comparison of the purely competitive firm's demand for labor and the monopolist's demand for labor is shown in Figure 12–5. The competitive firm's demand for labor is $MRP_c$ in Figure 12–5c, obtained, as before, by multiplying $MR_c$ from part a by $MP_L$, the marginal product of workers in part b. The monopolist's demand for labor, $MRP_m$, is obtained by multiplying $MR_m$ in part a by $MP_L$ in part b. The marginal product of workers, $MP_L$, falls at the same rate whether they are working for monopolists or for purely competitive firms. But the monopolist's MRP diminishes more quickly than the pure competitor's because not only is the marginal product of labor falling, but the monopolist's marginal revenue is falling as well.

A firm that enjoys a monopoly in its output market does not necessarily have any advantage in resource markets. For example, your local electric company may have exclusive rights as the only seller of electricity in your area. But this monopoly position does not imply that the electric company is the only employer of electricians, computer programmers, or bookkeepers. For these occupations, the monopoly is a purely competitive buyer of labor and must pay a competitive market wage for its labor. The marginal factor cost of labor is equal to the wage rate because the supply of labor to the firm is infinitely elastic.

In the end, the monopolist's decisions in hiring labor require the same

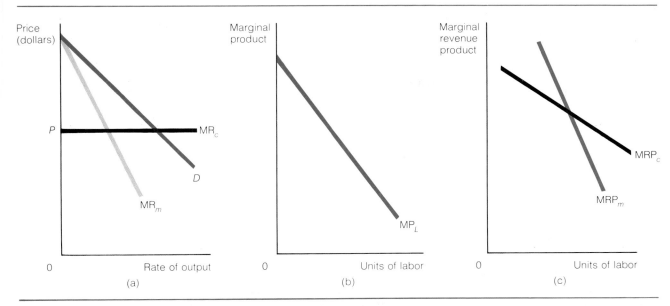

**FIGURE 12-5** The Monopolist's MRP versus the Competitive Firm's MRP

(a) The marginal revenue curves for a monopoly ($MR_m$) and for a competitive firm ($MR_c$). (b) The marginal product of labor for both types of firm. (c) The marginal revenue product for each firm, obtained by multiplying marginal revenue by marginal product. The monopolist's marginal revenue curve is downward sloping, and the competitive firm's is equal to a given and constant product price; therefore, the monopolist's demand for labor, $MRP_m$, diminishes more rapidly than the competitive firm's $MRP_c$.

marginal analysis as for any other firm. The firm will hire any unit of labor when its marginal revenue product is greater than its marginal factor cost and will continue to hire labor up to the point where the $MRP_L$ is equal to $MRC_L$. The only difference for a monopoly firm is that its marginal revenue falls along with the labor's marginal product. Focus, "The Marginal Revenue Product of Professional Baseball Players," discusses these ideas as they relate to another market.

## The Supply of Labor

Now that we have some insight into the demand for labor, the next logical category is labor supply. We know that labor supply in competitive markets is highly elastic. But this fact does not answer two important questions: What determines the overall, or aggregate, supply of labor and how are wage rates and labor supply interrelated?

### Individual Labor Supply

Individuals in free markets are able to choose whether to offer their labor services in the market, and individuals' labor market decisions are based on the principle of utility maximization. The individual must choose between time spent in the market earning the wage that he or she may command and time spent in **nonmarket activities,** such as going to school, keeping house, tending a garden, or watching television.

**Nonmarket activities:** Anything that an individual does while not earning income from working.

Fortunately, the individual does not have to make an all-or-nothing

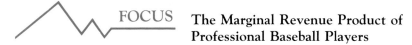

## FOCUS: The Marginal Revenue Product of Professional Baseball Players

The extra revenue that a firm obtains from hiring a worker is the marginal revenue product; if this is higher than the extra cost of the worker's wages, then the firm adds to its profits by employing the worker. Baseball players' salaries have increased sharply in the last few years, especially for superstars. Are they worth it? Are their MRPs greater than their wages?

To answer this question, Gerald Scully calculated the extra revenues that team owners received from hitters and pitchers. He first found that a player's contribution to his team's win-loss record was best measured for pitchers by the percent each added to the team's strikeout-to-walk ratios and for batters to the contribution of each to the team's batting average. Using this information, Scully then estimated the total team revenue where revenue is simply the number of tickets sold per season times the average ticket price. After adjusting for the size of the city in which the team played its home games, the age of the stadium, the team's league, and other factors affecting revenue, he calculated that every point added to the team batting average by a batter increased the team's revenue by $9,504. Similarly, every .01 point added to the pitching staff's strikeout-to-walk ratio increased revenue $9,297.

Once he estimated player MRPs, it was a straightforward calculation for Scully to determine whether the salary paid to a particular player was commensurate with the amount the player added to team revenue. For example, star hitters added from $250,000 to $383,700

*Source:* Gerald W. Scully, "Pay and Performance in Major League Baseball," *American Economic Review* 64 (December 1974), pp. 915–30.

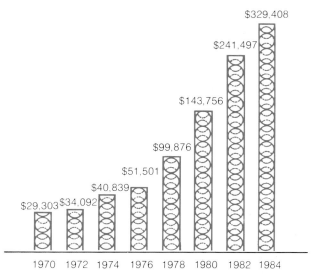

*The average salary of major league baseball players has increased from $29,303 in 1970 to $329,408 in 1984.*
Source: Major League Baseball Players Association.

to team revenues. Star pitchers contributed even more, from $321,700 to $479,000. However, during the 1968 and 1969 seasons, no player in Scully's sample was paid more than $125,000. He concluded, therefore, that players' salaries were well below their MRPs, with the greatest deficiency being the superstars' salaries. Scully's results explain in part free agency and the huge increases in players' salaries in the 1970s and 1980s.

---

choice. No one chooses to work twenty-four hours a day, seven days a week. People can divide their time between market and nonmarket activities. How many hours would one choose to work during an average week? This question cannot be easily answered, but we can speculate about the effect that the wage rate may have on an individual's labor supply choices.

A supply curve relates the quantity supplied of a commodity to its price. Applied to labor supply, it is a schedule that relates the quantity of work offered and different wage rates. The labor supply curve of an individual is given in Figure 12–6. The first thing to observe about this person's labor supply curve, $S_i$, is that below $w_1$ she would prefer not to work at all; she chooses all nonmarket activity. This cutoff occurs because wages below $w_1$ do not meet her opportunity cost; she receives more utility from nonmarket activities than she would from the income received from working. Wage $w_1$ may not be high enough to encourage her to forgo fishing, painting the house, or attending school, for example. At wages immediately above $w_1$, the hours of labor she supplies respond positively to increases in pay. As her

**FIGURE 12–6**

**Individual Labor Supply**

The individual labor supply curve, the number of hours offered at various wages, may be positively or negatively sloped. If the substitution effect is greater than the income effect, the curve is upward sloping, as shown from $w_1$ to $w_3$. The individual substitutes hours of work for hours of nonmarket activities. As the individual's wage rises beyond $w_3$, his or her demand for nonmarket time increases. The income effect is greater than the substitution effect, and the curve is negatively sloped above wage rate $w_3$.

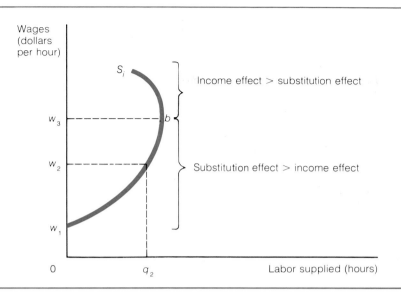

wage rate rises from $w_1$ to $w_2$, the hours she is willing to supply increase accordingly, to $q_2$.

The value of one hour of market time (one hour spent working) is equal to the wage rate. This means that for each hour an individual chooses to enjoy nonmarket activities, he or she is forgoing an hour of pay. In other words, the opportunity cost or the price an individual must pay for an hour of leisure time is the wage rate. As the wage rate rises, the price of nonmarket time also rises. An individual's demand for nonmarket time is downward sloping, just as it is for any other good. If the wage rate rises, the individual is encouraged to purchase less nonmarket time; that is, he or she is encouraged to substitute hours of work for hours of leisure.

The substitution effect, however, may be offset by an income effect. As the wage rate rises, the individual's income rises. If nonmarket time is a normal good, as we might expect, then the demand for nonmarket time will increase as income rises. This type of income effect will encourage the individual to enjoy more nonmarket time and less work.

As the wage rate rises, the two effects pull the worker in opposite directions. The effect that dominates will determine whether higher wages encourage more or fewer hours of work and determine whether the labor supply curve is positively or negatively sloped. If the substitution effect is greater than the income effect, higher wages will bring more hours of labor. In Figure 12–6, the substitution effect is greater than the income effect from $w_1$ to $w_3$, so the slope is positive. However, the income effect is greater than the substitution effect at wages above $w_3$. In this range, higher wages bring fewer hours of labor, so the curve is negatively sloped. At point $b$, where the income and substitution effects are equal, the supply curve begins to bend backward.

The actual shape of any individual's labor supply curve cannot be determined theoretically. It can be positively sloped, negatively sloped, or vertical through any range of wages. The wage level that initially entices people to enter the labor market also varies.

Since the shapes of individual labor supply curves are difficult to deter-

**FIGURE 12–7**

**The Aggregate Supply of Labor**

The aggregate labor supply curve, $S_A$, representing the numbers of hours of labor supplied by all individuals, is likely to be upward sloping. As wages rise, more people join the labor force, and many people work more hours.

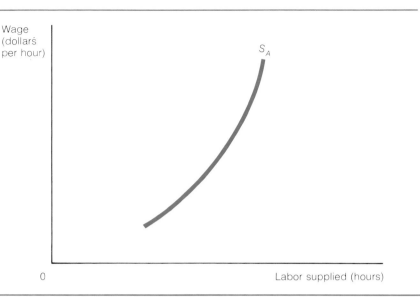

mine, the slope of the total or aggregate labor supply curve is difficult to predict with accuracy. The aggregate labor supply curve is theoretically obtained by summing individuals' labor supply curves horizontally. The aggregate curve shows the total amount of labor supplied at all the different wages. Even though some individuals have backward-bending labor supply curves, we expect the aggregate labor supply curve to be positively sloped, in Figure 12–7, for two reasons. First, as wages rise, many workers will work more hours. Second, as wages rise, more people will enter the labor force.

Although the wage rate has an influence on the number of hours people may choose to work, individuals may also choose their wage rate, within limits. They do so by choosing an occupation. Some occupations must pay higher wages than others to induce people to join.

### Human Capital

Some occupations, such as medicine, law, and nuclear physics, require many years of training. The fact that the wages in these skilled professions are higher than the wages of unskilled labor is not coincidental. Before people are willing to endure the years of training to become highly skilled professionals, they must be reasonably sure that their investment of time and other resources will pay off in the long run. People frequently choose to invest in some form of training to make themselves more productive and to enhance their income-earning potential. While a person is going to college, attending trade school, or gaining on-the-job training, he or she is building **human capital.**

**Human capital:** Any nontransferable quality an individual acquires that enhances productivity, such as education, experience, and skills.

Gaining human capital requires an investment period, such as four years in college. This investment involves a large opportunity cost. A college student loses the next-best alternative when attending college. For many students, the lost opportunity is the income they would have earned if working. Figure 12–8 portrays a simplified version of two alternative lifetime income streams for an individual.

After graduating from high school at age eighteen, individuals have a

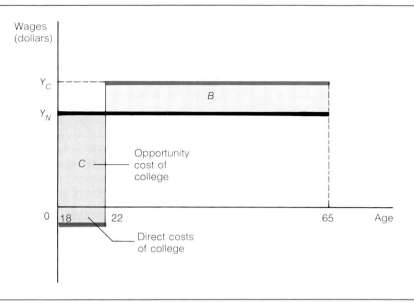

**FIGURE 12-8  The Economic Costs and Benefits of College**

An individual may choose to earn income $Y_N$ from age eighteen to retirement at age sixty-five. Or an individual may attend college, lose income from age eighteen to age twenty-two, pay the direct costs of college, and then earn income $Y_C$ from age twenty-two to retirement. The total costs of college are represented by the shaded area C, the lost income and direct costs; the benefits of college are shaded area B, the extra income earned with a college degree. In monetary terms it pays to go to college if B is greater than C.

choice. They may enter the labor market immediately and earn income stream $Y_N$ from that time until retirement at age sixty-five. This income stream could rise through the years as the individuals acquire skills, but for simplicity we let it remain constant.

Some individuals may go to college. From age eighteen to twenty-two, they do not earn income. The opportunity cost of going to college—that is, the income the individuals forgo by choosing not to work—is represented by the shaded area above the x-axis between ages eighteen and twenty-two. In addition, they must pay the direct costs of college—tuition, books, and so on. The shaded area below the x-axis between ages eighteen and twenty-two represents the direct costs of college, a negative income. The total cost of going to college is the sum of both the opportunity cost and the direct costs of college, the entire shaded area C. After graduation, these individuals enter the labor market and begin earning positive income. Figure 12-8 indicates that their starting pay at age twenty-two after attending college is higher than it would have been had they not gone to college. They earn negative income for four years and then $Y_C$ as a lifetime average from age twenty-two until retirement.

Which is the best choice? If the deciding factor is money, then one determines the relative values of the two income streams and chooses the higher. One method of doing so is to compare the area C (the investment period) with area B. Area C represents the total cost of going to college: the direct costs plus the lost income. Area B represents the benefits of going to college; it is the extra income earned with a college degree. If area B is greater than area C, then it pays to go to college.[1]

Some investments in human capital require more time and some require less time than others. But regardless of the length of the investment period, if there is pure competition in the market and if enough individuals choose the higher income streams, then in the long run the income streams will all be equal. (This equilibrium requires that the wage rates for occupations with long investment periods be greater than the wage rates for occupations with short investment periods. The difference in wages is what equalizes the lifetime income streams.) Focus, "Does It Pay to Go to College?" elaborates on the economic decision-making process of whether to go to college.

### Other Equalizing Differences in Wages

A long investment period is not all that discourages individuals from entering a particular occupation. Other characteristics make it necessary for some occupations to offer higher wages to induce workers to enter. People will evaluate their alternative work possibilities in both monetary and nonmonetary terms. Pay will be important, to be sure, but so too will be working conditions, location, co-workers, risk, length of contract, personality of the boss, and myriad other factors. Although some of these factors are nonmonetary or psychological, this does not mean that they are not income. They are a part of workers' pay. The **total compensation** of a worker in a given occupation consists of the wage plus any nonmonetary aspects of the job.

The principle of **equalizing differences in wages** works in the following way. In a competitive labor market, laborers choose the occupation with the highest total compensation. This brings the wage down until the total compensation for the occupation is on a par with that of other occupations. If total compensation is too low in some occupation, then people leave that occupation until the wage rises enough to equalize the total compensation. When total compensation is equalized across occupations, the competitive equilibrium is achieved. If this condition is not met, workers will move around and change jobs until it is met.

One student who spends the summer in an air-conditioned office may make a lower wage but more nonwage pay per hour than another student who spends the summer on a construction crew. As students compete for summer jobs, total pay between jobs will be equalized. This is the principle of equalizing differences at work in a competitive labor market.

The basic point is that observed wage differences across occupations may reflect differences in nonmonetary aspects of employment. Of course, labor markets may not always be competitive as we have assumed here. Some jobs may pay more and have more attractive working conditions if there are barriers to entry into the occupation. However, under competition, wage differences can exist. The following are just a few of the many reasons for wage differences other than those caused by human capital differences.

*Wages will vary directly with the disagreeableness of a job.* The more uncomfortable the job, the higher the pay will be. To induce workers to accept jobs that create discomfort—such as tarring roofs in the heat of summer—a higher monetary reward must be offered. Workers will not offer their labor below a certain wage, given that they have alternatives.

*The more seasonal or irregular a job, the higher the pay will be.* To induce

**Total compensation:** The lifetime income that an individual receives from employment in a particular occupation, including all monetary and nonmonetary pay.

**Equalizing differences in wages:** The differences in wages across all occupations that result in equality in total compensation.

---

[1] The values of C and B must be discounted for time, as is the case for any investment; this point is discussed in Chapter 14.

## FOCUS  Does It Pay to Go to College?

The decision to go to college is a decision to invest in human capital with alternative payoffs for different levels of investment (see Table 12–3). As with all investments, there are costs and benefits. The monetary costs are the forgone income one could earn by working and the direct costs such as tuition and books. The monetary benefits are the years of extra income that one may earn with a degree. Is college a good investment? Is the rate of return on an investment in education higher than that of other investments?

Richard Freeman calculated the rate of return on a college degree over an individual's life cycle. Freeman's findings were:

| Year | Rate of Return |
|---|---|
| 1959 | 11.0% |
| 1969 | 11.5% |
| 1972 | 10.5% |
| 1974 | 8.5% |

These rates of return represent the interest rates that individuals receive on their investments in education. From 1969 to 1974 there was clearly a downward trend.

Freeman investigated whether the trend has continued into the 1980s. He found that through the 1970s the rate of return to college education fell considerably, suggesting that Americans are overeducated. However, the trend seems to have bottomed out. Freeman predicts that during the 1980s the market for college graduates will improve and increase the rate of return on investment in a college education.

**TABLE 12–3  Estimated Lifetime Earnings for Men and Women by Educational Level (1981 dollars)**

Expected earnings from additional education rise for both men and women with more education, but less than is commonly supposed. A dramatic increase for both men and women comes with the high school diploma and with a college degree (versus only some college work). However, the figures probably underestimate actual earnings since they do not include human productivity changes over lifetimes. Also note the obvious disparity in estimated future income for men and for women, a difference due partially to sex discrimination which we fully discuss in Chapter 16 on income distribution.

| Category | Men | Women |
|---|---|---|
| Less than 12 years education | $ 845,000 | $500,000 |
| Completed 12 years education | 1,041,000 | 634,000 |
| Completed 1 to 3 years college | 1,155,000 | 716,000 |
| Completed 4 years college | 1,392,000 | 846,000 |
| Some graduate work | 1,503,000 | 955,000 |

*Source:* U.S. Department of Commerce, Bureau of the Census, *Consumer Income* (February 1983).

*Sources:* Richard Freeman, "Overinvestment in College Training," *Journal of Human Resources* 10 (Summer 1975); and "The Overeducated American in the 1980s: A Report to the National Commission on Student Financial Assistance," *Higher Education Marketing Journal* 11:9 (Summer 1983).

---

individuals to supply labor to irregular employments, the wage must compensate them for the likelihood of being laid off frequently. For example, housing construction workers have frequent periods of unemployment between jobs. To have readily available workers, construction employers must pay a higher wage.

*Jobs that require trustworthiness carry a higher wage.* Take the jobs of bank teller, armored truck driver, and blackjack dealer. One reason that such occupations pay more is that being trustworthy is both an extra burden and a form of human capital; employers must compensate such workers for their extra efforts. Since workers have a choice between stealing and thereby losing future wage income and not stealing but keeping the job, the wage must be high enough to induce continued honesty in supplying labor.

*Jobs with greater risk to health will have higher pay.* Occupations with higher than average probabilities of death or disability must compensate employees for taking a risk. People are not willing to risk lost future income because of injuries unless they can receive higher income in the present. Steelworkers who construct frames for one-hundred-story skyscrapers must be paid more than those who work on the ground.

*Jobs that carry the possibility of tremendous success will have a lower wage.* Acting is a good example of such a profession. Many young people aspire to be actors and are arguably attracted into acting by the sumptuous lifestyles of movie stars. Most never make it that big; the average salary of all actors is quite low. Yet some people are willing to take the plunge, feeling that they are good enough to be big winners. Attitudes toward risks are therefore important in determining relative wages. The wages for most actors will be lower because some successful actors make startlingly high pay, a possibility that induces many young people to enroll in acting school.

Regardless of the equilibrium wage, the supply of labor to each occupation is highly elastic. In the short run, wages may deviate from equilibrium because of changes in demand, but in the long run new people entering the labor force and competing for jobs will force the difference in wages that yields equal total compensation across occupations.

## Marginal Productivity Theory in Income Distribution

The profit-maximizing tendency of competitive firms to set marginal revenue product (MRP) equal to marginal factor cost (MFC) leads to a wage that equals the value of the marginal product. This result applies to the prices of all factors of production in competitive markets. Each resource—be it labor, land, or capital—is paid the value of its marginal product.

The marginal productivity theory of factor prices is a positive, not a normative, theory. It is not meant to be an ethical theory of income distribution even though it is sometimes attacked on such terms. It is meant as a demand and supply theory that explains the behavior of input prices and the allocation of factors of production to different employments.

However, the marginal productivity theory has been criticized even as a positive theory. Labor markets are not perfect (as discussed in detail in the next chapter), and resources do not seem to flow around the economy in a smooth, frictionless way to equalize rates of return in factor markets. Labor and physical capital are particularly difficult to characterize as easily movable. People and machines and buildings generally like to stay in the same place. In this sense the marginal productivity theory may not always offer a realistic explanation of short-run factor prices. At any time we might observe different wage rates in a market for equally skilled workers, which marginal productivity theory may have difficulty in rationalizing.

Over a longer period of time, marginal productivity theory does offer a good explanation of factor prices. In the long run, resources will relocate to take advantage of higher rates of return. For this reason, marginal productivity theory is alive and well as a scientific explanation of factor prices. Given proper time, it is the best explanation we have of factor prices.

Another criticism of marginal productivity analysis is that business managers could not possibly estimate the marginal revenue products of their inputs. How hard it is to estimate MRP is an interesting question, but the fact is that business managers are forced to make some sort of estimate of what their inputs are capable of doing on an ongoing basis. For example, firms must decide whether to promote from within or hire managers from other firms. Why do firms typically promote from within? The answer is that it is less costly to form an estimate of the MRP of inside candidates. The firm has observed the performance of the inside candidate but must take letters of recommendation and other types of indirect evidence about the outside candidate.

It should now be clear that the theory of marginal productivity implies many things about income distribution. Where there is competitive voluntary contracting for labor services, labor's share of national income will be directly related to its marginal product. "To each according to what he or she produces" perhaps best describes what happens in marginal productivity theory. Of course, factor markets are not perfect, and various other forces affect income distribution in a society. These issues will be carefully addressed over the next several chapters. Our argument in this chapter is not that the marginal productivity theory is necessarily a good normative theory of how income should be distributed (although you might believe it so) but rather that it provides an objective basis for understanding the economic behavior of demanders and suppliers of inputs.

## Summary

1. The demand for labor is a derived demand; it is derived from the demand for the product that labor produces.
2. The demand curve for labor is a function of its marginal product and the demand for the product it produces. In the short run, demand is equivalent to marginal revenue product: $MRP_L = MP_L \times MR$.
3. The competitive firm hires the profit-maximizing amount of labor at the point where MRP equals the wage rate. The short-run market demand curve for labor shows the amount of labor that firms hire when labor is the only variable input and the number of firms is constant, but it allows for changes in product price as wages change.
4. The long-run market demand curve for labor shows the amount of labor hired at different wages when all inputs are allowed to vary and the firms are in long-run competitive equilibrium.
5. The demand for labor changes when the demand for the product changes or when the marginal product of labor changes. The elasticity of demand for labor is directly related to the elasticity of demand for the product it produces, the percentage of total cost accounted for by labor, and the number and availability of substitute inputs for labor.
6. The relation between the wage rate and the number of hours of labor an individual supplies depends on the substitution and income effects. If the substitution effect is greater than the income effect, then the individual's labor supply curve is positively sloped; the curve is negatively sloped if the income effect is greater than the substitution effect. The aggregate labor supply curve is the sum of individual labor supply curves and is usually positively sloped.
7. Human capital is anything an individual acquires that increases productivity. In an occupation that requires human capital, the longer and more costly the investment period, the higher the wages.
8. The total compensation in all occupations in competitive markets is equalized by differences in wages. The long-run supply of labor to each occupation ensures the equilibrium difference in wages.
9. In competitive markets, resources are paid a price that is equal to their marginal product. Labor's share of the overall income distribution is related to its marginal product.

## Key Terms

factor market
derived demand
competitive labor market
market demand for labor
market supply of labor
marginal product of labor
marginal revenue product
marginal factor cost
firm's long-run demand for labor
complementary inputs
substitute inputs
elasticity of demand for labor
nonmarket activities
human capital
total compensation
equalizing differences in wages

## Questions for Review and Discussion

1. Suppose that the demand for each of the following goods and services increases. What derived demands will increase as a result? (a) Automobiles, (b) candy, (c) government regulation, (d) physical fitness, (e) education.
2. Explain why an individual's supply of labor curve may bend backward at some sufficiently high wage level.
3. What does the law of diminishing marginal productivity have to do with the demand curve for labor?
4. What is the difference between the demand curve for labor of a monopolist and of a perfectly competitive firm in the output market?
5. Apply the principle of equalizing differences to explain why the relative wage of the following occupations is high or low: (a) politicians, (b) schoolteachers, (c) morticians, (d) actors, (e) brain surgeons.
6. In what sense is the marginal productivity theory of wages a positive theory? In what sense is it a normative theory?
7. What activities will frustrate the equalization of total wages across jobs in the labor market? Does this mean that there is such a thing as nonequalizing wage differences?
8. Are tractor drivers substitutes for human cotton pickers? What happens to the demand for tractor drivers if cotton pickers' wages fall?
9. Is the demand for toothpicks elastic or inelastic? What about elasticity of demand for a machine that makes toothpicks?

## ECONOMICS IN ACTION   Regional Wage Differences

In the long run, wages may differ across occupations for many reasons. Some occupations require large investments in education or training, some are risky, and others have random periods of unemployment. Economists in the past have also noticed that geographical factors influence wage differences within the same occupations. Why do we observe higher wages in the North than in the South or higher wages in San Francisco than in Oklahoma City? Can we expect these wage differences to persist in the long run?

Wages in any region are determined by supply and demand. However, if wages in market A exceed the wages in market B, then we expect to see labor flow from B to A. The shifting short-run supply brings wages into equilibrium. Wages fall in A and rise in B.

The equality of wages is the result of individuals' actions. Furthermore, an individual's decision to move is based on the costs and benefits of doing so. Relocation requires resource expenditures that must be offset by extra income. Indeed, the decision to move to a higher-paying location within the same occupation represents an investment in human capital.

Suppose a worker could stay in location B and earn wage $W_B$ or could move to location A and earn a higher wage $W_A$. The direct costs of moving are the expenses incurred in transporting the worker, her family members, and their personal belongings. There are also additional opportunity costs. The act of moving requires time, and so might finding a job in A. This is time that could have been spent earning income in B. The lost income is a cost of relocation, as shown in Figure 12–9.

At time 0 the individual must choose between earning wage $W_B$ from T to retirement and spending time moving so as to earn wage $W_A$. The time $0T$ is the amount of time spent moving from B to A plus the amount of time spent seeking employment in A. The total cost of moving is the shaded area between 0 and T above and below the x-axis, which represents the income forgone plus direct moving expenses. The benefit of moving—represented by the shaded area between T and retirement—is the extra income that may be earned after the move. It pays to move if the area of extra benefits is greater than the area of costs.[a]

If benefits outweigh costs, then many people will choose to move to A—the move is a good investment in human capital. However, the shifting labor supplies will increase wages in B, where workers are now scarcer, and decrease them in A. The maximum amount by which $W_A$ can exceed $W_B$ in the long run is the amount that forces the area of benefits to just equal the area of costs. Therefore, the lower the costs of moving, the smaller the difference in wages needs to be.

In fact, there are many reasons why we expect little variation in wages across the country within the same occupation in today's markets. First, the duration of lost income for most movers is very small. Indeed, it is quite possible to work until 5:00 P.M. on Friday in New York City and start work at 9:00 A.M. on Monday in Los Angeles. Many people will not move unless they have already found employment in the new location. If

---

[a] As in the earlier example of the economic costs and benefits of a college education, these values must be discounted for time, as explained in Chapter 14.

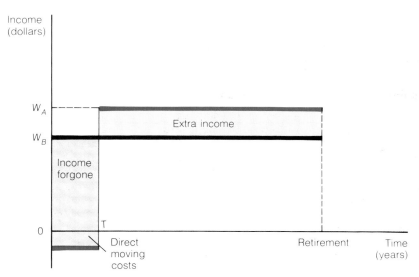

**FIGURE 12–9  Wage Differences Across Regions**

An individual must choose whether to remain in location B and earn wage $W_B$ from time 0 to retirement or move to location A and earn wage $W_A$ from time T to retirement. The decision to move to A necessitates costs in terms of the time 0T spent moving and income forgone while making the move, represented by the shaded area from 0 to T above the x-axis. After the move, the individual earns extra income represented by the shaded area from T to retirement. If this area of extra income is greater than the area of total costs of moving, the individual will choose to move.

the amount of time it takes to travel across the country today were as great as it was two hundred years ago, then wage differences could be very large.

Another reason that regional differences in wages are small is that young workers have small costs and large benefits in moving. Young workers typically have small families or no families and few material possessions to move. Also, the years of receiving extra benefits are longer for young people. Thus, the difference in wages can be smaller to induce them to move.

Another important aspect of regional wage differences involves the demand side of the labor market. Firms will be encouraged to expand operations in areas where there are low costs of production and decrease operations where costs are high. Firms will invest more capital in low-wage area B and decrease investment in high-wage area A. Capital moves around the country just as labor does, and this movement tends to equalize wages. In fact, all resources that are mobile move to the locations that yield the resource owners the highest income.

In spite of this equalizing tendency, we observe that wages in the North are about 10 percent higher than in the South in the same occupations. Don Bellante suggests that the difference occurs only in money wages and not in real wages.[b] The cost of living in the South is about 10 percent lower than in the North. This difference in costs equalizes the real wage, just as we expect in the long run.

### Question

The City of Houston recently advertised for police men and women on Atlanta TV stations offering 30 percent (on average) higher wages than those paid by the Atlanta police department. Some personnel moved to Houston. What effect would this movement have on regional wage differences for police services between Texas and Georgia? What might we expect the average age of the migratory police workers to be?

[b]D. Bellante, "The North-South Differential and the Migration of Heterogeneous Labor," *American Economic Review* 69 (March 1979), pp. 166–175.

# 13

# Labor Unions

The previous chapter discussed demand and supply conditions in a competitive labor market, where the price of labor tends to equality. Although we see a great deal of wage discrepancies in competitive markets, many of these can be attributed to circumstances such as the amount of risk a particular job involves or the amount of training it requires. We naturally expect brain surgeons to make more than dishwashers in a competitive environment because of the investment a brain surgeon must make in order to perform services. The mobility of a work force also affects its wage. The wages of workers will often vary because many of us are simply unable to move to fill available jobs.

All labor markets, of course, are not strictly competitive. In general, factor markets can assume the same kinds of imperfect structures that we have already studied in Chapter 11. The imperfection in labor markets can occur on the supply side, the demand side, or both. On the supply side, labor unions account for the largest source of imperfect competition.

In this chapter we will look at how imperfectly competitive labor markets work. Specifically, we will look at the effects of labor unions.

The imperfection in labor markets can occur on the supply side, the demand side, or both. On the supply side, labor unions account for the largest source of imperfect competition.

## Types of Labor Unions

**Labor union:** A group of workers who organize to act as a unit in an attempt to affect labor market conditions.

A **labor union** is essentially a group of workers who organize collectively in an effort to increase their market power. By acting as a collective unit, they are able to exert a greater influence over working conditions or wages. In this sense, a labor union is very similar to a cartel (discussed in Chapter 11). The sellers of labor agree not to compete among themselves but to act as a single seller of labor.

## Types of Labor Unions

The first labor unions in the United States started as workers' guilds in the late 1700s and early 1800s. These organizations of workers within the same trade—carpenters, cordwainers (shoemakers), hatters—began meeting to set standards and prices for their output. These loosely organized trade unions were typically short-lived but were the beginnings of the American labor movement.

The Industrial Revolution brought about new opportunities for labor unions. Large manufacturing plants employed many workers with similar skills and interests who could organize to pursue their common goals. The early stages of the new industrial growth also brought the buyer of labor—the factory owner—considerable market power, the effects of which further encouraged workers to unify to enhance their own power. Today there are three major types of labor unions: craft unions, industrial unions, and public employees' unions.

### Craft Unions

Craft union: Workers with a common skill who unify to obtain market power and restrict the supply of labor in their trade; also called trade union.

In 1886, the American Federation of Labor (AFL) was started in an effort to organize craftspeople into local unions. **Craft unions** organize workers according to particular skills—such as electricians, carpenters, or plumbers—regardless of the industry in which they work. The main function of the guilds is to advance their members' economic well-being. When workers act as a unit, their power in the market can limit competition and raise total compensation above the competitive level.

Trade or craft unions increase members' total compensation by decreasing the supply of skilled workers. They do so by excluding potential workers from membership. Frequently, trade unions require high initiation fees, monthly dues, and long apprenticeship programs in an effort to discourage potential entrants. Existing members enjoy the higher wage brought about by the limited supply of workers in their trade. Figure 13-1 shows the impact of this decreased supply on wages: Wages rise from the competitive level, $W_c$, to the unionized level, $W_u$.

Trade unions are very much like cartels in the sense that supply is artifi-

**FIGURE 13-1**
**A Craft Union's Effect in the Labor Market**

Craft unions decrease the supply of workers in a particular craft by limiting the number of people who may join the union. When a union organizes, the supply of labor decreases to $S_u$ from the competitive supply, $S_c$. Wages rise from the competitive level $W_c$ to the unionized level $W_u$. Employment falls from $L_c$ to $L_u$.

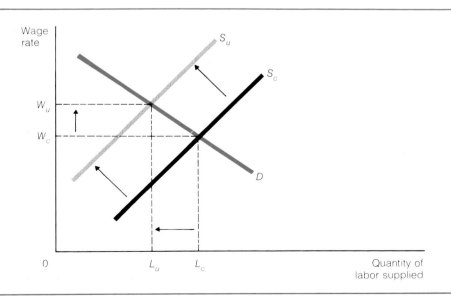

cially restricted so as to command a higher price. As such, craft unions face the same problems as cartels. Just as cartel members have an incentive to cheat on the cartel price, workers outside the union who offer their labor at a wage lower than the union wage can obtain jobs at the expense of union members.

### Industrial Unions

Large manufacturing plants first appeared during the late 1800s and early 1900s. A large firm in the textile, automobile, steel, or coal industry would frequently be the only employer in a small, isolated town. As a single buyer of labor, such a firm would be able to keep wages below the competitive level. Under these circumstances, workers often found it in their interest to organize. Rather than organize individual craft unions according to different skills, all workers within the same industry would organize a single industry-wide union. In 1938, John L. Lewis formally organized the Congress of Industrial Organizations (CIO), which unified workers first at the firm level and then at the level of the industry as a whole. Contemporary examples of these **industrial unions** include autoworkers' unions and steelworkers' unions.

For an industrial union to be effective, it must unionize all firms in an industry. Otherwise, the lower-cost, nonunionized firms would prosper while the higher-cost, unionized firms would dwindle. Thus, industrial unions often encourage membership rather than restrict it.

**Industrial union:** Workers within a single industry who organize regardless of skill in an effort to obtain market power.

### Public Employees' Unions

**Public employees' unions** are organizations of government workers. Such unions cover a wide variety of jobs—such as firefighting, police work, teaching, and clerical work—and include both blue-collar and white-collar workers. This sector of the union movement has been one of the fastest growing in recent years. The American Federation of State, County, and Municipal Employees (AFSCME) is now among the ten largest unions in the country, as indicated in Table 13–1. Indeed, membership in AFSCME almost tripled over the 1968–1980 period.

**Public employees' union:** Workers who are employed by the federal, state, or local government and who organize in an effort to obtain market power.

TABLE 13–1  The Ten Largest Unions, 1980

The larger unions are industrial unions, which usually encourage membership. The smaller craft unions (electricians, carpenters, and so on) typically restrict new membership.

| Type of Union | Membership |
| --- | --- |
| Teamsters | 1,891,000 |
| Autoworkers (UAW) | 1,357,000 |
| Food and commercial workers | 1,300,000 |
| Steelworkers | 1,238,000 |
| State, county, and municipal employees (AFSCME) | 1,098,000 |
| Electrical workers (IBEW) | 1,041,000 |
| Carpenters | 832,000 |
| Machinists | 745,000 |
| Service employees (SEIU) | 650,000 |
| Laborers (LIUNA) | 608,000 |

Source: Statistical Abstract of the United States, 1984, p. 440.

# Union Activities

Regardless of the means of organization, when workers successfully unite to act as a single seller of labor, the union gains monopoly power in its labor market. The union and its members are therefore no longer wage rate takers. There is a downward-sloping demand curve for the members' labor, and the union may seek any wage along this curve. However, the level of employment for union members is inversely related to the wage rate. Higher wages are gained at the expense of fewer jobs. In the face of this fact, a union's activities will depend on its choices among conflicting objectives, the elasticity of demand for its workers' product, and the union's effect on demand for its workers' labor.

## Union Goals

The ultimate goals of the union may depend on many competing objectives. Understandably, the union's elected, policy-making officials suggest that members want improved economic well-being: higher wages, more jobs, greater job security, safer jobs, more retirement pay, more fringe benefits, and so on. But relating union policy to a particular objective is difficult. For example, the union objective may be to maximize any number of options, some of which may be inconsistent, such as the utility of the union officials, the utility of members, the wage rate, the level of employment, the level of membership, the total wage income of members, the total wage income of senior members, and so on. The ramifications of three of these objectives are discussed below.

**Employment for All Members.** One objective of a union may be to achieve full employment for all members. To do so would require a wage rate that ensures that the quantity of labor demanded by firms is equal to the quantity of labor supplied by existing members. Suppose that the demand for the union's labor can be represented by curve $D_L$ in Figure 13–2. If the amount of labor offered by the union is $L_1$, then wage $W_1$ will ensure full employment of union members.

**Maximizing the Wage Bill.** Although wage $W_1$ means that everyone in the union is working, full employment may not be the union's goal. Unions frequently have members sitting on the unemployment bench. Any wage higher than $W_1$ will leave some members unemployed, but there may also be some benefits to a higher wage. For example, it may increase the total wage bill. The **total wage bill** to firms hiring union workers is the wage rate times the total amount (in person-hours) of labor hired. This total wage bill is not only the total cost of labor to the firms; it is also the collective wage income of all the union members. Since workers prefer more income to less, all things being equal, the union may choose to maximize the wage bill. If the full employment wage occurs where the elasticity of demand is less than 1 (inelastic demand), then an increase in the wage will increase total union income ($W \times L$).

To understand this possibility, recall from Chapter 5 the relation between price, total revenue, and elasticity. Total revenue is maximized at the price where the elasticity of demand is equal to 1. If elasticity is less than 1, then an increase in price increases total revenue; if elasticity is greater than 1, it takes a price decrease to increase total revenue. In analyzing labor demand, the total wage bill, the wage rate, and the elasticity of demand have the

*Total wage bill:* The total cost of labor to firms, equal to wage times total quantity of labor employed; the total income of all workers.

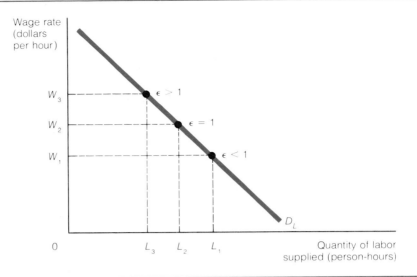

**FIGURE 13–2  The Union's Wage Goals**
In setting its wage rate, a union must choose among many goals. If there is $L_1$ amount of laborers in a union, then wage $W_1$ must be chosen to achieve full employment of all members. If the elasticity of demand at $W_1$ is less than 1, then an increase in wages up to $W_2$ will increase the total wage bill while bringing about union unemployment in the amount $L_1 - L_2$. Still higher wages such as $W_3$ will decrease total union income, although those union members who are working enjoy greater incomes. In trying to maximize total wages, the union will face some degree of unemployment, according to this model.

same relation. If the union's objective is to maximize the total union income, then the wage is set where the elasticity of demand is equal to 1.

In Figure 13–2, wage $W_2$ maximizes the wage bill. However, with $L_1$ amount of labor in the union, unemployment in the amount $L_1 - L_2$ results. This may not be a problem because the total union income ($W_2 \times L_2$) is now higher than at $W_1$. All members can still be better off at a higher wage if the working members provide unemployment compensation to unemployed members. On the other hand, a trade union may choose to limit union membership to $L_2$.

**Maximizing Income for Limited Members.** While wage $W_2$ would maximize income for the union as a whole, a higher wage such as $W_3$ in Figure 13–2 would increase income only for those who remain fully employed. $L_3$ workers would be hired, and their incomes would be higher so long as they are working the same number of hours. The workers on the unemployment bench would receive no income. For example, senior members usually retain employment under a seniority system. Thus, wages above $W_2$ would probably increase senior members' incomes at the expense of younger members. These unemployed members could be phased out of the union so that remaining members could retain higher incomes.

## Union Objectives and the Elasticity of Demand

A union's ability to accomplish its goals is limited by the elasticity of demand for its members' labor. Increases in wages decrease the level of employment, but the percent decrease in employment is determined by the elasticity of demand. Unions would like to see large increases in wages with minor effects on employment; in other words, they would like the demand for their members' labor to be as inelastic as possible. The elasticity of demand for union labor is determined by many factors (discussed in Chapter 12). Unions themselves often engage in activities that decrease the elasticity of demand for union members. Such a change in elasticity of demand is shown in Figure 13–3 as a rotation of the demand curve from $D_1$ to $D_2$. Wages are increased without causing so much unemployment. Following are some familiar examples of this process.

**Elasticity of Demand for the Product.** The more inelastic the demand for the good or service that the union produces, the more inelastic the demand for the union's labor. While a union would like the demand for the good or service its workers produce to be inelastic, it cannot do much to decrease this elasticity. Union workers are mostly at the mercy of the market, although in some circumstances they can and do affect elasticity of demand. For example, an item's elasticity of demand is determined in part by the number and availability of substitutes. If a union can limit the amount of substitutes for the product they produce then, the elasticity of demand for the union workers' labor decreases. We see this happening in the automobile industry. Through the political process (see Focus, "The Political Power of Unions") U.S. autoworkers encourage import quotas, which have the effect of limiting the supply of foreign autos and decreasing the elasticity of demand for U.S.-made cars. With import quotas, increases in autoworkers' wages have less effect on employment than the same increases in wages if quotas are not imposed.

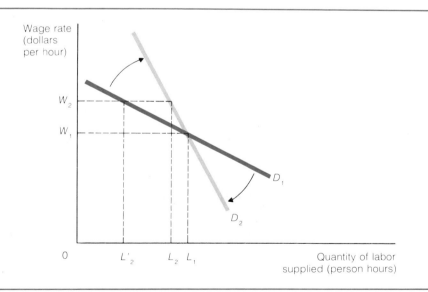

**FIGURE 13–3  Decreasing the Elasticity of Demand for Union Labor**
Unions typically attempt to decrease the elasticity of demand for their members' labor. For example, unions support immigration laws that retard or block the flow of competitive foreign labor into this country. The effect of such activities is represented by a rotation of the demand curve from $D_1$ to $D_2$. This rotation of the demand curve indicates that future wage increases result in smaller employment decreases. If wages rise from $W_1$ to $W_2$, employment falls only to $L_2$ rather than $L'_2$.

> **FOCUS** The Political Power of Unions
>
> One way to increase the demand for union labor is to lobby for various types of protective legislation, such as tariffs and minimum-wage laws. Over the years, unions have been quite successful in garnering political support for legislation favorable to their membership. The unions' political effectiveness results primarily from their ability to control a supply of votes.
>
> Unions often attempt to persuade members to work in campaigns and to vote in a bloc for candidates who support union positions. The incentive for a union member to vote can be stronger than that of the average voter because the union member knows that a lot of other members will vote the same way. In other words, individual union members know that their votes will count, unlike average voters, who sometimes feel that their votes are insignificant. Union leaders, in turn, broker union votes among political candidates to win votes for union positions in legislative struggles.
>
> Labor union institutions such as the union hall and the union boss are designed to mitigate the problem of free riding—the tendency of individuals to let others work in campaigns and then reap the benefits of their efforts without bearing any of the costs. Unions overcome this problem through their organizing and monitoring of members' activities.

**Availability of Substitute Inputs.** The fewer substitute inputs for union labor, the lower the elasticity of demand for union labor. Unions would prefer that there be no substitute inputs for their members' labor. If this were true, then wage increases would have minimal effects on employment. However, there are almost always some inputs that may be substituted for union labor. Typically, nonunion labor is a good substitute. For this reason, labor unions usually form contracts with employing firms in which the firms agree to hire only union labor. Unions also take direct actions to limit the amount of nonunion labor available. For example, laws that restrict immigration to the United States are supported by unions in an effort to decrease the availability of immigrant workers who could substitute for their members in the work force.

**Union Labor as a Proportion of Total Costs.** The smaller the percentage of firms' total costs accounted for by unions, the more inelastic the demand for union labor. If labor represents a very small percentage of total costs and if wages increase substantially, then the increase in total cost and thus product price will be very small. For this reason, the level of employment will fall by only a very small percent. In such a situation, union goals conflict. If unions attempt to maximize the total wage bill and increase the number of unionized occupations, then their ability to affect wages might decrease.

### Increasing the Demand for Union Labor

Regardless of the goals of unions, the members are always in favor of an increase in the demand for their labor. With increased demand, wages, employment, or both are increased. The curve representing demand for labor shifts to the right, as shown in Figure 13–4, when one of two basic things happens: an increase in the demand for the product or an increase in the marginal product of workers.

Unions can and do engage in activities that affect the demand for their members. Some of these demand-increasing activities also decrease elasticity of demand for labor.

## FIGURE 13–4
### Increases in Demand for Union Labor

Labor unions encourage increases in the demand for their members' labor. Increasing the demand from $D_1$ to $D_2$ allows an increase in wages from $W_1$ to $W_2$ with a constant level of employment at $L_1$. Alternatively, employment can increase from $L_1$ to $L_2$ with wages constant at $W_1$. Or there can be an increase in both wages and employment to $W_3$ and $L_3$.

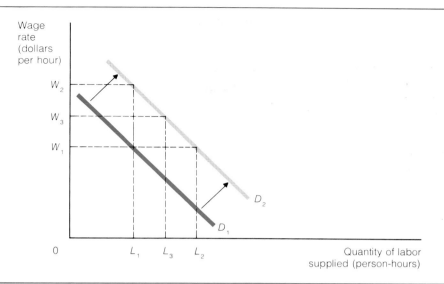

**Increasing Product Demand.** Unions frequently attempt to increase the demand for the product they produce. Garment workers advertise the union label and encourage consumers, as well as union members, to buy union-made clothes. Autoworkers strongly encourage union members to buy domestically produced cars. As we have seen, unions sometimes try to influence lawmakers to decrease the supply of foreign products that are substitutes for union-made goods. Import tariffs or quotas increase the demand for U.S.-made products.

**Increasing Substitute Input Prices.** If unions can increase the relative price of inputs that are substitutes for union labor, then the demand for union labor will increase. Unskilled nonunion workers using machines can substitute for skilled union labor, for example. Unions have therefore supported increases in the minimum wage. As the relative price of unskilled nonunion labor rises, the demand for union labor increases.

**Increasing Productivity of Members.** Unions prefer that their members be very productive. If the marginal product of new members joining the union is relatively high, then the demand for all members increases. Apprenticeship programs offered by unions train new entrants in an effort to increase overall productivity.

## Monopsony: A Single Employer of Labor

Having examined unions' attempts to build the economic power of laborers, we now turn to the nature of the firms they face as employers. Before the Industrial Revolution, large manufacturing plants were not common. In most towns or regions there were several potential employers for most workers. As a matter of fact, most skilled craftspeople were self-employed. However, as technology changed, the benefits of mass production increased. Large manufacturing plants such as steel mills, textile mills, and coal mines became

**FIGURE 13–5**

**Monopsony in the Labor Market**

A monopsonist faces an upward-sloping supply curve of labor, $S_L$. It maximizes profit by hiring the amount of labor that equates its marginal factor cost (MFC) to its marginal revenue product for labor ($MRP_L$). The monopsonist's levels of employment and wage, $L_m$ and $W_m$, fall below the competitive levels, $L_c$ and $W_c$.

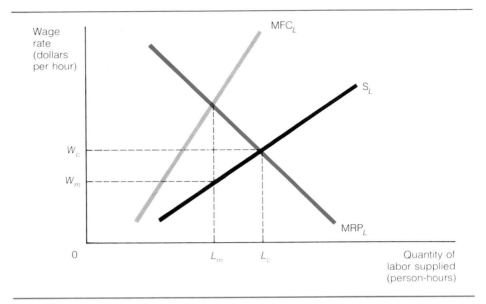

**Monopsony:** A single buyer of a resource or product in a market.

the principal employers of labor in some regions. In some areas only one firm employed workers.

A single buyer of a resource in a market is known as a **monopsony.** The monopsonist buyer has market power because rather than purchase all that it is willing and able to purchase at a market-determined price, the monopsony faces an upward-sloping supply curve. From along this curve it seeks the price that yields the profit-maximizing quantity of resource employment.

Consider Figure 13–5. The firm sells its product in a competitive market; its marginal revenue product curve for labor is $MRP_L$. Since it is the only buyer of labor in the market, it faces an upward-sloping supply curve of labor, $S_L$. To hire more labor, the firm must offer higher wages. Under these circumstances, the marginal factor cost of labor is no longer equal to the wage rate. The increase in total cost that results from hiring one more unit of labor is greater than the price paid for one more unit. And to attract one more unit of labor, the firm must offer a higher wage to all units of labor. For this reason, the $MFC_L$ curve has the same relation to the upward-sloping supply curve as the marginal cost curve has to the average cost curve (discussed in Chapter 8). Indeed, the supply curve of labor represents the average factor cost curve.

The firm will hire all units of labor that add more to revenue than to cost. Thus the monopsonist firm will maximize profit by hiring labor up to the point where $MFC_L$ equals $MRP_L$. (The monopsonist makes the same marginal analysis as the competitive buyer of labor.) The monopsonist hires $L_m$ amount of labor and must pay wage $W_m$ to attract that amount of labor.

The monopsonist wage and level of employment may be compared to those of pure competition under the same circumstances. If $MRP_L$ in Figure 13–5 had been the summation of many firms' demand curves for labor, then competitive forces would have forced wages to $W_c$ and employment to $L_c$. Under the same supply and $MRP_L$ conditions, the presence of monopsony results in lower wages and lower employment.

The fact that the monopsonist pays a wage rate less than the $MRP_L$ is known as **exploitation of labor.** If individuals are not being paid the value of their marginal product, then the monopsonist is extracting value from them.

**Exploitation of labor:** A situation in which the wage rate is less than the marginal revenue product of labor.

In the small regional markets of early days, monopsonists were actually exploiting the immobility of labor. As single employers, firms were able to pay lower-than-competitive wages in isolated areas because the cost of transportation was very high. It was very difficult for individuals to move around the country to work for other employers.

The more mobile an employee, the greater the number of potential employers and the more elastic the labor supply. Today, in spite of the seemingly high price of gasoline, the costs of transportation are relatively low. The opportunity cost of moving from one side of the country to another or commuting to all firms within a sixty-mile radius is lower today than it was a hundred years ago. It is extremely rare to find an individual who has only one potential employer.

### Monopsony and the Minimum Wage

The monopsony model represents one of the few cases where a minimum-wage law may have a productive impact on the economy. To see why, ask yourself what would happen in Figure 13–5 if the government set a minimum wage at the level of $W_c$. In this case, the firm would be a wage rate taker when hiring labor up to $L_c$ units. The supply of labor would be infinitely elastic at $w_c$ from zero to $L_c$; the supply would be upward sloping from $L_c$ on. With infinitely elastic supply, $MFC_L$ equals the wage. Thus the firm would hire labor up to the point where $MRP_L$ equals the wage. The monopsonist would increase employment to $L_c$ and pay the minimum wage. A minimum wage can increase employment in a monopsonistic market as long as it is not higher than the competitive wage.

### Bilateral Monopoly and the Need for Bargaining

If there is a monopoly in a market, then the single seller has the power to determine price. If there is monopsony in a market, then the single buyer has the power to determine price. But in some markets, **bilateral monopoly**—a single seller and a single buyer—could exist. In such a case, what determines the price?

**Bilateral monopoly:** A market in which there is only one buyer and one seller of a resource or product.

Theoretically, the price in a bilateral monopoly market cannot be determined. No competitive forces determine a single price, and the seller has a preferred price that is higher than the buyer's preferred price. However, if the maximum price the buyer is willing to pay is greater than the minimum price the seller is willing to accept, then a bargain can be reached. Bargaining between two parties is the process by which price is determined.

Consider a labor market with bilateral monopoly. Suppose a monopsony exists in an isolated mining town with only one employer, a mining firm. The firm must rely on the town's supply of labor. As Figure 13–6 illustrates, the mining firm maximizes profit by hiring $L_m$ workers and by paying wage $W_m$. The workers, though, are unhappy with wage $W_m$ and decide to form a union. They manage to sign up workers of all skill levels throughout the community to the union, maximizing its bargaining strength.

The wage most preferred by the union is not clear. If the union's goal is to maximize employment, then $W_c$, which coincides with the competitive wage, would be most preferred. However, a higher wage such as $W_u$ may result in the maximum total wage bill.

There is a maximum wage that the union can obtain. Any wage above the maximum would result in no union employment at all because of the high production costs that would be imposed. If wages are too high, then the monopsony may actually go out of business, turn entirely to substitutes

## FIGURE 13–6
### Bilateral Monopoly

When a monopsonist chooses lower-than-competitive wage $W_m$ and a monopoly union chooses higher-than-competitive wage $W_u$, the wage theoretically cannot be determined. But it can be determined by bargaining, and it will be between $W_u$ and $W_m$.

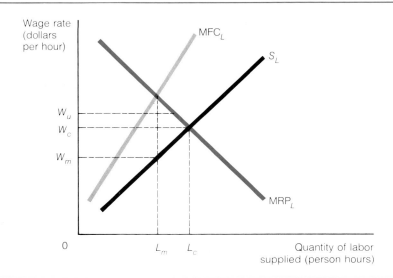

for the union labor, or relocate in a lower-cost area. The union's ability to extract value from the firm is limited by the mobility of the firm, just as the firm's ability to extract value from the workers is limited by the mobility of the workers.

### Collective Bargaining

When labor bargains collectively with a monopsonist, the final wage will be somewhere between the wage chosen by the monopsonist ($W_m$ in Figure 13–6), and the wage most preferred by the union ($W_u$ in the figure). The party that comes closest to establishing its goal is the one with the best bargaining strength. In **collective bargaining,** all buyers or all sellers of labor act as a unit to enhance their bargaining power.

Bilateral monopoly brings forth the need to negotiate. The absence of competition on both sides of the market creates a situation in which neither employers nor workers are able to dictate a wage rate. Indeed, there may be no need for bargaining unless a bilateral monopoly exists. Yet each year collective bargaining contracts are negotiated that cover the pay and working conditions for six to nine million American workers. Unions definitely limit wage competition among their members in many markets, but are there that many monopsonies?

In some industrial unions, such as the autoworkers' union, the union representatives bargain with a single firm. Under these circumstances, a firm that employs a very large percentage of a union's members does have some monopsony power. Wages cannot be dictated by the union. On the other hand, craft unions' members are employed by many firms. Hundreds of firms in different sectors of the United States employ carpenters, for example. It might seem that a union would not have to negotiate with an employer that has no monopsony power. However, the potential employers of craft union members may form a coalition, a buyer's cartel, to negotiate with unions as a single buyer of labor. Thus, when labor bargains as a unit to gain market power, previously competitive employers would do the same. Any time that wage contracts are determined by bargaining negotiations there is, in effect, a bilateral monopoly in the market.

**Collective bargaining:** The determination of a market wage through a process in which sellers or buyers act as bargaining units rather than competing individually.

### Strikes

**Strike:** A refusal to work at the current wage or under current conditions.

The ability of a union to achieve higher wages depends on its bargaining strength, and its strongest weapon is the strike. When firms fail to meet labor's demands for a higher wage or any other demand, the union can withhold the labor services of its members. Such a **strike,** or refusal to work, can severely limit or even eliminate the firm's ability to continue production. By striking, a union withholds not only its members' labor but also that of other unions who respect the picket line. Under some circumstances, members of nonnegotiating unions also withhold their labor from the negotiating firm. A firm's failure to concede to labor's demands can therefore be very costly. Of course, conceding to higher wages is costly as well.

Yet the strike is a double-edged sword. Not only does it cost firms income, but union members lose income as well. A strike is not attractive to either party, but it is the best weapon labor has to force firms to accept the union wage. The union must decide if the lost income is worth its members' increased future income.

Similarly, firms have the power to reject employment at the union's desired wage. By doing so, a firm in effect refuses to hire labor at a particular wage or any higher wage. In a sense, the firm strikes against employing labor at too high a wage rate.

The potential costliness of a strike leads to serious bargaining by both management and labor because both sides have significant incentives to arrive at a new labor contract without a strike. In fact, negotiation is the general pattern of collective bargaining in the United States. Each year, some 120,000 labor contracts are renegotiated in collective bargaining processes. In more than 96 percent of these cases, labor and management arrive at a new agreement without recourse to a strike.

Table 13–2 provides some data about strikes since 1950. Perhaps the most relevant statistic in the table is the percentage of working time lost to strikes.

**TABLE 13–2    Strikes: 1950–1983**

While there are thousands of strikes each year, the percentage of working time lost is relatively small.

| Year | Number of Strikes | Number of Workers Involved (thousands) | Number of Worker-Days Idle (thousands) | Percentage of Working Time Lost |
|---|---|---|---|---|
| 1950 | 4,843 | 2,410 | 38,800 | 0.33 |
| 1955 | 4,320 | 2,650 | 28,200 | 0.22 |
| 1960 | 3,333 | 1,320 | 19,100 | 0.14 |
| 1965 | 3,963 | 1,550 | 23,300 | 0.15 |
| 1969 | 5,700 | 2,481 | 42,869 | 0.28 |
| 1973 | 5,353 | 2,251 | 27,948 | 0.14 |
| 1975 | 5,031 | 1,746 | 31,237 | 0.16 |
| 1977 | 5,600 | 2,300 | 35,822 | 0.17 |
| 1979 | 4,827 | 1,727 | 34,754 | 0.15 |
| 1980 | 3,885 | 1,366 | 33,289 | 0.14 |
| 1981 | 2,568 | 1,081 | 24,730 | 0.11 |
| 1982[a] | 96 | 656 | 9,061 | 0.04 |
| 1983[a] | 81 | 909 | 17,461 | 0.08 |

[a]After 1981, data on work stoppages involving less than one thousand workers ceased to be collected.
*Sources:* U.S. Bureau of Labor Statistics, *Handbook of Labor Statistics,* 1983, p. 380; and *Monthly Labor Review* (November 1984), p. 103.

As the data make clear, this figure has fallen consistently over time (1969 is an exception) and generally constitutes less than 0.5 percent of total working time, a figure far less than the time lost to worker absenteeism each year. The fact that not much working time is lost to strikes does not mean that strikes are not a powerful union weapon. The threat of a strike may be sufficient to make collective bargaining work smoothly.

Sometimes, if negotiations are not successful, both parties can agree to **binding arbitration.** Such an agreement brings in a third party acceptable to labor and management to make a decision that both sides must abide by. Binding arbitration is often used for public sector union disputes because public employees usually are forbidden by law to strike. Public sector strikes in basic service activities—such as commuter transportation, garbage collection, teaching, and fire and police protection—have the potential to bring the economy to a halt. If essential services provided by government cease to be performed because of a strike, much private economic activity will cease as well.

In private sector strikes, lost output can be made up to some extent if workers work overtime after the strike is settled. Moreover, substitute products and services are available to consumers during a private sector strike. Purchasers of new cars can buy used cars or drive their present cars until an autoworkers' strike is over, for example, and trips planned in taxis can be made by rental cars or postponed.

### Political Influence in Bargaining

The bargaining power of unions and management is affected by legislation, which in turn is affected by the political environment. Legislation can favor either management or labor in the negotiating process. Until the 1930s, the struggle between management and labor in the United States was tilted toward management. The strike was not a strong labor weapon because employers could obtain a court order, called an injunction, against a strike. Employees suspected of union sympathies could be fired or roughed up by company thugs. New employees could be required to sign a "yellow-dog" contract, which required that a potential employee not join the union to get the job.

All this changed in the 1930s with the New Deal. The Norris–La Guardia Act (1932) and Wagner Act (1935) denied management the use of antiunion tactics and granted labor the right to organize and engage in collective bargaining in a legal framework regulated by the National Labor Relations Board (NLRB). The NLRB has five members appointed by the president for five-year terms. Its functions include ruling on unfair labor practices, settling jurisdictional disputes among unions over bargaining rights, and calling for plant elections at the request of workers. These legal changes lent a tremendous impetus to the growth of the union movement and gave unions a favorable environment for pursuing their ends.

The post–World War II era saw much industrial unrest. Unions showed little self-restraint or discipline, so Congress tilted the balance between labor and management back toward management. The Taft-Hartley Act, passed in 1947, allowed states to enact **right-to-work laws,** which forbid unions from coercing workers into their ranks. (Focus, "The Right-to-Work Controversy," discusses these laws in more detail.) It added several other constraints on union behavior, such as outlawing strikes by government workers and secondary boycotts, by which the union would set up picket lines against other suppliers of the company being struck. After the Taft-Hartley Act,

---

**Binding arbitration:** An agreement between employers and labor to allow a third party to determine the conditions of a work contract.

**Right-to-work law:** A law that prevents unions from forcing individuals to join a union as a prerequisite to employment in a particular firm.

charges of union corruption and ties to organized crime led to further legislation to regulate union behavior. The Landrum-Griffin Act, passed in 1959, contained various provisions to ensure that unions were honestly managed. Among other things, this law required the filing of financial reports by union officers and the auditing of union finances.

Even in the face of such restraining legislation, the labor union is important and powerful in American life. Unions are now an accepted fact in our economy. No longer are unions viewed as weak associations of workers fighting for a better standard of living for their members. In some respects, unions are best thought of as strong special-interest groups that are proficient at obtaining special legislative favors such as protective tariffs. On the other hand, union political power is sometimes used to promote general-interest legislation. Civil rights and welfare laws, Social Security, and similar programs have been strongly supported by organized labor.

## Union Power Over Wages: What Does the Evidence Show?

As we have seen, unions may have an effect on the relative wages of their members, depending on a number of conditions. But what effect have the unions actually had? Several studies have attempted to measure the influence of unions using data on wages and employment in the economy.

H. Gregg Lewis, who made the pioneering study of this issue, estimated that the average wage of union members was 10 to 15 percent higher than that of nonunion members who had about the same marginal productivity as the union members.[1] Some unions made more than this, on average, while others made less. Strong unions, such as the airline pilots' union, were able to raise members' wages by more than 25 percent compared to the wages of nonunion pilots with the same skill level. Weaker unions included those in the textile and retail sales industries, where unionization has had little perceptible influence on members' wages. Craft unions exhibited more strength than industrial unions. Lewis's results square with the economic theory of union power as we have presented it.

Lewis based his estimates on data for the 1940s and 1950s. Similar empirical research on more recent data suggests that unions' power to raise relative wages may be increasing. Holding the characteristics of employees constant, Frank Stafford estimated that the earnings of unionized craft and semiskilled workers were 25 percent higher than that of their nonunion counterparts. Michael Boskin estimated that the union-nonunion wage difference was 15 to 25 percent on average in 1967.[2] Other estimates have shown even higher differentials between union and nonunion members. It is not completely clear why union power has grown over time, but one factor is the increasing political power of the union movement. As unions have become more important in politics—for instance, as major suppliers of votes to political parties—their power to increase the demand for union labor and to restrict competing labor supply has increased.

---

[1] H. Gregg Lewis, *Unionism and Relative Wages in the United States* (Chicago: University of Chicago Press, 1963).

[2] Frank P. Stafford, "Concentration and Labor Earnings: Comment," *American Economic Review* 58 (March 1968), pp. 174–181; and Michael J. Boskin, "Unions and Relative Wages," *American Economic Review* 62 (June 1972), pp. 466–472.

## FOCUS  The Right-to-Work Controversy

A union shop provision in a contract between labor and management makes it mandatory for a new employee to join the union after a short probationary period. However, Section 14b of the Taft-Hartley Act allows states to enact right-to-work laws. These laws ban the union shop provision and let individuals decide voluntarily whether to join a union. Such laws are a thorn in the side of organized labor. Union leaders feel it is unjust for workers to refuse to join a union and pay dues to support it while at the same time benefiting from the bargaining gains achieved by the union. Opponents of compulsory unionism perceive a need to check the growth of union power and its negative impact on economic growth and employment. Periodically, labor tries to have Section 14b of the Taft-Hartley Act repealed.

As of 1982, twenty states had right-to-work laws on their books. These states are listed in Table 13–3. Historically, there have been as many as twenty-four states with such laws, but over time some states have repealed their right-to-work laws and others have added them. The thirty states in which labor union membership was compulsory in 1982 are listed in Table 13–4.

TABLE 13–3  Net Change in Manufacturing Employment in Right-to-Work States: 1970–1982

| State | Employment Gain |
|---|---|
| Texas | 325,900 |
| Florida | 79,300 |
| North Carolina | 65,100 |
| Arizona | 61,000 |
| Louisiana | 37,100 |
| Virginia | 31,700 |
| Georgia | 29,900 |
| Kansas | 29,000 |
| Arkansas | 27,200 |
| South Carolina | 21,900 |
| Mississippi | 21,400 |
| Alabama | 9,800 |
| South Dakota | 9,200 |
| Georgia | 6,400 |
| North Dakota | 5,000 |
| Tennessee | 4,200 |
| Nebraska | 2,000 |
| Wyoming | 1,800 |
| Nevada | −2,400 |
| Iowa | −8,300 |

Source: U.S. Bureau of Labor Statistics, Handbook of Labor Statistics, 1983, pp. 175–176.

TABLE 13–4  Net Change in Manufacturing Employment in Non-Right-to-Work States: 1970–1982

| State | Employment Gain (+) or Loss (−) |
|---|---|
| California | 370,000 |
| Colorado | 62,000 |
| Washington | 50,400 |
| Oklahoma | 49,500 |
| Minnesota | 27,600 |
| New Hampshire | 20,500 |
| Oregon | 13,300 |
| New Mexico | 12,600 |
| Vermont | 8,600 |
| Idaho | 7,500 |
| Alaska | 2,500 |
| Maine | −3,000 |
| Hawaii | −3,100 |
| Delaware | −3,200 |
| Montana | −3,800 |
| Rhode Island | −3,800 |
| Wisconsin | −4,600 |
| Kentucky | −8,400 |
| Massachusetts[a] | −18,500 |
| Connecticut | −25,700 |
| West Virginia | −27,600 |
| Missouri | −42,100 |
| Maryland | −59,200 |
| Indiana | −126,900 |
| New Jersey | −133,000 |
| Michigan | −197,400 |
| Illinois[a] | −223,700 |
| Ohio | −306,300 |
| Pennsylvania | −361,700 |
| New York | −398,800 |

[a]Massachusetts and Illinois data cover 1970–1981.
Source: U.S. Bureau of Labor Statistics, Handbook of Labor Statistics, 1983, pp. 175–176.

Also listed in Tables 13–3 and 13–4 are some data on the change in the level of employment in right-to-work and non-right-to-work states between 1970 and 1982. In general, employment has grown in right-to-work states and decreased in non-right-to-work states. These data would seem to indicate that compulsory unionism retards state and local economic growth and employment. Such a conclusion must be drawn with care, however. The level and rate of change in employment in a state is a function of many factors, including the levels of state and local taxation, the abundance and availability of natural resources, and the quality of the local labor force. Many states that have right-to-work laws are also attractive to industry for other such reasons. It is the sum of these factors that determines employment changes across states. In this context we can say that while the weakness of unions in right-to-work states helps to promote employment and growth in these states, it is not the whole, and maybe not even the most important, part of the story of economic growth in the states.

## The Impact of Unions on Labor's Share of Total Income

The studies just mentioned measure the impact of unions on relative wages; that is, the wages of union members versus those of nonunion members. A broader question concerns the impact of unions on the share of national income that goes to all labor, not just union labor.

Primarily, a general increase in wages requires an increase in the productivity of workers. An increase in productivity can occur in a variety of ways. The quality of tools and other physical capital used by workers can increase, workers' skill levels can be raised through education and experience, innovation or better management can increase worker productivity, and so on. Higher real wages for all workers can be realized only if the output of goods and services in the economy is increased. It would seem to follow that unions can increase the wages of all workers only if they increase the general level of productivity in the economy. While such an impact of unions is not out of the question, it is hard to imagine how it might happen.

Who pays for the benefits that unions obtain through bargaining power, legislation, and other means? The naive answer is that, of course, employers pay, but the evidence suggests otherwise. In fact, economic analysis indicates that consumers and nonunion employees pay for the benefits that unions obtain for their members.

Consumers pay because higher union wages mean higher costs for firms that employ union labor. If these firms produce in a competitive market for their output, output prices will rise to reflect the higher cost of union labor. Basically, the same result holds if the firm that hires union labor has monopoly power, although the monopolist may bargain with the union over the level of monopoly profits earned. In either case, higher costs caused by union wages lead to higher prices for consumers. Consumers thus pay part of the tab for higher union wages.

Nonunion employees pay because higher union wages lead firms to produce less output and to substitute nonhuman capital for union labor. Both these effects cause the level of employment in unionized industries to fall. In other words, as we saw in the models of how unions increase the relative wages of their members, these are cases where union bargaining power leads to a lower employment level for labor in the unionized industry. The workers who do not make it into the union must seek employment elsewhere in the economy. This increases the supply of labor to alternative, nonunion employments and drives down nonunion wages. Thus, nonunion employees also pay part of the tab for higher union wages.

It can be argued on the basis of economic theory that employers do not generally pay for all union gains. Consumers and nonunion employees share the burden because the price of union-made goods rises and the wage rate for nonunion labor falls. The latter effect can be substantial. Lewis estimated

TABLE 13-5  Labor's Share of National Income

The percentage of the nation's real output that labor receives has increased slightly since World War II. This increase is attributed to a decrease in small owner-managed firms rather than to union activities.

| Year | Total Employee Compensation (Including Employer Contribution for Social Insurance) (percent of national income) | Total Employee Compensation plus Self-Employment Income (percent of national income) |
|---|---|---|
| 1935 | 71 | 84 |
| 1940 | 68 | 83 |
| 1945 | 68.2 | 82.4 |
| 1950 | 71.8 | 82 |
| 1955 | 72.2 | 82.6 |
| 1960 | 74 | 81.8 |
| 1965 | 74.3 | 82.3 |
| 1970 | 74.2 | 82.8 |
| 1975 | 74.3 | 83.4 |
| 1980 | 73.9 | 82.2 |
| 1981 | 74.6 | 81.4 |
| 1982 | 76.1 | 82.5 |
| 1983 | 74.9 | 79.6 |
| 1984[a] | 73.4 | 78.6 |

[a]provisional
Source: U.S. Department of Commerce, Survey of Current Business, various issues.

that nonunion wages are 3 to 4 percent lower than they would be without unions.[3] Since this effect on nonunion labor covers about three-fourths of the labor force, it is clear that the resource allocation effects of unions can be substantial in the aggregate.

Table 13-5 provides some data about the behavior over time of the share of national income that goes to labor. Two ways of measuring labor's share are shown. The second column measures total employee compensation (wages and salaries) plus the employer's share of the Social Security tax as a percentage of national income. This series shows a slight upward trend since World War II, a rise that has been attributed to a decline in the number of self-employed workers over this period, primarily in agriculture.

The third column in the table adds the income of self-employed persons—such as business proprietors, lawyers, and accountants—to employee compensation. Self-employment compensation is clearly income that goes to labor. When we look at this more inclusive concept of labor's share of national income, we see that it has been virtually constant for about fifty years. Thus, during an era when labor union strength grew in the United States, we do not find any evidence that labor's share of national income rose. While labor unions have clearly increased the wages of their members relative to the wages of nonunion workers, there is no evidence to suggest that they have made all workers better off in terms of their share of national income.

The fact is that higher real wages for workers can come about only through increases in worker productivity. Workers in the United States earn high wages because their productivity is high. Their wages can rise only if

[3]Lewis, Unionism and Relative Wages, pp. 1-308.

the output of goods and services in the American economy is increased. All things being equal, it is the level of productivity that drives real wages.

## Conclusion

The goals of unions are easy to understand. The members want more income, greater income security, and more pleasant working conditions than the market provides. Their ability to achieve these goals is limited by many factors: the supplies of competing inputs, the monopsony power of employers, and the political power of the unions. It is clear that unions have provided their members with higher wages and better job security, but they also provide something more.

The union's role extends beyond an economic wage and employment analysis to the plant, where the union provides workers with a set of rules and representation to settle disputes with employers. This presence of the union on the job cannot be discounted as a primary source of nonpecuniary benefits to union members. Such benefits accrue to workers in the form of greater job security, a feeling of power and dignity at work, and less alienation from their work. These are important aspects of work in modern factories and offices, and the union's role in providing workplace representation is a powerful force for creating greater worker satisfaction and thus greater productivity. (See Economics in Action, "Toward a New Theory of Labor Unions," at the end of this chapter for more on the productivity-enhancing aspects of labor unions.)

There is much concern today over the meaningfulness of work. Workers are said to be alienated and bored by assembly-line types of jobs. In this respect, the role of the union as an arbitrator of workplace rules and procedures will probably become increasingly important. Workers may be willing to trade off some wage gains to obtain changes in the way their work is done. And if workers seek more meaningful jobs and factory arrangements, the union will surely be at the forefront in helping promote such arrangements in future labor contracts.

## Summary

1. A labor union is a group of workers in a craft, industry, or government job who organize to gain market power.
2. Labor unions attempt to improve the economic welfare of their members by increasing wages, decreasing the elasticity of demand for their services, and increasing the demand for their members' labor through a variety of activities.
3. Monopsony exists when there is a lack of competition in the employment of labor. A single buyer may force wages below the competitive level.
4. In a bilateral monopoly in a labor market, the wage is determined by the relative strength of the bargainers.
5. A union's ability to achieve its goals is in part determined by its bargaining strength. A union's ability to strike is its strongest negotiating tool.
6. It is estimated that unions have increased the wages of their members relative to nonunion workers in similar occupations.
7. Unions have not increased all labor's share of total income. Gains made by union labor have generally been at the expense of nonunion labor and total production in the economy.

## Key Terms

labor union
craft union
industrial union
public employees' union
total wage bill
monopsony
exploitation of labor
bilateral monopoly
collective bargaining
strike
binding arbitration
right-to-work law

## Questions for Review and Discussion

1. How do craft unions increase wages for their members? What happens to the people who are not allowed into a craft union?
2. If industrial unions allow anyone who wishes to enter the union, how can it increase wages? Could an industrial union in an isolated mining town increase both wages and employment?
3. Suppose the United States exports beef to Japan and imports cars from Japan. If autoworkers persuade Congress to impose a tariff on Japanese cars, what happens to the incomes of cattle ranchers and ranch hands?
4. Why do unions want to decrease the elasticity of demand for their members? How do they do so?
5. What is a monopsony? Does the existence of labor unions promote the existence of monopsonies?
6. It is frequently suggested that firms exploit women in the labor force. Monopsonists' power to exploit is based on the immobility of labor. Are women less mobile than men?
7. In some countries such as the Soviet Union, there is only one potential employer—the government. Do these countries have the power to exploit their workers? How can their workers avoid exploitation?
8. Why are unions and union employers forced to negotiate wages while nonunion wages are determined without negotiation?
9. Are union wages closer to the monopsony wage or to the wage that maximizes the wage bill?
10. Suppose that a union is successful at organizing only part of the laborers in an industry. Do you predict that the union will be strong or weak? Why?
11. Why do you think unions favor minimum-wage laws?

## ECONOMICS IN ACTION

### Toward a New Theory of Labor Unions

Two economists from Harvard University have developed a new theory of the role of labor unions in the U.S. economy. In *What Do Unions Do?* Richard B. Freeman and James L. Medoff argue that two conflicting views of unions dominate the thinking of economists. On one side is the standard or orthodox view that unions are basically monopolies in the labor market whose main effect is to raise the wages of members, which in turn has the effect of decreasing the efficiency of the economy. On the other side are those who defend unions as increasing labor productivity by improving worker morale, assisting in the development and retention of skills, pressuring management to improve its efficiency, and protecting workers against arbitrary management decisions through collective bargaining.

In their study, Freeman and Medoff argue that the truth is somewhere in between. They present detailed findings in their book, but five basic results stand out:

1. Although unions provide their members with a significantly higher-than-competitive wage, there is no single union/nonunion wage differential among all socioeconomic groups. The effect of a union on wages is greatest for less-educated, younger, and low-seniority workers in heavily organized and regulated industries. Further, Freeman and Medoff state that "the social costs of the monopoly wage gains of unionism appear to be relatively modest, on the order of .3 percent of gross national product, or less" (p. 2).
2. Unions reduce the cost of their members' pensions as well as the cost of their life, accident, and health insurance.
3. Unions on balance increase the equality of income distribution among workers.
4. Unions increase the stability of the work force by providing various services such as grievance and arbitration proceedings and seniority clauses.
5. In many industries, unionized establishments actually have a higher rate of productivity than nonunion establishments.

The main contention of Freeman and Medoff's book—that unions provide services to the work force that are essentially unrelated to monopoly wage differentials and that, taken by themselves, increase economic efficiency—would receive the assent of most economists. Critics of their work have emphasized that Freeman and Medoff underestimate the social costs of union monopoly and that the efficiency-increasing services offered by unions are basically unrelated to their efficiency-decreasing monopoly aspects. In other words, it is possible to have unions provide benefits to workers without simultaneously engaging in restrictive labor market practices such as the closed shop.

*Source:* Richard B. Freeman and James L. Medoff, *What Do Unions Do?* (New York: Basic Books, 1984).

## Question

If we assume that unions actually produce higher than average wage rates for younger or less educated workers in certain fields, is there reason to believe that future unemployment will increase in these areas? Will businesses faced with such unions attempt to substitute relatively lower-priced machinery for the now relatively higher-priced labor? Comment.

# 14

# Capital and Interest

Wealthy societies such as our own consume, or use, goods and services at ever-increasing rates. Think, for instance, of the rate at which automobiles have become available to nearly every American. Little more than sixty years ago, relatively few Americans owned cars, and most of these proud owners were happy to keep the same car for several years. Today autos are bought and sold in the millions. Many of us own more than one car; few of us are happy to hold onto the same car for more than four or five years. Since consumption is possible only through production, the basis of a wealthy society is not its ability to consume but its ability to produce, to transform raw materials into more useful goods and services through the application of human skills and physical capital such as tools and factories. Societies that are better at this transformation are able to reach higher levels of wealth and to consume more. This chapter is about the role of capital in determining the ability to produce and the level of wealth in a society.

## Roundabout Production and Capital Formation

Economic behavior is forward-looking. The availability of the food we eat today is due to the foresight of farmers in the past who undertook the appropriate productive actions. The food we will eat tomorrow depends on actions farmers take now. A simple analogy of a shipwrecked sailor illustrates this point.

Imagine Robinson Crusoe trying to survive alone on a windswept island in the middle of the ocean. Crusoe's economy is primitive. He lives by fishing with a simple wooden spear that he found on the beach. With his spear Crusoe can catch 5 pounds of fish in about 10 hours of fishing. Suppose that one day Crusoe decides to improve his fishing techniques by weaving a fishing net.

Crusoe takes a day off to build the net. His opportunity cost of building the net is the fish that he would have caught during this day by fishing with his spear. Assuming that it takes 10 hours to build the net, the opportunity cost of the net to Crusoe is 5 pounds of fish.

Crusoe's purpose in building the net is obviously to catch more fish in the future. That is, he hopes the net will enable him to catch more than 5 pounds in 10 hours. It is the anticipation of more production and therefore more consumption in the future that drives Crusoe to the more **roundabout** method of catching fish.

**Roundabout production:** The production and use of capital goods to produce greater amounts of consumption goods in the future.

Of course, Crusoe must have some way to support himself during the period that he stops fishing to weave the net. He provides for his needs during this time by **saving.** If he consumes 5 pounds of fish every 10 hours and it takes 10 hours to make a net, Crusoe must set aside 5 pounds of fish to support himself while making the net. To do so, he could fish extra hours for several days, or he could eat only 4 pounds each day for 5 days. The hallmark of saving is some form of current sacrifice—extra hours worked mean less hours for current leisure, or fewer fish eaten in the present mean fewer calories consumed—in anticipation of a higher level of future consumption. However Crusoe manages to save fish, it is clearly the act of current abstinence that leads to saving and ultimately to the greater yield of fish.

**Saving:** The act of forgoing present consumption in an effort to increase future consumption.

What is Crusoe's incentive to abstain? Suppose that with the net he can catch 25 pounds of fish in 10 hours and that the net will last indefinitely. By abstaining from spear fishing for 10 hours, at a cost of 5 pounds of fish, Crusoe is able to raise his daily catch by 20 pounds. The increase is even more significant in long-range terms: For an investment of a mere 5 pounds of fish Crusoe raises his weekly catch from 35 pounds to 175 pounds, an increase of 140 pounds per week. This particular act of saving and investment of resources to produce more capital is especially profitable. Indeed, even if the net wears out every so often, Crusoe will be able to afford to take 10 hours to make a new net in anticipation of similar rates of return in the future.

This simple example has relevance to a modern economy. In a modern economy individuals specialize in their activities. They don't spend part of the day fishing and part of the day making rods and reels. But this specialization does not change any of the essential features of what is called capital formation.

**Capital formation:** The use of roundabout production to increase capital stock.

The term **capital formation** refers to the process of building capital goods and adding to the capital stock in the economy. Capital goods are things such as machines and implements that are used to produce final goods and services. The capital stock of an economy is the amount of capital goods that exists at a given point in time. To increase capital stock, producers must resort to methods of roundabout production, the use and production of capital goods as a means of achieving greater output in the future. Saving, or the abstinence from present consumption, is required for capital formation to take place. Moreover, the act of saving requires entrepreneurship, a concept involving risk taking, which we discuss in Chapter 15. In the case of capital production, the future is uncertain. Crusoe did not know whether a net would increase his daily catch. He took a chance that it would, and it paid off.

Another important feature of capital is that it does not automatically re-

**Depreciation:** The wearing out of capital goods that occurs over a period of time.

produce itself. Instead, capital goods are automatically **depreciating;** they decrease in productivity or value over time. This wearing out of capital goods cannot be stopped. If Crusoe does not repair or replace his net, he will ultimately return to spear fishing and to a lower standard of living. The capital goods or capital stock of a society must be repaired and replaced over time if the society is to continue to grow and experience high levels of consumption. This is the primary problem of economics: maintaining and increasing the ability to produce.

**Capital consumption:** The loss of capital that occurs because the rate of depreciation is greater than the rate of capital formation.

**Capital consumption** is the opposite of capital formation. Present consumption is increased temporarily at the cost of a reduction in the future rate of consumption. Suppose that a lumber company owns trees ranging in age from 1 to 25 years. Assume that it cuts down one thousand 25-year-old trees and plants one thousand seedlings each year. As long as external circumstances do not change, this forest can yield one thousand 25-year-old trees annually. For a while, however, the yield from the forest can be increased by harvesting some of the 24-year-old trees in addition to the regular harvest, then dipping into the stock of 23-year-old trees the next year, and so on. Cut in this manner, the amount of timber harvested will increase for a while. Still, the time will come when the total output of the forest will necessarily decline. Capital in the form of younger trees has been used up to increase current output, so the future output of timber must fall. This process is called capital consumption.

## The Rate of Interest

Saving and capital formation depend on the willingness of individuals to pass up current consumption to achieve greater consumption in the future. In other words, individuals must abstain from current consumption to provide the flow of saving that is used to produce capital goods. Economists stress that an individual's willingness to abstain from current consumption is related to his or her **rate of time preference.** Time preference is the degree of patience that an individual has in forgoing present consumption in order to save. A person with a high rate of time preference has a strong preference for current rather than future consumption. Most individuals have positive rates of time preference; that is, they prefer present to future consumption. Since people prefer to consume now rather than later, they must be paid a price for waiting. This price is called **interest.**

**Rate of time preference:** The percent increase in future consumption that is necessary to induce an individual to forgo some amount of present consumption.

**Interest:** The price a borrower pays for a loan or a lender receives for saving, measured as a percentage of the amount; the price of not consuming now but waiting to consume in the future.

The concept of interest has two related meanings. In the first sense, interest refers to returns on investments. A person who saves $100 in a passbook account earns 5¾ percent interest on saving. A firm that invests $1000 in a new piece of machinery earns $200 more in revenue because of the resulting improvement in productivity. The firm's investment thus yields 20 percent interest, the firm's reward for roundabout production. In these terms, interest is the amount individuals are willing to *receive* in order to sacrifice current consumption; it is a payment for their abstinence or waiting to consume later.

Interest is also the price that individuals are willing to *pay* to obtain a good or service now rather than later. A consumer who wants a car today might pay 13 percent interest for a loan from the bank. A firm that needs a new plant might pay 10 percent interest to get the necessary financing.

Keep in mind that interest is not just a monetary phenomenon. Whether paid or received, interest is based on the fact that people prefer to consume and invest now rather than later. Moreover, the rate of interest reflects the

## Demand and Supply of Loanable Funds

rate of time preference in the economy. The more impatient people are, the higher the rate of interest must be to induce them to save and to create capital goods.

### Demand and Supply of Loanable Funds

The rate of interest is determined in the market for loanable funds. It is, in effect, the price of loanable funds. Like other markets, the market for loanable funds has a demand side and a supply side.

**Demand for loanable funds:** A curve or schedule that shows the various amounts of money that people are willing and able to borrow at all interest rates.

The **demand for loanable funds** arises from two sources. Consumers demand loanable funds because they want to consume more now than their current incomes will permit. Loans to these individuals are called *consumption loans*. They are made for myriad reasons. Individuals may want to take a vacation now and pay for it on time, borrow to tide themselves over a temporary decline in income, or borrow to buy cars or household appliances.

The second source of demand for loanable funds is desire for *investment loans*. Investors borrow in order to finance the construction and use of capital goods and roundabout methods of production that are expected to be productive. Firms borrow funds to build new plants and to purchase new equipment in the expectation that such investments will increase profits.

The sum of the demand for consumption loans and investment loans equals the total demand for loanable funds. The law of demand applies to the demand for loanable funds just as it does to any other commodity (see Figure 14–1). The price one must pay to borrow funds for consumption or investment loans is the interest rate. According to the law of demand, the amount of funds demanded is inversely related to the interest rate. As the interest rate falls, the cost of borrowing money to finance the earlier availability of consumption and capital goods falls. Other things being equal, we thus expect to see more borrowing when the interest rate falls. At a lower rate of

**FIGURE 14–1**

**Determination of Interest Rates**

D and S are the demand and supply curves for loanable funds. The market for loanable funds reaches equilibrium at *i* and Q, where the plans of borrowers and lenders are compatible. That is, for a given interest rate, the amount that borrowers want to borrow and the amount that lenders want to lend are equal.

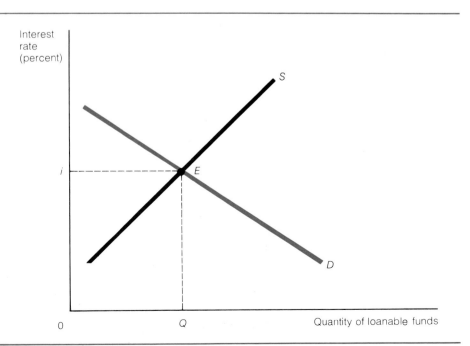

interest, consumers will expand current consumption, and more investment projects will appear to be profitable.

In Figure 14–1, the **supply of loanable funds,** S, is provided by savers. These are individuals or firms who are willing to consume less than their present earnings in order to set aside something for the future. As the interest rate—in this case, the return on investment—rises, more and more individuals and firms will be enticed to save; that is, to forgo current consumption in order to consume even more in the future. Thus, the quantity supplied of loanable funds varies positively with the interest rate offered to savers.

The intersection of the demand and supply curves for loanable funds in Figure 14–1 yields an equilibrium interest rate and the equilibrium level of loanable funds. At $i$ and $Q$, the market for loanable funds is in equilibrium, and the plans of borrowers are compatible with the plans of lenders. At interest rates above $i$, there will be an excess supply of saving, putting pressure on the interest rate to fall. Below $i$, there will be an excess demand for loanable funds, putting pressure on the interest rate to rise.

> **Supply of loanable funds:** A curve or schedule that shows the various amounts of money that people are willing and able to lend (save) at all interest rates.

### Variations Among Interest Rates

The previous discussion of the supply and demand for loanable funds might suggest that there is one interest rate in the economy. In reality there is a multiplicity of interest rates: the prime interest rate given to businesses with excellent credit ratings, mortgage rates offered to home buyers, credit card rates, consumer loan rates, and so on. These interest rates tend to be different. For instance, the interest rate on government bonds is generally lower than the interest rate on corporate bonds. What are some of the reasons that interest rates differ?

**Risk.** The **risk** associated with particular borrowers—their likelihood to default on repaying a loan—is an important source of the differences in interest rates. Creditors go to great lengths to ascertain the degree of this risk, and they adjust the interest rate they charge accordingly. Loans to a government agency will carry a low interest rate because the risk of default is very low. Government can use its power to tax to repay loans. On the other hand, loans to unemployed workers will carry high interest rates because the risk of default is high.

> **Risk:** The probability of a default or a failure of repayment of a loan.

**Cost of Making Loans.** The cost of making loans differs. Large loans and small loans may require the same amount of accounting and bookkeeping work. The large loan will therefore be less costly to process per dollar loaned; for this reason it will carry a lower interest rate than the small loan. This distinction means, for example, that loans to large companies will carry a lower interest rate than loans to small companies.

**Time.** The length of time over which a loan is made will affect the rate of interest charged. The longer the term of the loan, the more things that can go wrong for the borrower. Because the risk of default rises with the length of the loan, longer-term borrowers must pay a premium for this rise. Long-term loans will carry higher interest rates than short-term loans, other things being equal. In addition to the risk factor, borrowers are willing to pay a premium for the longer availability of funds.

**Nominal rate of interest:** The price of loanable funds measured as a percentage of the dollar or nominal amount of the loan.

**Nominal and Real Rates of Interest.** The **nominal rate of interest** is the interest rate set in the market for loanable funds. With inflation in the economy, the nominal rate of interest can be a misleading measure of how much borrowers pay for consumption and investment loans. Suppose that the inflation rate, which is a rise in the average level of all prices in the economy, is 12 percent per year and that the nominal rate of interest is 17 percent. A borrower who borrows $100 will have to repay $117 in a year. However, during the year the average level of prices increases by 12 percent. The $117 paid to the lender at the end of the year will not buy the same amount of goods as the $100 made available to the borrower one year earlier. Since prices have risen by 12 percent, the $117 will buy only 5 percent more goods for the lender after a year. Thus, when the rate of inflation is factored in, the **real rate of interest** is 5 percent, or 17 percent minus 12 percent.

**Real rate of interest:** The nominal interest rate minus the rate of inflation; the price of loanable funds measured as a percentage of the real buying power of the amount loaned.

Lenders and borrowers will not generally be fooled by inflation. The real rate of interest will adjust to account for the expected inflation rate. The nominal rate of interest will include a premium to compensate lenders for the expected depreciation of the purchasing power of their principal and interest. Lenders will have to be compensated for expected inflation, or they will reduce the amount they are willing to lend. Borrowers will also recognize that they will be repaying loans with dollars of less purchasing power and will adjust the amount of interest they are willing to pay to obtain loans. In the above example, if both borrowers and lenders fully anticipate a 12 percent inflation rate, the nominal rate of interest will adjust to 17 percent. As the inflation rate changes, the nominal rate of interest will also change to reflect the level of expected inflation.

The real rate of interest is the nominal rate of interest minus the expected inflation rate. In the latter part of the 1970s, the nominal rate of interest rose to over 20 percent on some types of loans, but since the expected inflation rate was on the order of 15 percent, the real rate of interest was only 5 percent.

## Interest as the Return to Capital

So far, we have discussed interest as the price of loanable funds. Interest may also be viewed as the return that goes to capital as a factor of production. In other words, interest is the return to capital as a productive input in the circular flow model of the economy.

If an entrepreneur buys a machine for $100 and makes $25 a year by using it in production, the entrepreneur earns 25 percent interest on the investment in the machine. Using this concept of interest, we can show how the market for capital equipment works in Figure 14-2.

The demand curve, $D$, is the marginal revenue product curve of capital. It represents the entrepreneur's estimate of how much each unit of capital equipment will add to the firm's revenue. In effect, $D$ is the derived demand curve for capital equipment. We assume that the short-run supply of physical capital—machines, building, tools, and the like—is fixed and cannot be augmented except over the long run. $S_1$ and $S_2$ are therefore drawn as vertical straight lines in Figure 14-2. The point where demand and supply intersect determines the rate of interest that entrepreneurs earn on each level of fixed capital investment.

What is the link between the market for loanable funds and the market for physical capital? If the rate of interest being earned on physical capital,

## FIGURE 14–2
### Return on Investment in Capital Goods

D is the derived demand curve for capital goods. $S_1$ and $S_2$ are short-run fixed supply curves for capital goods. Investment in new plant and equipment will take place if the interest earned on such investment ($i_1$ and $i_2$) exceeds the interest rate on loanable funds. New investment shifts the supply curve of capital goods from $S_1$ to $S_2$.

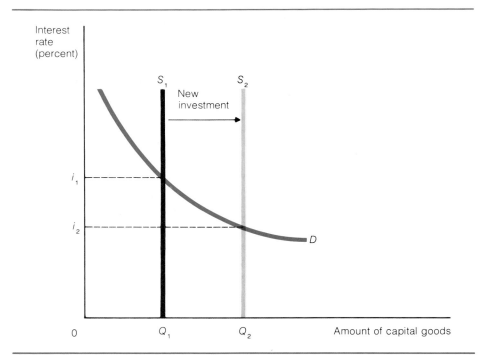

such as $i_1$, is higher than the rate of interest on loanable funds, it will clearly pay entrepreneurs to borrow funds at the lower rate of interest to purchase additional physical capital. Over time, the given supply of capital will shift to the right as entrepreneurs invest in plant and equipment. The movement from $S_1$ to $S_2$ represents such a shift. If the rate of return on physical capital falls below the loanable funds rate, the opposite process will take place, with the level of capital investment falling over time and the supply of physical capital shifting to the left.

Generally, there is a tendency for the rate of return on capital goods to fall over time and for an associated decline to exist in the demand for loanable funds. Offsetting this tendency in the economy are forces such as technological change that increase the productivity of capital goods and shift the demand curve for capital goods to the right.

## The Nature of Returns to Owners of Capital

We have seen that individuals invest in capital goods because they expect to make a rate of return in excess of the cost of capital. The elements of this rate of return can be broken down into three categories: pure interest, risk, and profits and losses.

### Pure Interest

Capital investment yields a return called **pure interest,** the interest rate that must be paid to induce saving when the lender bears little or no risk. Think of pure interest this way: You have a sum of money, and you're trying to decide whether to invest in a piece of capital equipment or a savings account

**Pure interest:** The interest obtained from a risk-free loan.

at the bank with a virtually guaranteed return of 5 percent. The 5 percent is a pure interest yield on your money; in a year you will have earned a 5 percent return on your investment in the savings account. Unless you are confident that the alternative investment in the equipment will yield at least 5 percent, you have no incentive not to put your money in the bank. Investments in capital equipment must pay at least the pure rate of interest, or investors will not provide funds for investments. Again, the pure rate of interest is the rate that must be paid to induce individuals to abstain from current consumption and to save.

### Risk

Suppose that you are confronted with the following offer: You can have $100 for sure or you can have a 50-50 chance of receiving either $200 or nothing. In fact, the expected value of each alternative is the same. In the second alternative, the probability of receiving $200 is 0.50, and the probability of receiving no money is 0.50. Thus, 0.50 ($200) + 0.50 ($0) = $100. Yet the individual decision maker is not indifferent between these alternatives. The latter alternative involves risk, while the former does not. Individuals generally view the bearing of risk as a cost and have to be compensated to bear risk.

Risk has a lot to do with capital investment. There is no guarantee of a handsome rate of return on investments in capital goods. Market conditions can change, making the capital goods obsolete or making the things that the capital goods produce less popular with consumers. In other words, capital investments involve risks, and to induce individual investors to bear risks, they must be paid a premium on their investments. This risk premium is above and beyond the pure rate of interest, and it is paid to induce investors to bear the risk of capital investment.

### Profits and Losses

Investors in capital goods can earn economic profits or losses. The opportunity cost of fixed capital is the price that must be paid to keep it committed to its present use. This cost consists of the pure interest and risk premium discussed above. Yet something unexpected can happen in the economy to make the fixed capital more valuable. For example, the demand for the product produced by fixed capital may increase many times over what was expected. In such a case the owners of capital earn economic profits on their investment in addition to pure interest and a risk premium. In a competitive market, these returns will be competed away as other investors become aware of the excess returns and make capital investments of their own in the industry. Keep in mind too that economic profits can be negative; that is, investors in capital goods can incur large and unexpected losses.

These profits and losses are analogous to capital gains and losses to investors in the stock market. A capital gain, for example, derives from an unexpected increase in the value of a firm, its capital stock, and what the capital stock produces.

Summarizing, the rate of return to capital investment embodies three components: a pure interest return, a premium for bearing risk, and a residual component reflecting unexpected changes in the value of capital goods. Each of these components has an important allocative function in the economy. Pure interest induces people to save. The risk premium leads individuals to invest in risky but valuable ventures. Economic profits are the spur

## The Present Value of Future Income

**Present value:** Today's value of a payment received in the future; future income discounted by the rate of interest.

Investment in capital goods and saving balances the needs of tomorrow with the needs of today. Economists find it useful to look at this balancing act in a precise mathematical formula called present value. **Present value** refers to the value today of some payment that will be received in the future. This value depends on the rate of interest and the length of time between the present and future.

How, for instance, would the prospect of making $100 in one year or in two years compare with the prospect of making $100 now, if the rate of interest is 10 percent? The $100 made now would by definition have a present value of $100. But what about the $100 made in one year? What would be the equivalent amount of money made now? That is, what amount invested now at 10 percent would be able to generate a fund of $100 in one year? The answer is $90.91 with simple interest, computed by the formula

$$PV = \frac{\text{Receipts in one year}}{1 + \text{Interest rate}} = \frac{\$100}{(1 + 0.10)} = \$90.91.$$

What sum invested now at 10 percent and compounded annually would produce $100 after two years? The answer is $82.64, which is the present value of $100 in two years at 10 percent compounded annually. This amount can be computed from the formula

$$PV = \frac{\$100}{(1 + 0.10)^2} = \$82.64.$$

The logic of this formula is that $82.64 will increase to $90.91 after one year, which in turn will increase to $100 after two years.

By extension of this procedure, the present value of any amount at any time in the future can be computed. The formula for doing this is

$$PV = \frac{A_n}{(1 + r)^n},$$

where $A_n$ is the actual amount anticipated in a particular year in the future, $r$ is the rate of interest, and $n$ refers to the particular year in the future.

Note two aspects of present value computations. First, the lower the rate of interest used in making these computations, the higher will be the present value attributed to any given amount of income to be made in the future. Table 14–1 illustrates this point clearly. The present value of $100 one year from now is $97.10 at a 3 percent interest rate and $89.30 at a 12 percent interest rate. Second, the present value of a given amount to be received in the future declines as the date of receipt advances further into the future. In Table 14–1, for any given interest rate the present value of $100 is less the further it is to be received in the future. To see this relation, simply read down any column in the table. The point is that the present value of future revenues or costs is inversely related to the rate of interest and the distance of the date in the future when payment will be received. Focus, "Present Value and Calculating the Loss from Injury or Death," presents an application of the concept of present value.

## TABLE 14–1  The Present Value of $100 at Various Years in the Future

The further in the future the $100 is received, reading vertically down the table at any interest rate, the less it is worth now. Reading horizontally across the table, the higher the interest rate the less is the value of $100 at any year in the future.

| Years in the Future | 3% | 5% | 7% | 10% | 12% |
|---|---|---|---|---|---|
| 1 | $97.10 | $95.20 | $93.50 | $90.90 | $89.30 |
| 2 | 94.30 | 90.70 | 87.30 | 82.60 | 79.70 |
| 3 | 91.50 | 86.40 | 81.60 | 75.10 | 71.10 |
| 4 | 88.80 | 82.30 | 76.30 | 68.30 | 63.60 |
| 5 | 86.30 | 78.40 | 71.30 | 62.00 | 56.70 |
| 6 | 83.70 | 74.60 | 66.60 | 56.40 | 50.70 |
| 7 | 81.30 | 71.10 | 62.30 | 51.30 | 45.20 |
| 8 | 78.90 | 67.70 | 58.20 | 46.60 | 40.40 |
| 9 | 76.60 | 64.50 | 54.40 | 42.40 | 36.00 |
| 10 | 74.40 | 61.40 | 50.80 | 38.50 | 32.20 |
| ... | ... | ... | ... | ... | ... |
| 15 | 64.20 | 48.10 | 36.20 | 23.90 | 18.30 |
| ... | ... | ... | ... | ... | ... |
| 20 | 55.40 | 37.70 | 25.80 | 14.80 | 10.40 |
| ... | ... | ... | ... | ... | ... |
| 30 | 41.20 | 23.10 | 13.10 | 5.73 | 3.34 |
| ... | ... | ... | ... | ... | ... |
| 40 | 30.70 | 14.20 | 6.70 | 2.21 | 1.07 |
| ... | ... | ... | ... | ... | ... |
| 50 | 22.80 | 8.70 | 2.13 | .85 | .35 |

## Investment Decisions

The estimation of present value is important to firms because investment decisions involve current expenditures for plant and equipment in the expectation of future revenues from the goods and services produced by the plant and equipment. Two aspects of investment decision making are paramount. First, the future revenues that an investment project will yield are estimated and converted to present value terms. This process provides the firm with an estimate of how much a project is worth now. Second, the firm must know what its cost of loanable funds is; that is, it must know on what terms it can borrow or lend money. This cost is the interest rate that the firm will apply to investment decisions. A simple decision rule follows from these two steps: If the estimated present value of a project exceeds the cost of loanable funds, a profit-maximizing firm will undertake the project. If not, it will turn down the project. For example, if a firm estimates that a project will yield an 8 percent return but money market certificates are currently paying 15 percent, the firm will reject the project and place its funds in the money market.

## FOCUS: Present Value and Calculating the Loss from Injury or Death

In many lawsuits, the formula for present value is used to estimate the loss of earnings that an individual suffers from an incapacitating injury or premature death. Economists are retained by law firms across the country to make such estimates and to present them to judges and juries.

Consider an instance of premature death. The cost of premature death is treated as the loss of income that would otherwise have resulted over the remaining working lifetime of the deceased. For someone who earned $50 a day and worked 240 days per year, the loss of income would be $12,000 per year. If this person died at age sixty but had expected to work five more years before retiring, the total loss of income for the missing five-year period is $60,000. This amount would not be the cost of the premature death. Rather, the cost would be some lesser amount, reflecting the fact that a loss of $100 in the future is not equivalent to a loss of $100 now.

The formula for present value (PV) given in the text is $PV = A_n/(1 + r)^n$. In the case of a person who earned $12,000 per year, A would be $12,000. At a 10 percent rate of interest, what would be the present value of the premature death at age sixty of someone who earned $12,000? Applying the general formula for present value, the answer would be found by solving

$$PV = \frac{\$12,000}{(1.1)^1} + \frac{\$12,000}{(1.1)^2} + \frac{\$12,000}{(1.1)^3}$$
$$+ \frac{\$12,000}{(1.1)^4} + \frac{\$12,000}{(1.1)^5} = \$45,521.$$

While a total of $60,000 of income is lost, the present value of this loss is $45,521. For a person who dies prematurely at age sixty, the cost of this death, labeled as premature because it occurred before age sixty-five, would therefore be estimated at $45,521.

The concept of present value thus plays a role in the courtroom. It helps lawyers, judges, and juries establish reasonable bounds for settling lawsuits based on the loss of future income through injury or death.

---

Here is an example of this reasoning. A firm is contemplating an investment in a $50,000 machine. Its managers estimate that after expenses for maintenance and repair the machine will add $10,000 per year for the next six years to firm revenues. After six years the machine no longer has any value; it has no scrap value. Assuming that the firm can borrow the funds to purchase the machine at a 12 percent rate of interest, it will discount the estimated future revenues at a 12 percent rate. The present value calculations are shown in Table 14–2.

**Discount rate:** The interest rate that a firm uses to determine the present value of an investment in a capital good; the best interest rate that a firm can obtain on its savings.

By taking the appropriate rate of interest, commonly called the **discount rate** when used in investment planning, from Table 14–1 and applying it to the estimated future revenue at the end of each year, we derive in the last column of Table 14–2 how much the flow of future income from the machine for each year is worth today. As you can see, the total present value of the estimated future income produced by the piece of equipment is $41,200 (the sum of the figures in the last column). This total present value is considerably less than the cost of the machine ($50,000), and so the firm will rationally reject this investment proposal.

Whether an investment project will be undertaken is obviously sensitive to the interest rate that the firm faces. As we have seen, lower interest rates lead to higher present values. If the discount rate in this example had been 5 percent, the present value calculations would have indicated that estimated revenues would be greater than $50,000. At a 5 percent interest rate, the firm would have undertaken the project.

Will the rates of return on profitable investments differ greatly in a competitive economy? Ignoring differences in risk of different investments, competition will tend to equalize the rate of return to investment projects everywhere in the economy. Where there are profits from investment projects,

TABLE 14-2  The Discounted Present Value of $10,000 for Six Years

| Year | Estimated Future Revenue (received at year-end) | Discounted Value (12% rate) | Present Value of Income |
|---|---|---|---|
| 1 | $10,000 | 0.89 | $8,900 |
| 2 | 10,000 | 0.80 | 8,000 |
| 3 | 10,000 | 0.71 | 7,100 |
| 4 | 10,000 | 0.64 | 6,400 |
| 5 | 10,000 | 0.57 | 5,700 |
| 6 | 10,000 | 0.51 | 5,100 |
|  | Total $60,000 |  | Total $41,200 |

entry by competing investors in similar projects will occur. For example, in the 1950s Ray Kroc pioneered the concept of a fast-food restaurant by opening McDonald's restaurants across the country. The profits that he earned on this investment attracted many competitors—Burger King, Wendy's, and so on—to make similar investments. Such entry (or exit in the case of losses on investment projects) will drive the rate of return on capital investment to equal its cost. In simple terms, competition will tend to equate the expected future revenues from capital investment to the current cost of the investment. In such a way competition equalizes the rate of return on capital investments in a competitive economy.

## The Benefits of Capital Formation

Saving means abstaining from consumption now in the expectation of being able to consume more in the future. Saving thus provides the resources needed to increase the stock of capital goods in the economy. A large stock of capital goods raises the rate of consumption that can be sustained in the future. Saving clearly benefits the saver, but it also benefits those who don't save. Indeed, the benefits from saving are diffused throughout the whole economy. Most of us benefit from advances in computer technology, though few of us contributed to the saving that made such advances possible.

Saving and capital formation also lead to higher wages and incomes for workers. In a competitive economy wages are a function of the productivity of workers, and this productivity is in turn a function of the amount and quality of the equipment and tools that workers use. A worker with a tractor is far more productive than a worker with a mule-drawn plow. Since more productive workers are paid more, some of the benefits from capital formation are spread to workers who did not necessarily contribute to the saving that made the capital formation possible.

Indeed, those who save and who build capital goods are among the chief benefactors of society. As Ludwig von Mises put it:

> Every single performance in this ceaseless pursuit of wealth production is based upon the saving and the preparatory work of earlier generations. We are the lucky heirs of our father and forefathers whose saving has accumulated the capital goods with the aid of which we are working today. We favorite children of the age of electricity still derive advantage from the original saving of the primitive fishermen who, in producing the first nets and canoes, devoted a part of their working time to provision for a remoter future. If the sons of these legendary fishermen had worn out these intermediary products—nets and canoes—without replacing them by new ones, they would have consumed capital

and the process of saving and capital accumulation would have had to start afresh. We are better off than earlier generations because we are equipped with the capital goods they have accumulated for us.[1]

The reverse is also true: Policies that reduce the incentive to save and to produce capital goods, while appearing to harm the suppliers of saving and the owners of capital goods, actually hurt all of us.

## Summary

1. Wealth can be defined as a high level of sustainable consumption in an economy. The elements of a wealthy society are production, resources, knowledge, capital, technology, and institutions.
2. Roundabout production is the process of saving and investing in capital goods production in order to produce more in the future.
3. Time preference is a measure of an individual's concern for present versus future consumption.
4. Interest is the price that individuals must be paid for waiting to consume later, or the price that individuals are willing to pay in order to consume now rather than later.
5. The rate of interest is determined in the market for loanable funds by the demand and supply of loanable funds.
6. In reality there are a multiplicity of interest rates—such as prime rates, government rates, and credit card rates—determined by the degree of risk, cost, and time allowed for payment. Taking the rate of inflation into account, the real rate of interest is lower than the nominal rate of interest.
7. The returns to capital are composed of pure interest, a risk premium, and economic profits or losses.
8. Present value is the amount of money in the present that is equivalent to a certain amount of money in the future.
9. The discount rate is the interest rate used to discount future amounts of revenues and costs in order to obtain present values.

## Key Terms

roundabout production
saving
capital formation
depreciation
capital consumption
rate of time preference
interest
demand for loanable funds
supply of loanable funds
risk
nominal rate of interest
real rate of interest
pure interest
present value
discount rate

## Questions for Review and Discussion

1. Give some examples of roundabout production. Describe why going to college is a roundabout method of production.
2. Suppose that you are thinking about investing in a state of the art typewriter. The typewriter costs $2000 and will last for four years. You expect to earn $800 a year typing papers part time with the typewriter. The bank will lend you the money to buy the typewriter at 19 percent interest. Should you buy the typewriter? Suppose the bank's interest rate was 12 percent?
3. Why is the concept of time preference important in the discussion of interest rate? Explain.
4. How would the following events affect the rate of interest in the United States?
   (a) The threat of war in Central America
   (b) An increase in the inflation rate
   (c) The discovery of massive domestic oil resources
   (d) Greater impatience in the general population
   (e) More capricious government intervention in the economy
5. If the interest return that entrepreneurs can earn on capital investments is less than the interest rate paid for loanable funds, how will entrepreneurs behave? What happens to the level of investment in the economy?
6. J. M. Keynes in *General Theory* suggests that the marginal product of capital can be driven to zero in one generation by increasing the production of capital. Would this ever happen in a free market? If it did, what would this imply about time preference? What would happen to the nation's output? What would be the interest rate?
7. What is the formula for computing present value?

---

[1] Ludwig von Mises, *Human Action*, 3rd ed. (Chicago: Henry Regnery, 1966), p. 492.

How is present value affected by changes in the interest rate used for discounting?
8. What is the present value of your entire future income? Would you consider yourself a millionaire? How much of your future income would a bank be willing to lend you now?

## ECONOMICS IN ACTION

## Stocks and Bonds

At the back of most daily newspapers are tables listing the current prices of stocks and bonds. Such tables are good evidence of how the market for loanable funds works, coordinating the supply of funds, primarily held by investors and investor groups, with the demand for funds, primarily the desire by corporations to raise capital for future production.

Corporations can raise capital by two means: They can sell shares of stock or issue bonds. By selling shares of stock, a corporation is effectively selling property rights to itself. The shareholders of a corporation are its owners; shares usually convey a voting right: Each share of stock permits its holder one vote at meetings that determine the firm's future management. In this way corporate policy is directed by the will of the holders of the majority of the firm's shares—whether one or many individuals. Shareholders do not make operating decisions for the firm, but they are legally responsible for hiring the managers—the board of directors—who do. Usually when shareholders are pleased with the existing management of a firm, they assign their votes to the management, who then act as a proxy for the shareholders in voting on corporate policy.

The stock market is an organized exchange where corporate shares are bought and sold. In the stock market the prices of shares issued by different firms reflect the market's estimate of the future profitability of the firms in question. From the shareholders' standpoint, investing in the stock market means bearing the risk resulting from the uncertain prospects of any given firm. A share of the stock's future value is not guaranteed and can rise or fall with the firm's fortunes, or because of general market conditions such as interest rates.

Corporations and governments also issue another type of obligation called a bond. Unlike shares of stock, bonds do not confer ownership rights; they simply represent IOUs. The firm or government body agrees to pay the bondholder a fixed sum (the principal that represents the loan to the firm) either on a specific date or in installments over a specified period. Most bonds pay, in addition, a fixed return per year, usually expressed as a percentage of the face value of the bond. For example, a corporation may issue a $10,000 bond (which, if sold, represents a $10,000 loan to the corporation for some specified period) at an interest rate of 5 percent. This bond would pay the holder $500 per year until maturity, when the entire $10,000 is returned to the investor.

| 52 Weeks High | Low | Stock | Div. | Yld % | P-E Ratio | Sales 100s | High | low | Close | Net Chg. |
|---|---|---|---|---|---|---|---|---|---|---|
| | | – A–A–A – | | | | | | | | |
| 39⅞ | 30¾ | Alcoa | 1.20 | 3.8 | 16 | x2283 | 32 | 31⅜ | 31⅜ | – ½ |
| 25⅝ | 15½ | Amax | .20 | 1.1 | .. | 523 | 17⅝ | 17⅜ | 17½ | – ⅛ |
| 42¼ | 32½ | Amax pf | 3 | 8.8 | .. | 6 | 34 | 34 | 34 | – ½ |
| 52½ | 22¾ | AmHes | 1.10 | 3.5 | 16 | 3877 | 32¼ | 31½ | 31⅝ | + ½ |
| 2¾ | 1¼ | AmAgr | .. | .. | .. | 70 | 2⅛ | 2 | 2⅛ | .... |
| 19⅝ | 15⅛ | ABakr | .. | .. | 8 | 10 | 18½ | 18¼ | 18½ | + ⅛ |
| 70 | 53 | ABrand | 3.90 | 5.9 | 9 | 709 | 67⅞ | 66⅛ | 66⅜ | –1½ |
| 27¾ | 24⅝ | ABrd pf2.75 | | 9.9 | .. | 166 | 27¾ | 27⅜ | 27¾ | + ⅜ |
| 115 | 55¾ | ABdcst | 1.60 | 1.5 | 16 | 899 | 108⅜ | 108 | 108 | .... |
| 26¼ | 19½ | ABldM | .86 | 3.4 | 13 | 12 | 25½ | 25½ | 25½ | .... |
| 27⅛ | 20⅛ | ABusPr | .64 | 2.5 | 15 | 6 | 26 | 25¾ | 25¾ | – ⅜ |
| 55¾ | 40⅛ | AmCan | 2.90 | 5.5 | 11 | 805 | 53½ | 53 | 53⅛ | – ⅜ |
| 48 | 36 | ACan pf | 3 | 6.5 | .. | 54 | 46⅝ | 46¼ | 46¼ | – ¼ |
| 110 | 103 | ACan pf13.75 | | 12. | .. | 16 | u110¾ | 109½ | 110¾ | +1¾ |
| 19⅞ | 16¾ | ACapBd | 2.20 | 12. | .. | 85 | 19⅜ | 18⅞ | 19 | – ⅛ |
| 33 | 25⅛ | ACapCv | 2.51e | 9.0 | .. | 13 | 27⅞ | 27¾ | 27¾ | – ⅛ |
| 11 | 6½ | ACentC | .. | .. | 11 | 32 | 8⅝ | 8½ | 8½ | – ¼ |

*The photo above shows a listing of stocks, their prices, and other information from the New York Stock Exchange in the Wall Street Journal. To the right of the first two columns of numbers, an abbreviation of the stock's name is found. ABdcst, for example, stands for the American Broadcasting Company. In columns one and two, the high and low price of the stock in the previous year (ending at the close of trading the day before) is given. Price information from the latest day is in the last four columns on the right. The latest day's high and low price is reported along with the last (closing) price paid for it. If the high or low price paid for a stock on that day is also a 52-week high or low, a footnote (u for up, d for down) is attached to the day's high or low price and will be reflected in the following trading day's data. The final column gives the net change in a stock's value from the previous closing price. Thus the price of Alcoa (the Aluminum Company of America) stock fluctuated between a high of $32 and a low of $31⅜, also closing at that value on this day. The net change (from the value of the previous day's closing price) was down by 50¢ a share. Other important stock data is also given. The annual cash dividend is given in the column right of the stock's name. It is based on the rate of the last quarterly payout, and extra dividends are indicated by footnotes following the cash amount. The next column gives yield of the stock by dividing the cash dividend by the closing price. This valuable information permits you to compare the yield with other kinds of financial instruments. The P-E, Price-Earnings ratio, is found by dividing latest closing price of the stock by latest available earnings per share, which is (for the most recent four quarters) the profit of the company divided by the number of shares outstanding. The P-E ratio is an indicator of the stock's performance. The "Sales 100s" column lists the number of shares traded. Thus the number 6 means that 600 shares were traded that day.*

Although bonds tend to be issued for long terms (thirty-eight years in the case of some U.S. Treasury bonds), bonds are commonly bought and sold prior to the date of maturity. The bond market is like the stock market in that it represents an organized exchange specializing in the buying and selling of bonds.

Interest rates affect the prices of shares of stock indirectly by their impact on the activities and hence the profitability of firms, but interest rates bear a direct relation to the price of bonds. Take, for example, a bond issued with a face value of $10,000 that pays an annual rate of interest of 6 percent. Until maturity, whoever holds the bond is entitled to $600 annually. Suppose, however, that the market interest rate rises to 10 percent. This means that purchasers of newly issued $10,000 bonds can earn $1,000 annually, other things equal. What happens to the price of the 6 percent bonds? Their market value must fall to reflect the new, higher interest rate. No one, in other words, would invest in a bond paying less than 10 percent since they can earn 10 percent on newly issued bonds. The price of the 6 percent bond must fall to $6,000 to reflect the new, higher rate of 10 percent. Individuals will not buy the original bonds unless their price falls so as to produce a 10 percent return on a given investment ($600/$6,000). The purchaser of the original bond (assuming he or she still held it after the interest rate increase) experiences a capital loss of $4,000 ($10,000 − $6,000). The bond will still pay $10,000 at maturity (the firm must pay back what it borrowed), but the sales value of the bond right now is $6,000.

The general point to remember about bond price is that the price of a bond varies inversely with the interest rate. When the interest rate rises, the price of a bond falls; when the interest rate falls, the price of a bond rises.

*This U.S. Treasury bond, with a face value of $100,000 and 3½% interest, is dated October 30, 1960, and will mature on November 15, 1998, a term of thirty-eight years. The present sale value of a bond like this is now below $100,000 because interest rates have risen considerably above 3½% in recent years. Current sale prices of bonds are reported in daily newspapers.*

## Question

In what ways are stocks and bonds similar? In what ways are they different? How does the sale of old, previously issued securities (both stocks and bonds) contrast to the issuance of new securities?

# 15

# Rents, Profits, and Entrepreneurship

Thus far in our study of the factors of production, we have looked at labor and capital and their corresponding factor payments, wages and interest. In this chapter we look at two more factors: land and the special form of labor called entrepreneurship. In discussing these two factors, we will present the economic theories of rents and profits.

The concept of rent has a broader economic meaning than simply the payment for use of land. In theory, any factor of production, including labor, can earn rents. Therefore, after we have examined the relation of land supply to rent, we will look at other types of rents and discuss the concept of rent seeking, which refers to the competition for rents among individuals and firms.

In economic terms, rent and profit are closely allied. Profits are the residual payment to capitalist entrepreneurs after all other factor payments—including rents, salaries, and interest—have been made. Every firm and every individual seeks to maximize profits and minimize losses. Competition for profits, or risk taking, involves decision making under uncertainty. We will explore how this process takes place.

## Types of Rent

In everyday language, rent is simply payment for the use of something. A farmer pays rent to a landowner for the use of an acre of land. A student leases an apartment and pays rent to the landlord each month. A sales representative rents a car to call on customers in a distant city. These types of rents are all represented in the circular flow of income as the income to property and landowners, a broad category that includes all types of rent payments made according to lease agreements. In 1984, such payments to-

**Rent:** A payment to a factor of production in excess of its opportunity cost.

taled $62.4 billion, or about 2.1 percent of national income. In economic analysis the concept of rent has a broader meaning than the payment for the lease of a factor such as land. For the economist, **rent** refers to payment to any factor of production—land, labor, or capital—beyond its opportunity cost, or the forgone value of its next most productive use. Nearly every factor you can name has alternative uses. An acre of farmland could be used as the site of an industrial park. An engineer could be put to use typing correspondence. The economic theory of rents, therefore, has a much broader application than simply the payment for lease.

**Pure economic rent:** The payment to a factor of production that is perfectly inelastic in supply.

Economists distinguish among various types of economic rent. A **pure economic rent** is the return to a factor of production that has a perfectly inelastic supply curve. In such a case the price of the resource is determined solely by the level of demand for its services because supply is fixed and does not change as price changes. Land in the aggregate is the classic example of a resource whose supply is given and whose return is characterized as a pure economic rent.

**Inframarginal rent:** A type of rent that accrues to specialized factors of production.

Another type of economic rent is paid to factors of production with rising supply curves. These rents are called **inframarginal rents**, and they represent payments to a factor of production above and beyond its opportunity cost as reflected in its supply price. Virtually all factors of production earn inframarginal rents.

Finally, there are categories of rents that accrue to firms. Quasi-rents are returns to a firm owner in a competitive industry above the opportunity cost of the owner's invested capital. Quasi-rents are thus the short-run economic profits of competitive firms, which we analyzed in Chapter 9. Monopoly rents are the returns that accrue to the owner of a monopoly firm when the firm restricts output and raises price. Monopoly rents are the profits of the monopolist, which we studied in Chapter 10. Competitive and monopoly profits are recast in this chapter as quasi-rents and monopoly rents to illustrate the relation between economic rent and economic profits.

## Land Rents

How is land priced? The obvious answer is that the price of land is determined where its demand and supply curves intersect. While this is correct, there are some special aspects of the price-determination process in the case of land. These aspects concern how the supply curve of land is defined. Although the focus in our discussion will be on the supply of land, keep in mind the role of the demand curve for land. As we learned in Chapter 12, this demand curve is based on the marginal productivity of land. In this regard the demand curve for land is the same as the demand curve for any factor of production—it is based on the expected marginal revenue product of land as a factor of production. When land is rented, rent is paid to the landowner. The greater the demand for the land, the larger the rent will be. In equilibrium, where the demand and supply of land are equal, the rent to the landowner is equal to the marginal revenue product of the land to the renter.

### Pure Economic Rents

To pursue the relation of the supply of land to its price, assume that land is leased by the acre per year. We are concerned with the rent that must be paid to use the flow of services yielded by land for a year. We are not concerned with the related question of the price of a given stock or amount of

**Aggregate supply of land:** The amount of land available to the entire economy at various rental rates.

land, although the two ways of looking at the pricing of the services of land amount to practically the same thing. Land is a resource that yields a flow of services over time. The price of a stock of land will reflect the estimated present value of this future flow of services. Alternatively, the rental price of land will reflect the estimated value of the flow of services for a given period of time. Since our interest is in land rent, we speak in terms of the price of the flow of services from land for a given period of time.

A distinction must also be made between the fixed supply of land and its potential allocation to varying uses. In Figure 15–1a the supply curve of land is drawn for all land, or the **aggregate supply of land.** The curve is drawn as a vertical line, indicating that land in the aggregate is in fixed supply. In general, the amount of land on earth is given by nature and cannot be changed. Actually, common sense tells us that the assumption of a perfectly inelastic supply of land is not completely true. The supply of usable land can be increased through reclamation of swamps and marshes or reduced through erosion. Nonetheless, it is useful to think of the aggregate supply curve of land as perfectly inelastic, as drawn in Figure 15–1.

What are the economic implications of this vertical supply curve for all land? First, the perfectly inelastic nature of land in the aggregate means that the supply of land is unresponsive to price. Whether land is priced at $1 per acre per year or $1 million per acre per year, the supply of land in the aggregate will not change. In economic terms this fixed quantity, $Q$ in Figure 15–1a, means that the rental price of land will be determined solely by the level of demand. Thus, the level of rent on the vertical axis depends on where the demand curve for land intersects the fixed supply curve.

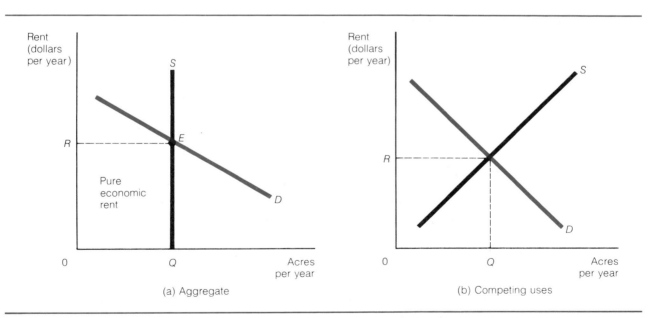

**FIGURE 15–1    Pure Economic Rent and the Supply of Land**

(a) The supply curve for land in the aggregate is perfectly inelastic with respect to rent, which means that the return to land in the aggregate is a pure economic rent.
(b) The supply curve of land to competing uses appears as it is normally drawn—sloping upward to the right. This slope reflects the fact that land has competing uses and hence that the price of using land or rent plays an important role in determining the uses to which land is put.

The return to land in the aggregate is labeled a pure economic rent. To understand this concept, recall that economic rent is defined as a payment in excess of the opportunity cost of a factor of production. Recall also that a factor's opportunity cost is reflected in its supply schedule because a supply schedule indicates what it takes to bid resources away from their next-best alternative use. A vertical supply curve such as in Figure 15–1a means that land in the aggregate has no opportunity cost; nothing is given up for its supply, for it has no alternative uses. Very simply, this means that the whole return to land in the aggregate is a pure economic rent. In Figure 15–1a, land garners $REQ0$ in pure rent.

The concept of pure economic rent is normally associated with the supply curve of all land. However, it is possible that other resources will exhibit a range of their supply curve that is perfectly inelastic and that they therefore stand to earn pure economic rents. This characteristic may be true of uniquely talented performers in the world of sport and art, a subject to which we will return later in the chapter.

### The Supply of Land to Alternative Uses

Since the return to land in the aggregate is a pure economic rent, a price does not have to be paid to call forth a supply of land. The supply of land is fixed and given at $Q$, and it is not affected by the level of rent, $R$. Since the supply of all land is independent of its price, we say that price has no allocative function in this case.

If rent does not have to be paid to call forth a supply of land, then why pay it? Or why not tax all rents away as an unearned surplus to landowners? The attractiveness of such a tax is apparent—it raises revenue without affecting the available supply of land. In Figure 15–1a, for example, a tax equal to the area of pure rents, $REQ0$, would leave the supply of land unaffected at $Q$.

In 1879, in his immensely popular book *Progress and Poverty*, Henry George proposed a single tax of 10 percent on land rent on just these grounds. George argued for a single tax on land because he felt that landowners contributed nothing to the land's productivity. Rising land values were determined by general economic growth and by increases in the demand for land. Thus, a tax on land would not affect the amount of land available to the economy and would prevent landowners from getting rich from windfall profits while nonlandowners remained poor. Moreover, George thought that a single tax on land rents could finance all the government of his day, eliminating the need for other types of taxes.

George's proposal rests on the assumption of a perfectly inelastic supply curve of land. He was essentially assuming that land has no alternative uses. But this assumption concerns land only in the aggregate. While land in the aggregate has a vertical supply curve, the supply of land to alternative uses does not. Consider what happens, for example, as a city expands. Farmland is converted to use as sites for houses and shopping centers as people bid for the right to use the land in other ways. The market process reallocates the land from less valued to more highly valued uses, from farmland to city land.

The supply curve of land for particular uses slopes upward to the right, as shown in Figure 15–1b. To rent land for a particular use, one must bid it away from competing uses. As more desirable land is rented, the rental rate rises. With a normally sloped supply curve, the rental price paid for land serves to allocate land to competing uses. Those who value the use of a parcel of land more highly will offer higher rents than others. Focus, "Land

## FOCUS: Land Rent, Location, and the Price of Corn

The rental rate of land varies greatly from one location to another depending on the land's productivity. While the fertility of the soil may be identical in two different locations, the rental rates may not be equal because the physical locations of two plots may account for differences in productivity. To understand this point, consider a straight highway that leaves a large city and extends far into a rural area. Along this highway are farms of equal size that produce the same product, say corn.

Each farm transports its corn to the city and incurs the costs of transporting each bushel. The distance from each farm to the city is of course different, and the cost of transportation increases as the distance increases. Thus, farms nearer the city have lower average and marginal costs than farms farther from the city.

Figure 15–2 shows cost curves of four farms at 50-mile intervals from a city. Farm A is closer to town than Farm B, B is closer than C, and C is closer than D. The costs of production are lower at locations closer to the city; this land can be described as more productive.

Since each farm is a price taker in a competitive market in town, all farms face the same price, P. Profits are shown by the shaded area in each panel. The land closer to town is more profitable because of its locational advantage and thus is in greater demand. The demand for land decreases as the location is farther from town. Farm D makes no profits; there would be no demand for land by farmers beyond this point. The varying profits bring forth varying demand and thus varying rental rates. The rental rates are bid up by potential farm owners until a normal profit is obtained at each location. The shaded area of profits at each location actually becomes the rental rate. The advantages of a productive location yield income in the form of rents to the suppliers of the land.

What happens to rents if the demand for corn, and thus its price, increases? The higher price for corn will increase profits. Even the land at greater distances than farm D will now be brought into production. With higher profits, the demand for land will increase, and thus rental rates will again rise. After rental rates have risen, normal profits will result at each location, given competitive markets. A decrease in the demand for corn and a reduction in its price will lead to the opposite results. Profits will fall, some distant farms will go out of business, and rental rates will fall, restoring a normal rate of return to farming.

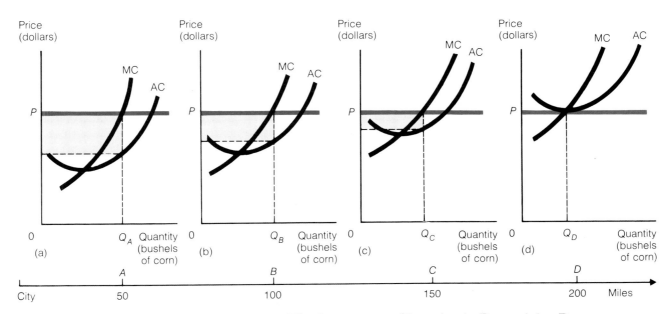

**FIGURE 15–2** **The Importance of Location in Determining Rents**
Graphs a–d show the cost curves for four different farms at four different locations. Farm A is closest to the city, farm D is farthest away. Since costs of transportation increase the farther a farm is from the city, farm A will enjoy lower costs and higher productivity than any of its three competitors.

Rent, Location, and the Price of Corn," discusses this process in more detail. In this sense, rent functions like any other price in signaling the intentions of buyers and sellers in the market for land. Rent for land is not a pure surplus in this case. It serves a useful purpose in the price system, and to tax it away could impose unnecessary shortages and other costs on the economy.

The proper distinction to keep in mind is that while the supply curve of land in the aggregate is approximately vertical, the supply curve of land to alternative uses is positively sloped. Since the latter concept of the supply of land is far more relevant to real-world situations than the former, a proposal such as Henry George's could do much damage to the economy. It was never implemented. Although George became a popular politician and ran for mayor of New York City twice as the candidate of the Labor and Socialist parties, he lost his first race and died during the second, in 1897. Whether land is viewed in the aggregate or as a resource of varying desirability and alternative uses, rent does serve a function: It rations land among available bidders.

### Land and Marginal Revenue Product

Like the demand curve for any other factor of production, the demand curve for the services of land is a derived demand curve. It is derived from the value of the output that the land, used in conjunction with other factors of production, can produce. Entrepreneurs will therefore employ land in their production processes as long as the marginal revenue product of land exceeds its rental rate. Where the demand and supply for land are equal, as shown in Figure 15–1b, rental payments equal the marginal revenue product of the land. As we stressed in Chapter 12, the principles of marginal productivity theory apply to any factor of production, including land.

## Specialization of Resources and Inframarginal Rents

The concept of economic rent is more general than a rent payment in a lease agreement. An economic rent is defined as a payment to any factor of production—land, labor, or capital—above and beyond its opportunity cost. Rent, in other words, is an excess payment to a resource owner. This concept can easily be applied to resources other than land. For example, a worker provides labor services to a firm. The firm pays the worker the prevailing market wage. The opportunity cost of the worker—her evaluation of her next-best employment prospect—is less than the wage she is paid. The worker thus earns an economic rent in her present job, for she is paid more than it takes to induce her to continue to work in her present employment.

Note that rents are caused by the specialization of resources. **Resource specialization** simply means that a resource or factor of production is relatively better at working in one occupation or industry than in others. A worker, for example, may be specialized as a welder because he went to technical college to develop his welding skills. As a welder, he will supply his specialized labor to those industries that employ welders.

In general, we can detect the degree of resource specialization by observing the supply curve of the resource. The supply curve of a factor of production measures the willingness of its owner to supply its services at various prices. It therefore gives a measure of the opportunity cost of the resource at different prices.

**Resource specialization:** The devotion of a resource to one particular occupation that is based on comparative advantage.

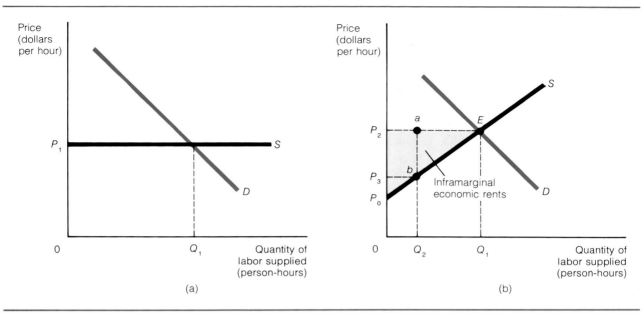

**FIGURE 15-3  Unspecialized and Specialized Resources**
(a) The supply curve of an unspecialized resource. Such resources earn no economic rents. (b) The normal case of economic specialization and an upward-sloping resource supply curve. Inframarginal rents accrue to the resource owners in the amount represented by $P_2EP_0$.

Examples are given in Figure 15-3. In part a, the supply curve of the resource is flat or perfectly elastic. This means that the resource is completely unspecialized with respect to its use in this industry. Any amount of the resource can be purchased at the same price, $P_1$. The resource's price is determined solely by the level of the supply function. This is the opposite extreme from the case of the perfectly inelastic supply curve of land given in Figure 15-1a. Whereas the return to land in the aggregate was a pure rent, the resource depicted in Figure 15-3a earns no rents. Completely unspecialized resources earn no rents.

Many categories of manual labor and jobs that do not require extensive training or experience can be characterized as unspecialized. The supply, for example, of paper carriers, yard workers, and baby-sitters is probably a flat line at the prevailing wage for these services. That is, given the level of demand for such services, the supply of labor to fulfill them is perfectly elastic.

In Figure 15-3b we have drawn a more normal-looking supply curve for a factor of production. In this case, we can say that the upward-sloping supply curve indicates degrees of economic specialization. If the factor is word processors, for instance, some people will be highly specialized at word processing but untrained for anything else. Their next-best alternative would be some low-paying job that requires little skill or training. Such people theoretically would be available as word processors at a relatively low wage—such as $P_3$ in Figure 15-3 because their opportunity cost is low. At the upper end of the supply curve are people trained as word processors who are also highly skilled in other areas, such as data programming or legal secretarial work. They can be bid into word processing jobs only by relatively high wages such

## FIGURE 15-4
### Inframarginal Rents for Volunteer Soldiers

S and D are the demand and supply curves for volunteer soldiers. Because of the upward-sloping nature of the supply curve, volunteer soldiers earn inframarginal economic rents in the amount represented by WEA.

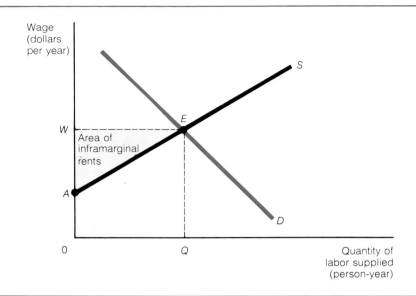

as $P_2$ or an even higher wage because the cost of their forgone opportunities—jobs as legal secretaries or data programmers—is high.

Although workers with varying degrees of specialization would thus accept different wages, interaction of supply and demand tends to set an equilibrium wage, $P_2$, paid to all people or other resources doing the same job. At this price, the more specialized resources on the lower end of the supply curve—those who are relatively better at the work for which they are hired than at anything else—will earn economic rents. Look at the word processors or other resources represented by $Q_2$ in Figure 15-3, for example. Their opportunity cost is $P_3$, which is the price just sufficient to attract them into this industry. Yet because of this market's equilibrium, all units of the resource receive the same price, $P_2$. The difference between $P_2$ and $P_3$, or $ab$, is an economic rent that accrues to the resource depicted by $Q_2$. It is a payment above and beyond the opportunity cost of the resource. In fact, the whole area below $P_2$ and above the supply curve, $P_2EP_0$, represents economic rents earned by suppliers of this resource. The last unit, $Q_1$, is paid just its opportunity cost, $P_2$, so we say that $Q_1$ is the marginal unit of the resource that earns no rents. All the previous units hired earn economic rents by the amount of the difference between their supply prices and their market rate of compensation. Economists call this kind of economic rent an inframarginal rent.

As another example of inframarginal rents, examine the supply curve of volunteer soldiers given in Figure 15-4. We assume that the army wants to hire $Q$ soldiers per year to maintain its force structure. To do so it finds that it must offer a wage of $W$. All soldiers hired are paid $W$, but it is clear from the upward slope of the supply curve that some would have volunteered for lower wages. For example, the very first volunteers would have signed up for a wage slightly above $A$. The volunteer army therefore leads to the receipt of inframarginal economic rents by volunteer soldiers. In Figure 15-4, these rents are given by the area $WEA$.

The concept of inframarginal rents has general applicability to economic life. Factors of production by their very nature are different, and these differences lead to economic specialization, which in turn leads to inframarginal rents. Yet the prevalence of rents does not suggest that there is something unwholesome about the market process. While it is true that these are "excess returns" that do not have to be paid in order to allocate resources to their most valued uses, it would be a virtual administrative impossibility to tax such returns away. Think of the complexity of determining the opportunity cost of each resource. Moreover, it is the ability to capture such returns that leads to the process of resource specialization in the first place. These rents provide incentives for resource owners to behave in certain ways that are good for the economy as a whole.

## The Economics of Accountants and Superstars

It is possible that the same resource can attract both inframarginal and pure economic rents. Such a situation may be created temporarily by a sudden surge in demand for a resource. Consider what happens to the rents earned by accountants if the government unexpectedly passes a complicated tax law that can be understood and applied only by those with special accounting skills. Prior to the passage of this law, the market for tax accountants would be in equilibrium at a wage of $W_1$ and a quantity of $Q_1$ accountants, as indicated in Figure 15–5. The supply curve of accountants slopes upward to the right, and at the usual wage $W_1$ accountants earn inframarginal rents represented by $W_1BA$.

After the passage of the new tax law, the demand for the services of tax accountants increases to $D_2$. Yet at this point in time, the supply of tax accountants is fixed at $Q_1$. There are just so many tax accountants and their supply cannot be increased overnight. The supply curve of accountants therefore becomes vertical at B for some specified time period. As in the case of aggregate land supply, a vertical supply curve means that the accountants

**FIGURE 15–5**
**Two Kinds of Economic Rent Earned Simultaneously**

When the demand curve for accountants shifts from $D_1$ to $D_2$, the supply of accountants remains constant in the short run at $Q_1$. Over this period accountants earn both pure economic rents of $W_2CBW_1$ and inframarginal rents of $W_1BA$.

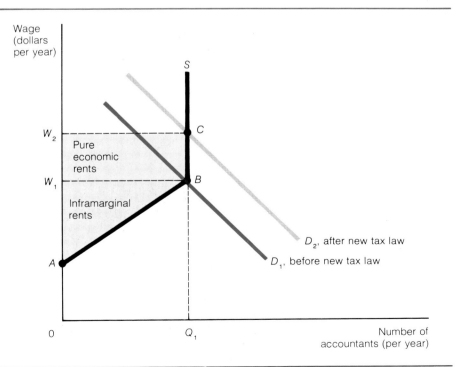

will earn pure economic rents as the demand for their services increases. In Figure 15–5 these rents are given by the area $W_2CBW_1$. In such a situation, accountants thus earn two types of economic rent: pure economic rents equal to $W_2CBW_1$ and inframarginal economic rents equal to $W_1BA$.

The pure economic rents in this example are only temporary. Over time, these returns in excess of opportunity cost will attract entry into the accounting profession. Entry of new accountants will shift the supply curve for accountants to the right, reducing the price of accounting services and the economic rents earned by accountants. During the period in which entry does not occur and the accountants earn pure economic rents, a rising wage for accountants serves to ration their services to the highest bidders.

A supply schedule like that drawn for accountants in Figure 15–5 can also apply to the income of superstars, such as Barbra Streisand or basketball star Julius Erving. Figure 15–6 shows a possible supply curve for a superstar. In this case the next-best alternative to being a superstar is reflected in a price for the star's services of $P_1$. The job earning $P_1$ as an alternative to superstardom might be something as ordinary as working in a bank. The supply curve of the superstar becomes vertical at $P_1$ because he or she possesses unusual abilities that cannot be copied by anyone else. Only Dr. J has certain moves on a basketball court; only Streisand can offer her unique musical sound. Thus the supply curve of the superstar above $P_1$ indicates that he or she is earning pure economic rent. As in other cases of vertical supply curves, price here serves to ration the services of the superstar among the competing demands of music producers or team owners.

We must be careful, however, in defining economic rents in the case of a superstar. If we define the next-best alternative of a star as being a truck driver, the economic rents that he or she earns will be large. Only if the price for his or her superstar services falls below $P_1$ in Figure 15–6 would the

## FIGURE 15–6
### A Superstar's Supply Curve

The superstar's unique talent cannot be replicated, so it is represented by the vertical portion of the supply curve. The next-best alternative of the superstar is given by $P_1$. The difference between $P_2$ and $P_1$ means that the superstar's income includes a pure economic rent in the amount of $P_2ABP_1$ as well as an inframarginal rent of $P_1BC$.

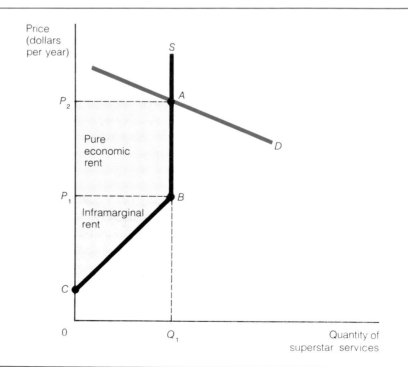

star switch to such an alternative job. Suppose, however, that the next-best alternative of the basketball star who earns $1 million per year is to play for another team who would pay him $900,000. In this case economic rent would be $100,000:

$$\underset{\substack{\text{Resource owner's}\\\text{income}}}{\$1,000,000} - \underset{\substack{\text{Resource owner's}\\\text{opportunity cost}}}{\$900,000} = \underset{\substack{\text{Economic}\\\text{rent}}}{\$100,000.}$$

The measurement of economic rents hinges on how the next best alternative or the opportunity cost of the resource is defined.

Do the rents that superstars earn affect the future supply of talent in their professions? Are their pure economic rents transitory, like those of the accountants, as new stars rise to compete with them and capture some of their rents? No one knows the answers to these questions exactly, but we can surmise that the answers are yes. Surely, it is the lifestyle and general level of ability and wealth of superstars that provide the incentive for thousands of youngsters to practice hard and to develop extraordinary talents. Over a sufficient period of time, new entry may erode the pure economic rents of superstars.

## Rents and Firms

Firms may also earn rents, and these rents are analogous to the firms' profits. In fact, the concept of economic rent as a return in excess of the opportunity cost of a factor of production is just another way to think about the profits earned by competitive and monopolistic firms. The purely competitive firm can earn quasi-rents, and the pure monopolist can earn monopoly rents.

### Quasi-Rents

**Quasi-rent:** The short-run payments to owners of capital in a competitive industry that exceed the opportunity cost of capital.

**Quasi-rents** are the return to the owner of a competitive firm in the short run. They are called quasi-rents because they are a short-run or temporary return. Figure 15–7 illustrates a short-run equilibrium for the competitive firm and industry. The firm earns an economic profit in the short run, which we have labeled as quasi-rent. This return is a rent because it exceeds the firm owner's opportunity cost (which by definition is included in the average total cost curve). Quasi-rents are temporary; entry of new firms into the industry will erode these returns over the long run. Keep in mind that quasi-rents can also be negative instead of positive, for the firm can suffer economic losses in the short run. The rents themselves are a return to the owner of the firm for decision making under conditions of risk and uncertainty.

### Monopoly Rents

**Monopoly rent:** The payments to owners of capital in a monopolized industry that exceed the opportunity cost of capital.

Figure 15–8 shows the case of a monopolist who makes profits of *PABC*. These profits are sometimes called **monopoly rents** because they are a return in excess of the monopolist's opportunity cost, which is reflected in the average total cost curve. Unlike the quasi-rents earned by the competitive firm, the monopoly rents earned by the monopolist will persist over time. They are permanent, short of an event that would threaten the monopolist's control of an industry. But what happens if the monopolist sells the monopoly? Primarily, the monopolist will sell for the present value of future monopoly rents, as discussed in Chapter 10. Monopoly rents are therefore captured by the original owner of the monopoly; buyers of the monopoly right may earn only a competitive rate of return on their purchase of the monopoly.

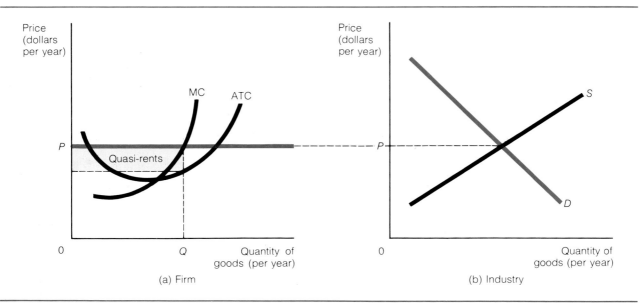

**FIGURE 15–7  Quasi-Rents**

Quasi-rents are illustrated as the economic profits earned by the owner of a competitive firm in the short run. They are rents because they are a return in excess of the owner's opportunity cost, included in the ATC curve. They are called quasi-rents because they are temporary and will be eroded by the entry of new firms in the long run.

**FIGURE 15–8**

**Monopoly Rents**

Monopoly profits can be called monopoly rents. They are returns in excess of the monopolist's opportunity costs, which are reflected in the ATC curve. Unlike quasi-rents, they persist over time because of the monopolist's dominant position in the market.

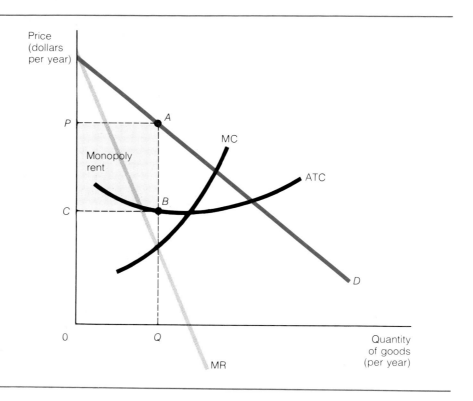

## Rent Seeking

When competition is viewed as a dynamic, value-creating, evolutionary process, economic rents play a crucial role in stimulating entrepreneurial decisions and in prompting an efficient allocation of resources. Profit seeking is a normal feature of economic life in a competitive market. The returns of resource owners will be driven to normal levels by competitive profit seeking as some resource owners earn positive rents that promote entry into the industry and others earn negative rents that cause exit. In this basic sense, profit seeking in a competitive setting is socially beneficial.

When the setting in which profits are pursued is changed, sometimes through government intervention, problems can arise. Suppose the government wishes to grant a monopoly right in the production of playing cards. In this case an artificial scarcity is created by the state. As a consequence, monopoly profits or rents are present to be captured by monopolists who seek the government's favor. In a naturally occurring monopoly these rents are thought of as transfers from playing card consumers to the monopolist. Yet this can be the case only if the aspiring monopolists employ no resources to compete for the monopoly rents. To the extent that resources are used to capture monopoly rents by lobbying the government, these expenditures create no value from a social point of view. This activity of wasting resources in competing for artificially contrived transfers is called **rent seeking**. If a potential monopolist hires a lawyer to lobby the government for the monopoly right, the opportunity cost of the lawyer (such as the contracts that he or she does not write while engaged in lobbying) is a social cost of the monopolization process.

**Rent seeking:** The process of spending resources in an effort to obtain an economic transfer.

The economic wastefulness of rent-seeking activity is difficult to escape once an artificial scarcity has been created. At one level the government can allow individuals to compete for the playing card monopoly and waste resources through such activities as bribery. Such outright corruption is perhaps the simplest and most readily understood level of rent seeking. At a second level the state could sell the monopoly right to the highest bidder and put the proceeds at the disposal of government officials. In such a case the monopoly rents may show up in the wages of state officials; to capture rents at this level individuals will compete to become civil servants. At still another level, the monopoly right may be sold to the highest bidder and the resources dispersed through the state budget in terms of expenditure increases and/or tax reductions. Even in this case, rent-seeking costs will be incurred as individuals seek to become members of the groups favored by the tax-expenditure program. Rent-seeking costs are incurred in any of these options, and the form that such costs take is influenced by how the government transacts its business in creating artificial scarcity.

## Profits and the Allocation of Resources

As we saw in the discussion of quasi-rents and monopoly rents, rents and profits are closely related concepts. The word *profits*, however, is the term that we usually apply to the earnings of owners of firms. Profits are the last category of input returns in the circular flow process that we analyze. As we will see, profits are a residual return to those individuals in the economy who provide economic foresight and leadership in an uncertain environment. Though we discuss the role of profits in the resource allocation process last, this does not mean that their role is small. To the contrary, profits are a very important reason that the competitive price system works to allocate resources to their most highly valued uses.

### Profit Data

For individual firms, industries, or the economy as a whole, profits are not computed directly but are determined by what is left over from revenues after payments have been made to land, labor, and capital. It is in this sense that the profit earner, or the productive factor to which profits are imputed, is referred to as the *residual claimant*. This view of profits as a residual contrasts sharply with the incorrect notion that employees and other input suppliers are paid only after the entrepreneur has secured profits.

Have levels of profits earned in the economy as a whole grown over the years? Table 15–1 traces after-tax corporate profits and proprietors' income for selected years between 1929 and 1984. Even when taking inflation into account, the absolute level of corporate profits has risen quite steadily over this period, but when measured as a percent of GNP the corporate profit level has remained remarkably stable, ranging between 3 and 5.5 percent

TABLE 15–1  Corporate Profits and Proprietors' Income, 1929–1984

| | Corporate Profits | | Proprietors' Income | |
|---|---|---|---|---|
| Year | After Tax (billions of 1972 dollars) | Percent of GNP | After Tax (billions of 1972 dollars) | Percent of GNP |
| 1929 | 22.3 | 7.1 | 45.8 | 14.5 |
| 1933 | 0.4 | .2 | 23.5 | 10.6 |
| 1939 | 16.5 | 5.2 | 41.5 | 13.0 |
| 1940 | 21.0 | 6.1 | 44.7 | 13.0 |
| 1945 | 20.0 | 3.6 | 83.9 | 15.0 |
| 1950 | 40.3 | 7.5 | 72.3 | 13.5 |
| 1955 | 35.8 | 5.4 | 70.5 | 10.7 |
| 1960 | 29.8 | 4.0 | 68.7 | 9.3 |
| 1965 | 51.1 | 5.5 | 76.5 | 8.2 |
| 1966 | 53.2 | 5.4 | 78.8 | 8.0 |
| 1967 | 48.8 | 4.8 | 77.4 | 7.7 |
| 1968 | 47.9 | 4.5 | 77.5 | 7.3 |
| 1969 | 41.7 | 3.8 | 77.2 | 7.1 |
| 1970 | 32.6 | 3.0 | 72.4 | 6.7 |
| 1971 | 37.1 | 3.3 | 72.3 | 6.4 |
| 1972 | 43.0 | 3.6 | 76.9 | 6.5 |
| 1973 | 53.0 | 4.2 | 88.7 | 7.1 |
| 1974 | 55.1 | 4.4 | 77.2 | 6.2 |
| 1975 | 52.6 | 4.3 | 71.7 | 5.8 |
| 1976 | 62.3 | 4.8 | 71.2 | 5.5 |
| 1977 | 67.3 | 4.9 | 74.0 | 5.4 |
| 1978 | 71.3 | 5.0 | 78.0 | 5.4 |
| 1979 | 76.0 | 5.1 | 80.8 | 5.4 |
| 1980 | 68.0 | 5.0 | 73.6 | 5.0 |
| 1981 | 57.98 | 3.8 | 69.4 | 4.6 |
| 1982 | 35.15 | 2.6 | 53.57 | 3.6 |
| 1983 | 42.17 | 2.8 | 56.52 | 3.7 |
| 1984[a] | 50.45 | 3.1 | 69.25 | 4.2 |

[a]preliminary

Source: *Economic Report of the President* (Washington, D.C.: U.S. Government Printing Office, 1985), pp. 246, 256.

since the mid-1950s. In contrast, proprietors' income exhibits a strong and continuous downward trend as a percent of GNP, while showing little change from year to year in constant dollar terms. Overall, the Table 15–1 data fail to confirm the widely held perception that profits have been growing steadily over time.

### Rate of Return on Capital

The profit data given in Table 15–1 are not relevant to the decisions of firms concerning investments in plant and equipment. What matters for such purposes is not the absolute level of profits but the economic **rate of return on invested capital**. That is, at the margin it will pay to buy an additional dollar's worth of equipment only if the firm expects to earn more than a dollar from the sale of the resulting output from the equipment.

The economic rate of return on capital is approximated in Table 15–2 with accounting data by dividing the amount or stock of capital valued at its current replacement cost into profits earned during the year. The data indicate that after-tax rates of return on capital have fallen over time. The average annual rate of return between 1955 and 1969 was 10.1 percent, with

> **Rate of return on invested capital:** Profits that are measured as a percentage of the costs of capital.

**TABLE 15–2** Rates of Return on Depreciable Assets, Nonfinancial Corporations, 1955–1982

| | Rate of Return (percent) | |
|---|---|---|
| Year | Before Tax | After Tax |
| 1955 | 19.8 | 9.8 |
| 1956 | 16.8 | 7.9 |
| 1957 | 15.2 | 7.4 |
| 1958 | 12.8 | 6.5 |
| 1959 | 16.4 | 8.5 |
| 1960 | 15.0 | 8.0 |
| 1961 | 15.1 | 8.2 |
| 1962 | 17.4 | 10.3 |
| 1963 | 18.8 | 11.2 |
| 1964 | 20.2 | 12.5 |
| 1965 | 22.1 | 14.0 |
| 1966 | 21.8 | 13.7 |
| 1967 | 19.3 | 12.4 |
| 1968 | 18.9 | 11.3 |
| 1969 | 16.5 | 9.7 |
| 1970 | 12.8 | 7.9 |
| 1971 | 13.5 | 8.5 |
| 1972 | 14.3 | 9.1 |
| 1973 | 14.3 | 8.7 |
| 1974 | 11.0 | 6.1 |
| 1975 | 11.9 | 7.7 |
| 1976 | 12.9 | 7.9 |
| 1977 | 13.7 | 8.6 |
| 1978 | 13.3 | 8.2 |
| 1979 | 12.3 | 7.6 |
| 1980 | 10.7 | 6.9 |
| 1981 | 11.0 | 7.7 |
| 1982[a] | 9.5 | 7.5 |

[a] preliminary
Source: *Economic Report of the President* (Washington, D.C.: U.S. Government Printing Office, 1983).

the highest rates occurring during the 1960s. By contrast, the average annual rate of return on capital between 1970 and 1981 was 7.9 percent. This decline in rates of return has two important implications. The first involves incentives to make capital investments. For investments in capital to be profitable, not only must the rate of return be positive, but the rate of return on capital must exceed the expected return on alternative investments. Since a larger number of competing uses of funds—buying bonds, for example—become attractive as the rate of return on capital falls, the data in Table 15–2 suggest that incentives to invest in plant and equipment during the 1970s were lower than in earlier years. Second, the data provide further evidence that the popular view of large and ever-increasing corporate profitability is mistaken.

### Accounting Versus Economic Profits

**Accounting profit:** Total revenue minus total explicit money expenditures.

**Economic profit:** Total revenue minus total opportunity costs.

It is important to note that **accounting profits** such as those in Table 15–2 tend to overstate the level of **economic profits** because accountants by necessity take a less theoretical view of costs than do economists. Specifically, accountants count as costs only those business expenditures involving direct money outlays—wages and salaries, raw materials purchased, rental payments on land and buildings, borrowing costs, advertising expenses, and so forth. To economists, costs are opportunities forgone, whether explicit or implicit. For example, what is the economic cost of the company-owned land on which corporate headquarters sits? It is the revenue forgone from putting the land to its next-best use, perhaps by leasing it to a shopping center developer. But the accounting cost of company-owned land will generally be zero because no explicit money payments are presently made for the land by the firm. Thus, because the accounting concept of costs tends to understate economic costs, accounting profits overstate economic profits.

Accounting profits also tend to be misleading because cost is a subjective, forward-looking concept. That is, insofar as they influence economic decisions, costs are based on anticipations about the future that may or may not materialize. The cost of a particular decision will therefore vary depending on when in time cost is measured. Moreover, what the decision maker thinks is being sacrificed by a choice cannot be directly observed by someone other than the decision maker. We are left with the conclusion that accounting profits, based as they are on a restricted concept of costs, provide only a limited guide to the actual level of economic profits in the economy.

### Profits and Entrepreneurship

**Entrepreneur:** An individual who organizes resources into productive ventures and assumes the uncertain status of a residual claimant in the resulting economic outcome.

Assuming that economic profits are capable of being measured, what role do they play in the economic process? Economic profits are the return to a particular factor of production, **entrepreneurship.** As such, profits are the residual income accruing to the entrepreneur as a return on the services he or she brings to the productive process. These services include technical abilities in organizing the other factors of production into combinations appropriate for the efficient manufacture of goods. More important, though, is the entrepreneur's alertness to the existence of potential profit opportunities in the economy. In this sense entrepreneurship consists of linking markets by perceiving the opportunity to buy resources at a lower total cost than the revenue obtainable from the sale of output or by recognizing that sellers in one market are offering to sell output at a price lower than buyers in some other market are willing to pay.

Put another way, entrepreneurship consists of offering the most attractive opportunities to other market participants—perceiving unfulfilled demands, offering higher prices to sellers or lower prices to buyers, improving existing goods or making them more cheaply, finding more effective means of communicating to consumers the availability and attributes of goods. In all of these activities the entrepreneur takes advantage of opportunities that exist because of the initial ignorance of other market participants. But as the result of the entrepreneur's actions, markets move closer to the prices and quantities emerging in equilibrium, the plans of buyers and sellers more closely dovetail, and the knowledge of economic data held by market participants is increased.[1]

In addition to alertness and organizational skills, the entrepreneur brings to the production process a willingness to act in the presence of uncertainty. (We distinguish risk from uncertainty in that risk involves events that occur with known probabilities—the toss of a coin, for example—while uncertainty entails outcomes whose probabilities cannot be specified with precision—for example, war or peace, long-run weather forecasts, and so on.) There is no guarantee that perceived profit opportunities will in fact materialize. Because it takes time between the purchase of inputs and the sale of output—between the expenditures on resources and the receipt of revenue—intervening events may cause the entrepreneur's plans to be either overambitious or underambitious when evaluated with the advantage of hindsight. Entrepreneurship thus entails the bearing of responsibility for incorrect anticipations. Profits can be viewed as a return to uncertainty-bearing.

In summary, the entrepreneur is the prime actor in the economic process, and entrepreneurship is characterized by technical skills in organizing production, alertness to the appearance of profit opportunities, and a willingness to act under uncertain conditions. Profits are just the payment to another factor of production, but they are different in that the entrepreneur is "paid" last, claiming the residual after the other factors have been compensated. Profits are the incentive spurring entrepreneurial activity in the economy. Focus, "Venture Capitalism," discusses a specific type of entrepreneurship.

## General Equilibrium

Our study of the essential parts of microeconomic theory is now complete. The basic tools of economic analysis—demand and supply, elasticity, marginal utility, and costs of production—were introduced in Chapters 1 through 8. The pricing and production of goods and services in a variety of market structures ranging from pure competition to pure monopoly were covered in Chapters 9 through 11. In Chapters 12 through 15, the elements of the theory of input markets were presented. We have analyzed each part of the economy in isolation from the other parts. This type of economic analysis is called *partial equilibrium analysis*. Changes in other parts of the economy are held constant while the functioning of a particular market or industry is analyzed. We are now in a position to gather some understanding of how all the parts work together. This study of the whole economy, where no part of the system is held in isolation from the others, is called *general equilibrium analysis*. The appendix to this chapter is devoted to a simple exposition of general equilibrium theory.

[1]The entrepreneur's role in the economic process is detailed in Israel Kirzner, *Competition and Entrepreneurship* (Chicago: University of Chicago Press, 1973).

## FOCUS  Venture Capitalism

Entrepreneurs are people with new ideas about how to make money. They see profit opportunities where others do not—in new products, different locations, and an almost endless number of other ways. An entrepreneur, in essence, makes conjectures about the economic future, and these conjectures typically entail very risky undertakings. Since not all entrepreneurs can finance their dreams, who provides the financial backing for their proposals?

The answer is a venture capitalist, a person who supplies capital to fund new investment ideas by entrepreneurs. Some venture capitalists are wealthy individuals, and others are subsidiaries or divisions of large firms or financial institutions. Each year, billions of dollars are invested by venture capitalists in the new ideas of entrepreneurs who cannot find support for their projects through alternative channels.

Venture capitalists must exercise great care and caution in evaluating the proposals that come to them. For

*Source: Wall Street Journal, December 19, 1983, p. 27.*

example, Citicorp Venture Capital Ltd. has $100 million in capital to invest and each year receives about 3,000 proposals. The proposals range from the interesting (a new concept in toilet seat sanitizing) to the vague (a proposal with no details except the address to which to send $150,000 to open a supermarket). Obviously, most venture projects are quite risky, and the venture capitalist seeks a high rate of return on the investment in such projects as well as a large measure of control in the undertaking.

There are no systematic data on the type of projects that tend to be funded, but it appears that a ripe source of entrepreneurial proposals come from operating executives in established firms who are seeking to establish their own businesses. These individuals can typically provide some of the finance for their proposals, and they have credibility with the venture capitalist because of their industrial experience.

Venture capitalism is an important part of the economy. In a way, it finances the progress that the economy makes over time.

## Summary

1. An economic rent is a return in excess of the opportunity cost of an owner of a resource.
2. Land in the aggregate is fixed and has a perfectly inelastic supply curve. Its entire return is a pure economic rent determined by the level of demand for land as a factor of production.
3. The supply of land to competing uses is not fixed. Land must be bid away from competing uses, so the supply curve of land to a specific use is upward sloping.
4. Proposals to tax land rents away would be costly to the economy. Land is like any other factor of production. It has alternative uses, and it is used in the production process by entrepreneurs and firms up to the point where its marginal revenue product equals its marginal cost.
5. Specialized resources earn economic rents. The degree to which the supply curve of a factor of production is upward sloping reflects the degree to which it is specialized. Where all units of a resource earn the same market price, the more specialized units will earn economic rents. These rents are called inframarginal rents because all units of the resource except the very last, or marginal, unit earn rents.
6. Superstars earn economic rents because their talents are rare gifts that cannot easily be duplicated. The supply curve of superstar services is vertical, and a great deal of the income of a superstar will consist of pure economic rents.
7. Another way to look at the profits of competitive and monopolistic firms is through the concept of economic rent. Competitive firms earn quasi-rents, or returns in excess of the entrepreneur's opportunity cost. Monopolists can earn monopoly rents, which are equivalent to monopoly profits.
8. Rent seeking is the process of competing for artificially contrived rents normally created through government intervention in the economy. From society's point of view, expenditures to capture such transfers are wasted.
9. Economic profits are a residual that the entrepreneur receives after the other factors of production have been paid. Profits can be positive or negative (or zero), and the receivers of profits are called residual claimants. Although the absolute level of profits in the economy has grown over time, the rate of return on capital has not. It has declined in recent years. The rate of return on capital is the crucial determinant of firm profitability.
10. Accounting profits differ from economic profits. Typically, accounting procedures understate economic costs and thus overstate economic profits.
11. Profits are the force in the economy that leads to

entrepreneurship. They are a return to such activities as decision making under conditions of risk and uncertainty. The concept of entrepreneurship has many dimensions, but perhaps the most basic characteristic is that of being alert to economic opportunities.

## Key Terms

rent
pure economic rent
inframarginal rent
aggregate supply of land
resource specialization
quasi-rent
monopoly rent
rent seeking
rate of return on invested capital
accounting profit
economic profit
entrepreneur

## Questions for Review and Discussion

1. Distinguish between pure and inframarginal economic rents.
2. At what point in the production process do residual claimants receive their return? Can their return be negative?
3. How does rent seeking lead to the waste of resources?
4. What are the essential things that an entrepreneur does?
5. There is a fixed supply of oil in the world. Is the return to oil producers therefore a pure economic rent? Explain carefully.
6. "Land rent should be completely taxed away." Evaluate.
7. Suppose that your whole wage was a rent. What implications would this situation have for your behavior where you work? Would you be cantankerous with your superiors or docile and easy to get along with? Why?
8. Explain why a firm making zero economic profit will probably stay in business.
9. Are you earning economic rents as a college student? For example, suppose your college raised tuition. Would you drop out of school or transfer to another school?
10. Suppose several cable television companies were each attempting to obtain a franchise from your city government. Only one company would obtain the exclusive rights to provide cable services. How much would each company be willing to spend in an effort to obtain the franchise? What is the term that describes their activity, and what is the term that describes the winning firm's profits?
11. Suppose a firm is earning economic profits that are greater than the profits of most other firms in the industry because it has a superior business manager. In the long run what happens to the manager's salary? Also, what happens to the firm's production costs and profits?

## ECONOMICS IN ACTION  Computer Entrepreneurs

In the view of many economists, entrepreneurs play a vital role in the growth of an economy. By their willingness to take risks, their inner drive to succeed, and their ability to identify new products and markets, entrepreneurs often spark sudden breakthroughs in technology and reap large fortunes as a result.

The computer revolution of the 1970s and 1980s is a case in point of the role of the entrepreneur. Many of the men and women who saw early on the potential of the personal computer for use in the home, office, or classroom are now managing corporations that provide goods and services undreamed of fifteen years ago. But not everyone who took risks in developing computer products succeeded. As sizeable profits attracted new entrants into computer markets, competition intensified and pricing and marketing strategies became the only means of survival for many. By the early 1980s, news of bankruptcies in America's "silicon valleys" became as common as the new products themselves.

Steven Jobs and Jack Tramiel are two of the most celebrated computer entrepreneurs. Jobs, a 1972 college dropout who spent several years tinkering with computer technology in his parents' garage with his friend and fellow "hacker" Stephen Wozniak, found the means to raise money and market the Apple computer, the first successful personal computer. Jobs's first capital—$1300—came from the sale of his Volkswagen bus. Today Jobs is the Chairman of the Macintosh division of Apple Computer, Inc., whose revenues top $1 billion. He and Wozniak have both become multimillionaires.

Jack Tramiel's career spans many more years than

*Steven P. Jobs*

*Jack Tramiel*

Jobs's, and he has overcome many more obstacles to success. Often described as a somewhat ruthless, intensely driven man, Tramiel survived the Nazi holocaust in Poland, emigrated to the United States, opened a typewriter repair shop, and built it into Commodore International, another billion dollar company that today dominates the home computer market. (Apple's more expensive models have been much more successful with small businesses and schools.) Tramiel has seldom been afraid of taking risks. In 1968 he became one of the first marketers of hand-held calculators, but he soon lost most of his business to Texas Instruments, as prices for the calculators dropped almost tenfold in less than 5 years. He retaliated by introducing the PET tabletop computer (later the VIC-20) and became so aggressive in marketing and pricing his products that he drove several competitors, including Texas Instruments, out of these markets. In 1984 Tramiel abruptly resigned his post at Commodore and purchased the moribund Atari company, another computer manufacturer that had suffered huge losses in competition with Commodore. He is now Chairman of the Board at Atari. Many observers believe that Tramiel is bent upon removing his former company, Commodore, from its profitable niche.

Strong egos and intense competitive desire are common traits of entrepreneurs such as Jobs and Tramiel.

(Tramiel is often quoted as saying, "Business is not a sport. It's a war.") Aside from their fascinating psychological profiles, however, entrepreneurs are important in a strictly economic sense. In *The Spirit of Enterprise*, a book about entrepreneurial behavior, author George Gilder writes,

> The capitalist is not merely a dependent of capital, labor, and land; he defines and creates capital, lends value to land, and offers his own labor. . . . He is not chiefly a tool of markets but a maker of markets; not a scout of opportunity, but a developer of opportunity; not an optimizer of resources, but an inventor of them; not a respondent to existing demands but an innovator who evokes demand; not chiefly a user of technology but a producer of it. He does not operate within a limited sphere of market disequilibria, marginal options, and incremental advances.[a]

## Question

How, specifically, would you characterize the role of the entrepreneur in society? How is the role of the entrepreneur related to risk taking? Does the entrepreneur provide only economic benefits for herself or himself?

[a]George Gilder, *The Spirit of Enterprise* (New York: Simon and Schuster, 1984), pp. 16–17.

# APPENDIX
## Putting the Pieces Together: General Equilibrium in Competitive Markets

At this point each of the many parts of microeconomics has been introduced and explored. Studying each individually can add a great deal to our understanding of many events that occur in the economy. However, bringing all the parts together can create an overall picture that may help answer the "what," "how," and "for whom" questions introduced in the first chapter of the book. To illustrate, we can describe a situation of general equilibrium and then create a change in an economic variable, tracing its effects throughout the economy.

Consider an economy that has only two goods, wheat and coal. All citizens spend their income on these two goods in a manner that maximizes utility. With limited incomes and diminishing marginal utility, downward-sloping demand curves result. In Figure 15–9a and 15–9b, which represents supply and demand conditions in the output markets for wheat and coal, $D_{w1}$ and $D_{c1}$ are the current demand curves for wheat and coal, respectively.

Firms that maximize profits and have diminishing returns will have upward-sloping supply curves for the two goods. $S_{w1}$ and $S_{c1}$, reflecting the current supply curves for wheat and coal.

The supplies of farm workers and coal miners are determined in part by each individual's maximizing of utility in occupational choice. In Figure 15–9c and 15–9d, the equilibrium wage rates $W_{f1}$ and $W_{m1}$ for farm workers and coal miners are determined by competitive forces. There may be a difference in these two wage rates for a variety of reasons. For example, if coal mining is a skilled craft that requires a long training period or if coal mining is especially dangerous, in the long run the wages of coal miners must be greater than the wages of farm workers. The current quantities of labor employed in each industry are $L_{f1}$ and $L_{m1}$.

$P_{w1}$, $W_1$, $P_{c1}$, and $C_1$ represent the equilibrium prices and quantities of wheat and coal. In this long-run competitive equilibrium, profits are zero, each firm is forced to operate at minimum average cost, and price is equal to marginal cost. In other words, there is an optimal allocation of resources.

From this initial equilibrium, we can examine a disturbance in the economy. Suppose that consumers permanently decide to spend a larger portion of their income on wheat and less on coal. The demand for wheat shifts to the right to $D_{w2}$ in Figure 15–9a, and the demand for coal shifts to the left to $D_{c2}$ in Figure 15–9b. This action will lead to a series of short-run changes and eventually to a new long-run equilibrium.

In the short run the price of wheat will rise to $P_{w2}$ and the price of coal will fall to $P_{c2}$. These shifts will generate short-run profits for firms in wheat production and losses for firms in coal production. The demand for farm workers increases to $D_{f2}$, and the demand for coal miners decreases to $D_{m2}$. The wages for farm workers rise to $W_{f2}$, while the wages of coal miners fall to $W_{m2}$. If wages do not adjust to these new levels, there will be a shortage of farm workers and unemployment for coal miners.

As time goes by, firms and resources adjust to the changed conditions.

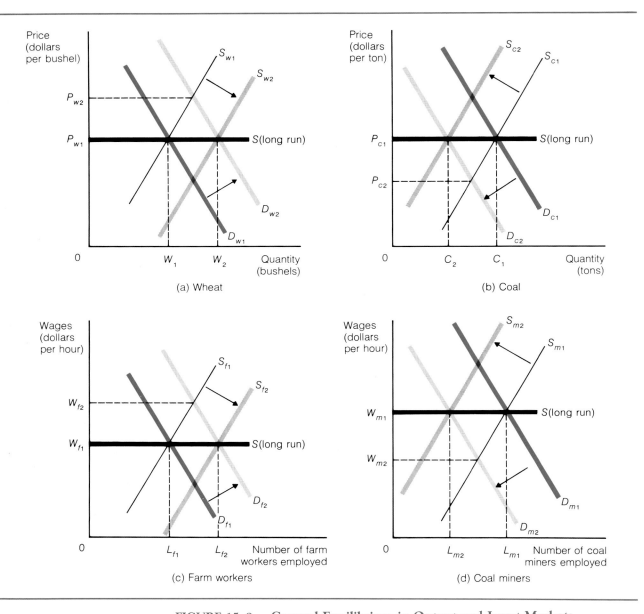

**FIGURE 15–9  General Equilibrium in Output and Input Markets**

The current demand curves for wheat and coal are given in (a) and (b) by $D_{w1}$ and $D_{c1}$. If consumers decide to spend more of their income on wheat and less on coal, the demand curve for wheat shifts to the right, from $D_{w1}$ to $D_{w2}$, and that for coal shifts to the left, from $D_{c1}$ to $D_{c2}$. The short-run price of wheat will rise from $P_{w1}$ to $P_{w2}$, and the short-run price of coal will fall from $P_{c1}$ to $P_{c2}$. In the long run the economy adjusts, and the equilibrium price and quantities become $P_{w1}$, $P_{c1}$, $W_2$, and $C_2$. The equilibrium wage rates $W_{f1}$ and $W_{m1}$ for farm workers (c) and coal miners (d), respectively, are determined by competitive forces when the demands for wheat and coal shift as in (a) and (b). The demand for farm workers increases, and the demand curve shifts from $D_{f1}$ to $D_{f2}$. The demand for coal miners decreases, and the demand curve shifts from $D_{m1}$ to $D_{m2}$. The wages of farm workers rise from $W_{f1}$ to $W_{f2}$ and those of coal miners fall from $W_{m1}$ to $W_{m2}$. When general equilibrium is reached in the economy, as in (a) and (b), the equilibrium quantities of farm workers and coal miners are $L_{f2}$, $W_{f1}$, $L_{m2}$, and $W_{m1}$.

The higher economic profits in the wheat industry will attract new firms. The losses in the coal industry will encourage firms to leave the industry. The supply of wheat will begin to increase, while the supply of coal decreases.

In the resource markets, shown in Figure 15–9c and 15–9d, let there be only one input, labor. Workers may choose to enter the wheat industry as farm workers or the coal industry as coal miners. The demands for farm workers and for coal miners are derived from the demands for wheat and coal. The higher wages in the wheat industry will attract workers, and the lower wages in the coal industry will discourage the entry of workers. The supply of farm workers increases while the supply of coal miners decreases.

If resources are able to shift, as they are in this case, the new long-run equilibrium will result in the former prices and wages. Profits will return to zero in both industries. The only permanent changes are in the quantities of wheat, coal, farm workers, and coal miners, $W_2$, $C_2$, $L_{f2}$, and $L_{m2}$, respectively. The long-run supplies of wheat, coal, and labor are infinitely elastic.

This example demonstrates the way in which resources are demand-directed. As consumers' tastes change from coal to wheat, resources shift from coal to wheat. Thus it is said that resources flow to their most highly valued use.

An understanding of what makes this process work is important. Self-interest is the motivating force of individuals; it is also the impelling force that drives the competitive market to the desired result. Self-interest is what attracts firms to profitable industries and leads them away from unprofitable industries. Self-interest is what attracts workers to occupations with high wages and leads them away from those with low wages. Without self-interest the market system would not produce the products that are most highly valued by society.

In actuality, there are many obstacles to long-run competitive equilibrium. Imperfect competition, externalities, or government intervention may inhibit the free flow of resources to the most desired goods.

In the previous example, what would be the result if there had been a monopoly in the wheat industry or a labor union for farm workers? Or what would have happened if government had taxed away the profits of wheat-producing firms and subsidized the coal industry? The most desired quantities of wheat and coal would not have been produced. While both natural and artificial barriers to optimal resource allocation do exist, this example should provide the basis for an understanding of the overall workings of the microeconomic system.

# 16

# The Distribution of Income

Are you rich or poor? In what sense? Diane, who is passionately interested in art, may be devoting a long and happy life to painting seascapes, earning no more than five or six hundred dollars a month. John, a stockbroker, may regard himself as one of the walking dead in spite of a six-figure income. Is Diane rich? Is John poor? Surely the question of whether one is rich or poor in an economic sense is different from whether one feels rich or poor in emotional well-being.

While most people, including economists, might argue that it is far better to be rich and unhappy than poor and unhappy, happiness cannot be measured in objective terms or easily be related to income. The equation of rich with happy or poor with unhappy, though tempting, is inadmissible in any scientific sense. Economists cannot measure happiness, but they can measure and analyze differences in income.

In studying income distribution, economists look at how and why personal revenue is measured and divided among members of a society—without any implications for the members' relative happiness. Economists seek to explain a given pattern of income distribution as the product of such factors as individual choice, socioeconomic discrimination, and government policies. More important, they go beyond the facts and look to the consequences of policy changes. What are the economic effects of a distribution of income determined by free-market forces? If society, working through the political process, determines that distribution to be unjust, what are the effects on incentives and economic growth of altering the distribution produced through market forces? These and similar important questions are discussed in this chapter. Before turning to these larger issues, consider the simpler question, Why are individuals rich or poor?

# The Individual and Income Distribution

**Individual income:** The sum of labor income, asset income, and government subsidies minus tax payments.

The economic question of why one is rich or poor has many facets. Individuals accumulate income from their labor and entrepreneurial skills, from the ownership of assets such as land, or from the receipt of cash or government subsidies called transfer payments. At the same time, most individuals lose a certain portion of income through taxes. We will analyze each factor of **individual income** briefly.

## Labor Income

**Labor income:** The payments an individual receives from supplying labor, equal to the individual's wage rate times the number of hours of labor supplied.

**Labor income**—income received from supplying labor—is the wage rate received multiplied by the number of hours worked. The supply of labor is by far the largest source of individual income. The supply of labor ordinarily varies according to wage rates, but individuals also differ in their choice of work hours over leisure hours, and wages themselves vary for a number of reasons.

**Choices of Work Versus Leisure.** Different people with the same skills will often supply different quantities of labor at identical wage rates. The same wage rate affects people differently because they can make choices in the trade-off between work and leisure. Harpo, who has the same skills as Gummo, chooses to play his harp six hours a day for no pay and to work two hours a day for pay. Gummo chooses work and income over leisure, working twelve or fourteen hours a day. The terms *lazy* and *industrious* are certainly relative, but these personality traits do not totally explain different attitudes toward work among those with identical skills.

What else accounts for differences in attitudes toward work? No human being is a machine who can supply labor without limit. At higher and higher wage rates and longer and longer hours worked, we all become leisure lovers (see Chapter 12 on this point). As we choose more work, the costs of losing more and more leisure rise and the rewards of earning more income fall.

The explanation for income-leisure trade-offs also relates to differences in incentives to work. Incentives to work—willingness to supply labor at a given wage rate—are ordinarily thought to result from different doses of the Protestant ethic of industriousness. The economist must accept this ethic as given and point out that, for some, leisure is freely chosen over income. Unfortunately, the choice of income or leisure is not freely open to all. Disabled persons, the sick, the old, and the unskilled may have few or no choices. Society often compensates by making transfer payments to such individuals, providing them with an income. We will have more to say about such government activities later in the chapter.

**Differences in Wage Rates.** Wage rates vary widely among occupations. The wage rate you earn is largely a function of the human capital you have built up through education and investments in other skills. Human capital is a major reason why you are ultimately rich or poor. Level of schooling attained (a college degree, a high school diploma) is a ticket to enter some occupations. Given government enforced regulations of the American Medical Association, entry into the physicians' market without a medical degree would almost certainly land you in jail. It is worth noting that if education level and achievement are related to the income class one initially comes from, income differences may be somewhat self-perpetuating over time.

Luck also pays a role in the wage rate. At the beginning of the energy crisis in the early 1970s, those entering the labor market with degrees in petroleum engineering received much higher wages than under previous conditions. (The situation was short-lived, of course, since more and more college students were attracted to petroleum engineering by high wages, and, as usual, supply expanded and wages fell.) Others are naturally endowed with rare talents or physical attributes and are rewarded accordingly. P. T. Barnum's midget Tom Thumb was paid a handsome income simply for being abnormally small. For all of us, it is far better to be "lucky than good."

Individuals' or families' incomes are also determined by their stage in the life cycle. Law students at Harvard, Stanford, or the University of Michigan may be poor—students everywhere tend to be poor—but they will not always be poor. Income or earnings tend to rise through the life cycle, ordinarily reaching a peak between the mid-40s and the mid-50s and falling off thereafter. Other factors also influence the wage rate an individual receives. Without question, race and sex discrimination have been ugly practices affecting wage rates. We look at this matter in more detail later in the chapter.

Broadly speaking, then, an individual's income is determined by his or her choices in the trade-off between work and leisure and by the individual's stock of human capital, current life cycle period, special skills, luck, and degree of discrimination. These factors are major determinants of whether one is rich or poor, but there are other factors related to income from savings and government benefits or taxation.

### Income from Savings and Other Assets

The other major source of income is the return from savings or other assets one owns or has accumulated. We may save as an ingrained habit. Our parents or culture may instill the ethic of frugality in us to "save for a rainy day" or to accumulate wealth or assets for other reasons. Motives for the sacrifice of current consumption are numerous: security, expected future gaps in income, education, retirement, large consumer expenditures, and so forth. In any case, the sacrifice of current consumption for future consumption is rewarded by an interest return or **asset income.**

*Asset income:* The income received from savings, capital investment, and land, all of which require forgone present consumption.

Many people save for future generations. Some current income earners set aside assets for their children's later use, which brings us to another major source of income. Inheritance of money or other assets is related to the luck of having had wealthy parents. Historically, free societies have permitted inheritance and protected the rights of income owners to bequeath gifts at death or while still alive. Most societies have nevertheless taxed the lucky recipients of inheritance on the grounds that such wealth was not due to their own productivity.

The returns from both savings and other wealth and asset accumulations also depend on one's skill at investing. Interest returns on all sorts of investments—monetary, real capital, or land—are determined by the costs of investing and the incentives to maximize income from investments. Small savers and investors often receive lower returns than large-scale investors partly because of the high costs of gathering information about markets.

### Taxes and Transfer Payments

*Ex ante distribution:* The distribution of income before the government transfer payments and taxes are taken into account.

A final source of income is government tax collections and benefit distributions. Economists distinguish between the **ex ante distribution** of income, the before-tax and transfer payments distribution, and the **ex post,** or after-tax and transfer payments, **distribution.** We all receive goods, services, and benefits from all levels of government, and we all pay taxes in one form or

*Ex post* **distribution:** The distribution of income after the government influences the disposable income of individuals with taxes and transfer payments.

**Transfer payment:** The transfer by government of income from one individual to another; it may take the form of cash or goods and services such as education, housing, health care, or transportation.

another. National defense, roads, and the local public swimming pool are some of the benefits; property taxes, sales taxes, and income taxes are some of the costs.

Most individuals are either net taxpayers or net benefit receivers. The determination of how rich or poor we are *ex post*, in other words, depends partly on our position with regard to taxes and benefits. The receipt of government services and **transfer payments**—such as direct welfare payments, food stamps, subsidized public housing, Social Security, and unemployment compensation—must be added to privately earned income. Transfer payments are all money or real goods or services transferred by government from one group in society (taxpayers) to other groups (transfer recipients). Tax payments of all kinds must likewise be subtracted to help explain *ex post* why we, as individuals, are rich or poor. We will see that most government statistics on income distribution reveal only *ex ante* money distributions of family or household income, but economists have attempted to make some *ex post* calculations as well.

To summarize, we are rich or poor as individuals because of three major factors: our wage income, our income from savings and other assets, and our position with respect to government taxes and benefits. Part of our position in income distribution depends on luck (health, genetic and financial inheritance, the socioeconomic position of our family), part depends on choice (incentives, investments in human capital and skills), and part depends on political decisions—taxes and public benefits—about the justice of *ex ante* income distribution. With these important distinctions in mind, we turn to the issue of how the government measures income and to the economic tools used to explain income distribution and how it changes over time.

## How Is Income Inequality Measured?

**Family income:** The sum of incomes earned by all members of a household.

The money income of families (often termed *households*), or **family income,** is calculated in *Current Population Reports* published by the U.S. Bureau of the Census. Table 16–1 shows family income in the United States for selected years between 1960 and 1982. Total family income is broken down into the percentage received by each fifth of the total number of families over these years, ranked from the lowest, or poorest, fifth to the highest, or

TABLE 16–1  Money Income of Families, Percent of Aggregate Income: 1960–1982

These data show an amazing uniformity in pretax income distribution over a twenty-year plus period. Special care must be used in interpreting the data, however.

| | Percent of Aggregate Income | | | | | |
|---|---|---|---|---|---|---|
| Families by Quintile | 1960 | 1965 | 1970 | 1975 | 1979 | 1982 |
| 1. Lowest fifth | 4.8 | 5.2 | 5.4 | 5.4 | 5.3 | 4.7 |
| 2. Second fifth | 12.2 | 12.2 | 12.2 | 11.8 | 11.6 | 11.2 |
| 3. Middle fifth | 17.8 | 17.8 | 17.6 | 17.6 | 17.5 | 17.1 |
| 4. Fourth fifth | 24.0 | 23.9 | 23.8 | 24.1 | 24.1 | 24.3 |
| 5. Highest fifth | 41.3 | 40.9 | 40.9 | 41.1 | 41.6 | 42.7 |
| Highest 5% | 15.9 | 15.5 | 15.6 | 15.5 | 15.7 | 16.0 |

*Source:* U.S. Bureau of the Census, *Current Population Reports*, in *Statistical Abstract of the United States*, various issues.

richest, fifth. (A fifth of a total distribution is called a *quintile*.) In addition, Table 16–1 gives the percentage of total money income earned by the top 5 percent of income earners between 1960 and 1982.

### Interpretation of Money Distribution Data

We are struck immediately by a feature of Table 16–1: *The distribution of reported money income did not change significantly between 1960 and 1982.* Reading horizontally across the table, we see that the shares for each quintile have stayed roughly the same. The lowest quintile still receives little more than 5 percent of aggregate income, while the highest quintile consistently gets the lion's share: over 40 percent. Before accepting these figures at face value, we should note some limitations of the statistics in Table 16–1:

1. Money income statistics reported by the Bureau of the Census do not conform to the economist's definition of *income*. Income, in the economist's sense, is what an individual accumulates from labor, assets, or transfer payments minus what is paid in taxes. By contrast, the *Current Population* statistics on income omit all transfer payments, all assets such as capital gains, and all income and payroll taxes.
2. The government statistics report family, not individual, income. Therefore, an increase in the number of family units, income remaining the same, will alter the reported income distribution. In the period 1960–1982, the number of family units increased significantly through later marriages, high divorce rates, and increasing independence of the elderly. The government's statistics do not take this increase into account.

These two deficiencies in the Census data require great care in drawing conclusions about income distribution from the raw statistics. Refinements must be made before concluding anything about the distribution of economic welfare. Before actually making some of these refinements, we turn to the tools with which economists deal with income distribution.

### Economists' Measures of Income Inequality

Economists have developed useful methods for describing and analyzing income distribution. The primary tool for measuring income distribution is called a Lorenz curve, developed in 1905 by M. O. Lorenz. A second device derived from the Lorenz curve is also named for its inventor: the Gini coefficient, developed by Corrado Gini in 1936.

Lorenz curve: A graph that shows the cumulative distribution of family income by comparing the actual distribution to the line of perfect equality.

**The Lorenz Curve.** A **Lorenz curve** plots the relation between the percentage of families receiving income and the cumulative percentage of aggregate family income. To understand the Lorenz curve, consider a hypothetical income distribution for some country in the year 2020, as shown in Table 16–2. Table 16–2 tells us that the lowest 20 percent of family income earners receive only 5 percent of total income. The next 20 percent of families get 15 percent, and the lowest 40 percent receive a cumulative percentage share of 20 percent. The middle 20 percent get 20 percent of total income, indicating that the lowest 60 percent of income earners receive only 40 percent of total income. Obviously, 100 percent of income recipients receive a cumulative total of 100 percent of income.

The data of Table 16–2 may be translated into graphic form. The vertical axis of Figure 16–1 represents the cumulative percentage of family income, and the horizontal axis shows the percentage of income-earning families. If all families earned equal incomes, the Lorenz curve would be a straight line.

## How Is Income Inequality Measured? 349

**TABLE 16–2  Hypothetical Income Data for the Year 2020**

The second column shows the percentage share of total income received by each quintile of families in this hypothetical country; the third column adds these percentage shares cumulatively. For instance, since the lowest fifth receives only 5 percent of total income and the second fifth receives 15 percent, together these lowest two fifths receive a cumulative share of only 20 percent of total income. The lowest four fifths receive a cumulative share of 65 percent. Data from the third column are plotted on a Lorenz curve in Figure 16–1.

| | Aggregate Income | |
|---|---|---|
| All Families by Quintile | Percent Share (year 2020) | Cumulative Percent Share (year 2020) |
| A. Lowest fifth | 5 | 5    20% |
| B. Second fifth | 15 | 20   40% |
| C. Middle fifth | 20 | 40   60% |
| D. Fourth fifth | 25 | 65   80% |
| E. Highest fifth | 35 | 100  100% |

Ten percent of all families would earn 10 percent of income; 80 percent of families would earn 80 percent; and so on. Such a curve would actually be a diagonal cutting the square in half, as represented by the line of perfect equality in Figure 16–1.

A Lorenz curve is constructed from the hypothetical data of Table 16–2. Point A in Figure 16–1 corresponds to line A in Table 16–2—it is the

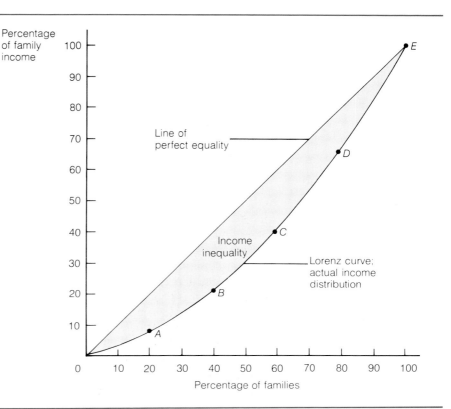

**FIGURE 16–1**

**A Hypothetical Lorenz Curve for the Year 2020**

The Lorenz curve shows the cumulative percentage of family income earned by percentages of families. The Lorenz curve indicates that 40 percent of families earned only 20 percent of total income, while 80 percent earned a cumulative 65 percent. This distribution is unequal and falls short of the line of perfect equality. The shaded area in the diagram indicates the degree of income inequality. The farther the Lorenz curve moves away from the line of perfect equality, the less equal the income distribution.

intersection of the lowest 20 percent of families and their cumulative percentage share of total income. Point *D* in Figure 16–1 shows that 80 percent of families (the fourth fifth in Table 16–2) received 65 percent of total income. The combination of points *A*, *B*, *C*, *D*, and *E* forms a Lorenz curve.

The distance of the Lorenz curve from the line of perfect equality is a measure of income inequality. The shaded area of Figure 16–1 is a measure of the degree of income inequality for our hypothetical society. Should the shaded area become larger or more bowed, money income would be more unevenly distributed. A smaller area or a flatter curve would signal a more equal income distribution.

**The Gini Coefficient.** An economic tool related to the Lorenz curve is the Gini coefficient. A **Gini coefficient** is the ratio of the shaded area of Figure 16–1 to the total area under the line of perfect equality. The Gini coefficient is simply a number between zero and one. The smaller the number—the closer it is to zero—the more equal the income distribution. A larger Gini coefficient, resulting from a more bowed Lorenz curve, means that income is less evenly distributed. A Gini coefficient of 0.30 for a certain year would indicate the presence of greater income inequality than a year with a coefficient of 0.20. Care must be taken in interpreting both Lorenz curves and Gini coefficients.

> Gini coefficient: A numerical estimation of the degree of inequality of the distribution of family income.

## Problems with Lorenz and Gini Measures

We have already seen some of the limitations associated with government statistics of money income distribution. Such data are not adjusted for taxes, capital gains, or many kinds of income transfers from government to families. Since the Lorenz and Gini measures are based on these statistics, we must be especially careful in using these instruments. In addition, the Lorenz and Gini measures have limitations of their own:

1. The Lorenz and Gini income distribution measures are simply mechanical devices and not precise indicators of the relative level of wealth in society. Economists are particularly interested in the effects of policies and programs on income distribution. Yet a change in the Gini coefficient from, say, 0.30 in 1990 to 0.28 in 1995 may not give an accurate enough indication of how distribution has changed. While the lower number does mean more equality overall, it may be that the poorest 20 percent of families received a lower percentage of income and the richest 20 percent received a greater percentage. If economists are concerned with policies to raise the economic welfare at the lower end of the distribution, the mechanically calculated Gini number is a very misleading indicator of social welfare changes. The actual Gini ratios of 0.376 in 1947 and 0.364 in 1977 tell us little of what the economist wants to know about the component shares in income distribution.

2. Important microeconomic effects may be hidden or neglected by a measure such as the Lorenz curve. Suppose, for example, that policies designed to reduce income inequalities at the top 20 percent are instituted through higher taxes for the rich. The initial *ex ante* effect would be a reduction in incomes of executives and others earning very high salaries and wages. The Lorenz curve would move closer to the line of perfect equality. But *ex post* factors must also be considered. High income earners might decide to work less, and the labor supply in these areas of the economy would diminish. A reduced supply, given demand conditions,

yields higher future returns to workers. The new high returns also signal new entrants into these areas of the economy. These factors may create a new *ex post* bowing out of the Lorenz curve, meaning that the original redistribution policies did not have the ultimate effect of reducing inequalities in income distribution at the upper end.

While changes in both the Lorenz and the Gini measures are suspect for reasons such as these, they remain two of the essential economic tools to measure economic well-being. They are useful so long as we keep their limitations in mind. More important, some of their limitations have been lessened by refinements attached to them by economists interested in problems of economic welfare. We turn to some of these refinements in the following section.

## Income Distribution in the United States

We have already looked at family income distribution for selected years between 1960 and 1982. We now construct Lorenz curves with the data for the years 1960 and 1982. In Table 16–3, the cumulative percent share of all families by quintiles is calculated from the simple shares for 1960 and 1982. In 1960, for example, the lowest 60 percent of families received 34.8 percent of total income and in 1982 received 33.0 percent. In 1960, the lowest 20 percent received 4.8 percent of money income, but 4.7 percent in 1982, and so on.

Lorenz curves are calculated for these two years in Figure 16–2. Two important facts may be concluded from these Lorenz curves. First, it appears that money income distribution has changed very little over the twenty-year period, as shown by the minute differences between the Lorenz curves. For the lowest, middle, and fourth highest quintiles of income receivers, the distribution of income remained about the same. The second fifth lost one percent and the richest 20 percent gained one and a half percent, although this movement may be insignificant in a statistical sense. (Focus, "Do the Very, Very Rich Get Richer?" analyzes income distribution further.)

What can we conclude from this money income data? Has overall distribution become more unequal over the twenty-year period 1960–1982? Have

TABLE 16–3 Share and Cumulative Share in Income Distribution of All Families, 1960 and 1982

The data show insignificant changes in the lowest, middle, and fourth highest quintiles in income distribution, with a loss of one percent in the second fifth and a gain of about one and a half percent in the highest 20 percent of income receivers.

| All Families by Quintile | Percent Share 1960 | Percent Cumulative Share 1960 | Percent Share 1982 | Percent Cumulative Share 1982 |
|---|---|---|---|---|
| Lowest fifth | 4.8 | 4.8 | 4.7 | 4.7 |
| Second fifth | 12.2 | 17.0 | 11.2 | 15.9 |
| Middle fifth | 17.8 | 34.8 | 17.1 | 33.0 |
| Fourth fifth | 24.0 | 58.8 | 24.3 | 57.3 |
| Highest fifth | 41.3 | 100.0 | 42.7 | 100.0 |

*Source:* U.S. Bureau of the Census, *Current Population Reports*, in *Statistical Abstract of the United States*, 1984.

**FIGURE 16–2**

**Lorenz Curves for 1960 and 1982**

The Lorenz curves of income distribution for 1960 and 1982 show little change in before-tax income distribution in the United States.

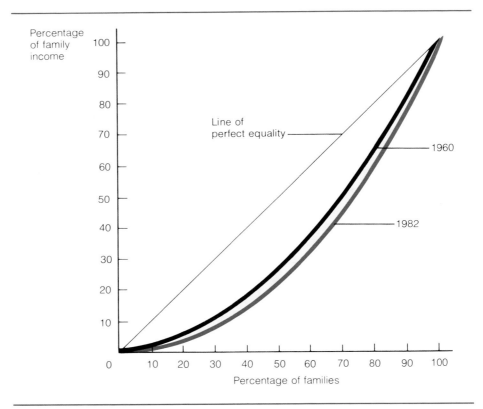

**Adjusted income:** The income of an individual after taxes are subtracted, transfer payments are added, and other items are accounted for.

the lowest 60 percent of income recipients on balance become worse off? Have the rich gotten slightly richer?

The answers to these questions lie in the taxes and transfers that are not reported in the *ex ante* money income statistics from which the Lorenz curves of Figure 16–2 were constructed. We have already discussed some of the problems associated with the construction of simple Lorenz curves and Gini coefficients. Consider what happens to income distribution when adjustments are made for such things as income and payroll taxes, capital gains, in-kind transfers, money transfer or welfare payments, and other factors.

### Adjustments to Money Income

Two important factors related to money income data must be considered before any conclusions are drawn about recent trends in income distribution in the United States. First, while the reported money income data do reflect the impact of indirect business taxes—excise taxes, property taxes, and corporate income taxes—since these items are paid before money factor payments are distributed, the data in Tables 16–1 and 16–3 do not account for payroll and income taxes. Since the U.S. income tax system is progressive, meaning that the higher-income recipients pay higher tax rates, **adjusted income** distribution is affected in the direction of greater equality.

A second and increasingly important set of items that is not included in the raw money income statistics is transfer payments. These include **in-kind transfers** of benefits other than money—such as food stamps, public housing, Medicare, Medicaid, and other subsidies received largely by lower-income groups—in addition to money transfers or welfare payments. Such in-kind subsidies also tend to equalize income distribution across the various classes.

## FOCUS  Do the Very, Very Rich Get Richer?

F. Scott Fitzgerald, the Jazz Age American writer, once remarked to Ernest Hemingway, "You know, Ernest, the very very rich are different than you and I." Hemingway's reply: "I know, they have more money." How rich are the very rich?

The rich in America are not as rich as they were sixty or a hundred years ago. Gone are the good old days when Alva Vanderbilt could spend $250,000 1883 dollars on a fancy dress ball or when door prizes consisted of five minutes with a shovel in a sandbox filled with diamonds, emeralds, and rubies. Today, we do not seem to observe huge family accumulations such as those of the Rockefellers, Mellons, Carnegies, or Vanderbilts, although there are some exceptions (such as the Hunts of Dallas). But the rich are still among us, as the following table shows.

| Year | Percentage of Personal Wealth Held by Top 1% |
|---|---|
| 1922 | 31.6 |
| 1929 | 36.3 |
| 1933 | 28.3 |
| 1939 | 30.6 |
| 1945 | 23.3 |
| 1949 | 20.8 |
| 1953 | 24.3 |
| 1954 | 24.0 |
| 1956 | 26.0 |
| 1958 | 23.8 |
| 1962 | 22.0 |
| 1965 | 23.4 |
| 1969 | 20.1 |
| 1972 | 20.7 |

*Source:* 1922–1956: Robert J. Lampman, *The Share of Top Wealth-Holders in National Wealth,* National Bureau of Economic Research, 1962; 1958: James D. Smith and Staunten K. Calvert, "Estimating the Wealth of Top Wealth-Holders from Estate Tax Returns," *Proceedings of the American Statistical Association,* Philadelphia, 1965 (copyright); 1962–1972: James D. Smith, unpublished estimates, the Urban Institute, Washington, D.C., and the Pennsylvania State University (copyright).

The percent share of total U.S. wealth (not just income) held by the wealthiest 1 percent of all Americans has changed a good deal since the 1920s. Over the 1920s, wealth holdings concentrated at the top, reaching a high of 36.3 percent in 1929. Since World War II, however, wealth holdings have in general become less concentrated. Tax laws, especially inheritance laws, have probably reduced the concentrations of income in the hands of a few.

Composition of the wealthiest class may have changed, too. Gross statistics on wealth holdings give no indication of who the wealth holders are. With more economic and social mobility in this century than in the last one, membership in the top 1 percent was probably different in 1949, 1972, and 1984. *Forbes* magazine's 1983 poll of the four hundred richest Americans ($125 million and above) is very revealing in this regard.[a] Of a list of the thirty richest Americans compiled in 1981 only seven or eight names from the same families made it in 1983. A full 18 percent of the names on the 1983 list were different from 1982's roster. Rather than being frozen in inheritance or traditional economic power, wealth originated in a large variety of occupations, settings, and backgrounds. Women make up one-third of the 1983 list. It is also the case that high-tech society provides a ripe setting for rags to riches—and riches to rags—experiences. Given the large amount of economic mobility, tax laws, and incentives of inheritors, the likelihood of families going from "shirtsleeves to shirtsleeves in three generations" increases.

Ambition, incentive, and intelligence are still winning attributes in U.S. society, in spite of tax laws tending to reduce wealth concentrations. Americans still believe, in the sentiments of the late Duchess of Windsor, that there are two things a person can never be—"too thin or too rich."

[a] "The Richest People in America: The Forbes Four Hundred, 1983 Edition," *Forbes* (Fall 1983).

---

**In-kind transfer payments:** Transfers of benefits other than money from government to citizens, such as food stamps, public housing, and Medicare.

Economist Edgar K. Browning has attempted to develop a measure of income that conforms more closely to economists' definition. Browning estimated the effects of compensation for taxes, welfare transfers, in-kind transfers, and other factors for the year 1972, which we reproduce as Table 16–4. The table shows the distribution of money income in 1972 in billions of dollars with adjustments for in-kind transfers and other factors added in and with income and payroll taxes subtracted from the raw money income data.

Consider the additions to money income shown in Table 16–4. In-kind

### TABLE 16–4 Distribution of Adjusted Income in 1972

Adjusted income distribution statistics for any single year require the addition of in-kind transfer benefits and other items and the subtraction of income and payroll taxes. In 1972, a more equal distribution resulted from these adjustments, as is evident in the comparison between the adjusted and unadjusted distribution figures at the bottom of the table.

| Income Item | Families by Quintile (shown in billions of dollars) | | | | | |
|---|---|---|---|---|---|---|
| | Lowest | Second | Third | Fourth | Highest | Total |
| Unadjusted money income | 37.1 | 81.7 | 120.1 | 164.1 | 284.2 | 687.2 |
| *Plus* Benefits in kind, capital gains, potential additional earnings, and other adjustments | 58.6 | 54.8 | 54.1 | 50.5 | 102.9 | 320.9 |
| *Minus* Income and payroll taxes | 1.3 | 5.8 | 15.5 | 26.6 | 67.5 | 116.6 |
| *Equals* Adjusted money income total | 94.4 | 130.7 | 158.7 | 188.0 | 319.6 | 891.4 |
| Adjusted percent distribution | 12.5% | 15.8% | 17.9% | 20.4% | 33.3% | 100.0% |
| Unadjusted percent distribution | 5.4% | 11.9% | 17.5% | 23.9% | 41.4% | 100.0% |

*Source:* Edgar K. Browning, "The Trend Toward Equality in the Distribution of Net Income," *Southern Economic Journal* 43 (July 1976), p. 914.

benefits are received largely by the poorest classes in society, whereas capital gains on stocks, bonds, and houses—which are not reflected in money income data—are six to seven times greater for the highest quintile of income recipients than for the lowest.

After these items are added to money income, taxes must be subtracted to obtain an estimate of net income. Since the United States employs a progressive income tax system, the highest burden of income taxes—in both percentage and total amount—falls on the highest 20 percent of income recipients. In 1972, as the income and payroll taxes line in Table 16–4 shows, this group paid about $20 billion more in income taxes than all other income earners combined.

What is the net result of these adjustments to money income? Table 16–4 reports the results in terms of total income adjustment and in unadjusted and adjusted percent distribution. Total income rises for each class after additions to and subtractions from money income are made. But the important point is that the percent distribution among the quintiles is affected. The poorest 60 percent of income recipients all receive increased shares, with the lowest 20 percent receiving a full 7 percent increase in adjusted income.

The effects indicated in the data of Table 16–4 are summarized in Figure 16–3 as Lorenz curves. Figure 16–3 shows a set of three Lorenz curves for

**FIGURE 16–3**

**Lorenz Curve Including Income Adjustments, 1972**

When taxes are subtracted from before-tax money income and in-kind transfers and other factors are added, the Lorenz curve moves closer to the line of perfect equality.

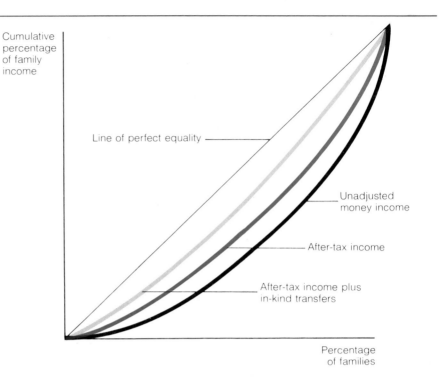

Source: Figure based on conclusions of Browning, "The Trend Toward Equality in the Distribution of Net Income," *Southern Economic Journal* 43 (July 1976), pp. 912–923.

1972. The most bowed curve is the unadjusted distribution of money income. Figure 16–3 also shows a Lorenz curve that is adjusted for payroll and income taxes. The tax adjustment makes the curve less bowed and moves income distribution closer to the line of perfect equality. This shift indicates that a progressive tax system does effect greater equality in income distribution. The inclusion of in-kind transfers and other adjustments pushes income distribution to still greater equality, as the third Lorenz curve shows. These curves indicate the importance of social programs such as Medicare, food stamps, and public housing in creating greater income equality across the various segments of society.

### Historical Changes in Adjusted Income

While a study of adjustments to money income over a given year leads to the conclusion that after-adjustment income is distributed more equally, what about long-range trends? Edgar K. Browning has calculated adjusted net income for 1952, 1962, and 1972 to answer this question. His results are reported in Table 16–5.

Table 16–5 indicates that a significant change took place in income distribution over the twenty-year period 1952–1972. Gains were concentrated in the bottom two quintiles, largely due to the growing importance of in-kind transfers. These benefits grew in importance over the period, and the benefits were concentrated in the poorest 40 percent of families. The growth in educational expenditures also had an equalizing effect.

The relative positions of the richest two quintiles also changed. A loss of 5.8 percentage points for these groups balanced the gain of 6.1 in the bottom two quintiles. Middle-income families did not fare significantly better or

**TABLE 16–5  Adjusted Net Income for 1952, 1962, and 1972**

The data spanning two decades from 1952 to 1972 show gains in the poorest 40 percent of income earners and losses in the highest 40 percent. Middle-income families (the third quintile) received about the same proportion of total income.

| Year | Families by Quintile (percent share) | | | | |
|------|--------|--------|-------|--------|---------|
|      | Lowest | Second | Third | Fourth | Highest |
| 1952 | 7.8    | 14.8   | 18.8  | 23.3   | 35.3    |
| 1962 | 9.0    | 15.1   | 19.1  | 22.9   | 34.0    |
| 1972 | 12.6   | 16.1   | 18.4  | 20.9   | 31.9    |

Source: Browning, "The Trend Toward Equality in the Distribution of Net Income," *Southern Economic Journal* 43 (July 1976), p. 919.

worse over the twenty-year period. These data suggest that a redistribution did in fact take place from richer to poorer families, creating a more equal income distribution between 1952 and 1972. The growth of in-kind transfer payments, moreover, appears to have been the principal means of the redistribution from rich to poor.

## Age Distribution and Income

Another useful factor in understanding income distribution is the age structure of the income-receiving population. Take your own situation as an example. Perhaps you are a college student with a part-time job. If so, you are now probably part of the poorest quintile of the population in calculated annual income distribution. But you are not always likely to be poor. Education and training alone will give you an edge in earning income later. The relative mean income of heads of households tends to peak between ages forty-five and fifty-five and perhaps even later. Therefore, annual income figures will either overstate or understate multiyear inequality in income distribution, depending on the age structure of the population.

If the population is heavily populated by young people in the college age group, annual data will understate actual lifetime or multiyear inequality in income distribution. There is in fact less inequality over multiyear distributions in the United States than in any given year. Economists' estimates vary on how much more equal distribution is over the long run, but all agree that there is more equality when age distribution is considered.[1]

## Rags to Riches Mobility

**Economic mobility**—the opportunity for movements up or down in the quintiles of income distribution—also influences the question of how seriously annual data should be taken. How, for example, do we know that a family in the poorest quintile of income receivers in 1960 was not in the richest quintile in 1979 or vice versa? The answer is that we do not know. To the extent that there is upward (and downward) mobility among income classes, greater equality is achieved over time.

The United States, as a nation of rugged individualists, has historically championed the idea that a hard-working and frugal boy or girl, though poor,

**Economic mobility:** The ability or ease with which an individual may move from one income range to another.

---

[1] See Alan S. Blinder, "Inequality and Mobility in the Distribution of Wealth," *Kyklos* 29 (1976), pp. 607–638. Also see an estimate by Morton Paglin, "The Measurement and Trend of Inequality: A Basic Revision," *American Economic Review* 65 (September 1975), pp. 598–609.

could climb to the top. Such economic mobility can promote equality over the long run. Most studies indicate that there is a high degree of mobility in the United States, especially in the lower quintiles, but there are important statistical problems with identifying mobility over long-run periods.[2] Such problems also extend to measuring age distribution factors and multiyear inequality. While it is safe to say that age distribution factors and economic mobility undoubtedly reduce income distribution inequality over time, economists and policymakers do not yet have the data or the wherewithal to judge the exact amount of the reduction. Annual statistics are nonetheless useful as indicators of certain important trends and characteristics of income distribution. Let us consider some of these important characteristics, especially those related to poverty in the United States.

## Some Characteristics of Income Distribution

The data collected in population surveys and census reports, though imperfect, reveal a great deal of information about ourselves as a nation.

### Family Income Characteristics

Consider Table 16–6, for example, which shows the total, average or mean, and per capita incomes of all families, black families and white families, by certain characteristics for one year, 1981. The data of Table 16–6 illustrate a number of the points made earlier about income distribution and also show some persistent and disturbing trends.

One important fact that glares out of the data is that the **per capita** (per person) **income** of white families is almost $4,000 greater than that of black families. Such income differences may result from differences in educational attainment, skills, and training, but it is a sad and undeniable fact that some of the differences in black and white income result from economic discrimination. **Mean**—or average—black **incomes** are persistently only 60 percent of white incomes, and, as we will analyze later, discrimination is a large part of the reason.

Another look at Table 16–6 gives us other important information. **Age-distributed income** (statistics broken down by ten-year periods, such as 15 to 24 years, 35 to 44, and so on) rises through all age categories. Maximum per capita income is not achieved until the decade 55–64 years (in 1981 at least) for all groups in society, black or white. Note, however, the persistent per capita income differentials between black and white families at all age levels.

Table 16–6 also tells us something about income distribution by sex. Note the differences in average and per capita incomes of families with a male married householder (head of family) and those with a female married householder. There is little difference between incomes of male householders whether the wife is absent or present, but for female householders, incomes are three times as great when the husband is present than when he is absent. Furthermore, widowed, separated, divorced, and single men receive more than twice the average and per capita income as their female counterparts. In other words, males did better than females on practically all counts. The

**Per capita income:** The income per individual, found by dividing total income by the total number of people.

**Mean income:** The income of the average income earner, found by dividing the total income by the number of income earners.

**Age-distributed income:** The distribution of income over the various age brackets of the population, such as ten-year intervals.

---

[2]See Alan S. Blinder, "The Level and Distribution of Economic Well-Being," in *The American Economy in Transition*, ed. Martin Feldstein (Chicago: University of Chicago Press, 1980), pp. 450–454. This entire essay is an excellent nontechnical and modern introduction to the many problems of income distribution.

TABLE 16-6 The Money Income of Families by Family Characteristics, 1981

The table shows total number of families, their aggregate income, average income, and per capita income by certain characteristics such as race, age, occupation, and family size.

| Characteristic | All Races | | | | White | | | | Black | | | |
|---|---|---|---|---|---|---|---|---|---|---|---|---|
| | All Families (1000) | Aggregate Income (billions of dollars) | Mean Income (dollars) | Per Capita Income (dollars) | All Families (1000) | Aggregate Income (billions of dollars) | Mean Income (dollars) | Per Capita Income (dollars) | All Families (1000) | Aggregate Income (billions of dollars) | Mean Income (dollars) | Per Capita Income (dollars) |
| *Age of householder* | | | | | | | | | | | | |
| 15 to 24 years | 3,621 | 56.2 | 15,528 | 5,632 | 3,030 | 49.7 | 16,399 | 6,075 | 522 | 5.5 | 10,487 | 3,446 |
| 25 to 34 years | 14,449 | 330.2 | 22,855 | 6,757 | 12,205 | 290.9 | 23,837 | 7,114 | 1,846 | 30.0 | 16,256 | 4,616 |
| 35 to 44 years | 13,083 | 380.9 | 29,114 | 7,332 | 11,359 | 344.9 | 30,361 | 7,707 | 1,399 | 26.8 | 19,172 | 4,617 |
| 45 to 54 years | 10,710 | 343.5 | 32,070 | 8,861 | 9,363 | 313.6 | 33,495 | 9,445 | 1,089 | 21.6 | 19,812 | 4,906 |
| 55 to 64 years | 9,752 | 287.6 | 29,492 | 10,585 | 8,802 | 268.9 | 30,547 | 11,336 | 794 | 14.4 | 18,093 | 5,005 |
| 65 years and over | 9,403 | 178.1 | 18,945 | 8,171 | 8,511 | 166.8 | 19,599 | 8,696 | 763 | 8.8 | 11,566 | 3,913 |
| *Marital status of householder* | | | | | | | | | | | | |
| Male householder, total | 49,513 | 1,388.3 | 28,039 | 8,507 | 44,968 | 1,283.5 | 28,542 | 8,792 | 3,442 | 74.1 | 21,531 | 5,788 |
| Married: | | | | | | | | | | | | |
|   Wife present | 47,527 | 1,343.4 | 28,266 | 8,509 | 43,326 | 1,244.8 | 28,732 | 8,784 | 3,168 | 69.3 | 21,872 | 5,777 |
|   Wife absent | 274 | 6.6 | 24,108 | 8,691 | 236 | 5.9 | 24,863 | 9,283 | 32 | .6 | (B) | (B) |
|     Separated | 227 | 5.2 | 23,063 | 8,234 | 194 | 4.6 | 23,928 | 8,851 | 29 | .5 | (B) | (B) |
| Widowed | 364 | 8.7 | 23,819 | 8,020 | 295 | 7.7 | 26,018 | 9,002 | 59 | .8 | (B) | (B) |
| Divorced | 681 | 15.9 | 23,298 | 8,846 | 588 | 14.1 | 23,980 | 9,300 | 78 | 1.5 | 19,288 | 6,379 |
| Single | 666 | 13.7 | 20,621 | 8,218 | 524 | 11.0 | 21,010 | 8,722 | 104 | 1.9 | 18,394 | 6,759 |
| Female householder, total | 11,506 | 188.3 | 16,364 | 5,327 | 8,301 | 151.3 | 18,226 | 6,320 | 2,971 | 33.0 | 11,096 | 3,105 |
| Married: | | | | | | | | | | | | |
|   Husband present | 2,103 | 58.8 | 27,947 | 8,699 | 1,680 | 49.7 | 29,591 | 9,783 | 366 | 7.6 | 20,640 | 5,191 |
|   Husband absent | 1,859 | 18.7 | 10,078 | 2,934 | 1,161 | 12.2 | 10,498 | 3,276 | 668 | 6.2 | 9,345 | 2,442 |
|     Separated | 1,651 | 16.2 | 9,808 | 2,847 | 996 | 10.0 | 10,066 | 3,139 | 628 | 5.9 | 9,382 | 2,452 |
| Widowed | 2,462 | 43.4 | 17,621 | 5,996 | 1,895 | 36.4 | 19,212 | 7,009 | 514 | 5.9 | 11,051 | 3,174 |
| Divorced | 3,478 | 50.3 | 14,475 | 4,890 | 2,838 | 42.7 | 15,049 | 5,264 | 576 | 6.8 | 11,875 | 3,451 |
| Single | 1,604 | 17.1 | 10,629 | 3,651 | 726 | 10.3 | 14,127 | 5,617 | 846 | 6.4 | 7,569 | 2,321 |
| *Occupation of householder* | | | | | | | | | | | | |
| White-collar workers | 21,107 | 751.6 | 35,609 | 10,710 | 19,345 | 703.2 | 36,353 | 10,994 | 1,285 | 31.2 | 24,310 | 7,192 |
| Blue-collar workers | 16,383 | 421.4 | 25,724 | 7,272 | 14,489 | 378.1 | 26,097 | 7,493 | 1,565 | 35.1 | 22,414 | 5,663 |
| Farm workers | 1,411 | 22.4 | 15,904 | 4,631 | 1,318 | 21.3 | 16,146 | 4,784 | 74 | .8 | (B) | (B) |
| Service workers | 3,784 | 74.2 | 19,603 | 5,744 | 2,856 | 58.6 | 20,534 | 6,258 | 812 | 12.9 | 15,936 | 4,182 |
| *Size of family* | | | | | | | | | | | | |
| Two persons | 24,426 | 552.3 | 22,612 | 11,251 | 22,072 | 519.4 | 23,530 | 11,726 | 2,024 | 26.7 | 13,203 | 6,460 |
| Three persons | 14,079 | 370.3 | 26,305 | 8,682 | 12,248 | 337.4 | 27,545 | 9,093 | 1,550 | 26.2 | 16,886 | 5,565 |
| Four persons | 12,594 | 365.0 | 28,978 | 7,232 | 10,998 | 331.9 | 30,177 | 7,530 | 1,264 | 23.3 | 18,407 | 4,604 |
| Five persons | 5,971 | 173.7 | 29,092 | 5,802 | 5,020 | 153.7 | 30,617 | 6,108 | 757 | 14.7 | 19,450 | 3,866 |
| Six persons | 2,409 | 71.6 | 29,723 | 4,948 | 1,900 | 59.8 | 31,456 | 5,221 | 406 | 8.3 | 20,535 | 3,468 |
| Seven persons or more | 1,539 | 43.7 | 28,359 | 3,662 | 1,031 | 32.7 | 31,698 | 4,169 | 412 | 7.8 | 19,048 | 2,365 |
| Total families | 61,019 | 1,576.6 | 25,836 | 7,941 | 53,269 | 1,434.8 | 26,934 | 8,444 | 6,413 | 107.1 | 16,696 | 4,571 |

Source: U.S. Bureau of the Census, *Current Population Survey*, in *Statistical Abstract of the United States*, 1984.

numbers strongly suggest that sex discrimination might be at work, a subject for later discussion.

Not surprisingly, families of two persons enjoy higher per capita income in all categories than larger families. White-collar workers earn significantly more than blue-collar workers on the average, again in all categories. Black farm workers earn the lowest per capita income of all, and white blue-collar workers earn more on average and per capita than black white-collar workers.

In summary, these numbers suggest some features about income distribution in 1981 that are fairly characteristic of all years surveyed: (1) that income is very much distributed along racial and sex lines and (2) that poverty, in a relative sense and perhaps also in an absolute sense, is still very much a feature of generally affluent American life. In the following two sections these interrelated matters are analyzed together with some proposed economic interpretations and solutions.

## Poverty

Poverty and economic discrimination are interrelated within our economic system. One cannot be understood without understanding the other. First we examine poverty, its definition, and some of the connected reasons why some families are poor.

**Poverty:** A term describing family income below a defined level when other things such as size of family, location, and age are considered.

**Poverty,** in the relative sense that some people have lower incomes than others, is always going to be with us. As long as the perfectly adjusted Lorenz curve is not diagonal, some families or persons will be better off relative to others. Moreover, an equitable distribution of income does not necessarily mean perfect equality. Clearly, however, some degree of absolute poverty persists in the United States: A number of families in the lower quintiles of income distribution are even worse off than the lowest living standards considered acceptable in this society.

Government measures of poverty are based on a poverty index first devised by the Social Security Administration in 1964 and modified by a Federal Interagency Committee in 1969. The index defines a "poverty level" that varies according to such factors as the size of family, inflation, age and sex of the head of family, number of children under age eighteen, and farm or nonfarm residence. Accepting the government's definition of poverty—income below the poverty level—we may observe aspects of poverty over time and in different groups.

Table 16–7 provides an overall poverty profile for persons by race, family status, and sex of householder for selected years between 1959 and 1981. In 1959, almost 40 million people in the United States were poor by government standards, but that number fell to barely 32 million in 1981. More important, the number of poor in percentage terms fell significantly over the twenty-year period. Clearly progress has been made, but disturbing problems remain.

The percentage of black persons below the poverty level, though declining since 1959, remained at 34 percent in 1981. Likewise, the total number of poor persons in families with female householders rose by 5 million between 1959 and 1981, though the percentage dropped from 50.2 to 35.2. The 1981 comparison between poor persons in families with female heads and the percentage of poor in all other families is stark (35.2 percent versus 8.8 percent).

These gross statistics tell us that the low end of income distribution is made up, disproportionately, of blacks and women householders. Other groups in our society, not shown in Table 16–7, fare poorly also. The num-

TABLE 16–7 Poverty by Family Status, Race, and Sex of Householder, 1959–1981

While poverty has declined in absolute numbers and in percentage terms over time, the relative position of blacks and female householders is still a matter of concern.

| Family Status, Race, and Sex of Householder | Number Below Poverty Level (millions) | | | | | Percentage of Persons Below Poverty Level | | | | |
|---|---|---|---|---|---|---|---|---|---|---|
| | 1959 | 1966 | 1969 | 1975 | 1981 | 1959 | 1966 | 1969 | 1975 | 1981 |
| All Persons | 39.5 | 28.5 | 24.1 | 25.9 | 31.8 | 22.4 | 14.7 | 12.1 | 12.3 | 14.0 |
| White | 28.5 | 19.3 | 16.7 | 17.8 | 21.6 | 18.1 | 11.3 | 9.5 | 9.7 | 11.1 |
| Black | 11.0 | 8.9 | 7.1 | 7.5 | 9.2 | 56.2 | 41.8 | 32.2 | 31.3 | 34.2 |
| Families with female householder, no husband present | 10.4 | 10.3 | 10.4 | 12.3 | 15.7 | 50.2 | 41.0 | 38.4 | 34.6 | 35.2 |
| All other families | 29.1 | 18.3 | 13.7 | 13.6 | 16.1 | 18.7 | 10.8 | 8.0 | 7.8 | 8.8 |

*Source:* U.S. Bureau of the Census, *Current Population Reports,* in *Statistical Abstract of the United States,* various issues.

ber of Hispanic poor tends to fall between that of poor whites and poor blacks. The over-sixty-five age group, as a whole, contained about 3.5 million persons below the poverty level in 1979, for example. Some 35.5 percent of this number were black and 26.1 percent were of Spanish origin, but only 13.2 percent were white. Improvements have been made since 1960 in the poverty status of the elderly, but the improvements have been among the white over-sixty-five age group.

Poverty also has a geographic dimension. The highest percentage of persons below the poverty level is found in the southern states (15.3 percent in 1975) with the lowest in the northeastern section of the United States (8.9 percent in 1975). Alaska and Connecticut tied in 1975 for the lowest percentage of poor (6.7 percent), while 26.1 percent of persons in Mississippi were defined as poor in government statistics.

## Poverty During Recession

One important factor in analyzing poverty is that poverty levels can change drastically when the total real output and unemployment in society increase or decrease because of changing total demand. A recent and dramatic example of this feature of poverty is revealed in the recessionary conditions of declining production and employment that took hold of the U.S. economy in 1981–1982.

Table 16–8 focuses on the Census Bureau's estimate of the poverty experience of alternative groups—Hispanics, blacks, the elderly, and so on—between 1979 and 1982. Table 16–8, moreover, gives two definitions of poverty: an official definition counting only the cash income of families in the various groups and an alternative definition that counts cash income plus the market value of noncash benefits such as food stamps, school lunches, public housing, Medicaid, and Medicare. A family of four was classified as poor if it had cash income of less than $7,386 in 1979 or less than $9,862 in 1982. The official poverty level, in other words, is adjusted for inflation, as are the market values of noncash benefits.

According to either measure, poverty was on a serious upswing between 1979 and 1982. Counting only cash income, there were 26.1 million poor people in 1979 but 34.4 million in 1982. But if noncash benefits are included, there were 15.1 million poor in 1979 and 22.9 million in 1982. The

**TABLE 16–8  Poverty Rates and Recession, 1979–1982**

Reductions in total output and employment, especially between 1981 and 1982, created increases in poverty rates for many groups in society. When cash income only (the official definition) is considered or when the market value of noncash benefits is added to cash income (alternative definition), poverty rose between 1979 and 1982.

| | 1979 | | 1982 | |
|---|---|---|---|---|
| Category | Official Poverty Definition | Alternative Definition | Official Poverty Definition | Alternative Definition |
| Total | 11.7% | 6.8% | 15.0% | 10.0% |
| White | 9.0 | 5.6 | 12.0 | 8.3 |
| Black | 31.0 | 14.9 | 35.6 | 21.5 |
| Hispanic | 21.8 | 12.0 | 29.9 | 20.5 |
| Children under six | 18.2 | 11.3 | 23.8 | 17.2 |
| Elderly | 15.2 | 4.3 | 14.6 | 3.5 |
| Married couples | 6.1 | 3.9 | 8.9 | 6.4 |
| Families headed by women | 34.9 | 16.6 | 40.6 | 24.8 |

*Source:* U.S. Bureau of the Census, *Statistical Abstract of the United States*, 1984.

inclusion of noncash benefits significantly reduces the absolute number of poor people, but note that, in relative terms, there was a higher percent increase in poverty using the noncash benefits calculation than using the official cash income definition.

With the exception of the elderly, recessionary conditions have increased the poverty level of all other groups considered by the Census Bureau, using either definition of poverty. Declining expenditures on all goods and services and the unemployment conditions created by the recession are clearly the major reasons for the observed increase, but there are several other possible explanations. The rate of increase in some noncash benefits was reduced by Congress in 1981. More important, there was a 33 percent increase in the consumer price index between 1979 and 1982 (primarily felt in 1979 and 1980), an increase that reduced the average market value of the noncash benefits received by the poor by more than 10 percent. In sum, poverty levels undoubtedly increased over the recession, but it is important to remember that decreases in poverty levels, both absolute and in percentage terms, are reduced during periods of economic expansion and recovery.

The overall statistics on poverty since 1960 do show improvement, but the undeniable fact is that there are poor among us and that poverty is often related to race, sex, and family status. What are the reasons for the facts of poverty? Factors are many and often interwoven. As noted earlier, some people choose to be poor because they choose more leisure. Others are unlucky. Still others are not able-bodied; they are simply unable to work. But these factors cannot explain the existence of all poverty.

Lower incomes also exist for clearly identifiable economic reasons. The poor tend to have lower skills and lower stocks of human capital. These lower stocks of human capital are the result of less education, inferior education, and lower job-related training. Lower productivity means unemployment and lower wages, lower wages mean lower incomes, and lower incomes mean poverty. Environmental factors may make poverty a vicious cycle: Children of the poor often receive less education and training. Moreover, many wage and income differences are the result of outright race and sex discrimination. Indeed, differences in training and education may be related to past and present economic discrimination.

### Race and Sex Discrimination in Wages

One might object that poverty, race, and sex discrimination are not necessarily related. After all, a black woman may earn $75,000 per year in legal practice but still endure discrimination with respect to her white male counterparts. It is more than coincidence, however, to find such groups as blacks, Hispanics, and women ranking consistently high in poverty statistics.

We must be careful to distinguish between social discrimination and economic discrimination that leads to significantly lower wages and income, although the two are often related. Groups such as Jews, gays, and the handicapped may meet with social discrimination in seeking housing and club memberships but be little affected by economic discrimination.

**Economic wage discrimination** exists when individuals of equal ability and productivity in the same occupation earn different wage rates. The bald economic facts, already observed in the statistics discussed earlier in this chapter, are that the median income of blacks is only 60 percent of that of whites (in all income classes) and that females earn only about 60 percent of median male income for full-time work. Empirical studies show that about half of this difference is related to educational, productivity, and job-training differences.[3] The remaining half is the apparent result of economic discrimination or other unidentified factors.

**The Practice of Discrimination.** Economists do not pretend to know the causes of discrimination. Some observers have argued that sex discrimination may originate from historical roles of women that are matters of cultural tradition. Women's biologically unique potential for childbearing may also be a factor in differences in wage rates. Employers may not want to invest in as much job training or pay high wages for women whose work is to be interrupted or terminated by childbirth. Both women and blacks receive less formal education and on-the-job training than men and whites, a factor that may be related to past discrimination.

Economic discrimination involves costs to those who discriminate as well as to those against whom discrimination is practiced. An employer who refuses to hire women or blacks of equal productivity to men or whites has what economists call a "taste" for discrimination. If the wages of the preferred groups are driven up by these actions, employers must pay a premium for men or whites. In a competitive system, discriminating firms will be at a disadvantage in the marketplace. Employers who do not discriminate in hiring can acquire equally productive labor at lower wage rates and, in the end, will tend to drive discriminating competitors out of business.

Even if employers do not discriminate, consumers may. In a competitive

*Economic wage discrimination:* A situation in which an employer pays individuals in the same occupation different wages, the wage difference based on race, sex, religion, or national origin rather than productivity differences.

---

[3]See Dwight R. Lee and Robert F. McNown, *Economics in Our Time: Concepts and Issues* (Chicago: Science Research Associates, 1983), chapter 10, for an excellent analysis of economic issues related to discrimination. Some of the present discussion is drawn from their work.

**Consumer-initiated discrimination:** A circumstance in which people prefer to purchase a good or service produced or sold by individuals of a particular sex, race, religion, or national origin; such consumers are willing to pay a premium to indulge their taste for discrimination.

system, **consumer-initiated discrimination** is costly to the discriminator as well as to those discriminated against. Consumers who will deal only with whites or men (avoiding, say, black or female lawyers, doctors, or interior decorators) must pay a premium for their prejudice. Prices of goods and services of favored sellers will be higher than prices for the same goods or services available to those who have no taste for discrimination. In sum, the competitive system makes it costly for employers or consumers to discriminate. This economic deterrent has failed to eradicate discrimination in our economy, however, as have government regulations explicitly forbidding discrimination.

**Economic Restrictions and Discrimination.** The law is clear on one point—employers are to give equal pay for equal work without regard to race, creed, or sex. But employers may simply avoid the hiring of women or blacks. Laws establishing quotas (minimum numbers) of females or racial minorities in employment or educational situations may not have the fully intended results. Reverse discrimination cases have resulted in some instances of quota imposition. In other cases, employers may subtly hide discrimination by reducing the amount of on-the-job training while paying the same money wages. The full wage—both money and training—may not be paid to those against whom discrimination is practiced.

Although there are encouraging statistics on the entry of women and blacks into male- and white-dominated occupations in the past twenty years, a number of factors remain that make the rapid and total elimination of discrimination unlikely. As suggested above, a fully competitive system discourages discrimination, placing extra costs on discriminating employers and consumers. Where, then, do we observe the most discrimination? In markets where regulations or restrictions prevent the competitive process from working. According to the research of economist Thomas Sowell, high degrees of economic discrimination have traditionally been found within labor unions, in regulated industries, and in the professions where legal and government-sanctioned entry restrictions are permitted.[4] In such industries and occupations, nonmarket characteristics—such as color and sex—may be used in deciding whom to employ. Such discrimination may be detected in medicine, law, and dentistry, for example, as well as in those historically regulated areas of our economy such as railroads and the postal service. Government itself has practiced a good deal of discrimination in this century as, for example, in the U.S. military up to World War II and the Korean conflict.

In sum, discrimination appears to be most pronounced and most long-lived where government-granted regulations and restrictions substitute for freely competitive market forces. Many economists feel that a movement toward a deregulated system of competition would go far in removing much sex and race discrimination. Where economic incentives are placed before employers and consumers to avoid discrimination, less discrimination may be confidently predicted.

## Programs to Alleviate Poverty

Poverty is the result of a number of factors—lower productivity, inferior educational opportunities, and economic discrimination. Until full equal economic opportunity for all is a reality, some type of redistribution of wealth is likely to take place.

---

[4]See Thomas Sowell, *Markets and Minorities* (New York: Basic Books, 1981), pp. 34–51.

Contemporary welfare programs (see Table 16–9) center on the relative needs of the poor. In-kind benefits such as food stamps, Medicare, housing subsidies, and outright money transfers (such as Aid to Families with Dependent Children) are allotted to the poor on the basis of a government agency's determination of need. Benefits, moreover, vary from state to state and among local governments. Without questioning need or the existence of poverty, economists might analyze the costs of fairly and accurately administering such programs. But more important, economists are concerned with the effects of such programs on the incentives of the poor.

While there will always be poor unfortunates, who, for health or other reasons, cannot work, it is clear that the contemporary welfare system discriminates against most of the poor in one respect: It discourages any effort to better themselves by earning income additional to their subsidy and perhaps acquiring job training in the process. Take a highly stylized example: Rita, a widow with three small children, could earn $4,000 a year to supplement her modest welfare subsidy of $7,000 per year. Such work would surely not be in her interest, however, since under the welfare system in some circumstances her subsidy would be taxed at a rate of 100 percent. By "taxed" we mean that Rita would lose her welfare benefits if she works. In other words Rita would have less incentive to work if her subsidy were reduced by an amount equal to what she earned and still less incentive if she were to lose the entire subsidy by working.

Consider another possibility, one supported by a large number of economists interested in public policy relating to poverty: the **negative income tax.** The negative income tax originates from the observation that welfare recipients such as Rita will react to economic incentives if such incentives are provided. Under a negative income tax system Rita would be taxed for working but not at a rate of 100 percent.

Suppose that Rita and her children are guaranteed a minimum income subsidy of $7,000; an income of $14,000 is determined by some agency calculation to be adequate for her situation—a widow with three children to support (the figures given here are wholly hypothetical). Further suppose that

**Negative income tax:** A progressive income tax that allows for a negative tax rate (income subsidy) for income below a particular level. As income rises, the subsidy gradually diminishes to zero.

TABLE 16–9  Summary of Principal Welfare Programs

U.S. welfare programs include both cash subsidy and in-kind benefits. Welfare programs, moreover, are aimed at various segments of society and are administered at all levels of government.

| Program | Level of Administration | Purpose | Beneficiaries |
| --- | --- | --- | --- |
| Aid to Families with Dependent Children (AFDC) | Federally mandated; administered by state and local governments | Income maintenance, often used in conjunction with in-kind subsidies | Poor families with unemployed heads of household |
| Unemployment Compensation | State governments | To aid the unemployed for limited periods | Directly benefits the unemployed and their families |
| In-Kind Benefits, Housing Subsidies, Food Stamps | Federal, state, and local governments | Supplements to nutrition and to the quality of housing of the poor | Poor families and the children of poor families |
| Social Security, including Medicare and Medicaid | Federal government | To provide retirement insurance | Insurance for survivors, welfare for the aged poor, medical care for the elderly (Medicare) and the poor (Medicaid) |

**TABLE 16–10** Negative Income Tax for a Hypothetical Welfare Recipient

If Rita could earn $4,000 per year from working, she would be taxed only at a rate of 50 percent, giving her a total income of $9,000. At a marginal or additional negative income tax rate of 50 percent, she would have an incentive to work and earn additional income. After she earns an income greater than $14,000, Rita becomes a net taxpayer.

| Rita's Income from Working or Other Sources | Rita's Subsidy (Negative Income Tax) | Rita's Total Income (Private Sources plus Subsidy) |
|---|---|---|
| $ 0 | $7,000 | $ 7,000 |
| 4,000 | 5,000 | 9,000 |
| 8,000 | 3,000 | 11,000 |
| 14,000 | 0 | 14,000 |
| 16,000 | −1,000 | 15,000 |
| 20,000 | −2,000 | 18,000 |

rather than being taxed at a rate of 100 percent, as in the current welfare system, Rita is taxed at a lower rate, say 50 percent of all additional private earnings. What happens when Rita works and earns income?

As Table 16–10 shows, Rita now has an incentive to enter the marketplace. If she earns $4,000, she does not lose $4,000 in subsidy but instead gains $2,000. In other words, if Rita earns $4,000, she pays only 50 percent of it in taxes, which, in effect, reduces her subsidy by $2,000. Her total income would climb to $9,000. Should Rita be able to earn $8,000 by job advancement, she would pay in taxes only $2,000 of the additonal $4,000 earned. Her subsidy would be reduced, in effect, to $3,000, giving her a total income of $11,000. Should she be able to earn $14,000, her minimum acceptable income, Rita would receive no subsidy (paying, in effect, an additional tax of $3,000). If Rita earns an income over $14,000 per year she would become a net taxpayer.

The point is that Rita would have positive incentives in this system to work and be self-supporting. While the numbers in this example—such as a 50 percent tax rate—are completely hypothetical (progressivity and regressivity may be built into the plan), the principle is clear. If the poor could improve their lot by working, economists would predict that such incentives would encourage work effort. While a negative income tax plan has not received much political support, many economists have strongly urged its implementation in preference to much of the contemporary welfare system. In an economic view at least, it would help make the welfare system a tonic to the poor rather than a sedative, as the present system can be regarded.

## The Justice of Income Distribution

Economists, who have been writing about income distribution for more than two hundred years, have attempted—not always successfully—to avoid advocating any given distribution of income or any redistribution of income. In other words, economists have tried to avoid normative statements about what should be and have taken a more positive stance—what income distribution is and how it might be changed to produce greater incentives to work and produce while maintaining some politically derived concept of justice. The proposal that a negative income tax plan be instituted in place of the

> ### FOCUS Pareto Movements and Income Distribution
>
> Are there any rules governing the ability of an economist to make statements concerning income distribution? Practically speaking, can an economist say anything about whether a housing subsidy or a new dam would make people better off, all things considered?
>
> A famous turn-of-the-century economist, Vilfredo Pareto (1848–1923), had something to say on these matters. His view, sometimes called Pareto optimality, was that the economist could pronounce one distribution of income or wealth as preferable to another if and only if the change made one group better off without leaving another group worse off. "Better off" is expressed as satisfaction or utility, so judgment depends on the ability of the economist to measure welfare.
>
> It is almost impossible to think of distribution changes that do not help one group without simultaneously making another group worse off. The provision of food stamps obviously aids the recipients, but it reduces the welfare of those who pay taxes to support such expenditures. While it is tempting to argue that a dollar's worth of something provides more satisfaction to a poor person than it takes away from a rich person, the positive economist cannot make this leap because utility cannot be measured. Support for a progressive income tax on this basis is invalid for the same reason. It would involve what economists call an interpersonal utility comparison—a judgment about the relative utility of, say, a dollar to a rich person and to a poor person. As noted in this chapter, the economist can measure rich or poor but cannot measure happy or unhappy. Thus the economist depends on the political process to decide on a particular distribution to be made through taxes and expenditures. The economist is then concerned with the effects of specific taxes and specific expenditures on economic efficiency.

contemporary system of welfare transfers is an example of a positive economic solution related to income distribution. Focus, "Pareto Movements and Income Distribution," considers further the role of economists.

Justice itself is not part of the positive economist's vocabulary, but the effects of some concept of justice are within the economist's purview. For example, some concepts of justice could create extreme income redistributions that would make income distribution more equal at the expense of working and producing members of society. This notion of extreme redistribution is condemned in some popular literature as putting too many people in the wagon with too few pulling the wagon. In economic terms, such redistributions and the high tax rates involved would probably create disincentives to work. If so, total real output of goods and services available to all citizens would decline. As a result, a shrinking segment of a smaller pie would be available to all classes of economic society, including the poor.

More than a hundred years ago, the classical economist John Stuart Mill observed that two kinds of equality are involved in questions of distribution: *ex ante* equality and *ex post* equality. *Ex ante* equality is a situation where "all start fair"—where, through educational opportunities, the absence of discrimination, social and economic mobility, and so on, every individual is given an opportunity to maximize his or her potential. *Ex post* equality is a guarantee that through redistributive policies all end up at the same place. While Mill championed *ex ante* equality, he believed that differing incentives, innate talents, luck, and so on meant there would and must be differences in income distribution in the end. To guarantee all citizens *ex post* equality would stultify incentives, which are the very mode of economic progress. Enforced *ex post* equality would surely shrink the total economic pie.

Modern economists concerned with income distribution face the same problem. The trick is to evaluate the contemporary system and to propose means that provide that all start fair (akin to golfing or bowling handicaps) but that do not force redistributions that significantly reduce incentives to produce. The modern economist is as much on a razor's edge as were past writers dealing with the questions of income distribution. The issues of fairness, justice, and equality will always be matters of dispute. The economist, as a maker of positive statements and proposals, will always be a contributor to this important debate.

## Summary

1. Economists deal with positive and not normative aspects of income distribution, for the economist's view of "who should receive what" is only as good as anyone else's. The economist therefore does not deal with questions of justice or fairness but focuses instead on the facts of income distribution and on the possible effects of alternative welfare programs.
2. Money income distributions differ for a number of reasons: Some persons choose leisure over work and income while others earn lower incomes because of bad luck, poor health, lower educational opportunities and productivity, and race or sex discrimination.
3. A Lorenz curve and the related Gini coefficient are measures of income inequality. Ordinarily, the more bowed the Lorenz curve or the larger the Gini coefficient, the more inequality there is in a society's income distribution.
4. Lorenz curves showing money income distribution for a given year overstate actual inequality when in-kind benefits, income and payroll taxes, and other factors are considered. Age distribution of the population and income mobility also create greater long-term equality in income distribution.
5. Poverty and economic discrimination are interrelated features of our economic system. Economic discrimination exists when individuals of equal productivity are paid different wages on the basis of sex or racial differences.
6. Economic discrimination exists against blacks and women in our society in the sense that only half of the difference between black-white and male-female earnings is explainable on the basis of productivity differences.
7. Economic discrimination cannot be practiced by employers or consumers without costs in a competitive environment. Government-sanctioned regulations and restrictions on competitors appear to account for a large amount of observed economic discrimination in the United States.
8. Contemporary welfare programs based on need do not provide positive incentives for the poor. Many economists support a negative income tax structure in place of the current system to provide work incentives to the poor.

## Key Terms

individual income
labor income
asset income
*ex ante* distribution
*ex post* distribution
transfer payment
family income
Lorenz curve
Gini coefficient
adjusted income
in-kind transfer payments
economic mobility
per capita income
mean income
age-distributed income
poverty
economic wage discrimination
consumer-initiated discrimination
negative income tax

## Questions for Review and Discussion

1. What are the components of an individual's income? Which of these are matters of choice by the individual?
2. How does an individual obtain asset income?
3. What does a Lorenz curve show? What does it mean if the Lorenz curve moves closer through time to the line of perfect equality?
4. What is the difference between income as measured by government and adjusted income as reported by some economists?
5. What is economic wage discrimination? Give an example of wage discrimination that fits the description.
6. What are the problems associated with the use of Gini coefficients to report income distribution?
7. Why do young people have lower incomes than older people? If average lifetime incomes rather than annual incomes could be used to construct Lorenz curves, would the Gini coefficient be lower?
8. If every married couple in the United States obtained a divorce, would the Lorenz curve move away from the line of perfect equality? Would this alteration change the actual distribution of income?
9. Suppose all people were born with the same

amount of natural abilities and the same amount of inherited money wealth. Under these circumstances, what would the sources of income differences be?

10. Do income redistribution programs affect the supply of labor? How does a negative income tax compare to other transfer programs in its effect on work effort?

## ECONOMICS IN ACTION

### Who Are the Economic Minorities in the United States?

Who are the economic minorities in the United States, and on what does their relative position in income distribution depend? Economist Thomas Sowell has suggested answers to these questions in an extensive investigation of the characteristics of a number of minorities in America.[a] Sowell's thesis is that discrimination is not a good single explanation for why some minority incomes are above or below the national average. Other factors—some of them brought out in this chapter—must figure prominently in any explanation of income differences.

A part of Sowell's intriguing explanation is revealed in the following table. The table ranks ethnic origin (or ethnicity) with the ethnic group's average income compared to the average per capita income of the U.S. population as a whole. WASPs' incomes—the average income of white, Anglo-Saxon, Protestant Americans—is not reported in the table but amounts to 104 percent of national average income.

| Ethnicity | Relative Income (percent of national average) | Median Age | Children per Woman |
|---|---|---|---|
| Jewish | 172 | 46 | 2.4 |
| Japanese | 132 | 32 | 2.2 |
| Polish | 115 | 40 | 2.5 |
| Chinese | 112 | 27 | 2.9 |
| Italian | 112 | 36 | 2.4 |
| Irish | 102 | 37 | 3.1 |
| Mexican | 76 | 18 | 4.4 |
| Puerto Rican | 63 | 18 | 3.5 |
| Black | 62 | 22 | 3.7 |
| American Indian | 60 | 20 | 4.4 |

Source: Adapted from Thomas Sowell, *Markets and Minorities* (New York: Basic Books, 1981), Tables 1.1, 1.2, and 1.3, pp. 8, 11, and 16.

What factors explain the differences? According to Sowell, genetic color differences are not much of an answer. The income of the average black West Indian American (also not reported in the table) is 94 percent of the mean U.S. income, and the average Japanese American earns 132 percent as much as the average American. Contrast these relatively high earnings for some dark-skinned minorities with the earnings of light-skinned Puerto Ricans: only 63 percent of the mean income. Color alone cannot be a big factor explaining ethnic income differences.

According to Sowell, at least three factors explain differences in average incomes: (1) median age in the ethnic group, (2) locational concentrations, and (3) the fecundity of women (children per woman) within the ethnic group. The table above shows, in general, that the older the median age and the fewer children per woman in the ethnic group, the higher the median income of that group. We have seen in this chapter that in all groups, income earned varies with age differences. Income earnings generally peak between the ages of forty-five and fifty-five and perhaps at higher ages. We would thus expect those ethnic groups with the lowest median ages to earn less income than those with higher median ages. This age differential in part explains, for example, why Mexican Americans—with a median age of eighteen—earn 63 percent of average national income and why Jewish Americans—with a median age of forty-six—earn 172 percent.

In these age distribution statistics nevertheless lie hidden signs of hope for the condition of ethnic minorities. Attitudes toward minorities are clearly changing, but the effects of attitude changes on equal education and job opportunities are being felt largely by the younger members of ethnic minorities. As Sowell reports, blacks over forty-five years of age earn less than 60 percent of the sum earned by their age-peers in the population, but blacks aged eighteen to twenty-four earn 83 percent of their age-peers' income.

The number of children per woman is also a factor explaining the relative economic position of minorities. In general, those with lower relative incomes have a higher fecundity rate. The number of children can also have an impact on differences between family and per capita incomes among minorities. In family income terms, for example, Mexican Americans earn significantly more than black families. But in per capita terms, when the number of children per woman becomes a factor, the more fecund Mexicans earn lower incomes than blacks. Age and fecundity factors lead to other surprising statistics. For example, half of all Mexicans and Puerto Ricans in the United States are below the age of eighteen.

[a] Thomas Sowell, *Markets and Minorities* (New York: Basic Books, 1981).

Geographic differences also account for relative income differences. As we saw in this chapter, greater poverty exists in the southern part of the United States. The same fact pertains to comparisons of ethnic incomes and poverty. As Sowell notes and as indicated in the table, Mexican Americans and Puerto Ricans earn more than blacks nationally, but blacks outside the South earn more than both those groups. The place one lives also makes a difference. American Indians living in Chicago or New York earn more than twice as much as those living on reservations. Mexican Americans living in Detroit earn more than twice the amount of those living in cities in the Rio Grande Valley of Texas such as Brownsville.

With arguments such as these, Sowell questions the extent to which discrimination is a function of others' sins. A number of factors explain the relative position of minorities, including age, number of children, and geographic differences. Discrimination cannot explain all differences. How, for example, could we explain Jewish and Japanese relative incomes in spite of rampant discrimination against these groups in the past and to a certain extent in the present as well?

Question

Explain why most generalizations about income distribution and minority incomes must be carefully analyzed and dissected.

## POINT-COUNTERPOINT

## Thomas Malthus and Gary Becker: The Economics of Population Growth

Thomas Malthus

Gary Becker

THOMAS MALTHUS (1766–1834) was born in Surrey, England, during the Industrial Revolution. It was widely believed at this time that the effects of industrialization, such as increased trade and specialization of labor, would eventually improve the quality of human life. When anarchist and pamphleteer William Godwin published his utopian outlook for society in his book, *Political Justice*, in 1793, an argument developed between Malthus and his father. Malthus's father was inclined to agree with Godwin's views, but Malthus believed that society was caught in a trap in which population would increase more rapidly than the food supply, leaving the standard of living at a subsistence level at best. In 1798, the same year he became a minister of the Church of England, Malthus published his views of population and the economy in his treatise, *An Essay on the Principle of Population*. It was this pessimistic treatise on the future of humanity that led essayist Thomas Carlyle to label economics "the dismal science."

Malthus went on to become the first professor of political economy in England at the East India College in 1805. His work on population influenced his friend and critic David Ricardo as well as Charles Darwin.

GARY BECKER (b. 1930), a current University Professor of Economics at the University of Chicago, gained international recognition for his work in applying microeconomic theory to areas such as marriage, crime, and prejudice. His book, *The Economics of Discrimination* (1971), introduced prejudice and discrimination as forces that can be analyzed and measured in their effect on the economy. In *The Economics of Human Behavior* (1976), Becker argues that all human behavior, even selection of a marriage partner, is based on economics.

In 1965, Becker authored his famous treatise on the concept of human capital. In *Human Capital*, Becker treats the individual as a "firm" that makes investment decisions (such as education or on-the-job training) on the basis of rate of return. Soon after the publication of *Human Capital*, Becker was awarded the John Bates Clark medal by the American Economic Association for excellence in research by an economist under the age of forty.

Viewed simply in economic terms, population growth represents sustained demand for children on the part of parents. Both Malthus and Becker have analyzed the decision to have or refrain from having children on the basis of rational self-interest. Their widely different conclusions offer insight into this timely issue.

### The Check of Misery

Malthus's theory of population growth is strongly pessimistic. In *An Essay on the Principle of Population*, Malthus wrote that the world's population will increase at a rate that will ultimately test the limits of our available food supply and other subsistence goods. As the supply of food increases through additional labor and agriculture, population will increase as well, with the result that per capita income—the fruits of labor—will never exceed bare subsistence standards for the population as a whole. The threat of widespread famine will persist indefinitely. Only the lucky will survive.

An alternative way of stating Malthus's proposition is that children are what economists call normal goods, demand for which rises and falls in response to changes in income. Any increase in income to parents will increase their demand for children, other things being equal. At the point when an additional child will actually reduce living standards below subsistence—a point Malthus called the "check of misery"—parents will cease reproducing.

Given this gloomy scenario, it is not surprising that Malthus, a parson, would urge "moral restraints" on parents. Such restraints included abstinence and postponing marriage. It is interesting to speculate about whether Malthus would urge modern forms of birth control. The morality of contraceptive use, forced sterilization, and abortion is, of course, one of the burning issues of our times.

Malthus's theory of population growth seems today most applicable to the exploding populations of nations such as China, Mexico, Bangladesh, and other poor, developing countries. In the industrialized nations, including the United States, the last century has seen actual declines in average family size, while average family incomes have risen well above the level of bare subsistence. Many economists, including Gary Becker, have sought reasons why the theory has not, fortunately, become generally true.

### The Price of Children

Becker's theory of human capital helps explain both the power and the limits of the Malthusian theory. For Becker, the choice of whether to reproduce has a second rationale, one beyond a simple change in income. Parents' demand for children will be influenced not only by changes in income but also by the relative "price" of children. This price is partly the direct costs of raising the child, but it also includes the opportunity costs that additional children represent. As the parents' income increases, the opportunity cost of children will also tend to increase. In other words, the sacrifices borne by increasing the size of a family will increase with income. To take one example, if an executive forsakes a high-paying job to stay home with the children so that his or her spouse can pursue a career, the cost of that decision is the wages and income the executive receives now and the higher income he or she could expect to receive in the future.

As the costs of children increase, the quantity demanded will decrease, other things being equal. As a result, parents may substitute quality for quantity in their decisions to raise children. Instead of feeding an additional child, parents might decide to spend additional income on housing, education, or any of a host of goods that improve living standards for their children.

Becker's approach to the costs of children helps explain why parents in more developed countries, with relatively high incomes, have fewer, better-educated children than parents in undeveloped countries with relatively low incomes. In undeveloped, largely subsistence agricultural countries, children represent direct labor inputs—even small children can do useful work on a subsistence farm—thus lowering the price of children. Where the price of children is lower, we expect to see more children produced, other things being equal. Thus we see the law of demand at work in a novel way.

According to Becker, then, Malthus went wrong by failing to take into account relative prices in his theory of population. While rats or horses may naturally tend to breed up to the limit imposed by the available food supply, the behavior of human beings will also reflect the influence of opportunity costs and the rational choices such costs inspire.

# IV

# Microeconomics and Public Policy

# 17

# Market Structure and Public Policy

Our market economy has grown in ways that Adam Smith could never have imagined. Smith, you may recall, believed that market forces, operating under a laissez-faire form of government, would eventually bring us to a point of perfect competition among firms, what Smith termed "a nation of shopkeepers." Our economy, however, has followed a different path, and its market structures range from nearly pure competition to nearly complete monopoly. In the process, a few of our "shopkeepers" have managed to capture huge shares of their markets. In 1882, for example, John D. Rockefeller's Standard Oil Trust controlled 95 percent of the oil refineries in the United States, virtually eliminating all competition. Such amassing of market power may temporarily allow lower prices through economies of scale and coordination of production—characteristics often necessary to undermine competitors. But, once such a monopoly is created, some believe that it can dominate the industry at the consumers' expense. Responding to public alarm over this potential for abuse, the U.S. government began late in the nineteenth century to pass laws designed to thwart the growth of monopolies. These laws have slowed, but not eliminated, the trend toward monoply power. When U.S. authorities intervened and broke up Rockefeller's Standard Oil Trust, it created several firms in its place. Today one of those firms, Exxon, is still the largest industrial corporation in the world, with income in 1984 of over $5½ billion.

Is big business bad for the economy? Should the government intervene in private markets to regulate the size and monopoly power of firms? How does the govenment determine whether a firm has monopoly power? Which government policies make sense from an economic point of view? All of these questions form the focus of this chapter. The overall question of mo-

## Measuring Industrial Concentration

Before it can intervene to break up, regulate, or even personally operate a monopoly in the public interest, the government must be able to determine whether a monopoly exists. The number of firms in an industry is a major indicator of the degree of competition. We examine two rough estimates of the degree of concentration of market power within an industry—aggregate concentration and industry concentration ratios. We also question why some industries exhibit high concentration ratios while others don't, noting that changing demands or costs will change the number of firms in an industry and thus the potential for monopoly concentration.

### Aggregate Concentration

One method of gauging the extent to which production assets are dispersed among different firms is to examine the **aggregate concentration** of firms in the economy as a whole. These statistics measure the percentage of all business assets that are owned by the largest one hundred or two hundred firms in the country. This approach ignores the percentage of individual industries' assets that firms own and instead focuses on the percentage of the nation's total assets held by the largest enterprises. Instead of determining the degree to which General Motors controls the assets of the automobile industry, aggregate concentration is a measure of the degree to which General Motors, Allied Chemical, and other manufacturing giants control assets among all manufacturing firms.

Table 17–1 shows trends in aggregate concentration for firms in the manufacturing sector between 1925 and 1977. The share of total manufacturing assets held by the top one hundred and top two hundred companies increased by about ten percentage points over the past fifty years, but the aggregate concentration has remained remarkably stable since 1958. Table 17–1 suggests that there is little cause for concern that the productive assets of this country are becoming increasingly concentrated in the hands of fewer and fewer firms.

### Industry Concentration Ratios

An alternative method of looking at business structure is to calculate **concentration ratios** (CR) for various industries. Concentration ratios are determined by ranking firms within an industry from largest to smallest and then calculating the share of some aggregate factor such as sales, employment, or assets held by the largest four or eight firms. Sales is the most common factor used in these calculations.

Figured by this method, the concentration ratio will lie between zero and one hundred. If an industry is totally monopolized, one firm will have 100 percent of industry sales, CR = 100. If an industry has several firms but the four largest have 50 percent of the market sales, then CR = 50. If a very large number of companies each have a very small market share, then CR will be close to zero, suggesting the existence of pure competition.

Table 17–2 displays four-firm and eight-firm concentration ratios for a small sample of manufacturing industries. A brief look at the concentration ratio data indicates that quite a bit of variation in business structure was hidden within the aggregate concentration percentages shown in Table

---

**Aggregate concentration:** A measure of the percentage of total productive assets held by the largest one hundred or two hundred firms in the economy.

**Industry concentration ratios:** An estimate of the degree to which assets, sales, or some other factor is controlled by the largest firms in an industry.

TABLE 17–1  **Trends in Aggregate Concentration in the Manufacturing Sector, 1925–1977**

To determine aggregate concentration, we measure the percentage of assets owned by the largest firms in the country. Since 1958, the percentage of assets held by the top one hundred and top two hundred firms in the U.S. manufacturing sector has remained remarkably stable.

| | Percentage of Assets Held by the Largest Firms | |
|---|---|---|
| Year | 100 Largest | 200 Largest |
| 1925 | 34.5% | — |
| 1929 | 38.2 | 45.8% |
| 1935 | 40.8 | 47.7 |
| 1939 | 41.9 | 48.7 |
| 1948 | 38.6 | 46.3 |
| 1950 | 38.4 | 46.1 |
| 1958 | 46.0 | 55.2 |
| 1963 | 45.7 | 55.5 |
| 1967 | 47.6 | 58.7 |
| 1972 | 45.3 | 56.5 |
| 1974 | 46.1 | 57.6 |
| 1975 | 46.0 | 57.3 |
| 1976 | 46.1 | 57.3 |
| 1977 | 45.6 | 56.6 |

*Source:* Richard Duke, "Trends in Aggregate Concentration," *FTC Working Paper No. 61,* (June 1982), p. 18.

17–1. Some industries are apparently very highly concentrated. In 1977, for example, the top eight producers of flat glass accounted for 99 percent of industry shipments, indicating that the remaining flat-glass manufacturers were very small businesses indeed. In contrast, the eight largest firms supplying ready-mix concrete had only an 8 percent share of total sales in 1977, suggesting that this industry is made up of a large number of relatively small companies.

Table 17–2 also reveals trends in concentration. Concentration ratios that have risen over the years indicate that the leading firms have grown relative to other industry members, either by internal expansion, merger, or decreases in the number of other firms. Declining concentration ratios imply that more and more competitors have made inroads into the industry and thereby reduced the dominance by leading firms. While examples of each pattern appear in the table, concentration ratios remain fairly stable in general.

One problem with using concentration ratios to determine the degree of competition in an industry is that the data can obscure quite vigorous competition. Since the ratios do not identify individual firms, the concentration ratio can remain stable over time even though the leading firms rapidly turn over. For example, if the first and fourth companies in meat packing change places, the four-firm concentration ratio would remain unchanged.

A further difficulty with the concentration ratios in Table 17–2 is that they are based on the Commerce Department's Standard Industrial Classification (SIC) groups. The SIC codes group firms that produce similar products, but these codes do not account for substitutability by buyers. For example, plastic panes may frequently be substituted for flat glass, but the SIC codes compare glass manufacturers only with other glass manufacturers. Thus, the concentration ratios do not show the competition between groups

TABLE 17–2  Selected Concentration Ratios in Manufacturing

Concentration ratios measure the share of sales held by the largest four or eight firms in an industry. The data shown here represent a selected group of manufacturing industries for the years 1947, 1963, and 1977. In some industries—such as men's and boys' suits—the concentration ratio increased through the years, suggesting that the largest firms obtained a larger share of total industry output. In some cases—such as meat packing—the ratio fell, suggesting a more competitive situation.

| | Concentration Ratios | | | | | |
|---|---|---|---|---|---|---|
| | 1947 | | 1963 | | 1977 | |
| | Four-Firm | Eight-Firm | Four-Firm | Eight-Firm | Four-Firm | Eight-Firm |
| Meat packing | 41 | 54 | 31 | 42 | 19 | 37 |
| Fluid milk | – | – | 23 | 30 | 18 | 28 |
| Cereal breakfast foods | 79 | 91 | 86 | 96 | 89 | 98 |
| Distilled liquor | 75 | 86 | 58 | 74 | 52 | 71 |
| Roasted coffee | – | – | 52 | 68 | 61 | 73 |
| Cigarettes | 90 | 99 | 80 | 100 | – | – |
| Men's and boys' suits and coats | 9 | 15 | 14 | 23 | 21 | 32 |
| Women's and misses' dresses | – | – | 6 | 9 | 8 | 12 |
| Logging camps and contractors | – | – | 11 | 19 | 29 | 36 |
| Mobile homes | – | – | – | – | 24 | 37 |
| Pulp mills | – | – | 48 | 72 | 48 | 76 |
| Book publishing | 18 | 29 | 20 | 33 | 17 | 30 |
| Pharmaceutical preparations | 28 | 44 | 22 | 38 | 24 | 43 |
| Petroleum refining | 37 | 59 | 34 | 56 | 30 | 53 |
| Flat glass | – | – | 94 | 99+ | 90 | 99 |
| Ready-mix concrete | – | – | 4 | 7 | 5 | 8 |
| Blast furnaces and steel mills | 50 | 66 | 48 | 67 | 45 | 65 |
| Metal cans | 78 | 86 | 74 | 85 | 59 | 74 |
| Electric lamps | 92 | 96 | 92 | 96 | 90 | 95 |
| Radio and TV receiving sets | – | – | 41 | 62 | 51 | 65 |
| Motor vehicles, car bodies | – | – | – | – | 93 | 99 |
| Jewelry, precious metal | 13 | 20 | 26 | 33 | 18 | 26 |
| Pens and mechanical pencils | – | – | 48 | 60 | 50 | 64 |

Source: U.S. Department of Commerce, Bureau of the Census, *Census of Manufactures*, 1977.

that results from actual market structures. Moreover, the SIC categories include sales data for the entire nation and therefore do not give information about concentration levels in different geographic regions. The four- and eight-firm concentration ratios were replaced in 1982 by the Herfindahl index, described in Focus, "New Merger Guidelines," but the Herfindahl index is subject to the same problems.

### Differences in the Number of Firms

Why are there so few firms, sometimes only one, in some industries, while other industries have several hundred? The answer lies in the relation between the industry demand and the firm's average cost of production.

For simplicity's sake, consider a hypothetical industry in which all firms have identical products and costs of production. Let the curve $AC_1$ in Figure 17–1 represent one firm's average costs. The output that minimizes average costs, $q_1$, is the equilibrium output for each firm. In other words, as we recall from Chapter 8, the optimal size for the individual firm is the output level that allows the lowest possible per unit cost of production. On an industry-wide level, $D$ represents the entire demand in the industry. The amount

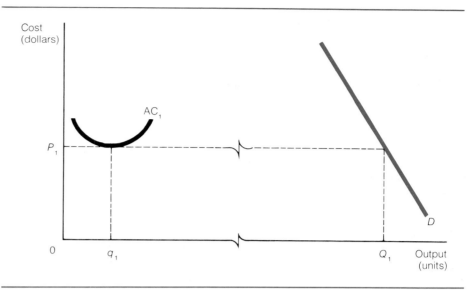

**FIGURE 17–1  Determination of the Number of Firms in an Industry**
The equilibrium number of firms is determined by the magnitude of industry demand relative to the firm's average cost of production. The average cost curve for one representative firm is given as $AC_1$. The minimum average cost determines the optimal rate of output and plant size for each firm. The total number of firms is equal to the output per firm, $q_1$, divided into the amount demanded in the industry, $Q_1$, at the price that is equal to minimum average cost. If $q_1 = 100$ and $Q_1 = 10{,}000$, then the equilibrium number of firms in this industry would be 100.

demanded in the industry at price $P_1$ is $Q_1$. The equilibrium number of firms in the industry is therefore $Q_1$ divided by $q_1$. If the optimal output for each firm is 100 units and the amount demanded in the market is 10,000 units, then 100 firms can exist in this industry.

The greater the industry demand relative to the firm's output at minimum average cost, the greater the number of firms will be. Highly concentrated industries result from industry demand that is insufficient to support a large number of firms of optimal size. In fact, a **natural monopoly** exists when industry demand is just great enough relative to a firm's average cost to support only one firm.

**Natural monopoly:** Market conditions that allow for the survival of only one firm in an industry.

Figure 17–2 shows two types of natural monopolies. In Figure 17–2a, a natural monopoly develops because demand is insufficient to support more than one firm. The monopolies held by small-town newspapers provide a familiar example: Demand for local news in small towns often is not great enough to support more than one newspaper. In such a case, $D_A$ represents the industry demand—the demand for local news in the small town; $Q_a$ represents the quantity of newspapers demanded at price $P_a$—a quantity that can be produced profitably by one firm. $AC_a$ and $MC_a$ represent this newspaper firm's average and marginal costs. In this industry, only one firm can survive because industry demand is not great enough to support two firms. If another firm enters the market, industry sales will be divided by two. For example, at $P_a$ the total industry output demanded is still $Q_a$, and thus each firm's sales are $Q_a/2$. Indeed, the marginal revenue curve represents each firm's demand curve when the industry demand is shared by the two firms. Since the individual firm's demand curve lies below $AC_a$ at all levels of

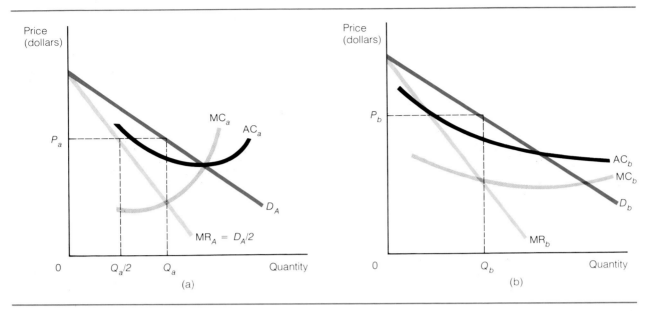

**FIGURE 17–2  Demand-Cost Relations of Natural Monopolies**
When industry demand is not great enough to support more than one firm of optimal size, then a natural monopoly results. (a) Only one firm can survive because half of industry demand, $MR_A$, the demand that each firm faces when there are two firms in the industry, does not rise above the firms' average cost. (b) A single firm survives because one large firm, through economies of scale, can achieve lower average cost than two smaller firms.

output, two local newspapers cannot profit and survive; only one firm can exist under these market conditions.

Figure 17–2b demonstrates a natural monopoly created by economies of scale. A single television company might dominate the market in a given area because it can supply the entire existing demand among customers at lower cost than two or more smaller firms. One large TV company serving many customers can therefore offer lower commercial prices than several firms serving fewer customers. The key feature of the model shown in Figure 17–2b is that the industry demand intersects the average cost curve at a point where average costs are still declining owing to economies of scale. Even though two firms could survive in this industry because their demand curves lie above $AC_b$ at some points, we expect that price competition would eliminate one firm. The reason is that a single firm with a larger output can produce the entire industry output at a lower cost than two smaller firms. The existence of economies of scale yields a cost advantage for a single large firm.

Comparing the industry demand to the firm's average cost is a useful approximation of the number of firms that can exist in an industry. This method does have some limitations, however, for its assumptions may be unrealistic. For one thing, not all firms have identical costs. Firms in different parts of the country are confronted with different input prices and a different economic environment, for example. Consequently, the optimal size of the firm varies from one location to another, and the number of firms becomes difficult to predict. The assumption of identical products may also be false. If firms in an industry produce products that are not perfect substi-

## FOCUS  New Merger Guidelines

The revised merger guidelines issued by the Department of Justice in 1982 list conditions under which the department's Antitrust Division will "more often than not" challenge a proposed merger or acquisition. The conditions, which focus on the pre- and post-merger concentration levels, are expressed in terms of the Herfindahl index.

In brief, the Herfindahl index seeks to summarize in one number the extent to which a market is "concentrated" or dominated by a few firms. The index is calculated by squaring and then summing the individual market shares of all firms in the market. That is,

$$H = s_1^2 + s_2^2 + \ldots + s_n^2,$$

where, for example, $s_1$ is firm 1's percentage market share (usually its percentage of industry sales) and there are $n$ firms in the market.

$H$ will always be between zero and 10,000. If the industry is monopolized, one firm will have 100 percent of the market, so $H = 100^2 = 10,000$. In contrast, if each of a very large number of firms has a very small market share, $H$ will be close to zero.

The advantage of the Herfindahl index over the more traditional four- and eight-firm concentration ratios is that it gives information about the dispersion of firm size within a market. The Herfindahl index does this by giving greater weight to large market shares than to small rather than treating all market shares sizes equally. A disadvantage of the Herfindahl index is that its calculation requires the gathering of information on the market shares of every firm in the industry.

The 1982 Justice Department merger guidelines label a market as concentrated if $H$ is equal to 1,800 or more. (This situation would occur, for instance, if four firms in an industry accounted for approximately 50 percent of sales.) In such cases a merger that causes $H$ to increase by 100 will generally be challenged by the Antitrust Division. In unconcentrated markets (where $H$ is 1,000 or less), mergers leading to a change in $H$ of 200 or more will usually trigger antitrust action.

Suppose that an industry is made up of three equally sized firms. The Herfindahl index would work out to

$$H = 33^2 + 33^2 + 33^2 = 3267.$$

A merger between any two of the firms would raise $H$ by 1178:

$$H = 66^2 + 33^2 = 5445.$$

Since the industry was already concentrated by Justice Department standards, the merger would undoubtedly draw legal action.

In contrast, suppose that one firm accounts for 25 percent of sales in an industry and that one hundred other sellers each have a 0.75 percent market share:

$$H = 25^2 + 0.75^2 + 0.75^2 + \ldots + 0.75^2$$
$$= 683.25.$$

If the large firm merges with one of the smaller firms, $H$ rises accordingly:

$$H = 25.75^2 + 0.75^2 + 0.75^2 + \ldots + 0.75^2$$
$$= 719.56.$$

This merger would generally go unchallenged.

Note that both the Herfindahl index and the merger guidelines are based on the assumption that firm size is an indication of market power. That is, concentration is considered bad because markets dominated by one or a few firms are thought to exhibit prices and profits exceeding the competitive level. Of course, firms can be large because they are efficient and serve consumers well. Thus, one should look beyond simple concentration levels before concluding that antitrust action is warranted.

---

tutes for one another, then industry demand and product price will be difficult to determine.

### Changes in the Number of Firms

Since the number of firms in an industry is determined by the magnitude of industry demand relative to each firm's average cost, any change in demand or cost can lead to a change in the equilibrium number of firms. If the demand in a particular industry increases, then we expect to see more firms enter the industry. As demand for computers has soared, IBM's previous dominance of the computer market has decreased. On the other hand, if the price of inputs changes such that the average cost curve shifts to the right, then the optimal plant size increases. Such a shift will lead to a decrease in the number of firms. Changes in technology can also affect the optimal plant size and thus the number of firms. Such changes have led to a decrease in the number of automobile manufacturers in the United States.

Since the economy is dynamic, we expect fairly frequent changes in production cost and industry demand. The number of firms in various industries will therefore change through time. New firms enter and old ones go out of business. The concentration ratio of an industry may thus change because of events external to the firms.

Figure 17–3 gives an indication of the turnover rate in U.S. markets between 1929 and 1980. The turnover rate is the rate at which new businesses open and older ones that are failing close their doors. The first point to note in Figure 17–3 is that many new enterprises are formed and quite a few fail in any given year. In 1980, for example, more than 500,000 new firms were incorporated, and nearly 12,000 businesses went bankrupt. The business failure rate during the Great Depression was very high. Since then the rate of failure has slowed considerably, but failure rates are still significant. Starting a new business continues to be a high-risk proposition. There appears to be no shortage of entrepreneurs willing to take those risks, however, in hopes of succeeding.

## Monopoly Profit Seeking

Entrepreneurs organize and operate businesses in an effort to obtain profits. The profit motivation is not socially undesirable, for businesses earn profits by producing products that are most highly valued by society while simultaneously using the fewest and least expensive resources. However, the level of profits is limited in purely competitive industries. When new firms are able to enter profitable industries and act independently in seeking profits, economic profits are forced to zero. On the other hand, firms that are natural monopolies may enjoy positive economic profits—a greater than normal return on their investments.

Even though profits in competitive industries are no greater than normal, there is a potential for obtaining greater profits by decreasing the amount of competition. If all the firms in a competitive industry could agree not to compete and simultaneously could prevent new firms from entering the market, monopoly-level profits could be obtained.

As explained in Chapter 10, converting a competitive industry into a monopoly results in loss of part of the consumer surplus. Consumer surplus is the amount that consumers would be willing to pay for a good above the price actually charged for it. This difference translates into real income to consumers. In Figure 17–4, a competitive market reaches equilibrium price and quantity for a certain good—perhaps lime juice—at price $P_c$ and quantity $Q_c$, with $D$ and $S$ indicating industrywide demand and supply conditions. At the competitive price and quantity, consumers enjoy a surplus equal to $FCP_c$, the area beneath the demand curve but above the supply curve.

If competition is eliminated in the lime juice industry, firms acting with monopoly power can create contrived scarcity by restricting output to $Q_m$ and raising the price of lime juice to $P_m$. At this level, consumer surplus shrinks to the area $FAP_m$, shifting the profits represented by $P_mABP_c$ to the pockets of the lime juice monopolists. The area $ACB$ is simply lost to everyone. It is neither transferred to monopolists in the form of profits nor re-

**FIGURE 17–3**

**Business Formation and Business Failures, 1929–1980**

During the Great Depression, the rate of business failures was quite high. The chances for survival have improved considerably since that time, yet business formation remains risky. The number of new incorporations has increased steadily since 1950, indicating that many individuals are still willing to take the risk.

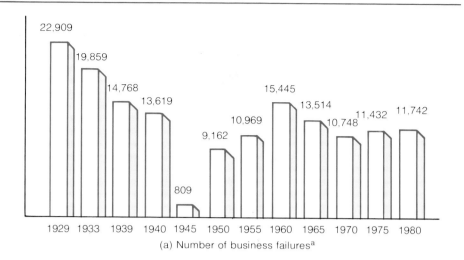
(a) Number of business failures[a]

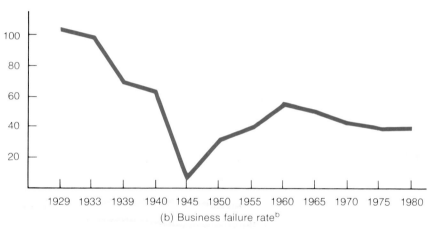
(b) Business failure rate[b]

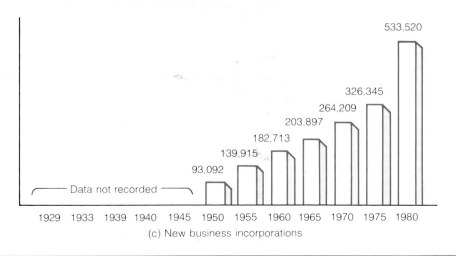
(c) New business incorporations

[a]Commercial and industrial failures only. Excludes failures of banks and railroads and, beginning in 1933, of real estate, insurance, holding and financial companies, steamship lines, travel agencies, and the like.
[b]Failure rate per 10,000 listed enterprises.
Source: *Economic Report of the President* (Washington, D.C.: U.S. Government Printing Office, 1982), p. 338.

## FIGURE 17–4
### Welfare Loss Resulting from Monopoly

With pure competition, the demand $D$ and supply $S$ intersect at $C$ and the price and quantity at $P_c$ and $Q_c$. The consumer surplus is $FCP_c$. If this industry is monopolized, then the profit-maximizing price and quantity are $P_m$ and $Q_m$. Profits are the area $P_mABP_c$, and consumer surplus is decreased to area $FAP_m$. The welfare loss is the lost consumer surplus ($ACB$) that is not transferred to the monopoly in the form of profits.

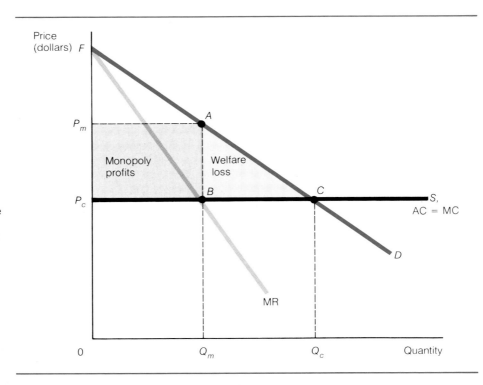

**Welfare loss:** The consumer surplus that is lost to consumers but not transferred to the monopoly in profits when a competitive industry is monopolized.

tained by consumers as consumer surplus. This area is therefore treated by economists as a **welfare loss** to society—to both consumers and producers. It is the cost to society of having a monopolist produce a product rather than having the good supplied by a competitive industry.

To capture monopoly profits, firms may engage in a variety of activities designed to decrease competition. As described below, anticompetitive behaviors include the formation of cartels, mergers, and trusts, and collusive agreements and government franchises.

### Cartels

Cartels, as described in Chapter 11, represent a formal alliance among producers of a good. The firms agree not to engage in price competition and agree to restrict output in order to extract monopoly profits. To be successful, a cartel must prevent member firms from charging a price below the monopoly price. The cartel must also be able to prevent entry of new firms—otherwise, profits are dissipated by the entry of competitive firms.

OPEC (Organization of Petroleum Exporting Countries) was a fairly successful cartel. OPEC firms are prevented from cheating on the cartel by the governments of the countries in the organization. New firms are prevented from entering by the endowments of nature—new firms cannot produce oil unless they acquire an oil deposit. These features explain the longevity of OPEC. Most cartels that are not able to police cheating and limit entry are short-lived as OPEC's experience in the early 1980s proved.

The endowments of nature are unknown until exploration becomes profitable. Indeed, OPEC's initial success made exploration and development profitable to firms that formerly would not have entered. Consider the new production from the North Sea, Mexico, China, and the North Slope of

Alaska. These new "firms" have entered, and OPEC is now much weaker in part as a result of their entry.

### Trusts

In the late 1800s, during the Industrial Revolution, the optimal size of firms—the size allowing the lowest possible per unit cost of production—began to increase. This trend lowered the number of firms in some industries and increased concentration ratios. When there are only a few firms in an industry, organizing them to act as a single firm becomes less costly. Firms in some industries joined holding companies called **trusts** that monitored the activities of members. For example, John D. Rockefeller capped his concentration of power in the oil-refining industry by forming the Standard Oil Trust, an organization to which stockholders of almost all oil-refining companies turned over controlling amounts of their stock in return for dividend-paying trust certificates. Such unifying organizations allowed firms to act as a cartel. Antitrust legislation, to be discussed later, has since been instituted to prevent most such organizations.

**Trust:** An institution that organizes firms in the same industry in an effort to increase profits by decreasing competition.

### Collusion

When there are very few firms in an industry but antitrust laws prevent formal organization, businesses sometimes engage in informal agreements to restrict output or raise prices, a form of **collusion.** Price competition can be diminished through explicit or implicit agreements. Firms may simply agree to charge a monopoly price and restrict output. The success of simple collusion is particularly rare since price cheaters and new entrants cannot be controlled informally.

**Collusion:** An explicit or implicit agreement between firms in the same industry not to engage in competitive behavior.

### Mergers

One way that concentration ratios may rise is through **mergers.** If there are only a few firms in an industry, the combination of two or more firms into one may decrease the degree of competition. However, the types of mergers vary and thus their effect in the market varies. There are three basic types of mergers: horizontal, vertical, and conglomerate. Each type is distinguished by the relation between the products produced by the merging firms.

**Merger:** The joining of two or more firms' assets that results in a single firm.

A horizontal merger occurs when the merger partners produce the same product. Examples include mergers between two beer producers, two manufacturers of glass containers, two retail grocery stores, and so forth. Horizontal mergers usually decrease competition because they have the direct effect of raising industry concentration ratios and moving the affected market closer to monopoly.

Mergers are characterized as vertical if the product of one firm serves as an input for the other firm, that is, the merging companies are at different stages of the production process. A vertical merger would occur if a company producing crude oil acquired a petroleum refinery, if a steel producer purchased iron ore deposits, or if an automobile manufacturer acquired a string of retail car dealerships. Vertical mergers do not raise industry concentration levels and are usually motivated by the prospect of lower costs. Accordingly, vertical mergers do not normally decrease competition.

Conglomerate mergers occur when the products of the firms are unrelated: for example, when an insurance company buys a firm that produces bread or if a cigarette producer merges with a soft drink company. The usual explanation for such mergers is that they are motivated by a desire to spread the risks associated with economic ups and downs across products. However, spreading risks can be accomplished much more simply by, for example, purchasing shares of a variety of businesses. Rather, conglomerate mergers can better be explained by the desire to diffuse managerial expertise across firms. That is, if an efficient corporate management team believes that their executive talents can be profitably applied to some other firm, they will have an incentive to acquire another company regardless of the product produced by the other firm. Since conglomerate mergers do not raise industry concentration levels, they do not generally decrease competition.

### Government-Managed Cartels

Given the problems associated with organizing and policing a cartel in an industry with a large number of firms, it is not surprising that firms turn to the power of government when seeking profits. Sometimes firms are able to entice the government to enforce a higher-than-competitive price or a lower-than-competitive output or to prevent new firms from entering their otherwise competitive industry. Long before OPEC, for example, the Texas Railroad Commission limited the rate at which oil could be pumped from wells in Texas. This action bolstered price above the competitive level.

## Public Policy

Purely competitive industries represent the ideal in market performance. In industries characterized by imperfect competition, the ideal is not obtained—profits may be greater than normal, production costs are not necessarily minimized, and welfare losses can occur. If these losses to society are significant, then there is a call for social action in the form of government regulation of some businesses' activities.

Public policies regulating business performance take many forms. Variety in policies is desirable because the imperfections in markets are caused by many diverse conditions. For example, the market conditions that create a natural monopoly differ significantly from those that promote a trust company or cartel. Even though the conditions vary, the final consequences of these market structures are indeed the same—monopoly price, restricted output, and welfare loss.

Public solutions to the monopoly problem take four basic forms:

1. Antitrust laws basically designed to prevent the creation of monopolies.
2. Price regulations designed to allow the existence of monopolies but to control prices and profits.
3. Government ownership that removes the firm from the private sector and thus places its performance directly in the hands of the public.
4. Laissez-faire policy that allows the market to determine the fate and performance of firms with no government interference.

### Antitust Law

**Antitrust policy:** The laws and agencies created by legislation in an effort to preserve competition.

**Antitrust policies** are aimed at preventing firms from engaging in anticompetitive activities. During the 1870s and 1880s, an antitrust movement sprang up primarily among American farmers. Facing falling prices for their products and apparently stable prices for the goods they bought, they became fearful that monopolies or trusts were wielding unwarranted economic power. Through organizations such as the National Grange and the National Anti-Monopoly Cheap Freight League, the farmers were successful in influencing the two major political parties to add antimonopoly planks to their 1888 election platforms. Antitrust legislation was enacted soon after, with subsequent revisions to strengthen the government's ability to fight monopolistic practices in court.

**Sherman Antitrust Act.** The Sherman Antitrust Act of 1890, the first antimonopoly law passed by Congress, provided that "every contract, combination in the form of trust or otherwise, or conspiracy, in restraint of trade or commerce among the several states, or with foreign nations, is hereby declared to be illegal. . . ." The Sherman Act also declared "every person who shall monopolize, or attempt to monopolize . . . guilty of a misdemeanor, and subject to fine or imprisonment." The statute did not spell out what would constitute an illegal restraint of trade. Its main purpose was to place the antitrust issue under federal law and to provide a means of penalizing monopoly whenever it was discovered.

**Clayton Antitrust Act.** Believing that the Sherman Act was ineffective in preventing monopoly, Congress passed the Clayton Act in 1914. This act sought to limit a firm's acquisitions of the stock of another firm "where in any line of commerce in any section of the country, the effect of such acquisition may be substantially to lessen competition, or to tend to create a monopoly." Unfortunately, the Clayton Act was silent on the legality of mergers through the acquisition of physical assets, leaving a large loophole for avoiding antitrust prosecution.

The Clayton Act also enumerated some specific restraints of trade that could be challenged if their effects tended to lessen competition substantially or create a monopoly. These included (1) price discrimination—selling the same product to different customers at different prices not related to cost differences; (2) tying arrangements—making the sale of one good contingent on the purchase of some other goods; (3) interlocking directorates—the same person's serving on the boards of competing companies; and (4) exclusive dealing—selling to a retailer only on the condition that the firm not carry rival products.

**Federal Trade Commission Act.** Section 5 of the Federal Trade Commission (FTC) Act, also passed by Congress in 1914, contains the broadest statutory language, declaring illegal "unfair methods of competition in commerce." The act established a five-member commission independent of the executive branch, with the idea that the FTC would be a repository of economic and antitrust expertise not available to the federal courts.

**Other Antitrust Statutes.** The Clayton Act has been amended twice. In 1936, the Robinson-Patman Act revised the provisions relating to price discrimination and added language prohibiting predatory pricing, the act of selling goods below cost as a method of destroying competitors (see Focus, "Cutthroat Competition"). In 1950, the Celler-Kefauver Act closed the Clayton Act loophole relating to mergers through the acquisition of physical assets.

The FTC Act has been revised even more often. First, in 1938 the Wheeler-Lea Act added "unfair or deceptive acts or practices in commerce" to the behavior declared illegal. In the late 1970s, the Hart-Scott-Rodino Antitrust Improvement Act established a premerger notification system by which firms contemplating a merger would notify the FTC and the Department of Justice prior to consummating the acquisition.

## FOCUS: Cutthroat Competition

Predatory pricing was declared illegal in the Robinson-Patman amendment to the Clayton Act. Predatory pricing—sometimes known as "cutthroat competition"—is said to exist when a firm lowers its price below its competitors' cost of production with the intention of driving competitors out of business. If the firm is able to do this, it may eventually become the only firm in the industry. It can then raise prices, and its subsequent monopoly profits can compensate for the losses it incurred during its predatory pricing period.

This activity was made illegal to prevent monopolies from arising in industries where competition could presumably exist. However, economic theory suggests that predatory pricing will not result in a monopoly unless the market conditions for a natural monopoly already exist. The only way that a single large firm can survive with no entry of new firms is to have lower average cost of production than the potential entrants. Otherwise, it must continually incur losses to prevent new firms from entering.

When a natural monopoly evolves in a market, one firm becomes larger, its price falls, and smaller firms go out of business. The term "predatory pricing" seems to describe these circumstances fairly accurately, but making cutthroat competition illegal does not change the market conditions that create a natural monopoly. Antitrust laws do not change the cost of production or the demand for the product.

Laws prohibiting predatory pricing are at best temporary hindrances to the emergence of a natural monopoly. Such prohibitions may indeed postpone the emergence of monopoly long enough for market conditions to change. Otherwise, outlawing price competition can have undesirable effects. For example, if a firm is prevented by law from charging a price that is below its competitors' average cost, then the law in effect may be protecting inefficient firms. Price competition is what forces the survival of only low-cost firms and thus low prices for consumers. Preserving competition by protecting firms from low prices ensures high prices for consumers.

---

**Antitrust Agencies.** The United States has a dual system of antitrust enforcement whereby two government agencies—the Federal Trade Commission and the Justice Department's Antitrust Division—share responsibility for policing the antitrust laws. Both agencies can enforce the Clayton Act, and, since 1948, the FTC has had the power to bring charges against behavior that violates the Sherman Act.

In addition to the two government agencies, complaints charging Sherman and Clayton antitrust violations can be brought by private parties. Private plaintiffs can sue for treble damages; that is, guilty defendants can be required to pay penalties of up to three times the value of the actual injury caused by their illegal conduct. Because private suits offer such potentially large rewards for successful plaintiffs, in any given year the number of private antitrust actions is many times higher than the number of cases brought by government.

### The Value of Antitrust Policies

The antitrust laws and agencies have provided a means to control anticompetitive behavior in industries. Initially they were promulgated to prevent firms from organizing into trusts or cartels designed to increase members' profits by decreasing price competition. Antitrust policies have since been extended to prevent other activities that decrease consumer surplus, such as price discrimination and tying arrangements. Under most circumstances, the free market prevents these forms of competition. However, when concentration of market power is high, the market can fail to prevent such behaviors. Antitrust laws and agencies have therefore extended government control over industry with the goal of preventing increases in concentration.

Many factors external to firms can increase concentration, however. As we have seen, a fall in industry demand or a change in either input prices or technology that increases the optimal plant size can cause increases in concentration. These changes are beyond the control of the firms in the industry, and a decrease in the number of firms is indeed socially optimal under these circumstances. Thus, antitrust policies designed only to promote low concentration can be detrimental to economic efficiency.

### Enforcement of Antitrust Policies

Antitrust policies are carried out in the courtroom. Antitrust officials, policymakers, and court officials face difficulties in distinguishing between behavior that is a conscious intent of firms to decrease competition and behavior that is the result of firms' responding to external changes. For this reason, the FTC treats every market situation as unique. Before the FTC takes action in altering market activities, it examines the individual firms and their activities. Firms judged to be in violation of antitrust laws may be ordered to stop their anticompetitive practices or to divest themselves of some of their assets by splitting the monopoly into several smaller competing firms. Sound economic theory is not always well represented in these court decisions. We will consider some specific cases that have been tried and the consequences of the decisions.

**Cartel Policies.** In 1927, the Supreme Court in the Trenton Potteries Case[1] held that price-fixing conspiracies were per se illegal under the Sherman Act even though the act did not specifically prohibit price fixing. That is, attempts by competitors to fix prices would be found illegal regardless of whether the fixed price was actually above the competitive level.

The case involved the activities of the Sanitary Potters Association (SPA), a Trenton, New Jersey trade association whose twenty-three member firms produced 82 percent of the U.S. output of bathroom sinks, tubs, and commodes. The SPA published price lists that included suggested discounts and surcharges for six geographic regions; members following the SPA lists charged identical prices.

The question addressed by the Supreme Court was whether a lower court judge was correct in directing the jury not to consider the reasonableness of the particular prices charged. The Court held that "uniform price-fixing by those controlling in any substantial manner a trade or business in interstate commerce is prohibited by the Sherman Law, despite the reasonableness of the particular prices agreed upon."[2] It seems that economic theory would support the court's decision. If a large percentage of the firms in an industry agree on price, then it is likely that price competition has been abolished.

**Merger Policy.** As we noted earlier, the Clayton Act, as amended by the Celler-Kefauver Act, seeks to limit corporate acquisitions that "tend to create a monopoly." Under the Hart-Scott-Rodino premerger notification system, most large firms contemplating a merger are required to apprise both the FTC and the Justice Department's Antitrust Division of their intentions. The agency chosen to handle the merger must decide whether to challenge it by seeking a preliminary injunction in federal court or to allow the merger to proceed without opposition. In the case of relatively small mergers, the

---

[1] *United States v. Trenton Potteries Co.*, 273 U.S. 392 (1927).
[2] See Phillip Areeda, *Antitrust Analysis: Problems, Text, Cases*, 3rd ed. (Boston: Little, Brown, 1981), p. 165.

antitrust authorities may become involved after the fact. In particular, a firm that is later found to have acquired another company in violation of the Clayton Act may be required to divest itself of the acquired assets in order to restore the premerger status quo.

An important aspect of merger law enforcement is the use of guidelines to establish which particular acquisitions will be challenged. For example, vertical and conglomerate mergers are not often challenged, but horizontal mergers that significantly decrease competition are. The merger guidelines issued by the Justice Department contain numerical standards that identify markets thought to be "unconcentrated" or "concentrated" prior to a merger. The guidelines set limits on the amount that concentration will be allowed to rise in each case before antitrust action is initiated. In 1968, concentration ratios were established for this purpose; in 1982, these ratios were supplanted by the Herfindahl index (described in the Focus essay earlier in the chapter).

In 1966, before these indexes were in use, the Supreme Court considered the legality of a merger between two Los Angeles retail grocery chains, Von's Grocery Company and Shopping Bag Food Stores.[3] At the time of the acquisition, 1960, Von's was the third largest grocery chain in Los Angeles; Shopping Bag was the sixth largest. Together, the two grocery retailers accounted for only 7.5 percent of sales in the Los Angeles area, however. Despite the relatively small market shares involved in the merger, the Court held that Von's had violated the Clayton Act. The court decision placed great weight on the fact that the number of owners operating single grocery stores in Los Angeles had declined from 5,365 in 1950 to 3,813 in 1961. This decline, the Court alleged, indicated a trend toward concentration.

In his dissenting opinion, Justice Potter Stewart called the Court's decision "a requiem for the so-called 'Mom and Pop' grocery stores . . . that are now economically and technologically obsolete in many parts of the country." He went on to say that the Court,

> through a simple exercise in sums, . . . finds that the number of individual competitors in the market has decreased over the years, and, apparently on the theory the degree of competition is invariably proportional to the number of competitors, it holds that this historic reduction in the number of competing units is enough under Section 7 of the Clayton Act to invalidate a merger . . . with no need to examine the economic concentration . . . , the level of competition . . . , or the potential adverse effect of the merger.[4]

If the number of firms in an industry is declining because of technological changes, economic theory suggests that the optimal size of firms is increasing. This change will naturally lead to an increase in concentration. According to economic theory, therefore, the merger probably should not have been prevented, especially since the two firms constituted such a small percentage of the market.

**Price Discrimination.** Charges of price discrimination may also require subtle distinctions. In a private suit, the Utah Pie Company charged Continental Baking Company,[5] Carnation Company, and Pet Milk Company with price discrimination in the sale of frozen pies. Utah Pie, located in Salt Lake City, alleged that its three California-based competitors sold pies shipped to

---

[3]*United States v. Von's Grocery Co.*, 384 U.S. 270 (1966).
[4]Areeda, *Antitrust Analysis*, pp. 961–964.
[5]*Utah Pie Co. v. Continental Baking Co.*, 386 U.S. 685, 699 (1967).

Salt Lake City at prices below those charged for pies sold nearer their own plants.

The evidence before the Supreme Court pointed to a highly competitive pie market in Salt Lake City. The price discrimination evidence appeared superficially correct in the sense that, given the cost of transportation, one would not expect the prices of the California firms to be lower in Salt Lake City. Accordingly, the Supreme Court found in favor of Utah Pie.

The Court's decision does not square with some of the facts, however. For one thing, the prices charged by Utah Pie were consistently lower than the prices of its rivals throughout the price discrimination episode. Second, Pet Milk Company suffered substantial losses on its Salt Lake City sales. Third, Utah Pie consistently increased its sales volume and continued to make a profit while facing the alleged anticompetitive practices of its rivals. Finally, prior to the entry of Continental, Carnation, and Pet, Utah Pie enjoyed a 67 percent market share in Salt Lake City, and while the new competition cost Utah pie a portion of its market, at the end of the price discrimination period the Salt Lake City firm still accounted for 45 percent of the pies sold in the area.[6] It appears that Utah Pie was using the antitrust policy to eliminate price competition and to gain a larger share of the market.

**Natural Monopoly and Antitrust Cases.** Antitrust policies may also be inappropriately applied to natural monopolies. The Aluminum Company of America (Alcoa) was involved in antitrust litigation as early as 1912 on allegations that it had violated the Sherman Act by, among other charges, monopolizing deposits of bauxite (the crucial ore in aluminum production), conspiring with foreign aluminum firms to fix world prices, and entering into exclusive contracts with power companies guaranteeing that the power companies would not supply electricity to any other aluminum producers. Finally in 1945, the U.S. Circuit Court of Appeals rendered its decision on charges brought by the government that Alcoa had monopolized the production and sale of virgin aluminum ingot.

The appeals court decision, written by Judge Learned Hand, is one of the most celebrated in American antitrust history.[7] Judge Hand, in overturning a lower court decision favoring Alcoa, wrote that even though Alcoa's monopoly was "thrust upon it" and the firm "stimulated demand and opened new uses for the metal," the firm should not increase production in anticipation of increases in demand. He contended that Alcoa prevented new firms from entering by increasing its capacity as demand increased and that because of Alcoa's lower cost, no new firms could effectively compete.

Although the court found Alcoa guilty, it refused to dissolve "an aggregation which has for so long demonstrated its efficiency." The "problem" of lack of competition in the aluminum industry was solved at the end of World War II, when the government sold to Reynolds Metals and Kaiser Aluminum the aluminum production plants it had set up for the war effort.

It is clear from this evidence that Alcoa had a natural monopoly and that **divestiture**—breaking the company into several smaller firms—was undesirable. Indeed, antitrust policies are generally ineffective when a natural monopoly exists. Antitrust legislation is best suited to stopping cartels, trusts, and conspiracies from eliminating competition when competition is the natural order.

**Divestiture:** A legal action that breaks a single firm into two or more smaller independent firms.

[6]Areeda, pp. 1070–1075.
[7]*United States* v. *Aluminum Co. of America*, 148 F.2d (2nd Cir. 1945).

### An Economic View of Antitrust Cases

How well enforcement of the antitrust laws agrees with economic principles is a complex question with no simple answers. On the basis of our brief sketch of laws, agencies, and court interpretations, we can draw the rough conclusion that economic theory has not made a large impression on legal thinking. In the words of George Stigler, "Economists have their glories, but I do not believe that the body of American antitrust law is one of them."[8]

In particular, the courts seem to equate the degree of competition with the number of firms in an industry and to seek to protect competition by protecting competitors. This focus on the number of firms in an industry ignores the other dimensions of competition (number of buyers, entry and exit conditions, degree of product homogeneity, and so forth). More important, the numbers game neglects the possibility that firms can grow relative to their rivals not because they use unfair methods of competition or expand by buying out their competitors but because they can serve consumers better at lower cost.

### Regulation

Antitrust policies are basically designed to maintain competition when competition can exist. Therefore, the antitrust policies are not very effective when a natural monopoly exists. How does the public avoid the potential problems of monopoly when only one firm can survive in an industry? Price regulation is offered as a solution to some of the problems that can arise under these circumstances. Governments may attempt to regulate prices to keep them low for consumers' sake or may even establish agencies controlling the activities of specific industries. But like antitrust litigation, these regulatory efforts may have unwanted economic effects.

**Price Regulation.** Public utilities, such as firms that provide local electricity, telephone, gas, and water services, are often considered natural monopolies because their provision requires economies of scale. In this case, Figure 17-5 illustrates the relation between demand for and cost of these services.

With no regulation, a firm holding a natural monopoly chooses the profit-maximizing price and output $P_m$ and $Q_m$, which result in profits. If the local government chooses to regulate the firm to eliminate profits, then it may force the firm to charge a price that just covers its average cost of production.

This price regulation is achieved through an agreement between the firm and the local government—the government allows the firm to exist as the sole provider of services and, in return, the firm agrees not to allow profits to rise above a normal rate of return and also agrees to meet the total demand for the product. The resulting price and output, $P_{AC}$ and $Q_{AC}$, occur where the industry demand crosses the average cost curve. Economic profits are zero since price equals average cost.

This **average cost pricing** is indeed a very popular form of price regulation. It seems to eliminate one of the problems of monopoly. However, it does not result in the optimal level of output. Many economists believe that price should instead be set equal to marginal cost, in what is called **marginal cost pricing**. The price and output that result from the intersection of the demand curve and the marginal cost curve, $P_{MC}$ and $Q_{MC}$, are theoretically optimal. That is, when price equals marginal cost, the benefit to society of one more

**Average cost pricing:** A form of price regulation that forces price equal to average cost and thus economic profits equal to zero.

---

[8]George Stigler, "The Economists and the Problem of Monopoly," *American Economic Review* 72 (May 1982), pp. 1–11.

**FIGURE 17–5**

**Two Forms of Price Regulation**

An unregulated natural monopoly might charge the profit-maximizing price $P_m$, at which it would sell the quantity $Q_m$. With average cost price regulation, the price would be set equal to the firm's average cost, $P_{AC}$, increasing the quantity sold to $Q_{AC}$. Marginal cost pricing would result in even lower price and higher quantity: Price $P_{MC}$ and output $Q_{MC}$ result where the demand curve intersects the marginal cost curve. But such a policy would force the firm to operate at a loss, for the marginal cost is lower than the firm's average cost of production.

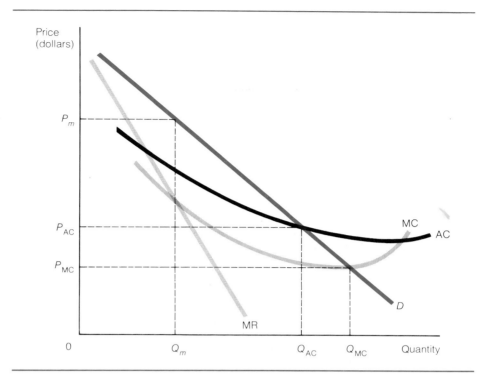

**Marginal cost pricing:** A form of price regulation that forces price equal to marginal cost and results in optimal allocation of resources.

unit of output is just equal to the extra cost of that unit. However, as in the case shown in Figure 17–5, marginal cost pricing may result in losses for the firm. One of the basic problems of price regulation is that monopolies cannot be forced into the ideal competitive solution where $P = AC = MC$. Since marginal cost pricing may result in losses, average cost pricing may be the preferred form of price regulation.

Another problem with price regulation is that when firms are forced to earn only a normal rate of return, their incentive to achieve efficiency is diminished. Maintaining low costs of production is not rewarded with higher profits. For this reason we may see public utilities allowing costs to rise.

In spite of the potential problems of price regulation, direct control of industry prices became popular in the late 1800s. It was believed that government regulation beyond antitrust legislation could promote more efficient behavior than could market forces.

**Government as a Cartel Manager.** Many government agencies have evolved through the years in an effort to regulate particular industries. In 1887 the Interstate Commerce Commission was established to regulate rates and routes of the U.S. railway industry, which was considered a monopoly. Since that time, the ICC has extended its control to the interstate trucking industry and the inland waterway transportation industry.

There are many more federal regulatory agencies intended to protect the public interest in specific industries. The Securities and Exchange Commission (SEC) regulates stock issues, brokers, and the major trading exchanges; the Federal Communications Commission (FCC) licenses radio and television broadcast rights; the Food and Drug Administration (FDA) controls the introduction of new drugs and regulates the purity of food and cosmetics; and the Federal Power Commission (FPC) regulates natural gas prices. Regulation is not confined to the federal level, however. State and local governments set standards and issue licenses for barbers, medical doctors, architects, teachers, and other professionals. Various state and federal agencies regulate public utilities (telephone, water, and electricity suppliers). The list could go on and on.

These agencies frequently were formed in an effort to control monopoly profits, but many have extended into competitive markets. In fact, we observe these agencies actually eliminating competition by forcing firms to charge a minimum (cartel) price. Their regulatory activities also prevent the entry of new firms into the industry. In a sense, the existing firms use the regulatory agency to police a cartel that is beyond the jurisdiction of antitrust legislation. Before the deregulation of the airline industry, for example, the Civil Aeronautics Board (CAB) prevented airlines from charging air fares lower (or higher) than those set by the agency.

Government regulation is often justified as a consumer-protection measure when, in fact, it usually ends up being a protection device for the producer. The prevalence of regulatory intervention in everyday life requires one to keep in mind that whether one believes regulation is good or bad from a social point of view, its economic effects are often to restrict output and raise prices.

### Government Ownership

An alternative to antitrust policy or regulation is government ownership, or **socialization.** When monopolies exist and their performance is less than socially optimal, government may choose to own and operate the firm in the interests of the public. We see this happening occasionally, more often at the state or local level than at the national level. For example, many public utilities such as electric services, water, and natural gas are owned and operated by local governments.

**Socialization:** Government ownership of a firm or industry.

Socialization of firms is a solution to the monopoly problem, but is it an efficient solution? Ideally, the firm should approximate the competitive equilibrium. It should minimize its costs of production and charge a price equal to marginal cost. However, the performance of the firm depends on its managers. The managers, of course, must operate within the bounds allowed by the politicians who employ them, the voters, and the buyers of their services.

The efficiency of publicly owned firms is questioned by many economists. Many suggest that there is a lack of incentive for efficiency because there are no residual claimants. That is, no one may receive any profits acquired by the socialized firm, so managers have no reason to ensure low costs of operation or product quality.

To judge the efficiency of publicly owned firms, it is possible under some circumstances to compare their price, costs, and quality of service to those of privately owned firms in the same industry. For example, we can compare a locally owned electric firm's price and service to a privately owned electric firm's price and service. Many industries have both privately owned and publicly owned firms, such as education, medical care, telecommunications, television, garbage services, and employment agencies. However, in most situations we must compare government-owned firms with government-regulated firms, as in the case of electricity.

The U.S. Postal Service is a socialized firm that we cannot compare directly to a private firm, for the government prohibits the delivery of first-class mail by anyone except the U.S. Postal Service. However, there are many firms wishing to enter the mail delivery market, and occasionally they do so illegally. The suggestion that private firms could make a profit at the same price whereas the U.S. Postal System does not is an indication that the Postal Service is run inefficiently. Indeed, many government-owned firms must be subsidized by tax dollars.

Despite its lack of incentives for efficiency, socialization is still considered by some supporters to be an appealing alternative to trust busting or regulation in monopoly-prone industries. The price of medical services has been rising, and socialized medicine has been offered as a solution. AT&T was recently divested through an antitrust case, but an alternative could have been to socialize telephone services. The possible consequences of such actions can be predicted by observing the prices and quality of services of similar industries that have been socialized in other countries.

### Laissez-Faire

Antitrust policies, regulations, and government ownership are not the only alternatives to the monopoly problem. Under many circumstances, the forces of the market are capable—given enough time—of eliminating the undesirable performances of monopolies, cartels, and trusts. Leaving an industry alone so that it can clear itself of monopolist inefficiencies and high prices is called **laissez-faire** (allow to do) policy.

**Laissez-faire:** A government policy of not interfering with market activities.

**Natural Shifts in Market Power.** To understand how markets can be self-correcting, let us examine the circumstances facing a natural monopoly.

Natural monopolies exist because of the special relation between industry demand and the firm's average cost. In an industry in which the demand curve intersects the firm's average cost curve at a point where average cost is still declining due to economies of scale, as in Figure 17–6, only one firm will survive, and profits will exist. These monopoly profits motivate competition from other firms.

New firms will want to reap the potential profits, and their profit-seeking activities can limit the welfare loss due to monopoly. For example, many entrepreneurs may attempt to discover and produce substitutes for the monopolist's product, or they may attempt to create a new technique of production that allows them to produce the same product on a smaller scale of production with equally low cost.

Without the threat of entry, the monopolist may charge price $P_M$, the typical profit-maximizing price. However, if new firms are on the verge of entry, then the monopolist may engage in **limit-entry pricing.** That is, the monopolist may drop to a price lower than $P_M$, such as $P_E$ in Figure 17–6. The entry-limiting price $P_E$ must be low enough so that new firms are not able to survive. Though designed to squeeze out potential rivals, this activity in effect allows the monopoly to maximize its long-run profits since demand is infinitely elastic at price $P_E$ in the long run. If the firm raises prices above $P_E$, it will lose customers through time to new entrants.

**Limit-entry pricing:** A pricing policy by a firm that discourages entry of new firms by selling at a price below the short-run profit-maximizing price.

The monopoly power of a firm may diminish in the long run anyway. The monopoly exists because of the particular demand-cost relation. Through time we expect demand and costs to change. New firms can enter and survive if industry demand increases or if prices of inputs and technology change in such a way that the optimal plant size decreases. Because the economy is truly dynamic, we expect to see the optimal number of firms and concentration ratios rise and fall in many industries, including monopolistic industries.

## FIGURE 17-6
### Limit-Entry Pricing

The monopoly with no threat of entry charges price $P_M$ and sells quantity $Q_M$. However, the threat of entry of new firms may be diminished if the monopoly charges a price, such as $P_E$, below the potential entrant's average cost. The entry-limiting price is still above the monopoly firm's average cost, which is kept low through economies of scale. Limit pricing in effect maximizes the firm's long-run profits and diminishes the adverse social effects of a monopoly.

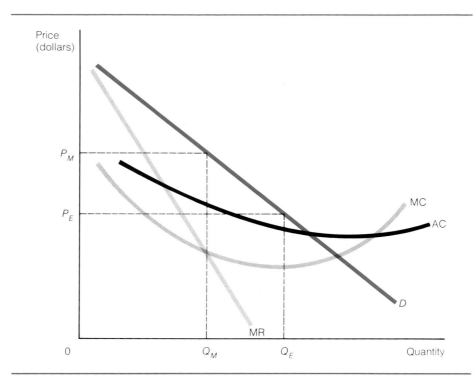

Just because a monopoly exists now does not imply that it will be there forever. According to laissez-faire thinking, there may be no need to intervene in certain monopolized industries because the monopolies may eventually disappear anyway. Focus, "What Happened to the IBM and GM Monopolies?" gives two recent examples of declining monopolies: General Motors and IBM.

The situation illustrated in Figure 17-6, with the demand curve intersecting the average cost curve where there are economies of scale (AC is falling), represents the usual natural monopoly. The figure also represents the situation believed to exist for public utilities such as electric companies. If there are economies of scale, then firms with low rates of output, such as electric companies in small towns, should have higher average cost and thus higher prices than electric companies in larger towns with high rates of output. If we do not observe this relation, then natural monopolies no longer exist, and the firms are monopolies only because the local authorities grant them monopoly rights. Indeed, it is difficult to find a free-market monopoly; that is, one that exists without monopoly rights granted by government.

**Deregulation.** In addition to simply leaving monopolized industries alone, a laissez-faire policy might dictate removing previous regulations—**deregulation** of the industries. As one would expect, firms in regulated industries often attempt to use the regulatory agencies to promote the firms' profits. Government regulation of industry often serves to protect the very companies that antitrust laws and regulations are intended to suppress. In particular, regulatory limits on the entry of new firms into an industry and on the prices that are charged by existing companies can give these firms a degree of market power they could not otherwise achieve.

In view of the fact that antitrust laws and regulations have been misused,

**Deregulation:** A situation in which government ceases to regulate a previously regulated industry in an effort to improve the performance of that industry.

## FOCUS: What Happened to the IBM and GM Monopolies?

In the 1960s IBM was essentially the only U.S. supplier of computers. For many years IBM enjoyed a lack of competition, which of course resulted in higher-than-normal profits and a call for antitrust action by IBM's customers and its few small competitors.

Similarly, in the 1960s, General Motors gained a disproportionately large share of the U.S. auto market. The firm was large and getting larger because it was making the most popular cars at the most popular prices. Through the 1950s and 1960s several smaller auto companies such as DeSoto and Packard went out of business. It seemed that GM was gaining a great deal of monopoly power. Antitrust authorities were contemplating drastic actions.

Customers, competitors, and antitrust lawyers suggested that IBM and GM should be divested of some of their market power by breaking each firm into several smaller firms. For example, an antitrust action could have made Chevrolet, Buick, Pontiac, Oldsmobile, Cadillac, and GMC Trucks all independently owned companies rather than divisions of General Motors. Such an action would have increased the number of firms and lowered concentration in the automobile industry. A similar action could have divided IBM into several smaller companies.

However, as we so frequently observe, market conditions changed. Under a laissez-faire government policy, both IBM and GM have lost their market power. Each now has many competitors. New firms were able to enter the market because demand or cost conditions changed. In the computer industry, the demand for computers increased along with tremendous changes in technology. IBM is no longer the only name in computers. The market thus solved the computer monopoly problem.

Competitors from abroad have of course decreased the market strength of General Motors. The expansion of production and technology in both Asia and Europe cured the domestic monopoly problem. Under some circumstances, free international trade is the ultimate weapon in destroying monopoly power. In fact, public policy has flip-flopped from an effort to diminish GM's profits to an effort to bolster them with import quotas.

The experience with IBM and GM leads us to believe that the market itself can decrease the monopoly power of firms through time. Industry concentration fluctuates through time and a "leave it be" policy can benefit the public in the long run.

---

it is not surprising that we have observed a trend toward deregulation of industry. Two recent examples of limited deregulation have occurred in the interstate trucking and airline industries. Until 1980 the Interstate Commerce Commission severely restricted entry of new firms into the trucking industry as well as price competition between firms. Since the Motor Carrier Act of 1980 decreased the ICC's ability to restrict new entry, the number of firms has increased by several thousand and price has fallen by 20 percent and more. In the airline industry, the Civil Aeronautics Board set passenger fares and restricted entry until the Airline Deregulation Act in 1978. From that time until the CAB's demise in late 1984, the CAB's ability to restrict price competition and entry was severely limited. The results of deregulation in the airline industry have been similar to those in trucking. The number of firms has increased and prices have fallen. Since deregulation, some trucking and airline firms have also gone bankrupt. One would expect that competition would eliminate the inefficient firms. Nevertheless, concentration has decreased in both industries.

### Conclusion

A purely competitive free market is impersonal and unforgiving. It forces efficient behavior by eliminating firms that are not able to achieve the minimum cost of production. No special-interest groups can reap profits because a competitive market acts as an impartial arbitrator in the task of resource allocation. However, as we have seen, certain market conditions can lead to the survival of only a few firms. In such cases, pure competition cannot exist, and firms are not forced into the socially optimal solution.

We have examined several forms of public policy that are aimed at preventing poor market performance. Each of these—antitrust, regulation, socialization, and laissez-faire—has its merits, but there is no public policy that can achieve the efficiency generated by a competitive market. When imperfect competition exists, our best solutions are themselves imperfect.

Not only are our best solutions theoretically imperfect, but actual government policies appear to be schizophrenic. One action attempts to increase competition and reduce profits while other actions promote monopolies and their profits. The reason for these contradictions is that public policies are formulated in a political arena and not by impartial economic theorists or forces. An understanding of public choice theory—to be explored in Chapter 20—can tell us more about public policies in monopolized industries.

## Summary

1. Aggregate concentration ratios have not increased significantly in the past several decades. This suggests that big businesses have not been gaining greater control over the nation's productive assets. Industry concentration ratios change through time because the optimal number of firms changes as demand and cost conditions change.
2. The number of firms in an industry is determined by the number of optimal-sized firms that industry demand is able to support. A natural monopoly exists when industry demand is not great enough to support two or more firms of optimal size.
3. Firms seek to increase profits by increasing their share of industry demand, by decreasing production costs, and by decreasing the number of competitors. They may achieve these goals in a variety of ways—cartels, trusts, mergers, collusion, and government grants of monopoly power.
4. The social welfare loss due to monopoly calls for public policy to end anticompetitive practices. Active public policies include antitrust legislation, regulation, and socialization.
5. Antitrust laws are basically designed to promote and maintain competition when competition is possible. Thus their main goal is to decrease concentration, a goal that is not always desirable.
6. Price regulation allows the existence of monopoly but attempts to decrease the social loss by forcing a particular price. However, price regulation may decrease the firm's incentive to maintain efficiency. Regulation of competitive industries has resulted in government-managed cartels.
7. Government ownership as an alternative to monopoly can prevent higher-than-normal profits. However, a government-owned business is not forced by the market to maintain efficiency.
8. A laissez-faire system has been offered as an alternative to active public policy. Under some circumstances, this policy to leave the market alone is clearly preferable. Under other circumstances, concentration could increase under laissez-faire policy.
9. All public policies have their shortcomings. When competition does not and cannot exist, then no public policy can force the competitive solution.

## Key Terms

| | | | |
|---|---|---|---|
| aggregate concentration | trust | divestiture | laissez-faire |
| industry concentration ratios | collusion | average cost pricing | limit-entry pricing |
| natural monopoly | merger | marginal cost pricing | deregulation |
| welfare loss | antitrust policy | socialization | |

## Questions for Review and Discussion

1. Suppose that the output of an entire industry were produced by one firm but there were no barriers to entry. Should the antitrust authorities be concerned about this industry?
2. What does market power mean in antitrust analysis? How can this concept be used to assist antitrust enforcers in deciding where to look for monopoly power in the economy?
3. What is the Herfindahl index of concentration? How is it used by government to screen mergers?

4. Do a few large firms control all the wealth in the economy? Does the identity of these firms remain constant over time?
5. In the United States there is dual antitrust enforcement by the Federal Trade Commission and the Department of Justice. Discuss the pros and cons of having two antitrust enforcement agencies.
6. What is a conglomerate merger? What is the rationale for conglomerate mergers?
7. Suppose that a merger reduces the costs and increases the market power of two firms. Can you show this shift graphically? How should the antitrust officials decide such a case?
8. Why does government regulation sometimes act to enforce a cartel? Why do voters and consumers put up with such programs and policies? What effect would deregulation have on such an industry?
9. Blue laws outlaw Sunday sales in a given locality. From the point of view of economic regulation, explain who gains and who loses from such legislation.
10. How does rate-of-return or profit regulation affect firm behavior? Does this regulation promote production efficiency? Do we achieve the optimal output with average cost pricing?
11. Are local television companies natural monopolies? How could the answer be determined? If they are, would they want to be regulated?
12. Can public policy change an industry characterized as a natural monopoly into a competitive industry that results in competitive equilibrium?

## ECONOMICS IN ACTION

## The Breakup of AT&T

For most of the twentieth century, the American Telephone and Telegraph Company (AT&T) was the regulated quasi-monopoly supplier of both local and long-distance telephone service in the United States. In 1974, for example, AT&T supplied almost 80 percent of U.S. telephone service.

In January 1982, AT&T struck a deal with the U.S. Justice Department ending an antitrust suit that the department filed in November 1974. By the terms of the agreement, the suit was dropped, and AT&T agreed to divest itself of its local telephone service subsidiaries. The twenty-two local Bell Telephone operating companies will continue much as before, organized into seven large regional holding companies. The "new" AT&T after January 1984 (when the divestiture took effect) has assets on the order of $35 billion and continues to supply long-distance telephone service, but in competition with other suppliers. Competing non–Bell long-distance services have been gradually entering the market over the last several years. The AT&T divestiture has eliminated some barriers to entry facing these firms, mostly by mandating "equal access" to local telephone facilities for all suppliers of long-distance service.

Meanwhile, the local telephone companies "spun off" from AT&T will still be subject to state and federal regulations, but their holding companies will be able to invest in areas previously prohibited. Some of these companies, for example, have entered the computer business.

AT&T voluntarily agreed to the divestiture, but one of its main goals in doing so may have been to end its expensive legal battle with the Justice Department. As

*An advertisement for one of AT&T's competitors in the long-distance phone market. Although AT&T still dominates the industry, many new companies have entered the market in recent years.*

things stand, it would be inaccurate to claim that divestiture has in and of itself significantly increased the degree of competition within the telephone industry. By the end of the 1960s certain forms of long-distance communication in addition to those supplied by AT&T had begun to be approved by the Federal Communications Commission. These developments accelerated during the 1970s, and it is not clear that divestiture itself significantly lowered entry barriers in the long-distance telephone service industry. Furthermore, local telephone service continues to be supplied by local companies that are effectively regulated monopoly suppliers, even though they are now independent concerns. Pacific Telephone in California and New England Telephone are now each a part of independent companies, but each has an effective monopoly in the supply of local telephone service in its respective operating area. In other words, divestiture has not introduced competition in the local telephone service market, which remains basically controlled by a set of regionally regulated monopoly firms.

It will be some time before a final assessment of the economic impact of forced divestiture in the case of AT&T can be made. One important fact is that the cost of a long-distance call has gone down. Beyond this, the aspects of the breakup likely to have the most significant positive effects are those relating to the relaxation in operating restrictions on both the "new" AT&T (now permitted to enter the fast-growing information processing industry) and the regional holding companies. It remains to be seen whether these consequences will outweigh any losses in efficiency the "old" AT&T may have enjoyed due to economies of scale.

Question

Why do you think AT&T went along with forced divestiture rather than continue to fight it in the courts? What will happen to local phone rates as a result of the divestiture? Why?

# 18

# Market Failure and Public Policy

**Market failure:** A situation in which a private market does not provide the optimal level of production of a particular good.

Chapter 17 addressed some of the ways that private markets can fail to achieve ideal economic results. Monopoly was the primary culprit in this discussion, with its attendant restriction of output and high prices causing a reduction in the economic welfare of society. Monopoly is a form of market failure, that is, a failure of private markets to provide goods and services in the right quantities at the right prices. It is not, however, the only type of market failure that is possible in an economy.

**Market failure** typically arises from problems of incomplete or nonexistent ownership rights to basic resources. Owners of private property bear directly the economic results of the use of property and are therefore motivated to use resources efficiently. In contrast, when property is held or used by everybody, users do not bear the full results of its use. Common ownership creates a conflict between the pursuit of self-interest and the common good.

Consider some examples. We are all familiar with the problem of litter in public parks or along city streets. Littering occurs in such places because each individual is not directly assigned property rights for the use of common property. We do not ordinarily allow litter to pile up indefinitely in our living rooms or automobiles because we are forced to bear the costs of our actions. We are not completely forced to bear the costs of public littering, however. As we know, fines and threat of arrest have failed, in many cases, to significantly reduce unsightly litter from our environment. Stream or water pollution by manufacturers or refiners, noise pollution by aircraft around airports and elsewhere, and overfishing in coastal waters are also examples of market failure. In some cases government regulation and intervention can be applied to correct market failures; other market failures may simply require a redefinition of who owns the rights to basic resources, after which the private market will work fine. Still other complex intermediate solutions emphasize public-private interaction to control market failures.

In this chapter we review all these approaches to market failures. We do not live in a perfect world, and to improve on situations of market failure we must carefully weigh the costs and benefits of alternative policies. Our discussion begins with an analysis of the problem of commonly owned resources and how this problem is approached in two concrete cases: the common stock of fish in international waters and the oil pool.

## The Economics of Common Property

**Common ownership:** Lack of clearly specified ownership rights of a resource for whose use more than one person competes.

The fundamental economic disadvantage of **common ownership**—the right to the use of a resource by anyone or by competing users—is the lack of incentive to invest in the productivity of the common property. Rather, each person with access to the resource has an incentive to exploit the resource and neglect the effects of his or her actions on productivity. This principle applies to both renewable resources such as fish and nonrenewable resources such as oil.

### Fishing in International Waters

In international waters anyone may acquire exclusive ownership of a fish by catching it, and no one has rights to the fish until it is caught. This system causes each fisher to ignore the effect of the well-known biological law that the current stock of a species of fish determines its reproduction rate. Thus, fish harvested today reduce today's stock and thereby affect the size of tomorrow's stock and tomorrow's harvesting costs and revenues. Overfishing results because no single fisher has an incentive to act on this bioeconomic relation. If all fishers would cooperate to restrain themselves now, tomorrow's stock would be larger and future harvesting cheaper and more profitable. However, under competitive conditions, each individual knows that if he or she abstains now, rivals will not, and much of the effect of abstention is lost. Further, the reduction in future costs would accrue to everyone, not just to the abstainer. Hence each person has little reason to abstain, since the major effect is to lower others' future costs at some immediate present cost to the abstainer.

Sole ownership removes this dilemma. If there is only one person fishing, he or she need not worry that rivals will not abstain. Alternatively, all the people involved could negotiate an agreement to abstain, and all would benefit. The fact that such agreements are not usually successfully negotiated is due to the costs of dealing with all current and potential people who will fish and the difficulty of ensuring that all abstain as agreed. A third method of dealing with this dilemma is government regulation.

Such regulation usually sets an annual catch quota. To allocate the quota, regulators rely on restrictions on technology or simply close the season when the quota is taken. Both means create difficult enforcement problems and potential economic waste. If the season is closed when the quota is taken, for example, excess profits are dissipated in competition among fishermen to buy bigger, faster boats and thus get a larger share of the quota. The season progressively shrinks, and resources stand idle or are devoted to inferior employments.

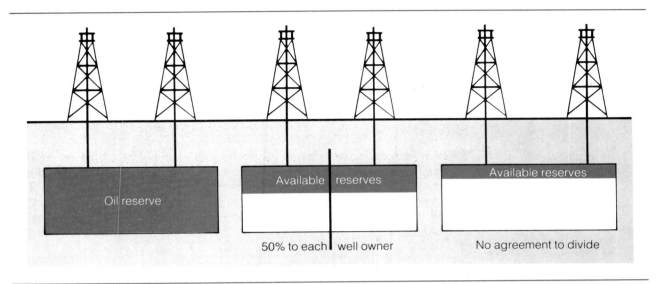

**FIGURE 18–1  The Problem of Common Ownership in Oil Drilling**
If a fixed reserve of oil is owned in common, and the co-owners cannot agree to share the reserves, then each owner will likely try to pump as much oil as possible from the reserve. Doing so reduces the available reserves at a rate disadvantageous to both owners.

## Drilling for Oil

Oil presents another version of the same common ownership dilemma (see Figure 18–1). Technically, oil, unlike fish, is nonrenewable. However, oil does move about, so pumping one well lowers the entire reserve and affects the pressure below other wells. These characteristics create a problem when more than one person has rights to the same oil reserve. For example, if the land above a reserve of oil is owned by two people, each has the right to drill for oil. But the only way to own the oil is to pump it to the surface. Oil in the ground belongs to the landowner only when that person pumps it. Herein lies the crucial lack of private ownership rights to oil reserves.

It may be more economically efficient to exploit an oil reserve with a single well than with two wells not only because it costs more to drill two wells than one but also because the existence of one well interferes with the other's technical ability to pump. If one person owned all the property rights, he or she would probably use only a single well. Likewise, if two landowners with rights to a common oil pool negotiate, both can be made better off by agreeing to split the proceeds of a single well rather than both drilling.

If they cannot negotiate, perhaps because of legal restrictions, or if a larger number of owners causes difficulties in communication, each may go ahead and drill. Once two wells are in production, there is an economically efficient rate of pumping for each well that a single owner would adopt to maximize his or her wealth. Nevertheless, if the two owners cannot agree on efficient exploitation, each will be tempted to pump oil at too high a rate, a tendency reinforced by the fear that the other is doing the same thing. In fact, even if negotiations are successful, it may be costly to monitor each other's pumping, and the agreement might dissolve in mutual cheating. The costs of negotiating and enforcement probably rise in proportion to the numbers of owners, as do the incentives for each to overpump because others are likely to do so.

The lessons to be learned about common property resources are straightforward: What belongs to everyone belongs to no one, and the deterioration and rapid exploitation of common property resources are clear implications of economic theory.

## Externality Theory

> **Externality:** Spillover costs or benefits.

When resources are not owned privately, a market failure called externality can occur. An **externality** is an economic situation in which an individual's pursuit of his or her self-interest results in costs or benefits to others. The costs and benefits are external to the individual who caused them. Obviously, any economic entity can create an externality, including firms, government agencies, and individuals.

### Negative and Positive Externalities

A common example of externality is a factory's emission of smoke as a by-product of production. If the factory is able to avoid responsibility for the consequences of its smoke—such as the expenses for painting soot-covered buildings nearby or laundering sooty clothes—the output of the factory will be excessive in that the value of additional output is less than the cost of the damage. If the factory were held responsible for the damage done by its smoke, its costs would rise and its product would become more expensive. This example illustrates a **negative externality,** a situation in which production entails costs for others, such as those living near a factory. Other potential examples of negative externalities are the exhaust emissions of automobiles, a neighbor who plays a stereo too loud, low-flying jet airplanes landing at an airport, cigarette smoke in a crowded elevator, drunk drivers, the contamination of underground water by toxic chemical dumps, and so on.

> **Negative externality:** A situation in which the social costs of producing or consuming a good are greater than the private costs.

> **Positive externality:** A situation in which the social benefits of producing or consuming a good are greater than the private benefits.

**Positive externalities** are social benefits accruing without costs to the beneficiaries. Suppose a beekeeper and the owner of an apple orchard have adjoining land. The beekeeper's bees fly into the orchard and pollinate the apple blossoms, making the orchard more productive. If the beekeeper were to reap the consequences of the increased value of the orchard, the yield on his investment in bees would rise, and he would raise more bees. But as it is, his bees' social service is provided free. For both illustrations the point is identical: Externalities arise when one person's activities affect the well-being of others, either positively or negatively. Some other examples of positive externalities are the impact of public education on literacy rates, personal hygiene and health care, inoculation against contagious disease, charity, invention, and so forth.

### Determining the Need for Government Intervention

To promote the general economic well-being, an ideal government would intervene to promote positive externalities and to discourage or prevent negative externalities. In the preceding examples, government might intervene to reduce the amount of smoke emitted by the factory or to increase the number of bees kept by the beekeeper. Intervention is not always called for, however. Individuals might settle such matters privately, because the costs of intervening might be prohibitive.

To enhance the bees' contribution to the orchard's productivity, the orchard owner could contract with the beekeeper to supply more bees.[1] Alternatively, the orchard owner could raise his own bees. As long as the beekeeper's marginal costs of raising additional bees are less than the value the bees contribute to the orchard owner, it will be possible for both parties to reach an agreement that will leave each better off. Such private contracting in effect eliminates the apple-bee externality and makes public intervention unnecessary.

In addition to the possibility of private settlements, a second consideration when determining the need for government intervention in externalities is whether the externality is irrelevant or relevant. Externalities are trivial when the situation is not worth anybody's effort to do anything about it. The world is full of trivial externalities. You may not like the way a friend dresses but not enough to say anything about it. Some people's unfailingly gracious and friendly behavior creates external benefits, and yet they are rarely complimented or rewarded for such behavior. The same goes for an attractive lawn. Neighbors reap external benefits from a well-kept lawn but not so much that they would offer a subsidy to support the yard work.

Externalities are termed irrelevant when the externality generates a demand, but not a sufficient demand, by the affected party to change the situation. Your neighbor's messy yard may disturb your sense of propriety. You would be willing to pay her $50 a year to clean up the mess, but her price to clean it up is $600 a year. In this event her yard will not be cleaned up, and the externality will persist. The messy yard is a nuisance but not enough of a nuisance to cause sufficient demand to get the yard cleaned up. In principle, the best thing to do with an irrelevant externality is to leave it alone because it costs more than it is worth to correct the problem.

A relevant externality creates sufficient demand on the part of those affected by it to change the situation. The neighbor's yard is messy, and the neighbor will supply a cleanup for $10. You are willing to pay as much as $50. In this case you can make an effective offer to correct the externality and reach a deal with your neighbor. With relevant externalities, it is worthwhile to correct the situation because the benefits of correction exceed the costs.

## Correcting Relevant Externalities

Economic theory suggests that correction of relevant externalities can be approached in a number of different ways. In general, these can be categorized as defining property rights, taxing negative externalities or subsidizing positive ones, selling rights to create an externality, and establishing regulatory controls.

**Establishing Ownership Rights.** Externality problems generally persist because of the presence of some element of common ownership. The beekeeper's and the apple grower's ownership rights are clearly established and easily transferable. In such a situation it will be in the interest of both parties to conclude an agreement that places resources in their most highly valued uses. But in the case of the polluting factory, does the factory have the right to use the air as it pleases, or do the people in the surrounding community have the right to clean air? Air-use rights are indefinite and nontransferable: It is generally impossible for people to buy and sell rights to the use of air. In this

---

[1] See Steven N. S. Cheung, "The Fable of the Bees: An Economic Investigation," *Journal of Law and Economics* 16 (April 1973), pp. 35–52.

case there is no reason to believe that the pursuit of individual interest will promote the use of air in its most highly valued manner.

The establishment of ownership, then, is one way of eliminating problems associated with externalities. Since nonownership is a source of market failure, the creation of ownership is a means of correcting market failure.

An important analysis of the economics of establishing ownership rights was first introduced by R. H. Coase.[2] Coase posed the question, Does the legal assignment of property rights to one party or the other in an externality relationship make any difference to the observed market outcome? To keep the analysis simple, Coase assumed that the costs of bargaining and transacting among the parties was zero.

We will apply the Coase analysis to an example of air pollution—a factory belching smoke over a nearby community.[3] Figure 18–2 illustrates the analysis. Firm output per unit of time is given along the horizontal axis. The marginal benefits (MB) to the firm of producing this output are given in dollar terms on the vertical axis. For the sake of this analysis, marginal benefits can be thought of as the net profits of producing additional units of output. The MB schedule therefore declines with increases in output because the rate of return on additional production generally declines (see Chapter 8). The marginal cost (MC) curve represents the externality caused by the firm's production and is also given in dollar terms along the vertical axis. The MC function measures the additional cost created at each output level by additional smoke. (Assuming that the amount of smoke the firm produces is directly related to its rate of output, more production will cause more smoke.) The MC curve rises as output increases.

Now we apply the Coase analysis. As an experiment, we give the ownership right to clean air to the homeowners in the surrounding neighborhood. Assuming that bargaining costs nothing, we can predict what will happen with the help of Figure 18–2. For rates of output up to $Q$, we observe that MB > MC. That is, the firm's profits on additional units are greater than the additional pollution costs borne by the neighborhood. This means that the firm would be willing to buy and the neighborhood willing to sell the right of using the air to produce these units. Since the externality caused by this output is represented by the area under the MC curve, $0EQ$, a bargain can be struck between the firm and the neighborhood to produce $Q$ by agreement on how to divide the surplus marginal benefit, $AE0$. Beyond $Q$, MC > MB. A bargain to produce these units cannot be reached between the firm and the neighborhood because the firm would not be willing to bid enough to obtain the right to produce these units. So $Q$, where MB = MC, is the equilibrium outcome when the neighborhood is given the transferable right to use of the air.

Suppose that a judge decided instead to award the firm the right to use the air as it chooses. What happens in this case? We go through the same analysis, but now we see that the neighborhood must pay the firm to produce less output. Up to $Q$ units of output, MB > MC, which means that the neighborhood cannot offer a large enough sum of money to induce the firm

---

[2] R. H. Coase, "The Problem of Social Cost," *Journal of Law and Economics* 3 (October 1960), pp. 1–44.

[3] More subtle, widespread damages such as the loss of fish and trees to acid rain and the loss of work time to health impairment are not dealt with here. For the sake of simplicity, we consider only obvious costs to the local community. But if claimants organize and the source of pollution can be clearly pinpointed, the same analysis can be applied to the more subtle costs of air pollution.

**FIGURE 18–2**

**The Coase Theorem**

In this graph depicting the Coase theorem, MC represents the externality imposed by the firm's pollution, and MB represents the profits to the firm from producing various levels of output (Q). The Coase theorem states that regardless of who is assigned ownership rights to the air, costless bargaining between the parties will result in the same market outcome at Q.

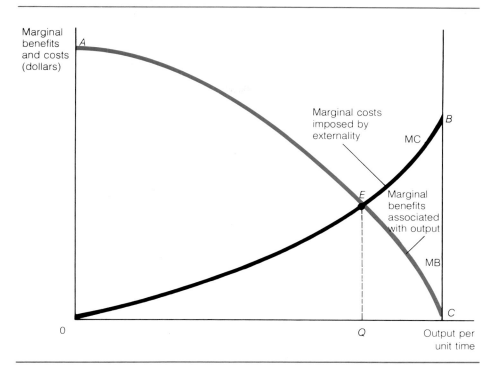

**Coase theorem:**
Externalities will adjust to the same level when ownership rights are assigned and when the costs of negotiation are nonexistent or trivial, regardless of which party receives the rights.

to reduce its output and its smoke. Beyond Q, the situation is changed: The firm would be willing to accept an offer of money to reduce its output. Beyond Q, the smoke causes EBCQ in damage to the neighborhood, and the benefits to the firm of this output are only ECQ. The neighborhood is therefore willing to make it worthwhile to the firm to reduce pollution to the point where MB = MC at Q.

Provided that we have been logical, we have shown a simple but powerful result: No matter who has the legal right to the use of the air, the amount of pollution is the same. When the firm must pay to pollute, it produces Q. When the firm has the unfettered right to pollute, it produces Q. This is the central insight offered by the **Coase theorem:** In a world in which bargaining costs nothing, the assignment of legal liability *does not matter*. A certain equilibrium level of output and its resulting level of pollution will exist regardless of whether firms or consumers "own" the air.[4] Under the stated conditions of the Coase theorem, government intervention—in the form of pollution guidelines, tax penalties, and the like—cannot improve upon a settlement negotiated by those parties who are directly involved with the externality problem.

The value of the Coase theorem is that it provides a benchmark for analyzing externality problems. It shows what would happen in a world of zero bargaining and transaction costs. Many real-world externality problems can be analyzed with the Coase theorem as long as the bargaining and transaction costs are low.

[4]The Coase theorem also marked the beginning of a new field in economics called law and economics, which studies the impact of legal rules and institutions on the economy; the Coase theorem, for example, analyzes the role of legal liability assignment in an externality problem. Many law schools have specialized fields in law and economics, and much of the literature in this area is published by the *Journal of Law and Economics*.

Nevertheless, using the establishment of ownership to correct externalities cannot always be relied upon to eliminate the difficulties of market failure because it may be exceedingly difficult if not impossible to establish ownership in some instances. Many externality problems do not fit the assumption of zero or low transaction and bargaining costs. In the factory smoke example, the neighbors in the area surrounding the plant are likely to be numerous, difficult to organize, and diffuse in their interests with respect to the air pollution problem. The widely dispersed sufferers of the smoke damage would face insuperable costs if they had to organize to purchase the agreement of the factory to reduce its emissions. If the firm had the right to pollute, it is not likely that the neighborhood would be able to overcome these problems, get organized, and offer the firm money to cut pollution back to $Q$ in Figure 18–2. Looked at the other way around, if the neighborhood owned the air space, the firm would find it costly to track down all homeowners and arrange for the purchase of their consent to produce up to output $Q$.

No matter who held the right to use of the air, bargaining costs would be high, and the likelihood is that nothing would be done to correct the situation. The costs of defining and enforcing a system of ownership can be so high that some system of taxation or administrative control would be more effective than reliance on the definition of property rights and contractual arrangements.

**Taxing or Subsidizing Externalities.** A seemingly simple way of dealing with externalities is to tax activities such as pollution that create negative externalities and to subsidize activities such as beekeeping that create positive externalities. Many complexities surround such a simple-sounding prescription, however. How could a negative externalities tax be levied? Against the product the factory produces? Against the amount of smoke it emits? Either solution might create burdens on the industry and on consumers that are not necessary in cleaning up the environment. Similarly, how would a subsidy for positive externalities be worked out? Based on the honey produced by the beekeeper? Based on the number of bees kept? Despite such complexities, the essential idea of taxing the negative and subsidizing the positive has a rich tradition in economic theory.[5]

To understand the effects of taxation, consider the pollution problem in terms of output decisions facing individual firms. We have learned that the competitive firm, in deciding what output to produce, compares marginal cost and marginal revenue. Marginal cost in this context means **marginal private costs** (MPC), the extra costs that the firm must pay to increase its output. These costs do not include externalities. By definition, an externality imposes costs on others that are not reflected in the private costs of the firm. The full costs to society of an increase in the firm's operations are summarized as **marginal social costs** (MSC). MSC are equal to MPC plus the externality costs (E): MSC = MPC + E. When there is no externality at the margin (E = 0), then MSC = MPC.

This analysis can be applied to the competitive industry as well. Consider the situation depicted in Figure 18–3. These are the same types of graphs that we used in Chapter 9. Figure 18–3b shows that an industry is initially in equilibrium at $E_1$ where industry demand (D) and supply ($S_1$) are equal. Under these conditions the firm (Figure 18–3a) reacts as a price taker to $P_1$

**Marginal private costs:** The increase in a firm's total costs resulting from producing one more unit.

**Marginal social costs:** The increase in total costs to society (the firm plus everyone else) resulting from producing one more unit.

---

[5] The tax-subsidy approach to externalities is often called *Pigovian*, after A. C. Pigou, author of *The Economics of Welfare* (London: Macmillan, 1932).

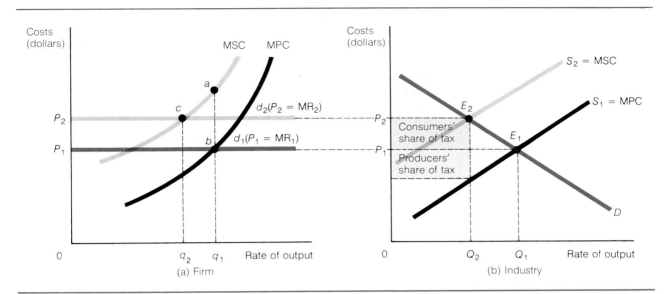

**FIGURE 18–3   The Effects of a Tax on an Externality**

(a) A pollution excise tax of *ab* per unit is placed on the output of the firm to correct the externality. The marginal private costs facing the firm are given by MPC, and marginal social costs are given by MSC. The tax raises the MPC curve to become identical with the MSC curve. In such a way the firm is made to account for the externality in its decision about how much output to produce. (b) At the industry level, the pollution tax reduces industry supply to $S_2$, and, according to our earlier analysis of the incidence of an excise tax in Chapter 9, the pollution tax is paid by consumers and producers in the proportions indicated in the diagram.

and sets its MPC curve equal to $P_1$ ($= d_1 = MR_1$). The competitive firm, acting on the basis of marginal private costs, produces $q_1$ units of output.

Notice, however, that the firm's MSC curve lies above the MPC curve at all levels of output; for example, it exceeds MPC by *ab* at $q_1$ units of output. The MSC curve traces the externality that the firm's operations impose, such as the amount of damage that the firm's air pollution causes in the surrounding neighborhood.

Our question is, What can government do to improve the situation? Basically, government must get the firm to behave as if its marginal cost curve is MSC, not MPC. In this way the competitive firm would equate MSC to price, and the externality would be "internalized" in the firm's output decision; that is, the firm will produce the output where MSC = P, which is the correct output from society's point of view. (As we saw with the Coase theorem, the optimal level of pollution is not zero. At some point, clean air costs more than it is worth.)

This solution can be obtained by imposing a tax on the industry's output that reflects the degree to which MSC and MPC diverge. Such a tax is illustrated in Figure 18–3 as an excise tax imposed on the firm of *ab* per unit. The impact of the pollution tax is to shift the marginal cost curve of the firm from MPC to MSC. This change is reflected at the industry level by a shift in the industry supply curve from $S_1$ to $S_2$. The tax forces the competitive firm to account for the externality in choosing its output level. It thus chooses the socially efficient output level where MSC = $P_2$, or where marginal social costs are equal to marginal private benefits.

Who pays the pollution tax? As we learned in Chapter 9, part of the tax will be paid by producers and part by consumers. Consumers will face a higher price for the industry's output, but the price will not rise by the amount of the tax. The producer thus absorbs some of the pollution tax.

**Selling Rights.** A simple alternative to the tax example is to establish a market for externality rights such as the right to pollute. In this approach government sets an allowable level of pollution and sells the right to pollute. In theory, this market works like any other market. Those who value the pollution rights most highly will buy and hold them. Firms will make decisions about whether it is less costly to install pollution-control equipment or to buy pollution rights. Government does not tell industry how to clean up the air. It sets the allowable level of pollution and lets firms decide how to control their pollution. This marketlike approach to pollution control has the virtue of ensuring that the acceptable level of pollution is reached at least cost.

Difficulties arise in such a scheme, however. Indeed, whether it chooses to tax pollution or sell pollution rights, government must somehow estimate what the "optimal" level of pollution is. The relevant knowledge in this case is economic rather than technical. The question is not, What is the technical damage of degrees of air pollution? but rather, How is this technical damage valued by individuals? Such knowledge is not easy to come by. Moreover, both a tax scheme and a pollution rights scheme must be monitored and enforced by government. Without such actions tax evasion and pollution without permit will be serious problems.

**Setting Regulations.** For such reasons government has most often not adopted an economic approach to control of externalities such as pollution. More typically, government has sought to control pollution and other externality problems through direct regulation of industry. In the case of pollution, regulation generally consists of detailed rules about the technology that firms must adopt to control pollution. Firms in certain pollution-prone industries are required to install special pollution-control equipment as a condition of being able to stay in the industry. This is a practical way to control pollution. Each firm must clean up its emissions in the prescribed manner. In contrast to the pollution rights approach, however, direct regulation does not allow firms to choose the most efficient means of staying below the allowable level of pollution. Firms cannot choose to pollute or not pollute or select among different types of pollution-control technologies. They must follow the rules laid down by government. Moreover, these rules can affect different firms in an industry in different ways. Large firms, for example, may be better able to adapt to the technology required by government than small firms. Some small firms may be driven out of the industry as a result, and the large firms will benefit through higher prices and profits.

Each approach to the problem of externality control that we have discussed—the establishment of property rights, taxation or subsidy, selling of rights, and direct regulation—has costs and benefits in different applications. In some cases, the establishment of property rights may resolve the externality. In other cases, like automobile pollution, some form of government taxation or regulation may be used to control the externality of pollution. No approach will yield a perfect solution to these problems. There will still be disputes among property owners about where one's property ends and another's begins. And government will not always act to resolve an externality in the most cost-effective way. There may, for example, be very little air pollution in Wyoming, which has fewer cars or industries, but new cars purchased there will nonetheless be required to have the same pollution control equipment as new cars purchased in smog-ridden Los Angeles or New York. Actual policies to control externalities are not perfect, as Focus, "Saving Endangered Predators," illustrates.

## Public Goods Theory

As we discussed in Chapter 3, government affects the economic well-being of a society in two essential ways. Government provides and maintains a system of laws to protect and enforce property rights and to permit the free flow of goods through markets. And government provides various goods and services that are not ordinarily provided through the free-market system.

These government-provided goods and services fall into the broader category of **public goods**—goods such as radio broadcasts and highways that, once provided to one person, are available to all on a noncompeting or nonrivalrous basis. It is difficult or impossible to exclude anyone from the use of a public good, and its use by one person does not prevent its use by another. A nuclear submarine, for example, provides national defense as a public good for all members of a society. One member's consumption of national defense does not preclude another's consumption. The use of a lighthouse beam by one ship does not stop other ships from using the same beam. The consumption of a public good by one consumer does not impose an opportunity cost on other consumers; all can consume the public good at the same time.

**Public good:** A product that is noncompetitive in consumption and nonexclusive; a good that, once produced for one individual, is available to everyone.

A private good is consumed exclusively by the person who buys it. Its consumption by one individual precludes consumption by others. When Benny eats a cheeseburger, Barbara cannot eat the same cheeseburger. The concern here is not with the fact that Benny bought the cheeseburger and owns it in the sense that he has legal title to it. The concern is with a technical characteristic of consumption—that when one person is consuming a private good, it cannot be simultaneously consumed by other people. The consumption of a private good carries an opportunity cost: Consumption of the same good is denied to other consumers.

A mixed good is a good that embodies attributes of both private and public goods. An outdoor circus, for example, would seem to be a pure public good. Two people can watch the circus at the same time without interfering with each other's consumption. If the crowd gets large enough, however, the circus can become more of a private good because congestion detracts from each person's view. In other words, rather than being equally available to all, consumption becomes rivalrous; some people's consumption of the circus detracts and may even prevent other people's consumption. We say then that the outdoor circus is a mixed good. It has characteristics of both private and public goods.

It is important to keep in mind the difference between public goods and goods that are publicly provided. The concept of a public good refers to the attributes of the good and not to whether it is produced by government. Indeed, the public sector produces many private goods such as first-class mail delivery, and the private sector produces many public goods such as beautiful architecture. In other words, the supply of public goods comes from both the public and the private sector. As we will see, it is largely the costs of exclud-

## FOCUS  Saving Endangered Predators

The preservation of endangered predators, such as the bald eagle, the mountain lion, and the timber wolf, is a pressing environmental issue. Because these predators cause an enormous amount of damage to ranchers (in 1979, approximately 1.3 million sheep were lost to predators), the wild animals and birds are hunted, trapped, and poisoned without regard to their endangered status. Environmentalists worry about the possible extinction of the predators themselves.

In the case of bald eagles, the present solution is a federal law that stipulates a $10,000 fine, a year in prison, or both for killing or possession of bald or golden eagles. This penalty appears to be ineffective. There is evidence that ranchers are willing to pay bounties of $25 for dead eagles, even in the face of a $10,000 fine and a year in jail. Why? People apparently do not take the fine and prison penalty seriously because they do not expect to get caught. The expected penalty (the probability of being caught times the fine and jail sentence) is very low; people rationally ignore the penalty and continue to kill eagles.

An obvious solution to the problem would be to raise the penalty. If the expected penalty were raised to equal the expected gain from killing an eagle, eagle killing would recede. Such a solution has been avoided out of concern that this fine would probably have to be so high that the unlucky rancher who got caught would have to go to jail for the rest of his life or pay a fine that would bankrupt him.

The economic theory of externalities suggests another solution: Establish property rights in eagles. Rather than considering eagles unowned as they are now, the "property rights" to eagles and other endangered predators could be transferred to conservation groups and their supporters. Then the damage done by the predators would become the liability of those who care about them. In assigning liability to conservation groups, it would be possible for a market for eagles to evolve. Farmers could take their damage claims to conservation groups (perhaps state by state), and conservation groups could finance whatever level of these claims they wanted to support out of their membership fees. As bargaining took place between farmers and conservationists, some idea of the "optimal" eagle population would emerge.

Needless to say, such a proposal would have many kinks to be ironed out. Farmers would seek to "overclaim" damages; conservationists would "underestimate" compensation. To get around such problems, independent arbitrators could perhaps be employed to settle damage claims.

Another possibility would be for the government to protect predators by compensating farmers for any livestock killed. Such a move would remove the burden of compensation from the few who care strongly and distribute it across the general population through taxation. In a modified version of this approach, the state of New Mexico is supporting a program that breeds Mediterranean sheepdogs and then leases the dogs to ranchers for $120 a year, the approximate price of one ewe. Subsidizing the breeding of sheepdogs is an alternative to outright compensation for lost sheep.

*Source:* R. C. Amacher, R. D. Tollison, and T. D. Willett, "The Economics of Fatal Mistakes: Fiscal Mechanisms for Preserving Endangered Predators," *Public Policy* (Summer 1972), pp. 411–441.

---

ing nonpayers that determine whether a public good is privately or publicly provided.

### Public Provision of Public Goods

Abraham Lincoln's dictum on government's proper place in the economic order still has merit: "The legitimate objective of government is to do for a community of people whatever they need to have done, but cannot do at all, or cannot do so well themselves, in their separate and individual capacities. In all that the people can individually do as well for themselves, government ought not to interfere."[6] The theory of public goods follows Lincoln's dictum. Just as Lincoln pointed out, there are some things that a free-market system fails to provide members of a society, usually because doing so would be difficult if not impossible.

[6]As quoted in Peter G. Sassone and William A. Schaffer, *Cost-Benefit Analysis: A Handbook* (New York: Academic Press, 1978), epigraph.

How, for example, can a free-market system provide national defense? Once a defense system is provided, it is equally available to all consumers or citizens. Since individuals cannot be excluded from the benefits of military defense, who would voluntarily contribute to its provision? Most individuals would hold back in the expectation that someone else would provide the necessary funds, and then they would reap the benefits of military defense for free.

Lighthouses provide another illustration of market failure to provide public goods. A lighthouse's beam is available to all ships passing by, regardless of whether a particular ship pays for the light. Since it is difficult or impossible to exclude nonpayers from using the beam, individual shipowners are likely to refrain from contributing to the lighthouse even though it provides an obviously valuable service for all shippers. To overcome this market failure, government assumes responsibility for the lighthouse. Through its power to tax, government can raise enough revenue to build and maintain the installation of the lighthouse.[7]

Public health services also illustrate the need for government to overcome market failure to provide a public good. At certain times when the public health seems in danger, such as the swine flu outbreak in the late 1970s, the government provides inoculations against disease. Individuals who are inoculated protect themselves against the disease. Their inoculation also protects those who are not inoculated. As a person becomes inoculated, the sources of contagion for the disease are reduced, so person $A$'s inoculation reduces the odds that $B$ and $C$, who have not been inoculated, will catch the disease. $B$ and $C$ theoretically owe $A$ something for the protection, but would they pay for it? Obviously not, since they cannot be excluded from $A$'s protection. Government's providing such a public health service is a way of overcoming the market failure in this case.

Not all goods provided to a number of people at once require government provision. In some cases the costs of excluding noncontributors are low enough that private firms will produce goods enjoyed on a nonrival basis by many people. For example, a movie, a football game, and a concert are goods that are jointly provided to many consumers at once. In this respect these goods are like the lighthouse and national defense. The difference is that it is feasible for the producers of these goods to exclude those who do not pay for their provision, and so they can be supplied by private producers within the free market.

## Free Riders

Sellers of private goods can exclude nonpayers from using the goods by charging a price for their use. Once a public good is produced, however, it is equally available to all consumers, regardless of whether they contribute to its production or not. Individuals who do not contribute to production and yet enjoy the benefits of goods are called **free riders.**

The problem facing producers of public goods is that free-riding behavior makes it difficult to discover the true preferences of consumers of a public good. Individuals will not reveal their true preferences for public goods because it is not in their self-interest to do so. For example, if a neighborhood tried to organize a crime-watch group, many neighbors might not contribute

**Free rider:** An individual who is able to receive the benefits of a good or service without paying for it.

---

[7]While the lighthouse has long served as an archetype of a public good, the private provision of lighthouses is a historical fact. See R. H. Coase, "The Lighthouse in Economics," *Journal of Law and Economics* 17 (October 1974), pp. 357–376.

but would still appreciate any protection provided, thus concealing their true preferences for the service. On the other hand, some who do not contribute voluntarily might be totally disinterested in the service. The possibility of free riding makes it difficult to determine the true level of demand for the public good.

If everyone free rides and no one contributes to the production of public goods, then everyone will be worse off as a result. No public goods will be produced, even though each individual would like some provision of public goods. Herein lies a rationale for government provision of public goods. Government has the power of coercion and taxation, and it can force consumers to contribute to the production of public goods. Citizens cannot refuse to pay taxes and thus contribute to the costs of public sector output. When the cost of excluding nonpayers is high, government can produce public goods through tax finance.

### Pricing Public Goods

Should a price be charged for use of public goods? In a sense, the correct price of a public good is zero. Take the example of a bridge. Once the bridge is built, the marginal cost of one more car's going over it is approximately zero. If we follow the rule that we learned in Chapter 9 for optimal pricing, we should set price equal to marginal cost. Since the marginal cost of an additional bridge crossing is effectively zero, price should be zero according to the $P = MC$ rule. According to this logic, public goods should be free.

This rationale is fine as far as it goes, but it does not go far enough. In rationing the use of the good among demanders, zero is the correct price. Except for cases of traffic congestion, a public good presents no rationing problem; by definition, additional individuals can consume a public good at zero marginal cost. But rationing is not the only function of price. A second important function of price is to achieve the right quantity of the public good by relating quantity to demand. Without information about what people are willing to pay to cross the bridge, authorities are left in the dark about the desired amount of bridge construction to undertake. After all, public goods involve the expenditure of scarce resources, and prices for public goods help public officials develop information about the demand for these goods.

How then should a price be set for a public good? Suppose a defense alliance is formed by two countries, A and B.[8] The alliance provides for a common defense policy through a treaty agreement. National defense is a public good for the alliance, and the problem facing the two countries is how to reach an agreement on the amount of national defense to be produced by the alliance. Figure 18–4 illustrates the problem.

We assume that each country faces the same marginal cost curve (MC) for providing defense. MC rises because additional scarce resources devoted to defense come at an increasing cost to the two countries.

Each country has a demand curve for national defense $D_A$ and $D_B$. Recall that the consumption of private goods is mutually exclusive. When A gets a unit of a private good, there is one less unit available for B to consume. In Chapter 4, this condition meant that to derive a market demand curve for a private good, the demand curves of individual consumers had to be summed

---

[8]See Mancur Olson and Richard Zeckhauser, "An Economic Theory of Alliances," *Review of Economics and Statistics* 47 (August 1966), pp. 266–279.

**FIGURE 18–4**
**Sharing the Costs of Defense**

MC is the marginal cost curve for production of military protection under an alliance. $D_A$ and $D_B$ are the individual country demand curves for national defense. The appropriate "prices" for national defense for each country are $P_B$ and $P_A$. $D_A + D_B$ is the vertical sum of $D_A$ and $D_B$. Optimal output for the alliance is where $D_A + D_B$ = MC, or at $Q_T$.

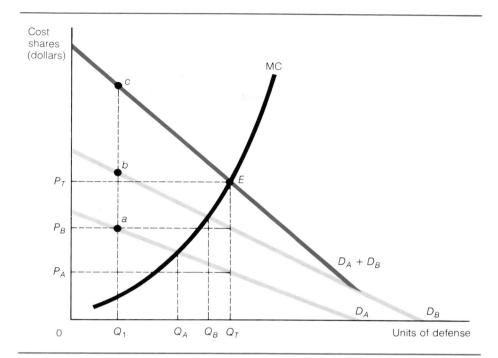

horizontally. National defense, however, is not a private good. It is a public good, and this makes the nature of the "demand curves" in Figure 18–4 quite different.

Consumption of a public good is not mutually exclusive among consumers. All the units along the horizontal axis in Figure 18–4 are equally available to both demanders. Both A and B receive benefits from, for example, $Q_1$ units of military protection. Thus, the total protection benefits offered by $Q_1$ to A and B are equal to the vertical sum of their demand prices for this quantity or output. At $Q_1$ the benefit that each country places on alliance protection is read vertically off its demand curve. A derives $aQ_1$, B derives $bQ_1$, and both countries together place a value of $cQ_1$ on $Q_1$ units of alliance output.

This point applies to all public goods. The market demand curve for a public good is obtained by vertically summing all individuals' demand curves. The demand curves are summed vertically rather than horizontally because all demanders can consume the same units of output along the horizontal axis. In Figure 18–4, then, market demand is given by $D_A + D_B$, which is the vertical sum of $D_A$ and $D_B$.

At points along the horizontal axis less than $Q_T$, marginal benefit exceeds marginal cost ($D_A + D_B$ > MC), and production will be expanded. At points beyond $Q_T$, marginal benefit is less than marginal cost ($D_A + D_B$ < MC), and production will be cut back. Equilibrium output is at E, where $D_A + D_B$ = MC. The equilibrium defense output in the alliance is $Q_T$, and the alliance defense budget is $P_T \times Q_T$.

How should this budget be divided between the two countries? One way to apportion the costs is in proportion to the benefits received. $P_B$ represents the benefit that accrues to country B, and $P_A$ is the benefit that goes to country A. Country B pays more because it benefits more from alliance output. $B$, for example, may be a larger country, with more income, capital, and population to defend.

This is, of course, an idealized presentation of alliance behavior. For example, what happened to the concept of free riding? Basically, we ignored it in order to depict a perfect outcome for the alliance. But free riding will be a thorn in the alliance's side. Look at the individual equilibrium positions of the two countries in Figure 18–4. The small country prefers $Q_A$ as its defense output; the large country prefers $Q_B$. When the treaty to form the alliance is signed, $A$ gains access to the benefits of $B$'s production. It is apparent that $B$'s production more than satisfies $A$'s demand for national defense. In group alliances, small countries like $A$ will therefore have strong incentives to reduce their defense spending and to free ride on the defense expenditures of larger countries such as $B$. This tendency will lead to much bargaining and many debates about contributions to alliance output with the large countries complaining about the efforts and contributions of the small countries. The problem of free riding will be more pronounced the more countries there are in the alliance.

## Cost-Benefit Analysis

Externalities and public goods require government intervention in the economy. As we saw, unless private ownership can be established, externalities call for government regulation or taxation of private activities that create externalities. The need for public goods calls for government provision of certain goods and services such as national defense. Up to this point we have assumed that government officials are able to obtain the necessary information about how to correct externalities and to provide public goods in more or less the correct quantities. This, of course, is a questionable assumption. Government decision makers face tremendous practical difficulties in obtaining information to guide their decisions. For example, since consumers will rationally not reveal their true preferences for a public good, how does government decide how much of the good to produce? This is not an easy question to answer in practice, but the technique of cost-benefit analysis has been applied to help government officials answer such questions.

**Cost-benefit analysis** is just what the term implies: It is a procedure by which the costs and benefits of government programs are weighed so that decisions can be made about which programs to undertake. Clearly, a wise use of public resources is to rank programs from highest to lowest in their benefit-cost ratios and undertake those programs with the highest ratios.

Cost-benefit analysis emerged as a result of efforts to find more systematic information on which to base government decisions about investment in physical resources. While cost-benefit analysis can be traced to the 1920s, its development truly got under way after World War II. In this postwar period various water resource investments under the aegis of the Army Corps of Engineers and the Bureau of Reclamation provided the subject matter for cost-benefit analysis. Military applications became important in the early 1960s in the Department of Defense, although these applications were typically referred to as systems analysis. Eventually, cost-benefit analysis came to be viewed as appropriate for application throughout the government budget process. Investments in human resources and proposed changes in legislation were added to investment in physical resources as reasonable subjects for cost-benefit analysis.

*Cost-benefit analysis:* A process used to estimate the net benefits of a good or project, particularly goods and services provided by government.

Government cost-benefit analysis can be divided into two main categories, according to its purpose. In the case of public investment, the goal of cost-benefit analysis is efficiency—to help choose more efficient over less efficient dams, for example. The other main category is the use of cost-benefit analysis to determine the merit of government intervention in private sector activity. The application of cost-benefit analysis in such areas as penalizing alcohol abuse, determining health effects of smoking, breaking up monopolies, regulating drugs, discouraging disposable containers, analyzing food additives, and promoting automobile safety are illustrations. The task of cost-benefit analysis in such instances is to provide evidence about the desirability of particular government interventions or public policies.

Although cost-benefit analysis is equally applicable to firms and governments, its development has been inspired mainly by an interest in promoting efficiency in government. Careful comparisons between the costs and benefits of a proposed dam, for example, help discipline government decisions, since market forces are not present to reward good or discourage bad government decision making. Estimating anticipated costs and benefits resulting from a particular government program helps decision makers select the most efficient programs from among those that could be adopted. Cost-benefit analysis is forward looking; it seeks to assess the consequences of various proposed courses of action.

Prior to conducting any cost-benefit analysis, planners must agree on a desired objective and array the various means of attaining the objective. Cost-benefit analysis does not propose a course of action. It is simply a technique for helping to evaluate proposed courses of action. The proposals themselves come from outside the framework of the analysis.

Once the goals of a government program are stated and the possible ways to reach these goals described, the primary task of cost-benefit analysis is to measure the anticipated benefits and costs of the project. The relevant benefits fall into two categories: direct benefits and indirect benefits. Direct benefits accrue to users of the project after its completion, such as the benefits that a new subway system provides to its users. Indirect benefits might occur if the subway system reduces congestion on the highways and highway users save travel time. Needless to say, estimating the benefits, even the simple direct benefits, of a program is a difficult task. It requires much judgment because hard evidence of real benefits is often impossible to develop.

In addition to measuring or estimating benefits, a cost-benefit analysis must measure or estimate the anticipated costs of a proposed project. The cost of a project is the value of the benefits from other potential uses of resources that must be sacrificed to undertake the project. If the costs of a dam are determined to be $10 million, this $10 million represents the value of the other services that could have been produced in place of the dam. This is why economists typically use the term *opportunity cost* as synonymous with *cost* since cost represents a sacrificed opportunity.

As with benefits, there will be direct and indirect costs associated with public programs. Moreover, the same inherent problems that plague benefit estimation plague cost estimation. As in the case of benefits, prices of programs often are not available to estimate the value of alternative uses of public resources. And what is forgone when specific alternatives are adopted is not easily observable. Cost and benefit estimations are therefore likely to be only crude approximations of real values even under the most favorable circumstances.

## A Role for Government

The presence of externalities and public goods means that government has an important and productive role to play in the economy. Externalities lead to a regulatory role for government. Public goods imply a role for government as a direct producer of some goods and services. In this case, in order to pay for such public goods and services, government levies taxes on citizens to raise revenue. It is this important area of government activity—taxation—that we turn to in the next chapter.

## Summary

1. Market failure occurs when the private market system fails to allocate resources in an optimal or ideal fashion. Externalities and public goods are two categories of market failure.
2. Owners of private property bear the economic results of its use. When property is held in common, users do not bear the full results of its use. Common or undefined ownership creates a conflict between the pursuit of self-interest and the common good. The implication is that common resources will be overused and exploited.
3. An externality occurs when an economic unit does not bear all the value consequences of its actions. Externalities can be positive or negative, trivial, irrelevant, or relevant. A relevant externality creates an effective demand to correct the externality.
4. The Coase theorem assumes that in a world of zero transactions and bargaining costs, the amount of an externality such as pollution is independent of who has legal liability for pollution damage. Under the stated conditions, government cannot improve upon what individuals negotiate.
5. Where transaction and bargaining costs are high, other routes to externality control become feasible. A per unit pollution tax would force a firm to act as if its marginal cost curve were identical to its marginal cost curve plus the cost of the externality, or the marginal social cost curve. Other approaches include establishment of a market for pollution rights and direct regulation of firm technology.
6. Public goods are goods that are nonrival in consumption, with high costs of excluding nonpayers. Whereas a private good is consumed exclusively by one individual, a public good can be simultaneously consumed by all consumers. A mixed good has both private and public characteristics.
7. To free ride means to reap the benefits of a public good without bearing its costs. If everyone free rides, no public good is produced by the private sector even though each individual values its production. Such behavior means that public goods typically have to be produced by government through its power to tax.
8. Public goods are not free; they cost real resources. In terms of rationing a public good among consumers, a zero price is the correct price because additional consumers can consume a public good at near-zero marginal cost. However, a zero price for a public good offers no guidance for public officials about how much of a public good to produce. There is thus a trade-off regarding whether public goods should be priced.
9. Individual demand curves for public goods are summed vertically. Optimal output of a public good occurs where the vertically summed individual demand curves intersect the marginal cost curve. Optimal prices for this output are read off the individual demand curves.
10. Cost-benefit analysis is an analytical tool for evaluating and promoting efficiency in government programs and policies. Cost-benefit analysis does not set goals; it evaluates the costs and benefits of attaining goals in alternative ways.

## Key Terms

market failure
common ownership
externality
negative externality
positive externality
Coase theorem
marginal private costs
marginal social cost
public good
free rider
cost-benefit analysis

## Questions for Review and Discussion

1. Are whales an example of a common property resource? If so, what sort of behavior would you predict among whale hunters?
2. What is an externality? Name the various types of externality. Why is an externality considered a market failure?
3. Suppose that pollution is controlled with a pollution tax on polluting firms. Who will pay the tax? Firms? Consumers?
4. Define a public good. What is the difference between a private and a public good? How will individuals behave when asked how much they will contribute toward payment for a public good? What does this mean for the nature of the demand curves for public goods?
5. "Museums should be free." Evaluate.
6. Describe how you would apply cost-benefit analysis to the evaluation of an addition to the football stadium at your college or university.

## ECONOMICS IN ACTION

### Is Urban Mass Transit Worth the Cost?

Several U.S. cities have new public transit systems in construction or in planning. Between 1984 and 1989 these cities (and the federal government) will spend roughly $14 billion for construction and repair in an effort to overcome traffic congestion, air pollution, and the economic decline of downtown metropolitan areas. Los Angeles, for example, will soon begin an 18.6-mile, $3.5-billion subway, and Washington, D.C., now operates a partially completed 47-mile subway system that will cost $6 billion by the time it is finished.

For the most part, urban mass transit systems represent a large subsidy to their ridership, who typically bear only a fraction of a system's true cost in the fares they pay. It was estimated that in 1982 fares in the New York City subway system failed to cover $1.3 billion in total operating expenses. But mass transit advocates argue that investment in these systems is nevertheless justified because of the positive externalities these systems offer, which more than make up for typical operating losses. The question faced by voters and politicians is quite straightforward: Is urban mass transit worth its cost?

Backers of mass transit point to three main categories of positive externalities generated by large-scale bus and subway systems. First, population density is increased in urban areas as a result of the availability of transit service because new residents are attracted by the reduced costs of transportation. This increase in population brings more tax income to hard-pressed local governments. The second positive externality is economic development. Businesses will concentrate in or near the denser population centers to reduce the costs of business transactions and to take advantage of

*Source:* David M. Stewart, "Rolling Nowhere," *Inquiry* (July 1984), pp. 18–23.

*The Jefferson Park station on the O'Hare line in Chicago.*

a large, accessible labor supply. The Bay Area Transit in San Francisco has been credited with stimulating $1.4 billion of construction since it opened. Third, mass transit may reduce traffic congestion and pollution levels as commuters switch from cars to buses or the subway. For these three broad reasons, mass transit proponents argue that the positive externalities generated by such systems are high enough to equal or exceed the costs of construction and operation.

Critics who oppose the large public investment in mass transit argue that the positive externalities are exaggerated and that the costs of construction and operation of these systems exceed their benefits. For example, some maintain that reduced business costs would tend to be canceled by increased property values and tax liabilities. Others point to the fact that in those areas with large mass transit systems, traffic congestion and pollution have remained approximately constant or have even increased. In Washington, D.C., for instance, rush-hour trips downtown by car actually increased after the transit system began operating. In short, the mass transit controversy hinges on weighing positive externalities against the costs of construction, maintenance, and operation.

## Question

Houston, Texas is a large and rapidly growing urban center. Voters in that city, however, have turned down several proposals for mass transit systems over the past decade. Is this proof that the costs of such systems outweigh the benefits to the city of Houston? Upon what economic considerations should such decisions be made?

# 19

# Taxation

When government finances the production of public goods or services—a local sewer system, a research program looking into the cause of acid rain, a nuclear warhead, a presidential limousine—it uses taxes to help pay the bill. In the United States about 82,000 governmental bodies at federal, state, county, city, school district, municipal, and township levels have the power to establish and collect taxes.

Unpopular as they may be, taxes are a necessary part of life in a mixed economy. Every tax has a purpose; directly or indirectly, tax revenues finance government expenditures. As we saw in the previous chapter, the market fails to provide the optimal amounts of public goods such as national defense, police and fire protection, and education or to deal with externalities through programs such as environmental protection. Traditionally, federal, state, and local governments have taken an active role in providing or influencing the production of such goods and services. **Public finance** is the study of government expenditures and revenue-collecting activities.

**Public finance:** The study of how governments at federal, state, and local levels tax and spend.

## Government Expenditures

In Chapter 3 we reviewed some of the major categories of government expenditures and gave some statistics on the rate of government growth. There are two basic types of government expenditures: direct purchases of goods and services and transfer payments, the redistribution of income from one group of people to another. Direct purchases involve spending on items such as defense, fire and police protection, and wages and salaries of government employees. Transfer payments include such expenditures as Social Security payments, Aid to Families with Dependent Children, and unemployment insurance.

## TABLE 19-1 Federal, State, and Local Government Expenditures

All government expenditures have increased in recent decades, with state and local expenditures increasing at about the same rate as federal expenditures. This increase was not simply a function of population growth; notice that the expenditures per person increased threefold between 1950 and 1984.

| Fiscal Year | Expenditures (billions of 1983 dollars) | | | Total Expenditures | |
|---|---|---|---|---|---|
| | All Governments | Federal | State and Local | Per Household | Per Capita |
| 1950 | 264.2 | 168.4 | 95.8 | 6,068 | 1,752 |
| 1960 | 511.6 | 329.0 | 182.6 | 9,692 | 2,864 |
| 1965 | 642.4 | 406.3 | 236.1 | 11,183 | 3,332 |
| 1970 | 837.9 | 523.9 | 314.0 | 13,215 | 4,134 |
| 1975 | 999.4 | 607.6 | 391.8 | 14,052 | 4,660 |
| 1976 | 1057.3 | 659.5 | 397.8 | 14,511 | 4,881 |
| 1977 | 1083.9 | 686.1 | 397.8 | 14,620 | 4,956 |
| 1978 | 1099.9 | 705.0 | 394.9 | 14,466 | 4,977 |
| 1979 | 1123.3 | 722.9 | 400.4 | 14,526 | 5,028 |
| 1980 | 1172.5 | 752.7 | 419.8 | 14,516 | 5,190 |
| 1981 | 1229.8 | 795.0 | 434.8 | 14,931 | 5,384 |
| 1982 | 1246.7 | 818.5 | 428.2 | 14,926 | 5,408 |
| 1983 | 1300.3 | 864.8 | 435.5 | 15,345 | 5,587 |
| 1984 | 1359.4 | 883.2 | 476.2 | 16,035 | 5,788 |

Source: *Tax Foundation's Monthly Tax Features*, (June/July 1984). Conversion to constant dollars provided by authors.

The size of all government expenditures is shown in Table 19-1 for selected years from 1950 to 1984. Total government expenditures have increased dramatically since 1950; they are expected to rise even higher in the future.

In Table 19-1, total expenditures are broken down into federal government expenditures and state and local government expenditures. Notice that federal expenditures (including grants-in-aid to state and local government) exceed state and local expenditures by a fairly constant proportion. This has not always been the case. Around the turn of the century, for example, local government expenditures exceeded the sum of federal and state expenditures. Also, the per household and per capita government expenditures have increased in the last few decades.

Not only has the absolute level of expenditures increased; the level of government expenditures as a percentage of GNP has increased. Government expenditures at all levels increased from 21.3 percent of GNP in 1950 to 26.9 percent in 1960, 31.3 percent in 1970, and 35.5 percent in 1982.

Federal government expenditures have increased most in the area of transfer payments. This trend began in the 1960s under the Great Society policies of the Lyndon Johnson administration (see Chapter 3 for a review of these trends). Direct purchases of goods and services by the federal government have also increased in absolute terms but have decreased as a percentage of total expenditures.

Direct purchases at state and local levels exceed those of the federal government. This may not be surprising when you consider the programs offered. Most streets, roads, and highways, police and fire protection, hospitals, education, and sewage and garbage disposal are provided by state and

## Tax Revenues

The two basic methods of financing public expenditures are taxing or borrowing. State and local governments usually borrow by selling bonds to finance special projects such as highways, schools, or hospitals. The federal government has come to use borrowing as a routine means to finance its burgeoning deficit. Nevertheless, the largest portion of government expenditures are still financed by taxes. Table 19–2 shows the estimated total tax receipts of the federal government from 1982 to 1986 from each source. The largest single source of federal revenues is the personal income tax, followed closely by Social Security taxes.

Property taxes and sales taxes are the two largest contributors to state and local government revenues, which are broken down in Table 19–3. The statistics reveal a recent trend in state and local financing: Property tax revenues have not been rising as rapidly as sales and income tax revenues. This indicates that property owners are being taxed relatively less over time than other groups. Focus, "Tax Freedom Day," discusses a little-known effect of taxation on individuals.

## Theories of Equitable Taxation

Given the government's need for tax revenues, how should taxes be levied among the people? Our society attempts to find an equitable way of distributing the tax burden, but there are no easy formulas for determining what is truly equitable. In theory, however, two basic ways are used to determine who should be taxed and how much: the benefit principle and the ability-to-pay principle.

TABLE 19–2  Actual and Estimated Tax Revenues of the Federal Government

The federal government obtains most of its tax dollars from personal income and Social Security taxes. Notice that total revenues are expected to rise by more than $200 billion between 1982 and 1986.

| | Tax Revenues (billions of dollars) | | | | |
|---|---|---|---|---|---|
| | Actual | Estimate | | | |
| Source | 1982 | 1983 | 1984 | 1985 | 1986 |
| Individual income taxes | 197.7 | 285.2 | 295.6 | 317.9 | 358.6 |
| Corporate income taxes | 49.2 | 35.3 | 51.8 | 60.5 | 74.0 |
| Social Security taxes and contributions | 201.5 | 210.3 | 242.9 | 275.5 | 304.9 |
| Excise taxes | 36.3 | 37.3 | 40.4 | 40.8 | 74.8 |
| Estate and gift taxes | 8.0 | 6.1 | 5.9 | 5.6 | 5.0 |
| Customs duties | 8.9 | 8.8 | 9.1 | 9.4 | 9.7 |
| Miscellaneous receipts | 16.2 | 14.5 | 14.0 | 14.5 | 14.8 |
| Total budget revenues | 617.8 | 597.5 | 659.7 | 724.3 | 841.9 |

Source: U.S. Office of Management and Budget.

### TABLE 19-3  State and Local Tax Revenues

Unlike the federal government's revenues, state and local governments' largest share of tax revenues is collected in the form of property taxes. Property taxes are falling in relative importance, while sales and income taxes are rising.

| | Tax Revenues (billions of dollars) | | | |
|---|---|---|---|---|
| | 1977 | | 1982 | |
| | Absolute | Percent | Absolute | Percent |
| *Level of Government* | | | | |
| State | 106,111 | 58 | 165,116 | 60.3 |
| Local | 76,661 | 42 | 108,814 | 39.7 |
| Total | 182,772 | 100 | 273,930 | 100 |
| *Type of Tax* | | | | |
| Individual income | 30,852 | 16.8 | 52,742 | 19.2 |
| Corporation net income | 9,709 | 5.3 | 13,494 | 4.9 |
| Property | 64,164 | 35.1 | 86,811 | 31.7 |
| General sales and gross receipts | 38,740 | 21.2 | 61,910 | 22.6 |
| Motor fuel sales | 9,365 | 5.1 | 10,597 | 3.8 |
| Tobacco product sales | 3,684 | 2.0 | 4,210 | 1.5 |
| Alcoholic beverage sales | 2,309 | 1.2 | 2,909 | 1.0 |
| Motor vehicle and operators' licenses | 4,961 | 2.7 | 6,712 | 2.4 |
| All other | 18,988 | 10.3 | 34,545 | 12.6 |

Source: Tax Foundation's *Monthly Tax Features.*

## Benefit Principle

According to the **benefit principle,** individuals who receive the most benefits from government-produced goods should pay the most for their production. Many taxes arise from this principle.

When nonbuyers can be excluded, government can directly tax the users of its goods and services. People who ride on public transportation facilities such as buses or subways are charged a tax in the form of a fare. Electrical power companies owned by local governments charge individuals according to the amount of electricity they use and thus the amount of the benefits they receive. Highway users may be taxed directly for their use of a road by a toll. Certain other taxes attempt to force users to pay indirectly for categories of benefits they enjoy. For example, the revenues collected from excise taxes on gasoline, tires, automobiles, and auto batteries are used to build highways, roads, and bridges.

The benefit principle is not easily applied to all taxes, however. For one thing, benefits that people receive are not always obvious. With some goods, such as highways and education, direct users receive the greatest benefits, but others receive indirect benefits. A Milwaukee resident who has no car may still buy fruits trucked across highways from California, Texas, and Florida. Families with no children to educate benefit from an educational system that increases worker productivity and raises the standard of living.

---

**Benefit principle:** The notion that people who receive the benefits of publicly provided goods should pay for their production.

## FOCUS  Tax Freedom Day

The average American worker spends a great deal of time earning income just to pay taxes. Not only does the federal government tax the worker directly in the form of income tax, but state and local governments collect taxes in many other forms. How much time does the average worker spend earning money to pay all these taxes?

In 1948, Dallas L. Hostetler of Florida decided to estimate how long the average worker spent each year working to pay taxes, beginning on January 1. Bringing his calculations up to date, the Tax Foundation, Inc. yearly estimates "tax freedom day," the day of the year that the average worker begins to earn take-home pay. Figure 19–1 represents the foundation's estimates.

In 1930, the average worker devoted only 43 days per year to pay all taxes. In 1985, 121 days—one-third of a year's work—were required to pay taxes.

The Tax Foundation also reports the number of hours per day that the average worker supports the public sector. Figure 19–2 breaks down the hours and minutes the average individual worked to support federal and state and local government.

Notice that the federal government now requires more of the worker's day than state and local governments combined. The total hours required for taxes represent one-third of a worker's day.

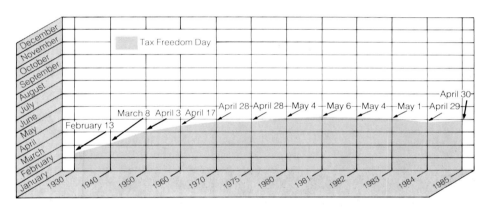

**FIGURE 19–1  Tax Freedom Day**
"Tax freedom day" is the day of the year on which the average worker is able to begin earning take-home pay.

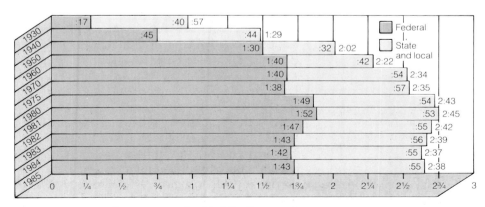

**FIGURE 19–2  Supporting the Public Sector**
The average individual worker spends a part of each work day supporting federal and state and local governments.

A second problem with the benefit principle is that individuals who receive government benefits cannot always afford to pay an equitable share of taxes. For example, welfare recipients are not likely to have the ability to pay for all the benefits they receive. For these reasons, many taxes are based on the ability-to-pay principle.

### Ability-to-Pay Principle

**Ability-to-pay principle:** The notion that tax bills should vary directly with income levels or the capacity of individuals to pay taxes.

According to the **ability-to-pay principle,** individuals who are more able to pay should pay more taxes than those less able. Under this principle, levels of an individual's income, wealth, or expenditures are measures frequently used for determining the level of tax obligations. The higher the level of income, wealth, or expenditures, the greater the tax liability, regardless of the benefits an individual receives. For example, income and property taxes result in higher tax payments for people with a greater ability to pay.

**Horizontal equity:** A tax structure under which people with equal incomes pay equal amounts of taxes.

**Vertical equity:** A tax structure under which people with unequal incomes pay unequal taxes; people with higher incomes pay more taxes than people with lower incomes.

When taxes are levied according to the ability-to-pay principle, the equity of taxation is theoretically measured in two dimensions. The ideal of **horizontal equity** suggests that people with equal abilities to pay should pay equal amounts of taxes. For example, all people who earn $10,000 income would pay the same amount in taxes—perhaps $2000. The ideal of **vertical equity** suggests that people with greater ability to pay must pay a higher absolute amount of taxes than people with less ability to pay. For example, if one individual earns $10,000 and pays $2000 in taxes, anyone who earns more than $10,000 must pay more than $2000 in taxes. Ideally, an equitable ability-to-pay tax has both horizontal and vertical equity.

This ideal is not easily met, unfortunately. One of the problems with establishing an equitable tax is defining ability to pay. Income is the most common measure used to determine ability to pay, but two families with the same income may have dramatically different abilities to pay. One family might have four children, huge medical bills, and no stored wealth. Another family with the same income might have no children, few medical bills, and a great store of wealth. Wealth itself is sometimes proposed as a better measure of ability to pay, but wealth is very hard to measure: Is it the value of a house, of a work of art, of a portfolio of stocks and bonds? A family with enormous wealth may find itself with very little income.

There is general agreement that the wealthy should pay more taxes than the poor, but the question of how much more is not easily resolved. Three ways have been used to relate taxes to ability to pay. The following paragraphs discuss these three ways with reference to an income tax, but they can also be applied to other taxes. Figure 19–3 graphically illustrates the effects of the three types of taxes.

**Proportional income tax:** A tax based on a fixed percentage of income for all levels of income.

**Proportional Tax.** A tax that requires individuals to pay a constant percentage of their income in taxes is a **proportional income tax.** If the proportion is 10 percent, a taxpayer earning $50,000 a year would pay $5000 in taxes. If a taxpayer earned $5000, his or her tax bill would be $500. A proportional income tax can result in both horizontal and vertical tax equity. Some states have a proportional income tax.

**Progressive income tax:** A tax based on a percentage of income that varies directly with the level of income.

**Progressive Tax.** A **progressive income tax** requires that a larger percentage of income be paid in taxes as income rises. If the tax structure is based on this principle, an individual with $15,000 income may pay only 5 percent of income in taxes while an individual with $30,000 income pays 15 percent. A progressive income tax is also capable of vertical and horizontal equity.

**FIGURE 19-3**

**Progressive, Proportional, and Regressive Taxes on Individual Income**

With a progressive tax, individuals pay a larger percentage of their income in taxes as their income rises. A proportional tax takes a constant percentage of income for taxes. A regressive tax takes a smaller percentage of income as income rises.

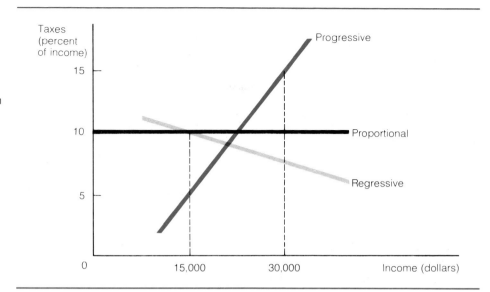

**Regressive income tax:** A tax based on a percentage of income that varies inversely with the level of income.

**Regressive Tax.** With a **regressive income tax** a lower percentage of income is paid in taxes as income rises. An individual with $15,000 income may pay 10 percent in taxes while an individual with $30,000 income may pay 8 percent in taxes. While a regressive tax results in a lower percentage of income paid in taxes as income rises, people with higher incomes may pay higher taxes in absolute amounts. According to our example, the individual with $15,000 income pays $1500 in taxes, while the individual with $30,000 income pays $2400 in taxes. For this reason, regressive taxes can be vertically equitable. Further, these taxes can be horizontally equitable so long as people with equal incomes pay equal taxes.

## Types of Taxes

Dozens of different taxes are imposed by national, state, and local governments on individuals and businesses. These taxes are usually based on income, wealth, or particular activities.

### Personal Income Tax

The federal government receives more tax revenues from personal income taxes than from any other single source. An individual's taxable income is found by subtracting tax exemptions and deductions from total income. The remainder is subject to a progressive tax rate that is used to determine total tax liability. Figure 19-4 illustrates some of the details for determining personal **taxable income.**

**Taxable income:** The amount of income that is subject to income taxes; total income minus deductions and exemptions.

Tax liability is usually determined by applying tax tables with a progressive tax rate to taxable income. Table 19-4 shows a sample tax table for an unmarried taxpayer. Taxable income in column (1) is divided into the tax obligation in column (2) to yield the **average tax rate** in column (3)—the percentage of income paid in taxes.

**Average tax rate:** The percentage of income that is paid in taxes; total tax liability divided by total income.

The **marginal tax rate** in column (6) is the percentage of a small increase in income that must be paid in taxes. It is the change in taxes (column 5) divided by the change in taxable income (column 4). For example, if a

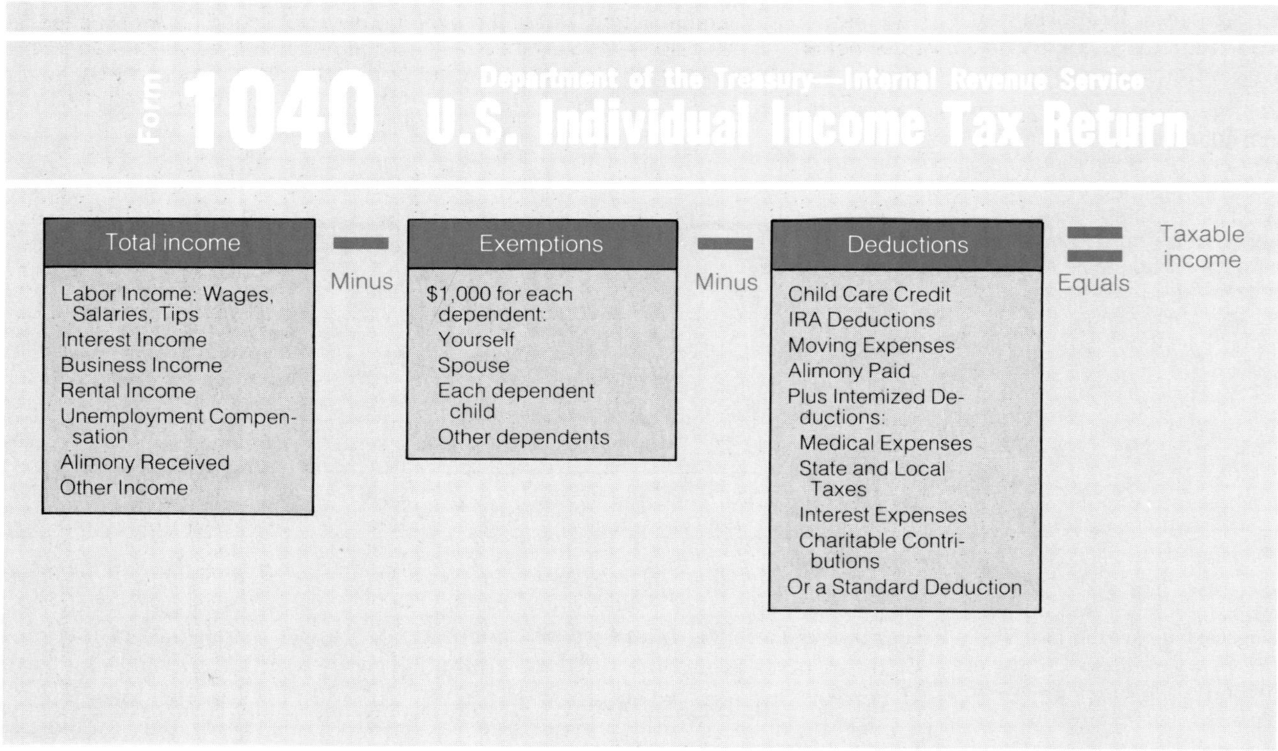

**FIGURE 19–4  Estimating Taxable Personal Income**

This chart demonstrates the basic method of determining personal income subject to federal income taxes. Notice that total income is not taxable. Individuals are allowed to exempt some income for each dependent. Also, certain other expenses are deducted from total income regardless of whether the taxpayer itemizes expenses.

*Source:* Abbreviated from IRS Form 1040, 1983.

**Form 1040 — U.S. Individual Income Tax Return — Department of the Treasury—Internal Revenue Service**

Total income (Labor Income: Wages, Salaries, Tips; Interest Income; Business Income; Rental Income; Unemployment Compensation; Alimony Received; Other Income) **Minus** Exemptions ($1,000 for each dependent: Yourself; Spouse; Each dependent child; Other dependents) **Minus** Deductions (Child Care Credit; IRA Deductions; Moving Expenses; Alimony Paid; Plus Itemized Deductions: Medical Expenses, State and Local Taxes, Interest Expenses, Charitable Contributions; Or a Standard Deduction) **Equals** Taxable income

**Marginal tax rate:** The percentage of an increase in income that must be paid in taxes; the change in tax liability divided by the change in taxable income.

worker normally earns $23,500 a year but has the opportunity to earn an extra $5300 in overtime pay, then 32 percent of that extra income would be paid in taxes; that is, $1696 of the $5300 would be used to pay taxes and only the remaining $3604 would be take-home pay. Notice that the marginal tax rate is greater than the average tax rate and that it rises as income rises. The maximum marginal tax rate is 50 percent. For every dollar earned over $55,300, one-half is taken in taxes.

Some economists believe that the federal income tax system does not result in vertical or horizontal equity. Some people with higher incomes pay lower taxes than others with lower incomes. Also, people with equal incomes do not always pay equal taxes.

Income tax inequities occur because many people engage in tax avoidance. This is a legal activity in which individuals are able to reduce their tax liability by using what are commonly called tax loopholes. One means of avoiding taxes is obtaining income from nontaxable sources. For example, the interest income received from municipal bonds is nontaxable, whereas the interest income received from bonds issued by businesses is taxable.

Individuals may also avoid taxes by spending their money in a particular way. For example, interest paid on borrowed money is deducted from total

## TABLE 19-4  Average and Marginal Income Tax Rates

As an individual's taxable income rises, the average tax rate rises. The average tax rate is equal to taxes paid divided by income. The marginal tax rate is the change in income from one tax bracket to the next divided by the change in taxes. With a progressive income tax, the marginal tax rate is greater than the average tax rate. The 50 percent bracket is the highest marginal tax rate.

| (1) Taxable Income (dollars) | (2) Taxes (dollars) | (3) Average Tax Rate (%) (2) ÷ (1) | (4) Change in Taxable Income (dollars) (1) | (5) Change in Taxes (dollars) (2) | (6) Marginal Tax Rate (%) (5) ÷ (4) |
|---|---|---|---|---|---|
| 2,300 | 0 | 0 | — | — | — |
| 3,400 | 121 | 3.5 | 1,100 | 121 | 11 |
| 4,400 | 251 | 5.7 | 1,000 | 130 | 13 |
| 8,500 | 866 | 10.2 | 4,100 | 615 | 15 |
| 10,800 | 1,257 | 11.6 | 2,300 | 391 | 17 |
| 12,900 | 1,656 | 12.8 | 2,100 | 399 | 19 |
| 15,000 | 2,097 | 14.0 | 2,100 | 441 | 21 |
| 18,200 | 2,865 | 15.7 | 3,200 | 768 | 24 |
| 23,500 | 4,349 | 18.5 | 5,300 | 1,484 | 28 |
| 28,800 | 6,045 | 21.0 | 5,300 | 1,696 | 32 |
| 34,100 | 7,953 | 23.3 | 5,300 | 1,908 | 36 |
| 41,500 | 10,913 | 26.3 | 7,400 | 2,960 | 40 |
| 55,300 | 17,123 | 31.0 | 13,800 | 6,210 | 45 |
| 69,100 | 24,023 | 34.8 | 13,800 | 6,900 | 50 |

Source: IRS Form 1040, Schedule X, 1983.

income. Therefore, mortgaging a house rather than renting can lower tax obligations. Deductions and exemptions such as these may result in tax inequities.

## Corporate Income Tax

The federal government charges incorporated businesses an income tax on their accounting profits. The corporate income tax is progressive, with a maximum marginal tax rate of 46 percent for profit income over $100,000. The after-tax profits can be distributed to the shareholders (owners) of a corporation in the form of dividend payments. The shareholders' income is also subject to personal income taxes. Thus corporate profits are taxed twice, once from corporations and once from shareholders.

## Social Security Taxes

Social Security taxes, the second largest source of federal tax revenues and the fastest-growing revenue source, represent taxpayers' contribution to federal programs that support the elderly, ill, and disabled. The Social Security tax is a payroll tax in which the employer and the employee each pay 6.7 percent of the employee's income up to $35,700 (as of 1984). The marginal tax rate is zero for income above $35,700. Nonlabor income is not subject to Social Security taxes. These features lead to a vertically and horizontally inequitable tax. For example, an individual with wage income of $35,700 pays $2,191.90 in Social Security taxes, and an individual who earns $50,000 in wage income pays the same amount because income over $35,700 is not taxed. Since unequals are not treated unequally, vertical inequity results. As currently structured, the Social Security tax is also horizontally inequitable: Equals are not treated equally because interest income is not taxed. An individual who earns $35,700 in interest from bonds pays no Social Security tax.

### Property Taxes

The largest source of state and local government tax revenues is the property tax. Owners of land and permanent structures are charged a tax based on the assessed value of their property. The assessed value is usually a percentage of the market value; that is, the selling price of the buildings and land.

A property tax is actually a tax on wealth rather than on income. Individuals who hold their wealth in the form of property pay more taxes than people who hold their wealth in other forms such as gold, art, or bonds. For example, people who hold bonds are taxed on the income from bonds—just as property owners are taxed on rental income—but they are not taxed on the value of the bonds, as property owners are taxed on the value of their property. From this perspective, the property tax may be regarded as inequitable.

### Sales and Excise Taxes

Sales and excise taxes also account for a large portion of state and local tax revenues. They are levied on the purchase of consumer goods. A sales tax is a specific percentage of the price of a good; an excise tax is a per unit tax. For example, a sales tax may be 5 percent of the price of a good regardless of the price. Under a sales tax, as the price of a good rises, the amount of the tax per unit rises. On the other hand, an excise tax may be 5 cents per unit of the good. The amount of the tax per unit does not change as the price rises or falls.

Sales taxes levied by state and local governments vary from state to state. In many locations, items such as food and medicine are exempt from sales taxes. Excise taxes are levied both by national and by state and local governments but only on selected items. For example, liquor, tobacco, gasoline, and certain sporting goods have excise taxes. Frequently these tax revenues are earmarked for specific government projects.

The purchase of nonconsumption goods such as stocks and bonds is not subject to a sales tax. This feature of sales taxes may create a horizontally inequitable tax distribution when people with the same income spend their money in different ways. An individual can pay less in sales taxes by spending more on investment goods than on consumption goods. Furthermore, if higher-income families invest more of their income than lower-income families, the sales tax is a regressive tax.

## The Effects of Taxes

The nature of the U.S. tax system effectively eliminates total tax avoidance—almost all goods and many resources are taxed. Indeed, this situation is desirable. Taxes are intended to eliminate free riders and thus should be unavoidable. Yet the burdens of taxes and the attempt to avoid them do influence individuals' activities. Altering economic behavior is sometimes intentional and desirable for society as a whole; other times it leads to undesirable economic inefficiency.

**Neutral tax:** A tax that has no effect on the production or consumption of goods.

Taxes that have no effect on behavior are called **neutral taxes.** Taxes are rarely neutral. More typically, by affecting price, they affect the quantity of goods supplied and demanded and the allocation of resources. Furthermore,

## Basic Effects on Price and Quantity

To understand how taxation affects price and quantity, consider a simple excise tax. Let $D$ and $S$ in Figure 19–5 represent, respectively, the demand and supply of gasoline in the United States with no taxes. $P_0$ and $Q_0$ represent the original price and quantity. Then assume that the federal government imposes an excise tax of $t$ cents per gallon for every gallon sold. This tax effectively increases the cost to the suppliers of producing and selling gasoline.

What happens? The supply of gasoline decreases. The supply curve shifts upward by the amount of the tax. Before the tax, producers were willing to offer quantity $Q_0$ at price $P_0$; after the tax, the producers are willing to offer $Q_0$ but only at a higher price, $P_0$ plus the tax $t$. The tax results in a new equilibrium $E_t$. The price increases, as we would expect, but probably not by the full amount of the tax. In this example, if the original price was $1.30 and the tax is 20 cents, then the price to consumers, $P_c$, rises by only 10 cents to $1.40. In this case, the price that producers receive after paying the tax is $1.40 minus 20 cents, or $1.20. The price the producers receive, $P_p$, may be found on the original supply curve at the new reduced quantity, $Q_t$.

The tax thus has two primary effects in the market for the taxed item. First, it changes the relative prices that sellers and buyers face. In fact, it lowers the price that sellers receive and raises the price that buyers pay. Second, as we would expect, the quantity exchanged falls—the lower price

**FIGURE 19–5**

**Effects of an Excise Tax**

An excise tax shifts the supply curve upward by the amount of the tax. This shift increases the price paid by consumers from $P_0$ to $P_c$ and lowers the net price received by the sellers from $P_0$ to $P_p$. The tax also lowers the amount produced and consumed from $Q_0$ to $Q_t$. The shaded areas represent tax revenues for the government and the welfare loss to society, discussed later in this chapter.

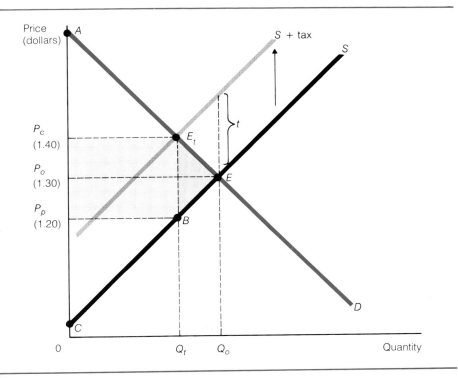

to sellers encourages them to offer fewer goods or services, and the higher price to buyers encourages them to buy fewer goods or services. The decrease in quantity depends on the size of the tax and on the elasticities of supply and demand.

If supply or demand for a good or service is perfectly inelastic, a tax will not affect quantity and is therefore neutral. This is rarely the case. As we have seen, elasticity varies at different price levels. The higher the tax-increased price, the more likely that the tax will affect quantity. With the Social Security tax, employers and employees turn over a fixed percentage of the employees' paychecks to the government. This tax increases the cost of labor to the firms and decreases the after-tax wage that workers take home. If the demand and supply of labor are less than perfectly inelastic, as is the case with most goods and services, the tax will encourage firms to hire fewer workers and discourage people from entering the labor force or encourage them to work fewer hours. If the supply of labor is relatively inelastic or if the percentage of wages paid in Social Security tax is lower, then the tax has relatively little effect on the level of employment. As the tax rises, more people seek ways to support themselves without earning heavily taxed wages, from fixing their own cars and growing their own food to investing in tax-exempt enterprises.

Although the burden of the Social Security tax is theoretically shared equally by employers and employees, who really pays taxes levied on items such as consumer goods? The distribution of the tax burden between buyers and sellers is known as tax incidence.

### Tax Incidence

**Tax incidence:** The burden of a tax or fiscal resting place.

At a basic level, we know that the buyers of goods in competitive markets cover the entire cost to producers—production costs plus any and all taxes. However, we can determine the actual burden of the tax—the **tax incidence**—by observing the relative change in price to producers and consumers. Notice that in the excise tax example in Figure 19–5, the price to producers fell by 10 cents ($P_0 - P_p$) and the price to consumers increased by 10 cents ($P_c - P_0$) after imposition of the tax. These changes indicate that the tax incidence was shared equally by the buyers and sellers—each faced a price change of 10 cents.

The incidence of a tax is not always shared equally by buyers and sellers—indeed, we suspect that an equal incidence is a rare event—but it *is* most often shared to some degree. It is most unusual for the entire burden of a tax to be shifted forward to the buyers or backward to the sellers. In other words, the price to consumers does not usually increase by the full amount of the tax or decrease to producers by the full amount.

Tax shifting depends on the elasticity of demand and supply. In general, the more elastic the demand for a product, the greater the incidence of tax shifted backward to the sellers. Conversely, the more elastic the supply, the greater the incidence shifted forward to the buyers. In some circumstances the tax incidence may be shifted entirely to one group or the other. In particular, if demand is perfectly inelastic or if supply is infinitely elastic, the entire burden is shifted to the buyers. The entire burden is shifted to the sellers if demand is infinitely elastic or if the supply is perfectly inelastic. Such extremes of elasticity are not common.

Regardless of whether a tax is shared or is borne entirely by the buyer or the seller, it drives a wedge between the price that buyers pay and the price that sellers receive. This chunk claimed by the government occurs whether the tax is an excise tax, a sales tax, a payroll tax, a property tax, or an income tax. For example, the income tax drives a wedge between the price that employers pay for labor and the price employees receive for their services.

### The Allocative Effect of Taxes

Unless taxes are neutral, they affect the allocation of resources. Taxes change relative prices, so the relative quantities of goods produced tend to change. In Figure 19-5, the imposition of the tax caused the equilibrium quantity to fall from $Q_0$ to $Q_t$. The quantity falls in such a case for two reasons: The price to consumers increases and the price to producers falls. The decrease in production and consumption represents a loss to society that can be measured in terms of consumer and producer surplus.

The consumer surplus at the original price is represented by the triangle $AEP_0$, and the producer surplus by $P_0EC$. When the tax causes an increase in the price to consumers, consumer surplus falls to $AE_tP_c$; when the tax lowers the price to producers, the producer's surplus falls to $P_pBC$. The loss to producers and consumers is represented by the area $P_cE_tEBP_p$. The area $P_cE_tBP_p$ represents tax revenues for the government ($t$ times $Q_t$), so revenue is not lost to society as a whole, but the area $E_tEB$ is lost. It is a deadweight loss to society that is similar to the welfare loss due to monopoly.

In some cases, a decrease in quantity is intentional because it is thought to be socially desirable. For example, if government wants people to consume less liquor or tobacco, an excise tax may discourage drinking and smoking. Or government may encourage the consumption and production of some goods by not taxing these goods and taxing all others. If government wants to increase consumption of agricultural products relative to other goods, for example, then agricultural products may be exempted from a general sales tax.

Tax policies may have unintended allocative effects. For example, property taxes may cause geographical shifts in population and industry. The aggregate supply of land, many people believe, is perfectly inelastic. However, the current property tax system is not neutral in its effects because demand for land is elastic. State and local governments tax land according to market value, which varies according to demand. Owners of land in the center of cities and other strategic locations must therefore pay higher taxes than owners of less valuable land. These high property tax rates may drive firms and homeowners out of such areas in search of lower taxes, other things being equal.

To summarize, the allocative effect of a tax depends on the size of the tax and the elasticity of supply and demand for the product. In general, the larger the tax, the greater the reduction in the production and consumption of the good. Also, the more elastic the supply or demand for the good, the greater the reduction in production due to the tax. Conversely, when demand or supply is perfectly inelastic, the tax has no effect on the quantity of the taxed good; the tax is said to be allocatively neutral.

### The Aggregate Effect of Taxes

By changing quantities demanded and supplied and thus reallocating resources, taxes have a strong impact on the economy. Some economists are concerned that the impact of taxes may stunt the growth of the economy. As we have indicated, for example, payroll taxes lower real wages to workers and thus may decrease the participation of workers in the labor force (a phenomenon discussed in Economics in Action at the end of this chapter). If this occurs, then the nation's total output will fall because of the tax.

Taxes on savings can also affect the overall performance of the economy. Interest obtained from bonds or savings accounts is subject to the personal income tax. The tax lowers the interest rate received from savings, so people tend to save less. With a decrease in savings, fewer funds are available for business purchases of capital. The tax on interest income can therefore diminish the rate of capital formation.

To encourage growth, the government sometimes cuts taxes on certain activities. The government may, for example, encourage productive behavior by lowering the tax rate for businesses or individuals engaged in activities that increase the supply of resources. For example, the investment tax credit lowers the tax liabilities of firms that invest in new production equipment. Moving expenses are also deductible when people relocate for new jobs. Such tax credits and deductions can improve the productive capacity of the nation and thus increase real output.

The overall effect of taxes on production in the United States is not certain. However, the burden of taxes is growing, and the notion that their effect is detrimental to economic performance is gaining popularity. Many people are calling for decreased taxes and for tax reform. Some economists even believe that when tax rates get too high, tax revenues to the government actually fall (see Focus, "The Laffer Curve"). The important issue is whether the net effect of taxes encourages or discourages productive efficiency, or, more specifically, whether we have the optimal amounts of public goods and whether our methods of taxation result in the most efficient use of resources.

## Tax Reform

The U.S. tax system has become so complex that the federal tax code, the textbook of federal tax regulation, is more than 40,000 pages long. Individual tax liabilities are difficult to determine without the help of specialists. Each taxpayer is also subject to a maze of sales, excise, and property taxes. And an unknown portion of the incidence of business taxes is passed on to taxpayers in the form of higher prices or lower wages.

Taxpayers are beginning to voice their displeasure with the current level and nature of taxes. People have always wanted a lower tax incidence. The current tax system itself may be inequitable and inefficient. If so, tax reform could improve the equity of the tax incidence and the overall productivity of the economy.

### The Need for Tax Reform

Much of the concern about our present tax system centers on the income tax. The nature of the current personal and corporate income tax tends to divert resources away from the production of desirable goods and services. For example, the personal and corporate income taxes induce people to devote resources to tax avoidance rather than to productive activity.

The income tax system allows a maze of deductions and exemptions that change through time. Households and businesses employ resources, both labor and capital, in an effort to minimize their tax liability. Keeping records and receipts throughout the year and staying abreast of new tax loopholes is not a trivial task. Several hundred million dollars are spent on professional

# FOCUS  The Laffer Curve

In the early 1980s, the Laffer curve, named for economist Arthur Laffer, became a favorite illustration of those pressing for cuts in the personal income tax. Essentially, the Laffer curve represents the relation between the tax rate and tax revenues. It predicts that as the tax rate rises, total tax revenues will rise at first and then begin to fall. The importance of the Laffer curve for tax cut advocates is its suggestion that the government may actually lose revenues by increasing tax rates too much (see Figure 19–6).

The Laffer relation can be demonstrated with a simple excise tax. In Figure 19–7, $S$ represents an infinitely elastic supply curve for a good produced by a constant-cost industry. When a simple, per unit excise tax is imposed on the good, the supply curve shifts upward by the amount of tax. Demand for the product is represented by $D$, and total tax revenues are shown in the shaded areas (the tax rate, $t_1$, $t_2$, and so on, times the amount sold, $Q$). For example, with excise tax $t_1$, the total tax revenues are $t_1$ times $Q_1$, or area $(P + t_1)AEP$. A higher tax rate will increase tax revenues. With excise tax $t_2$, tax revenues are represented by the shaded area $(P + t_2)BFP$. Notice that the excise tax $t_2$ maximizes tax revenues in this model. A higher tax, $t_3$, actually results in lower tax revenues, area $(P + t_3)CGP$. $(P + t_2)$ is identical to the price that would maximize profits if this industry could become a monopoly or a cartel. Indeed, with an excise tax the government can restrict output and raise price in an industry. The profits, however, are received by government in the form of tax revenues.

In this example, the elasticity of demand determines which tax will maximize revenues. Debate over the Laffer curve and the advantages of cutting income taxes has focused on this issue of elasticity. In theory, if the supply of or the demand for labor is perfectly inelastic, then higher tax rates would always yield higher tax revenues. If labor demand or supply is elastic, however, then raising taxes will not necessarily result in higher revenues, and cutting taxes may actually increase revenues.

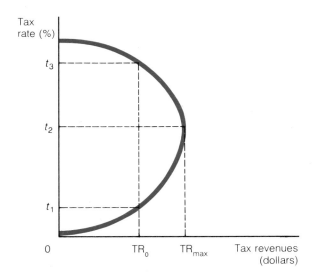

**FIGURE 19–6   The Laffer Curve**

When the tax rate is zero, total tax revenues are zero; but when the tax rate rises, revenues rise at first and then fall. This suggests that there is a tax rate that maximizes tax revenues.

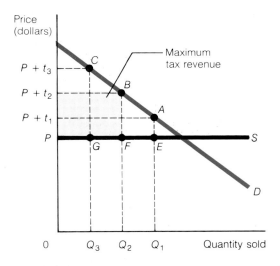

**FIGURE 19–7   Tax Revenues at Different Tax Rates**

Total tax revenues are a function of the tax rate and the elasticity of demand and supply. Excise tax $t_1$ results in tax revenues represented by the area $(P + t_1)AEP$. Maximum revenues occur at rate $t_2$. A higher rate such as $t_3$ results in lower tax revenues because of the high elasticity of demand.

## FIGURE 19–8
### The Effect of Tax Reform

An inefficient tax system can depress the supply of resources. Point A shows a low level of production with an inefficient tax system. If tax reform results in a more efficient tax, then there would be greater production of both public and private goods—point B on the shifted production possibilities curve.

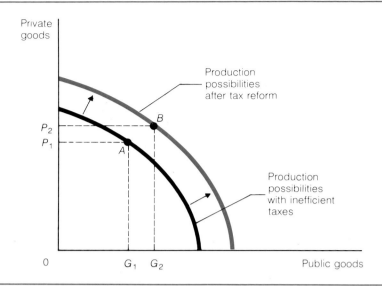

**Inefficient tax:** A tax that decreases the overall productive capacity of the country; a tax that decreases the supply of resources or decreases the efficiency of resource use.

tax services each year. Not only is labor employed in bookkeeping for tax purposes, but many lobbyists are employed by industry in an effort to obtain more loopholes from Congress. While these resources are employed in the production of a service that is valuable to the recipients of the tax reductions, they are not necessarily employed in the production of goods and services that benefit all of society.

Figure 19–8 shows the effects of an **inefficient tax** system on the productive capacity of the nation. Point A shows the level of output of publicly produced goods, $G_1$, and privately produced goods, $P_1$, with an inefficient tax system; that is, one that suppresses the supply of resources to the production of goods and services. Tax reform that encourages a more efficient tax system could increase the supply of resources and thus shift the production possibilities curve outward. More of all goods—public and private—could be produced with a more efficient tax system.

### Suggested Reforms

A tax system that is allocatively neutral, equitable, and efficient is certainly desirable. Although no known tax is both neutral and equitable, some economists believe there are more efficient and equitable tax systems than the current U.S. system. One radical reform would be a flat-rate tax on personal income. A milder version of the flat-tax concept, embodied in President Reagan's 1985 tax reform bill, is currently making its way through Congress.

**Flat-rate tax:** A proportional income tax with no exemptions or deductions.

**Flat-Rate Tax.** Some economists support the idea of replacing our present progressive tax with a **flat-rate tax,** a simple proportional tax with no exemptions or deductions. For example, a 20 percent tax on income for every level of income with no loopholes would generate an equitable tax. Everyone with the same income would pay the same amount in taxes, and people with higher incomes would pay more in taxes than people with lower incomes. The flat-rate tax would probably eliminate some of the costs of bookkeeping and the uncertainty of tax liability. Also, if Congress were not allowed to institute deductions or exemptions for special interests, lobbying efforts would probably decrease. Furthermore, the flat rate would eliminate the high marginal tax rates that can discourage work efforts. Problems emerge with implementing this proposal, however, including political opposition from those whose taxes would rise and the possible burden it would place on those with little ability to pay.

**Reagan's Tax Reform Plan of 1985.** The flat-tax concept has been modified by the Reagan administration into a sweeping legislative proposal to simplify the Federal tax code. Instead of a single tax rate for individuals, President Reagan proposes three tax brackets: a top rate of 35 percent (for all single incomes over $42,000), a middle rate of 25 percent (for single incomes between $18,000 and $42,000), and a bottom rate of 15 percent (for single incomes above $2,900). Anyone earning less than $2,900 would pay no income tax at all. Joint incomes would also be taxed in three brackets, but the income ranges would proportionately higher in each bracket.

The Reagan proposal eliminates many but by no means all of the deductions available to taxpayers. Deductions for interest payments on home mortgages, charitable contributions, and medical expenses would still be allowed. Wealthy taxpayers, however, would find fewer "loopholes" with which to reduce their tax bill. For their part, businesses would bear a higher percentage of the total tax bill. Corporate taxes would increase, in some estimates, by 23 percent.

It is too early to say whether the reform will achieve its primary purpose of reducing the enormous complexity and cost associated with the present system. If tax reform, as promised, results in a more efficient tax structure, then economic growth will be stimulated. At the same time, tax reform will have many effects on the incentives for individuals to invest. If, for example, homeowners can no longer claim credit on their tax bill for spending on energy-saving devices, the demand for solar heating panels and other such innovations will decline. If investors cannot deduct the interest for "second home" mortgages, the housing market will feel the effects.

As with any legislative proposal, the effort to simplify the tax code will depend as much upon political realities as upon economic costs and benefits. Who wins and who loses with tax reform? The myriad decisions involved rest with elected officials, whose behavior — the subject of our next chapter — is a matter of economic concern as well.

## Summary

1. Taxes are a means of financing public goods. Although some taxes are levied directly on the users of public goods, in many cases beneficiaries cannot be explicitly defined, so general taxes must therefore be used to diminish free ridership.
2. Taxes are levied according to two basic principles: the benefit principle and the ability-to-pay principle. The benefit principle suggests that the people who receive the greatest benefit from public programs should pay the most taxes. The ability-to-pay principle suggests that people with the greatest ability to pay taxes should pay the most.
3. The ideal of equity in taxation has both a horizontal and vertical dimension. An equitable tax takes greater revenues from higher-income groups than from lower-income groups and also taxes people with equal incomes equally.
4. In general, the relation of taxation to level of income falls into one of the following categories: A proportional tax takes a constant percentage of income regardless of the level of income. A progressive tax takes a larger percentage of income as income rises. A regressive tax takes a smaller percentage of income as income rises.
5. The marginal tax rate is the increase in tax liabilities that results from an increase in income. With a progressive income tax, the marginal tax rate is greater than the average tax rate.
6. The U.S. government imposes a progressive income tax. Because of loopholes—exemptions and deduc-

tions—the federal income tax has both horizontal and vertical inequities.
7. Most existing taxes effectively drive a wedge between the price buyers pay and the price sellers receive for a product or resource. The burden of a tax is usually shared by the buyers and sellers of the taxed item.
8. Taxes usually reallocate resources and may decrease the supply of resources. Taxes that decrease the productivity of the economy are inefficient. A tax that is neutral has no effect on the allocation of resources.
9. Tax reform such as the proposed flat-rate tax or a tax simplification scheme could lead to a more equitable tax incidence and to an increase in overall productivity and economic growth.

## Key Terms

public finance
benefit principle
ability-to-pay principle
horizontal equity
vertical equity
proportional income tax
progressive income tax
regressive income tax
taxable income
average tax rate
marginal tax rate
neutral tax
tax incidence
inefficient tax
flat-rate tax

## Questions for Review and Discussion

1. Why do we have taxes? What principle of taxation does a user tax follow? What principle does a property tax follow?
2. Which principle of taxation does the U.S. income tax follow? Is the U.S. income tax horizontally and vertically equitable? If a tax is not horizontally equitable, can it be vertically equitable?
3. When is a regressive income tax vertically equitable? With a regressive income tax, is the marginal tax rate greater than the average tax rate?
4. Draw the demand curve for a product with a supply curve that is perfectly inelastic. Show the effect of an excise tax. Who bears the burden of the tax? Is the tax neutral?
5. Would a property tax system that makes taxes identical at every location in the country be more efficient than the present property tax method?
6. Does the government use valuable resources to prevent tax avoidance? When state and federal agents attempt to detect and prevent bootlegging, are they attempting to protect a tax base? If there were no tax on liquor, would the government prevent bootlegging?
7. Does an income tax decrease the amount of labor supplied? If the supply of labor were perfectly inelastic, would an income tax decrease the level of employment? Who would bear the entire tax burden?
8. If consumption expenditures rather than income were taxed, would the after-tax wage still fall? Would a consumption tax decrease the amount of labor supplied? Would a consumption tax decrease the amount of capital supplied?
9. What would happen to the amount of capital in the United States if the interest income from savings were not taxed? Describe a tax that could increase the supply of capital.
10. Some states have a competitive market for alcoholic beverages in bottles, but they charge an excise tax. Other states have state-owned and -operated liquor stores that are profit-making monopolies. Under which of these two systems would a state maximize revenues? Could a state make more money by selling the monopoly right to the highest-bidding firm?
11. What are the benefits of eliminating tax loopholes? How can tax loopholes be used to improve the productivity of the economy?
12. How would the Reagan tax plan affect tax equity? How would it affect economic growth?

## ECONOMICS IN ACTION

### If You Are Taxed More, Will You Work Less?

Just about everyone has heard someone complain about high tax rates, especially in regard to earning extra income such as overtime pay. The percentage of extra income taken in taxes is determined by the marginal tax rate. With a progressive income tax, the marginal tax rate rises, so extra income is taxed at a higher rate than the rate at which the first dollars earned are taxed. As we have suggested, this system may make people who have the opportunity to earn extra income resent the higher tax rate, but does it actually affect their work decision? Specifically, does a progressive tax rate have a more detrimental effect on the supply of labor than a proportional tax rate, such as the proposed flat-rate tax?

Any tax based on labor income effectively decreases a worker's wage. A proportional tax decreases the wage by a constant amount regardless of the hours worked. For example, if a worker earns $10 per hour, a 20 percent income tax would lower his or her after-tax wage to $8 per hour regardless of the number of hours worked. A progressive income tax results in a lower hourly wage as the number of hours worked increases. For example, if the worker chooses to work 15 hours in a week, he or she may be taxed at a 10 percent rate, whereas 25 hours of work may be taxed at a 20 percent rate and 40 hours at a 30 percent rate. Figure 19–9 shows the relative effects of progressive and proportional tax rates. Both lower the after-tax income and thus lower the wage rate, but by different amounts at different numbers of hours worked.

What effects do these two patterns of taxation have on the amount of labor supplied? As we suggested in Chapter 12, a decrease in wages has two effects on labor supply decisions: an income effect and a substitution effect, which work in opposite directions. The income effect suggests that a wage decrease lowers real income and thus encourages individuals to work more. The substitution effect suggests that a wage decrease lowers the price of leisure and encourages individuals to buy more leisure; that is, to work less. Because these two effects work in opposite directions, the combined effect of a wage decrease on the amount of labor supplied cannot be determined. However, we may suggest the following: If per hour wages decrease because of the income tax, workers with an upward sloping supply curve of labor will work fewer hours, and workers with a negatively sloped supply curve will work more hours. Furthermore, if the aggregate supply of labor is upward sloping, an income tax will decrease the overall amount of labor supplied.

The relative effects of proportional and progressive tax rates on the supply of labor can also be determined. In Figure 19–9, the two tax rates result in equal tax revenues at 40 hours of work per week. At this level, the income effect of the two tax rates is therefore identical, for they both decrease income by the same amount. However, the substitution effects are not the same under both tax systems. The progressive income

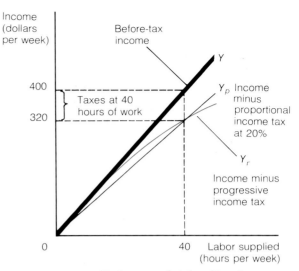

**FIGURE 19–9** Before- and After-Tax Income with Proportional and Progressive Tax Rates

The $Y$ curve represents the level of income earned before tax withholdings. The $Y_p$ curve shows after-tax income with a proportional tax rate, and $Y_r$ shows after-tax income with a progressive rate. As workers increase the number of hours they work, their after-tax hourly wage falls. The decline is steady with a proportional tax, but not with a progressive tax. In this example, both tax schedules result in identical tax revenues at 40 hours of work per week: $80 for the government, with $320 left as take-home pay.

tax has a stronger substitution effect than the proportional tax because it lowers the wage at the margin by a greater amount. This difference suggests that the progressive income tax will have a more detrimental effect on the amount of labor supplied than the proportional tax when both taxes bring in the same amount of tax revenues.

### Question

Consider that after graduation you can earn, over a period of years, from $25,000 to $80,000 per year. What factors would your income and substitution effects depend on? Given these factors, would your own work effort be reduced at any given annual income?

# 20

# The Theory of Public Choice

**Public choice:** The economic analysis of politics; the branch of economics concerned with the application of economic principles to political decision making.

The discussions of public goods and externalities in Chapter 18 and taxation in Chapter 19 were based on the concept of market failure. When or where markets do not produce ideal outcomes, government can and often does play a role in correcting the situation. In Chapter 18 we more or less assumed that government is capable of making these corrections. The overall capability of government to compensate for market failure, however, is itself an issue. Democratic governments act through a unique set of institutional arrangements, such as majority voting, representative democracy, political parties, bureaucracies, and special-interest groups. The theory of **public choice,** introduced in this chapter, is an attempt to apply economics to understand and to evaluate how democratic governments operate. A central precept of public choice theory is that government is not perfect; like markets, governments can fail to achieve ideal outcomes. Appropriate solutions to economic problems in the economy are frequently a choice between markets that are not perfect and government programs that are not perfect. Appropriate public policy cannot be decided simply by saying "Let the market work" or "Let government do it." Rather, the costs and benefits of alternative institutional arrangements must be carefully examined case by case.

The fundamental premise of public choice is that political decision makers (voters, politicians, bureaucrats) and private decision makers (consumers, brokers, producers) behave in a similar way: They all follow the dictates of rational self-interest. In fact, political and economic decision makers are often one and the same person—consumer and voter. The individual who buys the family groceries is the same individual who votes in an election. If the premise of public choice is correct, we can learn a lot about issues such as

Drawing by Mankoff; © 1984 The New Yorker Magazine, Inc.

why people take the time and trouble to vote by applying the same logic to the voting decision that we have applied to the grocery-buying decision. As we do so, we should keep in mind that the institutions and constraints facing political decision makers are different from those facing private decision makers. For example, the benefits of buying groceries are fairly obvious, but what are the benefits of voting? We will be careful to specify such differences between private and public choices throughout our analysis.

Public choice, like the rest of economics, has both a normative and positive side. **Normative public choice** involves value judgments about the desirability of certain political situations. In issues such as the design of voting procedures, for example, how does majority rule compare with other voting rules in reflecting the true preferences of voters? Is there a better voting rule or process through which to make political decisions? Normative public choice looks at the way political institutions work from the standpoint of how we might make them work better.

**Positive public choice** seeks to explain actual political behavior. Why do high-income individuals vote more often than low-income individuals? How do committees and legislatures work? What is the impact of special-interest groups on government? In this chapter we will examine both normative and positive aspects of public choice after first exploring how self-interest affects decision making.

**Normative public choice:** The study of shortcomings and possible improvements of political arrangements, such as voting rules.

**Positive public choice:** The analysis and explanation of political behavior.

## The Self-Interest Axiom

The public choice approach to politics is based on the idea that political actors are no different from anyone else: They behave in predictable ways. We do not seek to serve the public interest when we vote or the private interest when we buy a car. We seek our self-interest in both cases.

Nevertheless, market behavior and political behavior differ in the constraints facing decision makers. The market is a **proprietary** setting—one in which individuals bear the economic consequences of their decisions which either enhance or decrease their wealth. In the market economy, a firm that produces a new product stands to make profits or losses depending on the quality of its efforts. The political arena is a **nonproprietary** setting—one where individuals do not always bear the full economic consequences of their decisions. For example, the political entrepreneur who comes up with a new

**Proprietary:** Relating to private ownership and profit seeking by private owners.

**Nonproprietary:** Relating to public ownership.

political program does not bear the full costs of the program if it is a failure and does not reap all the benefits if it is a success. Behavior will therefore differ in market and in political settings, not so much because the goals of behavior are different but because the constraints on behavior are different.

We can distinguish the economic constraints at work in a market or proprietary setting and in a political or nonproprietary setting by analyzing the roles of agents and principals in each case. In both settings the agent, whether a firm or a politician, agrees to perform a service for the principal, whether a consumer or a voter. Because the agent and the principal are both self-interested, it is likely that the agent will not always act in the interest of the principal, particularly if the behavior of the agent is costly to monitor.

The agent-principal problem has been analyzed primarily in a private setting. For example, the corporate manager is an agent, and stockholders are principals. How do the stockholders get the manager to act in their interest? They do so primarily through economic incentives. Managers of private firms have incentives to control costs in their firms because increased costs cause a decrease in the firm's profitability. Stockholders, with such mechanisms as stock options and takeover bids, can discipline managerial behavior toward maximizing wealth.

Managers of political "firms," or bureaucracies, do not face a similar incentive to control costs. They cannot personally recoup any cost savings that they achieve for their agencies, and the means available to voters to curtail poor performance by political managers are minimal and costly to implement. This does not mean that public officials can do anything that they want to do. Like any other economic actors, they are constrained by costs and rewards.

The main point about the agent-principal problem is that political agents face different constraints on their behavior than do private agents because the principals in the two cases face different incentives to control the behavior of the agents. This difference is the focus of public choice analysis.

## Normative Public Choice Analysis

Normative public choice theory analyzes the performance of political institutions in serving voters. In this section we illustrate the usefulness of this approach by examining the process of voting by majority rule and by introducing the problem of constitutional choice.

### Majority Voting[1]

In a democracy, public choices are typically based on a majority vote: If candidate A wins one more vote than candidate B, then candidate A is elected to office. In some cases, particularly when more than two choices are available, majority rule does not always accurately reflect voter preferences.

Imagine a three-person nominating committee that must choose among three alternatives—Smith, Jones, and Tobin—as the nominee for club president. Each committee member ranks the three candidates in order of preference (first, second, third), and the winner is to be selected by majority vote (two out of three committee members in favor). Table 20–1 shows the rankings of the three candidates by the committee members (A, B, and C). Member A, for example, prefers Smith to Jones and Jones to Tobin. Note

---

[1]The inventor of this analysis was Kenneth Arrow, a Nobel Prize winner in economics. See his *Social Choice and Individual Values* (New York: Wiley, 1951).

## TABLE 20–1 A Voting Problem

The committee must choose among Smith, Jones, and Tobin. The candidate receiving a simple majority (2 votes) wins. When each candidate is paired against another (Smith versus Jones, Jones versus Tobin, Tobin versus Smith), no clear winner emerges. Smith beats Jones, Tobin beats Smith, but Jones beats Tobin. This is a problem of simple majority voting on more than two candidates two at a time. Individual voters can clearly rank candidates, but majorities cannot. Majority voting in this case leads to a repetitive cycle of winners.

| Committee Member | Smith | Jones | Tobin |
|---|---|---|---|
| A | 1 | 2 | 3 |
| B | 2 | 3 | 1 |
| C | 3 | 1 | 2 |

that each member has consistent preferences; that is, each member is able to rank the three candidates in order of preference.

Now let the voting begin. To find the preferred nominee, the committee pairs each candidate against one of the other two until a winner is found. Suppose they start with Smith against Jones. Who wins? A votes for Smith, B votes for Smith (B prefers Tobin overall but between Smith and Jones he prefers Smith), and C votes for Jones. Smith wins by a majority vote of 2 to 1. Can Smith now beat Tobin? A votes for Smith, B votes for Tobin, and C votes for Tobin. Tobin wins that round. Now, how does Tobin fare against Jones? Since Smith previously beat Jones, we would also expect Tobin to dominate Jones because Tobin dominated Smith. But look what happens: A votes for Jones, B votes for Tobin, and C votes for Jones. This time Jones wins! Thus in three rounds of voting, a different candidate wins each round.

The committee's stalemate points up a problem with majority voting. Under majority voting each individual voter can have consistent preferences among multiple candidates or issues, and yet voting on candidates two at a time can lead to inconsistent collective choices. In other words, individuals can make clear choices among Smith, Jones, and Tobin, but majorities cannot. Majorities will choose Smith over Jones, Tobin over Smith, and Jones over Tobin, and on and on in a repetitive voting cycle until a way is found to stop the voting. Either Smith, Jones, or Tobin will win when the voting is stopped, but only because the voting stopped at a particular point, not because one candidate is a clear winner over the other two.

This simple proof strikes at the rationality of majority voting procedures. It says that although individual choices will be clear and rational, majority choices will be inconsistent and cyclical. Where does such a paradoxical result lead us? First, it leads us to look for a better voting procedure than the simple majority rule when voting on two issues or candidates at a time. We explore this possibility in Focus, "A National Town Meeting," which discusses a proposal for national computer voting, and in Economics in Action at the end of the chapter, which examines how voting may be designed to protect minority interests. Second, we note that the problem of voting cycles is mitigated to some extent by the institutions of democracy. Our legislatures and committees do not cycle endlessly among alternative proposals without reaching a decision. Votes are taken and decisions are made. Attempts to reach agreement do not go on forever. Powerful committee chairpersons, for example, can call for votes on legislation and see to it that the legislature

## FOCUS  A National Town Meeting

Direct or participatory democracy, in which each citizen has the opportunity to cast his or her vote on all public issues, is often held up as the ideal method of reaching a political consensus. The use of such a voting process has been limited historically—the Athenian agora and the New England town meeting are the prime examples—chiefly because of the costs of organizing large groups. But the advent of small computers makes conceivable in the near future the direct voting system proposed by James C. Miller III.

In conjunction with cable television or other data transmission systems, Miller envisions personal computers being used to register and tally the votes of millions of individuals on legislative issues now decided by elected representatives. One can imagine that voters would spend a portion of each evening sitting in front of their computers examining the legislation that had been brought up for consideration, casting their votes, and soon after learning the results of the night's balloting.

Although the computer would vastly reduce the costs of registering and counting votes, it would not appreciably lower the cost to the public of becoming informed about issues. Miller's proposal consequently allows voters to assign proxies to experts or specialists. The proxy could be a general one, giving the proxy holder permission to cast a vote on all proposed legislation, or it could be confined to particular topics. For example, a voter could assign a proxy to an economist to vote on economic matters, to an expert in military affairs for votes on defense issues, and so forth. Voters would assign proxies to individuals who would be expected to cast ballots in the same way the voter would if the voter were informed about the proposed legislation. The assignment of a general proxy would be similar to electing a congressional representative or senator except that proxy holders would represent the views of each voter and not just the views of the majority. Moreover, the speed of computers would allow proxy assignments to be made for as long as each voter wished—perhaps only an hour—instead of for fixed terms of two to six years.

Under Miller's system a role would exist for a legislative assembly to propose bills or to debate important issues. He suggests that the "house of representatives" be composed of the 400 individuals wishing to serve there who have the largest number of proxies and that the "senate" be constituted from the 100 largest proxy holders. The possibility of quick recall would assure that each representative would exercise the proxy exactly as voters wish.

The three main elements of Miller's proposal—direct registering of citizen votes, proxy assignments, and a bicameral legislative body—are not far removed from the current system established by the U.S. Constitution. The main difference is that technological development makes it possible for the democratic ideal of a small New England town meeting to operate on a national scale.

*Source:* James C. Miller III, "A Program for Direct and Proxy Voting in the Legislative Process," *Public Choice* 6 (Fall 1969), pp. 107–113.

---

produces new laws. While such legislative procedures do not solve the problem of inconsistent choices in majority voting, they do keep the tendency toward voting cycles in check.

### Constitutional Choice

In a constitutional democracy there are two levels of political decision making: day-to-day decisions about running the government, which are made under given rules and procedures, and decisions about rules and procedures themselves. The first level of decision making takes place under given political constraints. The second involves the choice of the constraints themselves, which amounts to a constitution of the set of rules under which day-to-day government operates. For example, the choice of whether to employ simple majority voting or some other system would be a constitutional choice, while voting on current issues with the chosen voting system would be part of the day-to-day process of running the government.

Public choice theory can be applied to an analysis of constitutional deci-

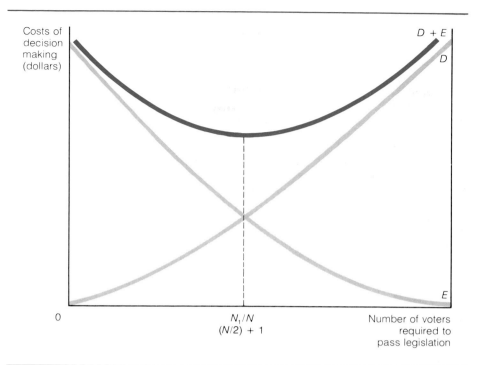

**FIGURE 20-1**

**Constitutional Choice**

Curves $D$ and $E$ represent the expected costs facing the constitutional decision maker. $D$ represents the costs of collective decision making; as the number of voters increases, $D$ increases at an increasing rate. $E$ is the cost of a decision rule requiring less than unanimity. Summing the two curves ($D + E$) and taking the minimum point gives the optimal voting rule chosen by the constitutional decision maker. $N_1/N$ represents the optimal number of individuals in the society of $N$ persons who must agree before a collective decision is made. $N_1/N$ may be any number. Here it is a simple majority, $(N/2) + 1$.

sion making.[2] As an illustration, we will analyze the choice of a voting rule for the legislature. We assume that the constitutional decision makers are unbiased and impartial (a large assumption). Figure 20–1 illustrates the analysis.

The horizontal axis gives the quantity of $N$, all the members of a hypothetical society; $N$ can also be thought of as all the voters in the society. The vertical axis is a measure in dollars of the expected costs of the various voting rules that could be selected for the society. Examples of these costs will be presented shortly.

Constitutional decision makers face the following problem: How do they choose the voting rule that minimizes the expected cost of collective decision making in this society? The answer to the question can be found by examining the $D$ curve and the $E$ curve. The $D$ curve measures the costs of collective decision making. As the voting rule is made more inclusive—requiring, for example, a two-thirds vote rather than a simple majority for approval or defeat of an issue—the costs of reaching collective decisions increases. In fact, as more people are required for agreement, the $D$ curve increases at an increasing rate. This is a logical result in any group decision process. More people require more discussion, larger meetings, more bargaining, and so on. Agreement is more costly to obtain in a larger group.

The $E$ curve reflects the potential costs that a collective decision can impose on the individual affected by the decision. If very few people are required to agree on a measure, agreement will occur more frequently and $E$-type costs will be high. In other words, a small number of people could get together and pass measures that benefit them at the expense of other voters.

[2]This analysis was first developed by J. M. Buchanan and G. Tullock, *The Calculus of Consent* (Ann Arbor: University of Michigan Press, 1962).

If a unanimity rule prevails, however, requiring that all members of the legislature must agree before a tax can be passed, then $E$-type costs to any individual will be zero. Under these conditions, no collective decision can be made unless every voter consents. Unanimity means that one person can block a collective decision.

We have, then, the two major categories of costs related to the choice of a voting rule by constitutional decision makers. The minimum of the two costs can be found by vertically summing the $D$ and $E$ functions and picking the minimum point on that curve. This occurs at $N_1/N$ in Figure 20–1. In this society, the two types of voting costs are minimized where $N_1$ persons are required to agree before a collective decision is made. $N_1/N$ may be any number. It may, for example, be a simple majority voting rule—$(N/2) + 1$ in Figure 20–1—or it may be a stricter voting rule in which more than a simple majority agreement is required before a collective decision is made. The analysis in Figure 20–1 leads to a more inclusive voting rule than simple majority. Though stricter voting rules involve higher $D$-type costs, the prospect of lower $E$-type costs could easily make stricter voting rules an important route to the improvement of collective decision procedures. What would happen, for example, if every proposal for a public expenditure had to carry with it a proposal for the taxes to finance it and, moreover, to pass the legislature with a majority of seven-eighths?[3]

## Positive Public Choice Analysis

Positive public choice analysis is like positive economics; it consists of the development of models of political behavior that can be subjected to empirical testing. A positive theory can be wrong; it can fail to be supported by the evidence. A positive theory can never be proved; it can just fail to be refuted by the evidence. In this case a positive theory will hold its ground against alternative theories until a theory comes along that offers a better explanation of the process. This is the spirit in which we outline positive public choice theory in this section. We present some positive propositions about government and political behavior that seem to be supported by evidence from the world around us.

### Political Competition

Have you ever wondered why by election time the positions of the two major parties' candidates sound like Tweedledum and Tweedledee, and you cannot see any difference in the views espoused by either? Perhaps more important, why do we have only two major political parties in the United States? If we had only two firms in every market, antitrust authorities would become alarmed and might intervene to increase competition. The following model helps us answer these and other questions about political competition in a democratic setting.

Assume that voter preferences on issues can be distributed along an imaginary spectrum running from radical left (perhaps the Socialist Workers) to radical right (perhaps the John Birch Society), as in Figure 20–2. Further assume that voters vote for the candidate closest to their ideological position. The normal distribution of voters holding different ideological positions is split down the middle by M. We call the voter at M the median voter. If

---

[3] A famous Swedish economist named Knut Wicksell proposed such a scheme more than ninety years ago.

**FIGURE 20–2**
**Ideological Distribution of Voters**

The distribution of voters in this example is single-peaked, ranging from radical left to radical right. *M* represents voters at the middle or median of the distribution. Two-party political competition leads to middle-of-the-road positions, similar to those held by voters at *M*.

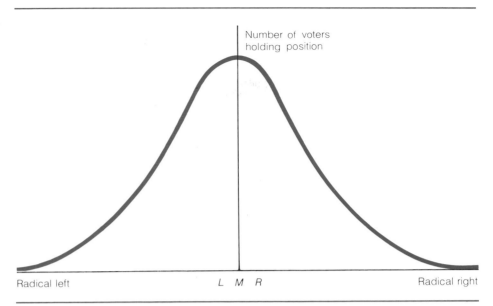

either candidate in a two-party race adopts a middle-of-the-road position such as M's position, he or she is guaranteed at least a tie in the election; he or she will receive at least 50 percent of the vote. If the other party's candidate adopts any other position, such as R, he or she will get less than 50 percent of the vote. (Remember: Voters follow the rule of voting for the candidate closest to their ideological position; fewer than half the voters in Figure 20–2 are closest to R.) The best strategy for each candidate is to adopt position M and hope that random error (pulling the wrong lever) by voters will make him or her a winner. M, the median position, is the vote-maximizing position for both candidates; by election time, therefore, both candidates are virtual carbon copies of each other.

What happens if a third candidate enters the race? If he or she adopts a position such as R while the other two are at M, he or she would get all votes to the right of R and half of those between M and R, thereby beating both of the look-alike candidates at position M. Such an entry would probably induce one of the other two candidates to move away from M, perhaps to L. The remaining candidate at M would now be isolated with a small fraction of the vote. This candidate at M would have an incentive to move outside of the *LR* segment, thereby trapping one of the other candidates, and so on. Starting from a position in which all candidates are at the center, at least one candidate in a three-party or larger system will have an incentive to move away from the center. But this spreading-out process has its limits. As long as the peak of the distribution of voters remains at M, a candidate can increase his or her percentage of the vote by moving toward M. With three candidates at L, M, and R, the two outside candidates can increase their votes by moving toward M.

This model of political competition is oversimplified, but it does offer some insights into political behavior. Where the distribution of voter preferences is single-peaked, as at M in Figure 20–2, there are strong tendencies for median outcomes to be produced. This holds true for virtually any type of collective decision process, from a committee consensus to a public election. Middle-of-the-road policies are vote-maximizing; therefore, we expect middle-of-the-road candidates and policies to be winners. In addition, the analysis underlying Figure 20–2 gives some clues as to why we have two parties in the United States. If parties can enter on either the left or right flank of parties at M, such entry jeopardizes the control of government by the center parties. We can expect that the M-type parties will do their best to place barriers to entry in the way of noncenter parties, such as making it very costly and difficult for a new party to get on the ballot for elections.

## Logrolling

**Logrolling:** The exchange of political favors, especially votes, to gain support for legislation.

Positive public choice also takes into account the behavior of officials once they are in office. A typical political behavior is **logrolling,** a term used to describe the process of vote trading. Representatives in a legislature are constantly making deals with one another. One wants a dam in his district; the other wants a new courthouse in hers. They agree to trade votes on issues: You vote for my dam; I'll vote for your courthouse.

There are many beneficial aspects of logrolling. Its main benefit is that it enables representatives to register their intensity of preferences across issues. For example, a minority representative who feels strongly about animal rights can trade his votes on issues about which he does not feel so intensely for support on the animal rights issue. Logrolling also helps mitigate the problem of indifferent majorities winning over intense minorities. In essence, logrolling or vote trading is a form of exchange, and economists usually view exchange as a productive and efficiency-enhancing process.

We must be careful, though, in assessing logrolling. Its general efficiency depends on the political setting. Most legislatures, for example, are formed on the basis of geographic representation, under which a representative's constituents have one thing in common: They all reside in the same area. Any bill the representative can get through the legislature that benefits the members of her district should win her votes at election time. In effect, a geographically based system provides the legislator with incentives to represent local interests in the national legislature at the expense of broader national issues.

Most individuals in a community share an interest in the vitality of the community's economy. In many instances a bloc of voters from a single voting district may receive income from a single firm or industry, as in a company town or in geographically concentrated industries like steel, automobiles, lumber, defense, and regional agricultural industries. When economic activity is concentrated, it is possible for representatives to win political support by serving the economic interests of their home districts. Tariffs, industry- and company-oriented tax concessions and subsidies, local public works projects, and defense contracts are all examples of issues that often are decided, in part, on the basis of their economic impact on certain regions.

Representatives therefore trade their support on national issues, such as air pollution control, for support on amendments or separate bills that serve their local interests. National legislation often becomes a vehicle for local support, and the result is the "pork barrel" type of legislation that carries rich rewards for specific locales. Logrolling, which attempts to redistribute income toward certain regions and industries, generally does not lead to a more productive economy. Rather it leads to legislation such as individual industry tariffs, unnecessary and costly public works legislation, and special-interest tax "reform" bills. This form of logrolling at the national level is unproductive not because it acts as a means for revealing relative intensities of preferences on national issues but because in its most blatant forms it is used to reveal relative intensities of preference on essentially local issues. When restricted to national issues, however, logrolling can be a beneficial means for revealing voter preferences. For example, one senator may prefer more foreign aid for Latin America and another a reform in the Social Security program. Through exchange of their votes on these two national issues, both may be passed as a result.

## Voting

In Chapter 6 we discussed whether voting was rational. Simply put, one voter out of many has very little impact on any election. Therefore, even if the marginal costs of voting—the time it takes to register and go to the polls—are very low, they may make it too costly for the average voter to vote. Of course, the voter is rational; he or she will vote when the marginal benefits exceed the marginal costs of voting. If an election is predicted to be close, the marginal benefit of voting will rise, and voter turnout will rise.

In this section we examine the act of voting in more detail by discussing some of the relevant differences between the market and voting as means to allocate resources. We compare consumer behavior within market institutions and voter behavior within political institutions.

*In markets, consumers make marginal choices; in politics, voters must evaluate package deals.* When you buy carrots or hot dogs, you can buy one more or one less; that is, you engage in marginal decision making. When you vote for a politician or a party, you vote not for a single issue but for a package or a platform—all-or-none decision making. Politicians do not offer voters a little more or a little less of this or that public program. They offer package deals. This feature of voter choice clearly makes voting complex and costly. Moreover, when people are forced to make package choices, they are likely to end up choosing a lot of items that they really do not want in order to get a few things that they do want.

*Elections generally involve the choice of numerous candidates and issues.* In addition to the package nature of political choice, the voter has to make a variety of choices at the same time; many candidates and issues are on the ballot. To some extent, of course, this is efficient. It is better to vote on many things at once than to vote on a lot of individual matters in separate elections. Clearly economies of scale accrue in voting. But having numerous candidates and issues on the ballot also leads to greater complexity and cost in voter choice.

*Consumption is frequent and repetitive; voting is infrequent and irregular.* The consumer goes grocery shopping weekly or daily; the voter votes every year or every two or four years. This infrequency makes it more difficult for the voter to find reliable policies and candidates. Imagine how the grocery-buying decision would be altered if you had to buy groceries once a year. Your ability to discard bad products and try new ones would be reduced dramatically.

*The primary difference between voting and the market, deriving from the above three factors, is that voters have little incentive to be informed.* Voting is more complicated than market choice. This means that voters have little incentive to gather information about their public choices. In market choice, as we have seen, consumers have direct and strong incentives to gather information and to search for useful, reliable products. In public choice, voters have difficulty evaluating candidates and issues. The costs of being informed are high. Suppose the Defense Department argues for an expensive new missile system on the grounds that it is needed to deter a Soviet threat. How can the average voter hope to obtain the information needed to make a rational choice? The same is true for proposals for dams, foreign aid, welfare, relations with China, money supply, jobs, and so on. Voters are quite rationally uninformed about such matters; they estimate that the costs exceed the benefits of being fully informed. Indeed, they may free ride by not gathering information, not voting, and letting those who do vote make the choices for them. If the choices of other voters happen to be beneficial to nonvoters, the nonvoters benefit without bearing any of the costs. This behavior leads to a fifth and final implication.

*Since voters are not well informed, the information transmitted to politicians by elections is not very useful in helping politicians determine what voters want.* Politicians predictably have great difficulty assessing the implications of an election with respect to what voters want them to do. Of course, politicians who are good at reading the public's pulse will survive and be reelected. Nevertheless, the transmission of information is not direct. In markets, if consumers want more carrots, they can effectively transmit this information to producers and get more carrots. In voting, if consumers want more butter and fewer guns, it is costly and hard to make this preference known to politicians through voting.

In sum, private and public choices differ in important ways. Political markets are more imperfect than private markets. This is a positive proposition derived from the manner in which our political institutions are designed.

### Interest Groups

Interest groups have received much attention in positive public choice analysis. The behavior of interest groups helps explain a significant amount of government activity. To see how interest groups affect government policy, consider the process by which a statewide group of barbers organizes to advance their interests.

The most fundamental problem that the barbers face is how to get organized. Specifically, each barber has incentives to stay outside the interest group, for if the interest group is successful in, say, obtaining higher prices for haircuts, a free-riding barber can benefit without having incurred any costs. Free-riding behavior makes the formation of interest groups difficult but by no means impossible. For such reasons, interest groups seek ways to limit the benefits of their activities to members only.

Assume that the barbers organize. The group's representatives then go to the state legislature and seek a legally sanctioned barbers' cartel. The purpose of this cartel is to protect barbers from new competition by erecting barriers to entry into barbering, thereby allowing barbers to raise the price of a haircut above the costs of providing a haircut. In other words, the barbers petition the state government to raise their wealth at the expense of the consumers of haircuts. Put another way, they are demanders of a wealth transfer from the state.

Who are the potential suppliers of wealth transfers? Clearly, the suppliers are those—in this case consumers of the barbers' product—who do not find it worthwhile to organize and to resist having their wealth taken away by the barbers. This is an unusual concept of supply, but it is a supply function nonetheless. The attitude that will prevail among suppliers is that it would cost them more to protest the barbers' proposal than any benefits they would derive from protesting. For example, the average consumer might have to spend $100 to defeat the barbers' proposal, the net result of which would be a saving of $10 in haircut costs. Why spend $100 to save $10?

This example describes a "demand" and a "supply" of wealth transfers in the case of barbers. How does the market for wealth transfers work? Politicians are brokers in this market and in the market for legislation generally. This means that politicians get paid for pairing demanders and suppliers of wealth transfers. If they transfer too much or too little wealth, they will be replaced at the next election by more efficient brokers.

The outcome for the barbers is shown in Figure 20–3. The barbers start out as a purely competitive industry, producing $Q_c$ haircuts per year at a price of $P_c$. When their interest group forms, they persuade the legislature to grant them cartel status. This persuasion is not very difficult because the consumers who are harmed by the cartel do not find it worthwhile to protest to the legislators. The barbers are the only ones to be heard, and perhaps by putting their arguments for a cartel in public-interest terms ("higher prices mean higher quality," "barbers must be licensed to ensure an orderly marketplace," and so on), they carry the day with the legislature.

The legislators assess the costs and benefits of creating a barber cartel. In Figure 20–3, the barbers seek $P_gABP_c$ in cartel profits at the expense of consumers' welfare; haircut consumers incur an additional loss equal to $ACB$. (Area $ACB$ represents the welfare loss due to monopoly discussed in Chapter 10.) The per capita gains to barbers, who are a relatively small number of producers, exceed the per capita costs to the more-numerous haircut consumers. The result: The legislature passes a law giving the barbers a cartel.

Note, however, that this discussion is highly stylized. Interest groups do not always work as producers versus consumers. Technically, any group can organize and act as an interest group, and we find in fact that interest groups represent many overlapping segments of society. There are labor unions, women's rights groups, religious groups, trade associations, organizations for

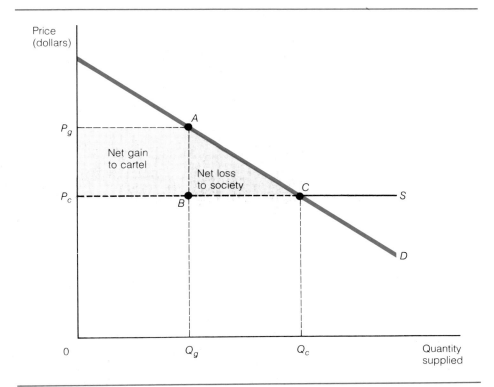

**FIGURE 20–3**

**How an Interest Group's Demands Are Met**

$P_c$ is the competitive price that the industry, in this case barbers, charges, and $P_g$ is the price established by the group after it is granted cartel status. After the cartel is formed, it stands to gain $P_gABP_c$ in profit from consumers, and consumers stand to lose $ACB$ in welfare. Generally, the former wealth transfer dominates the latter welfare loss, and a law is passed sanctioning the cartel.

doctors, lawyers, hunters, environmentalists, farmers, and so on. Moreover, not all of these groups are small. Many large groups, such as farmers, are effectively organized as an interest group. The basic point is that you begin explaining why a certain law is passed if you try to answer two questions: Who wins? and Who loses?[4] Focus, "Interest Groups and the British Factory Acts," presents a historical example of how the demands of interest groups result in legislation.

### Bureaucracy

While legislatures enact laws and authorize levels of public spending, the actual implementation of those laws is delegated to a variety of agencies, commonly referred to as bureaus. Bureaus are created by the legislature and, understandably, are generally supportive of the legislature's interests. Employees of bureaus generally fare better as the bureaus' budgets expand because, among other things, opportunities for promotion expand and, with them, salaries. With this expansion bureaus become more effective as vehicles for imposing their version of the public interest on the citizenry. At the same time legislators generally have particularly strong interests in the enactment of legislation that will be implemented by bureaus. Consequently, bureaus and the legislative committees with which they deal generally have similar interests.

As organizations engaged in the supply of services, public bureaus differ from private firms in two important respects. These differences in turn lead to important differences between the conduct of bureaus and the conduct of firms.[5]

1. Bureaus derive their revenue from legislative appropriations, which in turn come from tax collections. By contrast, firms derive their revenue from the voluntary buying decisions of customers.
2. There are no transferable ownership rights to public bureaus. In bureaus, there is no status comparable to that of stockholders, the residual recipients of differences between revenues and expenses. This difference does not mean that there are no profits in the operation of bureaus, but only that these profits do not show up as residual income to residual claimants or owners. Rather than being converted to personal use, any differences between revenue and expense disappear through added expenditure. For example, at the end of each fiscal year in Washington, bureaus with funds left over try to find ways to spend the funds. Many large research grants are made, for instance, at the end of the fiscal year.

The difference in incentives between private and public decision making yields many familiar results. It is often more difficult to register a car, get a passport, or buy alcoholic beverages in a state that has state-franchised stores than it is to buy automobiles, plane tickets, or wine from private suppliers. Under government operation, a bureau's hours of operation are usually shorter than those of private businesses and, hence, less convenient to customers. There is also a reluctance on the part of bureaus to accept checks and credit cards.

[4]For more analysis of interest groups see R. E. McCormick and R. D. Tollison, *Politicians, Legislation, and the Economy* (Boston: Martinus Nijhoff, 1981).

[5]See, for example, Ludwig von Mises, *Bureaucracy* (New Haven: Yale University Press, 1944); Gordon Tullock, *The Politics of Bureaucracy* (Washington, D.C.: Public Affairs Press, 1965); Anthony Downs, *Inside Bureaucracy* (Boston: Little, Brown, 1967); and William A. Niskanen, *Bureaucracy and Representative Government* (Chicago: Aldine-Atherton, 1971).

## FOCUS: Interest Groups and the British Factory Acts

The interest-group theory of government evaluates legislation in terms of who benefits and who loses. Not only is this framework a useful way to analyze present-day laws and regulations, but it is also helpful in the study of economic history. As an illustration, we can apply the theory to legislation passed in England more than a hundred years ago.

The British Factory Acts of 1833–1850 placed restrictions on the employment of children and women in the English textile industry. Most historians have analyzed this legislation by accepting at face value the rhetoric of those who agitated to get it enacted. This rhetoric was couched in humanitarian terms. Those who sought these laws said they were doing a great favor for women and children and for society in general; they said they were acting in the "public interest."

Suppose that we go behind what the reformers said and apply the interest-group theory to this legislation. Who won and who lost as a result of the Factory Acts?

Two primary groups stood to gain from the restrictions on children and women in the labor force. Male workers, especially skilled male workers, stood to gain because children and women were competitors for their jobs. If competing labor could be outlawed, male wages would rise. There is, in fact, evidence that male workers were the primary agitators for the factory legislation, as the interest-group theory suggests they should have been. Of course, they presented their lobbying efforts in terms of doing a favor for women and children and serving the public interest.

The other group that may have gained from the restrictions on women and children was the owners of steam textile mills. Mills in the 1830s were water- or steam-powered. Steam mills could run all the time, but water mills were subject to droughts. The water mills typically made up for time lost during droughts by operating for long hours when water power was available. The restrictions on women's and children's hours of work made this catching-up process more difficult. Textile prices rose as a consequence, and the wealth of the owners of the steam mills increased because their operations were less hampered by the law.

Who lost? Obviously, the women and children who lost jobs and income suffered to a degree from the legislation. One would have to count against this loss the extent to which the law led to better working conditions, especially for children. Women were particularly vocal about their loss, but since women could not vote, politicians could ignore their wishes at low cost. The other losers were the owners of water-powered mills.

So interest-group analysis goes beyond what people say they want and evaluates the costs and benefits that actually result from government action. Seen in this light, the Factory Acts appear to have been primarily an example of wealth transfer and sexism.

*Sources:* G. M. Anderson and R. D. Tollison, "A Rent-Seeking Explanation of the British Factory Acts," in David Colander, ed., *Neoclassical Political Economy* (Cambridge, Mass.: Ballinger, 1984), pp. 187–201; Howard P. Marvel, "Factory Regulation: A Reinterpretation of Early English Experience," *Journal of Law and Economics* 20 (October 1977), pp. 379–402.

*Carding, drawing, and roving in a Lancashire, England, cotton mill, 1834.*

In addition to these common observations, empirical studies have shown that public bureaus have higher costs of doing business than private firms. To cite a few examples, Roger Ahlbrandt has estimated that the cost of providing the same type of fire protection is about 88 percent higher when it is done by public bureaus than when done by private firms. David Davies has estimated that a private airline in Australia was 104 percent more productive in carrying freight and mail and 22 percent more productive in carrying passengers than a public airline. Robert Spann estimated that refuse collection is 43 percent more expensive when done by public bureaus than when done by private firms.[6]

Without the profit-maximizing incentives of private business, there are numerous ways that the indirect appropriation of excess funds through higher expenditures can take place within bureaus. Different public bureaus face different opportunities for appropriating profits. A public hospital, for instance, may overinvest in expensive equipment that is underutilized. A highway department may award contracts without competitive bidding and end up paying higher prices than necessary. Since excess funds cannot be appropriated directly, they are appropriated indirectly as the profits are dissipated through expenditure.

As we have noted, one main difference between a private firm and a public bureau lies in the identification of the customer. For private firms, the people who use the good or service are the customers whose continued favor is essential for the success of the firm. The public bureau's customers, however, are not the people who queue up for space at a public campground or who try to get efficient service at the motor vehicle department. Although the bureau provides services to those people, the public bureau's customer actually is the legislature because the legislature is responsible for the bureau's existence through legislation and funding. To remain in existence and to be successful, the bureau must please the legislature.

To some extent, of course, the legislature reflects the interests of citizens. But we must distinguish between the special interests of particular citizens and the interests of citizens in general. In cases where a bureau's performance pursues special interests, those special interests may represent a quite different set of people from those who use the bureau's service. For legislation to reward some special interests, it must penalize other, general interests. The often-poor performance of public bureaus is another feature of a special-interest approach to government, and this performance may not be so poor when it is viewed from the perspective of the bureau's true customers—the special interests that demanded the legislation.

## The Growth of Government

A particularly important problem for public choice analysis is to explain why government changes size over time. As Table 20–2 shows, the size of government in the United States has grown substantially over the period from 1929 to 1984. As the data clearly show, this growth has occurred in both absolute

---

[6]Roger Ahlbrandt, "Efficiency in the Provision of Fire Services," *Public Choice* 16 (Fall 1973), pp. 1–16; David G. Davies, "The Efficiency of Public Versus Private Firms: The Case of Australia's Two Airlines," *Journal of Law and Economics* 14 (April 1971), pp. 149–165; and Robert M. Spann, "Public Versus Private Provision of Governmental Services," in *Budgets and Bureaucrats*, ed. Thomas E. Borcherding (Durham, N.C.: Duke University Press, 1977), pp. 71–89.

TABLE 20–2  The Growth of Government Expenditures in the United States Since 1929[a]

| Year | GNP (billions of dollars) | Total Government Expenditures (billions of dollars) | Percentage of GNP |
|---|---|---|---|
| 1929 | 103.4 | 10.3 | 10.0 |
| 1934 | 65.3 | 12.9 | 19.7 |
| 1939 | 90.9 | 17.6 | 19.3 |
| 1944 | 210.6 | 103.0 | 48.9 |
| 1949 | 258.3 | 59.3 | 23.0 |
| 1954 | 366.8 | 97.0 | 26.4 |
| 1959 | 487.9 | 131.0 | 26.9 |
| 1964 | 637.7 | 176.3 | 27.6 |
| 1969 | 944.0 | 286.8 | 30.4 |
| 1974 | 1434.2 | 460.0 | 32.1 |
| 1976 | 1718.0 | 574.9 | 33.5 |
| 1978 | 2156.1 | 681.8 | 31.6 |
| 1980 | 2626.5 | 868.5 | 33.1 |
| 1982 | 3069.3 | 1090.1 | 35.5 |
| 1984 | 3661.3 | 1258.1 | 34.3 |

[a]Dollar amounts have not been adjusted for inflation.
Source: Adapted from Richard E. Wagner, *Public Finance* (Boston: Little, Brown, 1983), p. 13. Original data from *Facts and Figures on Government Finance*, 21st ed. (Washington, D.C.: Tax Foundation, Inc., 1981), p. 36.

and relative terms. The relative size of government at all levels rose from 10 percent of GNP in 1929 to 34.3 percent in 1984. Moreover, this growth continued under the Reagan administration, which came into power on a platform of reversing the trend.

Explanations for the growth of government are easy to find in some periods. Expenditures for World War II clearly explain the growth of government over the period 1939–1944. However, the general upward march of government expenditures in other periods has many complex causes. For public choice theorists, the growth of government poses both positive and normative issues.

On the positive side is the fundamental question, Why does government grow? Part of the answer rests with the various forces that we have discussed in this chapter. Those who pay for and elect government officials do not pay critical attention to what government is doing; interest groups and government expand in size and scope as a consequence. Politicians, of course, are happy to accommodate the interest groups and bureaus so long as voters remain uninterested. Such things can come to an end, however. Government and independent bureaus in government can become so large or controversial that they start to attract the voters' attention. In this case, the force of voter or citizen reaction can lead to a reduction in government or bureau size. It is also important to recognize that government growth is not inevitable. Our basic argument has been that government is a more imperfect mechanism for allocating resources than the market is. This does not mean that government is out of control; it just means that it takes longer to effect changes in the size of government.

On the normative side, scholars have addressed themselves to the issue of how to contain government expansion. The basic thrust of developments has been to suggest fiscal constraints on government. That is, in recognition of the incentive of politicians and bureaus to expand government, the way to check such expansion seems to be to impose a system of fiscal discipline on political decision makers. In this approach the Constitution would be amended to require that the annual budget be balanced or that government expenditure not exceed a given percentage of gross national product. This approach recognizes the incentives of politicians and bureaucrats, and it seeks to design constitutional rules that will enhance the degree to which the political process reflects the underlying desires of voters.

## Summary

1. Public choice analysis is the study of government with the tools of economics. It treats government decision makers like private decision makers—as rationally self-interested actors.
2. Government officials maximize their individual interests, not the "public interest." In the process they face different constraints than private decision makers, and so they behave differently.
3. Normative public choice analysis evaluates the effectiveness with which political institutions represent the preferences of voters, focusing on the evaluation of voting rules.
4. Voting on issues two at a time by majority rule can lead to a voting cycle where there is no clear winner even though each individual voter knows clearly who he or she prefers.
5. Constitutional decision-making analysis considers the choice of rules in the political arena. In other words, it asks how disinterested decision makers design such things as a voting rule for legislatures.
6. Positive public choice analysis seeks to present testable theories of government behavior. It is analogous to positive economics applied to politics.
7. In political competition the median voter is the controlling force. Parties and candidates compete by taking middle-of-the-road positions.
8. Logrolling, or vote trading, is useful in revealing relative intensities of preference across issues. When combined with geographic representation, however, logrolling can lead to an overexpansion of local public projects.
9. Voting and the market are two different means of allocating resources. For a variety of reasons, voters have little incentive to be informed in making voting decisions. As a result, voting is an imperfect mechanism for allocating resources.
10. Interest groups seek wealth transfers from government. Successful interest groups win transfers at the expense of general efficiency in the economy.
11. Bureaucracy is the production and management side of government. Unlike private production, government production takes place in a not-for-profit environment. As a consequence, bureaucrats behave differently from private decision makers.
12. Government grows because in general the costs of stopping growth appear to voters to exceed the benefits to them. There has been some interest in placing a constitutional constraint on the size and fiscal operations of government.

## Key Terms

public choice
normative public choice
positive public choice
proprietary
nonproprietary
logrolling

## Questions for Review and Discussion

1. What is a voting cycle? What conditions cause a voting cycle?
2. What is logrolling? What are the benefits and costs of logrolling?
3. Name three differences between voting and the market as mechanisms for allocating resources.
4. What is the difference between a public bureau and a private firm as a productive unit?
5. Discuss three experiences you have had in dealing with government that can be explained by the not-for-profit nature of government production.
6. Suppose that the size of government is declining relative to the size of the economy. Apply public choice theory to explain how this might happen.
7. Using the median voter model of voter preferences discussed in the chapter, explain why Democratic presidential candidates are more liberal in the primaries than in the general election.
8. Why do we not observe voting cycles in the U.S. House of Representatives?

# ECONOMICS IN ACTION

In this chapter we explored the problems of *majority* voting. What about the interests of *minorities*? How do they fare under majority rule?

Imagine the simplest of all democratic processes: a local referendum to increase the property tax by a relatively small amount and to use the funds to build a new school. This situation can easily lend itself to a problem of neglected minority interests. For example, a majority of voters may not have children and may be slightly opposed to the measure because of the small tax increase that accompanies it. The parents of school-age children may be intensely in favor of the tax-school package because of the poor condition of the existing school. Despite their strong concern, the parents lose to a relatively indifferent majority in the defeat of the referendum. Conversely, one could envisage a situation in which the parents are in the majority, the proposed tax increase is substantial, the present school is in good condition, and the nonparents feel tyrannized by the passage of the tax-school referendum because it promises more costs than benefits for them. In one person/one vote majority rule, unless all voters have an equal expected gain or loss from the outcome of an issue, voting may not accurately reflect the underlying intensities of voter preferences.

To clarify the example further, look at the distribution of potential gains for two voters over a set of issues, shown graphically in Figure 20–4. The height of each curve is a measure of the potential gain to a voter from a favorable outcome on a particular issue (a favorable outcome may represent either the passage of a desired bill or defeat of an undesired bill). To avoid the problem of minority interests under majority rule, each voter must have exactly the same utility curve. For example, all voters must experience a Y gain from the outcome on issue X. If 51 percent of the voters have a utility curve like A and 49 percent have a curve like B, the A's will win on every issue even if the B's feel more strongly in each case. Majority rule is disadvantageous to minorities in this case and does not lead to voting outcomes that reflect the preferences of all the voters.

The problem is that no mechanism exists by which minorities and majorities can vote according to their differing intensities of feelings on issues. When voters have only one vote per issue, this vote represents a different underlying amount of gain or loss to majority and minority voters.

*Source:* D. C. Mueller, R. D. Tollison, and T. D. Willett, "Solving the Intensity Problem in Representative Democracy," in R. C. Amacher, R. D. Tollison, and T. D. Willett, *The Economic Approach to Public Policy* (Ithaca, N.Y.: Cornell University Press, 1976), pp. 444–473.

## Voter Preferences and Majority Voting

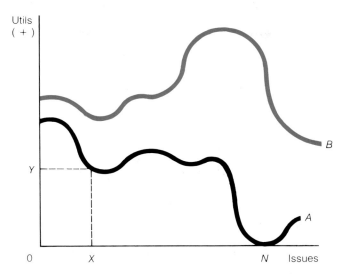

**FIGURE 20–4**  The Minority Voting Problem

Utility curves A and B measure the intensity of preference of two voters or two groups of voters across issues. The B group feels more strongly about each issue than the A group. A one person/one vote majority rule does not register this different intensity of feeling in voting outcomes. Point voting, when each voter is given a certain number of votes to allocate among all issues, may be a solution to this problem.

Suppose, however, that instead of having one vote on each issue, each voter is given 1000 votes and is asked to allocate them in proportion to his or her preferences over a wide number of issues. If issue J provides 7 times as much benefit as issue K, the voter allocates 7 times as many votes to it. Voting would consist of each voter's filling out a form indicating how the 1000 votes are to be assigned to each issue and whether the voter casts them for or against each issue. The issues would be decided by totaling the votes over all voters and passing all the issues that receive more yes votes than no votes.

This *point voting system* gives accurate information about the relative intensities of each voter's preferences over the various issues and makes it unnecessary for all voters to have the same utility curve. Stated in economic terms, point voting reveals individuals' marginal rates of substitution among issues, analogous to marginal rates of substitution among private goods. In other words, point voting for public issues is like the process of consumer equilibrium discussed in Chapter 6. The consumer allocates his or her fixed number of dollar votes in such a way as to establish equal marginal rates of substitution across goods. Point voting provides the voter with the opportunity to do the same with public issues.

Point voting is a possible solution to the problem that minority interests are not registered under one person/one vote majority rule. No voting system is perfect, however. The voting scheme that accurately reflects the preferences of all voters in final voting outcomes has yet to be invented. One of the key jobs of the public choice analyst is therefore to analyze voting systems and seek improvements in them.

## Question

It is well known that different cities provide different levels of public goods such as education, street repair, police and fire services, and so on. Individual citizens, however, do not value different "packages" of public goods in the same manner. One response to the situation is to move to other cities when those other cities offer a more satisfactory package. Describe conditions of local public goods supply under which you might prefer one city over another.

## POINT-COUNTERPOINT   A. C. Pigou and Ronald Coase: Solving the Problem of Social Costs

A. C. Pigou

Ronald Coase

ARTHUR CECIL PIGOU (1877–1959), an English economist, succeeded Alfred Marshall (see page 259) in the chair of political economy at Cambridge University in 1908. Like Marshall, Pigou had been drawn to economics because of its social importance. Both men were deeply interested in improving the living standards of the poor. Both had great faith in the power of economics to improve the welfare of all members of society. Pigou and Marshall disagreed, however, on the proper role of government. Like his neoclassical peers, Marshall was an advocate of laissez-faire policies, whereas Pigou felt that government's power to tax and redistribute wealth among all members of society was a crucial ingredient of general economic prosperity.

Pigou held his prestigious position at Cambridge for more than thirty-five years. During this period he wrote several books, including *Wealth and Welfare* (1912), *The Economics of Welfare* (1920), and *The Theory of Unemployment*. His students at Cambridge included Joan Robinson (see page 259) and John Maynard Keynes (see page 590). Pigou continued to be productive after his retirement from Cambridge in 1943. For many, his death in 1959 marked the end of the English neoclassical tradition in economics, a tradition begun in the 1870s.

RONALD COASE (b. 1910) has earned international recognition for his papers, "The Nature of the Firm" (*Economica*, November 1937) and "The Problem of Social Cost" (*Journal of Law and Economics*, October 1960). Modern theories of the firm and of externalities owe much to the work of this English-born economist. Coase's novel economic analysis of institutions, particularly the law of business, has helped spawn a new area of economic inquiry, much of which has challenged traditional views of how government and government regulations affect the behavior of firms. Coase himself has written extensively on utilities, broadcasting, and other government-regulated monopolies. He is generally critical of the effects of regulatory law, believing that solutions to market failure can and do exist outside of government intervention. Coase is currently retired Professor Emeritus of Economics at the University of Chicago. He has also held posts at the University of Buffalo and the University of Virginia.

For almost half a century, the theory of social cost was based on A. C. Pigou's original insights into the problem of market failure. Coase's challenge to Pigou's theory and his own approach to market failure offer a good illustration of how economics is evolving in our time.

### Taxing the Smoke

Pigou pioneered the modern microeconomic theories of externalities and market failure. In *The Economics of Welfare* he distinguished between private costs of production—the operating and maintenance costs of producers—and social costs of production—the expense or damage to society that results from the producer's activity. There are many possible illustrations of social costs. A nuclear power plant's social costs include the potential danger of radiation to the surrounding community. A railroad's social costs might include the danger of fire caused by sparks from its wheels. Pigou's own example of private and social costs is a smoke-belching factory. The factory owner's private costs derive mainly from the people and machines that make up the factory. The social costs of the factory derive from the smoke the factory produces as a by-product of its operation. The smoke from the factory soils clothing and property nearby, poses a health risk to the community, and creates a noxious odor.

Pigou pointed out that there is nothing in the economic forces of market supply and demand to prevent the factory owner from continuing and even increasing the pollution. Since the market, in essence, permits the owner to shift part of the costs (the smoke) onto others, the owner will likely protect this advantage.

Pigou also sought to prove that market failure resulted in inefficiency of resource allocation. To do so, he looked at the factory's costs and benefits in marginal terms. The owner's marginal private costs are the expenses to produce one more unit. The marginal social costs are the damages suffered as a by-product of producing one more unit. The marginal private benefit to the owner is simply the price of the product, or unit, and the marginal social benefit is whatever society gains from the product.

According to economic theory, the factory owner will try to manage production so that the marginal private costs of the factory equal the price of the unit. Pigou pointed out that by doing so, however, the owner is not responding to the true costs of production. Since the owner's private costs are less than the full (private plus social) costs of production, the owner will produce "too much" output. If, however, the owner managed production so that private plus social costs equaled price, the excess output would be eliminated, and all resources would be allocated according to their true opportunity cost.

To remedy the inequities and inefficiencies of market failure, Pigou proposed a tax on polluters and other sources of social costs. In the case of a smoke-belching factory, the tax would depend on the amount of smoke produced and would force the owner to restrict pollution to the point at which the benefits of production were equal to its true costs. Such a tax, of course, would also likely restrict output. In this way, both producers and consumers would share the costs of the tax.

### Assigning Liability

In his 1960 article, "The Problem of Social Costs," Ronald Coase challenged Pigou's major assumption that social costs always move in one direction—from producer to society. In fact, Coase argued, the issue of social cost poses potential harm to both parties. On the one hand, society may be harmed by the unwanted by-products of production, and on the other hand, producers may be harmed by society's attempt to correct the market failure through taxation. By the terms of Pigou's example, should the factory owner be allowed to pollute the local community, or should the local community be allowed to drive up the factory's costs of production by forcing it to cease polluting? The issue is especially pertinent today in the debate over acid rain and the problem of toxic waste.

Coase argued that an efficient solution to the problem of social costs would involve negotiation between the polluter and those affected by the pollution. If it were possible for both sides to negotiate on even terms, then a market might be created between them. The two sides could "trade" for the right to the pollution: Either the factory would pay the local community for the right to pollute or the community would pay the factory not to pollute. Coase showed that the actual outcome of such negotiations would be the same regardless of which side had the "right" to use the atmosphere as it saw fit.

In the real world, such negotiations between parties help solve a variety of potential social cost problems. For example, restaurant customers have influenced restaurants to segregate smokers from nonsmokers, eliminating the social cost of tobacco smoke. Home buyers shopping for a house near an airport can negotiate for a lower price to offset the social cost of airplane noise.

Often, however, one or both sides in the dispute face enormous transaction costs to enter such negotiation. A community, for instance, might find it very costly to organize into a bargaining unit with a nearby factory. The millions of people affected by acid rain and the hundreds of producers who might be responsible for the pollution could hardly transact freely. In such instances Coase suggested that the parties look to the courts, not the legislature, for a solution. Courts could determine the relative costs involved and could assign liability, or responsibility, to the party whose costs of adjusting to the social cost are lower.

For example, who should be responsible for injuries and death resulting from the use of defective products? For a long time, the rule of *caveat emptor,* or "let the buyer beware," prevailed among courts, and the consumer was fully responsible for the safe use of products. Such a rule worked reasonably well in times when products were relatively simple in design and use. Now, however, as products are becoming more complex, consumers may not be able to judge their safety and reliability. As a result, the courts have gradually increased the range of cases in which producers are held liable for shoddy and dangerous merchandise. This shift in the legal posture of the courts is an illustration of the Coase idea at work: The legal system alters liability assignment in response to the relative costs of adjusting to social costs.

# V

# International Trade and Economic Development

# 21

# International Trade

We do not often stop to ponder all the ways in which imported goods enrich our lives and improve the material well-being of all nations of the world. On a typical day, for example, an American student's consumption patterns are clearly global. She awakes on sheets made in England and prepares breakfast on a hot plate made in Taiwan. Donning a cotton dress made in India and shoes made in Mexico, she catches the morning news on a TV made in Japan and drives to school in a German-made Volkswagen. Her art history text was printed in the Netherlands, and the movie she sees that evening was imported from France.

In all but a few previous chapters and sections of this book, we have simplified our analyses by treating the national economy as closed—as isolated and with no international interchange. In the next three chapters, we focus exclusively on international trade, finance, and development, with attention to U.S. involvement.

Today's revolutions in technology, especially in telecommunications, are enabling nations to become more and more economically interdependent despite the often explosive political tensions that have divided nations for centuries. In a sense, the whole world is now an economic system. International economics is the study of this world economy with the tools of economic theory. The concepts of supply and demand and production possibilities introduced earlier in the book are quite useful in discussing worldwide trade, the international finance system, and the problems facing the less-developed economies of the world. In these chapters we also examine the organization and functioning of national economies such as that of the Soviet Union that are operated by means of central planning by government rather than by private property and markets.

International economics is composed of both microeconomic and macroeconomic elements. This chapter focuses essentially on the microeconomics of trade, including the reasons for and the advantages of specialization and trade. The effects of artificial interferences with trade, such as tariffs and quotas, are emphasized in an economic defense of free trade. Historical and contemporary trade policies in the United States are highlighted to provide a perspective on this issue of primary importance to our economic well-being.

## The Importance of International Trade

All nations have particular talents and resources; like individuals, whole nations are able to specialize in one or many activities. For example, the islands of the Caribbean have abundant sunshine and good weather year round, and so these islands specialize in tourism. Specialization enables nations to emphasize the activities at which they are most efficient and at the same time gain certain advantages through trade.

Examples of international specialization and trade abound. France, a country with a favorable climate and specialized land for wine growing, exports wine to Colombia and the United States and imports Colombian coffee and U.S. machinery. Likewise, both Colombia and the United States specialize and trade products that best utilize their qualities and quantities of resources.

Trade and specialization take place within a given economy as well as among economies. The former type of trade is called intranational or interregional trade; the latter is called international trade. In principle, the two types of trade are the same. As we saw in Chapter 2, Texas may specialize in beef production and Idaho may specialize in potato production. Each meets its need for the commodity it does not produce through intranational or interregional trade. This process is not fundamentally different from the trade of U.S. wheat for Japanese television sets.

In practice, though, some differences exist between interregional and international trade. Resources such as climate, fertile land, and work forces with specialized skills cannot easily be moved between countries. A nation typically trades with the products and services of its resources, not with the resources themselves. Beyond this basic point, countries have different currencies and political systems and values. For such reasons international trade is different from interregional trade and therefore merits special study by economists.

### National Involvement in International Trade

The overall magnitude of international trade and countries' shares in this trade can be measured in a variety of ways. The aggregate value of goods in international trade is given in Table 21–1 by major areas of the world. In 1984, world trade totaled more than $2 trillion, measured in exports or imports.

Table 21–2 lists the foreign trade of thirteen nations as a percentage of their gross national products for 1982. A wide range of difference occurs in the percentages. Exports account for almost 50 percent of Saudi Arabia's total output but less than 10 percent of Brazil's or the United States' output. Imports are about one-fourth to one-third of the real consumption of such countries as South Africa, West Germany, Sweden, and Switzerland.

Percentage figures do not tell the whole story, of course. In absolute terms the United States was the world's largest trader in both exports and imports in 1982. The value of exports totaled $261 billion and the value of imports

TABLE 21–1   World Trade: Exports and Imports, 1982

| Area | Exports (billions of dollars) | Imports (billions of dollars) |
| --- | --- | --- |
| Developed countries[a] | 1288.9 | 1405.0 |
| Developing countries[b] | 470.6 | 432.5 |
| Communist countries[c] | 244.6 | 220.2 |
| Total | 2004.1 | 2057.7 |

[a]Includes United States and other major developed nations such as Japan, West Germany, and so forth.
[b]Includes OPEC countries and other developing nations.
[c]Includes Soviet Union, China, and Eastern European countries.
Source: *Economic Report of the President* (Washington, D.C.: U.S. Government Printing Office, 1985), p. 352.

TABLE 21–2   Exports and Imports as a Percentage of GNP

| Country | Exports as a Percentage of GNP | Imports as a Percentage of GNP |
| --- | --- | --- |
| United States | 8.5 | 9.4 |
| United Kingdom | 27.0 | 24.8 |
| Switzerland | 33.4 | 33.5 |
| Saudi Arabia | 49.3 | 26.4 |
| Sweden | 33.3 | 34.1 |
| Mexico | 12.5 | 14.3 |
| Japan | 16.9 | 16.0 |
| Italy | 24.4 | 27.3 |
| West Germany | 33.5 | 31.2 |
| Canada | 27.6 | 23.4 |
| Brazil | 9.0 | 9.3 |
| South Africa | 28.8 | 28.9 |
| Venezuela | 25.5 | 27.8 |

Source: *International Financial Statistics*, International Monetary Fund (August 1983).

was over $290 billion. The United States trades with virtually every nation of the world, with the primary sources of imports being Canada (16 percent in 1981), Japan (8 percent), and Saudi Arabia and Mexico (5 percent each). The main destination of U.S. exports in 1981 was Canada (15 percent), Japan (8 percent), Mexico (6 percent), and the United Kingdom (5 percent). In terms of products, the three major U.S. imports are crude oil, motor vehicles, and food, while our three major exports are machinery, motor vehicles, and grain.

A more detailed account of U.S. exports and imports for 1982 is presented in Table 21–3. More than 25 percent of U.S imports consisted of petroleum and other fuels. Machinery and manufactured goods were also larger import items. While food and raw materials were only about 6 percent of total imports, food and raw materials constituted about 12 percent of U.S. exports in 1982. The export of machinery and transport equipment, the largest single export, was almost 40 percent of the total.

## Why Trade Is Important: Comparative Advantage

The reason for all this trading is that all nations benefit from specialization and trade. Suppose two nations produce computer components and food with an equal expenditure of time and resources. As we saw in Chapter 2, if

TABLE 21-3  U.S. Imports and Exports, 1982, by Major Classification

| Imports | Value (billions of dollars) | Percentage of Total | Exports | Value (billions of dollars) | Percentage of Total |
|---|---|---|---|---|---|
| Food and live animals | 14.5 | 5.9 | Food and live animals | 24.0 | 11.6 |
| Beverages and tobacco | 3.4 | 1.4 | Beverages and tobacco | 3.0 | 1.5 |
| Mineral fuels | 65.4 | 26.8 | Mineral fuels | 12.7 | 6.1 |
| Chemicals | 9.5 | 3.9 | Chemicals | 19.9 | 9.6 |
| Machinery and transportation equipment | 73.3 | 30.1 | Machinery and transportation equipment | 87.1 | 42.1 |
| Crude materials | 8.6 | 3.5 | Crude materials | 19.2 | 9.3 |
| Manufactured goods | 61.2 | 25.1 | Manufactured goods | 32.7 | 15.8 |
| Total imports | 244.0 | | Total exports | 207.2 | |

*Source:* U.S. Department of Commerce, *Statistical Abstract of the United States,* 1984 pp. 838–841.

**Absolute advantage:** The ability of a nation or a trading partner to produce a product with fewer resources than some other trading partner.

**Specialization:** Concentration in a single task in an effort to increase productivity.

**Comparative advantage:** The ability of a nation or a trading partner to produce a product at a lower opportunity cost than some other trading partner.

nation A can absolutely outproduce nation B in food and B can outproduce nation A in computer components, we say that A has an **absolute advantage**—the ability to produce a good at lower input cost—in producing food, and B has an absolute advantage in producing computer components. Trade between the two countries will likely emerge because each can specialize at what it does best—emphasizing the production at which it is most efficient—and trade with the other country for its requirements of the other good. As we will see, both countries will be better off because **specialization** and trade lead to increases in production and therefore to increases in the attainable consumption levels of both goods in both countries.

According to the principle of **comparative advantage,** countries will specialize in producing those goods and services in which they have lower opportunity costs than their trading partners. For example, a hilly, rocky country will not be able to raise as many sheep per acre as a country with fertile grasslands, but the rocky land cannot support any production other than sheep raising, whereas the grassland will support more lucrative cattle production. Even though the grassland is absolutely more efficient at producing both sheep and cattle, the rocky land has a comparative advantage in sheep growing because the opportunities forgone are nearly worthless. The rocky country will therefore tend to specialize in sheep, the grassy country in cattle.

Consider the simplicity and power of the idea that each country does what it can do best. Countries that have relatively lower opportunity costs of producing certain goods and services have a strong incentive to produce those goods and services. Production across the world will thus come to reflect the principle of comparative advantage at work. Indeed, the rough statistics in Table 21-3 provide a glimpse of America's comparative advantage. The large amounts of fertile U.S. farmland combined with advanced farm technology have made the American farmer the most productive in the world. America has a clear comparative advantage in farm products and raw materials. America's skilled labor force and high rate of technological advance through huge investments in private and public research and development have helped make the United States relatively efficient in producing highly sophisticated machinery and equipment. Trading partners of the United States, especially

Japan, have specialized in routinized productions such as automobiles and steel. Energy regulations and increased demands have forced the United States to import immense quantities of petroleum. The pattern of United States imports and exports therefore shows the process of economic specialization and comparative advantage at work.

**Production Possibilities Before Specialization.** To demonstrate more precisely how all trading partners benefit by exercising their comparative advantages, we must look at what happens to production possibilities before and after specialization and trade. Suppose that both the United States and Japan, in isolation, produce just two goods: computer components and food. Further assume that the pretrade production possibilities schedules facing the United States and Japan are depicted by the numbers shown in Table 21–4, which gives the alternative combinations of food and computer components that could be produced in the United States and Japan if resources are fully employed. The United States, for example, may choose to produce 60 units of food and no computer components, 40 units of food and 10 of computer components, 20 units of each, or 30 units of computer components and no food. With fully employed resources, Japan may choose between 45 units of computer components and no food at one end of the production spectrum and no computer components with 30 units of food at the other, or combinations between these extremes, as shown in Table 21–4.

These production alternatives can also be expressed in terms of the trade-offs among production possibilities. For the United States, the choice is between producing 60 units of food and 30 units of computer components. This equation of the possibilities can be expressed as 60F = 30CC, where F represents food and CC represents computer components. Within the United States, 1 unit of food therefore exchanges for ½ unit of computer components. Alternatively, 1 unit of computer components exchanges for 2 units of food. The equivalent expressions are as follows:

$$60F = 30CC,$$
$$1F = 1/2 CC,$$
$$2F = 1CC.$$

**TABLE 21–4  U.S. and Japanese Production Possibilities Schedules for Food and Computer Components**

Prior to trade between the United States and Japan, 2 units of food exchange for 1 unit of computer components within the United States. ⅔ unit of food trades for 1 unit of computer components within Japan. The relative opportunity costs of food production is lower in the United States, and the relative opportunity cost of computer component production is lower in Japan.

|  | Production Possibilities (at full employment) | | | |
|---|---|---|---|---|
|  | 1 | 2 | 3 | 4 |
| *United States* | | | | |
| Food | 0 | 20 | 40 | 60 |
| Computer components | 30 | 20 | 10 | 0 |
| *Japan* | | | | |
| Food | 0 | 10 | 20 | 30 |
| Computer components | 45 | 30 | 15 | 0 |

These ratios measure how many units of food must be given up to obtain a unit of computer components, or how many units of computer components must be given up to get an additional unit of food. In either case, what is being measured is the relative opportunity cost of producing food or computer components in the United States; that is, what must be given up to increase the production of food or computer components.

Exactly the same analysis can be applied to Japan. For Japan, the alternatives range from no food and 45 units of computer components to 30 units of food and no computer components. Japan's trade ratio is therefore 30F = 45CC, meaning that in Japan 1 unit of food exchanges for 1½ units of computer components, or that 1 unit of computer components exchanges for ⅔ unit of food. These alternative but equivalent expressions can be written as follows:

$$30F = 45CC,$$
$$1F = 1½CC,$$
$$\tfrac{2}{3}F = 1CC.$$

**Gains from Specialization and Trade.** The concepts of production possibilities and opportunity costs provide a perspective for understanding the benefits of trade. Continuing our example of trade between the United States and Japan, first look at the U.S. and Japanese production possibilities curves in Figures 21–1a and 21–1b. Notice that these production possibilities curves are straight lines. In practical terms, this means that resources can be transformed from food production to computer component production at constant opportunity cost, that resources are perfectly adaptable to one production or

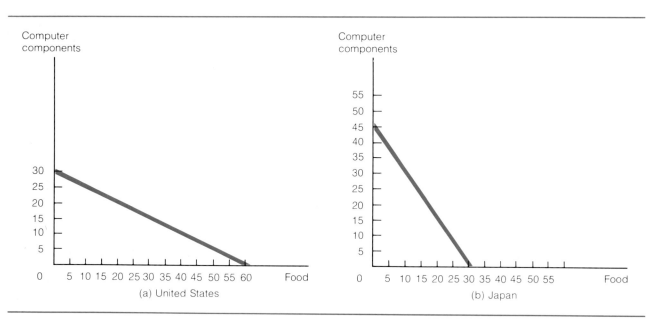

FIGURE 21–1  Prespecialization Production Possibilities Curves

The straight-line production possibilities curves for the United States and Japan indicate that resources can be transferred from computer component production into food production at constant opportunity cost. The differences between the two curves provide the basis for both countries to benefit from specialization and trade.

the other. This assumption is simplistic, but it does not change the argument for specialization and trade.

The important part of the argument for gains from specialization turns on the pretrade opportunity costs of producing the two goods that face Japan and the United States. Before trade, consumers in the United States must sacrifice 2 units of food to get 1 unit of computer components, while Japanese consumers must sacrifice only ⅔ unit of food to get the same amount of computer components. On the other side of the bargain, American consumers must sacrifice only ½ unit of computer components to obtain 1 unit of food, while their Japanese counterparts must give up 1½ units of computer components to get 1 unit of food.

Both countries can gain from trade. Since the United States produces food at an opportunity cost of ½CC and Japan produces computer components at an opportunity cost of ⅔F, we know that the Americans would gain if trade gave them more than ½CC for 1 unit of food, and the Japanese would be happy with anything more than ⅔F for a single unit of computer components. These differences in relative opportunity costs of producing food and computer components yield a basis for specialization and trade between the two countries.

At what price will trade take place? Clearly, the **terms of trade**—the ratio at which two goods can be traded for each other—will settle somewhere between the relative opportunity cost ratios for each country. For example, terms acceptable to both Japan and the United States will fall between ⅔F = 1CC (Japan) and 2F = 1CC (United States). We might say loosely that the two countries bargain to set the real exchange rate at which trade takes place, and the bargaining range is determined by each country's internal opportunity cost trade-off. For simplicity, let us assume that the rate settles at 1 unit of food for 1 unit of computer components, or 1F = 1CC.

Figures 21–2a and 21–2b illustrate the potential outcomes of this agreement in terms of the posttrade production possibilities curve. If the countries choose total specialization, the United States can completely specialize in food, producing 60 units, and Japan can completely specialize in computer components, producing 45 units. Given that each country can trade for the product it no longer produces at a price of 1F = 1CC, the production possibilities frontier for each country shifts outward to the right. In contrast to its domestic opportunity cost of 60F for 30CC, the United States now enjoys a situation in which production of 60F can be traded for production of 60CC. Japan enjoys a similar expansion of production possibilities through specialization and trade. By specializing and trading, both countries are made better off. Specialization does not even have to be complete for the two countries to gain from trade. We have used an example of complete specialization here just for illustration.

The production possibilities curves actually illustrate all possible combinations of both goods—food and computer components. If we assume that U.S. consumers originally purchased the bundle of food and computer components labeled US in Figure 21–2a, we can note the improvement in their well-being by considering the new possible consumption bundle $US_1$ on the posttrade production possibilities frontier. More of both commodities is available after specialization and trade. A similar consumption possibility exists for Japan if we compare hypothetical pre- and post-specialization and trade points J and $J_1$. Specialization and trade enlarge the consumption possibilities of the trading nations so that both countries can consume more of both commodities after trade.

> **Terms of trade:** The ratio of exchange between two countries, based on the relative opportunity costs of production in each country.

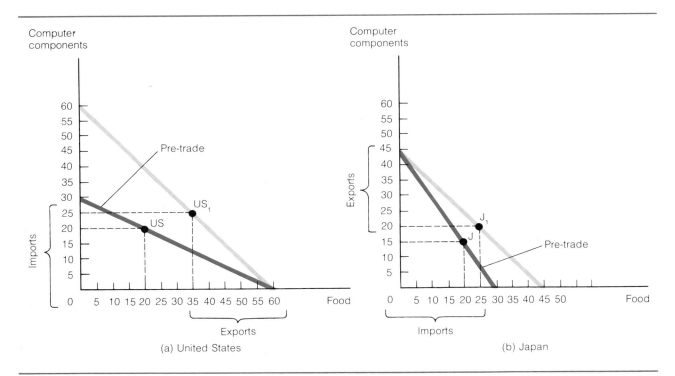

**FIGURE 21-2  Posttrade Production Possibilities Curves**
The pretrade production possibilities of the United States (a) and Japan (b) are given by the black straight-line curves. Specialization and trade permit the United States to specialize in food, producing 60 units, and Japan to specialize in computer components, producing 45 units. The production possibilities curve for each country shifts to the right (colored straight-line curve) because each country can trade 1F for 1CC: Japan can trade its 45 units of computer components for 45 units of food, whereas before specialization it could produce only 30 units of food for 45 units of computer components. Likewise, the United States can trade 60 units of food for 60 units of computer components instead of its pretrade production of 60 units of food for 30 units of computer components.

It is crucial to remember that the beneficiaries of trade are the consumers in all trading nations. American consumers benefit from the import of such goods as oil, automobiles, and shoes because they can be produced at lower real opportunity cost and therefore lower prices elsewhere. Likewise, consumers of other nations benefit from the United States comparative advantage—a lower real opportunity cost—in producing food, high-technology machinery, and products like Coca-Cola.

Our discussion has been based on a simple example of how the principle of comparative advantage works. In discussing two-good trade between Japan and the United States we assumed that each country was absolutely more efficient than the other at producing food or computer components. What if one of the countries has an absolute advantage at producing both goods? That is, for an equal expenditure of time and resources, one of the countries can outproduce the other in both industries. Is trade still possible? Will each nation stand to gain from specialization and trade? The answer to these critical questions is yes. The principle of comparative advantage still works in such a case. Although one country may be absolutely more efficient than the other at producing both goods, trade is possible so long as the relative opportunity costs of producing the two goods differ in the two countries.

### Barriers to Trade

Despite the great advantages of specialization and trade, there are certain barriers that make trade less than free. Some are natural costs of exchange; others are artificial barriers imposed by governments.

There are natural costs to all forms of exchange, from purchases at the local supermarket to international trade. These exchange costs decrease the possible gains from trade because they represent an increase in the real opportunity costs of trade. Contracting costs, negotiating costs, and transportation costs are examples of such natural trade barriers. We might think of them as reducing the size of the outward shift in the posttrade production possibilities curve. Means such as cheaper transportation and more efficient methods of trade negotiations are constantly evolving to reduce the magnitude of such barriers. Natural trade barriers, however, will always exist in one form or another.

Other barriers to the gain from trade are artificial in the sense that they are contrivances of governments designed to raise revenues or protect domestic industries from foreign competition. These artificial trade barriers fall into two major classes: tariffs and quotas. A **tariff,** or import duty, is simply a tax levied on particular imported goods. For example, the United States imposes a tariff on automobiles and steel imported into this country. A **quota** is a partial or absolute limitation on the quantity of a particular good that can be imported. For example, until 1972 the United States imposed a quota on the importation of foreign oil and permitted oil refiners to import only the allowed amount of oil. Both tariffs and quotas reduce specialization and the gains that consumers might obtain from trade. Quotas are generally more protective than tariffs because with tariffs, goods are at least admitted into the country and consumers can decide whether to pay the added amount imposed by the tariff. A more specific type of trade barrier within the United States is discussed in Focus, "State Protectionism."

**Tariff:** A tax on imported goods designed to maintain or encourage domestic production.

**Quota:** A limit on the quantity of an imported good.

## The Effects of Artificial Trade Barriers

Though tariffs and quotas are adopted by governments presumably out of national self-interest, they may not benefit most citizens in the long run. Tariff duties on imports were an essential source of U.S. government revenues from the Revolution until the late nineteenth century, and the use of tariffs to protect domestic industries is still being promoted. The effects of tariffs and quotas have long been debated, however. Adam Smith made the definitive economic statement on the matter in his *Wealth of Nations* in 1776:

> To give the monopoly of the home-market to the produce of domestic industry, in any particular art of manufacture, is in some measure to direct private people in what manner they ought to employ their capitals, and must, in almost all cases, be either a useless or a hurtful regulation. If the produce of domestic industry can be bought there as cheap as that of foreign industry, the regulation is evidently useless. If it cannot, it must generally be hurtful.[1]

[1] Adam Smith, *An Inquiry into the Nature and Causes of the Wealth of Nations*, ed. Edwin Cannan (New York: Modern Library, 1937), pp. 423–424.

## FOCUS  State Protectionism

The United States, like the European Common Market, can be regarded as a huge free-trade zone with no tariffs or other trade restrictions permitted in interstate commerce. Or can it?

Part of a state's police power is the authority to prevent agricultural pests from crossing the state's boundaries. But restrictions or absolute prohibitions imposed for safety or on "scientific grounds" also amount to partial prohibition quotas. For years, trucks and passenger cars entering California have had to submit to a fruit check conducted by the state Department of Agriculture, which restricts or confiscates incoming fruit. Regardless of whether these regulations prevent epidemics (inspections failed to prevent entry of the Mediterranean fruit fly in the late 1970s), producers benefit from reduced imports, and consumers lose.

The Kansas laws respecting the licensing and regulation-inspection of plant nursery operators provide another example of protectionist state quotas. The Kansas secretary of agriculture is empowered to license nursery operators within the state and to supervise inspections and certifications of nursery stock entering the state.[a] The instigators of such regulations would like the public to believe that they are sound and well intentioned—that, for example, such regulations improve the quality of the nursery industry in Kansas. But economists remain skeptical. A trade barrier is always a curious way to protect consumers.

[a]"Nursery Laws Governing Kansas," *Southern Florist and Nurseryman* (June 10, 1983), pp. 17–21.

---

Smith knew that individual traders, such as tailors and shoemakers, could gain from specialization and trade. But he went further and argued that the principle applied no less to nations:

> What is prudence in the conduct of every private family, can scarce be folly in that of a great kingdom. If a foreign country can supply us with a commodity cheaper than we ourselves can make it, better buy it of them with some part of the produce of our own industry, employed in a way in which we have some advantage. The general industry of the country, being always in proportion to the capital which employs it, will not thereby be diminished.[2]

**Free trade:** The free exchange of goods between countries without artificial barriers such as tariffs or quotas.

With very few exceptions (to be discussed later in the chapter), **free trade**—trade without artificial barriers—leads to maximum welfare among nations. Free trade and specialization expand production possibilities and deliver goods to consumers at the lowest possible costs. Yet given all the benefits of free trade, the world does not often seem to allow it to work, at least not completely. The reason is that the welfare of consumers is not always the single most important goal of a country. Other pressures are brought to bear on governments, such as the pressure by producer groups for protection from foreign competition. Sometimes these groups win political favors, such as tariffs or quotas, to reduce the competitive threat. In this section we use economic theory to show how these interferences with free trade reduce the economic welfare of a country.

### Trade and Tariffs: Who Gains? Who Loses?

To understand who really gains and loses from a tariff, consider Figure 35–3. It shows two possible prices facing domestic suppliers of automobiles: the free-market world price of autos, $P_w$, and the free-market world price plus some per unit tariff on autos imported to the United States, $P_{w+t}$.

Imposition of a tariff on imported autos clearly benefits domestic automobile manufacturers. Since imported cars are now more expensive, domestic

[2]Smith, *Wealth of Nations*, p. 424.

**FIGURE 21–3**

**Domestic Producers' Supply Schedule for Automobiles**

Imposition of an import tarriff on imported automobiles allows U. S. producers to raise the prices of their automobiles from the free-market world price, $P_w$, to $P_{w+t}$. The higher prices encourage them to increase the quantity of automobiles they produce from $Q_0$ to $Q_1$. Their profits are therefore increased by the shaded area $P_{w+t}E_1E_0P_w$.

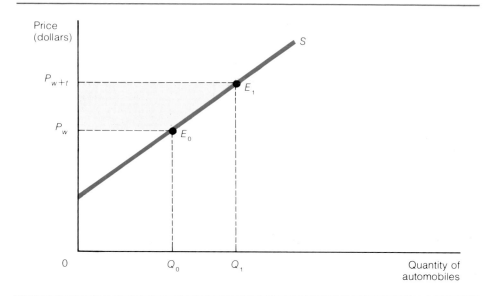

suppliers can raise their prices, without any increase in costs. The higher price $P_{w+t}$ encourages domestic auto producers to increase output from quantity $Q_0$ to quantity $Q_1$, increasing domestic suppliers' profits by the shaded area $P_{w+t}E_1E_0P_w$. Naturally, the tariff also increases domestic employment by an amount necessary to produce the additional cars.

What if the auto tariff were reduced? A reduction in the tariff would reduce producers' profits as well as domestic employment in the auto industry. But before we can present an economic assessment of tariffs, we must investigate their effects on consumers.

The effects of a tariff on consumers are shown in Figure 21–4. The demand curve for automobiles shows an initial price, $P_w$, and an initial quan-

**FIGURE 21–4**

**Effects of Tariff on Consumer Demand**

A tariff on imported automobiles causes the price to increase from $P_w$ to $P_{w+t}$, which in turn causes a reduction in the quantity of automobiles demanded by consumers from $Q_0$ to $Q_1$. Benefits to consumers are reduced by the amount represented by the shaded area $P_{w+t}E_1E_0P_w$.

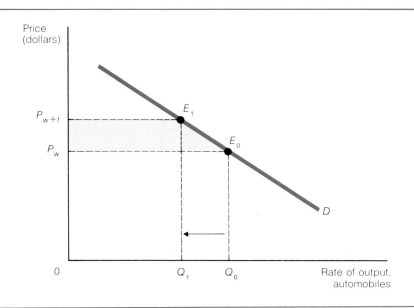

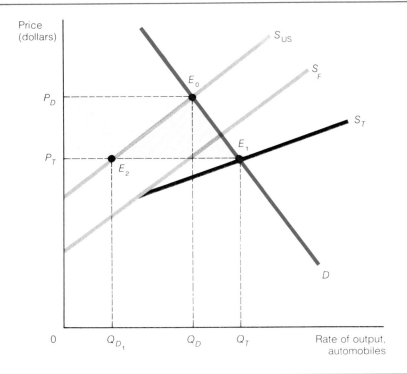

**FIGURE 21-5  The Benefits of Free Trade**
The total supply curve $S_T$ is the horizontal summation of domestic ($S_{US}$) and foreign ($S_F$) supplies of automobiles. In the pretrade situation, when consumers purchase only domestic automobiles, equilibrium $E_0$ is established at price $P_D$ and quantity produced $Q_D$. Under free trade, imports lower prices from $P_D$ to $P_T$ and increase quantity consumed in the domestic market from $Q_D$ to $Q_T$, with equilibrium established at $E_1$. Consumers benefit in the amount represented by $P_D E_0 E_1 P_T$, the amount they would be willing to pay over the amount they do pay, $P_T$.

tity of autos, $Q_0$, purchased by consumers. Imposition of a tariff on imported cars increases the price to $P_{w+t}$ and reduces consumption of autos from $Q_0$ to $Q_1$. A reduction in benefits to consumers accompanies the increase in price. (In earlier chapters, we have equated these benefits with consumers' surplus.) The shaded area $P_{w+t} E_1 E_0 P_w$ represents the reduction in benefits to consumers caused by the tariff on automobiles. This concept will be elaborated on in the following sections.

### The Benefits of Free Trade

Now we can see the benefits of free trade in the automobile industry. Consider Figure 21-5, which shows the U.S. domestic supply curve for automobiles, $S_{US}$, the foreign supply curve for automobiles that faces U.S. customers, $S_F$, and the American demand curve for automobiles. The domestic supply curve and the foreign supply curve for automobiles are added together horizontally to produce a total supply curve, $S_T$.

Equilibrium is established at the intersection of the demand curve for all automobiles, $D$, and the total supply curve, where the price to consumers is $P_T$ and the total quantity purchased is $Q_T$. If no trade took place, the price of automobiles in the United States would be higher, at $P_D$, and the quantity purchased would be lower, at $Q_D$. This is because the higher-priced domestic cars force some buyers out of the market. When free trade is introduced, a new price $P_T$ prevails, domestic production falls from $Q_D$ to $Q_{D_1}$ because part of the domestic supply of cars is replaced by lower-cost foreign production. U.S. imports of autos, as shown in Figure 21–5, now occur in the amount of $Q_{D_1}Q_T$. Total consumption rises from the pretrade quantity $Q_D$ to the posttrade quantity $Q_T$.

Who gains from free trade and who loses? In this simple example, domestic producers lose profits after trade in the amount represented by $P_D E_0 E_2 P_T$. U.S. consumers of automobiles gain benefits in the amount represented by $P_D E_0 E_1 P_T$. This is the consumers' surplus, the amount that consumers would be willing to pay for the additional automobiles they buy $(Q_D Q_T)$ over what they did pay for them, $P_T$. The net gain in U.S. benefits is equivalent to the difference between domestic producers' losses and domestic consumers' gains, or to the area $E_0 E_1 E_2$.

In assessing the effects of trade, it is also worth noting that imports of foreign automobiles release the resources formerly needed to produce $Q_D - Q_{D_1}$ cars in the domestic market. In other words, opening or expansion of imports creates temporary displacement or unemployment of labor and other inputs, a fact that helps explain some domestic opposition to free-trade policies.

### Welfare Loss from Tariffs

We have now seen how trade creates net benefits to domestic consumers. Let us now look at the effects of a protective tariff on the society as a whole, using a hypothetical market for television sets.

Figure 21–6 represents the domestic market for TV sets, including the domestic supply of the product and the U.S. demand. Prior to the imposition of a tariff, American consumers buy $Q_4$ units at the world price $P_w$. Of this quantity purchased, $Q_1$ are sold by domestic manufacturers and $Q_1 Q_4$ are imported from abroad.

Assume that a tariff is imposed in the amount $t$, causing the price of TVs to American consumers to rise to $P_{w+t}$. American TV producers gain additional profits in the amount represented by $P_{w+t} E_2 E_1 P_w$ because of the increased domestic output permitted by the higher price. Imports are reduced to $Q_2 Q_3$, and the total number of sets sold decreases to $Q_3$. Consumers lose total benefits in the amount of $P_{w+t} E_3 E_4 P_w$. At the same time, the tariff on imported TV sets creates government revenues. These revenues are composed of the per unit amount of the tariff multiplied by the units of TV sets imported after the tariff is imposed (area $E_2 E_3 AB$ in Figure 21–6).

Economists typically assume that the government revenue is used in a manner that produces benefits equivalent to those lost by consumers of TV sets. However, the tariff causes a net loss in benefits to society, of which TV consumers are a part, represented by the two areas $E_2 BE_1$ and $E_3 E_4 A$. The sum of these triangles represents a loss that is not counterbalanced by the sum of producers' gains and government revenue resulting from the tariff. Economists call this loss a welfare loss to society.

### Why Are Tariffs Imposed?

If tariffs create a welfare loss to society, why are they imposed? Certainly tariff revenue to the federal government is minuscule when compared to other sources of government revenue. More than two hundred years ago, Adam Smith identified the real cause of protective tariffs:

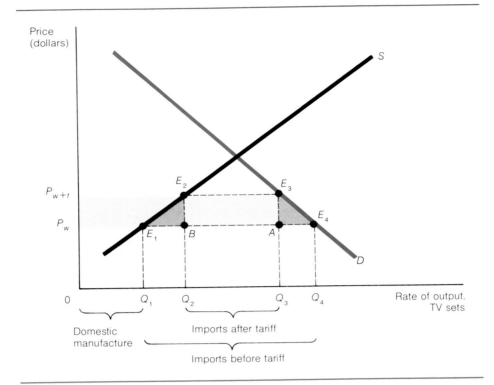

**FIGURE 21-6   The Effects of a Tariff**

Gains in producer profits and government revenue do not equal consumer losses from tariff imposition. Net consumer losses are shown in the shaded triangles in the figure. Before a tariff on TV sets, consumers purchase $Q_4$ units at price $P_w$. The tariff causes the price to rise to $P_{w+t}$, imports to decrease from $Q_1Q_4$ to $Q_2Q_3$, and number of TVs sold to decrease to $Q_3$. Consumers' total loss is represented by $P_{w+t}E_3E_4P_w$, but producers gain $P_{w+t}E_2E_1P_w$ and government revenues increase by $E_2E_3AB$. The net loss to society is therefore the sum of the areas $E_2BE_1$ and $E_3E_4A$.

> Merchants and manufacturers are the people who derive the greatest advantage from this monopoly of the home-market . . . . Manufactures, those of the finer kind especially, are more easily transported from one country to another than corn or cattle. It is in the fetching and carrying manufacturers, accordingly, that foreign trade is chiefly employed. In manufactures, a very small advantage will enable foreigners to undersell our own workmen, even in the home market . . . . They [merchants and manufacturers] accordingly seem to have been the original inventors of those restraints upon the importation of foreign goods, which secure to them the monopoly of the home-market.[3]

The source of tariff protection is to be found in the urgings of "merchants and manufacturers" today as it was in Smith's day. Consumers of TV sets and of any goods that actually bear a tariff could conceivably convince both producers and the government not to impose tariffs because, as we have shown, consumers lose more than producers and government gain from tariffs. Practically, however, the world does not work this way. Consumers, the gainers from free trade, are widely dispersed and costly to organize in any fight for free trade. Interest in fighting a tariff or quota is also apt to be low

---

[3]Smith, *Wealth of Nations*, pp. 426, 429.

among consumers because their pro rata share of losses from artificial barriers to trade is generally small. To an individual consumer, for example, the tariff is a small proportion of the total price of an automobile. The incentive to organize and fight protective tariffs is therefore small.

Not so for producers. The pro rata share of the effects of tariff protection is much higher among manufacturers, for they are far fewer in number than consumers. Manufacturers thus have stronger incentives than consumers to form an interest group, and their costs of organizing are lower since they are a smaller group. For such reasons, the consumer interest in free trade is often thwarted.

## The Case for Protection

Throughout history numerous arguments have emerged in defense of protection from free trade. Some arguments are well constructed; others are but thin veils for producers' interests. All the arguments—both the well-constructed and the questionable ones—deserve careful scrutiny.

### National Interest Arguments for Protection

The two oldest and best-formed arguments in defense of protection and against free trade are the national defense argument and the "infant industries" argument. Both arguments contain a grain of truth, but each should be closely scrutinized before serving as a base for protectionist policies.

**National Defense.** The oldest argument for protection—and the one with superficially the best justification—is that such restrictions are necessary for national defense. As military technology has changed, steel, gunpowder, manganese, uranium, and a host of other inputs have all been commodities essential to making war at various times, and they have been the subject of tariffs and quotas to keep domestic production of these commodities strong.

The logic of protection for defense is clear: Since the ultimate function of government is national defense, any possible threat to national defense, such as the unavailability of some resource during a crisis, must be avoided. A loss in consumers' benefits from protection is thus justified for a greater benefit—the availability of essential materials to be prepared for war. Artificial barriers to trade in these materials are used to protect domestic industries considered essential for national defense.

The national defense argument is plausible for some industries and some products, but consider that cheese, fruit, and watch manufacturers and other, non-defense-related industries have all resorted to the argument. History seems to prove that patriotism is the last and best refuge of a producer seeking protection from foreign competition. The national defense argument has seen double duty as an argument for maintaining unprofitable routes on railroads and protecting truck, air, and railroad companies by setting legalized cartel rates. The telephone industry was long protected on similar grounds.

**Subsidy:** A government cash grant to a favored industry.

Other ways can be devised to handle the national defense issue besides imposing tariffs and quotas. A tariff is nothing but a hidden **subsidy** or cash grant to domestic producers. Thus, if an industry is to be protected for national defense reasons, an explicit subsidy may be more straightforward than tariff protection. That way the national defense issue is made clear, and voters know how much defense actually costs them. If, for example, it is deemed vital to national security that the United States have a certain capacity to produce iron and steel, an alternative public policy would be to subsidize domestic iron and steel producers rather than to protect them against foreign competition with a tariff or quota.

**Protection of Infant Industries.** Possibly the most frequently head argument for protection, especially in less-developed nations, is the so-called **infant industries** argument. Discussed by Adam Smith and supported by Americans such as Alexander Hamilton, the infant industries argument reached its highest expression in the mid-nineteenth-century writing of German nationalist Friedrich List (1789–1846). List sought to unify and develop the German states against the incursions of British imports. He argued that protective tariffs were necessary in the transition from an agricultural-manufacturing to an agricultural-manufacturing-commercial stage of development. His reasoning was clear: Specific tariffs were necessary to maintain fledgling industries until they could compete with foreign imports on their own. After the "infancy" period of industrial development was over, protective tariffs would be lifted, forcing the businesses to meet the rigors of competition.

> Infant industry: A new or developing domestic industry whose average costs of production are typically higher than those of established industries in other nations.

A number of less-developed nations use this argument today in their quest for unilateral tariff protection. One-crop economies whose foreign earnings are heavily dependent on a single export seek to protect and diversify their economy with an umbrella of protectionist policies.

As much as one might be concerned for the economic plight of poor nations, great care must be exercised in applying the infant industry argument to them. At best, the argument is one for buying time, for allowing domestic industry to gain a foothold in international competition. Protection for any reason, as we have seen, means lost benefits for consumers. Obviously, any gains from ultimate independence must be set against the costs of lost consumer benefits over the period of protection.

A final question that should be applied to the infant industries argument concerns the vagueness of the goal of removing protection "when the industry grows up." It is hardly legitimate to apply the infant industries argument to the present-day steel industry in the United States and Western Europe, but the argument is still being used. Entrenched protectionist interests will always attempt to prolong the "infancy" of any industry. The dangers of giving protectionists a general legal foothold negate any merit the argument might have in limited and specific cases.

## Industry Arguments for Protection

A number of other arguments for protection crop up from time to time, usually put forward by domestic firms and industries seeking protection for protection's sake. We characterize these arguments as wrong because they are assaults on the very principle that gains may be realized from trade. They are all variations on the theme that special-interest groups deserve protection from international competition.

**The "Cheap Foreign Labor" Argument.** A common argument for tariff or quota protection is that labor or some other resource is cheaper abroad, enabling foreign manufacturers to sell goods at lower prices than U.S. manufacturers. The defenders of this argument are often producers or workers displaced by imports and foreign competition. During the early 1980s, for instance, U.S. producers and workers bitterly complained through their

## FOCUS: The Effects of Textile Quotas

Import quotas ordinarily are more damaging to consumer welfare than are tariffs. Tariffs permit consumers to import as much as they want so long as they pay the duty. Quotas place absolute limits on the foreign supply of a commodity. Quotas are imposed in two ways: (1) A limit on imports may be established, with profits going to the lucky sellers who are able to buy within the quotas on a first-come-first-served basis, or (2) the government may sell import licenses to sellers, with the profits going to government coffers. Either way, consumers lose because government has barred their access to foreign supplies.

In mid-1983, pressure from U.S. garment workers' unions, mills, and manufacturers led the U.S. Department of Commerce to provide "relief" to the industry through quota restrictions on eighty-nine categories of imported apparel and fabrics. Wool items, with American products of vastly inferior quality, were especially hard hit. Only about 20 percent of ordered imported wool skirts and men's and boys' wool coats got through to U.S. wholesalers and retailers. Prices rose on apparel and fabrics by 10 to 70 percent.

Domestic textile workers and manufacturers obviously gained, but who lost? Arnold Schmedock, executive director of the Ladies Apparel Contractors Association, said, "The only one that will lose out are the retail stores and importers because they won't have the markups and profits they've had."[a] Schmedock forgot to mention the other big losers: American consumers of wool and textile products. There have been other losers in the textile trade war as well. In a Chinese retaliation to the textile quota, U.S. exports to China in May 1983 fell to $75.2 million from $241.7 million in May 1982. The big losers were soybean and wheat producers: Retaliation punished American farmers as much as protection punished American buyers of textile products.

[a]Quoted in "We'll Pay More for Clothes," USA Today (July 21, 1983).

---

union representatives that imports were creating unemployment. (See Focus, "The Effects of Textile Quotas," for a discussion of quotas in one industry.)

While the argument may be correct—foreign producers may be more efficient combiners of resources—this truth in no way denies the benefits to free trade. Indeed, the request for protection from cheap labor turns the gains from trade position on its head: The reason *for* trade becomes a reason to limit or restrict trade. Open trade is beneficial precisely because resources may be cheaper or combined more efficiently elsewhere. The opening or extension of trade creates a temporary disruption of markets, including the unemployment of resources. To impose tariffs or quotas on the grounds that some groups of U.S. laborers are temporarily thrown out of work or that some U.S. stockholders are losing wealth is to subsidize special interests at a greater cost to all U.S. consumers of the subsidized product or service. Moreover, the domestic economy may actually gain from the movement of domestic resources into new fields of comparative advantage. For example, resources released from domestic steel production may be reallocated to computer production.

Side issues arise in the cheap foreign labor argument. One common complaint is that foreign governments subsidize the production of exported goods to shore up their domestic industries and prevent unemployment. As illustrated by Economics in Action at the end of this chapter, this complaint is used as a plea for protection from cheaper imports. But to argue for tariffs on these grounds is to look a gift horse in the mouth. The benefits from trade are independent of the reasons that imports are cheaper. If governments choose to subsidize their exports, they in effect tax their own citizens to benefit foreign consumers.

An activity called dumping is also related to the cheap foreign labor ar-

**Dumping:** The selling of a product in a foreign nation at a price lower than the domestic market price.

gument for protection. **Dumping** is simply the selling of goods abroad at a lower price than in the home market. The practice, an example of price discrimination (see Chapter 10), means that foreign buyers gain greater benefits from consuming the commodity or service than domestic consumers. Special interests in the favored country often complain that dumping constitutes "unfair competition," but the argument for free trade remains intact nonetheless. Consumers in the favored nation are able to purchase goods at lower prices and overall welfare in the consuming country is enhanced.

It should not matter why foreign prices are low as long as the foreign supplier who is dumping goods in U.S. markets has no monopoly power. If the producers of Japanese television sets undersell U.S. producers with an eye to putting them out of business and subsequently raising their price to a monopoly level, dumping does pose an issue for public concern. This is not usually the case, however. More often, dumping by foreign producers is simply a reflection of their lower costs of production and puts competitive pressure on U.S. producers.

**The "Buy American" Argument.** The "buy American" argument, a call to patriotism, means to keep money at home. More specifically, buy American means that imports should be restricted so that high costs or inefficient producers and their employees may be protected from foreign competition.

This well-known argument is fallacious on several counts. First, it asks consumers to pay higher prices for goods and services than are available to them through trade, thereby negating the potential expansion of trade benefits. Second, when money is kept at home, foreign consumers are unable to purchase domestic exports, reducing the welfare of domestic export producers and their workers and other input suppliers.

**"Terms of Trade" Advantage.** A final argument applies to countries that have monopoly or monopsony power in international markets. It suggests that a country employ its power to increase its share of the gains from international trade. Thus, if the United States is the world's largest supplier of computers, putting restrictions on computer exports will drive up the price of computers in the international market. Although the United States will then sell fewer computers, it will get a higher price per computer and possibly higher computer sales revenues. In effect, by restricting computer exports, the United States would shift the terms of trade, or the exchange rate, in international transactions in its favor.

**Terms of trade argument:** The use of export restrictions on goods in an effort to increase a country's monopoly or monopsony power in international markets.

The **terms of trade argument** therefore suggests that countries exercise monopoly and monopsony power where possible. Doing so is a means for a country to increase its revenues from international trade. The argument founders, however, on a simple point: the possibility of retaliation by other countries. If all countries seek terms of trade advantages, all countries will be worse off. It is thus hard for a country to win terms of trade advantages on a unilateral basis and have its trading partners sit idly by.

Perhaps the greatest flaw in all arguments for protection lies in the implicit assumption that other nations will lose export markets and will not retaliate. U.S. history is peppered with examples of tariff wars. One of the greatest tariff wars in history took place in the midst of the Great Depression of the 1930s, discussed in the next section. Tariffs or other forms of trade barrier retaliation can only create a reduction in worldwide economic welfare and massive and inefficient allocations of resources. Whatever the initial reasons for protective tariffs, consumers ultimately suffer the costs.

## U.S. Tariff Policy

U.S. tariff history has been punctuated by cycles of free trade and of protectionism, with a move toward free trade in the past fifty years that may be more apparent than real. The average tariff rates for the years between 1821 (the year of the first good statistics) and 1982 are shown in Figure 21–7 along with some of the highlights of our tariff history. Import duties as a percentage of the value of all imports subject to duty have ranged from almost 60 percent in the early 1930s to a mere 3.6 percent in 1982.

Why has the average tariff fluctuated so widely during U.S. history? Before turning to a specific analysis of changing conditions, one broad issue will help us answer the question. With the minor exception of tariffs imposed for revenues to fight wars or to apply foreign policy pressures, the average tariff has varied with business conditions, falling in periods of prosperity and rising during prolonged recessions or depressions. During periods of rising prosperity, manufacturers and workers displaced by free trade find little political support for the imposition of tariffs. When general business conditions turn downward, creating reduced demand for products and increased unemployment, the cry for protection grows louder in the political arena.

### Early Tariff Policy

Tariffs were a major source of revenue for the Republic in its early years, as they are for some less-developed countries today. Post-1820 tariffs tended to vary with business conditions, rising during recessions and falling during

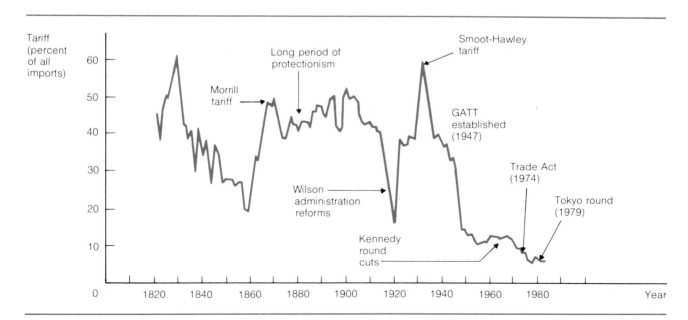

**FIGURE 21–7  Level of Tariffs**

The United States today has fewer tariffs than at any time in its history. Since the establishment of the General Agreement on Tariffs and Trade (GATT) in 1947, tariffs have followed a downward trend.

Sources: U.S. Bureau of the Census, *Historical Statistics of the U.S., 1976*; *Statistical Abstract of the United States*, 1984, p. 841.

more prosperous periods of business activity. On the eve of the Civil War, tariff revenues stood at about 19 percent of the value of imports, the lowest level in American history up to that time. In 1861, the Morrill tariff passed Congress under the pretext of generating much-needed revenue for the Union treasury. Whatever the initial intention, the Morrill tariff set off a wave of protectionism in U.S. policy that lasted until the second decade of the twentieth century. Tariffs declined dramatically during the Wilson administration, reaching an average of only 16 percent in 1920.

The dramatic downturn in the economy beginning in the late 1920s had a stark and lingering effect on protectionism in the United States. Manufacturing and commercial interests combined with a willing political climate to produce the Smoot-Hawley Act of 1930, which legislated the highest peacetime tariff in U.S. history. Average tariff levels were raised to almost 60 percent of the value of imports, setting off a frenzy of protectionist retaliations in other nations. The result of what was perceived as American self-interest was falling real incomes, sharply reduced consumer welfare, and a deepened and prolonged worldwide depression in the major world-trading countries.

### Modern Tariff Policies

Contemporary tariff policy may be viewed as an attempt to overcome the disastrous effects of protectionism embodied in the Smoot-Hawley Act. The first Roosevelt administration, under the aegis of Secretary of State Cordell Hull, acted swiftly to counteract the protectionism of the Smoot-Hawley tariff. In 1934, the Reciprocal Trade Agreements Act was passed, giving the president power to negotiate bilateral tariff reductions of up to 50 percent with other countries. This act was the first important step in establishing the character of modern U.S. tariff policy, for two reasons: (1) It established a framework within which free trade is envisioned as a goal of policy, and (2) it removed from Congress, to a large extent, the tariff-making power (but not the power to set nontariff barriers). While Congress could legally remove the tariff-setting powers of the president, these powers have been renewed and strengthened a number of times since 1934.

In the post-World War II era, the powers of the president have been expanded to include multilateral negotiations for tariff reductions, that is, negotiations with all nations simultaneously. A major manifestation of America's multilateral policy was its support of the multinational organization called GATT (General Agreement on Tariffs and Trade) in 1947. Originally a twenty-two-member body, GATT now includes more than eighty world nations representing about 80 percent of world trade. GATT sets rules and conditions for tariff reductions and oversees bargaining with all participating nations simultaneously.

Modern U.S. trade policies evolve through GATT, and tremendous gains and expansion of world trade have occurred since the formation of this multinational body. Moreover, Congress has expressed some willingness to move toward free trade. In 1962, a Trade Expansion Act was passed permitting the president to negotiate tariff reductions on all commodities simultaneously rather than commodity by commodity. This act led to the so-called Kennedy round of tariff reductions (1964–1967), which produced huge concessions on manufactured and industrial products with special concessions for poor or less-developed countries.

## Where Does the United States Stand in the Battle for Free Trade?

Protectionist interests in trading nations have not remained silent in the face of free trade. In fact, the era of lower tariffs coincided with the growth of nontariff trade restrictions such as quotas, outright prohibitions, and government subsidies to exporters. The requirement of import licensing is a feature of such restrictions. The expansion of these restrictions led to the Trade Act of 1974, giving the president expanded power over tariffs and new powers to deal with nontariff barriers. Negotiations under the auspices of GATT were conducted in Tokyo in 1979 (the Tokyo round) with promising results, especially on the issue of tariff reductions.

Economists are in virtually unanimous agreement that free trade is a critical goal, but some are doubtful about the viability of the move to free-trade policies. There are a number of reasons for skepticism, some revolving around the trade powers that Congress has retained and some relating to powers that Congress has delegated.

An escape clause is attached to the president's power to negotiate tariff reductions. The U.S. International Trade Commission is empowered by Congress through the Trade Act of 1974 to determine whether tariff reductions and imports would materially affect industries competing with the imports. The damage must be directly attributable to imports and not simply due to declining demand or inefficiency. This clause gives those opposed to free trade a wedge in the fight for protection, and a number of industries, such as color television manufacturers, have successfully invoked it.

The International Trade Commission is also charged with assessing damages done to workers in industries affected by tariff reductions. Trade adjustment—assistance payments to workers displaced by increased imports—is based on the fact that since government policies are responsible for the unemployment, government should address the problem with unemployment compensation and benefits. To date, workers in the automobile industry have been the major beneficiaries of trade adjustment, having received almost $3 billion in compensation. Some economists believe that such aid merely prolongs the inevitable and short-circuits the market adjustment process. In other words, this assistance may forestall movement of resources to industries in which the United States has a comparative advantage.

Most insidious of all to economists are the nontariff barriers and the hidden nontariff restrictions. Consent of Congress is still required of the president in negotiating reductions or elimination of nontariff barriers such as quotas, which Congress has the power to set. Protectionists therefore have simply changed the battlefield from tariffs to quotas. In periods of recession, protectionists lobby for trade restrictions other than tariffs, and they have been successful.

Conditions in the U.S. auto industry in the early 1980s are an example of this phenomenon. In the midst of a recession, reduced demand for U.S. automobiles, combined with competitive pressures from abroad, created pleas for protection: In the early 1980s, the U.S. automobile industry pressed for quota restrictions on Japanese imports. Fearing explicit and official congressionally imposed quotas and the retaliation they might bring, President Reagan in 1981 negotiated a three-year "self-imposed voluntary" lid on Japanese exports of automobiles in the amount of 1.76 million vehicles annually. In spite of Japan's adherence to the voluntary quota over the three-year period and notwithstanding a dramatic rise in U.S. automobile sales in 1982–1983, Senator Donald Reigle of Michigan demanded more protection. In 1983, when a Japanese minister indicated a possible unwillingness to renew the voluntary agreement, Reigle said, "The continuing Japanese attack on our basic industries is another Pearl Harbor. The time has come to close America's door to the flood of Japanese imported products."[4] The U.S. Congress threatened as much in early 1985.

As the senator's statement implied, we are a long way from free trade. Unfortunately, the U.S. auto industry is only one example of protectionist efforts. Any industry with enough power to affect political outcomes is in a position to gain some form of protection. In the battle, consumers are the inevitable losers.

Not only does free trade promote economic welfare; through the simultaneous creation of economic strength and interdependency, it also shores up the unity and strength of democratic countries. It is paradoxical that Western nations have long supported a goal of unity through multilateralism in foreign policy while simultaneously pursuing protectionist trade policies that limit economic growth and solidarity.

## Summary

1. All individuals possess talents that give them advantages in trade. Nations, like individuals, benefit from specialization and trade. Nations trade not resources but rather the products and services created with resources. Differing resource endowments are the basis for trade in products because products are ordinarily more mobile than the resources that produce them.
2. Specialization and trade may take place between countries of vastly differing degrees of economic development. The reason is the law of comparative advantage, which states that trade is possible when the relative opportunity cost of producing two goods differs between two countries.
3. Specialization according to the law of comparative advantage permits the production possibilities and hence the rate of sustainable consumption of two nations to expand. That is, the two parties to trade may obtain more of both traded goods after specialization and trade.
4. There are both natural and artificial barriers to trade. Natural barriers are all exchange costs including transportation costs of moving goods from one country to another. Artificial barriers include taxes on imports of goods, called tariffs, and limitations or prohibitions on imported items, called quotas.
5. Tariffs on imports increase the profits of producers and the revenue of government, but consumers lose more than producers and governments gain. This net welfare loss is the reason most economists oppose protectionist policies.
6. Though tariffs carry a net loss to society, they are imposed whenever domestic producer-competitors are strong enough to supply gains to politicians. Consumer groups may oppose tariffs but are seldom well enough organized or vocal enough to do so successfully.
7. Two substantive arguments are made for protection: the national defense and infant industries arguments. Most economists, however, question the adequacy of these arguments in actual operation. Other arguments—such as "cheap foreign labor" or "buy American"—are regarded by most economists as only thinly veiled protectionist fallacies.
8. U.S. trade policies have historically waxed and waned with prosperity and depression, becoming more protectionist during economic downturns. Although import tariffs have been reduced dramatically in the past fifty years, protectionists' interests have turned to a new battleground—the imposition of various types of quotas.

## Key Terms

| | | | |
|---|---|---|---|
| absolute advantage | terms of trade | free trade | dumping |
| specialization | tariff | subsidy | terms of trade argument |
| comparative advantage | quota | infant industry | |

[4]Quoted in "Uno's Surprise: Uncertainty About Auto Imports," *Time* (July 11, 1983), p. 19.

## Questions for Review and Discussion

1. Evaluate and discuss the following statement: "Trade across international boundaries is essentially the same as trade across interstate boundaries."
2. Nations Alpha and Beta have the following production possibilities for goods X and Y:

    | Alpha | X | 0 | 3 | 6 | 9 |
    |---|---|---|---|---|---|
    |  | Y | 12 | 8 | 4 | 0 |
    | Beta | X | 0 | 4 | 8 | 12 |
    |  | Y | 15 | 10 | 5 | 0 |

    a. Draw the production possibilities curves for both countries. What does 1 X cost in Alpha before trade? What does 1 Y cost in Beta? If these countries traded, which would export X and which would export Y?
    b. If Alpha can produce 100 units of X with all its resources or 60 units of Y, how much does 1 unit of Y cost? If Beta can produce 60 X or 40 Y, what does 1 unit of X cost in Beta? Which country has a comparative advantage in the production of X?
3. How can two countries simultaneously gain from trade? Under what circumstances can two countries not gain from trade?
4. What is the difference between natural barriers and artificial barriers to trade? Was the development of the Panama Canal an artificial encouragement of trade?
5. What does a tariff do to the terms of trade? Are consumers in both countries hurt by a tariff?
6. What are the differences and similarities between tariffs and quotas? Would consumers prefer one of these barriers to the other?
7. Who is hurt by a tariff? Who is helped? Who encourages government to impose tariffs?
8. What is the purpose of protective tariffs? Who or what is protected?
9. Evaluate the following statement: "Tariffs discourage the movement of goods between countries; therefore, they encourage the movement of resources such as capital and labor between countries."

## When Is a Rose Not a Rose? The Flower Industry Seeks Protection

Changes in technology and in resource prices will often alter comparative advantage in surprising ways. Consider conditions in the American flower industry, particularly among flower and plant growers.

Imported plants and cut flowers were a mere trickle in 1973 compared to what they were in recent years. Colombia, Mexico, the Netherlands, and Israel have made big incursions into U.S. wholesale and retail markets for fresh cut flowers (especially roses, carnations, and chrysanthemums), and Belgium, Denmark, the Netherlands, Israel, and Costa Rica have done the same in the plant market. Colombia, America's largest foreign source of cut flowers, exported almost 73 million roses to the United States in 1982 (an increase of almost 25 percent over the previous year) and half a billion carnations. In fact, cut flowers were second only to coffee as Colombia's leading legal export item in 1982. Israel has also come on strong as a rose supplier to the United States with the aid of Israeli government subsidies.

A number of reasons can be offered for the lower prices and increased supplies of cut flowers and plants from foreign sources. The energy crisis of the 1970s raised the price of heating greenhouses, putting a cost squeeze on many domestic growers, especially those in the colder climates of North America. (Rose growing is presently centered in California and other southwestern states for economic reasons.) Under such conditions, nations such as Colombia have a new source of comparative advantage in flower growing—no heating of greenhouses is necessary. A major Colombian flower grower said in 1983, "We work with what we have and only use varieties that will grow under our conditions. Even though OPEC prices are down, not heating is our only advantage."[a]

A second reason for the new strength of imports is the technical possibility of avoiding historical restrictions on the importation of plants potted in soil. Modern techniques permit the cultivation of plants in such nonsoil media as spagnum moss, unused peat, plastic particles, glass wool, and inorganic fibers. Dutch growers of such common plants as ferns and begonias have become especially adept at these new techniques for commercial propagation. Such plants (limited to seven varieties in mid-1983) are allowed into the United States under a U.S. Department of Argiculture (USDA) regulation called Quarantine 37. Under Quarantine 37, plants grown under very rigid specifications in foreign countries may be preinspected by USDA teams in the foreign country and admitted into U.S.

---

[a] Quoted in Barbara Bader, "How the Colombians Do It," *Florists' Review* 172 (April 14, 1983), p. 56.

markets. The plant-exporting nations (plus additional potential entrants) are asking for an expansion of permission to export more than forty additional varieties to the United States in nonsoil growing media.[b]

Predictably, the protectionists have drawn battle lines declaring "cheap foreign labor," "inferior quality products," "unemployment in the flower industry," "foreign government subsidies," and "health hazards through damaging insect pests." Leading the fight is the Washington-based lobby group of the industry called the Society of American Florists (SAF).

None of the reasons given by the flower industry, except the possible problem of the importation of damaging insects, holds a drop of water when compared to the benefits of free trade. If Colombian roses are of inferior quality, as U.S. growers allege, the market soon adjusts prices and quantities accordingly. (In fact, Colombia's initial stock of the Visa rose, the hearty mainstay of its export trade, had to be imported from France because of a refusal of American producers to trade.) If other nations wish to subsidize exports of flowers to the United States, American consumers gain.

How have the domestic flower growers fared? The International Trade Administration of the U.S. Department of Commerce directed that a tariff of about $11\frac{1}{2}$ percent of import value be imposed on Israeli roses to counteract the Israeli government's subsidy to promote flower exports. The lobbying efforts of SAF to impose further tariffs or quotas have concentrated on Congress and the USDA. The USDA is in a position to impose absolute and prohibitive quotas or restrictions on imported flowers and plants to prevent possible health hazards.

How much protection will be supplied in the flower industry? Politicians and government agencies have, in their own self-interest, been willing to grant protection to businesses before. But why should one small group of producers be subsidized at the larger costs of consumer welfare?

## Question

Suppose that the elimination of all quotas and tariffs on the United States imports of Japanese automobiles would reduce employment in the American automobile industry by 200,000 jobs. Would you support the immediate elimination of all trade restrictions? Would you support a gradual elimination? Would you support the implementation of re-training programs and subsidies to auto workers?

---

[b]Mike Branch, "Quarantine 37: What It Is, How It Works, What It Means," *Florida Foliage* (June 1983), pp. 9–12.

# 22

# The International Monetary System

When a car dealer in Virginia buys cars from a Detroit manufacturer to sell in Virginia, both use the same medium of exchange—U.S. dollars. Suppose, however, that the dealer is buying cars from Japanese or West German manufacturers to sell in Virginia. The Japanese firm will likely want payment in yen; the German firm, in marks. The car dealer thus confronts an international monetary problem. Are the prices stated in yen and marks fair? How are dollars converted into yen and marks? Is there a fixed rate for currency exchange, or does it vary from day to day or even from hour to hour? How would changes in exchange rates affect the price of the cars when they are sold in the United States? U.S. exporters face the same questions from a different perspective. They sell to citizens in foreign countries and seek payment for their goods and services in U.S. dollars. They must be careful, therefore, to pay attention to exchange rates because these rates can determine the prices and profits of their foreign sales.

In Chapter 21 we looked at the exchange of goods and services through foreign markets. The real, as opposed to monetary, side of the international economy concerns such topics as comparative advantage, resource specialization, the level of imports and exports between countries, and artificial barriers to free trade.

The monetary, or financial, side of international trade concerns the ways in which countries pay for the goods and services they exchange. In this chapter we examine many aspects of the relations among the currencies of the world, relations that reflect a kaleidoscope of changing conditions in each country, from interest rates and inflation rates to exports and imports of goods and capital. We look first at the surface—the relative values of national currencies—and then examine the many factors underlying these relative values.

## The Foreign Exchange Market

**Foreign exchange:** The currencies of other countries that are demanded and supplied to conduct international transactions.

**Foreign exchange markets:** The institutions through which foreign exchange is bought and sold.

International sellers and buyers usually prefer to deal in the currency of their own country. American sellers prefer U.S. dollars for their products, and Japanese sellers prefer yen. The currency of another country that is required to make payment in an international transaction is called **foreign exchange.** Foreign exchange is bought and sold in **foreign exchange markets,** usually made up of large brokers and banks around the world.

Sometimes consumers demand foreign exchange; for example, when travelers land in a foreign airport, they often exchange dollars for the local currency. However, large firms dealing in international transactions simply keep bank deposits in foreign currencies to cover their foreign transactions. An American importer of French wine will likely pay for it by writing checks on a large American bank that holds an account in a French bank. The American bank will use the importer's dollars to purchase the francs needed for payment.

The demand for foreign exchange arises because a country's residents want to buy foreign goods. Conversely, the supply of foreign exchange arises because foreign customers want to buy U.S. goods. For example, we have a demand for francs by the U.S. importer who wishes to purchase French wine and a supply of francs by French customers who wish to purchase U.S. personal computers.

### Rates of Exchange

**Foreign exchange rate:** The price of one country's currency stated in terms of another; for example, a dollar price of German marks of $0.37 means that an individual can buy one mark for $0.37.

The relation between dollars and francs is determined by the **foreign exchange rate:** the price of one currency in terms of another. These rates change daily and even hourly on the foreign exchange market. For example, on April 29, 1985, the value of the Philippine peso expressed in U.S. dollars was approximately $0.05, the Austrian schilling $0.05, the Canadian dollar $0.73, the British pound $1.23, the French franc $0.10, the Swedish krona $0.11, and the Mexican peso $0.004. (Figures have been rounded off.) On the same day, the exchange value of one U.S. dollar was approximately 252 Japanese yen, 1984 Italian lira, 952 Israeli shekels, 12 Indian rupees, and 4800 Brazilian cruzeiros.

For U.S. citizens, foreign exchange rates are used to convert foreign prices into U.S. prices and vice versa. Suppose that a bottle of French wine costs 1000 francs. Using the exchange rates just given, how expensive is the bottle of wine to the U.S. purchaser? The exchange rate of francs for dollars, expressed as the cost of 1 franc in U.S. dollars, is $0.10. The 1000-franc bottle of wine therefore costs the American connoisseur approximately $100.00.

**Floating exchange rate system:** An international monetary system in which exchange rates are set by the forces of demand and supply with minimal government intervention.

When the exchange rate is expressed in reverse, showing how much foreign currency can be purchased for a dollar, the foreign price is divided by the exchange rate to determine U.S. price. Suppose that a restaurant in Rome offers all the pasta you can eat for 10,000 lira. From the exchange rates given above, 1984 lira exchange for $1.00. The meal would thus cost you 10,000 ÷ 1984 = $5.04. As a final example of exchange rate conversion, suppose that a Swedish-made Volvo costs 100,000 krona if bought in the U.S. and 80,000 krona if bought at the plant in Sweden. How much do you save by buying the Volvo in Sweden? Since the price of the krona is $0.11, the Volvo would cost about $11,000 if purchased in the United States ($0.11 × 100,000 = $11,000) and $8800 if purchased in Sweden ($0.11 × 80,000 = $8800). You could save $2,200 by buying the Volvo in Sweden

**Fixed exchange rate system:** An international monetary system in which each country's currency is set at a fixed level relative to other currencies, and this fixed level is defended by government intervention in the foreign exchange market.

**Managed floating rate system:** A system in which a country's currency is allowed to float freely in the foreign exchange market, within certain bounds; drastic changes in the value of the currency are mitigated by central bank intervention in the foreign exchange market.

(although when transportation costs and tariffs are added, the car may be cheaper in the United States).

## How Are Exchange Rates Determined?

Exchange rates can be determined basically in two ways. The first is a **floating** or flexible **exchange rate system.** Under a floating exchange rate system, foreign exchange markets are allowed to operate without government intervention. A country's currency price is set by the forces of supply and demand in the foreign exchange market. The second case is a **fixed exchange rate system.** Under this system, a government intervenes in the foreign exchange market and tries to set, or fix, a price for its currency.

The present international monetary system is a blend of these two systems. Floating exchange rates are used in the United States and the other major industrial nations. Some smaller countries maintain fixed rates against such major currencies as the U.S. dollar and the English pound. No country, however, allows its exchange rate to float freely all the time. Wide swings in the exchange rate are controlled by government intervention in the currency market. For this reason the present international monetary system is called a **managed floating system.** In the next two sections we analyze how floating exchange rates and fixed exchange rates work.

## Floating Exchange Rates

As we pointed out earlier, the U.S. demand for foreign currency arises because U.S. citizens wish to buy foreign products, travel in foreign countries, invest in foreign companies, and carry on other international activities. The U.S. supply of foreign currencies arises because foreigners want to buy U.S. goods, travel in the United States, send their children to school in the United States, invest in this country, and so on. All of these forces of supply and demand affect the exchange rate between the U.S. dollar and other currencies. In effect, they form the basis of the supply and demand schedules for U.S. dollars.

### Exchange Rates Between Two Countries

We begin our analysis with a simple model of the determination of floating or flexible exchange rates. For simplicity we assume that foreign exchange takes place only between two countries, the United States and West Germany. The U.S. demand for foreign exchange is thus a demand for German marks; the supply of foreign exchange is a supply of German marks.

Figure 22–1 illustrates the U.S. demand and supply of marks. The vertical axis measures the exchange value of marks in U.S. dollars; the horizontal axis measures the quantity of marks supplied in the foreign exchange market. Like any other demand curve, the demand curve for marks is negatively sloped. Remember that Americans demand marks to buy West German goods. If the price that Americans pay for the mark falls, more marks can be purchased for a dollar. Suppose that the exchange value of a mark falls from \$0.37 to \$0.20 and American importers are buying German cameras that cost 100 marks. The old price of the cameras was $\$0.37 \times 100 = \$37.00$; when the value of the mark falls, the new price is $\$0.20 \times 100 = \$20.00$. Lower prices of the mark mean lower prices to U.S. purchasers of German goods. The lower the dollar price of marks, other things being equal, the greater the quantity demanded of German goods by U.S. consumers and the

### FIGURE 22–1
### A Two-Country Foreign Exchange Market

The vertical axis shows the exchange value of marks—how many dollars it takes to purchase one mark; the horizontal axis shows the quantity of marks available at various prices. The equilibrium price of marks is $0.37. At higher prices, such as $0.50, there is an excess supply of marks, which drives their price down. At lower prices, such as $0.20, there is an excess demand, which drives their prices up.

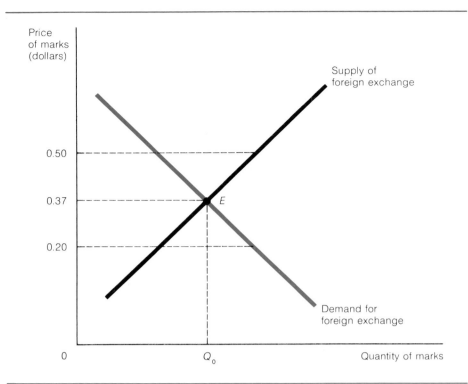

greater the quantity demanded of marks to make the additional purchases. The demand curve for foreign exchange is therefore negatively sloped, like the demand curve for any other economic good.

The supply curve of marks to the foreign exchange market is positively sloped, as drawn in Figure 22–1. As the price of the mark rises, the supply of marks on foreign exchange markets increases. Suppose the exchange value of marks rises from $0.37 to $0.50 and German consumers are buying American beef that costs $5.00 a pound. The old price of beef was $5.00 ÷ $0.37 = 13.51 marks; the new price is $5.00 ÷ $0.50 = 10 marks. As a result of the higher price of marks, German consumers face lower prices for U.S. goods such as beef. Germans will therefore increase their purchases of dollars to buy the now-cheaper U.S. goods. As they do so, the quantity supplied of marks is increased. The supply of foreign exchange is therefore positively sloped, like the supply curve of other economic goods.

The price of foreign exchange, or the exchange rate, reaches equilibrium where the demand and supply curves intersect. In Figure 22–1 the dollar price of marks is $0.37 at the point where the demand and supply curves of marks are equal, E, and a given quantity of marks are exchanged at this price. At prices below $0.37, such as $0.20, there is an excess demand for marks. Under these conditions pressure will be put on the dollar price of the mark to rise. There will be competition among U.S. importers of German goods over the relatively scarce supply of marks. As this competition causes the dollar price of marks to rise, the price that U.S. citizens pay for German goods will also rise, choking off U.S. imports, while the price Germans pay for U.S. goods will fall, stimulating U.S. exports and returning the exchange rate to equilibrium. Similarly, if the dollar price of marks is $0.50, there is an excess supply of marks, and the price of marks will be bid down to its equilibrium level of $0.37, just as it would be in any other competitive market.

### Changes in Floating Exchange Rates

As the dollar price of marks changes, the quantity demanded and quantity supplied of marks also change. A change in the price of a good or service, in this case the price of foreign exchange, leads to movements along given demand and supply curves. This is the familiar other things being equal experiment by which all demand and supply curves are determined. In addition to movements along given demand and supply curves, such as those analyzed in Figure 22–1, we can investigate changes in demand or supply that lead to shifts in the curves.

**Appreciation and Depreciation.** An **appreciation** of a country's currency is an increase in its foreign exchange value. If it took 2 German marks yesterday and 3 German marks today to purchase 1 U.S. dollar, we say that the dollar has appreciated relative to the mark. This means that the price of U.S. goods abroad has increased. A Mack truck priced at $50,000 cost 100,000 marks yesterday. Today, it costs 150,000 marks. A **depreciation** of a country's currency means that its foreign exchange value has fallen. If yesterday it took 2 German marks to purchase 1 dollar and today it took only 1½ German marks, the U.S. dollar has depreciated. Yesterday's Mack truck cost 100,000 marks in Germany; today it costs 75,000 marks.

We now turn to the analysis of why exchange rates change, that is, why currencies appreciate and depreciate in value. In economic terms this means that we are going to investigate factors that shift the supply and demand curves for foreign exchange.

**Inflation.** Inflation is an important factor causing exchange rates to change. Suppose that the United States goes through an inflationary period during which prices rise by 30 percent while prices in other countries are stable and unchanged. In this situation, U.S. goods become more expensive relative to foreign goods. U.S. consumers will increase their purchases of foreign goods, and foreign consumers will reduce their purchases of U.S. goods. In other words, U.S. imports will increase and exports will decrease.

Figure 22–2 shows what happens in the foreign exchange market as a consequence of high U.S. inflation. To purchase more imports, U.S. demand for foreign exchange (marks) shifts right, from $D$ to $D_1$. This shift causes the value of the dollar to depreciate and the value of the mark to appreciate. Moreover, these effects are reinforced as West Germans reduce their supply of foreign exchange from $S$ to $S_1$, reflecting their lower demand for dollars with which to buy more expensive U.S. goods. On both counts, inflation causes the value of the dollar to depreciate. Generally, if a country has inflation rates higher than its trading partner, the former country's currency will depreciate. Conversely, lower inflation will cause currency to appreciate.

According to the **purchasing power parity theory,** propounded by Gustav Cassel around 1917, the change in the exchange value of currency as a result of inflation allows international trade to go forward in the face of inflation. The depreciation of a currency is an adjustment in its purchasing power to account for the effect of inflation. Thus, a country with a high rate of inflation can still trade with other countries because its exchange rate depreciates to reflect the relative purchasing power of its currency. If the United States has an annual inflation rate of 100 percent and the value of the dollar in

---

**Appreciation:** The rise in the value of one currency relative to another.

**Depreciation:** The fall in the value of one currency relative to another.

**Purchasing power parity theory:** A theory of exchange rate determination that states that differential inflation rates across countries affect the level of exchange rates.

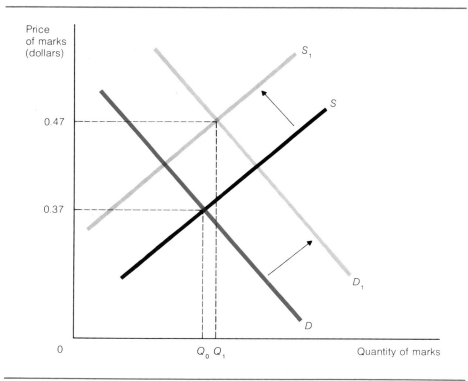

**FIGURE 22–2   Inflation and a Floating Exchange Rate System**

A higher inflation rate in the United States causes the demand for marks to shift from $D$ to $D_1$ and the supply of marks to shift from $S$ to $S_1$. These shifts drive the dollar price of marks up to $0.47. This change in the dollar price of marks reflects the desire of Americans to buy more German goods, which are now cheaper, and the desire of Germans to buy fewer American goods, which are now more expensive, in both cases because the United States has a higher inflation rate than West Germany.

exchange for marks falls by half, as the purchasing power parity theory predicts it will, then the prices of American goods will be unchanged to Germans. For example, as a result of inflation in the United States, the Mack truck that cost $50,000 rises to $100,000. The purchasing power parity theory predicts that the exchange value of the dollar will fall from 2 German marks to 1 German mark as a result of the same inflation. In this event the Mack truck will continue to cost Germans $50,000. The exchange rate adjusts to reflect the purchasing power of the dollar, leaving international trade undisturbed.

This theory works well in predicting the direction that exchange rates will take when there are large differences in inflation rates across countries. However, when inflation rates do not differ by much, movements of exchange rates are determined by other important factors such as the pattern of trade, capital movements, and interest rates.

**Changes in Exports and Imports.** Changes in the value of a country's exports or imports can lead to changes in its exchange rate. For example, the value of a nation's exports may increase relative to its imports because consumers' tastes shift to favor the former country's goods. For instance, U.S. tobacco products might become popular in Japan. Alternatively, a country might achieve comparative advantage in the production of certain products such as personal computers. For such reasons, the value of a country's exports can increase relative to the value of its imports.

Consequently, the country's exchange rate will appreciate, as Figure 22–3 illustrates. Suppose that West Germans increase their demand for personal computers produced in the United States. The supply of marks needed to get dollars with which to buy the computers will increase from S to $S_1$. The dollar therefore appreciates in value relative to the mark; that is, it now takes fewer dollars to purchase a mark ($0.20 versus $0.37). This appreciation of the dollar will raise U.S. prices, thereby stimulating U.S. imports, dampening U.S. exports, and restoring equilibrium in the foreign exchange market.

Obviously, this process can be reversed, and the value of a nation's imports can increase relative to its exports. Americans may choose to buy more foreign goods as their income rises. In this case, the U.S. demand for foreign exchange would increase, leading to a depreciation of the dollar. This depreciation would lead to a restoration of equilibrium in the foreign exchange market by stimulating U.S. exports and dampening U.S. imports.

**Changes in Interest Rates.** So far we have discussed foreign transactions as if only goods and services are exchanged among countries. In addition to these international transactions, financial capital flows between countries, and this flow can change. For example, short-run financial investments are sensitive to differences in interest rates across countries. Thus, financial capital will shift between countries as investors seek the highest rate of return on their money. How do these movements of money affect exchange rates?

Suppose that U.S. short-run interest rates rise above those in Western

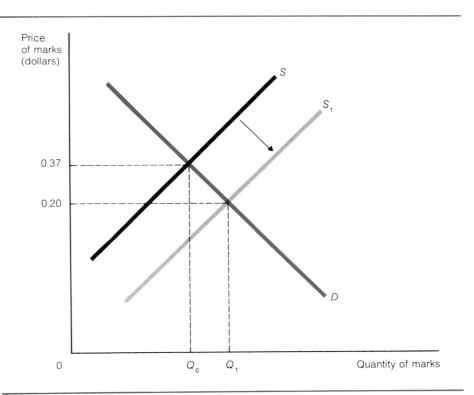

**FIGURE 22–3**

**Export and Import Changes**

The German demand for U.S. computers increases, and the value of U.S. exports rises relative to its imports. The supply of marks therefore increases to $S_1$, and the value of the dollar appreciates. That is, a dollar is now worth more marks.

## FIGURE 22-4

### A Shift in Foreign Investment

As Americans invest their money abroad, the demand curve for foreign exchange increases, shifting from $D$ to $D_1$. More marks are required to facilitate investments in West Germany. Foreign investment is thus associated with a depreciation of the U.S. dollar, which in turn will stimulate U.S. exports.

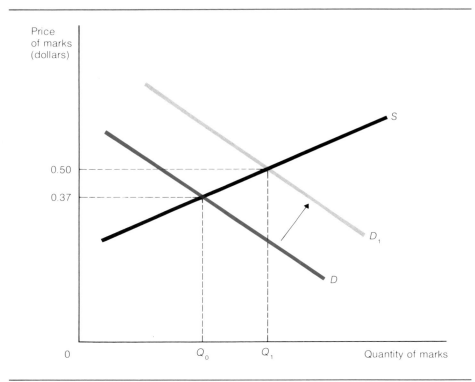

Europe. Then short-run financial investment will increase in the United States as Europeans seek to lend money in the United States where interest rates are higher. To make these loans, the Europeans will supply marks, francs, pounds, and so on to obtain dollars. This means that the dollar's exchange value will appreciate in response to higher U.S. interest rates.

The opposite analysis holds when U.S. interest rates are below those prevailing in other countries. Then short-run capital would flow out of the U.S. as investors sought higher returns in other countries. The demand for the currencies of these other countries would increase, and the dollar's exchange value would depreciate.

In Figure 22-4, the demand curve, $D$, represents the demand of U.S. citizens for foreign exchange to import foreign goods. When interest rates in foreign countries rise above those in the United States, demand for foreign exchange to make loans to those countries shifts from $D$ to $D_1$. As a result, the dollar price of foreign exchange rises, and the dollar depreciates in exchange value.

## Fixed Exchange Rates

As we have seen, a system of floating exchange rates allows the foreign exchange market to determine the prices at which currency will change hands. By contrast, a fixed exchange rate system does not allow exchange rates to float in a free market. Government intervenes in the foreign exchange market to fix the international price of its currency. Once fixed exchange rates are established, governments stand ready to protect the rates through intervention in the foreign exchange market.

## The Operation of a Fixed Exchange Rate System

Although the United States and West Germany no longer trade on the basis of a fixed exchange rate, imagine what would happen if they did rather than letting the rate float to accommodate shifts in the market. Figure 22–5 illustrates what happens when a fixed exchange rate between marks and dollars does not happen to correspond to ever-shifting market conditions. If the current market equilibrium rate is $0.37 per mark but the U.S. government seeks to maintain an exchange rate of $0.50 per mark, the official or fixed rate is above the equilibrium rate. In this case, the dollar is undervalued relative to its current equilibrium value, and the mark is overvalued.

Without intervention, this situation would lead to an excess supply of marks in the foreign exchange markets, creating pressure on the dollar price of marks to fall to the equilibrium value of $0.37. Instead, to maintain the fixed rate, the U.S. government would act to purchase the excess supply of marks. To do so, the central bank (the Federal Reserve System) would build up its holdings of marks. In terms of imports and exports, the undervalued dollar and the overvalued mark make U.S. goods a good buy and German goods a poor buy. This situation basically leads to a surplus of exports over imports for the United States in its trade with West Germany.

The fixed exchange rate can also be set below the point where the de-

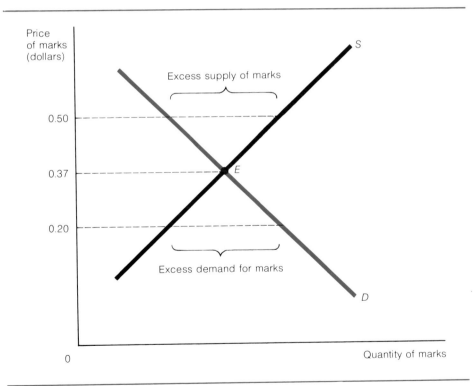

**FIGURE 22–5  Fixed Exchange Rates Above and Below Equilibrium**

Governments can interfere with floating foreign exchange markets and set official prices for their currencies. The equilibrium rate of dollars for marks is $0.37. An official fixed rate of $0.50 leads to an excess supply of marks, an undervalued dollar, and a U.S. surplus of exports over imports. An official rate of $0.20 leads to an excess demand for marks, an overvalued dollar, and a U.S. excess of imports over exports.

mand and supply of foreign exchange intersect. In this case the government would face the problem of keeping the value of the dollar from falling because there would be an excess demand for marks (an excess supply of dollars) in the foreign exchange market. The government would defend the fixed rate by increasing the supply of marks to the foreign exchange market from its holdings of official reserves, built up perhaps from its purchases when the official rate was above the equilibrium rate. This action would allow the U.S. dollars to be overvalued and the mark to be undervalued relative to their equilibrium values, and the imbalance would lead to an excess of imports over exports for the United States.

### Changes in Fixed Exchange Rates

The fixed exchange rate set by government can stray significantly from the equilibrium exchange rate. A country will therefore experience persistent problems in international trade; namely, it will confront persistent surpluses or deficits in its foreign trade. In the balance of trade between exports and imports, a country whose exports have greater value than its imports is said to have a balance of trade surplus; one whose imports' value exceed its exports' value has a balance of trade deficit. What are the options under a fixed-rate system for alleviating such situations?

**Changes in Domestic Macroeconomic Policy.** Consider first the example of Figure 22–5, where the dollar price of the mark is maintained at $0.50 while the equilibrium price is $0.37. The dollar is undervalued, and the United States experiences a balance of trade surplus, while West Germany runs a persistent deficit, importing more than it exports. The Germans will demand that something be done.

One option is to change domestic macroeconomic policies in the two countries. Suppose the United States allows its money supply to rise and West Germany lets its money supply fall. In the United States the increase in money supply will lead to higher prices and costs and to a higher level of GNP. In West Germany the opposite effects will occur: There will be lower prices and costs and a lower level of GNP. Thus, U.S. exports will fall because they are now relatively more expensive on world markets, and U.S. imports will rise because foreign goods are now cheaper and income is higher. In West Germany exports will rise because German goods are now relatively cheaper, and imports will decline because U.S. goods are more expensive and German GNP has fallen. The basic tendency in both countries, then, is toward adjustments that erase the surplus of exports over imports in the United States and the corresponding deficit in Germany.

An important point emerges from this discussion. In a fixed exchange rate system, where the government chooses to defend the fixed rate at all costs, imbalances in foreign trade must be resolved through macroeconomic adjustment of entire domestic economies. To keep one price—the foreign exchange rate—fixed, the United States and Germany manipulate the level of all the other prices in their economies. A floating-rate system, by contrast, changes one price—the foreign exchange rate—to resolve balance of trade problems. (The relation between macroeconomic policy and the foreign exchange market is explored in more detail in Focus, "Macroeconomic Policy and the Foreign Exchange Market.")

**Devaluation and Revaluation.** The second option facing two fixed-rate countries with a trade imbalance is to change the official exchange rate. In

## FOCUS: Macroeconomic Policy and the Foreign Exchange Market

Foreign trade can have dramatic effects on a government's macroeconomic policy. We consider briefly how monetary policy and fiscal policy are affected by foreign trade.

### Monetary Policy

The link between the foreign exchange market and monetary policy depends on whether trade takes place under a system of fixed or floating exchange rates. In the former case, the central bank fixes the exchange rate and allows the money supply to be determined by the economic system; in the latter case, the central bank lets the exchange rate be determined in the market for foreign exchange and manipulates the domestic supply of money.

**The Case of Fixed Exchange Rates.** To support a fixed exchange rate, the central bank must be prepared to buy and sell whatever foreign exchange is necessary to keep the exchange rate at the fixed level. Suppose there is a surplus in the balance of trade: At the current exchange rate, the quantity of foreign exchange supplied by exporters exceeds the amount of foreign exchange demanded by importers. In this case the central bank must purchase foreign exchange by an amount equal to the excess supply to eliminate the tendency for the exchange rate to fall. The central bank purchases foreign exchange by printing more domestic currency, and this increase in the supply of money sets forces in motion that tend to restore equilibrium. Other things being equal, domestic prices rise with a larger money supply, and these higher prices will discourage exports and encourage imports, thus diminishing the trade surplus. In the case of a trade deficit, the central bank will be selling foreign exchange (in effect decreasing domestic currency) to keep the exchange rate from rising. This action will lower the domestic money supply, and the accompanying lower domestic prices will tend to discourage imports and encourage exports. The important point is that to fix one price—the exchange rate—the central bank must give up other money supply controls and allow all other prices to adjust.

**The Case of Floating Exchange Rates.** If the government instead is using a floating exchange rate system, the central bank can use its domestic monetary policy tools, allowing the exchange rate to be determined by the market. Starting from equilibrium, an increase in the domestic money supply tends to raise prices, and these higher prices tend to encourage imports and discourage exports. Thus, there is a trade deficit (an excess demand for foreign exchange) at the old exchange rate. For individuals to be willing to supply the extra foreign exchange to finance these trade movements, the exchange rate must rise by whatever amount is necessary to eliminate the excess demand. The result of an increase in the domestic money supply is therefore a rise in the exchange rate by exactly the same proportion.

### Fiscal Policy

The effects of a government's fiscal policy initiatives on exchange rates depend in large part on how the changes in government spending are financed. If we assume, for simplicity, that all increases in government expenditures are financed by borrowing, then expansionary fiscal policies stimulate aggregate demand, including the demand for imports on the one hand, and raise government borrowing on the other. The increased demand puts upward pressure on the exchange rate because of the additional foreign currencies required to finance the increased imports. But the increased government borrowing tends to raise interest rates, attracting foreign capital and placing downward pressure on the exchange rate. The net change in the exchange rate owing to the expansionary policy depends on which of the two effects dominates.

Similarly, contractionary fiscal policies may raise or lower the exchange rate. A lower level of government spending will reduce aggregate demand, reduce the demand for imports, and, at the same time, lower government borrowing. The lower import demand will tend to lower the exchange rate by fostering an excess supply of foreign currencies. But with less government borrowing, interest rates tend to fall. Capital outflows increase, putting upward pressure on exchange rates. Again, the net change of the exchange rate depends on which event—the fall-off in imports or the decline in interest rates—has greater effect.

**Devaluation:** A change in the level of a fixed exchange rate in a downward direction; a depreciation of a fixed rate.

this way, the two governments can avoid costly manipulation of their domestic economies to restore foreign trade equilibrium. There are two possible changes in this regard.

One is **devaluation** of one country's currency. When the dollar is overvalued ($0.20 per mark in Figure 22-5), the United States suffers a trade deficit and West Germany a trade surplus. With West Germany's agreement, the United States can lower the official exchange value of its currency. In this case we say that a devaluation has occurred. In Figure 22-5 this means that the dollar price of marks is moved upward closer to its equilibrium value. This change in the exchange rate changes the relative prices of each country's imports and exports so as to help erase the U.S. deficit and the German surplus without each country's having to resort to costly deflation and inflation of its domestic economy.

Where the official rate is above the equilibrium rate (such as $0.50 per mark in Figure 22-5), the opposite process can be applied. In this case the United States has a trade surplus and West Germany a deficit. The two countries can support this exchange rate if the United States buys marks or if Germany sells dollars or some combination of both. However, to avoid inflation in the United States and deflation in Germany to correct the balance of trade situation, the two countries might agree to revalue the dollar. A **revaluation** occurs when the official price of a currency is raised. This means that the dollar price of marks would move downward in Figure 22-5, closer to its equilibrium value. The value of the dollar is thus raised, or revalued, and the corresponding value of the mark is lowered, or devalued.

**Revaluation:** A change in the level of a fixed exchange rate in an upward direction; an appreciation of a fixed rate.

Adjustments in official exchange rates seem less costly than the manipulation of whole economies. However, as we will see later, such adjustments are not always easy to accomplish. The Germans may reap advantages from a balance of trade surplus and resist efforts to devalue the dollar or to run their domestic economy so as to see this surplus dissipate. These are obviously complex matters of international negotiation that are inherent in a fixed-rate system.

**Changes in Trade Policy.** A third option for resolving balance of trade difficulties under a fixed-rate system is for a deficit country to erect barriers to international trade. Tariffs, quotas, and other barriers to the free movement of people and goods across international borders can be established to try to solve a trade problem. This is not an option that would find much favor among economists, however. As we learned in Chapter 21, such impediments to international trade cost the worldwide economy in terms of lost economic efficiency and specialization gains.

Both the floating and fixed exchange rate systems have been used in the past. The major countries presently operate essentially on a floating system and have done so since early 1973. Before that, starting shortly after World War II, a fixed exchange rate system governed international monetary relations. We will discuss the historical evolution of the international monetary system later in the chapter.

## The Gold Standard

The **gold standard** is a type of fixed exchange rate system with a long history in international trade. David Hume (1711-1776), a Scottish philosopher and friend of Adam Smith, was the first person to explain how an international gold standard works. The world economy functioned on a gold stan-

**Gold standard:** An international monetary system in which currencies are redeemable at fixed rates in terms of gold.

dard most recently from the 1870s until World War I. Many observers today wistfully call for a return to the gold standard.

Under a gold standard, all trading nations are willing to redeem their currencies for gold. With each currency linked to gold, the precious metal becomes, in effect, a world currency. Suppose that the U.S. dollar is defined as being worth 1/20 ounce of gold. That is, the U.S. Treasury is prepared to redeem the dollars of both domestic and foreign citizens at this fixed rate in gold. The exchange value of an ounce of gold is set at $20.00. Other currencies are also denominated in gold. Let us say that the German mark is fixed at 1/5 ounce of gold. This means that the mark is backed by four times as much gold as the dollar. The U.S. converts dollars at $1 = 1/20 ounce of gold; Germany converts marks at 1 mark = 1/5 ounce of gold. The fixed exchange rate between dollars and marks in this case is $4 = 1 mark. No one would ever pay more than $4 for a German mark under this system. Why? You could buy 1/5 ounce of gold for $4 from the U.S. Treasury and use this gold to buy a mark from the German Treasury. The fixed exchange rate between dollars and marks of $4 = 1 mark is thus guaranteed by such behavior on the part of individuals.[1] Since each country defines its currency in terms of gold, national currencies basically become just gold certificates.

A second key element of the gold standard is that each country links its money supply to its supply or holdings of gold. This link is established as discussed above: Both foreign and domestic citizens are allowed to redeem gold in a country at a fixed rate. This action creates a natural equilibrating mechanism (Hume called it the **price-specie flow mechanism**), which keeps the imports and exports of countries in balance. Suppose that while operating on a gold standard, the United States imports more than it exports over a given period of time. This differential of purchases has to be paid in gold. Since money supplies are tied to gold, the gold payment causes the money supply of the United States to fall, and the money supplies of surplus countries to rise. Prices and costs in the United States would thus fall, and prices and costs in the trade surplus countries would rise. Deflation in the trade deficit countries such as the United States would stimulate exports and dampen imports, and inflation in trade surplus countries would dampen exports and stimulate imports. Balance of trade equilibrium is thus restored in both types of countries under a gold standard.

**Price-specie flow mechanism:** The mechanism by which the gold standard equilibrates the balance of trade between countries.

The gold standard is very much like the fixed exchange rate system discussed in the last section. However, a gold-based standard is not necessary to operate a fixed exchange rate system. What is required is that countries act as if they are on the gold standard. When a country runs a trade surplus, it increases its money supply, and its domestic prices and costs rise; ultimately, exports decrease and imports increase. When a country has a trade deficit, it decreases its money supply, prices and costs fall, and thus exports increase and imports decrease. In both cases a country is required to manipulate its whole economy to maintain equilibrium in its balance of trade with a fixed exchange rate.

The crucial difference between the gold standard and modern fixed-rate systems managed by government intervention in the foreign exchange mar-

---

[1] Obviously, shipping gold between the United States and Germany costs real resources. Thus, the dollar price of marks could vary between, for example, $3.98 and $4.01 if it costs $0.02 to ship 1/5 ounce of gold between the two countries. The values $3.98 and $4.01 are called gold points, or values of the exchange rate below or above which gold will be shipped between the two countries.

ket is that under a gold standard the domestic authorities do not have a choice about changing the money supply, thereby inflating or deflating its currency. No international negotiations are required. The system works automatically. A modern fixed-rate system requires that countries agree to follow the appropriate domestic monetary and fiscal policies or to devalue or revalue the official rates when disequilibrium exchange rates lead to persistent deficits and surpluses. To reach such agreements requires complex and costly international negotiations. Modern fixed-rate systems are thus at the mercy of individual action by each country. Countries may or may not follow the rules of the game. The gold standard left no such discretion to political authorities.

## The Balance of Payments

**Balance of payments:** An official accounting record, following double-entry bookkeeping procedures, of all the foreign transactions of a country's citizens and firms. Exports are entered as credits and imports as debits.

Countries calculate their gross national product to keep track of the level of aggregate production and national income over time. Countries also keep track of the flow of international trade and periodically publish a report of their transactions, called the **balance of payments.** The balance of payments is essentially an accounting record of a nation's foreign business. It contains valuable information about the level of exports, imports, foreign investment, and the transactions of government such as purchases and sales of foreign currencies.

The balance of payments, as an accounting statement, is kept according to the principles of double-entry bookkeeping. The concept of double-entry bookkeeping is simple: Each entry on the credit side of the ledger implies an equal entry on the debit side.[2] Thus, each international transaction creates both a debit ($-$) and a credit ($+$) item in the balance of payments ledger.

There is a simple rule to follow in classifying debits and credits. Any foreign transaction that leads to a demand for foreign currencies (or a supply of dollars) is treated as a debit, or a minus, item. The act of importing is a debit entry in the balance of payments. Any foreign transaction that leads to a demand for dollars (or a supply of foreign currency) is entered as a credit, or a plus, item. The act of exporting is a credit entry in the balance of payments.

Suppose a U.S. firm exports computers to France, and the French importer of the computers pays for the purchase with an IOU. The French IOU is entered on the credit side in U.S. balance of payments bookkeeping because the French importer must ultimately demand dollars to pay off its IOU to the U.S. firm. The sale of computers is entered on the debit side as the offsetting part of the transaction. The U.S. firm must give up a computer in return for the French importer's payment. Since each transaction implies an equal debit and credit, the balance of payments must always balance. In this respect, the balance of payment is like any other accounting balance sheet.

Of course, although the total balance of payments of a country must always balance (debits minus credits equal zero), its component parts need not balance. For example, the imports and exports of specific merchandise such as automobiles do not have to balance. But overall, surpluses in one part of the balance of payments must be canceled out by deficits in other parts. This is not to say, of course, that countries do not experience balance of payments

[2] Double-entry bookkeeping is discussed in detail in Chapter 7.

## Exports and Imports: The Balance of Trade

Table 22–1 presents the data on the U.S. balance of payments for 1983. Items 1 and 2 represent exports and imports of goods and services. The difference between the sums of these aspects of the balance of payments—exports contrasted with imports—is the **balance of trade.** Note that trade in merchandise is only one facet of this balance. Merchandise trade figures are the result of a country's international trade in physical goods, such as books, cars, planes, and computers—the visible exports and imports. The data in Table 22–1 show that in 1983 merchandise exports (1a) were less than merchandise imports (2a) by $60.6 billion. This figure is sometimes called the merchandise trade balance and is equal to merchandise exports minus merchandise imports. If the figure is positive, there is a merchandise trade surplus. It was negative in 1983, so there was a merchandise trade deficit.

Many imports and exports are invisible. U.S. citizens and firms supply various services to foreigners, such as transportation, insurance, and loans.

> **Balance of trade:** Roughly, the balance of a country's imports and exports of goods and services with other countries. If imports are greater than exports, the balance of trade is in deficit; if exports exceed imports, the balance is in surplus.

problems. Problems arise, as we will see, from persistent surpluses or deficits in the component parts of the balance of payments.

### TABLE 22–1  U.S. Balance of Payments, 1983

U.S. exports enter the balance of payments with plus signs because buyers of these exports must pay U.S. firms with dollars. Dollars thus flow into the United States and are recorded as a plus, or credit, in the balance of payments. U.S. imports receive a minus sign, indicating that to pay for foreign goods U.S. buyers must supply dollars to foreign countries. Dollars thus flow out of the United States, and imports are then treated as a minus, or a debit, in the balance of payments. Similarly, outflows of U.S. capital and inflows of foreign capital are treated as debits and credits, respectively. To classify an item as a debit or a credit, think of whether dollars are leaving or entering the country. If they are leaving, the item is a debit; if they are entering, the item is a credit.

| Item | | Amount (billions of dollars) |
|---|---|---|
| 1. Exports of goods and services | | +332.2 |
|    a. Merchandise exports (including military sales) | +212.9 | |
|    b. Services | + 42.5 | |
|    c. Income from U.S. assets abroad | + 76.9 | |
| 2. Imports of goods and services | | −365.1 |
|    a. Merchandise imports (including military purchases) | −273.5 | |
|    b. Services | − 37.7 | |
|    c. Income from foreign assets in U.S. | − 53.4 | |
| 3. Net unilateral transfers abroad | | − 8.6 |
|    a. U.S. government grants and pensions | − 7.6 | |
|    b. Private remittances | − 1 | |
| 4. Current account balance | | − 41.5 |
| 5. Net capital movements | | + 32.3 |
|    a. Outflow of U.S. capital | − 49.4 | |
|    b. Inflow of foreign capital | + 81.7 | |
| 6. Statistical discrepancy | | + 9.3 |
| 7. Increase (−) in U.S. official reserve assets | | − 1.2 |
| 8. Increase (+) in foreign official assets in U.S. | | + 5.1 |
| 9. Total | | 0 |

*Source:* U.S. Commerce Department, *Survey of Current Business* (December 1984), p. 47. Some categories may not add exactly due to rounding error.

Payments are received for these services just as in the case of visible exports. Items 1b and 1c and 2b and 2c represent invisible exports and imports in the balance of payments. When both visible and invisible exports and imports are considered in Table 22-1, U.S. exports were substantially less than U.S. imports, yielding a trade deficit of $32.9 billion.

### Net Transfers Abroad

Net unilateral transfers abroad are one-way money payments from the United States to foreigners or United States citizens living abroad. Included in this category are such items as foreign aid and pension checks to retired U.S. citizens living abroad. Nothing tangible comes back to the United States for these transfers, but they do give rise to a demand for foreign exchange. They thus are entered as a debit item in the balance of payments. In 1983, the United States made $8.6 billion worth of such transfers.

### Current Account Balance

The balance on current account is equal to exports of goods and services minus imports of goods and services minus net unilateral transfers abroad. The United States had a deficit on current account in 1983 of $41.5 billion.

### Net Capital Movements

When U.S. citizens invest in foreign stocks and bonds, they hold a claim to foreign capital (buildings, factories, and so forth), and capital that would otherwise be invested in the United States is exported. When foreign citizens buy the stock and bonds of U.S. companies, capital is imported in the United States. In 1983, capital outflows were $49.4 billion.

Note what happens in bookkeeping terms when capital flows into or out of a country. Capital outflows are treated as a debit item in the balance of payments because they give rise to a demand for foreign exchange with which to make foreign investments. Capital inflows are a credit item because they increase the supply of foreign exchange traded for dollars with which to buy interests in U.S. firms.

Capital outflows exceeded capital inflows by $32.3 billion in 1983. This net outflow of capital helps explain why item 1c exceeds 2c in Table 22-1. The fact that the United States sends out more capital than it takes in is also reflected in the fact that the United States earns more on foreign investments than foreigners earn on investments in the United States.

### Statistical Discrepancy

Item 6 in Table 22-1 is called statistical discrepancy, an accounting fudge factor. When all the data were collected and debits and credits computed, the debits outweighed the credits by $9.3 billion. The statistical discrepancy is added to make the balance of payments balance by compensating for imperfections in data gathering. Some credits have apparently gone unrecorded in computing the balance of payments. These could include hidden exports or unrecorded capital inflows. Most experts think that the latter is the source of most of the discrepancy, which is clearly quite large.

### Transactions in Official Reserves

Items 7 and 8 reflect the activities of government in the foreign exchange market. The official reserve assets of a country are its holdings of gold, foreign exchange, and any other financial assets held by official agencies, such as the Federal Reserve System. Also included in reserve assets are special drawing rights (SDRs) with the International Monetary Fund, an international central bank located in Washington, D.C. SDRs are supplementary reserves of purchasing power that a country can draw on to pay international debts. SDRs are simply bookkeeping entries that member countries may draw against to settle an international payments deficit. They are not money in any physical sense, but SDRs may be used as a means of payment. Suppose that Mexico experiences a balance of payments deficit and is without dollars to repay Canada for its purchases of capital goods. Mexico, as a member of the IMF, may opt to settle the debt by transferring some of its special drawing rights credit to Canada. In this way international reserves and liquidity are increased without any actual transfer of funds. After the transfer, Canada may utilize the increased credit to settle some international obligations.

A government can finance an excess of imports over exports by drawing down its holdings of official reserves. Countries receiving surpluses may want to add to their official reserves. As we will see, during the 1947–1973 period of fixed exchange rates, these reserve accounts were quite important because countries used the reserves to stabilize their exchange rates. Deficit countries used reserves to defend the value of their currency, and surplus countries accumulated valuable international reserves.

Item 7 is the increase in U.S. holdings of official reserve assets. In 1983 these holdings increased by $1.2 billion. This item enters the balance of payments as a debit because an increase in official reserve holdings means an increase in the demand for foreign exchange. Item 7 is thus treated like a merchandise import.

Item 8 is the increase in investments of official agencies of foreign countries, such as the OPEC countries, in the United States. Since these investments represent demand for dollars, they are treated as a credit item in the balance of payments.

In 1983, the United States increased its net position in international reserves by $3.9 billion, the sum of items 7 and 8.

## The Balance of Payments and the Value of the Dollar

The last entry in Table 22–1 is zero, indicating that the balance of payments must balance. This is the way any double-entry accounting system works out.

The balance of payments, however, is an aggregate record of a country's international payments for one year. Behind this record are the myriad transactions of U.S. citizens and firms with foreign citizens and firms. These transactions determine how the U.S. dollar fares in the foreign exchange market. If, over time, the value of the dollar rises against other currencies, we say that the dollar is strong. This is precisely what happened over the period 1982–1985. The dollar has appreciated in the range of 10 to 15 percent against most major foreign currencies. Most observers feel that this appreciation of the dollar reflects high U.S. interest rates that have attracted an inflow of foreign capital into the United States. (The important role of short-run capital flows and interest rates in the balance of payments is treated in Economics in Action at the end of the chapter.) This capital inflow leads to a higher demand for the dollar and thus to its appreciation. There are obviously costs and benefits associated with large inflows of foreign capital into a country. Foreigners thereby own more U.S. assets and firms, but the United States has an expanded supply of capital as a consequence. The important point is that the dollar, for such reasons, appreciated in currency markets during the early 1980s, and though this fact is not directly reflected in the balance of payments, it indicates that the U.S. foreign economic position might loosely be described as a surplus. That is, the impact of all our foreign transactions over time has been to cause the dollar to rise in value.

# The Evolution of International Monetary Institutions

Major world trading nations today rely on floating exchange rates. These governments will occasionally intervene in the foreign exchange market when exchange rate changes are significant, but generally currency prices are free to adjust according to the forces of supply and demand. We briefly review the developments over the past hundred years that led the major trading nations to adopt floating exchange rates in 1973.

## The Decline of the Gold Standard

In the period before World War I, roughly dating back to the 1870s, most currencies in the world economy were tied to gold. This means that each country was prepared to redeem its currency at a fixed price in gold to both its own citizens and foreigners. The United States, for example, stood ready to exchange an ounce of gold for $20.67 over this period. In addition, countries on the gold standard linked their money supplies to their holdings of gold bullion.

We have seen how the gold standard works. Basically, it is a fixed exchange rate system in which balance of trade deficits and surpluses are settled by shipments of gold. These shipments cause domestic money supplies and hence domestic national income and employment to fluctuate as a function of the state of the country's trade balance. Deficit countries experience gold outflows and domestic deflation, and surplus countries face gold inflows and domestic inflation. These adjustments in domestic price levels lead to a restoration of equilibrium in the balance of trade by changing the relative prices of a country's imports and exports.

The gold standard worked reasonably well in eliminating balance of trade surpluses and deficits over the period 1870–1914. Nonetheless, for a variety of reasons the gold standard broke down after World War I. There were attempts to return to the gold standard in the 1920s, but the shock to the international economy caused by the Great Depression in the 1930s brought the gold system down once and for all. Most historians point to three basic reasons for the failure of the gold standard.

First, some countries, notably the United States, stopped adhering to gold standard etiquette after World War I. In particular, these countries did not continue to allow their money supplies to be linked to their balance of trade. That is, when gold left the country to settle a trade deficit, a country did not necessarily contract its money supply in proportion to its loss of gold. The gold standard cannot work to restore balance of trade equilibrium under such circumstances.

Second, when Great Britain and France restored the gold standard in the 1920s, they set the "wrong" exchange rates. Britain overvalued the pound, and France undervalued the franc. These actions led to trade surpluses in France and unemployment in England. The pressures on domestic economies caused by inappropriately set exchange rates made it clear to policymakers that it was not worth sacrificing economic stability to the gold standard.

Third, and most important, the Great Depression (1929–1939) ended the movement to return to the gold standard. With worldwide unemployment of resources, countries could no longer accept the discipline of the gold standard, particularly when a balance of trade deficit contributed to unemployment. The gold standard still has a great bit of luster in some circles, however. Focus, "Should We Go Back to the Gold Standard?" discusses the current debate on this issue.

### Bretton Woods and the Postwar System

It was not until after World War II that the problems of the international monetary system were addressed in a concerted way. Negotiators for the free world countries met in Bretton Woods, New Hampshire, in 1944 to develop a new international monetary order. The **Bretton Woods system,** which lasted almost thirty years, consisted of a system of fixed exchange rates and an international central bank, called the **International Monetary Fund (IMF),** to oversee the new system.

**Bretton Woods system:** The fixed-rate international monetary system established among Western countries after World War II.

**International Monetary Fund:** An international organization, established in the Bretton Woods system, designed to oversee the operations of the international monetary system.

We have already discussed the economics of fixed exchange rates. The basic idea of the Bretton Woods system was that over time countries would obtain or pay for their imports with their exports. However, there might be temporary periods over which a country might wish to run a trade deficit to obtain more imports than its current level of exports allowed it to obtain. Under the Bretton Woods system the country could do so by borrowing international reserves from the IMF. Over time, the country was expected to return to a trade surplus out of which the earlier loan of reserves could be repaid.

The IMF was the bank that held the reserves that allowed the system to operate in this way. When the IMF was formed, each member country was required to contribute reserves of its currency to the bank. The bank thus accumulated substantial holdings of dollars, marks, francs, pesos, and so on, and when member nations ran into balance of trade deficits, the bank would lend them reserves to resolve their difficulty. Each time a loan was made, the debtor nation was encouraged to reform its economic policies to avoid future deficits. Sound economic management might also lead to a trade surplus and a source of funds to repay the IMF loan.

The Bretton Woods system sounds fine in principle, and indeed it functioned tolerably well for a number of years. However, as we saw earlier, fixing the price of a currency is like fixing the price of any other good or service—it is very likely to cause surpluses and shortages in the currency market. As the conditions affecting imports and exports across countries change, the demand and supply of currencies will shift. Countries can find themselves with overvalued fixed exchange rates, which means that they will face persistent deficits in their balance of trade. Under the IMF system the deficit country could draw on its IMF reserves to settle the deficits, but it could not do this forever because it would exhaust its reserves. Chronic deficit countries with overvalued exchange rates were thus said to be in fundamental disequilibrium.

In our discussion of the theory of fixed exchange rates, we reviewed the various courses of action that a country could take in such a case. First, it could devalue its currency as a step toward restoring equilibrium in its balance of trade. Once its currency was devalued, the exchange rate would again be fixed and defended. Second, the country could attempt to improve its trade balance by imposing tariff and quota barriers to imports and perhaps subsidizing exports. In other words, the country could move away from free

## FOCUS  Should We Return to a Gold Standard?

The relatively high inflation rates experienced by the United States during the late 1970s and early 1980s led to a variety of proposals for a return to a gold standard. Congress established a U.S. Gold Commission to study the question, and the members not only debated the desirability of again tying the dollar to gold but also considered how to implement a new gold standard.

Proponents of a return to the gold standard argue that the Federal Reserve Board has neither the willingness nor the ability to follow monetary policies that promote price stability and that some constraint on excessive monetary expansionism is necessary. What are the advantages and disadvantages of gold-backed dollars?

### Advantages

Under a gold standard, currency is freely convertible into gold at some fixed rate, such as $35 per ounce, which was for many years the official U.S. price of gold. Thus, a nation's currency supply is directly related to its supply of gold. The supply of currency can be expanded relative to the supply of gold only by devaluing the dollar in terms of gold, that is, by raising gold's dollar price. A gold standard therefore enforces monetary discipline. In the absence of devaluation, the currency supply can be increased only at a rate equal to the increase in the output of gold, an expansion that has occurred historically at about 1.5 to 2 percent annually. Such a monetary policy would foster price stability and little or no inflation in the economy.

### Disadvantages

There are several major drawbacks to a gold standard. First, real resources are tied up in money production—gold must be mined, stored, and transported. Second, the supply of gold is relatively inelastic. Expansions in the demand for money therefore place downward pressure on the general level of prices. Third, even though the dollar price of gold is fixed, a variety of demand-side factors can cause the market price of gold to diverge from its official price. Indeed, it would be the sheerest coincidence for the two prices to be equal. Cooperation among nations is required to maintain the official price in such circumstances. For instance, during the late 1960s, a two-tier system evolved in which a world price of $35 per ounce was supported through sales among Western central banks and by an agreement between governments not to buy or sell gold in the open market. Prior to that time a group of countries had formed a gold pool for the purpose of using their gold reserves to intervene periodically to stabilize the open market price at the official price.

Last, incentives to engage in monetary expansionism are not completely eliminated by a gold standard. Increases in the stock of currency can be purchased for a time by a willingness to suffer a drain on gold reserves. That is, domestic inflation tends to create a trade deficit, which under a gold standard is balanced by shipments of gold to trade surplus nations. However, the inflating country's currency becomes overvalued, and pressures to devalue gradually become irresistible. It was in fact persistent monetary expansion by the United States that ultimately led to the collapse of the earlier world gold standard.

---

trade to balance its imports and exports. Third, the country could behave as if it were on the gold standard. This would mean adopting restrictive monetary and fiscal policies designed to promote domestic deflation and high interest rates in the hope that such changes would restore the balance of trade to equilibrium. This, of course, is a problematic course of action for any country. For example, if the country's unemployment rate were already high, it is hard to believe that the nation's leaders would be sufficiently disciplined to undertake a deflationary course of action.

**Speculation** by buyers and sellers of currencies also undermines a fixed exchange rate system. Suppose Great Britain is running chronic trade deficits, and everyone, including speculators, expects that the pound will be devalued (even though British central bankers will deny such rumors vehemently). In effect, speculators are in a no-lose position. Will they continue to hold pounds? Clearly, they will not; pounds are about to become less

**Speculation:** The buying and selling of currencies with an eye to turning a profit on exchange rate changes, predicted devaluations, and so on.

valuable relative to other currencies. Speculators will sell their pounds for other currencies, increasing the supply of pounds to the foreign exchange market and putting additional downward pressure on the pound. By selling the weak currency and buying a strong currency, speculators make it more difficult for authorities to find a new fixed value for the pound. This is why speculation in a fixed exchange rate system is often termed destabilizing.

### The Role of the Dollar in the Bretton Woods System

The U.S. dollar became the centerpiece of the Bretton Woods–IMF system. It assumed this role because it was the world's strongest currency after World Wall II and because the United States, which had a large stock of gold, stood ready to exchange the dollars of foreigners for gold at a rate of $35 an ounce.

The dollar thus became virtually an international currency in the Bretton Woods system. Many nations conducted their business in U.S. dollars and built up significant holdings of dollars. Indeed, a so-called Eurodollar emerged as firms, for various reasons, held dollar deposits in foreign banks and in foreign branches of U.S. banks. These dollars could be used to purchase goods and services not only in the United States but in many other nations as well. Moreover, they could be exchanged for gold. As international trade increased, the demand for dollars also increased. The use of the dollar as an international currency put the United States in a good position. Foreigners obtained dollars by supplying goods and services to U.S. citizens, so the United States was able to derive the benefits of imports such as cars and watches in exchange for dollars.

In the early period of the Bretton Woods system, the United States ran a large trade deficit, and this deficit provided a means of supplying dollar reserves to the rest of the world. It was thought that this deficit was a temporary problem that would soon go away to be replaced by a U.S. trade surplus. This was the premise of the Bretton Woods system. Yet the dollar was overvalued, and the U.S. deficits continued and grew larger into the 1950s and 1960s. The United States was in a difficult position. It was hard for it to devalue the dollar because the dollar was held by virtually every nation as an international reserve asset. A U.S. devaluation would have decreased the wealth of all those countries holding dollars. Several times, the United States tried to impose a restrictive macroeconomic policy to correct its balance of trade. However, when unemployment rose as a consequence, such policies were rapidly abandoned as pressures were brought to bear on policymakers. U.S. deficits continued to grow, and foreign holdings of dollars rose.

Speculators entered again. Many holders of dollars became concerned about what the United States was going to do. There were various runs on the U.S. gold stock as dollars were traded in by foreigners. In 1950, the United States had 509 million ounces of gold; by 1968, this stock had fallen to 296 million ounces. Confidence in the dollar fell further, and the stage was set for a drastic change in the international monetary system.

### The Current International Monetary System

The present international monetary system was born in 1971. In the face of continuing large trade deficits and mounting speculation against the dollar, President Nixon broke the link between the dollar and gold in August 1971. No longer would the United States stand ready to exchange gold for dollars at $35 an ounce. In effect, the dollar was set free to fluctuate and seek its own level in the foreign exchange market. As a result, the overvalued dollar depreciated substantially against other major currencies. There were interim attempts to fix the price of the dollar again, but they failed. By early 1973, all the major currencies were floating.

The international monetary system that has been in effect since 1973 is a managed floating rate system. Exchange rates are allowed to seek their free-market values so long as fluctuations are within an acceptable range. If fluctuations fall outside this range, governments may intervene with their reserve holdings to dampen the fluctuations in their currencies. Thus, exchange rates are not completely free; government intervention in the foreign exchange market will be forthcoming if a country's currency falls dramatically in value. In 1978, for example, the dollar fell sharply, and the Carter administration intervened to restrict this decline with international reserves and loans from West Germany and Japan.

Although the new system abandoned fixed exchange rates, it did not abandon the IMF, which still exists as an international monetary organization. The role of the IMF under the managed floating rate system is still evolving, and to this point it has consisted of helping countries that have persistent balance of payments problems by giving them loans and policy advice about how to conduct their domestic macroeconomic policy to overcome these problems. The World Bank, which is part of the IMF system set up by the Bretton Woods agreement, has also played an increasing role in the new system. This sister institution of the IMF makes long-run development loans to poor countries.

**Advantages of the Current System.** The consensus view seems to be that the new system has worked very well under difficult circumstances. The following are the major advantages of the floating-rate system:

1. It allows countries to pursue independent monetary policies. If a country wants to inflate its economy, it can do so by letting its exchange rate depreciate, thereby maintaining its position in international markets without sacrificing its preferred monetary policy.
2. Under the fixed-rate system, a country sometimes had to deflate its economy to solve a balance of trade deficit. With floating rates it only has to let its exchange rate depreciate. Clearly, it is easier to change one price than to change all prices to resolve balance of trade difficulties.
3. When the OPEC countries dramatically raised the price of oil in 1973–1974, the world economy experienced a tremendous shock. The greatest achievement of the floating-rate system is the way it handled this shock. Oil-importing countries developed large trade deficits; huge trade surpluses built up in OPEC countries. These deficits and surpluses were accommodated by floating rates. Moreover, the huge oil revenues of the OPEC countries were recycled into Western investments. Though exchange rates changed significantly over the period, the new system weathered the storm and got the job done.

**Disadvantages of the Current System.** The floating-rate system is not, however, without its critics. Some of its problems are as follows:

1. Some observers point out that floating rates can be very volatile, and this volatility creates considerable uncertainty for international trade. Thus, exchange rate flexibility leads to conditions that can retard the amount of international trade and therefore the degree of specialization in the world economy. Table 22–2 lists the exchange rates for several major currencies in 1973–1984. It is clear that some rates have changed signif-

TABLE 22–2  Foreign Exchange Rates, 1973–1984 (U.S. cents per unit of foreign currency)

| Year | French Franc | German Mark | Japanese Yen | British Pound | Swiss Franc |
|---|---|---|---|---|---|
| 1973 | 22.536 | 37.758 | 0.36915 | 245.10 | 31.700 |
| 1974 | 20.805 | 38.723 | 0.34302 | 234.03 | 33.688 |
| 1975 | 23.354 | 40.729 | 0.33705 | 222.16 | 38.743 |
| 1976 | 20.942 | 39.737 | 0.33741 | 180.48 | 40.013 |
| 1977 | 20.344 | 43.079 | 0.37342 | 174.49 | 41.714 |
| 1978 | 22.218 | 49.867 | 0.47981 | 191.84 | 56.283 |
| 1979 | 23.504 | 54.561 | 0.45834 | 212.24 | 60.121 |
| 1980 | 23.694 | 55.089 | 0.44311 | 232.58 | 59.697 |
| 1981 | 18.489 | 44.362 | 0.45432 | 202.43 | 51.025 |
| 1982 | 15.293 | 41.236 | 0.40284 | 174.80 | 49.373 |
| 1983 | 13.183 | 39.235 | 0.42128 | 151.59 | 47.660 |
| 1984 | 11.474 | 35.230 | 0.42139 | 133.56 | 42.676 |

*Source:* Council of Economic Advisers, *Economic Report of the President* (Washington, D.C.: U.S. Government Printing Office, 1985), p. 351.

icantly—the Swiss franc and the yen have appreciated strongly, while the value of the pound dropped considerably. Moreover, the strength of the dollar against all these currencies is evident since 1982. Do these changes impede international trade? Careful studies suggest that they have not. They have found basically that the volume of a country's trade is not very sensitive to fluctuations in its exchange rate.[3]

2. A second criticism of floating rates is that they promote increased world inflation rates. Under a fixed-rate system, countries experience a balance of trade deficit if they inflate their economy. This link is severed under a floating-rate system. Hence, domestic political authorities can more easily give in to those interests who benefit from inflation. Some proponents of the gold standard argue for its return to guard against domestic inflation.

3. Some see the present system of various international monies as an anachronism. These critics argue that it would be more beneficial for the world to adopt a single currency and to become a unified currency area. Such a system would allow various benefits to the world economy and would do away with the need to convert one currency to another in international trade. Although monetary unification has been tried on a modest scale in some areas of the world, notably the recent monetary cooperation of the European countries in seeking to establish a European currency unit, a single global currency seems to be some distance in the future. After all, the forces of national autonomy are still strong, and virtually no government would lightly give up its power to control its money supply.

In sum, there are pros and cons with respect to the present international monetary system, but in general the system seems to have worked well over a difficult period in the international economy. Growing problems in the international payments system, however, call for the prospect of still further reform of the system. Focus, "World Debt Crisis," details the growing problem less-developed countries have in repaying debts to lenders such as the World Bank and large multinational banks.

[3] See Leland B. Yeager, *International Monetary Relations* (New York: Harper and Row, 1976), Chapter 13.

## FOCUS: World Debt Crisis

In recent years, international debt in less-developed countries has become an increasing concern due to the apparent difficulties a number of governments have in meeting regular repayments. According to the International Monetary Fund, during 1983 about thirty developing countries completed or were engaging in debt refinancing—mostly consisting of postponing principal loan repayments due over the following year—involving a total debt of about $400 billion. Many Third World countries have become heavily in debt to international lending organizations such as the World Bank and to large private Western banks in recent years.

While most developed countries have a foreign debt of less than 20 percent of their GNP, the corresponding ratio in the case of many developing countries is much higher. Argentina, with a foreign debt of $45.3 billion, has a debt/GNP ratio of about 70.6 percent; Mexico's foreign debt is $89.8 billion, with a ratio of 60.5 percent; and Brazil's debt is $93.1 billion, with a ratio of 41.1 percent.

However, large indebtedness by itself does not indicate that a country is undergoing a debt crisis or that a country is likely to miss repayment. Indeed, some developed countries have high ratios of debt to GNP. Sweden's ratio is about 37 percent and Denmark's about 62 percent. For some developing countries, such as South Korea (with a debt/GNP ratio of about 53.5 percent), the comparatively large debt has represented foreign capital investment that has helped the countries develop rapidly and achieve impressive rates of economic growth while repaying their loans on schedule.

But some nations, such as Brazil and Mexico, have engaged in overambitious development projects based on projections of future revenues (oil in the case of Mexico) or costs (imported oil in the case of Brazil) that proved to be lower or higher, respectively. Other countries dependent on one major export (for example, Chile's dependence on copper) have experienced rapid declines in the prices of those exports that have further eroded their ability to meet their repayment schedules.

A large proportion of debt owed by less-developed nations represents loans from American banks, who are understandably concerned about the possibility of default on the part of debtor nations. For instance, Manufacturer's Hanover Trust in New York had $6.5 billion in loans outstanding to Latin American countries at the end of 1983. There are some hopeful signs, however. A long-awaited worldwide economic recovery may eventually help debtor nations meet regular loan payments. Also, the U.S. merchandise trade deficit represents good news for Third World debtor nations. In 1984, several developing countries, such as Mexico, Argentina, and Brazil, are expected to enjoy a trade surplus with the United States amounting to a total of over $9 billion, which will substantially assist those countries with especially heavy debt burdens to meet loan payments.

Various reforms have been proposed to provide short-run relief from current debt problems of Third World countries, including longer loan repayment periods and increased foreign private investment in developing countries. Over the longer run, however, successful economic development is the only viable solution to the problems Third World countries face.

## Summary

1. Foreign exchange is the currency of another country that is needed to make payment in an international transaction.
2. Foreign exchange can be obtained in the worldwide foreign exchange market. The demand for foreign exchange arises from the desire to buy foreign goods. The supply of foreign exchanges arises from the desire of citizens of one country to buy the goods of another country.
3. A foreign exchange rate is the price of one currency in terms of another. A currency's value appreciates when its purchasing power in terms of other currencies rises. It depreciates when its purchasing power falls.
4. Under a floating or flexible exchange rate system, a country allows its exchange rate to be set by the forces of supply and demand in the foreign exchange market. Floating rates change when inflation rates are different across countries, when the value of a country's imports or exports changes, or when there are differences in interest rates across countries. In a floating exchange rate system, rates change to maintain equilibrium in a country's balance of trade.
5. Under a fixed exchange rate system, governments intervene in the foreign exchange market to set and defend the value of their currency. When a country's currency is overvalued relative to the equilibrium rate, the country will run a balance of trade deficit. It can seek to cure this deficit by deflating

the domestic economy, establishing trade barriers, or devaluing its currency to a new, lower fixed rate of exchange. When the fixed exchange rate is below its equilibrium value, a country will experience a balance of trade surplus and will sometimes revalue its currency upward.
6. Under a gold standard, a historical form of a fixed exchange rate system, each country makes its currency redeemable in gold and ties its money supply to its stock of gold. When a surplus of imports over exports appears, gold is shipped to foreigners to settle the balance of trade deficit, causing deflation in the domestic money supply of the deficit country and inflation in the surplus country, restoring equilibrium.
7. An important difference between floating- and fixed-rate systems is that the former changes one price to maintain equilibrium in the balance of trade while the latter changes all prices in the economy to maintain a fixed exchange rate.
8. The balance of payments is a record of the international transactions of a country's economy. It follows the principles of double-entry bookkeeping and must therefore always balance in an accounting sense. Transactions that give rise to a demand for foreign exchange are entered as debits; transactions that give rise to a supply of foreign exchange are treated as credits.
9. The modern history of the international monetary system began with the gold standard, followed by a system of fixed exchange rates managed by the International Monetary Fund. The present international monetary system is a managed floating rate system. Exchange rates are set by supply and demand, but governments can and have intervened if rate movements are too large.

## Key Terms

foreign exchange
foreign exchange markets
foreign exchange rate
floating exchange rate system
fixed exchange rate system
managed floating rate system
appreciation
depreciation
purchasing power parity theory
devaluation
revaluation
gold standard
price-specie flow mechanism
balance of payments
balance of trade
Bretton Woods system
International Monetary Fund
speculation

## Questions for Review and Discussion

1. In September 1983, the British pound was worth $1.50, and $1 was worth about 8 Swedish krona. How much would a BMW cost in U.S. dollars if the British price was 10,000 pounds? How much would a Volvo cost in U.S. dollars if the Swedish price were 84,000 krona?
2. The following exchange rates came from a newspaper report for September 7, 1982.

|  | U.S. Dollar Equivalent | |
|---|---|---|
|  | Wednesday | Tuesday |
| Swiss franc | 0.4596 | 0.4591 |
| Italian lira | 0.000624 | 0.0006255 |

|  | Currency per U.S. Dollar | |
|---|---|---|
|  | Wednesday | Tuesday |
| Swiss franc | 2.1755 | 2.1780 |
| Italian lira | 1602 | 1598.50 |

  a. Did the Swiss franc rise or fall from Tuesday to Wednesday?
  b. What about the Italian lira?
  c. What happened to the dollar value of the franc?
  d. What about the dollar value of the lira?
3. Suppose the United States and Sweden are on a floating exchange rate system. Explain whether the following events would cause the Swedish krona to appreciate or depreciate.
  a. U.S. interest rates rise above Swedish interest rates.
  b. American tourism to Sweden increases.
  c. Americans fall in love with a newly designed Volvo, and Volvo sales in the United States skyrocket.
  d. The U.S. government puts a quota on Volvo imports.
  e. The Swedish inflation rate rises relative to the U.S. rate.
  f. The United States closes a military installation in Sweden at the urging of the Swedish government.
4. What happens to the floating exchange rate of a country that has a higher inflation rate than other countries? Show your answer graphically.
5. How would you treat the following items in the bal-

ance of payments, that is, would you enter them as debits or credits?
a. An American travels to Canada.
b. A U.S. wine importer buys wine from Italy.
c. A California wine grower sells wine to England.
d. An Austrian corporation pays dividends to American stockholders.
e. General Motors pay dividends to French investors.
f. Several Japanese visit Hawaii on an American cruise ship.
g. A Swedish citizen invests in a U.S. company in Houston.

6. Explain why capital inflows are a credit entry in the balance of payments and capital outflows are a debit entry.
7. When we say that the gold standard required that all prices but one be changed to overcome a balance of trade deficit, what do we mean?
8. What are the primary advantages and disadvantages of the present international monetary system?

## ECONOMICS IN ACTION

## International Capital Movements and the Dollar

When reports in the media emphasize U.S. merchandise trade deficits, the impression is that the dollar is weak on the world market. Yet in the early and mid-1980s the dollar was quite strong internationally despite the fact that this country's merchandise, or current account, deficit continued to increase. The reason for the dollar's strength can be traced to the international capital market. During the same period when the United States was running trade deficits, foreigners invested more in the United States than U.S. residents invested overseas: The United States was a net capital importer.

With international capital movements, it is no longer true that a country's domestic saving must equal its domestic investment, for some of what is saved is invested abroad and some of what is invested domestically has been saved by residents of other countries. Recall that GNP is composed of consumption (C), investment (I), government expenditures (G), and net exports (exports, X, minus imports, M). In addition, income (Y = GNP) equals consumption plus saving (S). If we assume that government spending and taxes are zero, it follows that

$$C + S = C + I + X - M,$$

or

$$S - I = X - M.$$

This expression shows that international capital flows are reflected in the balance of trade. If, for example, there is an excess of saving over investment in a country, this excess will be reflected in an excess of exports over imports. The excess saving becomes a net export of capital to other countries. The export of capital will be transferred into physical goods through a current account surplus (exports exceed imports). On the other hand, net capital imports will be associated with balance of trade deficits.

Capital movements take place as investors seek the highest possible rate of return on their investments. If interest rates are high in the United States compared to the rest of the world, foreign capital will tend to be drawn to the United States. If the United States has a relatively low interest rate, however, domestic capital will tend to flow to foreign investment opportunities. Such capital movements serve to raise interest rates in countries with low interest rates and to lower them in countries with high interest rates. In the long run, with perfect capital mobility and in a simple world with no risks, interest rates would be the same in all countries.

The international capital market does not display such perfection, however. Investments in some countries are more risky than in others. For example, governments can confiscate private property, and revolutions can change economic systems. International interest rate differentials can therefore persist.

If we look at interest rate data for recent years, shown in Figure 22–6, we see that during 1981–1982 both long-run and short-run rates in the United States were generally higher than comparable rates in Western Europe. The interest rates displayed in the figure are of course nominal rates (the real rate of interest plus the expected rate of inflation). The long-run rate differentials in Figure 22–6a therefore suggest that during most of the period, the rate of inflation was expected to be higher in the United States than in Europe. In Figure 22–6b, short-run interest rates in the United States exceeded a weighted average of rates in eleven foreign countries. (The rates on ninety-day certificates of deposit in both the United States and Europe track each other quite closely, however.) Overall, the data suggest that, other things being equal, investing in the United States was relatively attractive, and one would therefore expect this country to have been a net capital importer during 1981–1982.

When one looks at the capital movements during the same period, however, a different pattern emerges. For example, capital outflows of $118.3 billion were recorded for 1982. Comparing this figure with capital inflows of $84.5 billion results in a net capital outflow of $33.8 billion; that is, the United States was apparently

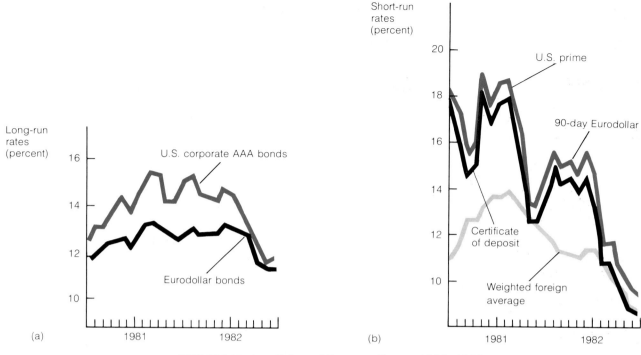

**FIGURE 22–6** Selected Interest Rates, 1981–1982

Source: Federal Reserve Board, U.S. Department of Commerce, Bureau of Economic Analysis, *Survey of Current Business* (December 1982), p. 42.

a net capital exporter at a time when interest rates were relatively high here. The discrepancy can be resolved by noting unrecorded inflows of $41.9 billion during 1982—the statistical discrepancy shown in Table 22–1. These unrecorded inflows are due to errors and omissions in reporting transactions, and most of these errors are thought to occur in the capital account. Thus, the United States was probably a net capital importer during the early 1980s, resulting in an appreciation of the dollar on the world market. That appreciation continued well into 1984 and early 1985.

## Question

In 1983 the United States trade deficit (balance on current account) reached $41 billion. In the first three quarters of 1984 the trade deficit had already doubled that amount. The American dollar, however, remained strong against other currencies. Should U.S. policy favor a strong dollar or a surplus in the balance of trade? Must we make a choice?

# 23

# Economic Systems and Economic Development

**Comparative economics:** The study of different economic systems, in particular the study of socialist versus capitalist economies.

Past chapters have focused on the operation of a mixed capitalist economy, in which a private market sector plays the primary role in the production and distribution of goods and services. The study of **comparative economics,** which we introduce in this chapter, analyzes the differences between capitalist systems and other types of economic systems, particularly socialist economies. Under socialism, the state (the public sector) plays the primary economic role in the economy. The state controls all the resources and makes the decisions about the allocation and distribution of resources. Although there are various forms of socialism, we focus our attention on the Soviet economy.

This chapter also introduces the study of economic development and the economic challenges facing the poorer economies of the world. For example, per capita income in 1984 in India and Burma was less than $200. Why are these countries so poor? How can they be made richer? What is the proper role of the richer countries, such as the United States, in helping the poor countries reach higher standards of living? The incentives created by the economic systems in these countries are a key to understanding their low levels of economic development.

The theme of the chapter is that the choice of economic system, whether a country follows a command-type or a market-type structure, is often a crucial factor in a country's economic development and performance.

## Comparative Economic Systems

What do we mean by an economic system? The economic system of a given country is based on institutions of ownership, incentives, and decision making, which underlie all economic activity.

The comparison of economic systems in different countries is made easier

**Pure capitalism:** An economic system in which all resources are owned and all relevant decisions are made by private individuals; the role of the state in such an idealized system is minimal or nonexistent.

**Pure socialism:** An economic system in which all basic means of production are owned by the state; the state operates the economy through a central plan.

by the classification of systems along a spectrum that ranges from pure capitalism to pure socialism. A **pure capitalist** economy is one in which all property is owned and all economic decisions are made by private individuals and in which government economic control is entirely absent. A pure capitalist economy would have no government at all. A **pure socialist** economy is one in which all property is owned by the state, which in turn makes all economic decisions and plans the economy's output in detail.

The real world is complex, however; there simply are no examples of either pure capitalist or pure socialist economic systems (although there are examples that come close to each). Moreover, it would be a mistake to draw a single line between pure capitalism and pure socialism and attempt to place various countries on either side of it. Some countries have economic systems in which most property is privately owned but in which government regulation and control is extensive; other countries have economic systems in which the opposite seems more nearly the case. Distinctions must therefore be made about the degree of capitalism and socialism in two dimensions—ownership and control of resources. Figure 23–1 illustrates the differences between existing economic systems in these two relevant dimensions. The

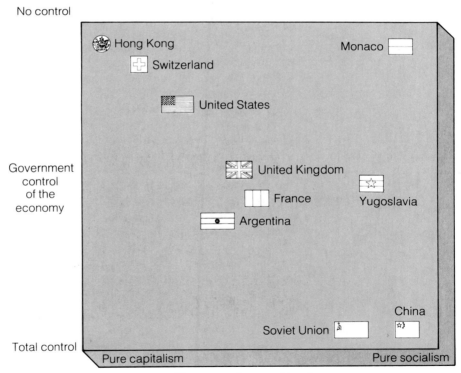

**FIGURE 23–1  Classification of Economic Systems**

The horizontal axis shows the degree of government ownership of productive property in an economy, ranging from pure capitalism on the left to pure socialism on the right. The vertical axis shows the degree of government involvement in economic decision making from no control at the top of the axis to total control at the bottom. Economies are classified by these two criteria. Thus, the economy of Hong Kong stands closest to the pure capitalism model; the Chinese economy is closest to the pure socialism model.

government ownership of property is along the horizontal axis, and the degree of government economic decision making is along the vertical axis.

Although there are many complex differences between any two economies, Figure 23–1 allows us to focus on two basic types of systems. To a varying degree, the economies of the United States, the United Kingdom, and France, for example, share the characteristics of **mixed capitalism.** In such a system the state is an important participant in the economy, but by and large the bulk of productive activity is undertaken by private firms and individuals. We need not dwell on the hallmarks of a capitalist economy here since this topic was covered in Chapter 2 and since this book so far has been based on the economic analysis of a mixed capitalist economy like that of the United States. (Note that one economy in Figure 23–1—that of Hong Kong—does come close to the pure capitalism model.)

In a large number of countries the dominant economic institutions are fundamentally different from those in capitalist societies. These are the socialist economies such as in China and the Soviet Union. Socialism is basically a system of economic organization in which resource allocation is determined by central planning rather than by market forces. Under socialism, at least in theory, resources are allocated and utilized according to a centrally determined and administered overall economic plan. The quantity and composition of output, the relative proportions of consumption and investment, the use of resources in production, and the allocation of final output are all decided by the central authorities. The central plan is also likely to include detailed decisions on quantities of raw materials and inputs, techniques of production, prices, wages, locations of plants and industries, and the pattern of employment of the labor force. The objectives of the central plan reflect the goals and preferences of the central planners. Consumers themselves have no direct input into the planning process; central planning effectively replaces the consumer sovereignty of capitalism.

Thus, capitalism and socialism are fundamentally different economic systems. Economists have been interested in how socialist economies function and how their performance compares to that of capitalist economies. (For a well-known debate among economists about socialism and how it works relative to capitalist markets, see Focus, "The Socialist Calculation Debate.") The Soviet economy is an example of socialism based on central planning. Soviet-style socialism is not the only type of socialism today, of course. Nonetheless, the Soviet economy can teach us a great deal about central planning and the performance of a socialist republic.

**Mixed capitalism:** An economic system in which most economic activities take place in a private sector but in which government also plays a substantial economic and regulatory role; an economy like that of the United States.

## The Generality of Economic Analysis

Before we begin our analysis of the Soviet economy, a general point needs to be made: The basic economic problem of scarcity confronts all economic systems. Scarcity means that choices have to be made. To have more of one thing means having less of another. If a socialist economy wants to have a larger army, the resources for the army must come from alternative uses in the economy, such as agricultural labor. Changing an economy to a socialist basis does not make such trade-offs go away. Scarcity prevades economic organization; it is not something that confronts capitalism but not socialism.

Scarcity is not the only economic concept that applies generally to all economies. We consider the relevance of the following economic principles

## FOCUS  The Socialist Calculation Debate

An important debate in this century among economists concerned whether socialism was possible at all, that is, whether a modern industrialized society could continue to exist if organized along socialist lines. This debate has enormous implications for the many economies in the world today that describe themselves as socialist.

The beginning of the controversy can be given as 1922, the year that Austrian economist Ludwig von Mises published a paper in German entitled "Economic Calculation in the Socialist Commonwealth." Mises attacked the position of contemporary socialist theorists that after a socialist state abolished money, the price system, and markets, it would be able to plan and direct all production. He argued that money prices determined across markets were necessary for rational economic calculation. The price system allowed resources to flow to their most highly valued uses in society. For example, although it would be technically feasible to build subway rails out of platinum, this would be an economically inefficient allocation if less-expensive substitutes were available for the rails. But only the price system, representing the competing bids of all potential users of platinum, allows for such judgments to be made. Without it, Mises argued, resources could not be allocated efficiently, and the economy could function only at a primitive level at best. A modern, technologically advanced, and complex economy would be impossible to achieve under socialism.

Socialists took the Mises challenge very seriously, with some of the more prominent writers (Oskar Lange and Abba Lerner in particular) acknowledging that Mises had identified an important weakness in the socialist position. Lange even half-seriously proposed that in the future socialist commonwealth a statue be erected in Mises' honor so that no one would forget that prices and markets would be essential under socialism too. In effect, the socialist counterattack took the form of a partial retreat from the original socialist position. Lange claimed that socialism might still work if central planning were abandoned in favor of a system wherein the state would set prices for goods and factors of production. Managers of state-owned firms would then produce until the marginal cost of their output equaled the assigned price of the good. These managers would simply requisition the inputs necessary to produce according to this rule; the state would adjust the prices it assigned in response to any shortages or surpluses of the factors of production.

While this plan seemed clever, Mises and his student, Friedrich von Hayek, countered that these "market socialist" schemes failed to solve the real problem of socialism. Although the socialist state could establish an arbitrary set of prices, it would not perform the function of a market price system unless these "prices" were able to convey an equal amount of information regarding the true opportunity costs associated with resource use. For the socialist state's prices to serve this purpose, they would have to reflect an enormous amount of information regarding the availability of resources, updated continually. If this task were possible at all, it would require high transaction costs. Further-

Ludwig von Mises (1881–1973)

Oskar Lange (1904–1965)

more, for such a system to approximate the efficiency of a market economy, incentives would have to be structured to ensure that individuals within the system would use information and resources efficiently. But this could be possible only where factors of production were privately owned, while the market socialists insisted that all resources be owned by the state.

In retrospect, was the socialist calculation debate relevant? Of existing socialist economies, only Yugoslavia resembles the market socialist proposals of Lange and Lerner. Although some of the other socialist countries—Hungary and Poland in particular—have introduced various reforms in recent years that involve a larger role for the private sector, most socialist economies are centrally planned in virtually all respects.

This observation would seem to imply that Mises and Hayek were wrong; socialism is indeed possible. However, two things are generally true about centrally planned economies: (1) Their economic performance is poor by comparison with capitalist market economies, and (2) the private sector in socialist economies, often in the form of illegal underground economies, is typically sizable and important. In these respects Mises and Hayek have been to some degree vindicated.

---

to the study of comparative economic systems: the law of demand, opportunity cost, diminishing marginal returns, and self-interest.

### The Law of Demand

The law of demand is a general proposition about human behavior. It is not simply an economic law that applies to capitalism and private markets. The law of demand says that, other things being equal, price and quantity demanded will vary inversely. That is, the lower the price, the greater the quantity demanded, and vice versa. If socialist planners want to encourage the use of public transit and discourage the purchase of private cars, they will set relatively low prices on transit tickets and relatively high prices on cars. Consumers will respond according to the law of demand. If meat is scarce, its price will be set high to reflect its relative scarcity, and consumers will reduce their purchase of meat in state stores. If the socialist planners set zero prices on basic services such as medical care, demand will be so great that queues will develop, and forms of nonprice rationing to decide who gets to receive medical care will emerge.

### Opportunity Cost

There is no such thing as a free lunch in any economic system. It may well be that a socialist republic offers "free" medical care, that is, medical care at zero price to recipients. But in no sense are the resources allocated to medical care free. They do not come out of thin air; they are directed away from alternative uses in the economy. These alternative uses of resources reflect their opportunity cost. If a socialist government wishes to increase its spending on space research, the resources must be reallocated from some other sector of the economy. The opportunity cost of a larger space program might be resources otherwise available for producing consumer goods.

### Diminishing Marginal Returns

The law of diminishing marginal returns states that the increased application of a resource will ultimately lead to smaller and smaller increments of output, other things being equal. This phenomenon can be found in all economic systems. Worldwide, agriculture exhibits the law of diminishing returns. Generally, the stock of land in an economy is fixed, and the increased application of capital, such as tractors or fertilizer, to the fixed stock of land will lead to diminishing marginal returns. Soviet agriculture as well as American agriculture is subject to diminishing marginal returns. The only way that societies overcome the law of diminishing marginal returns over time is by

### Self-Interest

Economic analysis is based on the idea that individuals weigh the costs and benefits of economic choices and normally respond according to their individual self-interest. Self-interest does not mean that individuals are totally selfish. Individuals can demand or want anything, including a better life for others. Economics simply says that the amount of things individuals want is determined by the costs and benefits of having those things.

Some socialist scholars have argued that self-interested behavior will disappear under socialism and that individuals will come to work for the common good instead. This is perhaps an appealing thought, but it does not seem to be true. The structure of personal incentives helps determine the actions of individuals in both capitalist and socialist economies. For example, in the Soviet Union managers of oil prospecting teams were once rewarded in a piece-rate fashion according to the number of meters they drilled. But drilling goes harder and slower as one drills deeper, with a greater chance that the pipe and drill bits will crack or break. With meters drilled rather than oil discovered as the indicator of success, drillers quickly concluded that they should drill only shallow holes, and lots of them, significantly retarding the discovery of new oil reserves. As this one small example indicates, even socialist producers and consumers weigh costs and benefits in terms of self-interest. Ignoring such tendencies often results in poor economic performance.

## The Soviet-Style Economy

With the understanding that the economic realities of the law of demand, opportunity cost, diminishing marginal returns, and self-interested behavior prevail in all economic systems, we turn now to an examination of how a socialist state—in particular, the Soviet Union—determines what gets produced, how, and for whom.

In 1917 the Bolsheviks, a group of revolutionary socialists, came into political power in Russia and formed a new state, the Soviet Union. The Bolsheviks, influenced by the writings of Karl Marx and Friedrich Engels and the leadership of Vladimir Ilich Lenin, sought to implement pure socialism—an economy in which all the means of production were owned and operated by the state and in which all economic activity would be centrally planned and controlled by the central government. Although Lenin, the revolutionary leader, did not live to see these goals accomplished, Joseph Stalin, his successor, was responsible for implementing them to a very great extent in the 1930s. Since that time, the Soviet economy has proven to be the model for most socialist economies founded in the twentieth century. In this section the basic organizational characteristics and performance of the Soviet economy are discussed.

### A Command Economy

The Soviet economy is sometimes loosely described as a **command economy.** Under authoritarian control, its economy is directed by a central planning system that oversees the production and distribution of most of its resources. The managers of individual state-owned enterprises follow orders imposed by

**Command economy:** An economy that is centrally planned and controlled; all economic decisions are made by a state central planning agency.

the central plan that tell them what to produce, how much to produce, and how to organize production. In effect, socialist managers are more like floor supervisors than the chief executive officers of capitalist firms; they have little independent authority.

The Soviet government basically owns and operates all industry, including the capital goods sector, transportation, communication, banking and financial institutions, and even the entire wholesale and retail network. In one sense, the Soviet economy is like a giant vertically and horizontally integrated firm. However, the analogy should not be taken too far. As we will see, the socialist state is not like a giant corporation, and its citizens are not its shareholders. Whereas corporate decisions are implemented by voluntary contractual agreements, the decisions of the central planning board have the force of law.

In the case of the Soviet Union, the Communist party plays an important role in the management of the planned economy. This is true of many other planned economies based on the Soviet model, such as those of Poland, East Germany, and China. The Communist party is responsible for the selection of enterprise managers and monitors the performance of both labor and management for the central planning board. This party monitoring is carried out through the various ministries of industry, which administer the central plan in individual enterprises.

### Central Planning

Although the details of the actual operation of a large economy with millions of participants are extremely complex, the basic pattern of operations of the centrally planned economy is briefly summarized in Figure 23–2.

The state central planning agency (called Gosplan in the Soviet Union) drafts an economic plan for the entire economy. This plan includes a long-range five-year plan as well as an annual plan. These plans have the force of law. The basic economic plan is directed to the more than 200,000 different Soviet enterprises, setting production targets and allocating inputs to each enterprise for all resources, including labor. Virtually all Soviet production is organized in the form of enterprises, which include everything from huge steel mills to small local baking plants. The manager of each enterprise is responsible for taking the resource inputs that have been allotted and using them to accomplish the goal assigned to the enterprise by the central plan. These targets are used to judge the performance of the firm over particular periods and to reward or penalize the responsible manager accordingly. For example, the plan might call for a steel mill to increase its production of rolled steel 5 percent over its previous year's output and reward the manager with a bonus for meeting the assigned quota (and a greater bonus for exceeding the quota).

**Materials balancing:** The method of central planning in the Soviet Union; the substitute for a price system in a capitalist economy; the attempt by Soviet planners to keep track of the availability of physical units of inputs and to program how these inputs are parceled out among state enterprises to produce final outputs.

Essentially, the central planning board attempts to ration raw materials throughout the entire economy without resort to a price system used in capitalist economies. It performs this task by the method of **materials balancing**—the maintaining of a balance sheet of all available supplies in the economy and of all sources of demand. In other words, the central planning board attempts to keep a comprehensive inventory of all raw materials and factor inputs—minerals, timber, labor, and so on—as well as a similar listing of the demand for these inputs by all the production units, given their assigned production goals. The method of materials balancing works out a consistent pattern of resource allocation throughout the economy that is compatible with the established planning priorities, that is, allocating necessary inputs

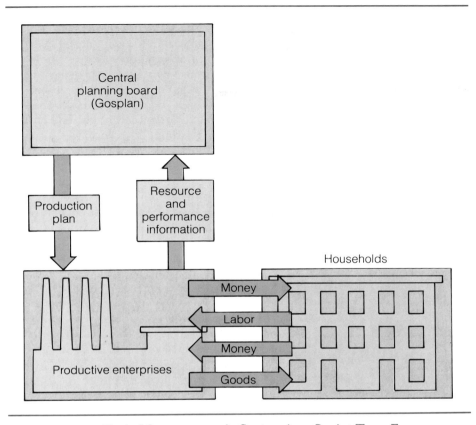

FIGURE 23–2   Basic Macroeconomic Sectors in a Soviet-Type Economy

This is a highly simplified model of the Soviet-style economy. Households supply labor inputs to enterprises in exchange for money. Labor inputs are assigned to firms by the planning process and not by a labor market. Money earned by labor is in turn exchanged for consumer goods and services produced by the enterprises. Exchange takes place in terms of money, although prices are fixed by the central planning board and remain fixed for long periods. In this simplified model, production takes place entirely within the state production establishment. Composition of output is determined not by consumers but by the central planning board.

perhaps first to military production, then to agricultural development, and so on.

By contrast with a capitalist economy, the socialist economy's central planners exercise essentially full control over production in the economy and can direct resources to whatever ends they choose. Production in the economic system is not directed by the final demand of consumers, who instead play a passive role. This is a fundamental difference in the two economies. In a capitalist economy the mix of goods provided reflects the demands of consumers. But in a centrally planned economy like the Soviet Union, the mix of goods provided reflects the decisions of the central planners. For example, if the Soviet government estimates that military spending should be 15 percent of GNP, it can reallocate resources accordingly to achieve this goal without having to convince consumers or voters to rearrange their priorities.

Central planning is not without its problems. The central planning board must coordinate millions of different inputs and potential outputs, an enormous task. Computer technology is available to make a coordinated overall plan feasible in principle. But the system must be constantly revised to accommodate unplanned-for circumstances. Droughts and other climatic events may cause crop failures, for example. When such events occur, agricultural plans must be revised. In recent years the planning system in agriculture has often failed to adjust to changed circumstances rapidly enough.

Another problem is production bottlenecks, which can develop if a supplier fails to deliver a critical input. The customer, who may in turn be some other enterprise's supplier, is not free to place an order with some other supplier of the necessary input. Enterprises have assigned suppliers, and securing official permission for shifting suppliers can be difficult and extremely time-consuming. Enterprise managers may be confronted with the choice of cutting corners where they can—perhaps radically reducing the quality of the goods they produce—or missing their assigned quota (or even failing to produce any output at all). An alternative is to make private, unofficial, and basically illegal private arrangements with other enterprise managers to overcome bottlenecks by switching suppliers. But such arrangements, however necessary, amount to cheating on the central planning process and to a partial introduction of the market process.

The Soviet planners must constantly monitor managerial performance to detect such cheating. To do so, the central planning board depends on information supplied by the managers themselves, whereas a capitalist economy would rely on a firm's success or failure in marketing its output. The incentives thus created can encourage managers to fabricate reports of available resources to avoid blame or to satisfy their output quota. While Soviet planners have long been aware of this problem and periodically undertake programs with great fanfare to overcome it, these reforms have often failed.

### The Legal Private Sector in the Soviet Economy

Although it seems inconsistent with the rhetoric of socialist planning, a legal private sector does exist in the Soviet Union and in most other centrally planned economies. Private economic activity is not entirely prohibited by Soviet law, but it is restricted in two major respects. First, a private individual cannot act as a trade middleman; second, no employee can be hired for the purpose of making a profit. These two restrictions obviously greatly limit the extent to which legally sanctioned private enterprise may operate. Effectively, the only areas in which private enterprise is permitted are personal services and farming. Professionals such as doctors and teachers and craft laborers like carpenters and shoemakers are free to sell their services to consumers. And peasants on state-owned collective farms are permitted to grow crops on small assigned plots and are free to sell whatever they produce. In both cases, production of personal services or of goods for private sale can legally take place only after the completion of the day's labor in the individual's state-sector job. In effect, only moonlighting is legally permissible in the private sector.

Despite these restrictions, activities in the legal private sector realize a considerably higher rate of productivity than do comparable activities in the state sector. The primary evidence for this productivity differential is the output record of the private agricultural plots. These private plots, which constitute only slightly more than 1 percent of land under cultivation in the Soviet Union, account for approximately 25 percent of the nation's total agricultural output, in spite of the fact that the private plots are typically far below the optimum size for efficient production of the crops grown on them.

The productivity of private plots is strong evidence for the crucial role of incentives in determining rates of output. The private plot owners gain the full value of any increase in the efficiency of resource use on their own plots. In marked contrast, the collectivized state agricultural sector is plagued by poor efficiency and poor productivity. It is true that relatively more land-intensive agricultural products—such as cotton and wheat—are grown on collective farms, partially accounting for the disparity in productivity because the value per acre of these crops tends to be relatively low. But this fact only explains a small portion of the difference. Incentives clearly play a large role in the productivity of private Soviet agriculture.

Although evidence concerning the relative performance of the privately provided personal services industry is scarce, the productivity advantages there are similar. Quality seems higher, too. Although in theory medical treatment is provided free to all Soviet citizens, the quality of such care often tends to be low. It is reported that some surgeons are able to command private fees of over 1000 rubles (about $1350) for major operations from individual patients concerned with the quality of their care.

In addition to the legally sanctioned private sector in the Soviet Union, there is reported to be a booming underground or illegal private economy, somewhat like the underground economy in the United States. Focus, "The Soviet Underground Economy," describes these activities in the Soviet Union.

## Comparative Performance

Despite the problems of centrally planned socialist economies, their performance relative to capitalist economies is worth noting. For example, a comparison is presented in Table 23–1 of the United States and the Soviet Union, the largest capitalist and socialist economies, with real GNPs of 2.9 and 1.4 trillion dollars, respectively, in 1980.

The total output of the U.S. economy is roughly twice that of the Soviet Union, although the Soviet Union has both a larger land area and a larger population. More food per person is produced in the United States by fewer agricultural workers; the United States also produces many more automobiles. However, the Soviet Union produces more oil and steel than the United States.

Looking at these two economies from a broader perspective, we must note that the investment rate in the Soviet Union is high. In 1980, for example, the Soviets allocated 29 percent of GNP to investment. The investment rate

TABLE 23–1  A Comparison of the U.S. and Soviet Production Characteristics, 1980

|  | United States | Soviet Union |
|---|---|---|
| Total output (per year) | $2.9 trillion | $1.4 trillion |
| Oil (barrels/day) | $8.57 million | $12.18 million |
| Steel (metric tons/year) | $100.8 million | $148.0 million |
| Automobiles (units/year) | $6.4 million | $1.3 million |
| Farm workers (percentage of population) | 3 percent | 24 percent |
| Meat (per person per year) | 256 pounds | 126 pounds |
| Grain (per person per year) | 2552 pounds | 1571 pounds |
| Population | 230 million | 248 million |
| Land area | 3.6 million square miles | 8.6 million square miles |

Source: U.S. News and World Report (March 1, 1982), pp. 34–35.

## FOCUS  The Soviet Underground Economy

The private sector in the Soviet economy is apparently much larger than what Soviet law permits. There is abundant evidence for the existence of a large Soviet underground economy, comprising private production activities that are officially prohibited or strongly discouraged by state authorities. In the Soviet Union the underground economy is called *na levo*—literally, "on the left."

Gur Ofer of Hebrew University, a specialist in the Soviet economy, calculates that up to 12 percent of the average Soviet citizen's income derives from the illegal private economy, and 18 percent of all consumer expenditures are made there. The size of the Soviet underground economy has been estimated to exceed 10 percent of the Soviet Union's GNP, or approximately $75 billion.[a] The underground economies in most Eastern European economies are of roughly comparable size, and significant underground sectors exist in other Communist countries like China and Vietnam.

Of course, some of this activity involves theft of resources owned by the state, but a large part of it seems to involve what might be termed capitalist acts between consenting adults. For example, records by the late John Lennon sell for up to 70 rubles (about $90) in Moscow. In Russia, buyers eagerly approach tourists on the streets, offering them up to $200 for the blue jeans they are wearing. It has been reported that between 20 and 33 percent of liquor consumed in the Soviet Union is illicit samogen (a vodka-like moonshine often distilled to over 100 proof), produced by 250,000 to 300,000 people devoting the equivalent of full-time work to the activity.[b] And all of this trade is quite illegal.

Since the 1930s, the Soviet government has used its criminal code to try to suppress underground trading, making acting as a private entrepreneur or speculator punishable by severe penalties, including the death sentence. Elizaveta Tyntareva, a lawyer in Lithuania, opened an underground business a few years ago in which she bought consumer goods (watches, cameras, umbrellas, and other things) in areas where they were in oversupply and sold them in areas where they were in short supply. Early in 1980, she was arrested and convicted under the tough antispeculation laws; she was sentenced to twelve years in prison.[c] In May 1981, the newspaper *Bakinsky Rabochii* reported the execution of three men by firing squad. They were convicted of conspiring to turn the No. 3 Knitwear Shop in Baku into an underground factory to make private profits through an illegal two-shift work schedule that signifi-

[a]*Time* (June 23, 1980), p. 50.
[b]*New York Times* (March 8, 1981), p. 11.
[c]*Time* (Jan. 23, 1980), p. 50.

*Black market activity in the Soviet Union.*

cantly increased production. By selling off the additional production to satisfied customers, they had accumulated more than 2 million rubles in three years.[d] But despite draconian penalties, the underground economy continues to flourish.

It would be a mistake to regard the underground economy as functioning solely to shift consumer goods to the highest bidder. The illicit private sector helps ease shortages and reduce inefficiency in the planned sector of the economy as well. Managers of collective farms admit that often the only way to meet their production targets is to buy supplies on the black market. In the Soviet economy (and apparently Eastern Europe as well) planned production depends on an extensive system of bribery to function; Russian enterprise managers employ *tolkachi*, professional "expediters" who basically bribe suppliers to provide necessary inputs.

In general, bribery (in the form of both actual payments and favors) performs a vital role in greasing the wheels of the planned economy. The underground private sector provides coordinating services that increase the efficiency of the aboveground planned sector.

[d]*U.S. News and World Report* (November 9, 1981), p. 42.

---

in the United States is only 19 percent. In effect, the Soviet Union emphasizes investment over consumption to a much greater extent than the United States. Centralized planning allows planners to stress industrial development and capital accumulation. In the United States, both investment and capital accumulation are dependent on the savings and investment decisions of individuals and, in the long run, on the decisions of consumers.

While the investment rate in the Soviet Union has been relatively high, the efficiency of investment has remained generally low. Productivity of both labor and capital lags appreciably behind that of Western economies. For example, while Soviet investment in agriculture has increased fivefold since World War II, the growth in Soviet agricultural output has been low, and negative per capita growth rates in annual grain production are common. The data in Table 23–2 illustrate this point.

In the early 1970s, the growth rate of the Soviet economy appeared to be relatively high, however. Table 23–3 presents data on the growth of real gross domestic product during the period 1970–1980 for a variety of countries, including the United States and the Soviet Union. Over the entire

**TABLE 23–2  Soviet Union per Capita Grain Production**

|  | Grain Production (millions of metric tons) | Population, Midyear (millions) | Kilograms of Grain (per capita) |
|---|---|---|---|
| 1970 | 186.8 | 242.8 | 769 |
| 1971 | 181.2 | 245.1 | 739 |
| 1972 | 168.2 | 247.5 | 680 |
| 1973 | 222.5 | 249.8 | 891 |
| 1974 | 195.7 | 252.1 | 776 |
| 1975 | 140.1 | 254.5 | 550 |
| 1976 | 223.8 | 256.8 | 871 |
| 1977 | 195.7 | 259.0 | 756 |
| 1978 | 237.4 | 261.3 | 909 |
| 1979 | 179.3 | 263.4 | 681 |
| 1980 | 189.1 | 265.5 | 712 |
| 1981 | 158.0[a] | 267.7 | 590 |

[a]Unofficially reported by Soviet Union.

Source: Hearings Before the Joint Economic Committee of the Congress of the United States, *Allocation of Resources in the Soviet Union and China, 1982*, 97th Cong., 2nd sess. (Washington, D.C.: U.S. Government Printing Office, 1982), p. 269.

TABLE 23-3  Average Annual Rate of Growth in Real Gross Domestic Product For Various Economies, 1970–1980

| Country | Rate of Growth, GDP | Rate per Capita |
|---|---|---|
| United States | 3.7 | 2.4 |
| Canada | 4.9 | 3.5 |
| West Germany | 3.5 | 3.3 |
| Sweden | 3.2 | 2.4 |
| United Kingdom | 2.5 | 2.2 |
| Australia | 4.4 | 2.8 |
| USSR[a] | 5.1 | 4.1 |
| Yugoslavia[a] | 5.9 | 5.4 |
| Japan | 8.0 | 6.9 |
| India | 3.4 | 1.4 |
| Brazil | 7.1 | 4.8 |

[a]Data are for net material product; this figure does not include public administration, defense, and professional services.
Source: World Development Report, 1982 (New York: Oxford University Press, 1982), Tables 1 and 2.

period, the Soviet growth rate was 5.1 percent. Only Japan's 8 percent exceeded this rate. The U.S. growth rate was 3.7 percent.

Soviet GNP growth has tended to be erratic, however. The rate of economic growth in 1961–1965 averaged 5 percent. This figure encompassed a high of 7.6 percent in 1964 and a low of 2.2 percent in 1963. Since 1975, Soviet growth rates have been generally falling, as illustrated in Figure 23-3. Analysts believe that Soviet GNP growth is gradually declining because of the increasing cost of raw materials as more accessible sources dry up. The Soviet Union is extraordinarily rich in raw materials, holding a large proportion of the world reserves of most economically important minerals, as well as oil and natural gas. But reserves of these resources are increasingly located in more remote areas, making exploitation more costly.

Finally, comparisons of Soviet and Western GNP do not take into ac-

FIGURE 23-3
Growth of Soviet GNP

Soviet economic growth in the late 1970s and early 1980s continued to be erratic and has generally declined over time. Data are constant-price data.

Source: Hearings Before the Joint Economic Committee of the Congress of the United States, *Allocation of Resources in the Soviet Union and China, 1982*, 97th Cong., 2nd sess. (Washington, D.C.; U.S. Government Printing Office, 1981), p. 4.

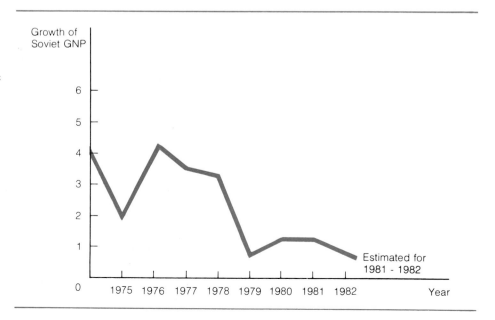

TABLE 23-4 Estimated Shares of Personal Income, by Quintile

| Income Quintile | Share of Income (percent) | | |
|---|---|---|---|
| | United States | Sweden | Soviet Union |
| Lowest | 6.9 | 7.7 | 7.5 |
| Second through fourth | 52.5 | 56.5 | 55.0 |
| Highest | 40.6 | 35.9 | 37.5 |
| Top 5 percent | 15.9 | 12.9 | 14.0 |

*Source:* Lowell Galloway, "The Folklore of Unemployment and Poverty," in S. Pejovich, *Governmental Controls and the Free Market* (College Station: Texas A & M University Press, 1976), pp. 41–72.

count the qualitative aspects of the goods produced. The quality of Soviet consumer goods is generally recognized to be quite poor relative to similar Western goods. As the planned economy does not function on the basis of consumer sovereignty, a considerable proportion of production represents low-quality goods—indeed, goods that consumers do not want at all.

### Income Distribution in the Soviet-Style Economy

One of the stated goals of socialist economies is to reduce income inequality among members of society. What can be said about the performance of socialist economies in this regard?

Data issued by the Soviet government are fragmentary, but they allow a rough comparison of income distribution in the Soviet Union with that in Western countries. One such comparison, involving the Soviet Union, the United States, and Sweden (a socialist democracy), is presented in Table 23-4. Interestingly, differences in the degree of inequality in the respective income distributions appear to be comparatively minor. The number of low-income individuals is about the same in the three economies, as is the number of individuals earning the highest incomes.

But this comparison may understate the degree of inequality in the Soviet economy. Special privileges are available to an enormous extent to Communist party members and bureaucrats, especially at the upper levels. For example, automobiles can usually be quickly purchased by members of the elite, while an ordinary buyer may have to place an order and wait a year or more. Soviet party officials and specially favored groups are permitted to shop in special stores stocked with foreign goods unavailable to ordinary citizens. Thus, the elite are often protected from the problems of shortages, waiting in long lines, and poor-quality goods that plague ordinary citizens. These special privileges mean that the effective degree of inequality in Soviet society is much higher than published figures suggest.

## The Problem of Economic Development

We turn now from comparisons of socialist and capitalist economic systems to a consideration of the disturbing differences between more-developed and less-developed countries.

In economic terms, development is not quite the same as growth. Economic growth usually refers to the increase in per capita GNP in an economy; **economic development** is the process of capital formation and the growth in economic productivity that causes rates of economic growth to rise. The history of the U.S. economy provides a model of successful eco-

**Economic development:** The processes through which a country attains long-run economic growth, such as by capital formulation and saving.

nomic development, leading to high and sustained rates of economic growth combined with a growing population. The U.S. economy in 1776 was in many respects an underdeveloped economy. The transition from a low to a high state of development occurred rapidly during the nineteenth century.

In the modern world, the developed nations are the relatively rich economies, and the less developed are the relatively poor economies. Many, though not all, less-developed countries are relatively stagnant economically. While their population is often growing rapidly, their per capita income grows slowly, if at all. As we will see, the form of economic institutions plays an important role in determining the extent to which economies develop.

### The Characteristics of Less-Developed Countries

Most countries of the world, including China, the most populous in the world, are defined as **less-developed countries,** and three-quarters of the world's population live in less-developed countries. But there is an amazing amount of diversity among these countries; the only thing they have in common is their less-developed status. Economically, what characteristics are associated with the low development rates of less-developed countries?

In describing less-developed countries, there is general agreement that four features stand out: low per capita income, a dominant agricultural-household sector, relatively low savings rates, and rapid population growth.

A less-developed country is a poor country in the sense that its per capita GNP is low. The poverty of less-developed countries compared with that of developed countries is in some cases quite severe. Table 23–5 presents some relevant data. In 1980, 49.3 percent of the world's population (that is, about 2.16 billion people) lived in countries where the average per capita GNP was less than $420 (the low-income group in Table 23–5). At the same time, these countries generated only 4.9 percent of the world's total income. In sharp contrast, the nineteen countries with a per capita GNP of $4800 or more accounted for only 16.3 percent of the world's population but generated about 65 percent of the world's total income. In many of the countries in the low-income group, the present low level of development has not changed dramatically in hundreds of years.

The second defining characteristic of less-developed countries is a domi-

**Less-developed countries:** Countries with extremely low levels of real GNP per capita, dependence on subsistence agriculture, extremely low rates of savings, and high rates of population growth.

TABLE 23–5  Population, Income, and Growth, 1980

| Country Group[a] | Population (millions) | (percentage of total population) | Per Capita Income (dollars) | Total Income (billions of dollars) | (percentage of total income) |
|---|---|---|---|---|---|
| 19 industrial countries | 714.4 | 16.3 | 10,320 | 7372.6 | 65.0 |
| 6 centrally planned economies | 353.3 | 8.1 | 4640 | 1639.3 | 14.4 |
| 100 less developed countries, including: | 3314.1 | 75.6 | 705.5 | 2338.1 | 20.6 |
| 4 high-income oil exporters | (14.4) | (0.3) | (12,630) | (181.9) | (1.6) |
| 63 middle-income countries | (1138.8) | (26.0) | (1400) | (1594.3) | (14.0) |
| 33 low-income countries | (2160.9) | (49.3) | (260) | (561.8) | (4.9) |
| Total (125 countries) | 4381.8 | 100 | 2585 | 11,350 | 100 |

[a]Industrial countries have per capita incomes of at least $4,800; middle-income countries have per capita incomes of $420–4800; low-income countries have per capita incomes below $420.
*Source:* World Bank, *World Development Report, 1982* (New York: Oxford University Press, 1982), Annex Table 1, pp. 110–111.

nant agricultural-household sector. In a way this is merely another way of saying that less-developed countries are poor—a large proportion of their population is engaged in subsistence agriculture. Nearly two-thirds of the labor force of the low-income countries of Asia, Africa, and South America are employed in agriculture. In contrast, only 2 percent of the U.S. labor force is employed in agriculture. The size of the household (or nonmarket) sector in less-developed countries is generally much larger than in developed nations. Most households in less-developed countries are engaged in subsistence production in the sense that they raise their own food, make their own clothes, and construct their own homes. The degree of specialization and exchange in these economies is very limited.

The other two characteristic features found in most less-developed countries (although there are exceptions) are low savings rates and high rates of population growth.

Many poor economies have low savings rates because saving is very difficult (or even impossible) when income is at or near the subsistence level. Low savings rates mean that very little income is set aside each year for investment in capital goods for increased production. This inability to save in turn contributes to continuing low rates of development. The term sometimes used for this problem is the **vicious circle of poverty.**

**Vicious circle of poverty:** The idea that countries are poor because they do not save and invest in capital goods and that they cannot save and invest because they are poor.

Rapid population growth is another condition commonly associated with less-developed countries. The population of the poor countries of Asia, Africa, and South America has been expanding at an average rate of about 2.5 percent. At this rate, the populations of these nations double every twenty-five or thirty years. In contrast, the populations of developed nations grow less than 1 percent per year on average. This difference is often taken to mean that rapid population growth is a major contributor to low rates of development because such growth imposes an increasing burden on the limited resources of less-developed countries. The more people, the greater the need to provide education, health care, and other basic services—which require resources that would otherwise be available for capital investment. To the extent that rapid population growth represents more mouths to feed, per capita income will decline as population grows, other things being equal.

## The Economic Gap Between Developed and Less-Developed Countries

Like wealth and poverty, development and underdevelopment are relative concepts. In 1750, England was probably the richest and most developed country in the world, but by modern standards mid-eighteenth-century England was relatively poor and undeveloped. Moreover, we cannot judge reliably whether any given country should be counted as a developed or less-developed economy on the basis of secondary characteristics—such things as the rate of population growth, degree of literacy, or degree of industrialization. There are, for example, some developed countries that have high rates of population growth and a lesser degree of industrialization than do some less-developed countries. A more reliable way to distinguish quantitatively between the developed and less-developed countries is by per capita income, that is, the level of a country's GNP divided by its population.

Comparing per capita income across countries involves some problems, however. Per capita income is a measure of the availability of goods and services to individuals, but it is obvious that we cannot draw meaningful comparisons between the incomes of Mexico, China, and Australia when GNP in each case is calculated in terms of a different national currency—

**Exchange rate conversion method:** A method of comparing economic well-being across countries by converting the currencies of different countries into a single currency, such as the U.S. dollar.

pesos, yuan, and pounds, respectively. The simplest means of overcoming this problem is to use the exchange rate between national currencies to convert the GNP of each nation into a common currency. Most international income comparisons follow this technique, with U.S. dollars usually constituting the common currency. The term for this technique is the **exchange rate conversion method.**

Unfortunately, even international income comparisons based on the exchange rate conversion method can be misleading. When international comparisons are made, the intention is to measure the differences in standards of living between different countries. Exchange rate conversion may reflect these differences poorly. While the exchange rate reflects differences in the purchasing power of currencies for goods that are traded across international markets, it may be a poor indicator of differences in the purchasing power of currencies with respect to goods and services that are not exchanged across international markets. For example, if 1 U.S. dollar will purchase 1½ Chinese yuan in the foreign exchange market, this exchange rate does not mean that the dollar will purchase exactly 1½ yuan of housing, dental care, or education in the United States (as 1½ yuan would in China). Differences in climate, culture, and tastes must be taken into account.

A technique for greatly improving international income comparisons involves expressing the conversion ratio between currencies in terms of their ability to purchase a typical bundle of goods and services in the countries where they are issued.[1] For example, the bundle of goods for the United Kingdom might include relatively expensive housing, heating, and food products, and the bundle for Thailand relatively cheap housing, heating, and food products. The United Nations International Comparison Project, a study begun in 1968 by the U.N. Statistical Office, has devised a workable purchasing power index for several currencies. In this **purchasing power parity method,** each category in the bundle of typical goods and services is weighted according to its contribution to GNP. The dollar cost of purchasing the typical bundle is then compared to the dollar cost of purchasing a similar bundle in the United States. After the purchasing power of the nation's currency is determined in terms of the typical bundle, this information is used to convert the GNP of the country in question to a common currency unit, the U.S. dollar.

**Purchasing power parity method:** A method of comparing economic well-being across countries based on the idea of what it costs (in dollars, for example) to purchase a typical assortment of goods in each country.

The size of the gap in per capita income between the developed and the less-developed countries has been calculated by the World Bank using both methods. The results are shown in Table 23–6. Using the exchange rate conversion method, per capita income in the industrial market economies in

TABLE 23–6 Measuring the Economic Gap Between Developed and Less-Developed Countries: Per Capita GNP, 1980

| Country Group | Exchange Rate Conversion Method (dollars) | Purchasing Power Parity Method (dollars) |
|---|---|---|
| Less-developed countries | 850 | 1790 |
| Industrial market economies | 10,660 | 8960 |

Source: World Bank, *World Bank Development Report 1981* (New York: Oxford University Press, 1981), p. 17.

[1] Dan Usher, *The Price Mechanism and the Meaning of National Income Statistics* (Oxford: Clarendon Press, 1968).

1980 was estimated to be $10,660, while that in less-developed countries was estimated to be $850. By this method of comparison the industrial market economies appeared to have a per capita income twelve times greater than that of the LDCs. However, the relative figures calculated in terms of the purchasing power parity method indicate that the more-developed countries' per capita GNP is only five times greater than that of the less-developed countries.

Purchasing power parity estimates of per capita income are a more accurate indicator of relative international performance. The problem is that they are more difficult to construct. However, even in purchasing power parity terms, there is still a very significant gap between the incomes of developed nations and less-developed nations.

## Causes and Cures for Underdevelopment

It is clear that some countries are much worse off than others economically. It is not altogether clear, however, what causes the difference. Various factors have been blamed as causes of underdevelopment, including the vicious circle of poverty, rapid population growth, exploitation by more-developed nations, and the lack of economic incentives. It is important to determine which circumstances are significant in each country because attempts to narrow the gap between the haves and the have-nots cannot be successful unless they address the true causes of underdevelopment.

### The Vicious Circle of Poverty

One idea that is often put forth to explain persistent underdevelopment is the vicious circle of poverty. Since incomes are low in underdeveloped countries, savings and investment rates tend to be low. A relatively large proportion of income is devoted to final consumption compared with that in developed countries. Low rates of investment retard capital formation, adoption of new technology, and the growth of future income. Thus, less-developed countries tend to remain less developed. Such countries may find themselves in a pit of underdevelopment from which they cannot escape.

Less-developed countries unquestionably could develop faster if they were able to allocate a higher proportion of their resources to economic growth. But the vicious circle theory contends that the relative poverty of poor countries actually prevents their economic development. Can this proposition be true? Clearly, all presently developed countries were at one time less developed. Yet these countries overcame their own dire conditions.

Whether the vicious circle of poverty is a cause of the lagging development of particular less-developed countries is basically an empirical question. We can determine a statistical answer to the question. Is much more income proportionately allocated to final consumption than to savings and investment in less-developed countries? The savings and investment rate in most countries of Africa and Southeast Asia is between 10 and 15 percent of GNP. In contrast, the savings and investment rate for industrial nations is generally between 20 and 25 percent of GNP. However, as Table 23–7 shows, some persistently underdeveloped countries have considerably lower rates of final consumption expenditure and therefore higher rates of savings and investment than do some highly developed economies. There evidently is no simple relation between the rate of savings and investment and the rate of economic development. While some less-developed countries, such as Bangladesh, have much higher rates of final consumption than developed

TABLE 23-7  Final Consumption Rates and per Capita GNP

|  | Final Consumption Expenditure, Government and Private (percentage of GNP) | Per Capita GNP (U.S. dollars) |
|---|---|---|
| *Developed Countries* | | |
| United States | 81 | 9869 |
| Japan | 68 | 8901 |
| Spain | 80 | 3968 |
| *Less-Developed Countries* | | |
| Bangladesh | 96 | 103 |
| Ethiopia | 91 | 116 |
| India | 79 | 169 |
| Indonesia | 70 | 347 |
| Mexico | 73 | 1481 |
| Sri Lanka | 88 | 196 |

*Source:* U.S. Bureau of the Census, *Statistical Abstract of the United States,* 1982–1983, charts 1524, 1525.

countries in general do, other less-developed countries, such as Indonesia, have significantly lower rates of final consumption than most developed countries.

### Population Growth and Economic Development

One of the characteristics shared by most developing countries is rapid population growth. The population of poor countries in Africa, Asia, and South America has been expanding in recent years at an annual rate of about 2.5 percent. By contrast, the population of the developed countries of Europe, Japan, and North America has been growing at a rate of around 1 percent per year. A cause-and-effect relation between rapid population growth and retarded economic growth of less-developed countries has often been claimed. Surely rapid population growth may impose substantial costs on less-developed countries, insofar as education and health care as well as elemental necessities like food and clothing must be provided from limited available resources. While this line of reasoning is plausible on the surface, however, there are reasons to believe that to blame underdevelopment on rapid population growth is an oversimplification that fails to consider that in economic terms both costs and benefits may be associated with population growth.

First, the high rate of population growth in the less-developed countries is fostered not by increasing birth rates but by stable birth rates combined with a sharp decline in mortality, or death, rates. Mortality rates in the less-developed world have fallen by 50 percent in the last thirty years because of advances in the control of diseases. Although infant mortality remains relatively high, life expectancy at birth in the less-developed world increased from about 35 years in 1950 to about 58 years in 1981. Surely, life expectancy at birth is an important measure of economic development. Hence, the population explosion in the less-developed countries is paradoxically evidence of economic advance because increased life expectancy is one result of economic progress.

High rates of fertility in less-developed countries—that is, a high demand for children—are also cited as a cause of underdevelopment. Both modern and traditional methods of birth control are generally available in less-developed countries, but this has not slackened demand for large families. High rates of fertility, other things being equal, do imply reduced per capita income. But the costs of having children that families in less-developed countries must face are lower than the costs faced by families in Western countries. In less-developed countries, children frequently represent valuable capital assets. It is common for quite young children—four or five years of age—to enter the labor force, especially in the agricultural sector. Only very young children are usually dependent in the sense that the marginal product of their labor does not substantially offset the cost to their families of clothing and feeding them. This usefulness of the young is reflected in the high labor force participation rates for many less-developed countries compared with developed countries. In short, children in less-developed countries apparently make a substantial contribution to aggregate output.

Thus, while population growth may explain underdevelopment, in whole or in part, in some cases it is clearly not a general explanation.

## International Wealth Distribution and Economic Development

Is economic exploitation by richer countries a cause of underdevelopment? If so, will financial help from more-developed countries solve the problem? In 1974, the U.N. General Assembly adopted a Declaration on the Establishment of a New International Economic Order (NIEO), in which it asserted that only large-scale redistribution of wealth from rich to poor countries can accelerate the latter's agonizingly slow development process.

Two ideas lie behind the NIEO: the vicious circle of poverty and the responsibility of developed countries for the underdevelopment of poor countries. In the minds of NIEO advocates, presently developed countries are all colonial powers who developed their own economies largely through a redistribution of resources taken from less-developed countries. Today the less-developed countries' circle of poverty is unbreakable unless the redistribution of wealth is reversed through trade preferences for less-developed countries, commodity agreements (in effect, long-run contracts for various agricultural and other goods), and, most important, cancellation of some of the enormous foreign debts owed by less-developed countries to foreign banks. An important clause of the declaration states that discrimination (that is, any restrictions on the use of funds) should not be involved in grants of aid to governments of less-developed countries. What those governments choose to do with the aid should be entirely left to them.

The claims of the advocates of the NIEO are understandable. Many presently developed nations long colonized many of the presently less-developed nations. International trade has continued among developed and less-developed countries on a large scale, yet the less-developed countries have progressed slowly. Finally, common sense would seem to suggest that additional aid could only help matters.

The NIEO proposal, however, has many opponents, who refute the claim that past and present commercial contacts between the Western developed countries and the less-developed countries is a source of the less-developed countries' persistent poverty. Opponents point out that the list of the least-developed countries consists almost entirely of those that have had little or no involvement with international trade, such as Burundi, Chad, Rwanda, and Bhutan. Those with a record of the most extensive contacts with the

West, including Mexico, Singapore, and Brazil, are among the most advanced poorer countries.

Opponents of NIEO also challenge the notion that trade between the developed nations and the less-developed nations has been a one-way street. They stress that there were and are gains to the less-developed countries from this trade. Opponents also disagree with NIEO's proposal to abolish economic discrimination against the less-developed countries, which would effectively allow them to follow any sort of economic policy. Opponents cite economic policies of the less-developed countries that have severely retarded their own economic development, including persecution and expulsion of productive minority groups (Asians in East Africa, Chinese in Southeast Asia, and many others), enforced collectivization of farming, the establishment of state export monopolies, and the confiscation of the property of productive groups for political purposes.

Thus, the economic policies pursued by governments in less-developed countries seem to play a crucial role in their own economic development. Many government policies seem to inhibit economic growth and reduce the performance of the economy. In this context, critics question unrestricted aid to less-developed countries. The same governments that promulgate policies detrimental to economic development are unlikely to invest foreign aid in an economically efficient manner. To opponents of the NIEO proposals, the theory of exploitation from without underestimates the importance of internal institutions and economic policies in the development process. These critics argue that development is not determined by the greed or generosity of already developed nations but to a large degree by economic policies in the less-developed countries themselves.

### Property Rights Arrangements and Economic Development

Can differences in economic institutions be pinpointed as a major cause of underdevelopment? It is often argued that economists are not able to bring societies into the laboratory and use the experimental method to test their theories. While this is certainly true, economists are able to make careful observations of how different economic institutions affect economic performance. There are contemporary examples of different economic systems that have a common basic cultural setting, a similar climate, and a common ethnic-religious origin but that have taken radically different routes to economic development. Comparison of such countries helps shed some light on the causes of underdevelopment. One relevant comparison can be made between North and South Korea.

**North and South Korea.** The two Koreas were one nation for hundreds of years, until 1950. When the two became separate nations, they adopted diametrically opposing paths to economic development. The People's Democratic Republic of (North) Korea has adhered strictly to the central planning model. In fact, by most accounts it has one of the world's most highly socialized and centrally planned economies. All industrial enterprises are either directly owned by the state or take the form of cooperatives, in which case they are owned indirectly by the state. Agriculture is carried out on either collective or state farms. Following the Soviet model, the central planning system has allocated priority to investment in heavy industry at the expense of the consumer and agricultural sectors. But despite this emphasis, the performance of the industrial sector has been disappointing. Productivity is low, and virtually all plant and equipment are obsolete, although North Korea has recently begun to purchase Western equipment and technology, including complete plants. Nominal per capita income in North Korea is only about $730 per year. There is little variety in consumer goods, and their quality is uniformly low. Shortages are common, and standing in long lines to purchase basic necessities is a way of life. Consumer durables like appliances are usually unavailable.

The land, climate, and people of South Korea are very similar to those of the North, but in other respects the differences between the two countries are dramatic. The Republic of (South) Korea has an economy based on private enterprise and a market economy. In little over twenty years South Korea transformed itself from an economy dominated by subsistence agriculture (characteristic of most poor nations) to a modern economy with an emphasis on light industry. A thriving export market has led South Korea's surge of development, with exports increasing at the impressive average annual rate of 40.4 percent between 1974 and 1977. This growth reflects in part the improving quality of its exports. The standard of living is among the highest in modern Asia; per capita GNP in 1978 was about $1200, with a diverse array of high-quality consumer goods available. Interestingly, South Korea has only 10 to 20 percent of the Korean peninsula's deposits of mineral resources, leaving the slower-developing north with the remainder.

**Mainland China, Hong Kong, and Taiwan.** Another example of the impact of different institutional settings on development is the comparative performance of mainland China and both Taiwan and Hong Kong. The same cultural and ethnic base has produced dramatically different results in terms of economic development.

The People's Republic of (Mainland) China is the world's most populous economy, with an estimated population in 1980 of about 1 billion (1,042,018,000). Unlike some other Communist countries, until recently China apparently did not have a legally sanctioned private sector in the

*A vendor in Jingzhon city, on the Yangtze River, Hubei Province, China.*

economy; economic activity was even more tightly controlled by the state than it is in the Soviet Union. Unlike the case of the Soviet economy, there has been little evidence in China of an underground economy, which in the Soviet case restores some degree of flexibility to the centrally planned system.

Since 1959, China has not issued economic statistics, so evaluation of the country's economic performance is based on outside estimates. But there is general agreement that despite promising reserves of oil, gas, coal, iron ore, and other natural resources, China is still one of the world's poorest countries. In 1979, the (estimated) GNP per capita was $516. However, even this low figure tends to overstate the standard of living in China, where quality consumer goods remain in scarce supply. Relative to average wages, goods tend in general to be more expensive than similar goods in the West. A worker can reasonably hope to save enough to buy a bicycle, but definitely not a car. Recently, the Chinese leadership, led by Communist Party Chairman Hua Guofeng and Deputy Chairman Deng Ziaoping, have begun relaxing the reliance on central planning by encouraging greater initiative on the part of enterprise management and encouraging private economic ventures to a limited extent.

In 1978, China introduced a contract responsibility system for some of its 800 million peasant farmers. The system allowed these farmers to sell their crops on the open market after they had turned over a certain percentage of their yield to the government. In 1984, China extended this free-market experiment to urban workers. In the past, urban workers had little incentive to produce; each job was guaranteed for a worker's lifetime, and both industrious and laggard workers were equally rewarded. Now, with the implementation of the new system, state-run plants are able to keep whatever profits they earn in excess of state taxes and distribute these profits to workers in the form of wage incentives. Managers are able to hire and fire workers and to set different wages for different jobs.

The most significant change in China's economy has been its adoption of a modified price system. The government has slowly relaxed its controls over prices, and the costs of many basic consumer items now fluctuate in response to supply and demand.

Although many Chinese are already enjoying greater prosperity because of these changes, the overall success of China's attempt to bring market principles into a command-type economy will depend on many factors, including the stability of the country's leadership. Eventually the world's most populous nation may become a powerful economic force in international trade.

Although the people of Taiwan and Hong Kong are ethnically identical to the Chinese people, Taiwan and Hong Kong are market economies, with relatively small public sectors. In fact, in both countries government regulation of the economy is considerably more limited than it is in the United States. Of the two, Taiwan has the greater endowment of natural resources. Hong Kong is very poor in natural resources. It has little level land for agriculture and must import both food and water from the mainland. Yet each of these economies has enjoyed successful economic development. Estimated 1979 GNP per capita for Taiwan was $1667; for Hong Kong, $2620. The standard of living in each economy is high and visibly improving. Largely unrestricted by government, individual entrepreneurship thrives. Whether Hong Kong's rapid economic growth will continue after it is returned to mainland China in 1999 is problematic, of course. The Chinese have assured

Hong Kong businessman of some freedom of action after the switch in regimes.

These real-world examples suggest that different systems of property rights and incentive structures play a vital role in determining a country's rate of economic development.

## Conclusion: Explaining Economic Development

Why have economies developed at different rates? More specifically, why do less-developed countries seem to have such difficulty in matching the levels of development typical of Europe, Japan, and North America? This question is the object of intense popular controversy; we have already discussed some of the major issues in the debate, such as the vicious circle of poverty. It appears that neither the vicious circle of poverty nor large populations have prevented development of economies such as Hong Kong, Taiwan, and South Korea, which have developed rapidly by attracting foreign investment rather than relying primarily on foreign aid.

Likewise, the level of technology in most less-developed countries tends to be low. But this is an effect, not a cause, of low levels of development. Poor countries are unable to afford equipment and techniques common in rich countries. The technology itself—inventions, new processes and techniques—is widely available to less-developed countries. The knowledge necessary to improve productivity is not lacking, but the economic development making that technical knowledge worthwhile is. The lack of entrepreneurial ability in less-developed countries is also a dubious explanation. They often have untapped reserves of entrepreneurial ability, a problem that countries such as China are only beginning to address.

Of all the explanations for poor growth, economists are generally satisfied that economic incentives matter. Property rights arrangements and institutions that reward economic efficiency and permit consumers to choose freely the kinds and amounts of goods and services produced tend to promote development.

For an in-depth look at the economic underpinnings of development, Economics in Action at the end of this chapter describes the phenomenal success of the modern Japanese economy. Despite its dense population and sparse resource base, Japan has risen rapidly from a poor agricultural country to a model for what a capitalist market structure and incentives to savings and investment can accomplish.

## Summary

1. The study of comparative economic systems focuses on the alternative institutional arrangements of different economies—such as who owns productive resources and who makes economic decisions—and the impact of these arrangements on economic performance.
2. Capitalism is an economic system in which a large proportion of economic activity is conducted by private individuals and firms. The private ownership of resources and the freedom to employ resources as the owner sees fit are a hallmark of capitalist economic systems. By contrast, socialism is an economic system in which the state owns and controls all the resources in the economy. Resource allocation under socialism is determined by central planning rather than by private initiative.
3. The basic economic concepts introduced in the earlier chapters of this book have general applicability to all economic systems. The law of demand, opportunity cost, diminishing marginal returns, and self-interest, for example, apply equally to capitalist and socialist economies.

4. The Soviet economy is a command economy in which virtually all economic decisions are made through a detailed system of central planning. The Soviet government owns and operates the whole economy.
5. There is a small legal private sector in the Soviet Union in agriculture and personal services. The private agricultural plots occupy a small percentage of cultivated land but are quite productive. There is also an extensive illegal private sector in the Soviet Union that eases the bottlenecks caused by central planning.
6. Comparisons of the performance of the Soviet economy and other economies are tricky. Data seem to indicate that the Soviet Union has achieved high but erratic growth rates over recent years. Moreover, when comparisons involving the quality of output are involved, Soviet performance in the consumer goods sector is generally poor.
7. Though an aim of Soviet policy is avowed to be greater equality in income distribution, the available evidence indicates that the effective distribution of income in the Soviet Union is very similar to that in the capitalist U.S. economy and the democratic socialist economy of Sweden.
8. Economic development is the study of how countries grow economically.
9. The less-developed countries of the world share several characteristics. They are characterized by low per capita income, large and growing populations, and low savings and investment rates.
10. Comparisons of per capita income between the industrial countries and the less-developed countries can be made by the exchange rate conversion method or the purchasing power parity method. However the computation is handled, the economic gap between the developed and the less-developed countries is large.
11. The vicious circle of poverty refers to the fact that there is little savings in a poor economy and so little chance for future development. Economies are thus said to be poor tomorrow because they are poor today. This idea has some merit in explaining underdevelopment, but it must be applied carefully. Some poor economies, for example, appear to have high savings and investment rates.
12. Population growth is a serious obstacle to economic development. However, much of the recent population growth in less-developed countries is due to a declining death rate and to the need for young children to augment the labor supply in the agricultural sector, so the phenomenon of population growth in the less-developed countries must be interpreted cautiously.
13. There have been various responses to the New International Economic Order (NIEO) adopted by the United Nations in 1974. Under the NIEO, the rich countries of the world would transfer wealth to the poor countries. This and similar proposals are predicated on the idea that the poverty of the poor countries is due to the policies of the rich countries and not to policies and institutions that prevail in the poor countries.
14. Comparative economic development suggests that property rights, incentives, personal freedom, and other similar institutions are important in sparking economic growth. The examples of Hong Kong, Taiwan, and Japan support the validity of this approach to development.

## Key Terms

comparative economics
pure capitalism
pure socialism
mixed capitalism
command economy
materials balancing
economic development
less-developed countries
vicious circle of poverty
exchange rate conversion method
purchasing power parity method

## Questions for Review and Discussion

1. Would you expect the underground economy or the illegal private sector to be larger in the United States or in the Soviet Union? Why?
2. A socialist economy does not recognize private property rights. The state "owns" all resources. Name three consequences of this lack of private property rights.
3. Is income distribution in the Soviet Union fundamentally different from that in Western economies?
4. Give three examples not discussed in the text of how basic economic principles apply to the Soviet economy.
5. What are the primary characteristics of a less-developed economy?
6. The vicious circle of poverty asserts that the less-developed countries are poor today because they were poor yesterday. Do you agree? Explain.
7. Evaluate this statement: "To achieve economic

growth an economy requires lots of resources, including land, and a low rate of population growth."
8. Explain why you agree or disagree with the following statement: "A major difference between a planned economy and a free market economy is that the supply of any particular good in a planned economy is perfectly inelastic while in a free market it is elastic."

## ECONOMICS IN ACTION  The Rising Japanese Economy

Industrial development in Japan began late by European standards. Japan's industrial revolution began in the early twentieth century, and by the time of World War II, Japan was a leading industrial power. But after the war, the Japanese economy was in shambles; its people were poor, and production processes were primitive. Japan was a less-developed country with a labor force employed predominantly in agriculture.

The growth and development of the Japanese economy since 1950 has been astonishing. Between 1950 and 1980 real GNP per capita in Japan rose elevenfold (from $843 to $9145). In 1980, Japanese GNP per capita was 80 percent of that of the United States; in 1950 it had been only 12.7 percent. Today the Japanese economy is the third largest in the world (after the United States and the Soviet Union).

To put the Japanese economy in perspective, it must be noted that Japan is 10 percent smaller than California, has only about half the population of the United States, and of the major industrial countries is the poorest in natural resources, importing virtually its entire supply of raw materials. Finally, Japan is one of the most densely populated countries on earth. Yet despite these obvious disadvantages, the Japanese development record is truly outstanding.

Several features of the Japanese economy are pertinent to an explanation of its success. The Japanese economy is primarily a capitalist market economy, but it is also characterized by extensive import restrictions. Virtually all productive resources are privately owned and operated. The labor market is different from that in the United States in that lifetime employment contracts are common in the case of larger firms (wherein workers can be dismissed only for misconduct prior to retirement at age fifty-five). There are no national labor unions. Local Japanese unions represent both blue- and white-collar employees and basically restrict their activities to establishing wage differentials between jobs. Assignment of workers to jobs is determined by individual firms without union involvement. The lifetime contract system seems to work well enough, and Japanese workers typically develop intense loyalty to

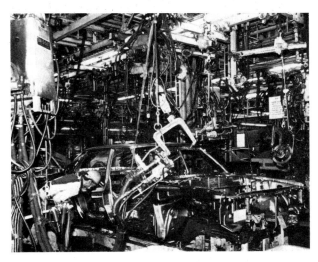

*Machines do most of the work with a minimum of human attention at this automobile production line at Yokosuka, south of Tokyo.*

their firms. The relative absence of labor market rigidity tends to increase economic efficiency.

Two features of Japanese economic affairs probably have played the most important role in Japan's rapid development: low marginal and absolute tax rates and high rates of savings and investment. Taxes consume a much smaller share of GNP in Japan than in Western economies. In 1978, taxes on productive effort (income, payroll, and profit taxes) consumed only 16.8 percent of GNP in Japan, compared to 21.4 percent in the United States, 20.5 percent in the United Kingdom, 23.8 percent in France, and more than 25 percent in both West Germany and Sweden. Moreover, tax rates on personal or business income have been steadily reduced by the Japanese government every year since 1950. In general, the tax rates are lower and the tax schedule is less progressive in Japan than in most Western countries.

The second significant influence on Japan's rapid development has been high rates of savings and investment. During the period from 1960 to 1979, almost one-third of the GNP of Japan was allocated to investment. This figure is considerably higher than the proportion of GNP invested in other developed countries.

*Sources:* The information in this feature was compiled from Steve Lohr, "The Japanese Challenge: Can They Achieve Technological Supremacy?" *New York Times Magazine* (July 8, 1984), pp. 18–41; and *Facts and Figures on Government Finance, 1981* (New York: Tax Foundation, 1981), Table 25.

The Japanese tax structure helps increase investment by taxing businesses and capital gains at much lower rates than in Western countries as well as exempting from taxation capital gains derived from sales of securities. Interest and dividends are taxed at a maximum rate of 25 percent, compared to marginal tax rates of up to 50 percent for similar income in the United States. Further, saving is encouraged by a system of tax credits. Unlike the case in centrally planned economies, where similarly high rates of investment by the state in certain areas of production are associated with shortages of consumer goods, Japan's market economy has induced high rates of voluntary investment to generate rapid economic growth and a high standard of living. In 1970, only 17 percent of Japanese families owned an automobile, compared with 62 percent in 1983; 26 percent owned TV sets in 1970, while 98.9 percent owned them in 1983.

Another feature of Japan's success is its openness to competition. At home, Japanese manufacturers must produce outstanding products at low prices in a competitive environment. Japan has nine auto manufacturers (the United States has only four), and its government has not intervened in the auto industry, allowing the extensive Japanese development of industrial robots. In exports, which constitute 13 percent of Japan's GNP, competition and price cutting are relentless. In 1977, Casio digital wristwatches cost $120; in 1984, they were selling for between $12 and $15. The Walkman portable stereos were introduced in 1979; by 1984, twelve firms were manufacturing them, with Sony offering eleven different models. Nine Japanese companies manufacture videocassette recorders (VCRs); in 1982, fifty-four new models were introduced.

The final ingredient of Japan's astonishing growth is its unabashed copying of technology developed in the West. Extremely successful in this area, it now is moving to achieve technological breakthroughs of its own by investing heavily in research and development. In 1984, 2.4 percent of its GNP was devoted to research—the same percentage as in the United States—and Japan plans to raise its expenditures in this area to 3 percent of its GNP. By 1990, the country plans to open the first of nineteen new centers of research and industrial development, each expected to cost at least $2.5 billion.

Japan is a model of market-directed growth and development. Private initiative and enlightened government policy have created perhaps the greatest success story of economic development in this century.

## Question

What are some of the dangers in assessing other cultures, economies, and nations on the basis of our own value systems? In what ways might ancient Japanese culture affect the performance and evaluation of the contemporary Japanese economy?

## POINT-COUNTERPOINT

## David Ricardo and Gunnar Myrdal: Does Free Trade Always Benefit the Traders?

David Ricardo

Gunnar Myrdal

DAVID RICARDO (1772–1823), contemporary and lifelong friend of Thomas Malthus, is widely regarded as the greatest of the early followers of Adam Smith. Ricardo initiated the use of economic models and developed a thorough and logical exposition of economic theory.

The study of economics began as an avocation for Ricardo. Extremely successful in securities investments and real estate, Ricardo accumulated a large fortune by his late twenties. He left school at the age of fourteen to work for his father, a Jewish immigrant from Holland, on the London Stock Exchange. Ricardo opened his own business at the age of twenty-two, became financially independent, and broke with his family and his religion in 1793 to marry a Quaker. After reading Smith's *Wealth of Nations* in 1799 while on vacation, Ricardo began to write on political economy. Despite his radicalism—he was a strong supporter of freedom of speech and an opponent of government corruption and religious persecution—his ideas gained rapid popularity. His articles were published in newspapers, and he retired from business in 1814 to concentrate on the study of economics.

It was through the publication of Ricardo's articles that Malthus became acquainted with Ricardo in 1809, and the two economists developed a strong and lasting friendship.

Ricardo agreed with Malthus's gloomy predictions about population and the economy and discussed this in his most famous work, *Principles of Political Economy*, published in 1817. In this book, Ricardo also set up a theoretical basis for the law of comparative advantage and developed the idea that free trade and exchange is beneficial to all parties involved. This book had a significant effect on the study of economics in England and has been influential to economists ever since its publication.

GUNNAR MYRDAL (b. 1898), sociologist and economist, was a joint recipient of the Nobel Memorial Prize in economics with Friedrich A. von Hayek in 1974. Myrdal is best known for his work on racial discrimination and social problems. His most famous work, *An American Dilemma: The Negro Problem and Modern Democracy*, which began as a study commissioned by the Carnegie Corporation in 1938, had an enormous impact on American attitudes toward integration. His later works explored trade and economic development: *The Political Element in the Development of Economic Theory* (1935) focused on the philosophical foundations of classical free-trade doctrine, and *Rich Lands and Poor: The Road to World Prosperity* (1957) discussed the application of the free-trade doctrine to the problems of economic development.

Myrdal was born in the Gustaf parish of Sweden. He studied law as an undergraduate at the University of Stockholm and went on to receive his Ph.D. in law from that university in 1927. In 1924, he married Alva Reimer, with whom he researched and wrote a study of Sweden's decreasing population, which was published in 1934. After graduation, Myrdal taught economics, traveled abroad, and served the government of Sweden both as a representative of the Social Democratic party and as cabinet minister. In 1947, he was appointed as executive secretary of the United Nations Economic Commission for Europe, a position he retained for the next ten years. In 1960, he became a professor of international economics at the University of Stockholm.

Is free trade always beneficial to the parties involved? Does the law of comparative advantage always apply? David Ricardo believed that the answer is yes. But Gunnar Myrdal, in his study of economic development among Third World nations, came to the conclusion that comparative advantage may not always work to the benefit of the poorer nation.

## The Advantages of Free Trade

Ricardo, in *Principles of Political Economy and Taxation*, put forth the notion of comparative advantage in international trade. Briefly stated, comparative advantage means that nations should specialize in whatever good or service they can produce at lower opportunity costs than their trading partners. The example used by Ricardo was the exchange of wool and wine between England and Portugal. He showed that both countries gain when England specializes in wool production and Portugal specializes in wine production.

The key to explaining this seeming paradox lies in understanding what is meant by lower opportunity costs. Even though Portugal, in Ricardo's example, was absolutely more efficient at producing both wool and wine with its resources than England, it was relatively more efficient at producing wine. This means that there was a lower opportunity cost for Portugal to produce wine than wool because Portugal's resources were more suited to harvesting grapes. England, on the other hand, was relatively more efficient at wool production and had a lower relative opportunity cost in doing so. If each country specialized by producing only those goods for which it had lower opportunity costs, trade could take place that was beneficial to both parties.

For comparative advantage to work, trade between nations must be free from tariffs and other artificial barriers. When tariffs, quotas, or other import restrictions are set up to protect domestic industries, domestic consumers are robbed of the ability to acquire desired products at the lowest possible prices. This means that the benefits from specialization and comparative advantage are either reduced or, in the extreme, eliminated entirely. For example, Japanese automobile import restrictions in the United States artificially raise the import price of automobiles. Although voluntary, these restrictions have the effect of protecting higher-priced or lower-quality automobiles produced domestically. The impact of these regulations is to deny all or part of the advantages of specialization due to lower opportunity costs of producing automobiles in Japan.

Without such barriers, will the poorer nations of the world overspecialize in one or two goods and rob themselves of the opportunity to strengthen and diversify their economies? The answer, according to those who advocate free trade, is no.

Trade restrictions imposed for any reason diminish the domestic welfare of consumers and protect inefficient, higher-cost domestic producers. Tariffs or quotas to encourage import substitution and diversification in underdeveloped nations, although sometimes well intentioned, have seldom been beneficial to the general welfare of these countries or to their trading partners. Suppose that Indonesia, instead of purchasing communication satellites from the United States, decided to put prohibitive tariffs on such technology in order to develop a new domestic industry. The opportunity cost of doing so would be enormous in terms of the resources forgone from other productions in Indonesia. Total output in Indonesia would fall drastically to produce a good that could be acquired far more cheaply from abroad. Defenders of free trade argue that internal development of human and nonhuman resources will encourage true economic diversification and new comparative advantages in underdeveloped nations.

## The Disadvantages of Free Trade

In Myrdal's view, the principle of comparative advantage and free trade works in favor of rich nations and keeps poor nations locked within a vicious cycle of poverty. In *Rich Lands and Poor: The Road to World Prosperity*, he writes, "The forces in the market tend to increase, rather than to decrease, the inequalities between regions."

To illustrate his argument, Myrdal points to the so-called banana republics, underdeveloped nations that virtually specialize in only one product. In Ricardo's view, such nations are protecting their self-interest because they can gain in trade what they could not gain by diversifying into other areas of production. But Myrdal believes that such overspecialization leaves a country extremely vulnerable to changes in demand. Should world demand for Costa Rica's bananas drop sharply, the country's economy would be in peril.

Myrdal characterizes the free-trade doctrine as outdated. "The English Classical economists did not set as their highest moral and political notion, directing their analysis, the welfare of mankind but rather the welfare of the British nation." In other words, Myrdal insists that free trade was relevant for nineteenth-century England and its self-interests, but not for the modern schism between developed nations and developing nations. He suggests that tariffs and subsidies would provide the impetus for developing nations to diversify production and to reduce dependence on one or a few exports. Ricardo and modern advocates of free trade argue that such restrictions are the certain path to welfare reductions and to lower economic growth and development.

# Glossary

**Ability-to-pay principle:** The notion that tax bills should vary directly with income levels or the capacity of individuals to pay taxes.

**Absolute advantage:** The ability of a nation or a trading partner to produce a product with fewer resources than some other trading partner.

**Absolute price:** The price of a product measured in terms of money.

**Accounting costs:** Direct costs of an activity measured in dollar terms; payments that a firm actually makes, in the form of bills or invoices; explicit costs.

**Accounting profit:** The amount by which total revenues exceed accounting costs; total revenue minus total explicit money expenditures.

**Adjusted income:** The income of an individual after taxes are subtracted, transfer payments are added, and other items are accounted for.

**Advertising:** Any communication that firms offer customers in an effort to increase demand for their product.

**Age distributed income:** The distribution of income over the various age brackets of the population, such as ten-year intervals.

**Aggregate concentration:** A measure of the percentage of total productive assets held by the largest one hundred or two hundred firms in the economy.

**Aggregate supply of land:** The amount of land available to the entire economy at various rental rates.

**Allocative efficiency:** A situation in which the socially optimal amount of a good or service is produced in an industry, given the tastes and preferences of society and the opportunity cost of production.

**Antitrust policy:** The laws and agencies created by legislation in an effort to preserve competition.

**Appreciation:** The rise in the value of one currency relative to another.

**Arc elasticity:** A measure of average elasticity across all intermediate points between two points along a demand curve.

**Artificial barriers to trade:** Any restrictions created by government that inhibit trade, including quotas and tariffs.

**Asset:** Anything of value owned by the firm that adds to the firm's net worth.

**Asset income:** The income received from savings, capital investment, and land, all of which require forgone present consumption.

**Average cost pricing:** A form of price regulation that forces price equal to average cost and thus economic profits equal to zero.

**Average fixed cost:** Fixed cost divided by the level of output.

**Average product:** The output per unit of a variable input; total output divided by the amount of variable input.

**Average tax rate:** The percentage of income that is paid in taxes; total tax liability divided by total income.

**Average total cost:** Total costs divided by the level of output, or average fixed cost plus average variable cost; unit cost.

**Average variable cost:** Variable costs divided by the level of output.

**Balance of payments:** An official accounting record, following double-entry bookkeeping procedures, of all the foreign transactions of a country's citizens and firms. Exports are entered as credits and imports as debits.

**Balance of trade:** Roughly, the balance of a country's imports and exports of goods and services with other countries. If imports are greater than exports, the balance of trade is in deficit; if exports exceed imports, the balance is in surplus.

**Balance sheet:** An accounting representation of the assets and liabilities of a firm.

**Barter:** The trading of goods for goods with no medium of exchange such as money.

**Benefit principle:** The notion that people who receive the benefits of publicly provided goods should pay for their production.

**Bilateral monopoly:** A market in which there is only one buyer and one seller of a resource or product.

**Binding arbitration:** An agreement between employers and labor to allow a third party to determine the conditions of a work contract.

**Bretton Woods system:** The fixed-rate international monetary system established among Western countries after World War II.

**Budget constraint:** A line that shows all the possible combinations of two goods that an individual is able to purchase given a particular money income and price level for the two goods; budget line or consumption opportunity line.

**Capital consumption:** The loss of capital that occurs because the rate of depreciation is greater than the rate of capital formation.

**Capital formation:** The use of roundabout production to increase capital stock.

**Capitalist economy:** An economic system in which the means of production are privately owned.

**Capital stock:** Supply of items used in the production of goods and services; these items include tools, machinery, plant and equpiment, and so on.

**Cartel:** A formal alliance of firms that reduces output and increases price in an industry in an effort to increase profits.

**Cartel enforcement:** An effort by the administrators of a cartel to prevent its members from secretly cutting price below the cartel price.

**Ceteris paribus:** All other things held constant.

**Change in demand:** A shift of the demand curve or a situation in which different quantities are purchased at all previous prices.

**Change in supply:** A shift in the supply curve or a situation in which different quantities are offered at all of the previous prices.

**Choices at the margin:** Decisions made by examining the benefits and costs of small, or one-unit, changes in a particular activity.

**Circular flow of income:** The flow of real goods and services, payments, and receipts between producers and suppliers.

**Coase theorem:** Externalities will adjust to the same level when ownership rights are assigned and when the costs of negotiation are nonexistent or trivial, regardless of which party receives the rights.

**Collective bargaining:** The determination of a market wage through a process in which sellers or buyers act as bargaining units rather than competing individually.

**Collusion:** An explicit or implicit agreement between firms in the same industry not to engage in competitive behavior.

**Command economy:** An economy that is centrally planned and controlled; all economic decisions are made by a state central planning agency.

**Command society:** An economic system in which the questions of "what," "how," and "for whom" are determined by a central authority.

**Common ownership:** Lack of clearly specified ownership rights of a resource for whose use more than one person competes.

**Comparative advantage:** The ability of a nation or a trading partner to produce a product at a lower opportunity cost than some other trading partner.

**Competition:** A market situation satisfying two conditions—a large number of buyers and sellers and free entry and exit in the market—and resulting in prices equal to the costs of production plus a normal profit for the sellers.

**Competitive labor market:** A labor market in which the wage rate of a particular type of labor is determined by the forces of supply by a large number of sellers of labor and demand by a large number of buyers of labor.

**Complementary inputs:** Two or more inputs with a relation such that increased employment of one increases the marginal product of the other.

**Complements:** Products that have a relation such that an increase in the price of one will decrease the demand for the other or a decrease in the price of one will increase the demand for the other: two goods whose cross elasticity of demand is negative.

**Constant-cost industry:** An industry in which the minimum average cost of producing a good or service does not change as the number of firms in the industry changes; an industry for which the supply of resources is perfectly elastic, resulting in a perfectly elastic industry supply.

**Constant returns to scale:** The relation that suggests that as plant size changes, the long-run average total cost does not change.

**Consumer equilibrium:** A situation in which a consumer maximizes total utility within a budget constraint; equilibrium implies that the marginal utility obtained from the last dollar spent on each good is the same.

**Consumer-initiated discrimination:** A circumstance in which people prefer to purchase a good or service produced or sold by individuals of a particular sex, race, religion, or national origin; such consumers are willing to pay a premium to indulge their taste for discrimination.

**Consumers' surplus:** The benefits that consumers receive from purchasing a particular quantity of a good at a particular price, measured by the area under the demand curve from the origin to the quantity purchased, minus price times quantity.

**Contrived scarcity:** The action of a monopoly that reduces output and increases price and profits above the competitive level.

**Corporation:** A firm that is owned by one or more individuals who hold shares of stock that indicate ownership and rights to residuals but who have limited liability.

**Cost-benefit analysis:** A process used to estimate the net benefits of a good or project, particularly goods and services provided by government.

**Costs:** The value of opportunities forgone in making choices among scarce goods.

**Costs of production:** Payments made to the owners of resources to ensure a continued supply of resources for production.

**Craft union:** Workers with a common skill who unify to obtain market power and restrict the supply of labor in their trade; also called trade union.

**Cross elasticity of demand:** A measure of buyers' responsiveness to a change in the price of one good in terms of the change in quantity demanded of another good. The percent change in the quantity demanded of one good divided by the percent change in the price of another good.

**Curve:** Any line, straight or curved, showing the correlation between two variables on a graph.

**Decreasing-cost industry:** An industry in which the minimum average cost of producing a good or service decreases as the number of firms in the industry increases; such an industry has a downward-sloping long-run supply curve.

**Demand curve:** A graphic representation of the quantities of a product that people are willing and able to purchase at all of the various prices.

**Demand elasticity coefficient:** The numerical representation of price elasticity of demand: $(\Delta Q/Q) \div (\Delta P/P)$.

**Demand for loanable funds:** A curve or schedule that shows the various amounts of money that people are willing and able to borrow at all interest rates.

**Depreciation:** The wearing out of capital goods that occurs over a period of time; the fall in the value of one currency relative to another.

**Deregulation:** A situation in which government ceases to regulate a previously regulated industry in an effort to improve the performance of that industry.

**Derived demand:** The demand for factors of production that is a direct function of the demand for the product that the factors produce.

**Devaluation:** A change in the level of a fixed exchange rate in a downward direction; a depreciation of a fixed rate.

**Differentiated products:** A group of products that are close substitutes, but each one has a feature that makes it unique and distinct from the others.

**Direct government purchases:** Real goods and services such as equipment, buildings, and consulting services purchased by the government.

**Discount rate:** The interest rate that a firm uses to determine the present value of an investment in a capital good; the best interest rate that a firm can obtain on its savings; an interest rate charged by the Federal Reserve to depository institutions for loans backed up by some form of collateral.

**Diseconomies of scale:** The relation between long-run average total cost and plant size that suggests that as plant size increases, the long-run average total cost curve increases.

**Divestiture:** A legal action that breaks a single firm into two or more smaller, independent firms.

**Division of labor:** Individual specialization in separate tasks involved in production of a good or service; increases overall productivity and economic efficiency.

**Dumping:** The selling of a product in a foreign nation at a price lower than the domestic market price.

**Dynamic efficiency:** A firm may at first glance impose welfare costs on the economy due to its monopoly power, but may on a closer look be a progressive, innovative firm. In other words, there may be a trade-off in analyzing real firms between static inefficiency and dynamic efficiency, between monopoly power and innovation, and so on.

**Economic development:** The process through which a country attains long-run economic growth, such as by capital formulation and saving.

**Economic efficiency:** Proper allocation of resources from the firm's perspective.

**Economic goods:** Scarce goods.

**Economic growth:** An increase in the sustainable productive capacity of society; increases in real GNP or real GNP per person through time.

**Economic mobility:** The ability or ease with which an individual may move from one income range to another.

**Economic profit:** The amount by which total revenues exceed total costs; total revenues minus total opportunity costs.

**Economics:** The study of how individuals and societies, experiencing limitless wants, choose to allocate scarce resources to satisfy their wants.

**Economic stabilization:** When aggregate variables such as the price level, the unemployment rate, and the economic growth rate are at acceptable levels, varying only slightly and temporarily from desired and achievable goals.

**Economic system:** A means of determining what, how, and for whom goods and services are produced.

**Economic wage discrimination:** A situation in which an employer pays individuals in the same occupation different wages, the wage difference based on race, sex, religion, or national origin rather than productivity differences.

**Economic welfare:** The situation in which products and services are offered to consumers at the minimum long-run average total cost of production.

**Economies of scale:** The relation between long-run average total cost and plant size that suggests that as plant size increases, the average cost of production decreases.

**Elastic demand:** A situation in which buyers are very responsive to price changes; the percent change in quantity demanded is greater than the percent change in price; $\epsilon_d > 1$.

**Elasticity:** A measure of the responsiveness of one variable caused by a change in another variable; the percent change in a dependent variable divided by the percent change in an independent variable.

**Elasticity of demand for labor:** A measure of the responsiveness of employment to changes in the wage rate; the percent change in the level of labor employed divided by the percent change in the wage rate.

**Elasticity of supply:** A measure of producers' or workers' responsiveness to price or wage changes; price elasticity of supply is the percent change in quantity supplied divided by the percent change in price.

**Entrepreneur:** An individual who organizes resources into productive ventures and assumes the uncertain status of a residual claimant in the resulting economic outcome.

**Equalizing differences in wages:** The differences in wages across all occupations that result in equality in total compensation.

**Equilibrium price:** The price at which quantity is equal to quantity supplied; when this price occurs there will be no tendency for it to change, other things being equal.

***Ex ante* distribution:** The distribution of income before government transfer payments and taxes are taken into account.

**Excess capacity:** A situation in which industry output is not produced at the lowest possible average total cost, the result of underutilized plant size.

**Exchange costs:** The value of resources used to make a trade; includes transportation costs, transaction costs, and artificial barriers to trade.

**Exchange rate conversion method:** A method of comparing economic well-being across countries by converting the currencies of different countries into a single currency, such as the U.S. dollar.

**Exploitation of labor:** A situation in which the wage rate is less than the marginal revenue product of labor.

***Ex post* distribution:** The distribution of income after the government influences the disposable income of individuals with taxes and transfer payments.

**External costs:** The cost of a firm's operation that it does not pay for.

**Externality:** Spillover costs or benefits; a situation in which the total costs or benefits to society of producing or consuming a good are greater than the costs or benefits to the individuals who produce or consume it.

**Factor market:** The market in which the prices of resources (factors of production, or inputs) are determined by the actions of businesses as the buyers of resources and households as the suppliers of resources; also called resource market.

**Factors affecting demand:** Anything other than price that determines the amount of a product that people are willing and able to purchase.

**Factors affecting supply:** Anything other than price that determines the amount of a product that producers are willing and able to offer.

**Fallacy of composition:** Generalization that what is true for a part is also true for the whole.

**Family income:** The sum of incomes earned by all members of a household.

**Firm:** An economic institution that purchases and organizes resources to produce desired goods and services.

**Firm coordination:** The process that directs the flow of resources into the production of a particular good or service through the forces of management organization within a firm.

**Firm's long-run demand for labor:** The various quantities of labor that a firm is willing to hire when all inputs are variable.

**Fixed costs:** Payments made to fixed inputs.

**Fixed exchange rate system:** An international monetary system in which each country's currency is set at a fixed level relative to other currencies, and this fixed level is defended by government intervention in the foreign exchange market.

**Fixed input:** Factors of production whose quantity cannot be changed as output changes in the short run.

**Fixed plant:** A situation in which the firm has a given size of plant and equipment to which it adds workers.

**Flat-rate tax:** A proportional income tax with no exemptions or deductions.

**Floating exchange rate system:** An international monetary system in which exchange rates are set by the forces of demand and supply with minimal government intervention.

**Foreign exchange:** The currencies of other countries that are demanded and supplied to conduct international transactions.

**Foreign exchange markets:** The institutions through which foreign exchange is bought and sold.

**Foreign exchange rate:** The price of one country's currency stated in terms of another; for example, a dollar price of German marks of $0.37 means that an individual can buy one mark for $0.37.

**Free enterprise:** Economic freedom to produce and sell or purchase and consume goods without government intervention.

**Free goods:** Things that are available in sufficient quantity to fill all desires.

**Free rider:** An individual who is able to receive the benefits of a good or service without paying for it.

**Free trade:** The free exchange of goods between countries without artificial barriers such as tariffs or quotas.

**Full price:** The total cost to an individual of obtaining a product, including money price and other costs such as transportation or waiting time.

**Gini coefficient:** A numerical estimation of the degree of inequality in the distribution of family income.

**Gold standard:** An international monetary system in which currencies are redeemable at fixed rates in terms of gold.

**Goods:** All tangible things that humans desire.

**Government license:** A right granted by state or federal government to enter certain occupations or industries.

**Government transfer payments:** Money transferred by government through taxes from one group to another, either directly or indirectly, also called income security transfers.

**Gross domestic product (GDP):** The dollar value of

all final goods and services produced in an economy in one year.

**Gross national product (GNP):** The market value of all final goods and services produced in an economy over a given period.

**Historical costs:** Costs of production from the past.

**Homogeneous product:** A good or service produced by many firms such that each firm's output is a perfect substitute for the other firms' output, with the result that buyers do not prefer one firm's product to another firm's.

**Horizontal equity:** A tax structure under which people with equal incomes pay equal amounts of taxes.

**Human capital:** Any nontransferable quality an individual acquires that enhances productivity, such as education, experience, and skills.

**Human resources:** All forms of labor used to produce goods and services.

**Imperfect competition:** A market model in which there is more than one firm but the necessary conditions for a purely competitive solution (homogeneous product, large number of firms, free entry) do not exist.

**Implicit costs:** The value of resources used in production for which no explicit payments are made; opportunity costs of resources owned by the firm.

**Income effect:** The change in quantity demanded of a particular good that results from a change in real income, which has resulted in turn from a change in price.

**Income elasticity of demand:** A measure of buyers' response to a change in income in terms of the change in quantity demanded; the percent change in quantity demanded divided by the percent change in income.

**Increasing-cost industry:** An industry in which the minimum average cost of producing a good or service increases as the number of firms in the industry increases; such an industry has an upward-sloping long-run supply curve.

**Increasing returns to scale:** The relation that suggests that the larger a firm becomes, the lower its costs are.

**Indifference curve:** A curve that shows all the possible combinations of two goods that yield the same total utility for a consumer.

**Indifference map:** A graph that shows two or more indifference curves for a consumer.

**Indifference set:** A group of combinations of two goods that yield the same total utility to a consumer.

**Individual income:** The sum of labor income, asset income, and government subsidies minus tax payments.

**Industrial union:** Workers within a single industry who organize regardless of skill in an effort to obtain market power.

**Industry concentration ratios:** An estimate of the degree to which assets, sales, or some other factor is controlled by the largest firms in an industry.

**Industry regulation:** Government rules to control the behavior of firms, particularly regarding prices and production techniques.

**Inefficient tax:** A tax that decreases the overall productive capacity of the country; a tax that decreases the supply of resources or decreases the efficiency of resource use.

**Inelastic demand:** A situation in which buyers are not very responsive to changes in price; the percent change in quantity demanded is less than the percent change in price: $\epsilon_d < 1$.

**Infant industry:** A new or developing domestic industry whose average costs of production are typically higher than those of established industries in other nations.

**Inferior good:** A product that an individual chooses to purchase in smaller amounts as income rises or larger amounts if income falls.

**Inflation:** A sustained increase in prices; a reduction in the purchasing power of money.

**Inframarginal rent:** A type of rent that accrues to specialized factors of production.

**In-kind transfer payments:** Transfers of benefits other than money from government to citizens such as food stamps, public housing, and Medicare.

**Interest:** The price a borrower pays for a loan or a lender receives for saving, measured as a percentage of the amount; the price of not consuming now but waiting to consume in the future.

**International Monetary Fund:** An international organization, established in the Bretton Woods system, designed to oversee the operations of the international monetary system.

**Joint unlimited liability:** The unlimited liability condition in a partnership that is shared by all partners.

**Kinked demand:** A curve that has a discontinuous slope, the result of two distinct price reactions of competitors to changes in price.

**Labor income:** The payments an individual receives from supplying labor, equal to the individual's wage rate times the number of hours of labor supplied.

**Labor-managed firm:** A firm that is owned and thus managed by the employees of the firm, who have the right to claim residuals.

**Labor union:** A group of workers who organize to act as a unit in an attempt to affect labor market conditions.

**Laissez-faire:** A government policy of not interfering with market activities.

**Laissez-faire economy:** A market economy that is allowed to operate according to competitive forces with little or no government intervention.

**Law of demand:** The price of a product and the amount purchased are inversely related. If price rises, the quantity demanded falls; if price falls, the quantity demanded increases.

**Law of diminishing marginal returns:** A relation that suggests that as more and more of a variable input is added to a fixed input, the resulting extra output decreases, eventually to zero.

**Law of increasing costs:** As more scarce resources are devoted to producing one good, the opportunity costs per unit of the good tend to rise.

**Law of one price:** Exists in a perfect market. After the market forces of supply and demand reach equilibrium, a single price for a commodity prevails.

**Law of supply:** The price of a product and the amount that producers are willing and able to offer are directly related. If price rises, than quantity supplied rises; if price falls, then quantity supplied falls.

**Legal barriers to entry:** A legal franchise, license, or patent granted by government that prohibits other firms or individuals from producing particular products or entering particular occupations or industries.

**Less-developed countries:** Countries with extremely low levels of real GNP per capita, dependence on subsistence agriculture, extremely low rates of savings, and high rates of population growth.

**Liability:** Anything that is owed as a debt by a firm and therefore takes away from the net worth of the firm

**Limited liability:** The legal term indicating that owners of corporations are not responsible for the debts of the firm except for the amount they have invested in shares of ownership.

**Limit-entry pricing:** A pricing policy by a firm that discourages entry of new firms by selling at a price below the short-run profit-maximizing price.

**Limit pricing:** The price behavior of an existing firm in which the firm charges a price lower than the current profit-maximizing price to discourage the entry of new firms and thus maximize its long-run profits.

**Logrolling:** The exchange of political favors, especially votes, to gain support for legislation.

**Long run:** An amount of time that is sufficient to allow all inputs to vary as the level of output varies.

**Long-run average total cost:** The lowest possible cost per unit of producing any level of output when all inputs can be varied.

**Long-run competitive equilibrium:** A situation in an industry in which economic profits are zero and each of the many firms is operating at minimum average total cost.

**Long-run industry supply:** The quantities of a product that all firms are willing and able to offer at all the various prices when the number of firms and scales of operation of each firm are allowed to adjust to the equilibrium level.

**Lorenz curve:** A graph that shows the cumulative distribution of family income by comparing the actual distribution to the line of perfect equality.

**Macroeconomics:** Analysis of aspects of the economy as a whole.

**Managed floating rate system:** A system in which a country's currency is allowed to float freely in the foreign exchange market, within certain bounds; drastic changes in the value of the currency are mitigated by central bank intervention in the foreign exchange market.

**Manager:** An individual or group of individuals that organize and monitor resources within a firm to produce a good or service.

**Margin:** The difference in costs or benefits between the existing situation and a proposed change.

**Marginal analysis:** Study of the difference in costs and benefits between the status quo and the production or consumption of an additional unit of a specific good or service. This, not the average cost of all goods produced or consumed, is the actual basis for rational economic choices.

**Marginal costs:** The extra costs of producing one more unit of output; the change in total costs divided by the change in output.

**Marginal cost pricing:** A form of price regulation that forces price equal to marginal cost and results in optimal allocation of resources.

**Marginal factor costs:** The change in total cost that results from employing one more unit of a variable input.

**Marginal opportunity costs:** The extra costs associated with the production of an additional unit of a product; these costs are the lost amounts of an alternative product.

**Marginal private costs:** The increase in a firm's total costs resulting from producing one more unit.

**Marginal product:** The extra output that results from employing one more unit of a variable input.

**Marginal product of labor:** The change in total output that results from employing one more unit of labor.

**Marginal rate of substitution:** The amount of one good that an individual is willing to give up to obtain one more unit of another good.

**Marginal revenue:** The change in total revenue that results from selling one additional unit of output; the change in total revenue divided by the change in amount sold.

**Marginal revenue product:** The change in total revenue that results from employing one more unit of a variable input.

**Marginal social costs:** The increase in total costs to society (the firm plus everyone else) resulting from producing one more unit.

**Marginal tax rate:** The percentage of an increase in income that must be paid in taxes; the change in tax liability divided by the change in taxable income.

**Marginal utility:** The change in total utility that results from the consumption of one more unit of a good; the change in total utility divided by the change in quantity consumed.

**Market:** An arrangement that brings together buyers and sellers of products and resources; any area in which prices of products or services tend toward equality through the continuous negotiations of buyers and sellers.

**Market coordination:** The process that directs the flow of resources into the production of desired goods and services through the forces of the price mechanism.

**Market demand:** The total demand for a product at each of various prices, obtained by summing all of the quantities demanded at each price for all buyers.

**Market demand for labor:** The sum of, or overall, demand for a particular type of labor by all firms employing that labor; the total level of employment of a particular type of labor at all the various wage rates.

**Market failure:** A situation in which a private market does not provide the optimal level of production of a particular good.

**Market power:** A situation characterized by barriers to entry of rival firms, giving an established firm control over price and, therefore, profit levels.

**Market society:** An economic system in which individuals acting in their self-interest determine what, how, and for whom goods and services are produced, with little or no government intervention.

**Market supply:** The total supply of a product, obtained by summing the amounts that firms offer at each of the various prices.

**Market supply of labor:** The total amount of labor that all individuals are willing to offer in a particular occupation at all the various wage rates.

**Materials balancing:** The method of central planning in the Soviet Union; the substitute for a price system in a capitalist economy; the attempt by Soviet planners to keep track of the availability of physical units of inputs and to program how these inputs are parceled out among state enterprises to produce final outputs.

**Mean income:** The income of the average income earner, found by dividing the total income by the number of income earners.

**Merger:** The joining of two or more firms' assets that results in a single firm.

**Microeconomics:** Analysis of the behavior of individual decision-making units within an economic system, from specific households to specific business firms.

**Mixed capitalism:** An economy in which both market forces and government forces determine the allocation of resources.

**Model:** An abstraction from real-world phenomena that approximates reality and makes it easier to deal with; a theory.

**Money:** A generally accepted medium of exchange.

**Money price:** The dollar price that sellers charge buyers.

**Monitor:** An individual who coordinates team production and discourages shirking.

**Monopolistic competition:** A market model with freedom of entry and number of firms that produce similar but slightly differentiated products.

**Monopoly rent:** The payments to owners of capital in a monopolized industry that exceed the opportunity cost of capital.

**Monopsony:** A single buyer of a resource or product in a market.

**Mutual interdependence:** A relation between firms in which the actions of one firm have significant effects on the actions and profits of other firms.

**Natural monopoly:** A monopoly that occurs because of a particular relation between industry demand and the firm's average total costs that makes it possible for only one firm to survive in the industry.

**Negative externality:** A cost of producing or consuming a good that is not paid entirely by the sellers or buyers but is imposed on a larger segment of society; a situation in which the social costs of producing or consuming a good are greater than the private costs.

**Negative income tax:** A progressive income tax that allows for a negative tax rate (income subsidy) for income below a particular level. As income rises, the subsidy gradually diminishes to zero.

**Negative, or inverse, relation:** A relation between variables in which the variables change in opposite directions. A negative relation has a downward-sloping curve.

**Net worth:** The value of a firm to the owners, determined by subtracting liabilities from assets; also called *equity*. For corporations, net worth is termed *capital stock*.

**Neutral tax:** A tax that has no effect on the production or consumption of goods.

**Nominal rate of interest:** The price of loanable funds measured as a percentage of the dollar or nominal amount of the loan.

**Nonhuman resources:** Inputs other than human labor involved in producing goods and services.

**Nonmarket activities:** Anything that an individual does while not earning income from working.

**Nonprice competition:** Any means that individual firms use to attract customers other than price cuts.

**Nonprofit firm:** A firm in which the costs of production and revenues must be equal and which does not have a residual claimant.

**Nonproprietary:** Relating to public ownership.

**Normal good:** A product that an individual chooses to

purchase in larger amounts as income rises or smaller amounts as income falls.

**Normative economics:** Value judgments about how economics should operate, based on certain moral principles or preferences.

**Normative public choice:** The study of shortcomings and possible improvements of political arrangements, such as voting rules.

**Oligopoly:** A market model characterized by a few firms that produce either a homogeneous product or differentiated products and entry of new firms is very difficult or is blocked.

**Opportunity cost:** The value placed on opportunities foregone in choosing to produce or consume scarce goods.

**Opportunity cost of capital:** The value of the payments that could be received from the next-best alternative investment; the normal rate of return.

**Ownership claims:** The legal titles that identify who owns the assets of a firm.

**Partnership:** A firm that has two or more owners who have unlimited liability for the firm's debts and who are residual claimants.

**Patent:** A monopoly granted by government to an inventor for a product or process, valid for seventeen years (in the United States).

**Per capita income:** The income per individual, found by dividing total income by the total number of people.

**Perfect information:** A condition in which information about prices and products is free to market participants; combined with conditions for pure competition, perfect information leads to perfect competition.

**Perfect market:** A market in which there are enough buyers and sellers so that no single buyer or seller can influence price.

**Positive economics:** Observations or predictions of the facts of economic life.

**Positive externality:** A benefit of producing or consuming a good that does not accrue to the sellers or buyers but can be realized by a larger segment of society; a situation in which the social benefits of producing or consuming a good are greater than the private benefits.

**Positive public choice:** The analysis and explanation of political behavior.

**Positive relation:** A direct relation between variables in which the variables change in the same direction. A positive relation has an upward-sloping curve.

**Post hoc fallacy:** From *post hoc, ergo propter hoc*, "after this, therefore because of this." The inaccurate linking of unrelated events as causes and effects.

**Poverty:** A term describing family income below a defined level when other things such as size of family, location, and age are considered.

**Present value:** Today's value of a payment received in the future; future income discounted by the rate of interest.

**Price ceiling:** A maximum legal price established by government to protect buyers.

**Price control:** Government intervention in the natural functioning of supply and demand.

**Price discrimination:** The practice of charging one buyer or group a different price than another group for the same product. The difference in price is not the result of differences in the costs of supplying the two groups.

**Price elasticity of demand:** A measurement of buyers' responsiveness to a price charge; the percent change in quantity demanded divided by the percent change in price.

**Price floor:** A minimum legal price established by government.

**Price leadership:** A pricing behavior in which a single firm determines industry price.

**Prices:** The opportunity costs established in markets for scarce goods, services, and resources.

**Price searcher:** A firm that must choose a price from a range of prices rather than have a single price imposed on it; such a firm has a downward-sloping demand curve for its product.

**Price-specie flow mechanism:** The mechanism by which the gold standard equilibrates the balance of trade between countries.

**Price taker:** An individual buyer or seller who faces a single market price and is able to buy or sell as much as desired at that price.

**Principle of diminishing marginal utility:** As more and more of a good is consumed, eventually its marginal utility to the consumer will fall, all things being equal.

**Private costs:** The total opportunity costs of production for which the owner of a firm is liable.

**Private sector:** All parts of the economy and activities that are not part of government.

**Production possibilities frontier:** The situation represented by a curve that shows all of the possible combinations of two goods that a country or an economic entity can produce when all resources and technology are fully utilized and fixed in supply.

**Productive efficiency:** A situation in which the total output of an industry is obtained at the lowest possible cost for resources.

**Products market:** The forces created by buyers and sellers that establish the prices and quantities of goods and services.

**Profit-maximizing price:** The price at which the difference between total revenue and total cost is greatest; the price at which marginal cost equals marginal revenue.

**Profits:** The amount by which total revenue exceeds total cost.

**Progressive income tax:** A tax based on a percentage

of income that varies directly with the level of income.

**Proportional income tax:** A tax based on a fixed percentage of income for all levels of income

**Proprietary:** Relating to private ownership and profit seeking by private owners.

**Proprietorship:** A firm that has a single owner who has unlimited liability for the firm's debts and who is the sole residual claimant.

**Public choice:** The economic analysis of politics; the branch of economics concerned with the application of economic principles to political decision making.

**Public employees' union:** Workers who are employed by the federal, state, or local government and who organize in an effort to obtain market power.

**Public finance:** The study of how governments at federal, state, and local levels tax and spend.

**Public franchise:** A right granted to a firm or industry allowing it to provide a good or service and excluding competitors from providing that good or service.

**Public good:** A good that no individual can be excluded from consuming, once it has been provided to another.

**Publicly owned firm:** A firm owned and operated by government.

**Purchasing power parity method:** A method of comparing economic well-being across countries based on the idea of what it costs (in dollars, for example) to purchase a typical assortment of goods in each country.

**Purchasing power parity theory:** A theory of exchange rate determination that states that differential inflation rates across countries affect the level of exchange rates.

**Pure capitalism:** An economic system in which all resources are owned and all relevant decisions are made by private individuals; the role of the state in such an idealized system is minimal or nonexistent.

**Pure economic rent:** The payment to a factor of production that is perfectly inelastic in supply.

**Pure interest:** The interest obtained from a risk-free loan.

**Pure monopoly:** An industry in which a single firm produces a product that has no close substitutes and in which entry of new firms cannot take place.

**Pure socialism:** An economic system in which all basic means of production are owned by the state; the state operates the economy through a central plan.

**Purely competitive market:** A coming together of a large number of buyers and sellers in a situation where entry is not restricted.

**Quantity demanded:** The amount of any good or service consumers are willing to purchase at some specific price.

**Quantity supplied:** The amount of any good or service that producers are willing to produce at some specific price.

**Quasi-rent:** The short-run payments to owners of capital in a competitive industry that exceed the opportunity cost of capital.

**Quota:** A limit on the quantity of an imported good.

**Rate of return on invested capital:** Profits that are measured as a percentage of the costs of capital.

**Rate of time preference:** The percent increase in future consumption that is necessary to induce an individual to forgo some amount of present consumption.

**Rational self-interest:** The view of human behavior espoused by economists. People will act to maximize the difference between benefits and costs as determined by their circumstances and their personal preferences.

**Rationing:** Prices are rationing devices; the equilibrium price rations out the limited amount of a product produced by the most willing and able suppliers or sellers to the most willing and able demanders, or buyers.

**Real income:** The purchasing power of income; the quantity of goods and services that an individual, a household, or a nation can consume with the money received from all sources of income; income adjusted for inflation or deflation.

**Real rate of interest:** The nominal interest rate minus the rate of inflation; the price of loanable funds measured as a percentage of the real buying power of the amount loaned.

**Relative price:** The price of a product related in terms of other goods that could be purchased rather than in money terms.

**Rent:** A payment to a factor of production in excess of its opportunity cost.

**Rent seeking:** The activity of individuals who spend resources in the pursuit of monopoly rights granted by government; the process of spending resources in an effort to obtain an economic transfer.

**Residual claimant:** The individual or group of individuals that share in the excess of revenues over costs, that is, profits.

**Resources:** Inputs necessary to supply goods and services. Such inputs include land, minerals, machines, energy, and human labor and ingenuity (called the factors of production).

**Resource specialization:** The devotion of a resource to one particular occupation that is based on comparative advantage.

**Resources market:** The forces created by buyers and sellers that establish the prices and quantities of resources such as land, labor, and capital.

**Revaluation:** A change in the level of a fixed exchange rate in an upward direction; an appreciation of a fixed rate.

**Right-to-work law:** A law that prevents unions from

forcing individuals to join a union as a prerequisite to employment in a particular firm.

**Risk:** The probability of a default or a failure of repayment of a loan.

**Roundabout production:** The production and use of capital goods to produce greater amounts of consumption goods in the future.

**Saving:** The act of forgoing present consumption in an effort to increase future consumption.

**Scale of production:** The relative size and rate of output of a physical plant that may be measured by the volume or value of firm capital.

**Scarcity:** Limitation of the amount of resources available to individuals and societies relative to their desires for the products that resources produce.

**Services:** All forms of work done for others—such as medical care and car washing—that do not result in production of tangible goods.

**Share:** The equal portions into which the ownership of a corporation is divided.

**Share transferability:** The power of an individual shareholder to sell his or her portion of ownership without the approval of other shareholders.

**Shirking:** A sometimes rational behavior of members of a team production process in which the individual exerts less than the normal productive effort.

**Shortage:** The amount by which quantity demanded exceeds quantity supplied when the price in a market is too low.

**Short run:** An amount of time that is not sufficient to allow all inputs to vary as the level of output varies.

**Short-run firm supply:** The portion of the marginal cost curve above the minimum average variable cost.

**Short-run industry supply:** A summation of all the existing firms' short-run supply curves.

**Shutdown:** A loss-minimizing option of a firm in which it halts production in the short run to eliminate its variable costs, although it must still pay its fixed costs.

**Slope:** The ratio of the change in ($\Delta$) the $x$ value to the change in ($\Delta$) the $y$ value; $\Delta y/\Delta x$.

**Social costs:** The total value of all resources used in the production of goods, including those used but not paid for by the firm.

**Socialist economy:** An economic system in which the means of production are owned and controlled by the government.

**Socialization:** Government ownership of a firm or industry.

**Specialization:** Performance of a single task in the production of a good or service to increase productivity.

**Speculation:** The buying and selling of currencies with an eye to turning a profit on exchange rate changes, predicted devaluations, and so on.

**Static inefficiency:** A condition, related to the concept of welfare loss due to monopoly power, which is summarized as the production of too little output at too high a price.

**Strike:** A refusal to work at the current wage or under current conditions.

**Subsidy:** A government cash grant to a favored industry.

**Substitute inputs:** Two or more inputs with a relation such that increasing the employment of one decreases the marginal product of the other.

**Substitutes:** Products that have a relation such that an increase in the price of one will increase the demand for the other or a decrease in the price of one will decrease the demand for the other; two goods whose cross elasticity of demand is positive.

**Substitution effect:** The change in the quantity demanded of a particular good that results from a change of its price relative to other goods.

**Sunk costs:** Past payment for a presently owned resource.

**Supply of loanable funds:** A curve or schedule that shows the various amounts of money that people are willing and able to lend (save) at all interest rates.

**Supply schedule:** A schedule or curve that shows the quantities of a product that producers are willing and able to offer at all prices.

**Surplus:** The amount by which quantity supplied exceeds quantity demanded when the price in a market is too high.

**Tangency solution:** A long-run situation in which the firm's downward-sloping demand curve is just tangent to the average total cost curve, necessarily implying zero economic profits.

**Tariff:** A tax on imported goods designed to maintain or encourage domestic production.

**Taxable income:** The amount of income that is subject to income taxes; total income minus deductions and exemptions.

**Tax incidence:** The burden of a tax or fiscal resting place.

**Team production:** An economic activity in which workers must cooperate, as team members, to accomplish a task.

**Terms of trade:** The ratio of exchange between two countries, based on the relative opportunity costs of production in each country; the price ratio or range of price ratios at which two entities are likely to trade.

**Terms of trade argument:** The use of export restrictions on goods in an effort to increase a country's monopoly or monopsony power in international markets.

**Total compensation:** The lifetime income that an individual receives from employment in a particular occupation, including all monetary and nonmonetary pay.

**Total cost:** All the costs of a firm's operations, including fixed and variable costs.

**Total cost of production:** The value of all resources used in production; explicit plus implicit costs.

**Total product:** The total amount of output that results from a specific amount of input.

**Total revenue:** The total amount of money received by a firm from selling its output in a given time period; price times quantity sold.

**Total utility:** The total amount of satisfaction obtained from the consumption of a particular quantity of a good; a summation of the marginal utility obtained from consuming each unit of a good.

**Total wage bill:** The total cost of labor to firms, equal to wage times total quantity of labor employed; the total income of all workers.

**Traditional society:** An economic system in which the "what," "how," and "for whom" questions are determined by customs and habits handed down from generation to generation.

**Transaction costs:** The value of resources used to make a purchase, including time, broker's fees, contract fees, and so on.

**Transfer payment:** The transfer by government of income from one individual to another; it may take the form of cash or goods and services such as education, housing, health care, or transportation.

**Transitivity of preferences:** A rational characteristic of consumers that suggests that if $A$ is preferred to $B$ and $B$ is preferred to $C$, then $A$ is preferred to $C$.

**Transportation costs:** The value of resources used in the transportation of goods.

**Trust:** An institution that organizes firms in the same industry in an effort to increase profits by decreasing competition.

**Unemployment of resources:** A situation in which human or nonhuman resources that can be used in production are not so used.

**Unit elasticity of demand:** A situation in which the percent change in quantity demanded is equal to the percent change in price: $\epsilon_d = 1$.

**Unlimited liability:** A legal term that indicates that the owner or owners of a firm are personally responsible for the debts of a firm up to the total value of their wealth.

**Utility:** The ability of a good to satisfy wants; the satisfaction obtained from the consumption of goods.

**Variable costs:** Payments made to variable inputs that necessarily change as output changes.

**Variable input:** Factors of production whose quantity may be changed as output changes in the short run.

**Vertical equity:** A tax structure under which people with unequal incomes pay unequal taxes; people with higher incomes pay more taxes than people with lower incomes.

**Vicious circle of poverty:** The idea that countries are poor because they do not save and invest in capital goods and that they cannot save and invest because they are poor.

**Welfare loss:** The consumer surplus that is lost to consumers but not transferred to the monopoly in profits when a competitive industry is monopolized.

**Welfare loss due to monopoly:** The lost consumers' surplus resulting from the restricted output of a monopoly firm.

**$x$-axis, $y$-axis:** Perpendicular lines in a coordinate grid system for mapping variables on a two-dimensional graph. The $x$-axis is the horizontal line; the $y$-axis is the vertical line. The intersection of the $x$- and $y$-axes is the origin.

**X-inefficiency:** The increase in costs of a monopoly resulting from the lack of competitive pressure to force costs to the minimum possible level.

**Zero economic profits:** The condition that faces the purely competitive firm in the long run; long-run equilibrium in a competitive industry leads to a condition where $P = MR = MC = LRATC$, which means that firms in the industry earn just a normal rate of return on their investment, or a zero economic profit.

of the publisher and author; Page 319: Listing of stocks from *The Wall Street Journal,* April 30, 1985. Reprinted by permission of *The Wall Street Journal,* © Dow Jones & Company, Inc. 1985. All Rights Reserved; Page 354: Table 16–4 from Edgar K. Browning, "The Trend Toward Equality in the Distribution of Net Income" *Southern Economic Journal* 43 (July 1976):914. Copyright © 1976 by the Southern Economic Association. Reprinted by permission of the publisher and author; Page 355: Figure 16–3 derived from Edgar K. Browning, "The Trend Toward Equality in the Distribution of Net Income," *Southern Economic Journal* 43 (July 1976):912–923. Copyright © 1976 by the Southern Economic Association. Reprinted by permission of the publisher and author; Page 356: Table 16–5 from Edgar K. Browning, "The Trend Toward Equality in the Distribution of Net Income," *Southern Economic Journal* 43 (July 1976):919. Copyright © 1976 by the Southern Economic Association. Reprinted by permission of the publisher and author; Page 368: Table adapted from *Markets and Minorities* by Thomas Sowell. Copyright © 1981 by the International Center for Economic Policy Studies. Reprinted by permission of Basic Books, Inc., New York. Data from U.S. Bureau of the Census; statistics from 1968, 1969, and 1970; Page 422: Data for Table 19–1 from *Monthly Tax Features* (Washington, D.C.: Tax Foundation, Inc., 1983); Page 424: Data for Table 19–3 from *Monthly Tax Features* (Washington, D.C.: Tax Foundation, Inc., 1980); Page 425: Data for tables in "Tax Freedom Day" from *Monthly Tax Features* Washington, D.C.: Tax Foundation, Inc., 1983); Page 455: Data for Table 20–2 from *Facts and Figures on Government Finance* (Washington, D.C.: Tax Foundation, Inc., 1981); Page 479: Arnold Schmedock quote from "We Pay More for Clothes," *USA Today,* July 21, 1983, p. 1. Copyright, 1983 USA TODAY. Reprinted with permission; Page 485: Quoted in Barbara Bader, "How the Colombians Do It," *Florists' Review,* April 14, 1983, p. 56. Reprinted by permission of Florists' Publishing Company; Page 526: Table 37–3 from World Bank, *World Development Report 1982* (New York: Oxford University Press, 1982), Annex Tables 1 and 2. Reprinted by permission; Page 527: Table 37–4 from Lowell Galloway, "The Folklore of Unemployment and Poverty," in Svetozar Pejovich (Ed.), *Governmental Controls and the Free Market: The U.S. Economy in the 1970's* (College Station: Texas A & M University Press, 1976), pp. 41–72. Reprinted by permission; Page 528: Table 37–5 from World Bank, *World Development Report 1982,* Annex Table 1. Reprinted by permission; Page 530: Table 37–6 from World Bank, *World Development Report 1981,* p. 17. Reprinted by permission.

# Index

Numbers in boldface indicate pages on which definitions of key terms appear.

Ability-to-pay principle, **426**–427
Absolute advantage, 37–39, **768**
Absolute price, **91**
Accounting costs **8**, 159
Accounting profit, 159–**160**, 336
Adjusted income, 352–357
Advertising, 231–232, 233–234, 247
Age-distributed income, 357
Aggregate concentration, **376**, 377
Aggregate supply of land, **323**
Agriculture. *See* Farming
Ahlbrandt, Roger, 454
Aid to Families with Dependent Children, 58
Airline Deregulation Act, 397
Allocative efficiency, **198**
Aluminum Company of America (Alcoa), 391
Amacher, R. C., 412n, 457n
Anderson, G. M. 453n
Antitrust policy, **386**–391, 392
Antitrust regulation, 55
Appreciation, **491**
Arbitration, binding, **298**
Arc elasticity, **105**
Areeda, Phillip, 389n, 390n
Artificial barriers to trade, **42**
Asset, **153**
Asset income, **346**
AT&T, 399–400
Average cost pricing, **392**
Average fixed cost (AFC), **162**
Average product, **165**–166
Average tax rate, **427**, 429
Average total cost (ATC), **162**
Average variable cost (AVC), **162**

Bader, Barbara, 485n
Balance of payments, **500**–504
Balance of trade, **501**–503
Balance sheet, **153**–155
Bargaining:
 bilateral monopoly and, 295–296
 collective, **296**, 298–299
 political influence in, 298–299
Barriers to entry:
 legal, **205**–206
 in monopolistic competition, 230–231
 in oligopoly, 241–242
Barter, **13**, 49
Barzel, Yoram, 133

Bazaar economy, 64–65
Becker, Gary, 370–371
Behrmann, Neil, 257n
Bell, Frederick W., 94
Bellante, D., 285n
Benefit principle, **424**–426
Bilateral monopoly, **295**, 296
Binding arbitration, **298**
Blinder, Alan S., 356n, 357n
Bolsheviks, 519
Bonds, 319–320
Boskin, Michael J., 299
Boulding, Kenneth, 4
Branch, Mike, 788n
Bretton Woods system, **505**–507
British Factory Acts, 453
Browning, Edgar K., 353, 355
Budget. *See* Federal budget
Budget constraint, **138**–139
Bureaucracy, 452–454
Business. *See* Firm
Business organization, 147–153

Calvert, Staunten, K., 353n
Capacity, excess, **238**
Capital, 6–7
 accumulation of, 150
 interest as return to, 311–312
 opportunity cost of, **159**
 rate of return on, 335–336
 venture, 338
Capital consumption, **308**
Capital formation, **307**
 benefits of, 317–318
 economic growth and, 450
Capitalism:
 mixed, 54–56, **516**
 pure, **515**
Capitalist economy, **50**
 American, 50–63
Capital stock, **34**, 155
Cartel, 204–205, **248**–252, 257–258
 characteristics of, 248–249
 formation of, 249, 250
 government management of, 386, 393–394
 profit seeking by, 384–385, 386
 public policy on, 389
Cartel enforcement, **250**–252
Cassel, Gustav, 491
Celler-Kefauver Act, 387, 389

Central planning, 520–522
*Ceteris paribus*, 73
Chamberlin, E. H., 288, 260
Change in demand, 73–**74**
Change in supply, **81**–82
Cheung, Steven N. S., 405n
China, 535–536
Choices at the margin, **30**, 32–33
Circular flow of income, **50**, 51
Civil Aeronautics Board (CAB), 394, 397
Clayton Antitrust Act, 387, 388, 389–390
Coase, Ronald H., 144, 176, 406, 459–460
Coase theorem, **407**
Cobweb effect, 201–202
Colander, David, 453n
Collective bargaining, **296**, 298–299
College, 278–280, 281
Collusion, **385**
Command economy, 821–822
Command society, **47**–48
Common ownership, **402**–404
Comparative advantage, 39–40, 44, **466**–471
Comparative economics, 514–516
Compensation, total, **280**
 *See also* Wage
Competition, 52–53, 54
 barriers to, 205–206
 cut-throat, 388
 demand curve and, 207
 excise tax and, 195
 in firm, 180–197
 imperfect, **228**
 in industry, 189–197
 in market, 52–53, 178–180, 181, 182, 198–200, 201–202
 monopolistic, **229**–239
 nonprice, **229**–230
 process of, 177–178
 pure, 178–200, 237–238
 *See also* Monopoly
Competitive equilibrium, long-run, **191**–194
Competitive labor market, **265**
Complement, **74**, 114
Complementary inputs, **272**
Composition, fallacy of, **15**–16
Computer entrepreneurs, 339–340
Concentration, aggregate, **376**, 377

555

Concentration ratios (CR), **376**–378
Constant-cost industry, **194**–**196**
Constant returns to scale, **172**
Constitutional choice, 444–446
Consumer choice, 123–141
Consumer equilibrium, 125–**127**, 139–140
Consumer-initiated discrimination, 363
Consumers' surplus, **130**, 131
Contrived scarcity, **220**–221
Corporate income tax, 429
Corporate takeovers, 152
Corporation, 149–151
Cost(s), **5**
   accounting, **8**, 159
   average fixed, **162**
   average total, **162**
   average variable, **162**
   constant, 194–196
   decreasing, 197
   effects of differing, 244
   exchange, 41–42
   explicit, 159
   external, **160**
   fixed, **161**, 167
   historical, **174**
   implicit, **159**
   increasing, 196–197
   law of increasing, **32**
   least, 144–145
   long-run, 170–173
   long-run average total, **170**–173
   marginal, **162**, 166–169
   marginal opportunity, **78**–79
   marginal social, **408**
   opportunity, **8**, **28**–33, 78–79, 159, 160, 173, 518
   pollution and, 176
   pollution and, 176
   private, **160**
   of production, 158, **159**–176
   short-run, 161–169
   social, **160**, 408, 459–460
   sunk, **174**
   supply decisions and, 174
   total, **161**, 167–168
   transaction, 41–42
   transportation, **42**
   variable, **161**, 167
Cost-benefit analysis, 416–417
Cost curves, 173
Cost pricing:
   average, **392**
   marginal, **393**
Craft union, 287–288
Cross elasticity of demand, **113**–114
Currency. *See* Foreign exchange rate; International monetary system; Money
Currency appreciation, **491**
Currency depreciation, **491**
Currency devaluation, **498**
Currency revaluation, **498**

Curve, **22**
   cost, 173
   indifference, **135**–141
   Lorenz, **348**–351, 352, 355
   slope of, 25–26
   supply, 80, 81–82, 134, 189–190, 194–197
   U-shaped, 24
   *See also* Demand curve

Davies, David, 454
De Beers diamond cartel, 257–258
Debt, world, 510
Decreasing-cost industry, **197**
Deficit. *See* Federal budget
Demand:
   change in, 73–**74**, 75
   competitive equilibrium and, 191–194
   cross elasticity of, **113**–114
   derived, **264**
   elastic, **101**
   factors of, 72–75
   graphic representation of, 84–85
   income elasticity of, **111**–113
   inelastic, **101**
   kinked, **244**–245
   for labor, 269–275
   labor unions and, 290–293
   law of, **71**–77, 128–130, 140–141, 518
   for loanable funds, 309–310
   marginal utility and, 128–130
   market, **76**–77
   price and, 85–97, 93–94
   price elasticity of, **101**–111, 210–211
   quantity and, 85–87
   tabular analysis of, 83–84
   time and, 110–111, 112
   unit elasticity of, **101**
Demand curve, 71–75
   advertising and, 232
   of competitive firm, 182–183
   elasticity along, 103–106, 108
   monopoly and, 206–208, 231
Demand elasticity coefficient, **101**
Demand schedule, 72
Depreciation, **308**, **491**
Depression. *See* Great Depression
Deregulation, 396–397
Derived demand, **264**
Devaluation, **498**
Differentiated products, 229–239, 247–248
Diminishing returns, 163–169, 518–519
Direct government purchases, **58**
Discount rate, **316**
Discrimination, 362–363
   consumer-initiated, 363
   economic wage, 362
Diseconomies of scale, **172**
Divestiture, 399–400

Division of labor, **37**, **142**–**143**
Dollar:
   balance of payments and, 503–504
   Bretton Woods system and, 507
   international capital movement and, 512–513
   *See also* Money
Downs, Anthony, 452n
Drilling rights, 403–404
Duke, Richard, 377n
Dumping, 479–**480**
Duopoly, 243
Dynamic analysis, 90–91
Dynamic efficiency, **224**

Economic analysis, 516–519
Economic development, **527**–537
Economic efficiency, **239**
Economic goods, **5**
Economic loss, 187–188
Economic mobility, **356**–357
Economic profit, 159–160, **336**
   zero, **191**
Economics:
   comparative, **514**
   defined, 4
   normative, **16**
   postiive, **16**
   theory in, 14–18
   *See also* Macroeconomics; Microeconomics
Economic stabilization, 14, **56**
Economic system, **46**
   classification of, 515
   command society, **47**–48
   comparative, 46–48, 514–516
   market society, **48**
   traditional society, **46**–47
Economic time, 161
Economic wage discrimination, 362–363
Economic welfare, **239**
Economies of scale, **171**–172, 206
Economy:
   bazaar, 64–65
   capitalist, **50**
   command, 519–520
   laissez-faire, **54**
   measurement of (*see* Gross national product)
   mixed, 54–56, 66–67
   socialist, **50**, 519–527
Efficiency:
   allocative, **198**
   dynamic, **224**
   economic, **239**
   productive, **198**
   *See also* Inefficiency
Elastic demand, **101**
Elasticity, **100**
   arc, **105**
   consumer expenditures and, 107
   of demand, 100–114, 210–211
   of demand for labor, **273**–274

farm problem and, 120–122
of supply, 114–119
Employment:
profit-maximizing level of, 266–269
*See also* Unemployment
Emshwiller, John R., 257*n*
Endangered predators, preservation of, 412
Engels, Friedrick, 95, 519
Entrepreneur, 6, **336–337**, 339–340
Equalizing differences in wages, **280–282**
Equilibrium analysis, 337, 341–343
Equilibrium price, **84**
Equity:
horizontal, **426**
vertical, **426**
*Ex ante* distribution of income, **346**
Excess capacity, **238**
Exchange costs, 41–42
Exchange rate. *See* Foreign exchange rate
Exchange rate conversion method, **530**
Excise tax, 117–118, 195, 430, 431
Exploitation of labor, 294–**295**
Export, 465, 436, 492–493, 501–502
*See also* Trade
*Ex post* distribution of income, **347**
External costs, **160**
Externality, **404**–411
correcting, 405–408
government and, 404–405, 408–411
irrelevant, 405
negative, **55, 404**
positive, **55, 404**
relevant, 405–408
taxing and subsidizing, 408–411
trivial, 405

Factor markets, **263**, 264–265
*See also* Resources market
Factor of demand, **72–75**
Factor of supply, **81**
Fallacy:
of composition, **15**–16
*post hoc*, 15, **16**
Family income, **347**, 357–359
Farming:
effects of strong dollar on, 123
income, 122
output, 121
productivity, 121
technology and, 121
Federal budget, 56–62
Federal Communications Commission (FCC), 393
Federal Power Commission (FPC), 394
Federal Trade Commission (FTC), 388, 389
Federal Trade Commission Act, 387
Firm(s), **142–157**

balance sheet of, 153–155
dominant, 246–247
labor managed, **151**–153
law, 156–157
management of, 6, 143, 146, 147
nonprofit, **153**
number of, in an industry, 378–382
publicly owned, **153**
purely competitive, 180–197
rents and, 331–332
Firm coordination, **143**–144
Firm organization, 147–153
corporation, **149**–151
labor-managed firms, **151**–153
nonprofit firms, **153**
partnership, **149**
proprietorship, **148**–149
publicly owned firms, **153**
Fiscal policy:
foreign exchange rates and, 497
Fishing rights, 402
Fixed costs, **161**, 167
average, **162**
Fixed exchange rate, **498**, 494–498
Fixed input, **161**
Fixed plant, **162**–163
Flat-rate tax, **436**–437
Floating exchange rate, **488**, 489–494
Flower industry, 485–486
Food and Drug Administration (FDA), 393
Food stamps, 58
Foreign exchange, **488**
Foreign exchange market, 488–498
Foreign exchange rate, **488**–498, 509
fixed, **489**, 494–498
floating, **488**, 489–494
Foreign exchange rate conversion method, **530**
Foreign trade. *See* Trade
Franchise, public, **205**
Free enterprise, **52**
Free goods, **5**
Freeman, Richard, 281, 304
Free riders, 413–414
Free trade, **472**, 474–475, 478–480, 483–484, 541–542
Friedman, Milton, 4
Full price, **91**, 92

Galbraith, John Kenneth, 225, 233
GATT (General Agreement on Tariffs and Trade), 482
GDP. *See* Gross domestic product
Geertz, Clifford, 64–65
General Agreement on Tariffs and Trade (GATT), 482
George, Henry, 324, 326
Gilder, George, 340
Gini coefficient, **350**–351
GNP. *See* Gross national product
Gold standard, 498, **499**–500, 504–505, 506
Good(s), **3**

consumer choices of, 126–128
distribution of, 12–13
economic, **5**
free, **5**
inferior, **74**
normal, **74**
public, 55, **411**–416
Gosplan, 520
Government:
antitrust policy of, 386–391
as cartel manager, 393–394
cost-benefit analysis by, 416–417
externalities and, 404–405, 408, 411
growth of, 56–63, 454–456
ownership of firms by, 394–397
public goods and, 411–416, 419–420
role of, 13–14, 53–56, 418
*See also* Federal budget; Regulation
Government expenditures, 421–423, 455
Government license, **205**
Government transfer payments, 58
Graphs, 20–26
bar, 21
drawing, 22–25
linear, 21, 23
purpose of, 20–22
slope, 25–26
Greenmail, 152
Gross domestic product (GDP), **61**–62
Gross national product (GNP), 20–21, 57–58
of Soviet Union, 526
*See also* Economic growth
Growth. *See* Economic growth

Hamilton, Alexander, 478
Hand, Learned, 391
Harberger, Arnold, 222
Hart-Scott-Rodino Antitrust Improvement Act, 387, 389
Hayek, Friedrich A., 517–518
Herfindahl index, 381
Homogeneous product, **179**
Hong Kong, 536
Horizontal equity, **426**
Hostetler, Dallas L., 425
Human capital, **278**–280
*See also* Labor
Human resources, **5**, 6
Hume, David, 498

Imperfect competition, **228**
Implicit costs, **159**
Import, 465, 466, 492–493, 501–502
*See also* Trade
Income, 278–283
adjusted, 352–357
age-distributed, 357
asset, **346**
circular flow of, **50**, 51

Income (*continued*)
    college education and, 278–280, 281
    family, **347**, 357–359
    individual, **345–347**
    labor, 345–346
    labor unions and, 289–290, 299–303
    marginal productivity theory and, 282, 283
    mean, 357
    per capita, 357
    real, **129**
    regional differences in, 284–285
    taxable, **427**, 428
Income distribution, 344–371
    characteristics of, 357–365
    *ex ante*, **346**
    *ex post*, **347**
    individual and, 345–347
    inequality of, 347–351
    interpretation of date, 348
    justice of, 365–367
    in United States, 351–357
Income effect, **129**
Income elasticity of demand, **111**–113
Income security transfers, 58
Income tax:
    corporate, 429
    negative, **364**–365
    personal, 427–429
    progressive, **426**, **427**, 438–439
    proportional, **426**, **427**, 438–439
    regressive, **427**
Increasing-cost industry, **196**–197
Increasing returns to scale, **173**
Indifference curve, **134**, **135**–141
Indifference map, **137**
Indifference set, **134**
Individual income, **345–347**
Industrial union, **288**
Industry:
    constant-cost, 194–**196**
    decreasing-cost, **197**
    increasing-cost, **196**–197
    infant, **478**
Industry concentration ratios, 376–378
Industry regulation, **55**
Industry supply:
    long-run, **194**–197
    short-run, **189**–190
Inefficiency:
    static, **224**
    x-, **223**
    *See also* Efficiency
Inefficient tax, **436**
Inelastic demand, **101**
Infant industry, **478**
Inferior good, **74**
Inflation, **13**
    foreign exchange rates and, 491–492
Inflationary gap. *See* Expansionary gap

Information, 7
    perfect, **180**
Inframarginal rent, **332**, 326–330
In-kind transfer payments, 352–**353**
Input:
    complementary, **272**
    fixed, **161**
    substitute, **272**, 292, 293
    variable, **161**
Interest, **308**
    pure, **312**–313
    as return on capital, 311–312
Interest groups, 450–452, 453
Interest rate(s), 308–311
    control of, 90
    foreign exchange and, 493–494
    nominal, **311**
    real, **311**
    United States vs. Western Europe, 513
    variations among, 310–311
International Monetary Fund (IMF), 503, **505**, 508
International monetary system, 487–513
    balance of payments, 500–504
    capital movements, 512–513
    debt crisis, 510
    evolution of, 504–509
    fixed exchange rates, 494–498
    floating exchange rates, 489–494
    foreign exchange market, 488–498
    gold standard, 498–500, 504–505, 506
International trade. *See* Trade
International Trade Commission, 483, 486
Interstate Commerce Commission (ICC), 393, 397
Inverse relation, **22**, 23
Investment:
    foreign exchange rates and, 494
Investment decisions, 315–317
Irrelevant externality, **405**

Japan, 467–470, 483–484, 539–540
Jewkes, John, 225
Jobs, Steven, 339–340
Johnson, Lyndon, B., 57
Joint unlimited liability, 149
Justice Department Antitrust Division, 388, 389

Keynes, John Maynard, 56, 259
Kinked demand, **244**–245
Kirzner, Israel M., 54n, 337n
Korea, 534–535

Labor:
    demand for, 269–275
    division of, **37**, 142–143
    elasticity of demand for, **273**–274
    exploitation of, 294–**295**
    foreign, 478–480

    long-run demand for, 270, **271**
    marginal factor cost of, **269**
    marginal product of, **266**–269
    marginal revenue product of, **267**
    market demand for, **265**
    market supply of, **265**
    monopoly and, 274–275
    supply of, 275–282
    *See also* Employment; Unemployment; Wage(s)
Labor income, 345–346
Labor-managed firm, **151**–153
Labor market, competitive, **265**
Labor supply, elasticity of, 114–115
Labor unions, **286**–304
    activities of, 289–293
    collective bargaining and, 296
    craft, **287**–288
    industrial, **288**
    new theory of, 304
    political influence and, 292, 298–299
    public employees', **288**
    strikes by, 297–298
    wages and, 289–290, 299–303
Laffer, Arthur, 435
Laffer curve, 435
Laissez-faire, **395**–397
Lampman, Robert J., 353n
Land:
    aggregate supply of, **323**
    supply to alternative uses, 324–326
Land rents, 322–326
Landrum-Griffin Act, 299
Lange, Oskar, 517–518
Law:
    antitrust, 386–391
    of demand, **71**–77, 128–130, 140–141, 518
    of diminishing marginal returns, 163–164
    of increasing costs, 32
    of one price, 82–83
    property rights and, 50–52
    of supply, 79–82
Law firms, 156–157
Lee, Dwight R., 362n
Legal barriers to entry, **205**–206
    *See also* Barriers to entry
Leibenstein, Harvey, 
Leibowitz, Arleen, 156n
Lenin, Vladimir Ilitch, 519
Less-developed countries, **528**–537
Lewis, H. Gregg, 299
Liability, 153–**154**
    joint unlimited, 149
    limited, 150
    unlimited, 148, 156
License, government, **205**
Limited liability, 150
Limit-entry pricing, **395**, 396
Limit pricing, 248
Lincoln, Abraham, 412
List, Friedrich, 478

Loanable funds:
  demand for, **309**–310
  supply of, **310**
Long run, **161**
Long-run average total cost (LRATC), 170–**173**
Long-run competitive equilibrium, 191–194
Long-run costs, 170–173
Long-run demand for labor, 270, **271**
Long-run industry supply, **194**–197
Lorenz curve, **348**–351, 352, 355
Loss, 187–188, 313

McCormick, R. E., 452n
McNaure, Robert F., 362n
Macroeconomic policy:
  foreign exchange rates and, 496–498
Macroeconomy, measurement of. *See* Gross national product
Malthus, Thomas, 120–121, 370–371
Managed floating rate system, **489**
Management, 6, 150–151
Manager, **143**, 146–147
Manufacturing:
  aggregate concentration in, 377
  concentration ratios in, 378
Margin, **9**–11
  choices at, **30**, 32–33
Marginal analysis, 9–11, 19–20
Marginal cost (MC), **162**, 166–169
Marginal cost pricing, **393**
Marginal factor cost, **268**–269
Marginal opportunity costs, **78**–79
Marginal private costs (MPC), **408**
Marginal product, **164**–165, 167
Marginal productivity theory, 266–269, 282–283
Marginal product of labor, **266**–269
Marginal rate of substitution, **136**
Marginal returns, diminishing, 163–169, 518–519
Marginal revenue (MR), **183**, 209–211
Marginal revenue product (MRP), **267**, 276
Marginal social costs (MSC), **408**
Marginal tax rate, 427–**428**, 429
Marginal utility, **124**, 125, 128–130
  consumers' surplus and, 131
  diminishing, **124**, 128–129
Market(s), **12**, **82**, **179**
  competitive, 52–53, 178–180, 181, 198–200, 201–202
  competitive labor, **265**
  factor, **263**, 264–265
  output vs. input, 264
  perfect, **82**–83
  products, **69**, 70
  purely competitive, 178, **179**–180, 181, 182, 198–200
  resources, **69**, 70
  society and, 95–96
  stock, 181
  United States, 48–50
  *See also* Monopolistic competition; Monopoly
Market coordination, **143**–144
Market demand, **76**–77
  for labor, **265**
Market failure, **401**–402
Market power, 241–**242**
Market society, **48**
  circular flow of income in, 50, 51
Market supply, **82**
  of labor, **265**
Marshall, Alfred, 4, 259–260
Marvel, Howard P., 453n
Marx, Karl, 4, 95–96, 821
Mass transit, 419–420
Materials balancing, **520**
Mean income, **357**
Medoff, James L., 304
Mercantilism, 204
Merger, **385**–386
  guidelines for, 381
  public policy on, 389–390
Microeconomics, **16**–17
Military service, 160
Mill, John Stuart, 366
Miller, James C., III, 444
Minorities, economic, 360, 362–363
Mises, Ludwig von, 4, 317–318, 452n, 517–518
Mixed capitalism, 54–56, **516**
Mixed economies, 54–56, 66–67
Model, **14**, 15, 30
Monetary policy:
  foreign exchange rates and, 497
Money, **13**, 49–50
  *See also* International monetary system
Money price, **91**
Monitor, **146**, 147
Monopolistic competition, **229**–239
  characteristics of, 229–232
  long-run equilibrium under, 235–236
  pure competition vs., 237–238
  resources and, 237–239
  short-run equilibrium under, 232–235
Monopoly, 203–227, 375–400
  bilateral, **295**, 296
  case against, 220–224
  case for, 224–225
  demand curve and, 206–208
  demand for labor and, 274–275
  in long run, 205–219
  mercantilism and, 204
  natural, **206**, 379–308, 391
  price and, 211, 213
  price discrimination and, 217–220
  profits of, 213–215, 216–217
  production costs of, 223
  profit seeking by, 382–386
  public policy and, 386–394
  pure, **203**
  revenues and, 208–211
  in short run, 206–215
  welfare loss due to, 220–**222**
Monopoly regulation, 55, 386–394
Monopoly rent, **331**–333
Monopsony, 293, **294**–299
Morocco, 64–65
Morrill tariff, 482
Motor Carrier Act of 1980, 397
Mueller, D. C., 457n
Mutual interdependence, **241**
Myrdal, Gunnar, 541–542

National income accounting. *See* Gross national product
National Labor Relations Board, 298
Natural monopoly, **206**, **379**–380, 391
Negative externality, **55**, **404**
Negative income tax, **364**–365
Negative relation, **22**, 23, 24
Net worth, **154**
Neutral tax, **430**
New International Economic Order, 533–534
Niskanen, William A., 452n
Nixon, Richard, 507
Nominal rate of interest, **311**
Nonhuman resources, **5**, 6–7
Nonmarket activities, **275**
Nonprice competition, **299**–230
Nonprofit firm, **153**
Nonproprietary setting, **441**
Normal good, **74**
Normative economics, **16**
Normative public choice, **441**, 442–446
Norris-La Guardia Act, 298
North Korea, 534–535

Oil drilling, 403–404
Oil supply, 112
Oligopoly, **241**–248
  characteristics of, 241–242
  pricing and, 243–248
Olson, Mancur, 414n
OPEC, 112, 249, 384–385, 508
Opportunity costs, **8**, **28**–33, 173, 518
  of capital, **159**
  of military service, 160
  production possibilities and, 30–33, 78
  supply and, 78–79
Ownership:
  common, **402**–404
  establishing, 405–408
Ownership categories, 148–153
  corporation, **149**–151
  labor-managed firms, 151–153
  nonprofit firms, **153**
  partnership, **149**

Ownership categories (*continued*)
 proprietorship, **148**–149
 publicly owned firms, **153**
Ownership claims, **154**

Paglin, Morton, 356n
Pareto, Vilfredo, 366
Parker, Dorothy, 4
Partial equilibrium analysis, 337
Partnership, **149**, 156–157
Patent, **205**–206
Per capita income, **357**
Perfect information, **180**
Perfect market, **82**–83
Personal income tax, 427–429
Personal saving. *See* Saving
Pickens, T. Boone, 152
Pigou, A. C., 176, 408n, 459–460
Plant:
 fixed, 162–**163**
 size of, 170–171
Policy. *See* Fiscal policy;
  Macroeconomic policy; Monetary
  policy; Regulation
Politics:
 labor unions and, 292, 298–299
Pollution, 176
Population growth:
 economic development and, 532–533
 economics of, 370–371
Positive economics, **16**
Positive externality, **55**, **404**
Positive public choice, **441**, 446–454
Positive relation, **22**, 23, 24
*Post hoc* fallacy, **15**, 16
Poverty, 359–362
 vicious circle of, **529**, 531–532
Preferences:
 revealed, **134**
 transitivity of, **137**
Present value, **314**–315, 316, 317
Price(s), 11–12, 82–88
 absolute, **91**
 demand and, 85–87, 93–94
 determination of, 83–85
 equilibrium, **84**
 full, **91**, 92
 money, **91**
 monopoly and, 211–213
 profit-maximizing, 212, **213**
 relative, **91**
 resource, 173
 supply and, 87–88
Price ceiling, **89**
Price control, **88**–90
Price discrimination, **217**–220, 390–391
Price elasticity of demand, **100**–111
 determinants of, 109–111
 formulation of, 100–101
Price floor, **89**
Price leadership, **246**
Price rationing, **85**

Price regulation, 392–393
Price searcher, **206**–207
Price-specie flow mechanism, 508
Price system, 69–71
Price taker, **180**
Pricing:
 average cost, **392**
 limit, **248**
 limit-entry, **395**, 396
 marginal, **393**
Principle of diminishing marginal
  utility, **124**, 128–129
Private costs, **160**
Private sector, **56**
Product:
 average, **165**–166
 differentiated, **229**–230, 247–248
 homogeneous, **179**
 marginal, **164**–165, 167
 total, **164**, 165
 *See also* Gross national product
Production:
 costs of, 158, **159**–176, 223
 roundabout, **307**
 scale of, **148**
 team, 145–147
 total cost of, **159**
Production possibilities, 30–33, 37, 78, 467–470
Production possibilities frontier, **31**
 shifts in, 33–35
Productive efficiency, **198**
Productivity:
 marginal, 266–269, 282–283
Products market, **69**, 70
Profit(s), **183**, 313, 333–337
 accounting, **159**–160, **336**
 economic, **159**–160, 191, **336**
 entrepreneurship and, 336–337
 maximization of, 183–187, 212, 213
 of monopoly, 213–215, 216–217
 as national income, 490, 491
 zero economic, **191**
Profit-maximizing level of
  employment, 266–269
Profit-maximizing price, 212, **213**
Profit seeking, 382–386
Progressive income tax, **426**, 427, 438–439
Property rights, 50–52
Property taxes, 430
Proportional income tax, **426**, 427, 438–439
Proprietary setting, **441**
Proprietorship, **148**–149
Protectionism, **472**, 477–480
Public choice, **440**–460
 constitutional, 444–446
 normative, **441**, 442–446
 positive, **441**, 446–454
Public employees' unions, **288**
Public finance, **421**–439
 government expenditures, 421–423

 taxation, 423–439
Public franchise, **205**
Public goods, **55**, **411**–416
 free riders, 413–414
 pricing, 414–416
 public provision of, 412–413
Publicly owned firm, **153**
Public policy, 386–394
 *See also* Government;
  Macroeconomic policy
Purchasing power parity method, **530**–531
Purchasing power parity theory, **491**–492
Pure capitalism, **515**
Pure economic rent, **322**–324, 329–331
Pure interest, 312–313
Purely competitive market, 178, **179**–180, 181, 182, 198–200
Pure monopoly, **203**
Pure socialism, **515**

Quantity demanded, **71**, 431–432
Quantity supplied, **79**, 81
Quasi-rents, **331**, 332
Quota, **471**, 479

Race:
 economic discrimination and, 362–363
 income and, 368–369
 poverty and, 360
Rate of return on invested capital, 335–336
Rate of time preference, **308**
Rational self-interest, **8**–11
Rationing, **85**
Real income, **129**
Real rate of interest, **311**
Recession:
 poverty during, 360–362
Reciprocal Trade Agreements Act, 482
Regressive income tax, **427**
Regulation, 392–394
 antitrust, 55, 386–391, 392
 costs and, 173
 of externalities, 410–411
 industry, **55**
 price, 392–393
Reigle, Donald, 484
Relative price, **91**
Relevant externality, **405**–408
Rent, **322**
 firms and, 331–332
 inframarginal, **322**, 329–330
 land, 322–326
 monopoly, 331–333
 pure economic, **322**–324, 329–331
 quasi-, **331**, 333
 types of, 321–322
Rent seeking, **223**, 333
Residual claimant, **147**, 334

Resources, **4**
  allocation of, 198–199
  costs of, 7–8
  human, **5,** 6
  monopolistic competition and, 237–239
  nonhuman, **5,** 6–7
  price of, 173
  scarce, 5–7
  unemployment of, **32**
Resources market, **69,** 70
  *See also* Factor markets
Resource specialization, **326**–329
Revaluation, **498**
Revealed preference, **134**
Revenue:
  marginal, **183,** 209–211
  monopoly and, 208–211
  tax, 423, 424, 435
  total, **106, 184**
Rhodes, Cecil, 257
Ricardo, David, 39, 541–542
Rights, selling, 402–404, 410
Right-to-work law, **298,** 300–301
Risk, **310,** 313
Robbins, Lionel, 4
Robertson, D. M., 9n
Robinson, Joan, 228, 259, 260
Robinson-Patman Act, 387
Rockefeller, John D., 385
Roundabout production, **307**

Sales tax, 430
Sassone, Peter G., 412n
Saving, **307,** 346
Sawers, David, 225
Scale:
  constant returns to, **172**
  diseconomies of, **172**
  increasing returns to, **173**
  of production, **148**
Scarcity, **5**–7, 516
  contrived, **220**–221
Schaffer, William A., 412n
Scherer, Fredric M., 225
Schumpeter, Joseph A., 224–225, 590–591, 429n
Scully, Gerald, 276
Securities and Exchange Commission (SEC), 393
Self-interest, 519
Senior, Nassau William, 4
Services, **3,** 12–13
Sex:
  economic discrimination and, 362–363
  poverty and, 360
Share, **149**
Shareholders, 150–151
Share transferability, **150**
Sherman Antitrust Act, 387, 388, 389
Shirking, **146**

Shortage, **83**–84
Short run, **161**
Short-run costs, 161–169
Short-run firm supply, **189**
Short-run industry supply, **189**–190
Shutdown, **188**
Silberberg, Eugene, 133
Slope, **25**–26
Smith, Adam, 4, 35–37, 53, 95–96, 128, 178
Smith, James D., 353n
Smoot-Hawley Act of 1930, 482
Social costs, **160,** 459–460
  marginal, 408
Socialism, 517–527
  pure, **515**
  Soviet, 519–527
Socialist economy, **50**
Socialization, **394**–395
Social Security, 58
Social Security taxes, 429–430
South Korea, 534–535
Soviet Union, 47, 519–527
Sowell, Thomas, 363, 368
Spann, Robert, 454
Special drawing rights (SDRs), 502–503
Specialization, 35–42, **466,** 468–471
Speculation, **506**–507
Stafford, Frank P., 299
Stalin, Joseph, 519
Standard Industrial Classification (SIC) groups, 377–378
Standard Oil Trust, 385
Static equilibrium analysis, 90–91
Static inefficiency, **224**
Stewart, Potter, 390
Stillerman, Richard, 225
Stock, 319
  capital, **34,** 155
  shares of, 149–150
Stock market, 181
Strikes, **297**–298
Subsidy, **477**–478
Substitute, **74,** 109, **113**
Substitute inputs, **272,** 292, 293
Substitution, marginal rate of, 136
Substitution effect, **129**
Supply:
  change in, **81**–82
  costs and, 174
  elasticity of, **114**–119
  factor of, **82**
  graphic representation of, 84–85
  of labor, 275–282
  law of, **79**–82
  of loanable funds, 310
  long-run industry, **194**–197
  market, **82**
  opportunity cost and, 78–79
  price and, 87–88
  quantity and, 87–88
  short-run firm, **189**
  short-run industry, **189**–190

tabular analysis of, 83–84
time and, 115–119
Supply curve, 80, 81–82, 87–88, 189–190, 194–197
Supply schedule, **79,** 80
Surplus, **83**
  consumers', **130,** 131
Sweden, 66–67
Sweezy, Paul M., 244, 245

Taft-Hartley Act, 298, 300
Taiwan, 536
Tangency solution, **235**–236
Tariff, **471**–477
  effects of, 472–477
  United States policy on, 481–484
Tax(es), 60–61, 421–439
  consumption, **437**
  costs and, 173
  effects of, 430–434
  equitable, 423–427
  excise, 117–118, 195, 430, 431
  externality and, 408–411
  flat-rate, **436**–437
  income distribution and, 346–347
  inefficient, 436
  negative, **364**–365
  neutral, **430**
  progressive, **426,** 427, 438–439
  property, 430
  proportional, **426,** 427, 438–439
  regressive, **427**
  revenues from, 423, 424, 435
  sales, 430
  Social Security, 429–430
  theories of, 423–427
  types of, 427–430
  *See also* Income tax
Taxable income, **427,** 428
Tax freedom day, 425
Tax incidence, **432**–433
Tax rate:
  average, **427,** 429
  marginal, **427**–428, 429
Tax reform, 434–437
Team production, **145**–147
Technology, 7, 173
Terms of trade, **39, 469**
Terms of trade argument, **480**
Textile quota, 479
Time:
  economic, 161
  elasticity of demand and, 110–111
  elasticity of supply and, 115–119
  as resource, 7
Time preference, rate of, **308**
Tollison, R. D., 156n, 412n, 452n, 453n, 457
Torrens, Robert, 39
Total compensation, 280
Total cost, **161,** 167–168
  average, **162**
Total product, **164,** 165

Total revenue, **106, 184**
Total utility, **124,** 125
Total wage bill, **289**
Town meeting, 444
Trade, 463–486
  artificial barriers to, **42**
  balance of, **501**–503
  barriers to, 42, 471–477
  comparative advantage in, 465–471
  free, **472,** 474–475, 478–480, 483–484, 541–544
  importance of, 464–471
  protection of, 472, 477–480
  specialization and, 35–42
  terms of, **39, 469,** 480
  United States tariff policy, 481–484
Trade Act of 1974, 483
Traditional society, **46**–47
Tramiel, Jack, 339–340
Transaction costs, 41–42
Transfer payment, **58, 347,** 421
  in-kind, 352–353
Transitivity of preferences, **137**
Transportation costs, 42
Trivial externality, 405
Trust, 385
Tullock, Gordon, 10, 222, 452n

Underdevelopment, 531–537
Unemployment:
  of resources, 32
  *See also* Employment
Unions. *See* Labor unions
United States:
  capitalist economy of, 50–63
  capitalist institutions in, 50–54
  farm problem in, 120–122
  government of, 13–14, 53–63
  market system of, 48–50
  tariff policy of, 481–484
*United States* v. *Trenton Potteries Co.,* 389
*United States* v. *Von's Grocery Co.,* 390
Unit elasticity of demand, **101**
Unlimited liability, **148,** 156
  joint, **149**
U.S. International Trade Commission, 483, 486
U.S. Postal Service, 394–395
U.S. Supreme Court, 389, 390
Usury, 90
*Utah Pie Co.* v. *Continental Banking Co.,* 390–391
Utility, **123**
  marginal, **124,** 125, 128–130, 131
  total, **124,** 125

Value:
  present, **314**–315, 316, 317
Variable costs, **161,** 167
  average, **162**
Variable input, **161**
Venture capitalism, 338
Vertical equity, **426**
Voters:
  ideological distribution of, 477
  preferences of, 457–458
Voting, 133, 449–450
  constitutional choice and, 444–446
  majority, 442–444, 457–458
  *See also* Public choice

Wage(s):
  differences in, 345–346
  equalizing differences in, **280**–282
  human capital and, 278–280
  labor unions and, 289–290, 229–303
  monopsony and, 295
  regional differences in, 284–285
Wage discrimination, 362–363
Wagner, Richard E., 455n
Wagner Act, 298
Welfare, 58, 347, 364–365
  economic, **239**
Welfare loss, **384**
Welfare loss due to monopoly, 220–222
West Germany, 39–40
Wheeler-Lea Act, 387
Willett, T. D., 412n, 457n

x-axis, **20**
X-inefficiency, **223**

y-axis, **20**
Yeager, Leland B., 811n

Zeckhauser, Richard, 414n
Zero economic growth, 457–458
Zero economic profits, **191**